The Upper Ocean

Editor-in-Chief

John H. Steele

Marine Policy Center, Woods Hole Oceanographic Institution, Woods Hole,
Massachusetts, USA

Editors

Steve A. Thorpe

National Oceanography Centre, University of Southampton,
Southampton, UK
and
School of Ocean Sciences, Bangor University, Menai Bridge, Anglesey, UK

Karl K. Turekian

Yale University, Department of Geology and Geophysics, New Haven,
Connecticut, USA

Subject Area Volumes from the Second Edition

Climate & Oceans edited by Karl K. Turekian
Elements of Physical Oceanography edited by Steve A. Thorpe
Marine Biology edited by John H. Steele
Marine Chemistry & Geochemistry edited by Karl K. Turekian
Marine Ecological Processes edited by John H. Steele
Marine Geology & Geophysics edited by Karl K. Turekian
Marine Policy & Economics guest edited by Porter Hoagland, Marine Policy Center,
Woods Hole Oceanographic Institution, Woods Hole, Massachusetts
Measurement Techniques, Sensors & Platforms edited by Steve A. Thorpe
Ocean Currents edited by Steve A. Thorpe
The Coastal Ocean edited by Karl K. Turekian
The Upper Ocean edited by Steve A. Thorpe

THE UPPER OCEAN

A DERIVATIVE OF ENCYCLOPEDIA OF OCEAN SCIENCES, 2ND EDITION

Editor

STEVE A. THORPE

ELSEVIER

AMSTERDAM • BOSTON • HEIDELBERG • LONDON • NEW YORK • OXFORD
PARIS • SAN DIEGO • SAN FRANCISCO • SINGAPORE • SYDNEY • TOKYO
Academic Press is an imprint of Elsevier

ACADEMIC
PRESS

Academic Press is an imprint of Elsevier
32 Jamestown Road, London NW1 7BY, UK
30 Corporate Drive, Suite 400, Burlington, MA 01803, USA
525 B Street, Suite 1900, San Diego, CA 92101-4495, USA

Material in the work originally appeared in *Encyclopedia of Ocean Sciences* (Elsevier Ltd., 2001) and *Encyclopedia of Ocean Sciences*, 2nd Edition (Elsevier Ltd., 2009), edited by John H. Steele, Steve A. Thorpe and Karl K. Turekian.

The following articles are US government works in the public domain and are not subject to copyright:

Satellite Oceanography, History, and Introductory Concepts
Satellite Passive-Microwave Measurements of Sea Ice
Wind- and Buoyancy-Forced Upper Ocean

Permissions may be sought directly from Elsevier's Science & Technology Rights Department in Oxford, UK: phone (+44)(0) 1865 843830; fax (+44)(0) 1865 853333; email: permissions@elsevier.com. Alternatively you can submit your request online by visiting the Elsevier website at (http://elsevier.com/locate/permissions), and selecting *Obtaining permissions to use Elsevier material*

Notice
No responsibility is assumed by the publisher for any injury and/or damage to persons or property as a matter of products liability, negligence or otherwise, or from any use or operation of any methods, products, instructions or ideas contained in the material herein, Because of rapid advances in the medical sciences, in particular, independent verification of diagnoses and drug dosages should be made

British Library Cataloguing in Publication Data
A catalogue record for this book is available from the British Library

Library of Congress Cataloging-in-Publication Data
A catalog record for this book is available from the Library of Congress

ISBN: 978-0-12-3813930

For information on all Academic Press publications
visit our website at www.elsevierdirect.com

CONTENTS

UPPER OCEAN CIRCULATION AND STRUCTURE

PLANKTON

ICE

MEASUREMENT TECHNIQUES INCLUDING REMOTE SENSING

INDEX

THE UPPER OCEAN: INTRODUCTION

There is no generally accepted definition of the region described in the title of this collection of articles as 'The Upper Ocean'. It is taken to be the part of the deep ocean that is closely involved in interaction with the atmosphere. As such it has immense importance, for through it passes those quantities that sustain and determine the life forms, dynamics and chemistry of the ocean: light, heat, momentum, freshwater, particulates and gases.

The oceans cover about 70.8% of the surface of the Earth, and of this about 7.5% overlies the continental shelves surrounding the continents. The surface of deep ocean therefore covers over 60% of the planet's surface, and it has a substantial effect on the overlying atmosphere. It is, for example, one source of cloud-forming aerosols and, having an albedo of 2–10%, generally lower than those of the soils, grasses and forests on land (6–20%) or of sandy deserts (35–45%), it has a substantial effect on the net planetary absorption of solar radiation.

It is, however, the phenomena that occur within the Upper Ocean, many driven by the atmosphere, that form the main content of this volume. The most obvious and visible are the wind-generated waves at the surface of the sea, a manifestation of the fluxes of momentum and energy from the atmosphere to the ocean. Currents are also generated, largely through the momentum dumped by the waves when they break. Precipitation is a visible transfer of freshwater to the ocean, evaporation an invisible transfer from the ocean to the atmosphere. Other fluxes are also invisible and less evident. Heat passes through the air-sea interface. Daytime solar radiation warms the surface layers, but in tropical regions at night, and more generally in extra-tropical regions in winter, long-wave radiation from the ocean result in a loss of heat, cooling and the formation of relatively dense water that sinks in convective plumes. Airborne particles of dust and pollutants contribute further to the air-sea fluxes. Invisible, but partly linked to the breaking of surface waves and the formation of clouds of subsurface, slowly dissolving bubbles, are the transfer of gases through the sea surface, vital in controlling their concentrations within the atmosphere.

Particles, heat and soluble gases transferred to the ocean through the sea surface are diffused downwards through a layer of relatively uniform density, the 'mixed layer' stirred by convection, the turbulence produced by breaking waves, Langmuir circulation and current shear. The properties and thickness of this layer are determined by the energy fluxes from the atmosphere. Beneath it temperature falls and the density of the sea increases relatively rapidly in a 'seasonal thermocline' ('seasonal' because its depth changes through the year). Sunlight within the near-surface euphotic zone allows photosynthesis to occur, mainly in the phytoplankton that forms the base of the marine foodweb. The resulting primary production results in the fixation of carbon and subsequently to the sinking of detrital material or 'marine snow' and carbon sequestration to the deep ocean through the 'biological pump'. This is very significant in limiting the atmospheric concentration of CO_2. Nutrients mixed upwards through the seasonal thermocline support production in the mixed layer.

But not all the ocean water surface is exposed to the atmosphere at all times; seasonally, some is covered by ice. The melting of the cryosphere in the Arctic (affecting living creatures, navigation and the Earth's albedo) is a clear indication of climate change and the subject of much present interest and concern. The ice-covered regions and some properties of the upper ocean can be observed and measured from space. Means are now developed for remote measurement of ice cover, surface temperature, salinity, and the colour of the sea, from which information can be obtained about its phytoplankton and sediment contents.

This is a selection of articles previously published within 'The Encyclopedia of Ocean Sciences' that relate to these aspects of the Upper Ocean: to the air-sea fluxes, to the physics of the sea surface and mixed layer, to plankton and ice, and to the means of measurement.

But what's the use of such a multidisciplinary collection of articles?

In the early 20th century there was, in oceanography, a strong emphasis on four separate component scientific disciplines, biology, chemistry, geology and physics. Research scientists were trained almost exclusively within a single discipline area. Contact with scientists of other disciplines occurred largely within research institutions such as the Woods Hole and Scripps Oceanographic Institutions in the USA and the National Institute of Oceanography in the UK, and, to a lesser extent, within fisheries laboratories and the laboratories of the Biological Associations, but rarely lead to joint research and scientific publications. Until the 1960s the expertise of the individual members of the scientific research community, although consequently narrow, was still generally fitted to address and solve the then important problems related to the ocean.

This is no longer true; the major present-day important – and most exciting – problems and issues (e.g., of climate, sustainability and pollution) demand interdisciplinary approaches, and a marine scientist's knowledge must span several of the 'old' disciplines. Take for example the International Programme, SOLAS, to study the climate-related exchanges between the Surface Ocean and the Lower Atmosphere. As recognised in the formation of that Programme, the upper ocean region is one of vital importance in the study of climate change. The ocean has been described as the flywheel of climate because of the relatively long times required for it to adjust to changes in temperature, hundreds of years compared to the few years of the atmosphere. Even more important than the exchange of heat between the ocean and the atmosphere – controlling temperature – is the exchange of gases, particularly the greenhouse gases – affecting radiative heat balance and climate. The absorption of such gases by the upper ocean or their release through the sea surface to the atmosphere involves many complex processes involving all the disciplines: bio-geo-physio-chemical understanding is required to address and lead in the solution of the problems faced, and consequently the Programme is largely supported by interdisciplinary research projects.

Reflecting the demand for informed scientists capable of addressing interdisciplinary problems, universities have established interdisciplinary schools and departments providing instruction and research in multidisciplinary areas. Their expertise often extends across the ocean surface to include atmospheric science as well as oceanography. This selection of articles from the Encyclopedia of Ocean Sciences provides, in a compact and focussed form, information of use in supporting such teaching and research, and reference suitable for young scientists at UG, Masters and PhD level, in an important area of the Planet where multidisciplinary knowledge is required.

The Editors are grateful to the authors who contributed the articles in the 'Encyclopedia of Ocean Sciences' that are included in this selection, and to the several members of the Editorial Advisory Board who helped to select the contents and writers of the Encyclopedia.

Steve A. Thorpe
Editor

AIR-SEA TRANSFERS

HEAT AND MOMENTUM FLUXES AT THE SEA SURFACE

P. K. Taylor, Southampton Oceanography Centre, Southampton, UK

Introduction

The maintenance of the earth's climate depends on a balance between the absorption of heat from the sun and the loss of heat through radiative cooling to space. For each 100 W of the sun's radiative energy entering the atmosphere nearly 40 W is absorbed by the ocean – about twice that adsorbed in the atmosphere and three times that falling on land surfaces. Much of this oceanic heat is transferred back to the atmosphere by the local sea to air heat flux. The geographical variation of this atmospheric heating drives the weather systems and their associated winds. The wind transfers momentum to the sea causing waves and the wind-driven currents. Major ocean currents transport heat polewards and at higher latitudes the sea to air heat flux significantly ameliorates the climate. Thus the heat and momentum fluxes through the ocean surface form a crucial component of the earth's climate system.

The total heat transfer through the ocean surface, the net heat flux, is a combination of several components. The heat from the sun is the short-wave radiative flux (wavelength 0.3–3 μm). Around noon on a sunny day this flux may reach about $1000 \, \text{W m}^{-2}$ but, when averaged over 24 h, a typical value is 100–$300 \, \text{W m}^{-2}$ varying with latitude and season. Part of this flux is reflected from the sea surface – about 6% depending on the solar elevation and the sea state. Most of the remaining short-wave flux is absorbed in the upper few meters of the ocean. In calm weather, with winds less than about $3 \, \text{m s}^{-1}$, a shallow layer may be formed during the day in which the sea is warmed by a few degrees Celsius (a 'diurnal thermocline'). However, under stronger winds or at night the absorbed heat becomes mixed down through several tens of metres. Thus, in contrast to land areas, the typical day to night variation in sea surface sea and air temperatures is small, $<1°C$. Both the sea and the sky emit and absorb long-wave radiative energy (wavelength 3–50 μm). Because, under most circumstances, the radiative temperature of the sky is colder than that of the sea, the downward long-wave flux is usually smaller than the upward flux. Hence the net long-wave flux acts to cool the surface, typically by 30–$80 \, \text{W m}^{-2}$ depending on cloud cover.

The turbulent fluxes of sensible and latent heat also typically transfer heat from sea to air. The sensible heat flux is the transfer of heat caused by difference in temperature between the sea and the air. Over much of the ocean this flux cools the sea by perhaps 10–$20 \, \text{W m}^{-2}$. However, where cold wintertime continental air flows over warm ocean currents, for example the Gulf Stream region off the eastern seaboard of North America, the sensible heat flux may reach $100 \, \text{W m}^{-2}$. Conversely warm winds blowing over a colder ocean region may result in a small sensible heat flux into the ocean – a frequent occurrence over the summertime North Pacific Ocean. The evaporation of water vapor from the sea surface causes the latent heat flux. This is the latent heat of vaporization which is carried by the water vapor and only released to warm the atmosphere when the vapor condenses to form clouds. Usually this flux is significantly greater than the sensible heat flux, being on average $100 \, \text{W m}^{-2}$ or more over large areas of the ocean. Over regions such as the Gulf Stream latent heat fluxes of several hundred W m^{-2} are observed. In foggy conditions with the air warmer than the sea, the latent heat flux can transfer heat from air to sea. In summertime over the infamous fog-shrouded Grand Banks off Newfoundland the mean monthly latent heat transfer is directed into the ocean, but this is an exceptional case.

Measuring the Fluxes

The standard instruments for determining the radiative fluxes measure the voltage generated by a thermopile which is exposed to the incident radiation. Typically the incoming short-wave radiation is measured by a pyranometer which is mounted in gimbals for use on a ship or buoy (**Figure 1**). For better accuracy the direct and scattered components should be determined separately but, apart from at the Baseline Surface Radiation Network stations which are predominantly situated on land, at present this is rarely done. The reflected short-wave radiation is normally determined from the sun's elevation

Figure 1 A pyranometer used for measuring short-wave radiation. The thermopile is covered by two transparent domes. (Photograph courtesy of Southampton Oceanography Centre.)

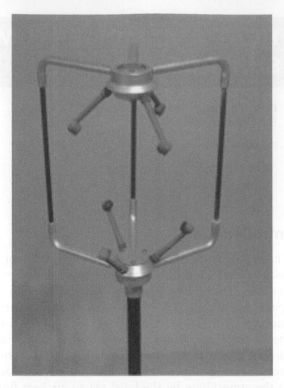

Figure 2 The sensing head of a three-component ultrasonic anemometer. The wind components are determined from the different times taken for sound pulses to travel in either direction between the six ceramic transducers. (Photograph courtesy of Southampton Oceanography Centre.)

and lookup tables based on the results of previous experiments. The pyrgeometer used to determine the long-wave radiation is similar to the pyranometer but uses a coated dome to filter out, as far as possible, the effects of the short-wave heating. Because the air close to the sea surface is normally near to the sea temperature, the use of gimbals is less important. However, a clear sky view is required and a number of correction terms have to be calculated for the temperature of the dome and any short-wave leakage. Again, only the downward component is normally measured; the upwards component is calculated from knowledge of the sea temperature and emissivity of the sea surface.

The turbulent fluxes may be measured in the near-surface atmosphere using the eddy correlation method. If upward moving air in an eddy is on average warmer and moister than the downward moving air, then there is an upwards flux of sensible heat and water vapor and hence also an upward latent heat flux. Similarly the momentum flux, or wind stress, may be determined from the correlation between the horizontal and vertical wind fluctuations. Since a large range of eddy sizes may contribute to the flux, fast response sensors capable of sampling at 10 Hz or more must be exposed for periods of the order of 30 min for each flux determination. Three-component ultrasonic anemometers (**Figure 2**) are relatively robust and, by also determining the speed of sound, can provide an estimate of the sonic temperature flux, a function of the heat and moisture fluxes. The sensors used for determining the fluctuations in temperature and humidity have previously tended to be fragile and prone to contamination by salt particles which are ever-present in the marine atmosphere. However, improved sonic thermometry, and new techniques for water vapor measurement, such as microwave

refractometry or differential infrared absorption instruments, are now becoming available.

Despite these improvements in instrumentation, obtaining accurate eddy correlation measurements over the sea remains very difficult. If the instrumentation is mounted on a buoy or ship the six components of the wave-induced motion of the measurement platform must be measured and removed from the signal. The distortion both of the turbulence and the mean wind by ship, buoy or fixed tower must be minimized and, as far as possible, corrected for. Thus eddy correlation measurements are not routinely obtained over the ocean, rather they are used in special air–sea interaction experiments to calibrate other less direct methods of flux estimation. For example, in the inertial dissipation method, fluctuations of the wind, temperature, or humidity at a few Hertz are measured and related (through turbulence theory) to the fluxes. This method is less sensitive to flow distortion or platform motion, but relies on various assumptions about the formation and dissipation of turbulent quantities, which may not be valid under some conditions. It has been implemented on a semi-routine basis on some research ships to increase the range of available flux data.

The most commonly used method of flux estimation is variously referred to as the bulk (aerodynamic) formulae. These formulae relate the difference between the value of temperature, humidity or wind ('x' in [1]) at some measurement height, z, and the value assumed to exist at the sea surface – respectively the sea surface temperature, 98% saturation humidity (to allow for salinity effects), and zero wind (or any nonwind-induced water current). Thus the flux F_x of some quantity x is:

$$F_x = \rho U_z C_{xz} (x_z - x_0) \qquad [1]$$

where ρ is the air density, and U_z the wind speed at the measurement height. While appearing intuitively correct (for example, blowing over a hot drink will cool it faster) these formulae can also be derived from turbulence theory. The value for the transfer coefficient, C_{xz}, characterizes both the surface roughness applicable to x and the relationship between F_x and the vertical profile of x. This varies with the atmospheric stability, which itself depends on the momentum, sensible heat, and water vapor fluxes, as well as the measurement height. Thus, although it may appear simple, Eqn [1] must be solved by iteration, initialized using the equivalent neutral value of C_{xz} at some standard height (normally 10 m), C_{x10n}. Typical neutral values (determined using eddy correlation or inertial dissipation data) are shown in **Table 1**. Many research problems remain. For example: C_{D10n} is expected to depend on the state of development of the wave field, but can this be successfully characterized by the ratio of the predominant wave speed to the wind speed (the wave age), or by the wave height and steepness, or is a spectral representation of the wave field required? What are the effects of waves propagating from other regions (i.e., swell waves)? What is the behavior of C_{D10n} in low wind speed conditions? Furthermore C_{E10n} and C_{H10n} are relatively poorly defined by the available experimental data, and recent bulk algorithms have used theoretical models of the ocean surface (known as surface renewal theory)

to predict these quantities from the momentum roughness length.

Sources of Flux Data

Until recent years the only source of data for flux calculation routinely available from widespread regions of the world's oceans was the weather reports from merchant ships. Organized as part of the World Weather Watch system of the World Meteorological Organisation, these 'Voluntary Observing Ships (VOS)' are asked to return coded weather messages at 00 00, 06 00, 12 00, and 18 00 h GMT daily, also recording the observation (with further details) in the ship's weather logbook. The very basic set of instruments provided will normally include a barometer and a means of measuring air temperature and humidity – typically wet and dry bulb thermometers mounted in a hand swung sling psychrometer or a fixed, louvered 'Stevenson' screen. Sea temperature is obtained using a thermometer and an insulated bucket, or by reading the temperature gauge for the engine cooling water intake. Depending on which country recruited the VOS an anemometer and wind vane might be provided, or the ship's officers might be asked to estimate the wind velocity from observations of the sea state using a tabulated 'Beaufort scale'. Because of the problems of adequately siting an anemometer and maintaining its calibration, these visual estimates are not necessarily inferior to anemometer-based values.

Thus the VOS weather reports include all the variables needed for calculating the turbulent fluxes using the bulk formulae. However, in many cases the accuracy of the data is limited both by the instrumentation and its siting. In particular, a large ship can induce significant changes in the local temperature and wind flow, since the VOS are not equipped with radiometers. The short-wave and long-wave fluxes must be estimated from the observer's estimate of the cloud amount plus (as appropriate) the solar elevation, or the sea and air temperature and

Table 1 Typical values (with estimated uncertainties) for the transfer coefficients[a]

Flux	Transfer coefficients	Typical values
Momentum	Drag coefficient $C_{D10n}(\times 1000)$	$= 0.61 \ (\pm 0.05) + 0.063 \ (\pm 0.005) \ U_{10n}$ $(U_{10n} > 3\,\mathrm{m\,s^{-1}}) = 0.61 + 0.57/U_{10n} < 3\,\mathrm{m\,s^{-1}}$
Sensible heat	Stanton no., U_{H10n}	$1.1 \ (\pm 0.2) \times 10^{-3}$
Latent heat	Dalton no., U_{E10n}	$1.2 \ (\pm 0.1) \times 10^{-3}$

[a]Neither the low wind speed formula for C_{D10n}, nor the wind speed below which it should be applied, are well defined by the available, very scattered, experimental data. It should be taken simply as an indication that, at low wind speeds, the surface roughness increases as the wind speed decreases due to the dominance of viscous effects.

humidity. The unavoidable observational errors and the crude form of the radiative flux formulae imply that large numbers of reports are needed, and correction schemes must be applied, before satisfactory flux estimates can be obtained. While there are presently nearly 7000 VOS, the ships tend to be concentrated in the main shipping lanes. Thus whilst coverage in most of the North Atlantic and North Pacific is adequate to provide monthly mean flux values, elsewhere data is mainly restricted to relatively narrow, major trade routes. For most of the southern hemisphere the VOS data is only capable of providing useful values if averaged over several years, and reports from the Southern Ocean are very few indeed. These shortcomings of VOS-derived fluxes must be borne in mind when studying the flux distribution maps presented below.

Satellite-borne sensors offer the potential to overcome these sampling problems. They are of two types, passive sensors which measure the radiation emitted from the sea surface and the intervening atmosphere at visible, infrared, or microwave frequencies, and active sensors which transmit microwave radiation and measure the returned signal. Unfortunately these remotely sensed data do not allow all of the flux components to be adequately estimated. Sea surface temperature has been routinely determined using visible and infrared radiometers since about 1980. Potential errors due, for example, to changes in atmospheric aerosols following volcanic eruptions, mean that these data must be continually checked against ship and buoy data. Algorithms have been developed to estimate the net surface short-wave radiation from top of the atmosphere values; those for estimating the net surface long wave are less successful. The surface wind velocity can be determined to good accuracy by active scatterometer sensors by measuring the microwave radiation backscattered from the sea surface. Unfortunately scatterometers are relatively costly to operate, since they demand significant power from the spacecraft and, to date, few have been flown. The determination of near-surface air temperature and humidity from satellite is hindered by the relatively coarse vertical resolution of the retrieved data. A problem is that the radiation emitted by the near-surface air is dominated by that originating from the sea surface. Statistically based algorithms for determining the near-surface humidity have been successfully demonstrated. More recently neural network techniques have been applied to retrieving both air temperature and humidity; however, at present there is no routinely available product. Thus the satellite flux products for which useful accuracy has been demonstrated are presently limited to momentum, short-wave radiation, and latent heat flux.

Numerical weather prediction (NWP) models (as used in weather forecasting centers) estimate values of the air–sea fluxes as a necessary part of their calculations. Since these models assimilate most of the available data from the World Weather Watch system, including satellite data, radiosonde profiles, and surface observations, it might be expected that NWP models represent the best source of flux data. However, there are other problems. The vertical resolution of these models is relatively poor and many of the near-surface processes which affect the fluxes have to be represented in terms of larger-scale parameters. Improvements to these models are normally judged on the resulting quality of the weather forecasts, not on the accuracy of the surface fluxes; sometimes these may become worse. Indeed, the continual introduction of model changes results in time discontinuities in the output variables. This makes the determination of interannual variations difficult. Because of this, NWP centres such as the European Centre for Medium Range Weather Forecasting (ECMWF) and the US National Centers for Environmental Prediction (NCEP) have reanalyzed the past weather and have gone back several decades. The surface fluxes from these reanalyses are receiving much study. Those presently available appear less accurate than fluxes derived from VOS data in regions where there are many VOS reports; in sparsely sampled regions the model fluxes may be more accurate. There are particular weaknesses in the short-wave radiation and latent heat fluxes. New reanalyses are planned and efforts are being made to improve the flux estimates; eventually these reanalyses will provide the best source of flux data for many purposes.

Regional and Seasonal Variation of the Momentum Flux

The main features of the wind regimes over the global oceans have long been recognized and descriptions are available in many books on marine meteorology (see Further Reading). The major features of the wind stress variability derived from ship observations from the period 1980–93 will be summarized here, using plots for January and July to illustrate the seasonal variation. The distribution of the heat fluxes will be discussed in the next section.

In northern hemisphere winter (**Figure 3A**) large wind stresses due to the strong midlatitude westerly winds are obvious in the North Atlantic and the North Pacific west of Japan. To the south of these regions the extratropical high pressure zones result in low wind stress values, south of these is the north-

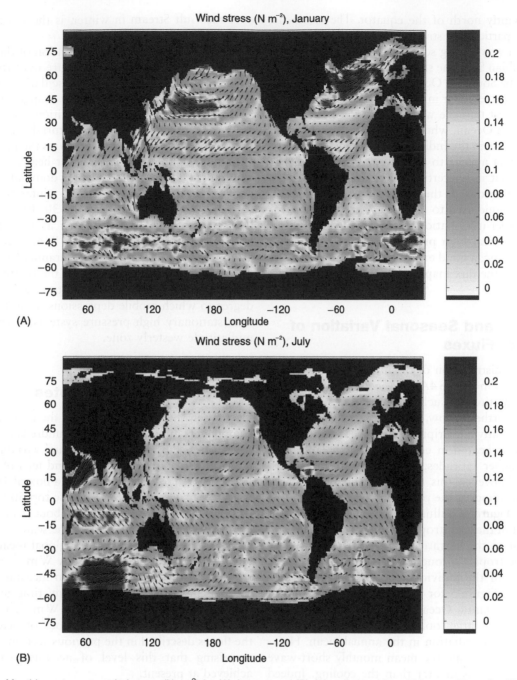

Figure 3 Monthly vector mean wind stress (N m⁻²) for (A) January and (B) July calculated from Voluntary Observing Ship weather reports for the period 1980–93. (Adapted with permission from Josey SA, Kent EC and Taylor PK (1998) *The Southampton Oceanography Centre (SOC) Ocean–Atmosphere Heat, Momentum and Freshwater Flux Atlas.* SOC Report no. 6.)

east trade wind belt. The Inter-Tropical Convergence Zone (ITCZ) with very light winds is close to the equator in both oceans. In the summertime southern hemisphere the south-east trade wind belt is less well marked. The extratropical high pressure regions are extensive but, despite it being summer, high winds and significant wind stress exist in the midlatitude southern ocean. The north-east monsoon dominates the wind patterns in the Indian Ocean and the South

China Sea (where it is particularly strong). The ITCZ is a diffuse region south of the equator with relatively strong south-east trade winds in the eastern Indian Ocean.

In northern hemisphere summer (**Figure 3B**) the wind stresses in the midlatitude westerlies are very much decreased. Both the north-east and the south-east trade wind zones are evident respectively to the north and south of the ITCZ. This is predominantly north of the equator. The south-east trades are particularly strong in the Indian Ocean and feed into a very strong south-westerly monsoon flow in the Arabian Sea. The ship data indicate very strong winds in the Southern Ocean south west of Australia. These are also evident in satellite scatterometer data, which suggest that the winds in the Pacific sector of the Southern Ocean, while still strong, are somewhat less than those in the Indian Ocean sector. In contrast the ship data appear to show very light winds. The reason is that in wintertime there are practically no VOS observations in the far south Pacific. The analysis technique used to fill in the data gaps has, for want of other information, spread the light winds of the extratropical high pressure region farther south than is realistic; a good example of the care needed in interpreting the flux maps available in many atlases.

Regional and Seasonal Variation of the Heat Fluxes

The global distribution of the mean annual net heat flux is shown in **Figure 4A**. The accuracy and method of determination of such flux distributions will be discussed further below, here they will be used to give a qualitative description. Averaged over the year the ocean is heated in equatorial regions and loses heat in higher latitudes, particularly in the North Atlantic. However, this mean distribution is somewhat misleading, as the plots for January (**Figure 4B**) and July (**Figure 4C**) illustrate. The ocean loses heat over most of the extratropical winter hemisphere and gains heat in the extratropical summer hemisphere and in the tropics throughout the year. The relative magnitude of the individual flux components is illustrated in **Figure 5** for three representative sites in the North Atlantic Ocean. At the Gulf Stream site (**Figure 5A**) the large cooling in winter dominates the incoming solar radiation in the annual mean. However, even at this site the mean monthly short-wave flux in summer is greater than the cooling. Indeed the effect of the longer daylight periods increases the mean short-wave radiation to values similar to or larger than those observed in equatorial regions (**Figure 5C**). The midlatitude site (**Figure 5B**) is typical of large areas of the ocean. The ocean cools in winter and warms in summer, in each case by around $100 \, W \, m^{-2}$. The annual mean flux is small – around $10 \, W \, m^{-2}$ – but cannot be neglected because of the very large ocean areas involved. At this site, and generally over the ocean, this annual balance is between the sum of the latent heat flux and net long-wave flux which cool the ocean, and the net short-wave heating. Only in very cold air flows, as

over the Gulf Stream in winter, is the sensible heat flux significant.

As regards the interannual variation of the surface fluxes, the major large-scale feature over the global ocean is the El Niño-Southern Oscillation system in the equatorial Pacific Ocean. The changes in the net heat flux under El Niño conditions are around $40 \, W \, m^{-2}$ in the eastern equatorial Pacific. For extratropical and midlatitude regions the interannual variability of the summertime net heat flux is typically about $20–30 \, W \, m^{-2}$, being dominated by the variations in latent heat flux. In winter the typical variability increases to about $30–40 \, W \, m^{-2}$, although in particular areas (such as over the Gulf Stream) variations of up to $100 \, W \, m^{-2}$ can occur. The major spatial pattern of interannual variability in the North Atlantic is known as the North Atlantic Oscillation (NAO). This represents a measure of the degree to which mobile depressions, or alternatively near stationary high pressure systems, occur in the midlatitude westerly zone.

Accuracy of Flux Estimates

It has been shown that, although the individual flux components are of the order of hundreds of $W \, m^{-2}$, the net heat flux and its interannual variability over much of the world ocean is around tens of $W \, m^{-2}$. Furthermore it can be shown that a flux of $10 \, W \, m^{-2}$ over 1 year would, if stored in the top $500 \, m$ of the ocean, heat that entire layer by about $0.15°C$. Temperature changes on a decadal time scale are at most a few tenths of a degree, so the global mean budget must balance to better than a few $W \, m^{-2}$. For these various reasons there is a need to measure the flux components, which vary on many time and space scales, to an accuracy of a few $W \, m^{-2}$. Given the available data sources and methods of determining the fluxes described in the previous sections, it is not surprising that this level of accuracy cannot be achieved at present.

To take an example, in calculating the flux maps shown in **Figure 4** from VOS data many corrections were applied to the VOS observations to attempt to remove biases caused by the methods of observation. For example, air temperature measurements were corrected for the heat island caused by the ship heating up in sunny, low wind conditions. The wind speeds were adjusted depending on the anemometer heights on different ships. Corrections were applied to sea temperatures calculated from engine room intake data. Despite these and other corrections, the global annual mean flux showed about $30 \, W \, m^{-2}$ excess heating of the ocean. Previous climatologies

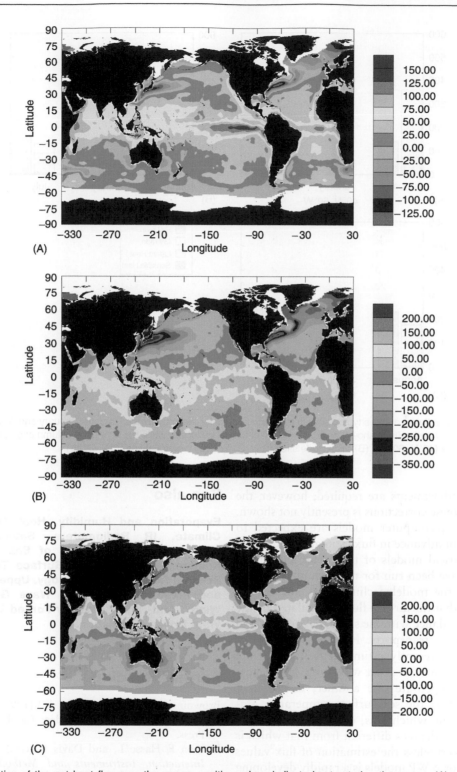

Figure 4 Variation of the net heat flux over the ocean, positive values indicate heat entering the ocean: (A) annual mean, (B) January monthly mean, (C) July monthly mean. (Adapted with permission from Josey SA, Kent EC and Taylor PK (1998) *The Southampton Oceanography Centre (SOC) Ocean–Atmosphere Heat, Momentum and Freshwater Flux Atlas.* SOC Report no. 6.)

calculated from ship data had shown similar biases and the fluxes had been adjusted to remove the bias, or to make the fluxes compatible with estimates of the meridional heat transport in the ocean. However, comparison of the unadjusted flux data with accurate data from air–sea interaction buoys showed good agreement between the two. This suggests that adjusting the fluxes globally is not correct and that

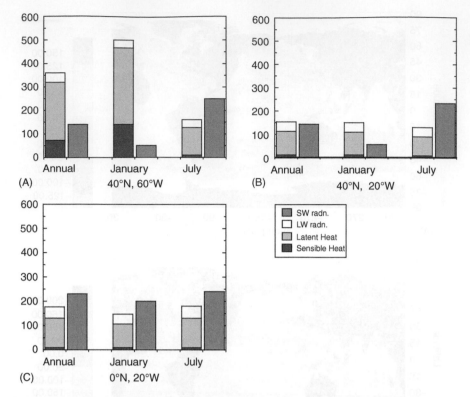

Figure 5 Mean heat fluxes at three typical sites in the North Atlantic for the annual mean, and the January and July monthly means. In each case the left-hand column shows the fluxes which act to cool the ocean while the right-hand column shows the solar heating. (A) Gulf Stream site (40°N, 60°W), (B) midlatitude site (40°N, 20°W), (C) equatorial site (0°N, 20°W).

regional flux adjustments are required; however, the exact form of these corrections is presently not shown.

In the future, computer models are expected to provide a major advance in flux estimation. Recently coupled numerical models of the ocean and of the atmosphere have been run for many simulated years during which the modeled climate has not drifted. This suggests that the air–sea fluxes calculated by the models are in balance with the simulated oceanic and atmospheric heat transports. However, it does not imply that the presently estimated flux values are realistic. Errors in the short-wave and latent heat fluxes may compensate one another; indeed in a typical simulation the sea surface temperature stabilized to a value which was, over large regions of the ocean, a few degrees different from that which is observed. Nevertheless the estimation of flux values using climate or NWP models is a rapidly developing field and improvements will doubtless occur in the next few years. There will be a continued need for *in situ* and satellite data for assimilation into the models and for model development and verification. However, it seems very likely that in future the most accurate routine source of the air–sea flux data will be from numerical models of the coupled ocean–atmosphere system.

See also

Evaporation and Humidity. Heat Transport and Climate. IR Radiometers. Satellite Passive-Microwave Measurements of Sea Ice. Satellite Remote Sensing of Sea Surface Temperatures. Sensors for Mean Meteorology. Upper Ocean Heat and Freshwater Budgets. Wave Generation by Wind. Wind- and Buoyancy-Forced Upper Ocean. Wind Driven Circulation.

Further Reading

Browning KA and Gurney RJ (eds.) (1999) *Global Energy and Water Cycles*. Cambridge: Cambridge University Press.

Dobson F, Hasse L, and Davis R (eds.) (1980) *Air–Sea Interaction, Instruments and Methods*. New York: Plenum Press.

Kraus EB and Businger JA (1994) *Atmosphere–Ocean Interaction*, 2nd edn. New York: Oxford University Press.

Meteorological Office (1978) *Meteorology for Mariners*, 3rd edn. London: HMSO.

Stull RB (1988) *An Introduction to Boundary Layer Meteorology*. Dordrecht: Kluwer Academic.

Wells N (1997) *The Atmosphere and Ocean: A Physical Introduction*, 2nd edn. London: Taylor and Francis.

SEA SURFACE EXCHANGES OF MOMENTUM, HEAT, AND FRESH WATER DETERMINED BY SATELLITE REMOTE SENSING

L. Yu, Woods Hole Oceanographic Institution, Woods Hole, MA, USA

Introduction

The ocean and the atmosphere communicate through the interfacial exchanges of heat, fresh water, and momentum. While the transfer of the momentum from the atmosphere to the ocean by wind stress is the most important forcing of the ocean circulation, the heat and water exchanges affect the horizontal and vertical temperature gradients of the lower atmosphere and the upper ocean, which, in turn, modify wind and ocean currents and maintain the equilibrium of the climate system. The sea surface exchanges are the fundamental processes of the coupled atmosphere–ocean system. An accurate knowledge of the flux variability is critical to our understanding and prediction of the changes of global weather and climate.

The heat exchanges include four processes: the short-wave radiation (Q_{SW}) from the sun, the outgoing long-wave radiation (Q_{LW}) from the sea surface, the sensible heat transfer (Q_{SH}) resulting from air–sea temperature differences, and the latent heat transfer (Q_{LH}) carried by evaporation of sea surface water. Evaporation releases both energy and water vapor to the atmosphere, and thus links the global energy cycle to the global water cycle. The oceans are the key element of the water cycle, because the oceans contain 96% of the Earth's water, experience 86% of planetary evaporation, and receive 78% of planetary precipitation.

The amount of air–sea exchange is called sea surface (or air–sea) flux. Direct flux measurements by ships and buoys are very limited. Our present knowledge of the global sea surface flux distribution stems primarily from bulk parametrizations of the fluxes as functions of surface meteorological variables that can be more easily measured (e.g., wind speed, temperature, humidity, cloud cover, precipitation, etc.). Before the advent of satellite remote sensing, marine surface weather reports collected from voluntary observing ships (VOSs) were the backbone for constructing the climatological state of the global flux fields. Over the past two decades, satellite remote sensing has become a mature technology for remotely sensing key air–sea variables. With continuous global spatial coverage, consistent quality, and high temporal sampling, satellite measurements not only allow the construction of air–sea fluxes at near-real time with unprecedented quality but most importantly, also offer the unique opportunity to view the global ocean synoptically as an entity.

Flux Estimation Using Satellite Observations

Sea Surface Wind Stress

The *Seasat-A* satellite scatterometer, launched in June 1978, was the first mission to demonstrate that ocean surface wind vectors (both speed and direction) could be remotely sensed by active radar backscatter from measuring surface roughness. Scatterometer detects the loss of intensity of transmitted microwave energy from that returned by the ocean surface. Microwaves are scattered by wind-driven capillary waves on the ocean surface, and the fraction of energy returned to the satellite (backscatter) depends on both the magnitude of the wind stress and the wind direction relative to the direction of the radar beam (azimuth angle). By using a transfer function or an empirical algorithm, the backscatter measurements are converted to wind vectors. It is true that scatterometers measure the effects of small-scale roughness caused by surface stress, but the retrieval algorithms produce surface wind, not wind stress, because there are no adequate surface-stress 'ground truths' to calibrate the algorithms. The wind retrievals are calibrated to the equivalent neutral-stability wind at a reference height of 10 m above the local-mean sea surface. This is the 10-m wind that would be associated with the observed surface stress if the atmospheric boundary layer were neutrally stratified. The 10-m equivalent neutral wind speeds differ from the 10-m wind speeds measured by anemometers, and these differences are a function of atmospheric stratification and are normally in the order of $0.2\,\text{m s}^{-1}$. To compute the surface wind stress, τ, the conventional bulk formulation is then employed:

$$\tau = (\tau_x, \tau_y) = \rho c_d W(u, v) \qquad [1]$$

where τ_x and τ_y are the zonal and meridional components of the wind stress; W, u, and v are the scatterometer-estimated wind speed at 10 m and its zonal component (eastward) and meridional component (northward), respectively. The density of surface air is given by ρ and is approximately equal to $1.225 \, \text{kg m}^{-3}$, and c_d is a neutral 10-m drag coefficient.

Scatterometer instruments are typically deployed on sun-synchronous near-polar-orbiting satellites that pass over the equator at approximately the same local times each day. These satellites orbit at an altitude of approximately 800 km and are commonly known as Polar Orbiting Environmental Satellites (POES). There have been six scatterometer sensors aboard POES since the early 1990s. The major characteristics of all scatterometers are summarized in **Table 1**. The first European Remote Sensing (ERS-1) satellite was launched by the European Space Agency (ESA) in August 1991. An identical instrument aboard the successor ERS-2 became operational in 1995, but failed in 2001. In August 1996, the National Aeronautics and Space Administration (NASA) began a joint mission with the National Space Development Agency (NASDA) of Japan to maintain continuous scatterometer missions beyond ERS satellites. The joint effort led to the launch of the NASA scatterometer (NSCAT) aboard the first Japanese Advanced Earth Observing Satellite (ADEOS-I). The ERA scatterometers differ from the NASA scatterometers in that the former operate on the C band ($\sim 5 \, \text{GHz}$), while the latter use the Ku band ($\sim 14 \, \text{GHz}$). For radio frequency band, rain attenuation increases as the signal frequency increases. Compared to C-band satellites, the higher frequencies of Ku band are more vulnerable to signal quality problems caused by rainfall. However, Ku-band satellites have the advantage of being more sensitive to wind variation at low winds and of covering more area.

Rain has three effects on backscatter measurements. It attenuates the radar signal, introduces volume scattering, and changes the properties of the sea surface and consequently the properties of microwave signals scattered from the sea surface. When the backscatter from the sea surface is low, the additional volume scattering from rain will lead to an overestimation of the low wind speed actually present. Conversely, when the backscatter is high,

attenuation by rain will reduce the signal causing an underestimation of the wind speed.

Under rain-free conditions, scatterometer-derived wind estimates are accurate within $1 \, \text{m s}^{-1}$ for speed and $20°$ for direction. For low (less than $3 \, \text{m s}^{-1}$) and high winds (greater than $20 \, \text{m s}^{-1}$), the uncertainties are generally larger. Most problems with low wind retrievals are due to the weak backscatter signal that is easily confounded by noise. The low signal/noise ratio complicates the ambiguity removal processing in selecting the best wind vector from the set of ambiguous wind vectors. Ambiguity removal is over 99% effective for wind speed of $8 \, \text{m s}^{-1}$ and higher. Extreme high winds are mostly associated with storm events. Scatterometer-derived high winds are found to be underestimated due largely to deficiencies of the empirical scatterometer algorithms. These algorithms are calibrated against a subset of ocean buoys – although the buoy winds are accurate and serve as surface wind truth, few of them have high-wind observations.

NSCAT worked flawlessly, but the spacecraft (ADEOS-I) that hosted it demised prematurely in June 1997 after only 9 months of operation. A replacement mission called QuikSCAT was rapidly developed and launched in July 1999. To date, QuikSCAT remains in operation, far outlasting the expected 2–3-year mission life expectancy. QuikSCAT carries a Ku-band scatterometer named SeaWinds, which has accuracy characteristics similar to NSCAT but with improved coverage. The instrument measures vector winds over a swath of 1800 km with a nominal spatial resolution of 25 km. The improved sampling size allows approximately 93% of the ocean surface to be sampled on a daily basis as opposed to 2 days by NSCAT and 4 days by the ERS instruments. A second similar-version SeaWinds instrument was placed on the ADEOS-II mission in December 2002. However, after only a few months of operation, it followed the unfortunate path of NSCAT and failed in October 2003 due – once again – to power loss.

The Advanced Scatterometer (ASCAT) launched by ESA/EUMETSAT in March 2007 is the most recent satellite designed primarily for the global measurement of sea surface wind vectors. ASCAT is flown on the first of three METOP satellites. Each METOP has a design lifetime of 5 years and thus, with overlap, the series has a planned duration of 14 years. ASCAT is similar to ERS-1/2 in configuration except that it has increased coverage, with two 500-km swaths (one on each side of the spacecraft nadir track).

The data collected by scatterometers on various missions have constituted a record of ocean vector winds for more than a decade, starting in August 1992. These satellite winds provide synoptic global

Table 1 Major characteristics of the spaceborne scatterometers

Characteristics	Scatterometer						
	SeaSat-A	ERS-1	ERS-2	NSCAT	SeaWinds on QuikSCAT	SeaWinds on ADEOS II	ASCAT
Operational frequency	Ku band	C band	C band	Ku band	Ku band	Ku band	C band
	14.6 GHz	5.255 GHz	5.255 GHz	13.995 GHz	13.402 GHz	13.402 GHz	5.255 GHz
Spatial resolution	50 km × 50 km with 100-km spacing	50 km × 50 km	50 km × 50 km	25 km × 25 km	25 km × 25 km	25 km × 6 km	25 km × 25 km
Scan characteristics	Two-sided, double 500-km swaths separated by a 450-km nadir gap	One-sided, single 500-km swath	One-sided, single 500-km swath	Two-sided, double 600-km swaths separated by a 329-km nadir gap	Conical scan, one wide swath of 1800 km	Conical scan, one wide swath of 1800 km	Two-sided, double 500-km swaths separated by a 700-km nadir gap
Daily coverage	Variable	41%	41%	77%	93%	93%	60%
Period in service	Jul. 1978–Oct. 1978	Aug. 1991–May. 1997	May. 1995–Jan. 2001	Sep. 1996–Jun. 1997	Jun. 1999–current	Dec. 2002–Oct. 2003	Mar. 2007–current

view from the vantage point of space, and provide excellent coverage in regions, such as the southern oceans, that are poorly sampled by the conventional observing network. Scatterometers have been shown to be the only means of delivering observations at adequate ranges of temporal and spatial scales and at adequate accuracy for understanding ocean–atmosphere interactions and global climate changes, and for improving climate predictions on synoptic, seasonal, and interannual timescales.

Surface Radiative Fluxes

Direct estimates of surface short-wave (SW) and long-wave (LW) fluxes that resolve synoptic to regional variability over the globe have only become possible with the advent of satellite in the past two decades. The surface radiation is a strong function of clouds. Low, thick clouds reflect large amounts of solar radiation and tend to cool the surface of the Earth. High, thin clouds transmit incoming solar radiation, but at the same time, they absorb the outgoing LW radiation emitted by the Earth and radiate it back downward. The portion of radiation, acting as an effective 'greenhouse gas', adds to the SW energy from the sun and causes an additional warming of the surface of the Earth. For a given cloud, its effect on the surface radiation depends on several factors, including the cloud's altitude, size, and the particles that form the cloud. At present, the radiative heat fluxes at the Earth's surface are estimated from top-of-the-atmosphere (TOA) SW and LW radiance measurements in conjunction with radiative transfer models.

Satellite radiance measurements are provided by two types of radiometers: scanning radiometers and nonscanning wide-field-of-view radiometers. Scanning radiometers view radiance from a single direction and must estimate the hemispheric emission or reflection. Nonscanning radiometers view the entire hemisphere of radiation with a roughly 1000-km field of view. The first flight of an Earth Radiation Budget Experiment (ERBE) instrument in 1984 included both a scanning radiometer and a set of nonscanning radiometers. These instruments obtain good measurements of TOA radiative variables including insolation, albedo, and absorbed radiation. To estimate surface radiation fluxes, however, more accurate information on clouds is needed.

To determine the physical properties of clouds from satellite measurements, the International Satellite Cloud Climatology Project (ISCCP) was established in 1983. ISCCP pioneered the cross-calibration, analysis, and merger of measurements from the international constellation of operational weather satellites. Using

geostationary satellite measurements with polar orbiter measurements as supplemental when there are no geostationary measurements, the ISCCP cloud-retrieval algorithm includes the conversion of radiance measurements to cloud scenes and the inference of cloud properties from the radiance values. Radiance thresholds are applied to obtain cloud fractions for low, middle, and high clouds based on radiance computed from models using observed temperature and climatological lapse rates.

In addition to the global cloud analysis, ISCCP also produces radiative fluxes (up, down, and net) at the Earth's surface that parallels the effort undertaken by the Global Energy and Water Cycle Experiment – Surface Radiation Budget (GEWEX-SRB) project. The two projects use the same ISCCP cloud information but different ancillary data sources and different radiative transfer codes. They both compute the radiation fluxes for clear and cloudy skies to estimate the cloud effect on radiative energy transfer. Both have a 3-h resolution, but ISCCP fluxes are produced on a 280-km equal-area (EQ) global grid while GEWEX-SRB fluxes are on a $1° \times 1°$ global grid. The two sets of fluxes have reasonable agreement with each other on the long-term mean basis, as suggested by the comparison of the global annual surface radiation budget in **Table 2**. The total net radiation differs by about $5 \, \text{W m}^{-2}$, due mostly to the SW component. However, when compared with ground-based observations, the uncertainty of these fluxes is about $10–15 \, \text{W m}^{-2}$. The main cause is the uncertainties in surface and near-surface atmospheric properties such as surface skin temperature, surface air and near-surface-layer temperatures and humidity, aerosols, etc. Further improvement requires improved retrievals of these properties.

In the late 1990s, the Clouds and the Earth's Radiant Energy System (CERES) experiment was developed by NASA's Earth Observing System (EOS)

Table 2 Annual surface radiation budget (in W m^{-2}) over global oceans. Uncertainty estimates are based on the standard error of monthly anomalies

Data 21-year mean 1984–2004	Parameter		
	SW Net downward	LW Net downward	SW+LW Net downward
ISCCP (Zhang et al., 2004)	173.2±9.2	−46.9±9.2	126.3±11.0
GEWEX-SRB (Gupta et al., 2006)	167.2±13.9	−46.3±5.5	120.9±11.9

not only to measure TOA radiative fluxes but also to determine radiative fluxes within the atmosphere and at the surface, by using simultaneous measurements of complete cloud properties from other EOS instruments such as the moderate-resolution imaging spectroradiometer (MODIS). CERES instruments were launched aboard the Tropical Rainfall Measuring Mission (TRMM) in November 1997, on the EOS *Terra* satellite in December 1999, and on the EOS *Aqua* spacecraft in 2002. There is no doubt that the EOS era satellite observations will lead to great improvement in estimating cloud properties and surface radiation budget with sufficient simultaneity and accuracy.

Sea Surface Turbulent Heat Fluxes

Latent and sensible heat fluxes are the primary mechanism by which the ocean transfers much of the absorbed solar radiation back to the atmosphere. The two fluxes cannot be directly observed by space sensors, but can be estimated from wind speed and sea–air humidity/temperature differences using the following bulk parametrizations:

$$Q_{LH} = \rho L_e c_e W (q_s - q_a) \qquad [2]$$

$$Q_{SH} = \rho c_p c_h W (T_s - T_a) \qquad [3]$$

where L_e is the latent heat of vaporization and is a function of sea surface temperature (SST, T_s) expressed as $L_e = (2.501 - 0.002\,37 \times T_s) \times 1.0^6$. c_p is the specific heat capacity of air at constant pressure; c_e and c_h are the stability- and height-dependent turbulent exchange coefficients for latent and sensible heat, respectively. T_a/q_a are the temperature/specific humidity at a reference height of 2 m above the sea surface. q_s is the saturation humidity at T_s, and is multiplied by 0.98 to take into account the reduction in vapor pressure caused by salt water.

The two variables, T_s and W, in eqns [2] and [3] are retrieved from satellites, and so q_s is known. The remote sensing of T_s is based on techniques by which spaceborne infrared and microwave radiometers detect thermally emitted radiation from the ocean surface. Infrared radiometers like the five-channel advanced very high resolution radiometer (AVHRR) utilize the wavelength bands at 3.5–4 and 10–12 μm that have a high transmission of the cloud-free atmosphere. The disadvantage is that clouds are opaque to infrared radiation and can effectively mask radiation from the ocean surface, and this affects the temporal resolution. Although the AVHRR satellite orbits the Earth 14 times each day from 833 km above its surface and each pass of the satellite

provides a 2399-km-wide swath, it usually takes 1 or 2 weeks, depending on the actual cloud coverage, to obtain a complete global coverage. Clouds, on the other hand, have little effect on the microwave radiometers so that microwave T_s retrievals can be made under complete cloud cover except for raining conditions. The TRMM microwave imager (TMI) launched in 1997 has a full suite of channels ranging from 10.7 to 85 GHz and was the first satellite sensor capable of accurately measuring SST through clouds. The low-inclination equitorial orbit, however, limits the TMI's coverage only up to *c.* 38° latitude. Following TMI, the first polar-orbiting microwave radiometer capable of measuring global through-cloud SST was made possible by the NASDA's advanced microwave scanning radiometer (AMSR) flown aboard the NASA's EOS Aqua mission in 2002.

While SST can be measured in both infrared and microwave regions, the near-surface wind speed can only be retrieved in the microwave region. The reason is that the emissivity of the ocean's surface at wavelengths of around 11 μm is so high that it is not sensitive to changes in the wind-induced sea surface roughness or humidity fluctuations in the lower atmosphere. Microwave wind speed retrievals are provided by the special sensor microwave/imager (SSM/I) that has been flown on a series of polar-orbiting operational spacecrafts of the Defense Meteorological Space Program (DMSP) since July 1987. SSM/I has a wide swath (∼ 1400 km) and a coverage of 82% of the Earth's surface within 1 day. But unlike scatterometers, SSM/I is a passive microwave sensor and cannot provide information on the wind direction. This is not a problem for the computation in eqns [2] and [3] that requires only wind speed observations. In fact, the high space-time resolution and good global coverage of SSM/I has made it serving as a primary database for computing the climate mean and variability of the oceanic latent and sensible heat fluxes over the past ∼20-year period. At present, wind speed measurements with good accuracy are also available from several NASA satellite platforms, including TMI and AMSR.

The most difficult problem for the satellite-based flux estimation is the retrieval of the air humidity and temperature, q_a and T_a, at a level of several meters above the surface. This problem is inherent to all spaceborne passive radiometers, because the measured radiation emanates from relatively thick atmospheric layers rather than from single levels. One common practice to extract satellite q_a is to relate q_a to the observed column integrated water vapor (IWV, also referred to as the total precipitable water) from SSM/I. Using IWV as a proxy for q_a is based on several observational findings that on monthly timescales the vertical distribution of water vapor is coherent throughout the entire atmospheric column. The approach, however, produces large systematic biases of over $2 \, \mathrm{g \, kg}^{-1}$ in the Tropics, as well as in the mid- and high latitudes during summertime. This is caused by the effect of the water vapor convergence that is difficult to assess in regions where the surface air is nearly saturated but the total IWV is small. Under such situations, the IWV cannot reflect the actual vertical and horizontal humidity variations in the atmosphere. Various remedies have been proposed to improve the q_a–IWV relation and to make it applicable on synoptic and shorter timescales. There are methods of including additional geophysical variables, replacing IWV with the IWV in the lower 500 m of the planetary layer, and/or using empirical orthogonal functions (EOFs). Although overall improvements were achieved, the accuracy remains poor due to the lack of detailed information on the atmospheric humidity profiles.

Retrieving T_a from satellite observations is even more challenging. Unlike humidity, there is no coherent vertical structure of temperature in the atmosphere. Satellite temperature sounding radiometers offer little help, as they generally are designed for retrieval in broad vertical layers. The sounder's low information content in the lower atmosphere does not enable the retrieval of near-surface air temperature with sufficient accuracy. Different methods have been tested to derive T_a from the inferred q_a, but all showed limited success. Because of the difficulties in determining q_a and T_a, latent and sensible fluxes estimated from satellite measurements have large uncertainties.

Three methods have been tested for obtaining better q_a and T_a to improve the estimates of latent and sensible fluxes. The first approach is to enhance the information on the temperature and moisture in the lower troposphere. This is achieved by combining SSM/I data with additional microwave sounder data that come from the instruments like the advanced microwave sounding unit (AMSU-A) and microwave humidity sounder (MHS) flown aboard the National Oceanic and Atmospheric Administration (NOAA) polar-orbiting satellites, and the special sensor microwave temperature sounder (SSM/T) and (SSM/T-2) on the DMSP satellites. Although the sounders do not directly provide shallow surface measurements, detailed profile information provided by the sounders can help to remove variability in total column measurements not associated with the surface. The second approach is to capitalize the progress made in numerical weather prediction models that assimilate sounder observations into the physically based system. The q_a and T_a estimates

from the models contain less ambiguity associated with the vertical integration and large spatial averaging of the various parameters, though they are subject to systematic bias due to model's subgrid parametrizations. The third approach is to obtain a better estimation of q_a and T_a through an optimal combination of satellite retrievals with the model outputs, which has been experimented by the Objectively Analyzed air–sea Fluxes (OAFlux) project at the Woods Hole Oceanographic Institution (WHOI). The effort has led to improved daily estimates of global air–sea latent and sensible fluxes.

Freshwater Flux

The freshwater flux is the difference between precipitation (rain) and evaporation. Evaporation releases both water vapor and latent heat to the atmosphere. Once latent heat fluxes are estimated, the sea surface evaporation (E) can be computed using the following relation:

$$E = Q_{\mathrm{LH}} / \rho_w L_e \qquad [4]$$

where Q_{LH} denotes latent heat flux and ρ_w is the density of seawater.

Spaceborne sensors cannot directly observe the actual precipitation reaching the Earth's surface, but they can measure other variables that may be highly correlated with surface rainfall. These include variations in infrared and microwave brightness temperatures, as well as visible and near-infrared albedo. Infrared techniques are based on the premise that rainfall at the surface is related to cloud-top properties observed from space. Visible/infrared observations supplement the infrared imagery with visible imagery during daytime to help eliminate thin cirrus clouds, which are cold in the infrared imagery and are sometimes misinterpreted as raining using infrared data alone. Visible/infrared sensors have the advantage of providing good space and time sampling, but have difficulty capturing the rain from warm-topped clouds. By comparison, microwave (MW) estimates are more physically based and more accurate although time and space resolutions are not as good. The principle of MW techniques is that rainfall at the surface is related to microwave emission from rain drops (low-frequency channels) and microwave scattering from ice (high-frequency channels). While the primary visible/infrared data sources are the operational geostationary satellites, microwave observations are available from SSM/I, the NOAA AMSU-B, and the TRMM spacecraft.

TRMM opened up a new era of estimating not only surface rainfall but also rain profiles. TRMM is equipped with the first spaceborne precipitation radar (PR) along with a microwave radiometer (TMI) and a visible/infrared radiometer (VIRS). Coincident measurements from the three sensors are complementary. PR provides detailed vertical rain profiles across a 215-km-wide strip. TMI (a five-frequency conical scanning radiometer) though has less vertical and horizontal fidelity in rain-resolving capability, and it features a swath width of 760 km. The VIRS on TRMM adds cloud-top temperatures and structures to complement the description of the two microwave sensors. While direct precipitation information from VIRS is less reliable than that obtained by the microwave sensors, VIRS serves an important role as a bridge between the high-quality but infrequent observations from TMI and PR and the more available data and longer time series data available from the geostationary visible/infrared satellite platforms.

The TRMM satellite focuses on the rain variability over the tropical and subtropical regions due to the low inclination. An improved instrument, AMSR, has extended TRMM rainfall measurements to higher latitudes. AMSR is currently aboard the *Aqua* satellite and is planned by the Global Precipitation Measurement (GPM) mission to be launched in 2009. Combining rainfall estimates from visible/infrared with microwave measurements is being undertaken by the Global Climatology Project (GPCP) to produce global precipitation analyses from 1979 and continuing.

Summary and Applications

The satellite sensor systems developed in the past two decades have provided unprecedented observations of geophysical parameters in the lower atmosphere and upper oceans. The combination of measurements from multiple satellite platforms has demonstrated the capability of estimating sea surface heat, fresh water, and momentum fluxes with sufficient accuracy and resolution. These air–sea flux data sets, together with satellite retrievals of ocean surface topography, temperature, and salinity (**Figure 1**), establish a complete satellite-based observational infrastructure for fully monitoring the ocean's response to the changes in air–sea physical forcing.

Atmosphere and the ocean are nonlinear turbulent fluids, and their interactions are nonlinear scale-dependent, with processes at one scale affecting processes at other scales. The synergy of various satellite-based products makes it especially advantageous to study the complex scale interactions between the atmosphere and the ocean. One clear example is the satellite monitoring of the development of the El Niño–Southern Oscillation (ENSO) in 1997–98. ENSO is the largest source of interannual variability in the global climate system. The phenomenon is characterized by the appearance of extensive warm surface water over the central and eastern tropical Pacific Ocean at a frequency of *c.* 3–7 years. The 1997–98 El Niño was one of the most severe events experienced during the twentieth century. During the

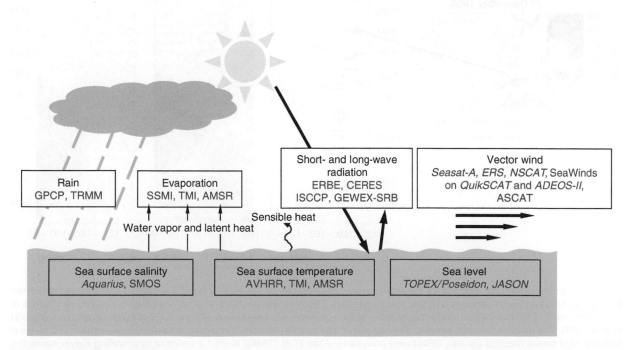

Figure 1 Schematic diagram of the physical exchange processes at the air–sea interface and the upper ocean responses, with corresponding sensor names shown in red.

peak of the event in December 1997 (**Figure 2**), the SST in the eastern equatorial Pacific was more than 5 °C above normal, and the warming was accompanied by excessive precipitation and large net heat transfer from the ocean to the atmosphere.

The 1997–98 event was also the best observed thanks largely to the expanded satellite-observing capability. One of the major observational findings was the role of synoptic westerly wind bursts (WWBs) in the onset of El Niño. **Figure 3** presents the evolution of zonal wind from NSCAT scatterometer combined with SSM/I-derived wind product, sea surface height (SSH) from TOPEX altimetry, and SST from AVHRR imagery in 1996–98. The appearance of the anomalous warming in the eastern basin in February 1997 coincided with the arrival of the downwelling Kelvin waves generated by the WWB of December 1996 in the western Pacific. A series of subsequent WWB-induced Kelvin waves further enhanced the eastern warming, and fueled

the El Niño development. The positive feedback between synoptic WWB and the interannual SST warming in making an El Niño is clearly indicated by satellite observations. On the other hand, the synoptic WWB events were the result of the development of equatorial twin cyclones under the influence of northerly cold surges from East Asia/western North Pacific. NSCAT made the first complete recording of the compelling connection between near-equatorial wind events and mid-latitude atmospheric transient forcing.

Clearly, the synergy of various satellite products offers consistent global patterns that facilitate the mapping of the correlations between various processes and the construction of the teleconnection pattern between weather and climate anomalies in one region and those in another. The satellite observing system will complement the *in situ* ground observations and play an increasingly important role in understanding the cause of global climate changes

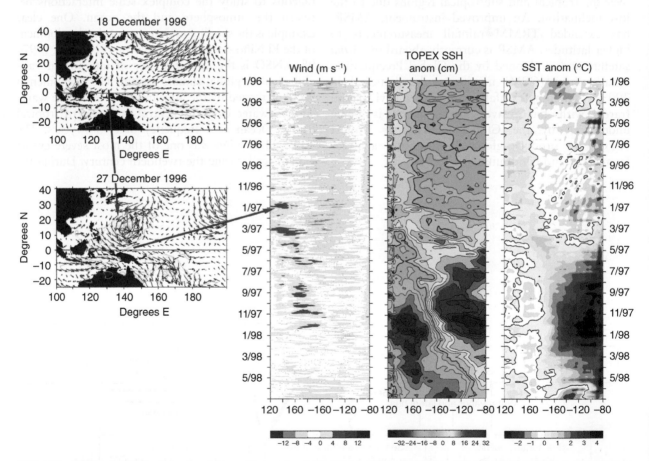

Figure 2 (First column) An example of the scatterometer observations of the generation of the tropical cyclones in the western tropical Pacific under the influence of northerly cold surges from East Asia/western North Pacific. The effect of westerly wind bursts on the development of El Niño is illustrated in the evolution of the equatorial sea level observed from *TOPEX/Poseidon* altimetry and SST from AVHRR. The second to fourth columns show longitude (horizontally) and time (vertically, increasing downwards). The series of westerly wind bursts (second column, SSM/I wind analysis by Atlas *et al.* (1996)) excited a series of downwelling Kelvin waves that propagated eastward along the equator (third column), suppressed the thermocline, and led to the sea surface warming in the eastern equatorial Pacific (fourth column).

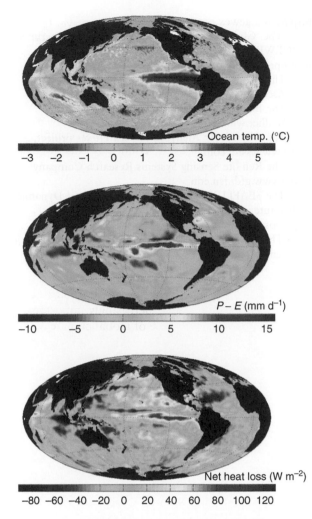

Figure 3 Satellite-derived global ocean temperature from AVHRR (top), precipitation minus evaporation from GPCP and WHOI OAFlux, respectively (middle), and net heat loss ($Q_{LH} + Q_{SH} + Q_{LW} - Q_{SW}$) from the ocean (bottom) during the El Niño in Dec. 1997. The latent and sensible heat fluxes $Q_{LH} + Q_{SH}$ are provided by WHOI OAFlux, and the short- and long-wave radiative fluxes are by ISCCP.

and in improving the model skills on predicting weather and climate variability.

Nomenclature

c_d	drag coefficient
c_e	turbulent exchange coefficient for latent heat
c_h	turbulent exchange coefficient for sensible heat
c_p	specific heat capacity of air at constant pressure
E	evaporation
L_e	latent heat of vaporization
q_a	specific humidity at a reference height above the sea surface
q_s	specific humidity at the sea surface
Q_{LH}	latent heat flux
Q_{SH}	sensible heat flux
T_a	temperature at a reference height above the sea surface
T_s	temperature at the sea surface
u	zonal component of the wind speed
v	meridional component of the wind speed
W	wind speed
ρ	density of surface air
ρ_w	density of sea water
τ	wind stress
τ_x	zonal component of the wind stress
τ_y	meridional component of the wind stress

See also

Air–Sea Gas Exchange. Evaporation and Humidity. Heat and Momentum Fluxes at the Sea Surface. Heat Transport and Climate. IR Radiometers. Satellite Oceanography, History, and Introductory Concepts. Satellite Remote Sensing of Sea Surface Temperatures. Satellite Remote Sensing: Salinity Measurements. Upper Ocean Heat and Freshwater Budgets. Wind- and Buoyancy-Forced Upper Ocean.

Further Reading

Adler RF, Huffman GJ, Chang A, *et al.* (2003) The Version 2 Global Precipitation Climatology Project (GPCP) monthly precipitation analysis (1979–present). *Journal of Hydrometeorology* 4: 1147–1167.

Atlas R, Hoffman RN, Bloom SC, Jusem JC, and Ardizzone J (1996) A multiyear global surface wind velocity dataset using SSM/I wind observations. *Bulletin of the American Meteorological Society* 77: 869–882.

Bentamy A, Katsaros KB, Mestas-Nuñez AM, *et al.* (2003) Satellite estimates of wind speed and latent heat flux over the global oceans. *Journal of Climate* 16: 637–656.

Chou S-H, Nelkin E, Ardizzone J, Atlas RM, and Shie C-L (2003) Surface turbulent heat and momentum fluxes over global oceans based on the Goddard satellite retrievals, version 2 (GSSTF2). *Journal of Climate* 16: 3256–3273.

Gupta SK, Ritchey NA, Wilber AC, Whitlock CH, Gibson GG, and Stackhouse RW, Jr. (1999) A climatology of surface radiation budget derived from satellite data. *Journal of Climate* 12: 2691–2710.

Liu WT and Katsaros KB (2001) Air–sea flux from satellite data. In: Siedler G, Church J, and Gould J (eds.) *Ocean Circulation and Climate*, pp. 173–179. New York: Academic Press.

Kubota M, Iwasaka N, Kizu S, Konda M, and Kutsuwada K (2002) Japanese Ocean Flux Data Sets with Use of

Remote Sensing Observations (J-OFURO). *Journal of Oceanography* 58: 213–225.

Wentz FJ, Gentemann C, Smith D, and Chelton D (2000) Satellite measurements of sea surface temperature through clouds. *Science* 288: 847–850.

Yu L and Weller RA (2007) Objectively analyzed air–sea heat fluxes (OAFlux) for the global oceans. *Bulletin of the American Meteorological Society* 88: 527–539.

Zhang Y-C, Rossow WB, Lacis AA, Oinas V, and Mishchenko MI (2004) Calculation of radiative fluxes from the surface to top of atmosphere based on ISCCP and other global data sets: Refinements of the radiative transfer model and the input data. *Journal of Geophysical Research* 109: D19105 (doi:10.1029/2003JD004457).

Relevant Websites

http://winds.jpl.nasa.gov
 – Measuring Ocean Winds from Space.

http://www.gewex.org
 – The Global Energy and Water Cycle Experiment (GEWEX).

http://precip.gsfc.nasa.gov
 – The Global Precipitation Climatology Project.

http://isccp.giss.nasa.gov
 – The International Cloud Climatology Project.

http://oaflux.whoi.edu
 – The Objectively Analyzed air–sea Fluxes project.

http://www.ssmi.com
 – The Remote Sensing Systems Research Company

http://www.gfdi.fsu.edu
 – The SEAFLUX Project, Geophysical Fluid Dynamics Institute.

http://eosweb.larc.nasa.gov
 – The Surface Radiation Budget Data, Atmospheric Science Data Center.

HEAT TRANSPORT AND CLIMATE

H. L. Bryden, University of Southampton,
Southampton, UK

Introduction: The Global Heat Budget

The Earth receives energy from the sun (**Figure 1**)
principally in the form of short-wave energy (sunlight). The amount of solar radiation is quantified by
the 'solar constant' which satellite radiometers have
measured since about 1985 to have a mean value of
$1366 \, \mathrm{W \, m^{-2}}$, an 11-year sunspot cycle of amplitude
$1.5 \, \mathrm{W \, m^{-2}}$, and a maximum at maximum sunspot
activity. A fraction of the sunlight is reflected directly
back into space and this fraction is termed the albedo. Brighter areas like snow in polar regions have
high albedo (0.8) reflecting most of the short-wave
radiation back to space, while darker areas like the
ocean have small albedo (0.05) and small reflection.
Averaged over the Earth's surface, the albedo is
about 0.3. Overall, the net incoming short-wave radiation (incoming minus reflected) peaks in equatorial regions and decreases to small values in polar
latitudes (**Figure 2**).

The Earth radiates energy back to space in the
form of long-wave, black-body radiation proportional to the fourth power of the absolute temperature at the top of the atmosphere. Because the
temperature at the top of the atmosphere is relatively
uniform with latitude varying only from 200 to
230 K, there is only a small latitudinal variation in
outgoing radiation (**Figure 1**). Over a year, the net
incoming radiation equals the net outgoing radiation
within our ability to measure the radiation, thus
maintaining the overall heat balance of the Earth.
For the radiation budget as a function of latitude,
however, there is more incoming short-wave radiation at equatorial and tropical latitudes and more
outgoing long-wave radiation at polar latitudes. To
maintain this heating–cooling distribution, the atmosphere and ocean must transport energy poleward
from the Tropics toward the Pole and the maximum
poleward energy transport in each hemisphere occurs
at a latitude of *c.* 35°, where there is a change
from net incoming to net outgoing radiation, and
the maximum has a magnitude of about 5.8 PW
(1 petawatt (PW) = 10^{15} W).

As recently as the mid-1990s, it was controversial
whether the ocean or the atmosphere was responsible
for the majority of the energy transport. Oceanographers found a maximum ocean heat transport of
about 2 PW at 25–30° N, while meteorologists reported a maximum atmospheric transport of about
2.5 PW from the analysis of global radiosonde network observations. Thus, there was a missing petawatt of energy transport in the Northern Hemisphere
for the combined ocean–atmosphere system. Recent
analyses combining observations and models suggest
that the atmosphere does carry the additional petawatt that observational analyses alone could not find

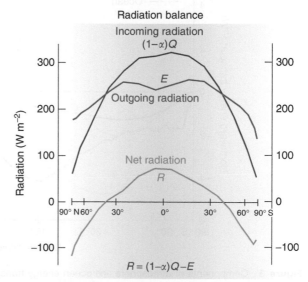

Figure 2 Latitudinal profiles of net incoming short-wave
radiation, outgoing long-wave radiation, and the net radiative
heating of the Earth. Note the latitudinal scale is stretched so that
it is proportional to the surface area of the Earth.

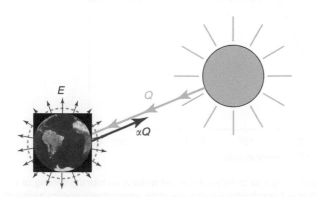

Figure 1 Schematic of the Earth's radiation budget. Q is
incoming, short-wave solar radiation; αQ is reflected solar
radiation where α is albedo, and E is outgoing, long-wave
black-body radiation.

due to a sparsity of radiosonde observations over the ocean. Now it is generally accepted that the atmosphere carries the majority of energy transport at 35° N, though some oceanographers point out that half of the maximum atmospheric transport is due to latent heat (water vapor) transport that can be considered to be a joint ocean–atmosphere process (**Figure 3**).

Here we will concentrate on ocean heat transport, especially on the mechanisms of ocean heat transport in the Northern Hemisphere. Conservation of energy in the ocean is effectively expressed as conservation of heat, where heat is defined to be $\rho C_p \Theta$, where ρ is seawater density, C_p is specific heat of seawater at constant pressure, and Θ is potential temperature, the temperature of a water parcel brought adiabatically (without heat exchange) to the sea surface from depth. Because ρC_p is nearly constant at about $4.08 \times 10^6 \, \mathrm{J\,m^{-3}\,{}^\circ C^{-1}}$, heat conservation is essentially expressed as conservation of potential temperature. There are many subtleties to the definitions that are described in entries for density, potential temperature, and heat, but here we use traditional definitions of potential temperature, density, and specific heat based on the internationally recognized equation of state for seawater.

Ocean heat transport is then the flow of heat through the ocean, $\rho C_p \Theta v$, where v is the water velocity. Such definition depends on the temperature scale and has little meaning until it is considered for a given volume of the ocean. Because of mass conservation, there is no net mass transport into or out of a fixed ocean volume over long timescales (neglecting the relatively miniscule contributions from evaporation minus precipitation), so it is the ocean heat transport convergence that is meaningful, that is the amount of heat transport into the volume minus the heat transport out of the volume. Since mass is conserved, the heat transport convergence is proportional to the mass transport times the difference in temperature between the inflow and the outflow across the boundaries of the volume.

For a complete latitude band like 25° N, where the Atlantic Ocean and Pacific Ocean volume north of 25° N can be considered to be an enclosed ocean, the heat transport convergence is commonly referred to as the ocean heat transport at 25° N. Individually the Atlantic and Pacific Oceans are nearly enclosed with only a small throughflow connecting them in Bering Straits, so the Atlantic heat transport at 25° N is commonly referred to, even though it is formally the heat transport convergence between 25° N and Bering Straits and similarly for Pacific heat transport at 25° N. Such definitions of heat transport are generally used throughout the Atlantic north of 30° S and the Pacific north of about 10° N, where each ocean

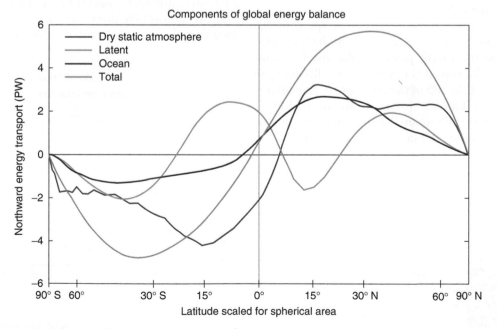

Figure 3 Components of atmosphere and ocean energy transports required to balance the net radiational heating/cooling of the Earth following **Figure 2**. The standard atmospheric energy transport is here divided into the dry static atmospheric energy transport and the latent heat transport. Latent heat transport is fundamentally a joint atmosphere–ocean process since the atmospheric water vapor transport is balanced by an opposing oceanic freshwater transport. The ocean heat transport is determined by integrating over the oceans the spatial distribution of atmosphere–surface heat exchange calculated by subtracting the atmospheric energy transport divergence from the radiative heating at the top of the atmosphere.

basin is closed except for the small Bering Straits transport.

Air–Sea Heat Exchange

Conservation of heat means that any convergence of ocean heat transport is balanced by heat loss to the atmosphere (the amount of heat gain or loss through the ocean bottom is small in comparison to exchanges with the atmosphere). Thus charts of air–sea energy exchange are primary sources for our understanding of ocean heat transport, where it occurs and how big it is. Estimates of air–sea energy exchange have long been made based on measurements of cloud cover, surface air and water temperatures, wind speed, humidity, and bulk formula exchange coefficients to calculate the size of the radiational heating and latent heat cooling of the ocean surface and the sensible heat exchange between the ocean and atmosphere. Combining such shipboard observations on a global scale produces air–sea flux climatologies giving air–sea exchange by month and region. From such climatologies, the global distribution of annual averaged air–sea energy exchange (**Figure 4**) shows that the ocean gains heat over much of the equatorial and tropical regions and gives up large amounts of heat over the warm poleward flowing western boundary currents like the Gulf Stream, Kuroshio or Agulhas Current, and over open-water subpolar and polar regions. Ocean heat transport convergence for any arbitrary ocean volume can technically be estimated by summing up the air–sea energy exchange over the surface area of the ocean volume. There is a problem, however, in that the air–sea energy exchange estimates have an uncertainty of about $30 \, \text{W m}^{-2}$. One way to determine this uncertainty is to sum the air–sea exchanges globally and to find that there is on average a heat gain by the ocean of $30 \, \text{W m}^{-2}$, in each of the two state-of-the-art air–sea exchange climatologies. It is of course possible to remove this bias, either uniformly, by region or by component (radiative, latent, or sensible heat exchange); despite careful comparison with buoy measurements with bulk formula estimates at several oceanic locations, there is no consensus on how to remove the $30 \, \text{W m}^{-2}$ uncertainty in air–sea energy exchange.

Distribution of Ocean Heat Transport

Estimating ocean heat transport convergence from *in situ* oceanographic measurements is a second method to quantify the role of the ocean in the global heat balance. The advantage of this direct approach is that the mechanisms of ocean heat transport are examined rather than just their overall effect in terms of the air–sea exchange. It was first reliably applied at 25°N in the Atlantic, a latitude where the warm

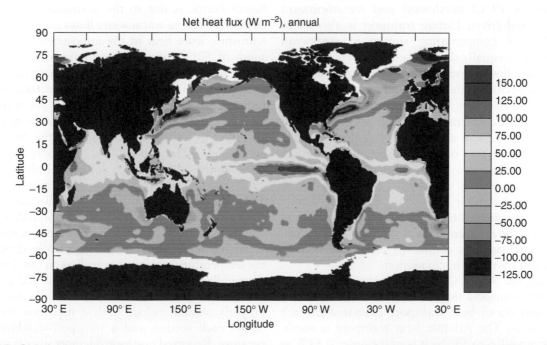

Net heat flux (W m⁻²), annual

Figure 4 Global distribution of the annual mean net heat gain by the ocean as determined from bulk formula calculations. Positive values indicate a gain of heat by the oceans. SOC climatology – Josey SA, Kent EC, and Taylor PK (1999) New insights into the ocean heat budget closure problem from analysis of the SOC air–sea flux climatology. *Journal of Climate* 12: 2856–2880.

northward Gulf Stream flow of about 30 Sv ($1\,\text{Sv} = 1 \times 10^6\,\text{m}^3\,\text{s}^{-1}$) through Florida Straits is regularly monitored by submarine telephone cable voltage (the varying flow of conducting seawater through a magnetic field produces varying voltage in the cable which is continuously measured). This is the latitude of the easterly trade winds whose westward wind stress generates a northward surface water transport of about 4 Sv in the Ekman layer. Because the Atlantic is closed to the north except for a small (1 Sv) flow from the Pacific through Bering Straits, a nearly equal amount of southward flow of 35 Sv must cross the mid-ocean section between the Bahamas and Africa. The vertical distribution of this southward return flow (and its temperature) has been estimated from transatlantic hydrographic sections where temperature and salinity profiles are used to derive geostrophic velocity profiles and the reference-level velocity for the geostrophic velocity profiles is set to make the southward geostrophic transport equal to the northward Gulf Stream plus Ekman transport.

Analysis of three hydrographic sections along 25° N made in 1957, 1981, and 1992 suggests that the ocean heat transport is 1.3 PW. Estimates of the error in heat transport using this direct method are about 0.3 PW, implying that direct ocean heat transport estimates are more accurate than air–sea flux estimates for areas larger than 30° latitude × 30° longitude (approximately, $30\,\text{W m}^{-2} \times 3100\,\text{km} \times 3100\,\text{km} = 0.3\,\text{PW}$). In terms of mechanisms, the Gulf Stream carries warm water (transport-weighted temperature of 19 °C) northward and the northward surface wind-driven Ekman transport is also warm (25 °C). The compensating southward mid-ocean flow is about equally divided between a recirculating thermocline flow above 800 m depth with an average temperature close to the 19 °C Gulf Stream flow and a cold deep-water flow with an average temperature less than 3 °C. Overall then, the northward heat transport of 1.3 PW across 25° N is due to a net northward transport of about 18 Sv of warm upper layer waters balanced by a net southward transport of cold deep waters. Deep-water formation in the Nordic and Labrador Seas of the northern Atlantic connects the northward flowing warm waters and southward flowing cold deep waters. This overall vertical circulation is commonly called the Atlantic meridional overturning circulation.

During the World Ocean Circulation Experiment 1990–99, transoceanic hydrographic sections were made and ocean heat transport estimated in each ocean basin. The Atlantic heat transport is northward from 40° S to 55° N; it is of the order 0.5 PW at the southern boundary of the Atlantic, increases in the tropical regions as heat is gained at the sea

surface, reaches a maximum near 25° N, and then decreases northward as the ocean gives up heat to the atmosphere (**Figure 5**). At all latitudes, the northward heat transport is due to the meridional overturning circulation where the cold deep-water transport of *c.* 15–20 Sv persists through the Atlantic and the compensating northward upper water flow changes temperature, warming in tropical regions and then cooling in northern subtropical and subpolar latitudes.

The North Pacific, north of 10° N where the basin is essentially closed, exhibits a similar pattern of northward heat transport, but the heat transport achieves a maximum of only 0.8 PW at 25° N, 50% less than the North Atlantic despite the Pacific being more than twice as wide as the Atlantic. The Pacific is different from the Atlantic in that there is no substantial deep meridional overturning circulation, for there is no deep-water formation in the North Pacific. The Pacific heat transport is due to a horizontal circulation where warm waters flow northward in the Kuroshio western boundary current and then recirculate southward over the vast zonal extent of the Pacific, all at depths shallower than 1000 m. The Kuroshio has about the same size and temperature structure as the Gulf Stream, but the zonal temperature distribution in the Pacific exhibits much colder upper water temperatures in mid-ocean and particularly along the eastern boundary of western North America than does the Atlantic along Europe and Africa. Thus, the northward heat transport in the North Pacific is due to the horizontal upper water circulation where warm water flows northward in the Kuroshio, loses heat to the atmosphere at latitudes south of 50° N, and then recirculates southward at colder temperatures in the mid- and eastern Pacific.

South of 10° N in the Pacific and throughout the Indian Ocean, ocean heat transports are somewhat ambiguous because of the throughflow from the Pacific to the Indian Ocean through the Indonesian archipelago. The throughflow transport is not yet well defined but there is a substantial observational effort now underway to quantify its mass transport and associated temperature structure. Because mass is not conserved for individual Pacific or Indian zonal sections south of 10° N, heat transports across such zonal sections in the literature generally depend on the temperature scale used as well as on the size of the throughflow assumed. These fluxes should properly be called temperature transports and to be interpreted properly should include both a net mass transport across each section and a transport-weighted temperature. Reported northward temperature transports in the South Pacific are mainly due to a net northward transport (to balance the throughflow) times

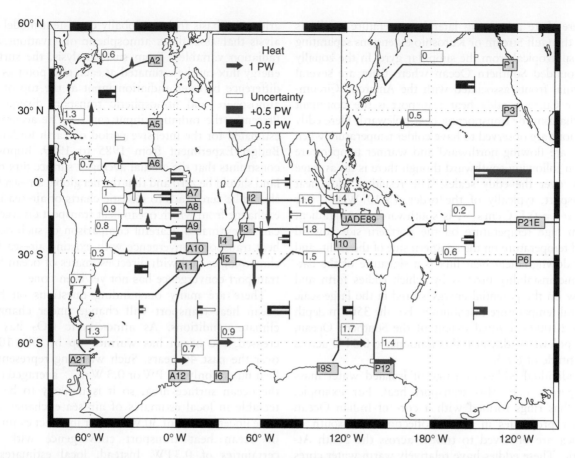

Figure 5 Heat transport and temperature transport across hydrographic sections taken during the World Ocean Circulation Experiment. For the Atlantic north of 40°S and the Pacific north of 10°N, heat transport values are presented where there is no net mass transport across the sections. For the remaining sections, temperature transports are presented where there is a net mass transport across the section and the temperature transport includes the net transport multiplied by temperature in degrees Celsius.

temperature in degrees Celsius. Similarly, southward temperature transports in the south Indian Ocean tend to be large because they include a substantial net southward mass transport (to balance the through-flow) times an average temperature. Careful ocean heat transport divergence estimates for the Indian Ocean taking into account the throughflow transport and temperature generally suggest that the Indian Ocean gains heat from the atmosphere between 0.4 and 1.2 PW depending on the size and temperature of the throughflow. For the South Pacific, the ocean heat transport convergence or divergence is ambiguous for a normal range of throughflow transport, so it is uncertain whether the South Pacific as a whole gains heat or loses heat to the atmosphere.

It is possible to combine the South Pacific and Indian sections to enclose a confined ocean volume north of say 32°S. For such combinations where mass is conserved, the Indo-Pacific Ocean heat transport at 32°S is meaningful and estimates are that it is southward with a magnitude of 0.4–1.2 PW. For the complete latitude band at about 30°S

combining Indo-Pacific and Atlantic Ocean heat transports, the ocean heat transport is southward or poleward but with only a maximum value of about 0.5 PW since the northward Atlantic heat transport partially cancels the southward Indo-Pacific heat transport. Thus, it appears that the ocean contributes much less than the atmosphere to the poleward heat transport in the Southern Hemisphere required to balance the Earth's radiation budget.

Eddy Heat Transport

In addition to heat transport by the steady ocean circulation, temporally and spatially varying currents with scales of 10–100 days and 10–100 km, which are called mesoscale features or eddies, can also transport heat. Correlations between time-varying velocity and temperature fluctuations can transport heat, $\rho C_p \langle \Theta' \upsilon' \rangle$ (where primes denote fluctuations and angular brackets indicate time averages), even when there is no net velocity or mass transport, that is, $\langle \upsilon \rangle$ is zero. Such eddy heat transport can be substantial in regions where there are strong lateral temperature gradients like the Gulf Stream or Kuroshio extensions separating the subtropical from the subpolar gyre. In the zonally unbounded Southern Ocean where there are several thermal fronts associated with the Antarctic Circumpolar Current, eddy heat transport is the dominant mechanism for transporting heat poleward. Here eddy motions are observed to have colder temperature when they are flowing northward and warmer temperature when following southward though there is no average flow over the eddy scales. The resulting eddy heat transport, typically of the order $4 \times 10^3 \, \mathrm{W \, m^{-2}}$ for $\langle \Theta' v' \rangle = 0.1 \, ^\circ \mathrm{C \, cm \, s^{-1}}$, is southward, downgradient from high temperature on the northern side toward cold temperature on the southern side of the front, and this downgradient heat flux is a signature of the baroclinic instability process by which eddies form and grow on the potential energy stored in the large-scale lateral temperature distribution. For the 3500 m depth and 20 000 km zonal extent of the Southern Ocean, this poleward eddy heat flux amounts to 0.3 PW across a latitude of 60° S.

Individual eddies or rings of isolated water mass properties may also transport heat. For example, Agulhas rings formed with a core of Indian Ocean water properties in the retroflection area south of Africa are observed to transit across the South Atlantic. These eddies have relatively warm water cores and their heat transport is often estimated by multiplying their heat content anomaly by an estimated number of how many such rings are formed each year. Such calculation is somewhat ambiguous because it is not clear how the mass is returned and what its temperature is. Similar estimates have been made with Gulf Stream and Kuroshio rings, both warm core and cold core, and with meddies formed from the outflow of Mediterranean water. While individual rings are impressive, it is not yet clear whether they carry a significant amount of heat compared with the annual averaged air–sea exchange in any region.

Future Developments

There is a third method, the residual method, for estimating ocean heat transport that takes the difference between the energy transport required to maintain the Earth's radiation budget and the atmospheric energy transport to define the ocean heat transport as a residual. In its original implementation, the residual method could only be applied to estimate zonally averaged ocean heat transport into the polar cap north of any given latitude because atmosphere energy transport could only be determined for a complete zonal integral. In a recent development, the atmospheric energy transport divergence is estimated on a

grid point basis from a globally consistent model analysis that assimilates atmospheric observations and radiation variables. From such analysis, the surface energy flux can be estimated at each grid point as the difference between radiation input at the top of the atmosphere and atmospheric transport divergence. Presently, the radiation input can only be accurately estimated for the intensive period of Earth Radiation Budget Experiment from 1985 to 1989. Imposing constraints that the annual averaged surface flux over land should be zero and that the net global air–sea flux should be zero leads to realistic charts of air–sea heat exchange from which ocean heat transport divergence can be estimated. Careful comparison of such ocean heat transport convergence with existing air–sea flux climatologies and with direct estimates of ocean heat transport convergence has not yet been done.

There are many outstanding questions on how ocean heat transport will change under changing climate conditions. As atmospheric CO_2 has increased, the ocean has warmed up by $14 \times 10^{22} \, \mathrm{J}$ over the past 40 years. Such warming represents a heat flux of only 0.1 PW or $0.3 \, \mathrm{W \, m^{-2}}$ averaged over the ocean surface area, so it is unlikely to be detectable in local estimates of air–sea exchange that have uncertainties of $30 \, \mathrm{W \, m^{-2}}$ or in direct estimates of ocean heat transport convergence with uncertainties of 0.3 PW. Instead, local estimates of ocean heat content change over decadal timescales provide a sensitive estimate of how the difference between ocean heat transport convergence and air–sea exchange is changing in a changing climate.

There may be changes in ocean circulation that will lead to measurable changes in air–sea heat exchange and ocean heat transport. For example, most coupled climate models predict that the Atlantic meridional overturning circulation will slow down by order of 50% over the next century as atmospheric CO_2 increases. Because Atlantic heat transport is presently closely related to the strength of the meridional overturning circulation, Atlantic ocean heat transport could reduce measurably. In addition, the absence of a meridional overturning circulation in coupled climate models leads to much colder (10 °C lower) temperatures in the northern Atlantic that greatly reduce the amount of heat given up by the ocean to the atmosphere in northern latitudes. In fact, there has been a recent suggestion that the Atlantic meridional circulation decreased by 30% since 1992. The heat transport decreased by only about 15% from 1.3 to 1.1 PW, however, as the horizontal gyre circulation increased to transport more heat northward. Thus, the Atlantic heat transport may not reduce proportionately with a decreased meridional overturning circulation, because in the absence of a meridional overturning circulation it is possible that the Atlantic will become more like the Pacific with a horizontal gyre circulation that still transports a substantial amount of heat northward.

If Atlantic ocean heat transport reduces under changing climate, will the atmospheric energy transport act to compensate with larger northward energy transport? Presently, the Atlantic Ocean circulation transports more than 20% of the maximum energy transport required to balance the Earth's radiation budget. The hypothesized Bjerkenes compensation mechanism suggests that a reduction in ocean heat transport will be compensated by increased atmospheric energy transport. For extratropical latitudes, atmospheric transport is primarily effected by eddies, cyclones, and anticyclones. Will a reduction in ocean heat transport then be accompanied by increased mid-latitude storminess and increased atmospheric energy transport? Or will the overall radiation budget for the Earth system be fundamentally altered?

Clearly, it is of interest to monitor the changes in ocean circulation and heat transport, most importantly in the Atlantic where there are concerns that increasing atmospheric CO_2 may lead relatively quickly to substantial changes in ocean circulation and heat transport. A program to monitor the Atlantic meridional overturning circulation and associated heat transport started in 2004 and such monitoring may provide the first evidence for changes in Atlantic ocean heat transport.

Further Reading

Bryden HL (1993) Ocean heat transport across 24° N latitude. In: McBean GA and Hantel M (eds.) *Geophysical Monograph Series, Vol. 75 Interactions between Global Climate Subsystems: The Legacy of Hann*, pp. 65–75. Washington, DC: American Geophysical Union.

Bryden HL and Beal LM (2001) Role of the Agulhas Current in Indian Ocean circulation and associated heat and freshwater fluxes. *Deep-Sea Research I* 48: 1821–1845.

Bryden HL and Imawaki S (2001) Ocean heat transport. In: Siedler G, Church J, and Gould J (eds.) *Ocean Circulation and Climate*, pp. 455–474. New York: Academic Press.

Bryden HL, Longworth HR, and Cunningham SA (2005) Slowing of the Atlantic meridional overturning circulation at 25° N. *Nature* 438: 655–657.

Cunningham SA, Kanzow T, Rayner D, *et al.* (2007) Temporal variability of the Atlantic meridional overturning circulation at 26.5° N. *Science* 317: 935–938.

Ganachaud A and Wunsch C (2000) Improved estimates of global ocean circulation, heat transport and mixing from hydrographic data. *Nature* 408: 453–457.

Josey SA, Kent EC, and Taylor PK (1999) New insights into the ocean heat budget closure problem from analysis of the SOC air–sea flux climatology. *Journal of Climate* 12: 2856–2880.

Lavín A, Bryden HL, and Parrilla G (1998) Meridional transport and heat flux variations in the subtropical North Atlantic. *Global Atmosphere and Ocean System* 6: 269–293.

Levitus S, Antonov J, and Boyer T (2005) Warming of the World Ocean, 1955–2003. *Geophysical Research Letters* 32 (doi:10.1029/2004GL021592).

Shaffrey L and Sutton R (2006) Bjerknes compensation and the decadal variability of the energy transports in a coupled climate model. *Journal of Climate* 19(7): 1167–1181.

Trenberth KE and Caron JM (2001) Estimates of meridional atmosphere and ocean heat transports. *Journal of Climate* 14: 3433–3443.

Trenberth KE, Caron JM, and Stepanaik DP (2001) The atmospheric energy budget and implications for surface fluxes and ocean heat transports. *Climate Dynamics* 17: 259–276.

Vellinga M and Wood RA (2002) Global climatic impacts of a collapse of the Atlantic thermohaline circulation. *Climatic Change* 54: 251–267.

EVAPORATION AND HUMIDITY

K. Katsaros, Atlantic Oceanographic and Meteorological Laboratory, NOAA, Miami, FL, USA

Introduction

Evaporation from the sea and humidity in the air above the surface are two important and related aspects of the phenomena of air–sea interaction. In fact, most subsections of the subject of air–sea interaction are related to evaporation. The processes that control the flux of water vapor from sea to air are similar to those for momentum and sensible heat; in many contexts, the energy transfer associated with evaporation, the latent heat flux, is of greatest interest. The latter is simply the internal energy carried from the sea to the air during evaporation by water molecules. The profile of water vapor content is logarithmic in the outer layer, from a few centimeters to approximately 30 m above the sea, as it is for wind speed and air temperature under neutrally stratified conditions. The molecular transfer rate of water vapor in air is slow and controls the flux only in the lowest millimeter. Turbulent eddies dominate the vertical exchange beyond this laminar layer. Modifications to the efficiency of the turbulent transfer occur due to positive and negative buoyancy forces. The relative importance of mechanical shear-generated turbulence and density-driven (buoyancy) fluxes was formulated in the 1940s, the Monin-Obukhov theory, and the field developed rapidly into the 1960s. New technologies, such as the sonic anemometer and Lyman-alpha hygrometer, were developed, which allowed direct measurements of turbulent fluxes. Furthermore, several collaborative international field experiments were undertaken. A famous one is the 'Kansas' experiment, whose data were used to formulate modern versions of the 'flux profile' relations, i.e., the relationship between the profile in the atmosphere of a variable such as humidity, and the associated turbulent flux of water vapor and its dependence on atmospheric stratification.

The density of air depends both on its temperature and on the concentration of water vapor. Recent improvements in measurement techniques and the ability to measure and correct for the motion of a ship or aircraft in three dimensions have allowed more direct measurements of evaporation over the ocean. The fundamentals of turbulent transfer in the atmosphere will not be discussed here, only the

special situations that are of interest for evaporation and humidity. As the water molecules leave the sea, they remove heat and leave behind an increase in the concentration of sea salts. Evaporation, therefore, changes the density of salt water, which has consequences for water mass formation and general oceanic circulation.

This article will focus on how humidity varies in the atmosphere, on the processes of evaporation, and how it is modified by the other phenomena discussed under the heading of air–sea interaction. All processes occurring at the air–sea interface interact and modify each other, so that none are simple and linear and most result in feedback on the phenomenon itself. The role of wind, temperature, humidity, wave breaking, spray, and bubbles will be broached and some fundamental concepts and equations presented. Methods of direct measurements and estimation using *in situ* mean measurements and satellite measurements will be discussed. Subjects requiring further research are also explored.

History/Definitions and Nomenclature

Many ways of measuring and defining the quantity of the invisible gas, water vapor, in the air have developed over the years. The common ones have been gathered together in **Table 1**, which gives their name, definition, SI units, and some further explanations. These quantitative definitions are all convertable one into another. The web-bulb temperature may seem rather anachronistic and is completely dependent on a rather crude measurement technique, but it is still a fundamental and dependable measure of the quantity of water vapor present in the air.

Evaporation or turbulent transfer of water vapor in the air was first modeled in analogy with down-gradient transfer by molecular conduction in solids. The conductivity was replaced by an 'Austaush' coefficient, A_e, or eddy diffusion coefficient, leading to the expression:

$$E = -A_e \rho \frac{\partial \bar{q}}{\partial z} \qquad [1]$$

where E is the evaporation rate, ρ the air density, $\bar{q}$ is mean atmospheric humidity, and z represents the vertical coordinate. Assuming no advection, steady state, and no accumulation of water vapor in the surface layer of the atmosphere (referred to as 'the constant flux layer'), the A_e is a function of z as

Table 1   Measures of humidity

Nomenclature	Units (SI)	Definition
Absolute humidity	kg m^{-3}	Amount of water vapor in the volume of associated moist air
Specific humidity	g kg^{-1}	The mass of water per unit mass of moist air (or equivalently in the same volume)
Mixing ratio	g kg^{-1}	The ratio of the mass of water as vapor to the mass of dry air in the same volume
Saturation humidity	Any of the above units	Can be given in terms of all three units and refers to the maximum amount the air can hold at its current temperature in terms of absolute or specific humidity, corresponds to 100% relative humidity
Relative humidity (RH)	%	Percent of saturation humidity that is actually in the air
Vapor pressure	hPa (or mb)	The partial pressure of the water vapor in the air
Dew point temperature	°K, °C	The temperature at which dew would form based on the actual amount of water vapor in the air. Dew point depression compared to actual temperature is a measure of the 'dryness' of the air
Wet-bulb temperature	°K, °C	This is a temperature obtained by the wetted thermometer of the pair of thermometers used in a psychrometer[a] (see Measurements chapter)

[a]A psychrometer is a measuring device consisting of two thermometers (mercury in glass or electronic), where one thermometer is covered with a wick wetted with distilled water. The device is aspirated with environmental air (at an air speed of at least 3 m s^{-1}). The evaporation of the distilled water cools the air passing over the wet wick, causing a lowering of the wet thermometer's temperature, which is dependent on the humidity in the air.

the turbulence scales increase away from the air–sea interface and the gradient is a decreasing function of height, z, as the distance from the source of water vapor, the sea surface, increases.

Determining E by measuring the gradient of q has not proved to be a good method because of the difficulties of obtaining differences of q accurately enough and in knowing the exact heights of the measurements well enough (say from a ship or a buoy on the ocean). The A_e must also be determined, which would require measurements of the intensity of the turbulent exchange in some fashion. The so-called direct method for evaluating the vapor flux in the atmosphere requires high frequency measurements. This method has been refined during the past 35 years or so, and has produced very good results for the turbulent flux of momentum (the wind stress). Fewer projects have been successful in measuring vapor flux over the ocean, because the humidity sensors are easily corrupted by the presence of spray or miniscule salt particles on the devices, which being hygroscopic, modify the local humidity.

Evaporation, E, can be measured directly today by obtaining the integration over all scales of the turbulent flux, namely, the correlation between the deviations from the mean of vertical velocity (w') and humidity (q') at height (z) within the constant flux layer. This correlation, resulting from the averaging of the vapor conservation equation (in analogy to the Reynolds stress term in the Navier–Stokes equation) can be measured directly, if sensors are available that resolve all relevant scales of fluctuations.

The correlation equation is

$$\overline{\rho w \cdot q} = \bar{\rho} \bar{w} \bar{q} + \bar{\rho} \overline{w' q'}, \qquad [2]$$

where w and q are the instantaneous values and the overbar indicates the time-averaged means. The product of the averages is zero since $\bar{w} = 0$. Much discussion and experimentation has gone into determining the time required to obtain a stable mean value of the eddy flux $\bar{\rho} \overline{q' w'}$. For the correlation term $\bar{\rho} \overline{w' q'}$ to represent the total vertical flux, there has to be a spectral gap between high and low frequencies of fluctuations, and the assumption of steady state and horizontal homogeneity must hold. The required averaging time is of the order of 20 min to 1 h.

Another commonly used method, the indirect or inertial dissipation method, also requires high frequency sensing devices, but relies on the balance between production and destruction of turbulence to be in steady state. The dissipation is related to the spectral amplitude of turbulent fluctuations in the inertial subrange, where the fluctuations are broken down from large-scale eddies to smaller and smaller scales, which happens in a similar fashion regardless of scale of the eddies responsible for the production of turbulence in the atmospheric boundary layer. The magnitude of the spectrum in the inertial subrange is, therefore, a measure of the total energy of the turbulence and can be interpreted in terms of the turbulent flux of water vapor. The advantage of this method over the eddy correlation method is that it is less dependent on the corrections for flow distortion and motion of the ship or the buoy platform, but it requires corrections for atmospheric stratification and other predetermined coefficients. It would not give the true flux if the production of turbulence was changing, as it does in changing sea states. Most of the time, the direct flux is not measured by either the direct or the indirect method; we resort to a parameterization of the flux in terms of so-called 'bulk' quantities.

The bulk formula has been found from field experiments where the total evaporation E has been measured directly together with mean values of q and wind speed, U, at one height, $z = a$ (usually referred to as 10 m by adjusting for the logarithmic vertical gradient), and the known sea surface temperature.

$$E = \overline{\rho w' q'} = \bar{\rho} \cdot C_{E_a} \overline{U_a} (\overline{q_s} - \overline{q_a}) \qquad [3]$$

where $\overline{q_s}$ is the saturation specific humidity at the air–sea interface, a function of sea surface temperature (SST). Air in contact with a water surface is assumed to be saturated. Above sea water the saturated air has 98% of the value of water vapor density at saturation over a freshwater surface, due to the effects of the dissolved salts in the sea. C_{E_a} is the exchange coefficient for water vapor evaluated for the height a. Experiments have shown C_{E_a} to be almost constant at $1.1–1.2 \times 10^{-3}$ for $U < 18$ m s^{-1}, for neutral stratification, i.e. no positive or negative buoyancy forces acting and at a height of 10 m, written as $C_{E_{10N}}$. However, measurements show large variability in $C_{E_{10N}}$ which may be due to the effects of sea state, such as sheltering in the wave troughs for large waves and increased evaporation due to spray droplets formed in highly forced seas with breaking waves. Results from a field experiment, the Humidity Exchange Over the Sea (HEXOS) experiment in the North Sea, are shown in **Figure 1**. Its purpose was to address the question of what happens to evaporation or water (vapor) flux at high wind speeds. However, the wind only reached 18 m s^{-1} and the measurements showed only weak, if any, effects of the spray. Theories suggest that the effects will be stronger above 25 m s^{-1}. More direct measurements are still required before these issues can be settled, especially for wind speeds >20 m s^{-1} (see Further Reading and the section on meteorological sensors for mean measurements for a discussion of the difficulties of making measurements over the sea at high wind speeds).

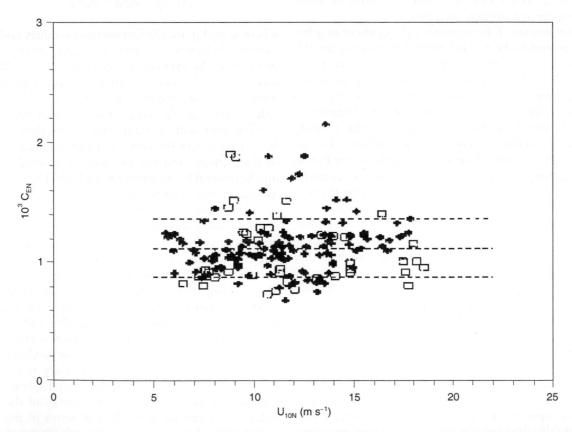

Figure 1 Vapor flux exchange coefficients from two simultaneous measurement sets: the University of Washington (crosses) and Bedford Institute of Oceanography (squares) data. Thick dashed line is the average value, 1.12×10^{-3}, for 170 data points. Thin dashed lines indicate standard deviations (from DeCosmo et al., 1996).

Clausius–Clapeyron Equation

The Clausius–Clapeyron equation relates the latent heat of evaporation to the work required to expand a unit mass of liquid water into a unit mass of water as vapor. The latent heat of evaporation is a function of absolute temperature. The Clausius–Clapeyron equation expresses the dependence of atmospheric saturation vapor pressure on temperature. It is a fundamental concept for understanding the role of evaporation in air–sea interaction on the large scale, as well as for gaining insight into the process of evaporation from the sea (or Earth's) surface on the small scale. Note first of all that the Clausius–Clapeyron equation is highly non-linear, viz:

$$\frac{d \ln \rho_v}{dT} = \frac{\Delta H_{vap}}{RT^2} \qquad [4]$$

where ρ_v is the vapor pressure, T is absolute temperature (°K), and ΔH_{vap} is the value of the latent heat of evaporation, R is the gas constant for water vapor $= 461.53 \, \mathrm{J \, kg^{-1} \, °K^{-1}}$. The dependence of vapor pressure on temperature is presented in a simplified form as:

$$e_s = 610.8 \exp \left[19.85 \left(1 - \frac{T_0}{T} \right) \right] (P_a) \qquad [5]$$

where e_s is vapor pressure in pascals, T_0 is a reference temperature set to $0°C = 273.16 \, °K$, and T is the actual temperature in °K which is accurate to 2% below 30°C (**Figure 2**).

Figure 2 displays the saturation vapor pressure and the pressure of atmospheric water vapor for 60% relative humidity. On the right-hand side of the figure, the ordinate gives the equivalent specific humidity values (for a near surface total atmospheric pressure of 1000 hpa). This figure illustrates that the atmosphere can hold vastly larger amounts of water as vapor at temperatures above 20°C than at temperatures below 10°C. For constant relative humidity, say 60%, the difference in specific humidity or vapor pressure in the air compared with the amount at the air–sea interface, if the sea is at the same temperature as the air, is about three times at 30°C what it would be at 10°C. Therefore, evaporation is driven much more strongly at tropical latitudes compared with high latitudes (cold sea and air) for the same mean wind and relative humidity as illustrated by eqn [4] and **Figure 2**.

Tropical Conditions of Humidity

By far, most of the water leaving the Earth's surface evaporates from the tropical oceans and jungles, providing the accompanying latent heat as the fuel that drives the atmospheric 'heat engines,' namely, thunderstorms and tropical cyclones. Such extreme and violent storms depend for their generation on the enormous release of latent heat in clouds to create the vertical motion and compensating horizontal

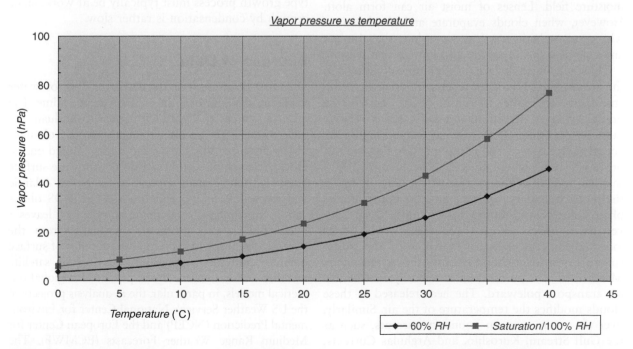

Vapor pressure vs temperature

Figure 2 Vapor pressure (hPa) as a function of temperature for two values of relative humidity, 60% and 100%.

accelerated inflows. Tropical cyclones do not form over oceanic regions with temperatures <26°C, and temperature increases of only 1° or 2°C sharply enhance the possibility of formation.

Latitudinal and Regional Variations

The Clausius–Clapeyron equation holds the secrets to the role of water vapor for both weather and climate. Warm moist air flowing north holds large quantities of water. As the air cools by vertical motion, contact with cold currents, and loss of heat by infrared radiation, the air reaches saturation and either clouds, storms and rain form, or fog (over cold surfaces) and stratus clouds. The warmer and moister the original air, the larger the possible rainfall and the larger the release of latent heat. Latitudinal, regional, and seasonal variations in evaporation and atmospheric humidity are all related to the source of heat for evaporation (upper ocean heat content) and the capacity of the air to hold water at its actual temperature. Many other processes such as the dynamics behind convergence patterns and the development of atmospheric frontal zones contribute to the variability of the associated weather.

Vertical Structure of Humidity

The fact that the source of moisture is the ocean, lakes, and moist ground explains the vertical structure of the moisture field. Lenses of moist air can form aloft. However, when clouds evaporate at high elevations where atmospheric temperature is low, the absolute amounts of water vapor are also low for that reason.

Thus, when the surface air is continually mixed in the atmospheric boundary layer with drier air, being entrained from the free atmosphere across the boundary layer inversion, it usually has a relative humidity less than 100% of what it could hold at its actual temperature. The exceptions are fog, clouds, or heavy rain, where the air has close to 100% relative humidity. The process of exchange between the moist boundary layer air and the upper atmosphere allows evaporation to continue. Deep convection in the inter-tropical convergence zone brings moist air up throughout the whole of the troposphere, even over-shooting into the stratosphere. Moisture that does not rain out locally is available for transport poleward. The heat released in these clouds modifies the temperature of the air. Similarly, over the warm western boundary currents, such as the Gulf Stream, Kuroshio, and Arghulas Currents, substantial evaporation and warming of the atmosphere takes place. Without the modifying effects of the hydrologic cycle of evaporation and precipitation on the atmosphere, the continents would have more extreme climates and be less habitable.

Sublimation–Deposition

The processes of water molecules leaving solid ice and condensing on it are called sublimation and deposition, respectively. These processes occur over the ice-covered polar regions of the ocean. In the cold regions, this flux is much less than that from open leads in the sea ice due to the warm liquid water, even at 0°C.

At an ice surface, water vapor saturation is less than over a water surface at the same temperature. This simple fact has consequences for the hydrologic cycle, because in a cloud consisting of a mixture of ice and liquid water particles, the vapor condenses on the ice crystals and the droplets evaporate. This process is important in the initial growth of ice particles in clouds until they become large enough to fall and grow by coalescence of droplets or other ice crystals encountered in their fall. Similar differences in water vapor occur for salty drops, and the vapor pressure over a droplet also depends on the curvature (radius) of the drop. Thus, particle size distribution in clouds and in spray over the ocean are always changing due to exchange of water vapor. For drops to become large enough to rain out, a coalescence-type growth process must typically be at work, since growth by condensation is rather slow.

Sources of Data

Very few direct measurements of the flux of water vapor are available over the ocean at any one time. The mean quantities ($\overline{U}$, $\overline{q_a}$, SST) needed to evaluate the bulk formula are reported regularly from voluntary observing ships (VOS) and from a few moored buoys. However, most of such buoys do not measure surface humidity, only a small number in the North Atlantic and tropical Pacific Oceans do so. The VOS observations are confined to shipping lanes, which leaves a huge void in the information available from the Southern Hemisphere. Alternative estimates of surface humidity and the water vapor flux include satellite methods and the surface fluxes produced in global numerical models, in particular, the re-analysis projects of the US Weather Service's National Center for Environmental Prediction (NCEP) and the European Center for Medium Range Weather Forecasts (ECMWF). The satellite method has large statistical uncertainty and, thus, requires weekly to monthly averages for obtaining reasonable accuracy ($\pm 30\,\mathrm{W\,m}^{-2}$ and $\pm 15\,\mathrm{W\,m}^{-2}$ for the weekly and monthly latent heat flux). Therefore, these data are most useful for climatological estimates and for checking the numerical models' results.

Estimation of Evaporation by Satellite Data

The estimation of evaporation/latent heat flux from the ocean using satellite data also relies on the bulk formula. The computation of latent heat flux by the bulk aerodynamic method requires SST, wind speed ($\overline{U}_{10_N}$), and humidity at a level within the surface layer $\overline{q_a}$, as seen in eqn [3]. Therefore, evaluation of the three variables from space is required. Over the ocean, $\overline{U}_{10_N}$ and SST have been directly retrieved from satellite data, but $\overline{q_a}$ has not. A method of estimating $\overline{q_a}$ and latent heat flux from the ocean using microwave radiometer data from satellites was proposed in the 1980s. It is based on an empirical relation between the integrated water vapor W (measured by spaceborne microwave radiometers) and $\overline{q_a}$ on a monthly timescale. The physical rationale is that the vertical distribution of water vapor through the whole depth of the atmosphere is coherent for periods longer than a week. The relation does not work well at synoptic and shorter timescales and also fails in some regions during summer. Modification of this method by including additional geophysical parameters has been proposed with some overall improvement, but the inherent limitation is the lack of information about the vertical distribution of q near the surface.

Two possible improvements in E retrieval include obtaining information on the vertical structures of humidity distribution and deriving a direct relation between E and the brightness temperatures (T_B) measured by a radiometer. Recent developments provide an algorithm for direct retrieval of boundary layer water vapor from radiances observed by the Special Sensor Microwave/Imager (SSM/I) on operational satellites in the Defense Meteorological Satellite Program since 1987. This sensor has four frequencies, 19.35, 22, 37, and 85.5 GHz, all except the 22 GHz operated at both horizontal and vertical polarizations. The 22 GHz channel at vertical polarization is in the center of a weak water vapor absorption line without saturation, even at high atmospheric humidity. The measurements are only possible over the oceans, because the oceans act as a relatively uniform reflecting background. Over land, the signals from the ground overwhelm the water vapor information.

Because all the three geophysical parameters, $\overline{U}_{10_N}$, W, and SST, can be retrieved from the radiances at the frequencies measured by the older microwave radiometer, launched in 1978 and operating to 1985 – the Scanning Multichannel Microwave Radiometer (SMMR) on Nimbus-7 (similar to SSM/I, but with 10.6 and 6.6 GHz channels as well, and no 85 GHz channels) – the feasibility of retrieving E directly from the measured radiances was also demonstrated. SMMR measures at 10 channels, but only six channels were identified as significantly useful in estimating E. SSM/I, the operational microwave radiometer that followed SMMR, lacks the low-frequency channels which are sensitive to SST, making direct retrieval of E from T_B unfeasible. The microwave imager (TMI) on the Tropical Rainfall Measuring Mission (TRMM), launched in 1998, includes low-frequency measurements sensitive to SST and could, therefore, allow direct estimates of evaporation rates. **Figure 3** gives an example of global monthly mean values of humidity obtained solely with satellite data from SSM/I.

To calculate q_s, gridded data of sea surface temperature can also be used, such as those provided operationally by the US National Weather Service based on infrared observations from the Advanced Very High Resolution Radiometer (AVHRR) on operational polar-orbiting satellites. The exact coincident timing is not so important for SST, since SST varies slowly due to the large heat capacity of water, and this method can only provide useful accuracies when averages are taken over 5 days to a week. Wind speed is best obtained from scatterometers, rather than from the microwave radiometer, in regions of heavy cloud or rain, since scatterometers (which are active radars) penetrate clouds more effectively. Scatterometers have been launched in recent times by the European Space Agency (ESA) and the US National Aeronautic and Space Administration (NASA) (the European Remote Sensing Satellites 1 and 2 in 1991 and 1995, the NASA scatterometer, NSCAT, on a Japanese short-lived satellite in 1996, and the QuikSCAT satellite in 1999).

Future Directions and Conclusions

Evaporation has been measured only up to wind speeds of $18\,\mathrm{m\,s}^{-1}$. The models appear to converge on the importance of the role of sea spray in evaporation, indicating that its significance grows beyond about $20\,\mathrm{m\,s}^{-1}$. However, the source function of spray droplets as a function of wind speed or wave breaking has not been measured, nor are techniques for measuring evaporation in the

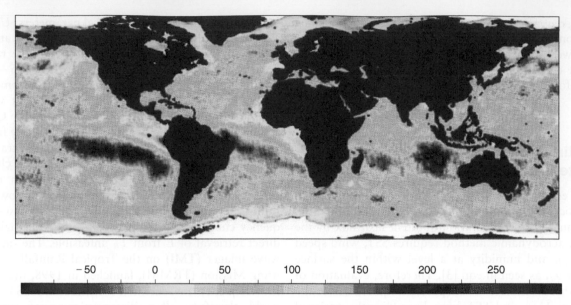

-50   0   50   100   150   200   250

Figure 3 Global distribution of monthly mean latent heat flux in W m^{-2} for September 1987. (Reproduced with permission from Schulz *et al.*, 1997.)

presence of droplets well–developed, whether for rain or sea spray. Since evaporation and the latent heat play such important roles in tropical cyclones and many other weather phenomena, as well as in oceanic circulation, there is great motivation for getting this important energy and mass flux term right. The bulk model is likely to be the main method used for estimating evaporation for some time to come. Development of more direct satellite methods and validating them should be an objective for climatological purposes. Progress in the past 30 years has brought the estimate of evaporation on a global scale to useful accuracy.

See also

IR Radiometers. Satellite Remote Sensing of Sea Surface Temperatures. Sensors for Mean Meteorology.

Further Reading

Bentamy A, Queffeulou P, Quilfen Y, and Katsaros KB (1999) Ocean surface wind fields estimated from satellite active and passive microwave instruments. *Institute of Electrical and Electronic Engineers, Transactions, Geoscience Remote Sensing* 37: 2469–2486.

Businger JA, Wyngaard JC, lzumi Y, and Bradley EF (1971) Flux-profile relationships in the atmospheric surface layer. *Journal of Atmospheric Science* 28: 181–189.

DeCosmo J, Katsaros KB, Smith SD, *et al.* (1996) Air–sea exchange of water vapor and sensible heat: The Humidity Exchange Over the Sea (HEXOS) results. *Journal of Geophysical Research* 101: 12001–12016.

Dobson F, Hasse L, and Davies R (eds.) (1980) *Instruments and Methods in Air–sea Interaction.* New York: Plenum Publishing.

Donelan MA (1990) Air–sea Interaction. In: LeMehaute B and Hanes DM (eds.) *The Sea*, Vol. 9, pp. 239–292. New York: John Wiley.

Esbensen SK, Chelton DB, Vickers D, and Sun J (1993) An analysis of errors in Special Sensor Microwave Imager evaporation estimates over the global oceans. *Journal of Geophysical Research* 98: 7081–7101.

Geernaert GL (ed.) (1999) *Air–sea Exchange Physics, Chemistry and Dynamics.* Dordrecht: Kluwer Academic Publishers.

Geernaert GL and Plant WJ (eds.) (1990) *Surface Waves and Fluxes*, Vol. 2. Dordrecht: Kluwer Academic Publishers.

Katsaros KB, Smith SD, and Oost WA (1987) HEXOS – Humidity Exchange Over the Sea: A program for research on water vapor and droplet fluxes from sea to air at moderate to high wind speeds. *Bulletin of the American Meteoroloical Society* 68: 466–476.

Kraus EB and Businger JA (eds.) (1994) *Atmosphere–Ocean Interaction* 2nd ed. New York: Oxford University Press.

Liu WT and Katsaros KB (2001) Air–sea fluxes from satellite data. In: Siedler G, Church J, and Gould J (eds.) *Ocean Circulation and Climate.* Academic Press

Liu WT, Tang W, and Wentz FJ (1992) Precipitable water and surface humidity over global oceans from SSM/I and ECMWF. *Journal of Geophysical Research* 97: 2251–2264.

Makin VK (1998) Air–sea exchange of heat in the presence of wind waves and spray. *Journal of Geophysical Research* 103: 1137–1152.

Schneider SH (ed.) (1996) *Encyclopedia of Climate and Weather.* New York: Oxford University Press.

Schulz J, Meywerk J, Ewald S, and Schlüssel P (1997) Evaluation of satellite-derived latent heat fluxes. *Journal of Climate* 10: 2782–2795.

Smith SD (1988) Coefficients for sea surface wind stress, heat flux, and wind profiles as a function of wind speed and temperature. *Journal of Geophysical Research* 93: 15467–15472.

RADIATIVE TRANSFER IN THE OCEAN

C. D. Mobley, Sequoia Scientific, Inc., WA, USA

Introduction

Understanding how light interacts with sea water is a fascinating problem in itself, as well as being fundamental to fields as diverse as biological primary production, mixed-layer thermodynamics, photochemistry, lidar bathymetry, ocean-color remote sensing, and visual searching for submerged objects. For these reasons, optics is one of the fastest growing oceanographic research areas.

Radiative transfer theory provides the theoretical framework for understanding light propagation in the ocean, just as hydrodynamics provides the framework for physical oceanography. The article begins with an overview of the definitions and terminology of radiative transfer as used in oceanography. Various ways of quantifying the optical properties of a water body and the light within the water are described. The chapter closes with examples of the absorption and scattering properties of two hypothetical water bodies, which are characteristic of the open ocean and a turbid estuary, and a comparison of their underwater light fields.

Terminology

The optical properties of sea water are sometimes grouped into inherent and apparent properties.

- Inherent optical properties (IOPs) are those properties that depend only upon the medium and therefore are independent of the ambient light field. The two fundamental IOPs are the absorption coefficient and the volume scattering function. (These quantities are defined below.)
- Apparent optical properties (AOPs) are those properties that depend both on the medium (the IOPs) and on the directional structure of the ambient light field, and that display enough regular features and stability to be useful descriptors of a water body. Commonly used AOPs are the irradiance reflectance, the remote-sensing reflectance, and various diffuse attenuation functions.

'Case 1 waters' are those in which the contribution by phytoplankton to the total absorption and scattering is high compared to that by other substances.

Absorption by chlorophyll and related pigments therefore plays the dominant role in determining the total absorption in such waters, although covarying detritus and dissolved organic matter derived from the phytoplankton also contribute to absorption and scattering in case 1 waters. Case 1 water can range from very clear (oligotrophic) to very productive (eutrophic) water, depending on the phytoplankton concentration.

'Case 2 waters' are 'everything else,' namely, waters where inorganic particles or dissolved organic matter from land drainage contribute significantly to the IOPs, so that absorption by pigments is relatively less important in determining the total absorption. Roughly 98% of the world's open ocean and coastal waters fall into the case 1 category, but near-shore and estuarine case 2 waters are disproportionately important to human interests such as recreation, fisheries, and military operations.

Table 1 summarizes the terms, units, and symbols for various quantities frequently used in optical oceanography.

Radiometric Quantities

Consider an amount ΔQ of radiant energy incident in a time interval Δt centered on time t, onto a surface of area ΔA located at position (x, y, z), and arriving through a set of directions contained in a solid angle $\Delta \Omega$ about the direction (θ, φ) normal to the area ΔA, as produced by photons in a wavelength interval $\Delta \lambda$ centered on wavelength λ. The geometry of this situation is illustrated in **Figure 1**. Then an operational definition of the spectral radiance is

$$L(x, y, z, t, \theta, \varphi, \lambda) \equiv \frac{\Delta Q}{\Delta t \, \Delta A \, \Delta \Omega \, \Delta \lambda} \quad [\text{Js}^{-1}\text{m}^{-2}\text{sr}^{-1}\text{nm}^{-1}] \quad [1]$$

In the conceptual limit of infinitesimal parameter intervals, the spectral radiance is defined as

$$L(x, y, z, t, \theta, \varphi, \lambda) \equiv \frac{\partial^4 Q}{\partial t \, \partial A \, \partial \Omega \, \partial \lambda} \quad [2]$$

Spectral radiance is the fundamental radiometric quantity of interest in optical oceanography: it completely specifies the positional (x, y, z), temporal (t), directional (θ, φ), and spectral (λ) structure of the light field. In many oceanic environments, horizontal variations (on a scale of tens to thousands of meters)

Table 1 Quantities commonly used in optical oceanography

Quantity	SI units	Symbol
Radiometric quantities		
Quantity of radiant energy	$J\,nm^{-1}$	Q
Power	$W\,nm^{-1}$	Φ
Intensity	$W\,sr^{-1}\,nm^{-1}$	I
Radiance	$W\,m^{-2}\,sr^{-1}\,nm^{-1}$	L
Downwelling plane irradiance	$W\,m^{-2}\,nm^{-1}$	E_d
Upwelling plane irradiance	$W\,m^{-2}\,nm^{-1}$	E_u
Net irradiance	$W\,m^{-2}\,nm^{-1}$	E
Scalar irradiance	$W\,m^{-2}\,nm^{-1}$	E_o
Downwelling scalar irradiance	$W\,m^{-2}\,nm^{-1}$	E_{ou}
Upwelling scalar irradiance	$W\,m^{-2}\,nm^{-1}$	E_{ou}
Photosynthetic available radiation	$Photons\,s^{-1}\,m^{-2}$	PAR
Inherent optical properties		
Absorption coefficient	m^{-1}	a
Volume scattering function	$m^{-1}\,sr^{-1}$	β
Scattering phase function	sr^{-1}	$\tilde{\beta}$
Scattering coefficient	m^{-1}	b
Backscatter coefficient	m^{-1}	b_b
Beam attenuation coefficient	m^{-1}	c
Single-scattering albedo	–	ω_o
Apparent optical properties		
Irradiance reflectance (ratio)	–	R
Remote-sensing reflectance	sr^{-1}	R_{rs}
Attenuation coefficients	m^{-1}	
of radiance $L(z, \theta, \varphi)$	m^{-1}	$K(\theta, \varphi)$
of downwelling irradiance $E_d(z)$	m^{-1}	K_d
of upwelling irradiance $E_u(z)$	m^{-1}	K_u
of PAR	m^{-1}	K_{PAR}

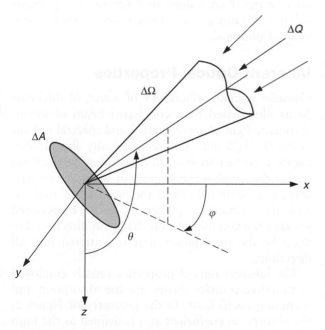

Figure 1 Geometry used to define radiance.

of the IOPs and the radiance are much less than variations with depth, in which case it can be assumed that these quantities vary only with depth z. (An exception would be the light field due to a single light source imbedded in the ocean; such a radiance distribution is inherently three-dimensional.) Moreover, since the timescales for changes in IOPs or in the environment (seconds to seasons) are much greater than the time required for the radiance to reach steady state (microseconds) after a change in IOPs or boundary conditions, time-independent radiative transfer theory is adequate for most oceanographic studies. (An exception is time-of-flight lidar bathymetry.) When the assumptions of horizontal homogeneity and time independence are valid, the spectral radiance can be written as $L(z, \theta, \varphi, \lambda)$.

Although the spectral radiance completely specifies the light field, it is seldom measured in all directions, both because of instrumental difficulties and because such complete information often is not needed. The most commonly measured radiometric quantities are various irradiances. Suppose the light detector is equally sensitive to photons of a given wavelength λ traveling in any direction (θ, φ) within a hemisphere of directions. If the detector is located at depth z and is oriented facing upward, so as to collect photons traveling downward, then the detector output is a measure of the spectral downwelling scalar irradiance at depth z, $E_{od}(z, \lambda)$. Such an instrument is summing radiance over all the directions (elements of solid angle) in the downward hemisphere; thus $E_{od}(z, \lambda)$ is related to $L(z, \theta, \varphi, \lambda)$ by

$$E_{od}(z, \lambda) = \int_{2\pi_d} L(z, \theta, \varphi, \lambda) \, d\Omega \qquad [\text{W m}^{-2} \text{nm}^{-1}] \quad [3]$$

Here $2\pi_d$ denotes the hemisphere of downward directions (i.e., the set of directions (θ, φ) such that $0 \leq \theta \leq \pi/2$ and $0 \leq \varphi < 2\pi$, if θ is measured from the $+z$ or nadir direction). The integral over $2\pi_d$ can be evaluated as a double integral over θ and φ after a specific coordinate system is chosen.

If the same instrument is oriented facing downward, so as to detect photons traveling upward, then the quantity measured is the spectral upwelling scalar irradiance $E_{ou}(z, \lambda)$. The spectral scalar irradiance $E_o(z, \lambda)$ is the sum of the downwelling and upwelling components:

$$E_O(z, \lambda) \equiv E_{od}(z, \lambda) + E_{ou}(z, \lambda)$$
$$= \int_{4\pi} L(z, \theta, \varphi, \lambda) \, d\Omega \qquad [4]$$

$E_o(z, \lambda)$ is proportional to the spectral radiant energy density $(\text{J m}^{-3} \text{nm}^{-1})$ and therefore quantifies how much radiant energy is available for photosynthesis or heating the water.

Now consider a detector designed so that its sensitivity is proportional to $|\cos \theta|$, where θ is the angle between the photon direction and the normal to the surface of the detector. This is the ideal response of a 'flat plate' collector of area ΔA, which when viewed at an angle θ to its normal appears to have an area of $\Delta A |\cos \theta|$. If such a detector is located at depth z and is oriented facing upward, so as to detect photons traveling downward, then its output is proportional to the spectral downwelling plane irradiance $E_d(z, \lambda)$. This instrument is summing the downwelling radiance weighted by the cosine of the photon direction, thus

$$E_d(z, \lambda) = \int_{2\pi_d} L(z, \theta, \varphi, \lambda) |\cos \theta| \, d\Omega$$
$$[\text{W m}^{-2} \text{nm}^{-1}] \qquad [5]$$

Turning this instrument upside down gives the spectral upwelling plane irradiance $E_u(z, \lambda)$. E_d and E_u are useful because they give the energy flux (power per unit area) across the horizontal surface at depth z owing to downwelling and upwelling photons, respectively. The difference $E_d - E_u$ is called the net (or vector) irradiance.

Photosynthesis is a quantum phenomenon, i.e., it is the number of available photons rather than the amount of radiant energy that is relevant to the chemical transformations. This is because a photon of, say, $\lambda = 400$ nm, if absorbed by a chlorophyll molecule, induces the same chemical change as does a photon of $\lambda = 600$ nm, even though the 400 nm photon has 50% more energy than the 600 nm photon. Only a part of the photon energy goes into photosynthesis; the excess is converted to heat or is re-radiated. Moreover, chlorophyll is equally able to absorb and utilize a photon regardless of the photon's direction of travel. Therefore, in studies of phytoplankton biology, the relevant measure of the light field is the photosynthetic available radiation, PAR, defined by

$$\text{PAR}(z) \equiv \int_{350 \, \text{nm}}^{700 \, \text{nm}} \frac{\lambda Z}{hc} E_o(z, \lambda) \, d\lambda$$
$$[\text{photons s}^{-1} \text{m}^{-2}] \qquad [6]$$

where $h = 6.6255 \times 10^{-34}$ J s is the Planck constant and $c = 3.0 \times 10^{17}$ nm s^{-1} is the speed of light. The factor λ/hc converts the energy units of E_o to quantum units (photons per second). Bio-optical literature often states PAR values in units of mol photons s^{-1} m^{-2} or einst s^{-1} m^{-2} (where one einstein is one mole of photons).

Inherent Optical Properties

Consider a small volume ΔV of water, of thickness Δr as illuminated by a collimated beam of monochromatic light of wavelength λ and spectral radiant power $\Phi_i(\lambda)$ (W nm^{-1}), as schematically illustrated in **Figure 2**. Some part $\Phi_a(\lambda)$ of the incident power $\Phi_i(\lambda)$ is absorbed within the volume of water. Some part $\Phi_i(\psi, \lambda)$ is scattered out of the beam at an angle ψ, and the remaining power $\Phi_t(\lambda)$ is transmitted through the volume with no change in direction. Let $\Phi_s(\lambda)$ be the total power that is scattered into all directions.

The inherent optical properties usually employed in radiative transfer theory are the absorption and scattering coefficients. In the geometry of **Figure 2**, the absorption coefficient $a(\lambda)$ is defined as the limit of the fraction of the incident power that is absorbed within the volume, as the thickness becomes small:

$$a(\lambda) \equiv \lim_{\Delta r \to 0} \frac{1}{\Phi_i(\lambda)} \frac{\Phi_a(\lambda)}{\Delta r} \, [\text{m}^{-1}] \qquad [7]$$

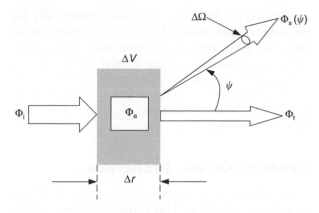

Figure 2 Geometry used to define inherent optical properties.

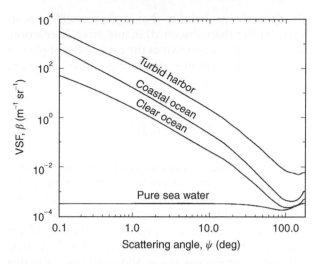

Figure 3 Volume scattering functions (VSF) measured in three different oceanic waters. The VSF of pure sea water is shown for comparison.

The scattering coefficient $b(\lambda)$ has a corresponding definition using $\Phi_s(\lambda)$. The beam attenuation coefficient $c(\lambda)$ is defined as $c(\lambda) = a(\lambda) + b(\lambda)$.

Now take into account the angular distribution of the scattered power, with $\Phi_s(\psi, \lambda)/\Phi_i(\lambda)$ being the fraction of incident power scattered out of the beam through an angle ψ into a solid angle $\Delta\Omega$ centered on ψ, as shown in **Figure 2**. Then the fraction of scattered power per unit distance and unit solid angle, $\beta(\psi, \lambda)$, is

$$\beta(\psi, \lambda) \equiv \lim_{\Delta r \to 0} \lim_{\Delta\Omega \to 0} \frac{\Phi_s(\psi, \lambda)}{\Phi_i(\lambda)\Delta r \Delta\Omega} \quad [\mathrm{m^{-1}sr^{-1}}] \quad [8]$$

The spectral power scattered into the given solid angle $\Delta\Omega$ is just the spectral radiant intensity scattered into direction ψ times the solid angle: $\Phi_s(\psi, \lambda) = I_s(\psi, \lambda)\,\Delta\Omega$. Moreover, if the incident power $\Phi_i(\lambda)$ falls on an area ΔA, then the corresponding incident irradiance is $E_i(\lambda) = \Phi_i(\lambda)/\Delta A$. Noting that $\Delta V = \Delta r \Delta A$ is the volume of water that is illuminated by the incident beam gives

$$\beta(\psi, \lambda) = \lim_{\Delta V \to 0} \frac{I_S(\psi, \lambda)}{E_i(\lambda)\Delta V} \quad [9]$$

This form of $\beta(\psi, \lambda)$ suggests the name volume scattering function (VSF) and the physical interpretation of scattered intensity per unit incident irradiance per unit volume of water. **Figure 3** shows measured VSFs (at 514 nm) from three greatly different water bodies; the VSF of pure water is shown for comparison. VSFs of sea water typically increase by five or six orders of magnitude in going from $\psi = 90°$ to $\psi = 0.1°$ for a given water sample, and scattering at a given angle ψ can vary by two orders of magnitude among water samples.

Integrating $\beta(\psi, \lambda)$ over all directions (solid angles) gives the total scattered power per unit incident irradiance and unit volume of water, in other words the spectral scattering coefficient:

$$b(\lambda) = \int_{4\pi} \beta(\psi, \lambda)\mathrm{d}\Omega = 2\pi \int_0^\pi \beta(\psi, \lambda) \sin\psi\,\mathrm{d}\psi \quad [10]$$

Eqn. [10] follows because scattering in natural waters is azimuthally symmetric about the incident direction (for unpolarized light sources and randomly oriented scatterers). This integration is often divided into forward scattering, $0 \le \psi \le \pi/2$, and backward scattering, $\pi/2 \le \psi \le \pi$, parts. Thus the backscatter coefficient is

$$b_\mathrm{b}(\lambda) \equiv 2\pi \int_{\pi/2}^\pi \beta(\psi, \lambda) \sin\psi\,\mathrm{d}\psi \quad [11]$$

The VSFs of **Figure 3** have b values ranging from 0.037 to 1.824 m^{-1} and backscatter fractions b_b/b of 0.013 to 0.044.

The preceding discussion assumed that no inelastic-scattering processes are present. However, inelastic scattering does occur owing to fluorescence by dissolved matter or chlorophyll, and to Raman scattering by the water molecules themselves. Power lost from wavelength λ by scattering into wavelength $\lambda' \ne \lambda$ appears as an increase in the absorption $a(\lambda)$. The gain in power at λ' appears as a source term in the radiative transfer equation.

Two more inherent optical properties are commonly used in optical oceanography. The single-scattering albedo is $\omega_0(\lambda) = b(\lambda)/c(\lambda)$. The single-scattering albedo is the probability that a photon will be scattered (rather than absorbed) in any given interaction, hence $\omega_o(\lambda)$ is also known as the probability of photon survival. The volume scattering phase function, $\tilde{\beta}(\psi, \lambda)$ is defined by

$$\tilde{\beta}(\psi, \lambda) \equiv \frac{\beta(\psi, \lambda)}{b(\lambda)} \quad [\text{sr}^{-1}] \qquad [12]$$

Writing the volume scattering function $\beta(\psi, \lambda)$ as the product of the scattering coefficient $b(\lambda)$ and the phase function $\tilde{\beta}(\psi, \lambda)$ partitions $\beta(\psi, \lambda)$ into a factor giving the strength of the scattering, $b(\lambda)$ with units of m^{-1}, and a factor giving the angular distribution of the scattered photons, $\tilde{\beta}(\psi, \lambda)$ with units of sr^{-1}. A striking feature of the sea water VSFs of **Figure 3** is that their phase functions are all similar in shape, with the main differences being in the detailed shape of the functions in the backscatter directions ($\psi > 90°$).

The IOPs are additive. This means, for example, that the total absorption coefficient of a water body is the sum of the absorption coefficients of water, phytoplankton, dissolved substances, mineral particles, etc. This additivity allows the development of separate models for the absorption and scattering properties of the various constituents of sea water.

The Radiative Transfer Equation

The equation that connects the IOPs and the radiance is called the radiative transfer equation (RTE). Even in the simplest situation of horizontally homogeneous water and time independence, the RTE is a formidable integro-differential equation:

$$\cos\theta \frac{\mathrm{d}L(z, \theta, \varphi, \lambda)}{\mathrm{d}z} = -c(z, \lambda)L(z, \theta, \varphi, \lambda)$$
$$+ \int_{4\pi} L(z, \theta', \varphi', \lambda)$$
$$\times \beta(z; \theta', \varphi' \to \theta, \varphi; \lambda)\, \mathrm{d}\Omega'$$
$$+ S(z, \theta, \varphi, \lambda) \qquad [13]$$

The scattering angle ψ in the VSF is the angle between the incident direction (θ', φ') and the scattered direction (θ, φ). The source term $S(z, \theta, \varphi, \lambda)$ can describe either an internal light source such as bioluminescence, or inelastically scattered light from other wavelengths. The physical environment of a water body – waves on its surface, the character of its bottom, the incident radiance from the sky – enters the theory via the boundary conditions necessary to solve the RTE. Given the IOPs and suitable boundary

conditions, the RTE can be solved numerically for the radiance distribution $L(z, \theta, \varphi, \lambda)$. Unfortunately, there are no shortcuts to computing other radiometric quantities. For example, it is not possible to write down an equation that can be solved directly for the irradiance E_d; one must first solve the RTE for the radiance and then compute E_d by integrating the radiance over direction.

Apparent Optical Properties

Apparent optical properties are always a ratio of two radiometric variables. This ratioing removes effects of the magnitude of the incident sky radiance onto the sea surface. For example, if the sun goes behind a cloud, the downwelling and upwelling irradiances within the water can change by an order of magnitude within a few seconds, but their ratio will be almost unchanged. (There will still be some change because the directional structure of the underwater radiance will change when the sun's direct beam is removed from the radiance incident onto the sea surface.)

The ratio just mentioned,

$$R(z, \lambda) \equiv \frac{E_u(z, \lambda)}{E_d(z, \lambda)} \qquad [14]$$

is called the irradiance reflectance (or irradiance ratio). The remote-sensing reflectance $R_{rs}(\theta, \varphi, \lambda)$ is defined as

$$R_{rs}(\theta, \varphi, \lambda) \equiv \frac{L_w(\theta, \varphi, \lambda)}{E_d(\lambda)} \quad [\text{sr}^{-1}] \qquad [15]$$

where L_w is the water-leaving radiance, i.e., the total upward radiance minus the sky and solar radiance that was reflected upward by the sea surface. L_w and E_d are evaluated just above the sea surface. Both $R_{rs}(\theta, \varphi, \lambda)$ and $R(z, \lambda)$ just beneath the sea surface are of great importance in remote sensing, and both can be regarded as a measure of 'ocean color.' R and R_{rs} are proportional (to a first-order approximation) to $b_b/(a + b_b)$, and measurements of R_{rs} above the surface or of R within the water can be used to estimate water quality parameters such as the chlorophyll concentration.

Under typical oceanic conditions, for which the incident lighting is provided by the sun and sky, the radiance and various irradiances all decrease approximately exponentially with depth, at least when far enough below the surface (and far enough above the bottom, in shallow water) to be free of boundary effects. It is therefore convenient to write the depth dependence of, say, $E_d(z, \lambda)$ as

$$E_d(z, \lambda) \equiv E_d(0, \lambda) \exp\left[-\int_0^z K_d(z', \lambda)\mathrm{d}z'\right] \quad [16]$$

where $K_d(z, \lambda)$ is the spectral diffuse attenuation coefficient for spectral downwelling plane irradiance. Solving for $K_d(z, \lambda)$ gives

$$K_d(z, \lambda) = -\frac{\mathrm{d}\ln E_d(z, \lambda)}{\mathrm{d}z}$$
$$= -\frac{1}{E_d(z, \lambda)}\frac{\mathrm{d}E_d(z, \lambda)}{\mathrm{d}z} \quad [\mathrm{m}^{-1}] \quad [17]$$

The beam attenuation coefficient $c(\lambda)$ is defined in terms of the radiant power lost from a collimated beam of photons. The diffuse attenuation coefficient $K_d(z, \lambda)$ is defined in terms of the decrease with depth of the ambient downwelling irradiance $E_d(z, \lambda)$, which comprises photons heading in all downward directions (a diffuse, or uncollimated, light field). $K_d(z, \lambda)$ clearly depends on the directional structure of the ambient light field, hence its classification as an apparent optical property. Other diffuse attenuation coefficients, e.g., K_u, K_{od}, or K_{PAR}, are defined in an analogous manner, using the corresponding radiometric quantities. In most waters, these K functions are strongly correlated with the absorption coefficient a and therefore can serve as convenient, if imperfect, descriptors of a water body. However, AOPs are not additive, which complicates their interpretation in terms of water constituents.

Optical Constituents of Seawater

Oceanic waters are a witch's brew of dissolved and particulate matter whose concentrations and optical properties vary by many orders of magnitude, so that ocean waters vary in color from the deep blue of the open ocean, where sunlight can penetrate to depths of several hundred meters, to yellowish-brown in a turbid estuary, where sunlight may penetrate less than a meter. The most important optical constituents of sea water can be briefly described as follows.

Sea Water

Water itself is highly absorbing at wavelengths below 250 nm and above 700 nm, which limits the wavelength range of interest in optical oceanography to the near-ultraviolet to the near infrared.

Dissolved Organic Compounds

These compounds are produced during the decay of plant matter. In sufficient concentrations these compounds can color the water yellowish brown; they are therefore generally called yellow matter or colored dissolved organic matter (CDOM). CDOM absorbs very little in the red, but absorption increases rapidly with decreasing wavelength, and CDOM can be the dominant absorber at the blue end of the spectrum, especially in coastal waters influenced by river runoff.

Organic Particles

Biogenic particles occur in many forms.

Bacteria Living bacteria in the size range 0.2–1.0 μm can be significant scatterers and absorbers of light, especially at blue wavelengths and in clean oceanic waters, where the larger phytoplankton are relatively scarce.

Phytoplankton These ubiquitous microscopic plants occur with incredible diversity of species, size (from less than 1 μm to more than 200 μm), shape, and concentration. Phytoplankton are responsible for determining the optical properties of most oceanic waters. Their chlorophyll and related pigments strongly absorb light in the blue and red and thus, when concentrations are high, determine the spectral absorption of sea water. Phytoplankton are generally much larger than the wavelength of visible light and can scatter light strongly.

Detritus Nonliving organic particles of various sizes are produced, for example, when phytoplankton die and their cells break apart, and when zooplankton graze on phytoplankton and leave cell fragments and fecal pellets. Detritus can be rapidly photooxidized and lose the characteristic absorption spectrum of living phytoplankton, leaving significant absorption only at blue wavelengths. However, detritus can contribute significantly to scattering, especially in the open ocean.

Inorganic Particles

Particles created by weathering of terrestrial rocks can enter the water as wind-blown dust settles on the sea surface, as rivers carry eroded soil to the sea, or as currents resuspend bottom sediments. Such particles range in size from less than 0.1 μm to tens of micrometers and can dominate water optical properties when present in sufficient concentrations.

Particulate matter is usually the major determinant of the absorption and scattering properties of sea water and is responsible for most of the temporal and spatial variability in these optical properties. A central goal of research in optical oceanography is to understand how the absorption and scattering properties of these various constituents relate to the particle type (e.g., microbial species or mineral composition), present conditions (e.g., the physiological state of a living microbe, which in turn depends on nutrient supply and ambient lighting), and history (e.g., photo-oxidation of pigments in dead cells). Bio-geo-optical models have been developed that attempt (with varying degrees of success) to predict the IOPs in terms of the chlorophyll concentration or other simplified measures of the composition of a water body.

Examples of Underwater Light Fields

Solving the radiative transfer equation requires mathematically sophisticated and computationally intensive numerical methods. *Hydrolight* is a widely used software package for numerical solution of oceanographic radiative transfer problems. The input to *Hydrolight* consists of the absorption and scattering coefficients of each constituent of the water body (microbial particles, dissolved substances, mineral particles, etc.) as functions of depth and wavelength, the corresponding scattering phase functions, the sea state, the sky radiance incident onto the sea surface, and the reflectance properties of the bottom boundary (if the water is not assumed infinitely deep). *Hydrolight* solves the one-dimensional, time-independent radiative transfer equation, including inelastic scattering effects, to

obtain the radiance distribution $L(z, \theta, \varphi, \lambda)$. Other quantities of interest such as irradiances or reflectances are then computed using their definitions and the solution radiance distribution.

To illustrate the range of behavior of underwater light fields, *Hydrolight* was run for two greatly different water bodies. The first simulation used a chlorophyll profile measured in the Atlantic Ocean north of the Azores in winter. The water was well mixed to a depth of over 100 m. The chlorophyll concentration *Chl* varied between 0.2 and 0.3 mg m^{-3} between the surface and 116 m depth; it then dropped to less than 0.05 mg m^{-3} below 150 m depth. The water was oligotrophic, case 1 water, and commonly used bio-optical models for case 1 water were used to convert the chlorophyll concentration to absorption and scattering coefficients (which were not measured). A scattering phase function similar in shape to those seen in **Figure 3** was used for the particles; this phase function had a backscatter fraction of $b_b/b = 0.018$. The second simulation was for an idealized, case 2 coastal water body containing 5 mg m^{-3} of chlorophyll and 2 g m^{-3} of brown-colored mineral particles representing resuspended sediments. Bio-optical models and measured mass-specific absorption and scattering coefficients were used to convert the chlorophyll and mineral concentrations to absorption and scattering coefficients. The large microbial particles of low index of refraction were assumed to have a phase function with $b_b/b = 0.005$, and the small mineral particles of high index of refraction had $b_b/b = 0.03$. The water was assumed to be well mixed and to have a brown mud bottom at a depth of 10 m. Both simulations used a

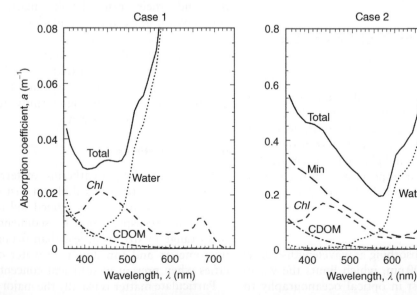

Figure 4 Absorption coefficients for the case 1 and case 2 water bodies. The contributions by the various components are labeled.

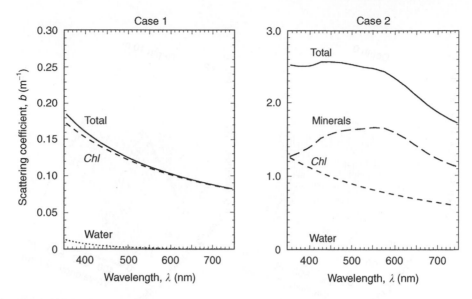

Figure 5 Scattering coefficients for the case 1 and case 2 water bodies. The contributions by the various components are labeled (CDOM is nonscattering).

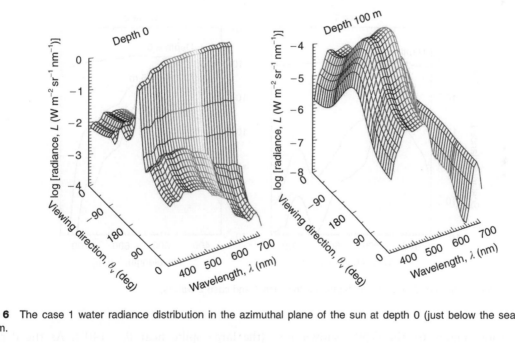

Figure 6 The case 1 water radiance distribution in the azimuthal plane of the sun at depth 0 (just below the sea surface) and at 100 m.

clear sky radiance distribution appropriate for midday in January at the Azores location. The sea surface was covered by capillary waves corresponding to a $5 \, \text{m s}^{-1}$ wind speed.

Figure 4 shows the component and total absorption coefficients just beneath the sea surface for these two hypothetical water bodies, and **Figure 5** shows the corresponding scattering coefficients. For the case 1 water, the total absorption is dominated by chlorophyll at blue wavelengths and by the water

itself at wavelengths greater than 500 nm. However, the water makes only a small contribution to the total scattering. In the case 2 water, absorption by the mineral particles is comparable to or greater than that by the chlorophyll-bearing particles, and water dominates only in the red. The mineral particles are the primary scatterers.

Figures 6 and **7** show the radiance in the azimuthal plane of the sun as a function of polar viewing direction and wavelength, for selected depths. For the

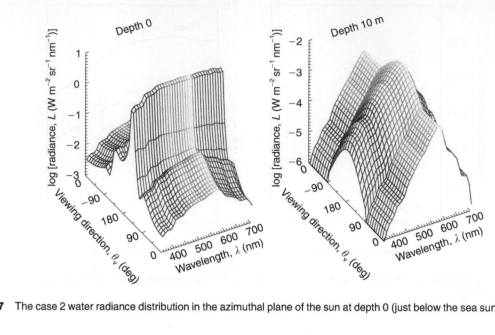

Figure 7 The case 2 water radiance distribution in the azimuthal plane of the sun at depth 0 (just below the sea surface) and at 10 m.

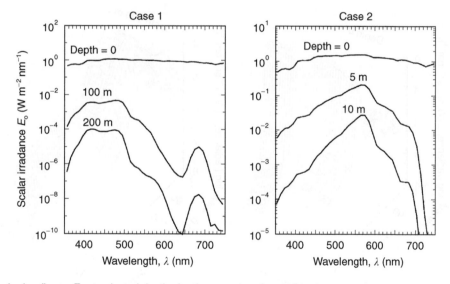

Figure 8 The scalar irradiance E_0 at selected depths for the case 1 and case 2 waters.

case 1 simulation (**Figure 6**), the depths shown are zero, just beneath the sea surface, and 100 m; for the case 2 simulation (**Figure 7**), the depths are zero and 10 m, which is at the bottom. Note that the radiance axis is logarithmic. A viewing direction of $\theta_v = 0$ corresponds to looking straight down and seeing the upwelling radiance (photons traveling straight up). Near the sea surface, the angular dependence of the radiance distribution is complicated because of boundary effects such as internal reflection (the bumps near $\theta_v = 90°$, which is radiance traveling horizontally) and refraction of the sun's direct beam

(the large spike near $\theta_v = 140°$). As the depth increases, the angular shape of the radiance distribution smooths out as a result of multiple scattering. By 100 m in the case 1 simulation, the shape of the radiance distribution is approaching its asymptotic shape, which is determined only by the IOPs. In the case 2 simulation, the upwelling radiance $(-90° \le \theta_v \le 90°)$ at the bottom is isotropic; this is a consequence of having assumed the mud bottom to be a Lambertian reflecting surface. As the depth increases, the color of the radiance becomes blue for the case 1 water and greenish-yellow for the case 2

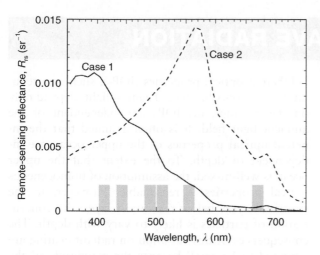

Figure 9 The remote-sensing reflectance R_{rs} for the case 1 and case 2 waters.

water. In the case 1 simulation at 100 m, there is a prominent peak in the radiance near 685 nm, even though the solar radiance has been filtered out by the strong absorption by water at red wavelengths. This peak is due to chlorophyll fluorescence, which is transferring energy from blue to red wavelengths, where it is emitted isotropically.

As already noted, the extensive information contained in the full radiance distribution is seldom needed. A biologist would probably be interested only in the scalar irradiance E_o, which is shown at selected depths in **Figure 8**. This irradiance was computed by integrating the radiance over all directions. Although the irradiances near the surface are almost identical, the decay of these irradiances with depth is much different in the case 1 and case 2 waters.

The remote-sensing reflectance R_{rs}, the quantity of interest for 'ocean color' remote sensing, is shown in **Figure 9** for the two water bodies. The shaded bars at the bottom of the figure show the nominal SeaWiFS sensor bands. The SeaWiFS algorithm for retrieval of the chlorophyll concentration uses a function of the ratio $R_{rs}(490 \, \text{nm})/R_{rs}(555 \, \text{nm})$. When applied to these R_{rs} spectra, the SeaWiFS algorithm retrieves a value of $Chl = 0.24 \, \text{mg m}^{-3}$ for the case 1 water, which is close to the average value of the measured profile over the upper few tens of meters of the water column. However, when applied to the case 2 spectrum, the

SeaWiFS algorithm gives $Chl = 8.88 \, \text{mg m}^{-3}$, which is almost twice the value of $5.0 \, \text{mg m}^{-3}$ used in the simulation. This error results from the presence of the mineral particles, which are not accounted for in the SeaWiFS chlorophyll retrieval algorithm.

These *Hydrolight* simulations highlight the fact that it is now possible to compute accurate underwater radiance distributions given the IOPs and boundary conditions. The difficult science lies in learning how to predict the IOPs for the incredible variety of water constituents and environmental conditions found in the world's oceans, and in learning how to interpret measurements such as R_{rs}. The development of bio-geo-optical models for case 2 waters, in particular, is a research topic for the next decades.

See also

IR Radiometers.

Further Reading

Bukata RP, Jerome JH, Kondratyev KY, and Pozdnyakov DV (1995) *Optical Properties and Remote Sensing of Inland and Coastal Waters*. New York: CRC Press.

Caimi FM (ed.) (1995) *Selected Papers on Underwater Optics*. SPIE Milestone Series, vol. MS 118. Bellingham, WA: SPIE Optical Engineering Press.

Jerlov NG (1976) *Marine Optics*. Amsterdam: Elsevier.

Kirk JTO (1994) *Light and Photosynthesis in Aquatic Ecosystems*, 2nd edn. New York: Cambridge University Press.

Mobley CD (1994) *Light and Water: Radiative Transfer in Natural Waters*. San Diego: Academic Press.

Mobley CD (1995) The optical properties of water. In Bass M (ed.) *Handbook of Optics*, 2nd edn, vol. I. New York: McGraw Hill.

Mobley CD and Sundman LK (2000) *Hydrolight 4.1 Users' Guide*. Redmond, WA: Sequoia Scientific. [See also www.sequoiasci.com/hydrolight.html]

Shifrin KS (1988) *Physical Optics of Ocean Water*. AIP Translation Series. New York: American Institute of Physics.

Spinrad RW, Carder KL, and Perry MJ (1994) *Ocean Optics*. New York: Oxford University Press.

Walker RE (1994) *Marine Light Field Statistics*. New York: Wiley.

PENETRATING SHORTWAVE RADIATION

C. A. Paulson and W. S. Pegau, Oregon State University, Corvallis, OR, USA

Introduction

The penetration of solar radiation into the upper ocean has important consequences for physical, chemical, and biological processes. The principal physical process is the heating of the upper layers by the absorption of solar radiation. To estimate the solar radiative heating rate, the net downward shortwave irradiance entering the ocean and the rate of absorption of this energy as a function of depth must be determined. Shortwave irradiance is the flux of solar energy incident on a plane surface ($W m^{-2}$).

Given the downward shortwave radiance field just above the sea surface, the rate of shortwave absorption as a function of depth is governed primarily by sea surface roughness, molecular structure of pure sea water, suspended particles, and dissolved organic compounds. The optical properties of pure sea water are considered a baseline; the addition of particles and dissolved compounds increases absorption and scattering of sunlight. The dissolved organic compounds are referred to as 'colored dissolved organic matter' (CDOM) or 'yellow matter' because they color the water yellowish-brown. The source of CDOM is decaying plants; concentrations are highest in coastal waters. Suspended particles may be of biological or geological origin. Biological (organic) particles are formed as the result of the growth of bacteria, phytoplankton, and zooplankton. The source of geological (inorganic) particles is primarily weathering of terrestrial soils and rocks that are carried to the ocean by the wind and rivers. Phytoplankton particles are the main determinant of optical properties in much of the ocean and the concentration of chlorophyll associated with these plants is used to quantify the effect of phytoplankton on optical properties. Case 1 waters are defined as waters in which the concentration of phytoplankton is high compared with inorganic particles and dissolved compounds; roughly 98% of the world ocean falls into this category. Case 2 waters are waters in which inorganic particles or CDOM are the dominant influence on optical properties. Case 2 waters are usually coastal, but not all coastal water is case 2.

Inherent optical properties (IOPs), such as attenuation of a monochromatic beam of light, depend only on the medium, i.e. IOPs are independent of the ambient light field. It is often assumed that the inherent optical properties of the upper ocean are independent of depth. To the extent that the upper ocean is well-mixed, the assumption of homogeneous optical properties is reasonable. However, in the stratified layers below the mixed layer, the concentration of particles is likely to vary with depth. The consequences of this variation on radiant heating are expected to be small because the magnitude of the downward irradiance decreases rapidly with depth.

Apparent optical properties (AOPs) depend both on the medium and on the directional properties of the ambient light field. Some AOPs, such as the ratio of downward irradiance in the ocean to the surface value, are sufficiently independent of directional properties of the light field to be useful for characterizing the optical properties of a water body.

Albedo

Albedo, A, is the ratio of upward to downward short-wave irradiance just above the sea surface and is defined by:

$$A \equiv \frac{E_u}{E_d}$$

where E_u and E_d are the upwelling and downward irradiances just above the sea surface, respectively. The upwelling irradiance is composed of two components: emergent irradiance due to back-scattered light from below the sea surface; and irradiance reflected from the sea surface. Emergent irradiance is typically <10% of reflected irradiance. The rate at which net short-wave irradiance penetrates the sea surface is the rate at which the sea absorbs solar energy and is given by:

$$(1 - A) E_d \ (W m^{-2})$$

R.E. Payne analyzed observations to represent albedo as a function of solar altitude θ and atmospheric transmittance Γ defined by:

$$\Gamma = E_d r^2 / S \sin\theta$$

where S is the solar constant ($1370 \ W^{-2}$) and r is the ratio of the actual to mean Earth–sun separation. The transmittance is a measure of the effect of the

Table 1 Mean albedos for the Atlantic Ocean by month and latitude

Latitude	Jan	Feb	Mar	Apr	May	Jun	Jul	Aug	Sep	Oct	Nov	Dec
80°N			0.33	0.14	0.10	0.09	0.08	0.08	0.12			
70°N		0.41	0.15	0.10	0.08	0.07	0.07	0.09	0.11	0.25		
60°N	0.28	0.12	0.09	0.07	0.07	0.07	0.06	0.07	0.07	0.10	0.16	0.44
50°N	0.11	0.10	0.08	0.07	0.06	0.06	0.06	0.07	0.07	0.08	0.11	0.12
40°N	0.10	0.09	0.07	0.07	0.06	0.06	0.06	0.06	0.07	0.08	0.10	0.11
30°N	0.09	0.07	0.06	0.06	0.06	0.06	0.06	0.06	0.06	0.07	0.08	0.09
20°N	0.07	0.06	0.06	0.06	0.06	0.06	0.06	0.06	0.06	0.06	0.07	0.07
10°N	0.07	0.06	0.06	0.06	0.06	0.06	0.06	0.06	0.06	0.06	0.06	0.07
0°	0.06	0.06	0.06	0.06	0.06	0.06	0.06	0.06	0.06	0.06	0.06	0.06
10°S	0.06	0.06	0.06	0.06	0.07	0.07	0.06	0.06	0.06	0.06	0.06	0.06
20°S	0.06	0.06	0.06	0.06	0.07	0.07	0.07	0.07	0.06	0.06	0.06	0.06
30°S	0.06	0.06	0.06	0.07	0.08	0.09	0.08	0.07	0.07	0.06	0.06	0.06
40°S	0.06	0.06	0.07	0.08	0.09	0.11	0.10	0.08	0.07	0.07	0.06	0.06
50°S	0.06	0.07	0.07	0.08	0.10	0.13	0.11	0.08	0.08	0.07	0.06	0.06
60°S	0.06	0.07	0.08	0.11	0.13		0.27	0.07	0.08	0.07	0.06	0.06

(Reproduced with permission from Payne, 1972.)

Earth's atmosphere, including clouds, on the radiance distribution at the Earth's surface. If there were no atmosphere, the transmittance would equal one and the radiance would be a direct beam from the sun. For very heavy overcast, the transmittance can be <0.1 and the downward radiance distribution may be approximately independent of direction.

Payne's observations were taken from a fixed platform off the coast of Massachusetts from 25 May to 28 September. Solar altitude ranged up to 72° and the mean wind speed was 3.7 m s^{-1}. The transmittance varied from near zero to about 0.75. Payne fitted smooth curves to the albedo as a function of transmittance for observations in intervals of 2° of solar altitude and 0.1 in transmittance. The smoothed albedos ranged from 0.03 to 0.5. Payne extrapolated the curves to values of solar altitude and transmittance for which there were no observations by use of theoretical calculations of reflectance for a sea surface roughened by a wind speed of 3.7 m s^{-1}. Albedo was obtained for the limiting case of $\Gamma = 1$ by adding 0.005 to the calculated reflectance to account for the irradiance emerging from beneath the surface.

Wind speed affects albedo through its influence on the surface roughness. The clear-sky reflectivity from a flat water surface is a strong function of solar altitude for altitudes <30°. As wind speed and roughness increase, reflectivity decreases because of the nonlinear relationship between reflectivity and the incidence angle. Payne investigated the variation of albedo with wind speed for solar elevations from 17° to 25° and found that albedo decreased at the rate of 2% per meter per second increase in wind speed.

Wind speed may also affect albedo by the generation of breaking waves that produce white caps. The albedo of whitecaps is higher, on average, than the whitecap-free sea surface. Hence the qualitative effect of whitecaps is to increase the albedo. Monahan and O'Muircheartaigh estimate an increase of 10% in albedo due to whitecaps for a wind speed of 15 m s^{-1} and of 20% for a wind speed of 20 m s^{-1}.

Payne estimated monthly climatological values of albedo at 10° latitude intervals for the Atlantic Ocean (**Table 1**). For the range of latitudes from 40°S to 40°N, mean albedo varies from a minimum of 0.06 at the equator in all months to a maximum of 0.11 at 40°S and 40°N for the months containing the winter solstice. The symmetry exhibited by mean albedo values for the winter solstice in the North and South Atlantic suggests that the values in **Table 1** may be reasonable estimates for the world ocean.

Spectrum of Downward Irradiance

The spectrum of downward short-wave irradiance at various depths in the ocean (**Figure 1**) illustrates the strong dependence of absorption on wavelength. The shape of the spectrum at the surface is determined primarily by the temperature of the sun (Wien displacement law) and wavelength-dependent absorption by the atmosphere. The area under the spectrum at each depth is proportional to the downward irradiance. The downward irradiance at a depth of 1 m is less than half the surface value because of preferential absorption at wavelengths in excess of 700 nm. The downward irradiance below 10 m depth is in a relatively narrow band centered

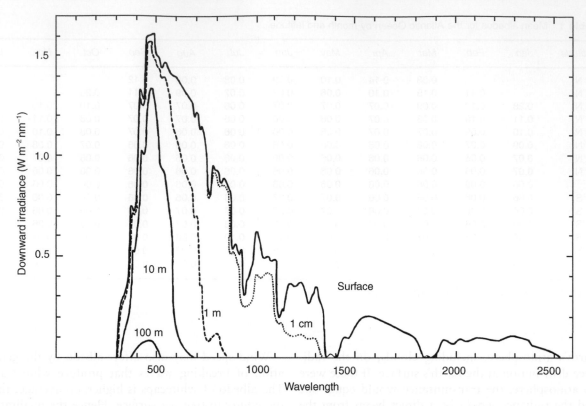

Figure 1   The spectrum of downward irradiance $E_d(z, \lambda)$ in the sea at different depths. (Adapted with permission from Jerlov, 1976.)

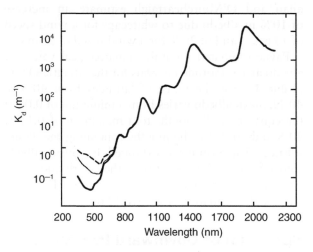

Figure 2   The diffuse attenuation coefficient for downward irradiance in sea water versus wavelength. Data for the wavelength band from 350 to 800 nm are from the Hydrolight radiative transfer model with the following conditions: depth 10 m, solar altitude 60°, cloudless sky, wind speed 2 m s^{-1}. The three lines (thin, thick, dashed) result from specified concentrations of chlorophyll (0.05, 1, 10 mg m^{-3}) and the beam attenuation coefficient at 440 nm for CDOM (0.01, 0.05, 0.1 m^{-1}). Data for the wavelength band from 800 to 2200 nm are for pure water. (Tabulations taken from Kuo et al., 1993.)

near 470 nm (blue-green). Pure sea water is most transparent near a wavelength of 450 nm which, by coincidence, is close to the peak in the downward irradiance spectrum at the surface.

Modeled Irradiance

Radiative transfer models are useful tools for investigating the characteristics of underwater light fields and their dependence on suspended particles and dissolved organic matter. The Hydrolight model, constructed by Mobley, is used to illustrate the diffuse attenuation coefficient for downward irradiance for three cases with different concentrations of chlorophyll and values of CDOM beam attenuation at 440 nm (**Figure 2**). The diffuse attenuation coefficient for downward irradiance K_d is defined by

$$K_d(z,\lambda) = \frac{d \ln E_d(z,\lambda)}{dz}$$

where z is the vertical space coordinate, zero at the surface and positive upward, and λ is the wavelength. If K_d is independent of z, monochromatic irradiance decreases exponentially with depth, consistent with Beer's law. K_d increases five orders of magnitude as wavelength increases from 500 to 2000 nm (**Figure 2**), consistent with the strong dependence of absorption on wavelength shown in **Figure 1**. In the wavelength band centered around 500 nm, K_d varies by a factor of 10 between the least absorbent and most absorbent cases.

The order of magnitude variation in K_d at 500 nm among the three cases (**Figure 2**) has a dramatic effect

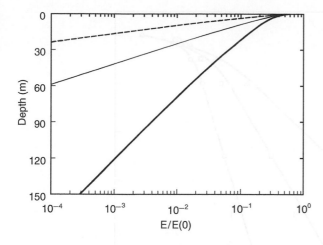

Figure 3 Net irradiance $E = (E_d - E_u)$ versus depth from the Hydrolight model for the conditions specified in the caption for **Figure 2**.

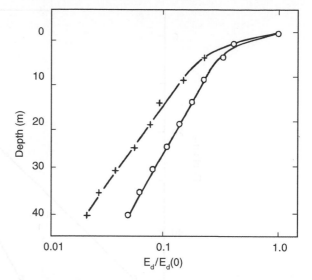

Figure 4 Measurements in the upper 40 m of downward irradiance normalized by downward irradiance just below the surface. The measurements were made in the North Pacific (35°N, 155°W) in February. The open circles are the average of five sets of observations with similar irradiance profiles; solar altitude ranged from 30° to 38° and the sky was overcast. The plus signs show one set of observations for which solar altitude was 16° and the sky was clear. The curves are the sum of two exponential terms fitted to the observations (see eqn [1]). (Adapted with permission from Paulson and Simpson, 1977.)

on net irradiance (**Figure 3**) because the peak of the solar spectrum is near 500 nm (**Figure 1**). Net irradiance, E, is the difference between net (wavelength integrated) downward and net upward irradiance. The difference between downward and net irradiance is negligible for most purposes because upward irradiance is typically 3% of downward irradiance within the ocean. At depths where the net downward irradiance is <10% of its surface value, the decay with z is approximately exponential because light at these depths is roughly monochromatic.

The three cases with different optical properties (**Figures 2** and **3**) can be characterized biologically as oligotrophic, mesotrophic, and eutrophic (ranging from least to most absorbent). Oligotrophic water has low biological production and low nutrients. Eutrophic water has high biological production and high nutrients and mesotrophic water is moderate in both respects. The oligotrophic case illustrated in **Figures 2** and **3** is typical of open-ocean water. Mesotrophic and eutrophic water are likely to be found near the coast.

Parameterized Irradiance versus Depth

Observations show that downward short-wave irradiance decreases exponentially with depth below a depth of about 10 m (**Figure 4**) as the result of absorption by the overlying sea water of all irradiance except for blue-green light. This suggests that E_d can be approximated by a sum of n exponential terms:

$$E_d/E_0 = \sum_{i=1}^{n} F_i \exp(K_i z) \qquad [1]$$

$$\sum_{i=1}^{n} F_i = 1$$

Table 2 Values of parameters determined by fitting the sum of two exponential terms (see eqn [1]) to values of downward irradiance which define *Jerlov's (1976)* water types

Water type	F_1	$K_1\ (m^{-1})$	$K_2\ (m^{-1})$	$C\ (mg\,m^{-1})$
I	0.32	0.036	0.8	0–0.01
IA	0.38	0.049	1.7	~0.05
IB	0.33	0.058	1.0	~0.1
II	0.23	0.069	0.7	~0.5
III	0.22	0.13	0.7	~1.5–2.0

Values of downward irradiance in the upper 100 m were used, except that values were limited to the upper 50 m for type I because of a change in slope below 50 m. F_2 is $1 - F_1$. (Adapted from Paulson and Simpson, 1977.) The column labeled C is the approximate chlorophyll concentration for each water type as determined by Morel (1988).

where F_i is the fraction of downward irradiance in a wavelength band i and K_i is the diffuse attenuation coefficient for the same band. The leading term in eqn [1] is defined as the short wavelength band that describes the exponential decay below 10 m (**Figure 4**). At least one additional term is required. A total of two terms fits the observations in **Figure 4** reasonably well, although accuracy in the upper few meters is lacking.

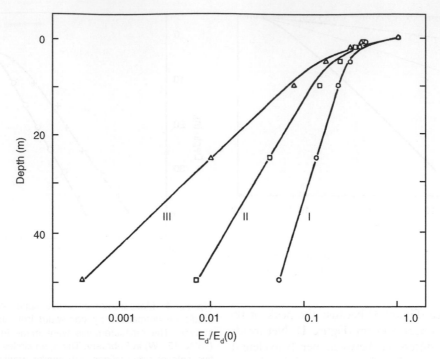

Figure 5 Normalized downward irradiance versus depth for water types I, II, and III. The data (open circles, squares, and triangles) are from Jerlov (1976) and the curves are a fit to the data with the parameters given in **Table 2**. (Adapted with permission from Paulson and Simpson, 1977.)

Jerlov has proposed a scheme for classifying oceanic waters according to their clarity. He defined five types (I, IA, IB, II, and III) ranging from the clearest open-ocean water (type I) to increasingly turbid water. Parameters for the sum of two exponential terms (eqn [1]) fit to the values of downward irradiance which define the Jerlov water types are given in **Table 2** and plots of values and fitted curves are shown in **Figure 5**. Apart from systematic disagreement in the upper few meters, the fit is good. Differences in the values of K_2 in **Table 2** are not significant. Water types IA and IB are not shown in **Figure 5**. However, the fits to the observations shown in **Figure 4** yield parameters very similar to those for types IA and IB (open circles and plus signs, respectively, in **Figure 4**). Most open-ocean water is intermediate between types I and II. The approximate chlorophyll concentration for each of the Jerlov water types is given in **Table 2**.

The Jerlov water types can be compared to the oligotrophic, mesotrophic, and eutrophic cases illustrated in **Figures 2** and **3**. The oligotrophic case is intermediate between types IA and IB. The mesotrophic case is similar to type III and the eutrophic case is similar to Jerlov's coastal type 5 water.

The sum of two exponential terms (eqn [1]) is adequate for modeling purposes when the required vertical resolution is a few meters or greater. For a vertical resolution of 1 m or less, additional terms are required. These additional terms can be constructed with knowledge of the surface irradiance spectrum (**Figure 1**) and the diffuse attenuation coefficient versus wavelength (**Figure 2**).

See also

Heat and Momentum Fluxes at the Sea Surface. Photochemical Processes. Radiative Transfer in the Ocean. Upper Ocean Heat and Freshwater Budgets. Upper Ocean Time and Space Variability. Wind- and Buoyancy-Forced Upper Ocean.

Further Reading

Dera J (1992) *Marine Physics*. Amsterdam: Elsevier.

Jerlov NG (1976) *Marine Optics*. Amsterdam: Elsevier.

Kou L, Labrie D, and Chylek P (1993) Refractive indices of water and ice in the 0.65- to 2.5-μm spectral range. *Applied Optics* 32: 3531–3540.

Kraus EB and Businger JA (1994) *Atmosphere–Ocean Interaction*, 2nd edn. New York: Oxford University Press.

Mobley CD (1994) *Light and Water*. San Diego: Academic Press.

Mobley CD and Sundman LK (2000) *Hydrolight 4.1 User's Guide*. Redmond, WA: Sequoia Scientific.

Monahan EC and O'Muircheartaigh IG (1987) Comments on glitter patterns of a wind-roughened sea surface. *Journal of Physical Oceanography* 17: 549–550.

Morel A (1988) Optical modeling of the upper ocean in relation to its biogenous matter content. *Journal of Geophysical Research* 93: 10749–10768.

Paulson CA and Simpson JJ (1977) Irradiance measurements in the upper ocean. *Journal of Physical Oceanography* 7: 952–956.

Payne RE (1972) Albedo of the sea surface. *Journal of Atmospheric Sciences* 29: 959–970.

Thomas GE and Stamnes K (1999) *Radiative Transfer in the Atmosphere and Ocean.* Cambridge: Cambridge University Press.

Tyler JE and Smith RC (1970) *Measurements of Spectral Irradiance Underwater.* New York: Gordon and Breach.

AEOLIAN INPUTS

R. Chester, Liverpool University, Liverpool, Merseyside, UK

Introduction

The oceans are an important reservoir in the global biogeochemical cycles of many elements, but until the recent past it was thought that material fluxes to the reservoir were dominated by fluvial inputs. Over the last two or three decades, however, it has become apparent that the atmosphere is a major transport pathway in the land–sea exchange of material. This atmospherically transported material differs from that introduced by fluvial inputs in two important ways. (i) It is delivered, albeit at different flux magnitudes, to all areas of the sea surface, whereas river inputs are initially delivered to the land–sea margins. (ii) It does not pass through the biogeochemically dynamic estuarine filter; a region of intense dissolved/particulate reactivity, which, under present day conditions, retains ~90% of fluvial particulate material. As a result, the atmosphere is the most important pathway for the long-range transport of much of the particulate material delivered directly to open-ocean regions. This material has an important influence on marine sedimentation; for example, in equatorial North Atlantic deep-sea sediments deposited to the east of the Mid-Atlantic Ridge and in central North Pacific deep-sea sediments, essentially all the land-derived components are aeolian in origin. Atmospheric aerosols can also exert an influence on climatic forcing by acting as cloud condensation nuclei and by processes such as the scattering of short-wave radiation by both anthropogenic and natural aerosols. Further, the presence of anthropogenic sulfate aerosols in the atmosphere can lead to an increase in albedo and so cool the planet; an effect, which, on a global scale, is comparable to that induced by the 'greenhouse' gases, but is opposite in sign.

The aeolian material delivered to the sea surface by the atmosphere is dispersed from the source regions via the major wind systems, such as the Trades and the Westerlies, within which relatively small-scale winds (e.g. the Sirocco and the Mistral in the Mediterranean) can be important locally. Large-scale meteorological phenomena can also affect aeolian transport; for example, long-term inter-annual variability in dust transport out of Africa to the Atlantic Ocean and the Mediterranean Sea has been linked to precipitation patterns induced by the North Atlantic Oscillation. Material in the marine atmosphere consists of gaseous and particulate components, both of which can originate from either natural or anthropogenic sources. Gas-to-particle conversions are important in the generation of particulate material, especially that derived from anthropogenic sources; however, air/sea gaseous exchange is covered in articles on air–sea interactions, and attention here is largely confined to the particulate aerosol.

The sea surface itself is a major source of particulate material to the marine atmosphere in the form of sea salt. However, these sea salts are re-cycled components, and the globally important terrestrial sources of material to the marine atmosphere, i.e. those supplying material involved in land–sea exchange, are (i) the Earth's crust (mineral dust), and (ii) anthropogenic processes (sulfates, nitrates, etc). Other terrestrial sources, which include volcanic activity and the biosphere (e.g. direct release from vegetation, biomass burning), can also supply components to the atmosphere. Material is removed from the atmosphere by a combination of two depositional modes; (i) the 'dry' mode, which does not involve an aqueous phase, and (ii) the 'wet' (precipitation scavenging) mode, either by cloud droplets (in-cloud processes) or by falling rain (below-cloud processes).

Land–sea Exchange of Individual Components

Marine Aerosol: Major Components, Sources, and Distribution

Data are available on the concentrations of aerosols over many marine regions, and it is now apparent that there is an 'aerosol veil' over all oceans. Concentrations of material in the aerosol veil, however, vary from $\sim 10^3$ ng m^{-3} of air close to continental sources to $\sim 10^{-2}$ ng m^{-3} of air over pristine oceanic regions. The aerosol veil is composed mainly of mineral dust and anthropogenic components. On a global scale, the anthropogenic material is dominated by sulfate aerosols, together with smaller amounts of nitrates. The mineral dust aerosol consists of a wide variety of minerals, with quartz, the clay minerals and feldspars usually being predominant. The signatures of the major clay minerals (chlorite, kaolinite, illite, and montmorillonite) can be used as tracers to identify the sources of the dusts, the extent to which the material has been transported over the oceans, and its contribution to marine sedimentation.

The principal continental sources of both the mineral dust aerosol and anthropogenic (mainly sulfate) aerosol are concentrated in specific latitudinal belts, predominantly in the northern hemisphere (see **Figure 1A**). Quantitatively, the crust-derived mineral dust, which is derived mainly from the arid and semi-arid desert regions of the world (**Figure 1A**), imposes the strongest fingerprints on the marine aerosol. Mineral dust fluxes to the world ocean are listed in **Table 1**, and are illustrated in **Figure 2**. From this figure it can be seen that the highest dust fluxes to the sea surface are found off the major deserts, e.g. the Sahara in the North Atlantic and the Asian deserts in the North Pacific. Much of the material injected into the atmosphere from these arid sources is transported in the form of

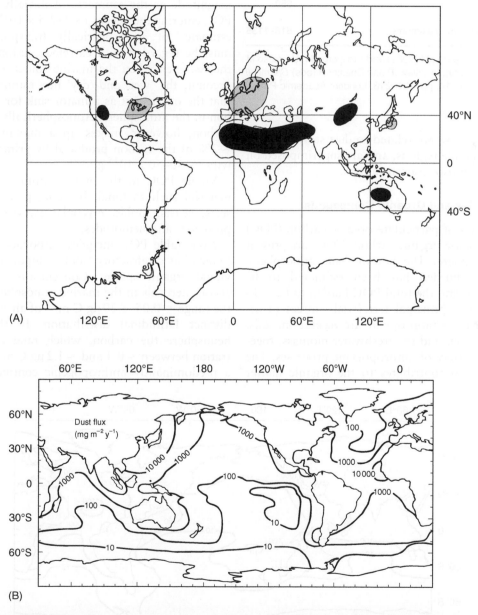

Figure 1 Aerosols: terrestrial sources and fluxes to the world ocean. (A) Terrestrial sources of aerosol production; light gray areas indicate regions of anthropogenic emissions, and dark gray areas indicate regions of mineral aerosol production. (Reproduced with permission from Gilman C and Garrett C (1994) Heat flux parameterizations for the Mediterranean Sea: The role of atmospheric aerosols and constraints from the water budget. *Journal of Geophysical Research* 99: 5119–5134.) (B) Mineral aerosol fluxes to the world ocean (units, mg m^{-2} y^{-1}). (Reproduced with permission from Duce RA *et al.* (1991)).

Table 1 Dust deposition rates to the world ocean

Ocean region	Deposition rate	
	10^{-6} g cm^2 y^{-1}	10^{12} g y^{-1}
North Atlantic, north of trades	82	12
North Atlantic trades	–	100–400
South Atlantic	85	18–37
Indian Ocean	450	336
North Pacific:		
western North Pacific	5000	300
central and eastern	11–62	30
South Pacific	5–64	18
Entire Pacific		350
All oceans (minimum–maximum)		816–1135

Adapted from Prospero JM (1981) and Prospero JM, Uematso M and Savoie DL (1989). In: Riley JP and Chester R (eds) *Chemical Oceanography*, vol. 10, pp. 137–218. London: Academic Press.

dust 'pulses'; these are related to dust storms on the continental source regions, and are superimposed on background aerosol concentrations.

Organic Matter and Organic Compounds

Various classes of particulate organic carbon (POC) and vapour-phase organic carbon (VOC) are present in the atmosphere. The non-methane atmospheric global VOC burden has been estimated to be $\sim 50 \times 10^{12}$ g, and the total POC burden to be ~ 1–5×10^{12} g. The principal terrestrial sources of organic matter to the atmosphere are vegetation, soils, biomass burning, and the freshwater biomass, together with a variety of anthropogenic processes. The sea surface also contributes to the organic matter

burden in the marine atmosphere, with $\sim 14 \times 10^{12}$ g y^{-1} of organic carbon being produced by the ocean surface; >90% being on particles >1 μm in diameter. Particulate organic matter (POM) is removed from the air via 'dry' and 'wet' deposition, and in addition the removal of VOM includes conversion to POM and transformation to inorganic gaseous products. Estimates of the 'wet' atmospheric depositional flux of carbon to the ocean surface range between $\sim 2.2 \times 10^{14}$ g y^{-1} and $\sim 10 \times 10^{14}$ g y^{-1}, and for the 'dry' flux a value of $\sim 6 \times 10^{12}$ g y^{-1} has been proposed. These are of the same order of magnitude as the estimates of fluvially transported POC entering the oceans (~ 1–2.5×10^{14} g y^{-1}). The estimates of atmospherically transported carbon must be regarded with extreme caution, but nonetheless, even when the marine source is taken into account, the 'wet' and 'dry' flux estimates indicate that the oceans act as a major sink for organic carbon in the atmosphere. Atmospherically transported carbon, however, makes up a maximum of only $\sim 2\%$ of the carbon produced by primary productivity (~ 30–50×10^{15} g y^{-1}).

Viable POC in the marine atmosphere includes material such as fungi, bacteria, pollen, algae, insects, yeasts, molds, mycoplasma, viruses, phages, protozoa, and nemotodes.

Non-viable POC includes carbonaceous material (which has a refractory 'soot' component) and individual organic species. Concentrations of carbonaceous aerosols in the marine atmosphere vary over the range ~ 0.05–1.20 μg C m^{-3} of air and display a distinct latitudinal distribution. In the northern hemisphere the carbon, which ranges in concentration between ~ 0.4 and ~ 1.2 μg C m^{-3} of air, has a predominantly anthropogenic continental source

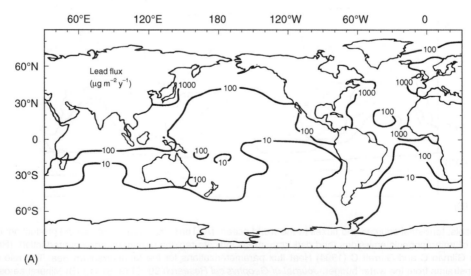

(A)

Figure 2 Fluxes of lead to the world ocean (units, μg m^{-2} y^{-1}). From Duce RA *et al.* (1991).

from combustion processes. In the southern hemisphere POC concentrations are lower (range ~ 0.05–$0.30\,\mu g\ C\,m^{-3}$ of air), and natural continental and marine sources are about equal.

A wide variety of individual species of both natural and synthetic organic compounds are transported from the continents to the sea surface via the atmosphere. The classes of organic compounds that have received particular attention include aliphatic hydrocarbons, wax esters, fatty alcohols, sterols, fatty acids, and long-chain unsaturated ketones. The most extensive investigation of atmospherically transported organic species on an ocean-wide basis was carried out in the Pacific as part of the SEXREX Program (Sea-Air Exchange Program). On the basis of data obtained from the Pacific it is apparent that in the remote marine atmosphere organic carbon accounts for $\sim 10\%$ of the total aerosol, although only $\sim 1\%$ of this has been characterized. The most abundant terrestrially derived components are the n-alkanes and the C_{21}–C_{36} fatty alcohols, which are common in the epicuticular waxes of vascular plants, and the most abundant marinederived species are the C_{13}–C_{18} fatty acid salts.

The atmosphere is a major pathway for the transport of a number of organic pollutants that enter sea water. These include the synthetic trace organics, such as high molecular-weight halogenated hydrocarbons of the following compounds, or compound classes; chlorobenzenes (e.g. hexachlorobenzene, HCB), chlorocyclohexanes (e.g. hexachlorocyclohexanes, HCHs), polychlorobiphenyls (PCBs), dichlorodiphenyltrichloroethanes (DDTs), and non-halogenated coumpounds (e.g. polynuclear aromatic hydrocarbons, PAHs). Global atmospheric fluxes of HCHs, HCBs, DDTs, and PCBs to the world ocean are listed in Table 2, and two important conclusions can be drawn from the data. (i) The dominant deposition of the organochlorines is to the North Atlantic and North Pacific, which is consistent with their source derivations; with HCH and DDT compounds having their highest deposition rates in the North Pacific and PCBs in the North Atlantic. (ii) The

atmospheric inputs of the organochlorines to the world ocean exceed those from fluvial inputs by 1 to 2 orders of magnitude.

Nutrients

These include nitrate, phosphate, and micro-nutrients such as iron. The main features in the atmospheric input of nitrogen nutrients to the world ocean can be summarized as follows. (i) The atmospheric input of total nitrogen (N) to the global ocean is $\sim 30.2 \times 10^{12}$ g N y^{-1}, made up of $\sim 13.4 \times 10^{12}$ g N y^{-1} oxidized nitrogen species and $\sim 16.8 \times 10^{12}$ g N y^{-1} reduced nitrogen species, which is similar in magnitude to the total (i.e. natural + anthropogenic) fluvial nitrogen flux (~ 21–49×10^{12} g N y^{-1}). (ii) The overall flux of nitrogen to the sea surface is $\sim 87\,mg$ N $m^{-2}\ y^{-1}$; the largest fluxes being to the North Atlantic and North Pacific. (iii) The 'wet' removal of reduced nitrogen species may be an important source of nutrients to the oceans. The atmospheric fluxes of nitrogen species usually include only inorganic forms, such as nitrate and ammonia; however, if dissolved organic nitrogen (DON), most of which has an anthropogenic source, is included in 'wet' deposition it would increase the anthropogenic input of fixed nitrogen to the oceans by a factor of ~ 1.5.

The atmospheric input of phosphate has been studied in detail in the Mediterranean Sea. However, the role played by atmospherically transported phosphorus in nutrient cycles is not clearly understood; for example, it has been proposed that Saharan dust may act as a *sink* for the removal of dissolved phosphorus in the water column by adsorption onto iron-rich particles, or as a *source* with up to $\sim 8\%$ of the phosphorus in the dusts being soluble in sea water. There is evidence that in summer months the atmosphere does provide a source of phosphorus to the Western Mediterranean and that this may account for new production, since the stratification of the surface waters prevents the input of nutrients from deep waters. For example, in the Ligurian Sea a strong summer desert dust transport

Table 2 The atmospheric input of some organochlorine compounds to the world oceans (units, 10^6 g y^{-1})

Compound[a]	North Atlantic	South Atlantic	North Pacific	South Pacific	Indian Ocean	Global atmospheric	Global fluvial
HCHs	850	97	2600	470	700	4800	40–80
HCB	17	10	20	19	11	77	4
DDTs	16	14	66	26	43	170	4
PCBs	100	14	36	29	52	240	40–80

[a]HCHs, hexachlorocyclohexanes; HCB, hexachlorobenzene; DDTs, dichlorodiphenyltrichloroethanes; PCBs, polychlorobiphenyls. Adapted from *Duce RA et al.* (1991).

episode was followed 10 days later by a significant increase in phytoplankton concentration, which could have resulted from the solubilization of phosphorus from the dusts.

In recent years there has been a renewed interest in the role played by iron as a limiting nutrient in primary production, especially in high-nitrogen, low-productivity (HNLP) regions in which there is sufficient light and nutrient concentrations but low productivity. The concentrations of iron in open-ocean waters are generally low, and it is thought that the element would run out before nitrate is exhausted. To provide sufficient iron it therefore has been suggested that it must be added to sea water from other sources, one of which is the long-range transport of atmospheric dust. However, iron oxyhydroxide particles and iron colloids are not directly available to phytoplankton, and the bioavailable forms of the metal are thought to be dissolved Fe(II), which is rapidly converted into Fe(III), and Fe(III) itself. As a result, iron must be dissolved from atmospheric dust before it becomes available to phytoplankton, and although adsorption onto particulate matter is the dominant control on the concentrations of dissolved iron in open-ocean waters, the deposition of mineral dust from high concentration episodic atmospheric events can result in a net addition of dissolved iron to surface waters. Further, during 'wet' deposition iron may undergo reductive dissolution to Fe(II), a form of iron that would be immediately available to phytoplankton. It also has been suggested that siderophores (compounds with a high affinity for ferric iron which are secreted by organisms) may play a role in the bioavailability of iron. Various small-scale laboratory simulations have shown that phytoplankton growth rates increase in response to the addition of iron to the system, but there is considerable disagreement over the interpretation of the results. Laboratory experiments also have been criticized on the grounds that they do not represent planktonic community response on an ocean-wide scale. To overcome this, the Fe-limitation hypothesis has been tested by large-scale intervention experiments, such as IronEx I and IronEx II, which were carried out in the equatorial Pacific. In these experiments patches of sea (defined by SF_6 tracer) were seeded with iron in the concentrations expected from natural events. Although there was a doubling of the plant biomass following iron addition, interpretation of the data from IronEx I was hampered as a result of the subduction of the seeded seawater patch below a layer of less dense water. During IronEx II, however, a massive phytoplankton bloom was triggered,

providing direct evidence that in these HNLP waters phytoplankton growth is iron limited.

Trace Metals

Particulate trace metals in the marine atmosphere that are involved in land–sea exchange are derived from two principal terrestrial sources. The first is the Earth's crust. Crustal weathering involves low-temperature generation processes, and crust-derived elements are found on particles with mass median diameters (MMDs) in the range ~ 1–$3\,\mu m$. Relative to other sources, crust-derived elements are referred to as the non-enriched elements (NEEs). The second source comprises a variety of anthropogenic processes, which often involve high temperatures (e.g. fuel combustion, ore smelting). Anthropogenically derived elements (e.g. Pb, Cu, Zn, Cd, As, Hg) are largely found on smaller particles with MMDs $< 0.5\,\mu m$, and are termed the anomalously enriched elements (AEEs). Less important terrestrial sources of trace metals to the marine atmosphere include volcanic activity and the biosphere. In addition, the sea surface supplies recycled elements to the marine atmosphere; this is a low-temperature source, and sea salt-associated elements are located on particles with MMDs in the range ~ 3–$7\,\mu m$. Global elemental atmospheric emission rates from natural and anthropogenic sources are listed in **Table 3**.

The overall trace metal composition of the marine aerosol is dependent on the extent to which material from the various sources are mixed together in the atmosphere, and trace metal concentrations are a function of factors such as the distance the air mass transporting them has travelled from the source, the

Table 3 Global elemental emission rates to the atmosphere (units, 10^9 g y^{-1})

Element	Natural	Anthropogenic
Al	20940	4000
As	12	19
Cd	1.4	7.7
Co	6.1–7.3	2.9
Cr	44	30.5
Cu	22–28	35–52
Fe	10370	6000
Hg	2.5	3.6
Mn	221–317	39–408
Pb	11.5–12	332–404
V	41–45	22–86
Zn	45–280	132–280

Adapted from Chester R (2000) (full references given in original table).

Table 4 Concentrations of particulate trace metals in the marine atmosphere (units, ng m^{-3} of air)

Trace metal	Coastal regions				Open-ocean regions			
	Close to anthropogenic sources		Close to crustal sources					
	North Sea	W. Black Sea	N. Atlantic north-east trades	N. Arabian Sea	Tropical N. Atlantic	Tropical Indian Ocean	Tropical N. Pacific	Tropical S. Pacific
Al	294	540	5925	1227	160	11	21	
Fe	353	420	3685	790	100	8.8	17	
Mn	14.5	17	65	17	2.2	0.16	0.29	
Ni	3.8	4.9	6.6	2.0	0.64	0.043	–	–
Cr	4.7	9.0	10	3.0	0.43	0.066	0.09	–
V	–	3.2	15	6.3	0.54	0.023	0.08	–
Cu	6.3	–	4.5	2.6	0.79	0.077	0.045	0.013
Zn	41	46	16	10	4.4	0.10	0.17	0.07
Pb	34.5	60	6.9	4.3	9.9	0.17	0.12	0.016

Adapted from Chester R (2000) (full references given in original table).

'aging' of the aerosol in the air, and the relative effectiveness of the processes that remove material from the atmosphere. As a result, trace metals in the marine atmosphere have concentrations ranging over several orders of magnitude – see **Table 4**. It is apparent from the data in this table that, in general, the trace metal concentrations decrease with increasing remoteness from continental sources in the general rank order: coastal seas > North Atlantic > North Pacific and tropical Indian Ocean > South Pacific. This is reflected in trace metal fluxes, which decrease in the same sequence; lead atmospheric fluxes to the world ocean are illustrated in **Figure 2**. Despite the fact that trace metal concentrations decrease towards more pristine oceanic environments, atmospheric inputs can be the dominant source of some trace metals to the mixed layer in open-ocean regions where inputs from other sources are minimal. This is especially the case for the 'scavenged-type' metals (e.g. Al, Mn, Pb), which have a surface source and a relatively short residence time in sea water.

Aerosol-associated trace metals are removed from the air by 'dry' or 'wet' depositional processes, and once deposited at the sea surface the initial constraint on the manner in which the metals enter the major marine biogeochemical cycles is a function of the extent to which they undergo solubilization in sea water. There is considerable size-dependent fractionation between the parent aerosol and the deposited material in both the 'dry' and the 'wet' depositional modes. However, with respect to the seawater solubility of trace metals, the major difference between the two depositional modes is that they follow separate geochemical routes. In 'dry'

deposition material is delivered directly to the sea surface and trace metal solubility is largely constrained by particle ↔ seawater reactivity; estimates of the seawater solubility of trace metals from aerosols are listed in **Table 5**. In contrast, in 'wet' deposition there is an initial particle ↔ rainwater reactivity; much of this is pH-dependent, and can involve the dissolution of some trace metals prior to the deposition of the scavenged aerosol at the sea surface. Data on the trace metal composition of marine rainwaters from a number of regions are now available, and a selection are listed in **Table 6**, from which it can be seen that trace metals in rainwaters, like those in aerosols, have their highest concentrations in coastal regions and decrease towards more pristine oceanic regions.

Estimates of particulate and dissolved atmospheric and fluvial trace metal fluxes to the world ocean are

Table 5 Seawater solubility of trace metals from particulate aerosols; % total element soluble

Trace metal	Solubility (%)
Al	~1–10
Fe	~1–50
Mn	~20–50
Ni	≤20–50
Cr	≤10–20
V	≤20–85
Cu	≤10–85
Zn	≤10–75
Pb	≤10–90

Data, which are from various sources, include both crustal and anthropogenic aerosols.

Table 6 Trace metal concentrations in marine-influenced rainwaters (units, ng l^{-1})

Trace metal	North Sea coast (VWM)	Irish Sea coast	Mediterranean Sea		North Atlantic		North Pacific	South Pacific
			South coast, France (VWM)	Sardinia (VWM)	Bermuda (VWM)	Bantry Bay, Ireland (VWM)		
Al	–	43	144	883	–	3.62	2.1	16
Fe	88	48	–	519	4.8	8.06	1.0	0.42
Mn	3.8	2.0	–	8.0	0.27	0.13	0.012	0.020
Cu	2.3	8.7	2.8	2.9	0.66	0.86	0.013	0.021
Zn	13	9.3	–	16	1.15	8.05	0.052	1.6
Pb	4.0	5.2	3.7	1.6	0.77	0.51	0.035	0.014

VWM, volume-weighted mean concentrations; this normalizes the trace metal concentration in a rain to the total amount of rainfall over the sampling period.
Adapted from Chester R (2000) (full references given in original table).

Table 7 Atmospheric and fluvial trace metal fluxes to the world ocean (units, 10^9 g y^{-1})

Element	Atmospheric input		Fluvial input	
	Dissolved	Particulate	Dissolved	Particulate
Fe	3.2×10^3	29×10^3	1.1×10^3	110×10^3
P	310	640	Total 300[a]	Total 300
Ni	8–11	14–17	11	1400
Cu	14–45	2–7	10	1500
Pb	80	10	2	1600
Zn	33–170	11–60	6	3900
Cd	1.9–3.3	0.4–0.7	0.3	15
As	2.3–5.0	1.3–2.9	10	80

Total phosphorus input to marine sediments.
Adapted from Duce RA et al. (1991).

listed in **Table 7**, from which a number of overall conclusions can be drawn. (i) Rivers are the principal source of particulate trace metals to the oceans; phosphorus being an exception. (ii) For iron, nickel, copper, and phosphorus the dissolved atmospheric and the dissolved fluvial inputs are the same order of magnitude. (iii) For lead, zinc, and cadmium the dissolved atmosphere fluxes are dominant; this will still be the case for lead even when allowance is made for the phasing out of leaded gasoline, which has accounted for a large fraction of the anthropogenic lead previously released into the atmosphere.

Conclusions

There is an 'aerosol veil' over all marine regions, and the atmosphere is a major transport route for the supply of mineral dust, organic matter, nutrients, and trace metals to the world ocean. Atmospheric fluxes decrease in strength away from the continental source regions, but in remote open-ocean areas they can be the dominant supply route for the deposition of land-derived particulate material and some trace metals to the ocean surface. Unlike fluvial inputs, which are delivered to the land–sea margins, the atmosphere supplies material to the 'mixed layer' over the whole ocean surface, and the material plays an important role in oceanic biogeochemical cycles and in the formation of marine sediments.

See also

Air–Sea Gas Exchange. Air–Sea Transfer: Dimethyl Sulfide, COS, CS$_2$, NH$_4$, Non-Methane Hydrocarbons, Organo-Halogens. Air–Sea Transfer: N$_2$O, NO, CH$_4$, CO; Atmospheric Input of Pollutants. Nitrogen Cycle. Phosphorus Cycle. Photochemical Processes.

Further Reading

Buat-Menard P (ed.) (1986) *The Role of Air-Sea Exchange in Geochemical Cycling*. Dordrecht: Kluwer Academic Publishers.

Charlson RJ and Heintzenberg (eds.) (1995) *Aerosol Forcing of Climate*. Berlin: Dahlem Workshop.

Chester R (2000) *Marine Geochemistry* 2nd edn. Oxford: Blackwell Science.

Duce RA, Mohnen VA, Zimmerman PR, et al. (1983) Organic material in the global troposphere. *Reviews of Geophysics and Space Physics* 21: 921–952.

Duce RA, Liss PS, Merrill JT, et al. (1991) The atmospheric input of trace species to the World Ocean. *Global Biogeochemical Cycles* 5: 193–529.

Guerzoni S and Chester R (eds.) (1996) *The Impact of Desert Dust Across the Mediterranean*. Dordrecht: Kluwer Academic Publishers.

Knap AH (ed.) (1990) *The Long Range Atmospheric Transport of Natural and Contaminant Substances.* Dordrecht: Kluwer Academic Publishers.

Peltzer ET and Gagosian RB (1989) Organic chemistry of aerosols over the Pacific Ocean. In: Riley JP and Chester R (eds.) *Chemical Oceanography*, vol. 10, pp. 282–338. London: Academic Press.

Prospero JM (1981) Eolian transport to the World Ocean. In: Emiliani C (ed.) *The Sea*, vol. 7, pp. 801–874. New York: John Wiley, Interscience.

Prospero JM (1996) The atmospheric transport of particles to the ocean. In: Ittekkot V, Schafer P, Honjo S, and Depetris PJ (eds.) *Particle Flux to the Oceans*, pp. 19–52. New York: John Wiley.

ATMOSPHERIC TRANSPORT AND DEPOSITION OF PARTICULATE MATERIAL TO THE OCEANS

J. M. Prospero, University of Miami, Miami, FL, USA
R. Arimoto, New Mexico State University, Carlsbad, NM, USA

Introduction

The atmosphere is the primary pathway for the transport of many geochemically important substances to the oceans. Although the magnitude of these wind-borne transports is not accurately known, there is growing evidence that atmospheric deposition significantly impacts chemical and biological processes in the oceans. It is only over the past several decades that marine scientists have come to appreciate the importance of atmospheric transport. Historically it had been assumed that the fluxes of continental materials to the oceans were dominated by rivers. But over time it was recognized that much of the riverine load was deposited in estuaries or on the continental shelves. In contrast, winds can rapidly span great distances to reach even the most remote ocean regions.

The transport and deposition of particulate matter (PM) to the oceans depends on many factors including the distribution of sources, physical and chemical properties of the particles, meteorological conditions, and removal mechanisms. Our interest here focuses on particles between about 0.1 and 10 µm in diameter which, because of their small size, have atmospheric lifetimes ranging from days to several weeks. These are commonly referred to as aerosol particles or aerosols. Larger particles are deposited close to their sources and do not contribute substantially to ocean deposition except in some coastal regions. Smaller aerosols carry little mass and, while they are important for other atmospheric issues, they are not particularly important for air/sea chemical exchange.

Winds carry billions of tons of PM to the ocean. Some of the PM is emitted by natural processes and some is produced anthropogenically, that is, as a result of human activities. Much PM is emitted directly as primary particles; this includes mineral (soil) dust, organic particles from plants, and emissions from anthropogenic combustion processes (e.g., from industry, homes, and vehicles) and biomass burning, which can be natural (e.g., started by lightning) or anthropogenic (e.g., in clearing land, burning agricultural waste). An important PM fraction – secondary PM – is that produced from gases, natural and anthropogenic, that react in the atmosphere to form particles.

One goal of marine scientists is to characterize atmospheric transport and chemical deposition to the ocean and to assess the impact of the air/sea exchange process. This is a difficult task which can only be achieved when we know the kinds of materials deposited and their temporal and spatial variability. Because of the patchy distribution of sources and the relatively short tropospheric residence times of aerosols, PM concentrations over the oceans vary by orders of magnitude in time and space. Here we review the sources and composition of aerosols and the removal mechanisms relevant to deposition issues. We then present estimates of deposition rates of some PM classes to the oceans.

Aerosol Sources, Composition, and Concentrations

The composition of PM over the oceans varies greatly due to the myriad sources and variations in their strengths. Soils emit fine mineral particles. Plants produce a wide range of organic particles, ranging from decayed leaf matter, to plant waxes, and condensed organic compounds. Volcanoes sporadically inject many tons of material into the atmosphere, and much of this is deposited in the oceans; but the total amount of PM deposited over time is relatively small compared with other sources. Smelters, power plants, and incinerators emit PM highly enriched with trace metal pollutants. Combustion sources, both natural (wild fires, biomass burning) and anthropogenic (power plants, vehicles, home heating) emit thousands of organic compounds. Combustion processes and the use of fertilizers contribute to the production of nitrogen-rich particles. Pesticides and other synthetic organic compounds are emitted from industrial and domestic sources.

Typically the dominant marine aerosol species by mass are: (a) sea salt, produced by breaking waves and bursting bubbles; (b) sulfate, including that from sea salt aerosol and non-sea-salt sulfate (nss-SO_4^{2-}), the latter of which is both transported from pollution sources on the continents and produced from gaseous precursors such as dimethyl sulfide (DMS) emitted from the oceans; (c) nitrate, originating from pollution sources on the continents and produced by lightning; (d) ammonium, derived mostly from continental sources but in some areas from ocean sources; (e) mineral dust, from arid lands; (f) organic carbon (OC), largely from anthropogenic and natural sources on the continents; and (g) black carbon (BC), from biomass burning and anthropogenic sources.

PM composition is strongly size dependent not only as a result of various production mechanisms but also because size-selective removal occurs during transport. Physical production mechanisms (grinding of rocks, bursting bubbles) normally produce large particles with most of the mass in PM greater than 1-µm diameter (coarse particles). For example, the mass median diameter (MMD: 50% of the mass is greater than the MMD and 50% is less) of dust particles over deserts can be extremely large, many tens or hundreds of micrometers, but over the oceans, it is typically only several micrometers. The MMD of sea salt particles is generally in the range of about 5–10 µm, depending on wind conditions. Other primary particles including soot emitted from smoke stacks, diesel exhaust, and particles shed by plants (e.g., plant waxes, fibers), tending to be in the size range of ~ 0.1–1.0 µm.

Gas-phase reactions produce secondary PM ranging in size range of 0.001–0.1 µm diameter. Examples are sulfate particles produced from SO_2 emitted from power plants or from the oxidation of DMS emitted from the oceans. Particles in this very fine particle mode are highly mobile. They can rapidly coagulate to form larger particles (typically 0.1–1 µm) or they can diffuse to the surface of cloud or fog droplets or to larger particles (e.g., sea salt, mineral dust).

Table 1 presents concentration data for the aerosols that make up most of the PM mass over the oceans; it also includes data for vanadium, which is included as an example of an element strongly affected by pollution sources. The column on the extreme right shows the total aerosol concentration less that of sea salt, so as to better illustrate the impact of transported PM. These data are the product of long-term measurements obtained from a global ocean network. In general, PM concentrations are much higher in the Northern Hemisphere relative to the Southern Hemisphere. Mineral dust shows an extremely wide range of concentrations over the oceans; the highest are over the tropical North Atlantic and the western North Pacific. These reflect the impact of dust transport from North Africa and China, respectively. Dust concentrations in the southern oceans tend to be extremely low due to the

Table 1 Annual mean aerosol concentrations measured at remote ocean stations

Station location[a]		Sea salt $(\mu g\,m^{-3})$	NO_3 $(\mu g\,m^{-3})$	nss-SO_4 $(\mu g\,m^{-3})$	NH_4 $(\mu g\,m^{-3})$	Dust $(\mu g\,m^{-3})$	V $(\mu g\,m^{-3})$	Total[b] $(\mu g\,m^{-3})$	Total SS[c] $(\mu g\,m^{-3})$
Lat°N	Lon°E								
North Pacific									
Western Pacific									
Cheju, Korea	33.5	126.5	19.8	4.1	7.2	3.0	15.5	4.1	49.5	29.8
Central Pacific									
Midway	28.2	−177.4	13.8	0.3	0.5	0.8	0.7	0.2	15.4	1.6
Oahu	21.36	−157.7	15.1	0.4	0.5	0.0	0.7	0.3	16.7	1.6
North Atlantic									
Mace Head	53.5	−9.9	14.1	1.5	2.0	0.9	0.5	0.9	19.0	4.9
Bermuda	32.3	−64.9	13.7	1.1	2.2	0.3	5.6	1.3	22.8	9.2
Barbados	13.2	−59.4	16.5	0.5	0.8	0.1	14.6	1.9	32.5	16.0
South Pacific									
American Samoa	−14.3	−170.6	16.7	0.1	0.4		0.0	0.1	17.2	0.5
Antarctic									
Mawson	−67.6	62.5	0.3	0.0	0.1	0.0			0.5	0.1

[a]Station Location: negative latitudes – Southern Hemisphere: negative longitudes – Western Hemisphere.
[b]Total aerosol: the sum of the major aerosol components – sea salt, soil dust, nss-SO_4, NO_3, and NH_4.
[c]Total SS: Total aerosol minus sea salt.

dearth of strong dust sources combined with the great distances to central ocean regions.

The impact of air pollution is evident over much of the Northern Hemisphere. Extremely high NO_3^- and $nss-SO_4^{2-}$ concentrations are seen in the western Pacific near Asia; these are attributable to continental outflow and exacerbated by limited emission controls. Moderately high pollutant levels are seen over the North Atlantic as well, a result of emissions from North America and Europe. In contrast, the concentrations of NO_3^- and $nss-SO_4^{2-}$ at American Samoa and the Antarctic stations Mawson and Palmer are extremely low; these represent conditions that one might expect when pollution impacts are minimal.

The larger-scale picture of PM distributions is provided by sensors such as the advanced very high resolution radiometer (AVHRR), which measures solar radiation backscattered to space by PM to estimate aerosol optical thickness (AOT) (**Figure 1**). There are three characteristics of the global distributions of AOT, all consistent with the data in **Table 1**. First, the highest AOT (i.e., the greatest column loadings of PM) is found close to the continents. This distribution affirms the fact that over most of the ocean PM is largely the result of material transported from the continents. Second, there are large seasonal differences in PM concentrations due to the seasonality of emissions and meteorology. Third, some continents emit more PM than others, illustrating the large-scale differences in production and transport.

Especially notable in satellite images is a large plume of AOT over the tropical Atlantic, extending from the coast of Africa to South America (December–February) and to the Caribbean (June–August). This plume is mainly African dust. A large region of high AOT over the Arabian Sea (June–August) is attributed to dust from Africa and the Middle East. In this same season, a large PM plume seen off the west coast of southern Africa is attributed to smoke from intense biomass burning. Substantial aerosol plumes are also seen over the North Atlantic; these are caused by pollutants from North America and Europe. Large regions of high AOT are seen along the coast of Asia; but the Asian plume is most prominent in spring when large quantities of soil dust mix with pollution aerosols. The attribution of these plumes to these dominant aerosol types is supported by evidence from field studies.

Aerosol Removal Mechanisms

Estimating Wet and Dry Deposition

PM is deposited to the ocean by two broadly characterized mechanisms: (1) dry deposition, in which a particle is transferred directly from the atmosphere to the surface; and (2) wet deposition, in which a particle is first incorporated into a cloud or rain droplet that subsequently falls to the surface. The relative efficiency of the removal processes is dependent on a number of factors, especially the particle-size distribution and the hygroscopic properties of the aerosol. In most ocean regions, wet removal is thought to dominate for most aerosol species.

Wet deposition is relatively easy to measure using precipitation collectors, such as automatic bucket systems that open only when precipitation falls. Dry deposition, on the other hand, is much more difficult to collect because this process is affected by a variety of factors, all highly variable: the properties of the aerosol and the water surface, atmospheric stability, relative humidity, wind velocity, etc. Furthermore, while there is a vast quantity of data on wet deposition to continental areas, there is very little for ocean environments. There are some long-term records for selected species in precipitation at a few island stations but there are no matching data for dry deposition.

Wet Deposition

Long-term studies show that on average the wet deposition rates of many species are related to their concentrations in the atmospheric aerosol and to rainfall rates. This relationship is expressed in terms of a dimensionless scavenging ratio, S:

$$S = C_p \rho C_a^{-1} \quad [1]$$

where C_p is the concentration of the substance in precipitation ($g\,kg^{-1}$), ρ the density of air ($\sim 1.2\,kg\,m^{-3}$), and C_a the aerosol concentration of the species of interest ($g\,m^{-3}$).

Wet deposition rates depend on the vertical distribution of PM and the type of precipitation event (e.g., frontal, cumulus, and stratus). In practice, comprehensive, long-term, aerosol data are only available from surface sites; consequently one must assume that over the longer term the PM concentrations in surface-level air are correlated with their vertical distributions. Therefore, S is appropriately calculated only when data have been obtained over periods of a year or more. That is, the variability in the aerosol and precipitation events must be smoothed by the averaging process. Typically used values for S fall in the range of 200–1000.

Scavenging ratios can be applied to regions where no precipitation data exist using the following

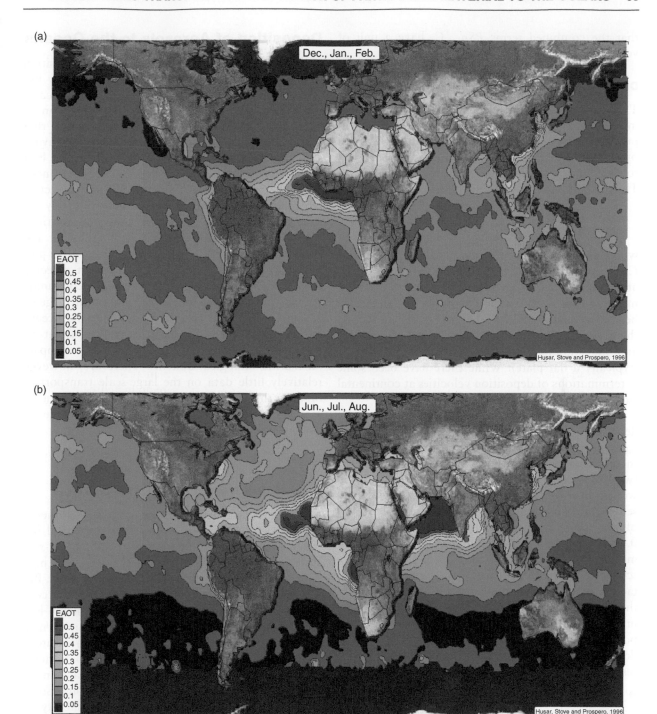

Figure 1 The distribution of aerosols over the oceans inferred from aerosol optical thickness estimated (EAOT) by National Oceanic and Atmospheric Administration (NOAA) AVHRR. Aerosol optical thickness is a measure of the attenuation of direct solar radiation at a specific wavelength due to the scattering and absorption caused by aerosols. Large values of optical thickness suggest high concentrations of aerosols. Distributions are shown for the months (a) Dec.–Feb. and (b) Jun.–Aug. Adapted by permission of American Geophysical Union from Husar RB, Prospero JM, and Stowe LL, Characterization of tropospheric aerosols over the oceans with the NOAA advanced very high resolution radiometer optical thickness operational product, *Journal of Geophysical Research*, vol. 102(D14), pp. 16889–16909, 1997. Copyright 1997 American Geophysical Union.

expression:

$$F_p = PC_p = PS\rho^{-1}C_a \qquad [2]$$

where F_p is the wet deposition flux ($g\,m^{-2}\,yr^{-1}$) and P is the precipitation rate ($m\,yr^{-1}$), using conversion factors to translate rainfall amounts to the mass of water deposited per unit area. Note that the combined terms $PS\rho^{-1}C_a$ have a unit of velocity.

Dry Deposition

There are no widely accepted methods for directly measuring PM dry deposition to water surfaces. In practice, dry deposition is almost always calculated by assuming that the deposition rate is proportional to PM concentration times a 'deposition velocity', v_d. The dry PM flux, F_d ($g\,m^{-2}s^{-1}$), is given by

$$F_d = v_d C_a \qquad [3]$$

where v_d is the dry deposition velocity ($m\,s^{-1}$) and C_a is the mass concentration of the substance in the atmosphere ($g\,m^{-3}$). Deposition velocities have been modeled based on physical principles, and they have been empirically derived by concurrently measuring the concentration of PM species in the atmosphere and the amount deposited to a surrogate surface (typically a flat plate). While there have been some determinations of deposition velocities at continental sites, the data for ocean regions are scant.

In eqn [3], v_d incorporates all the processes of dry deposition, but it is difficult to accurately parametrize v_d for the ambient aerosol because the importance of these processes varies with particle size. For PM between 0.1 and 1.0 μm, dry deposition is inefficient, and wet removal is normally the major sink. Gravitational settling and surface impaction control the dry removal of PM larger than about 1 μm while below about 0.1 μm, Brownian diffusion dominates. Each of these mechanisms depends on wind speed, aerosol hygroscopicity, relative humidity near the surface, and other factors which are poorly characterized. Unfortunately, there is no general agreement on how to resolve these uncertainties. Nonetheless, many estimates of dry deposition make the following assumptions about the dependence on particle size and the uncertainties in the resulting estimated deposition rate:

- submicrometer aerosol particles: $0.001\,m\,s^{-1} \pm$ a factor of 3,
- supramicrometer crustal particles not associated with sea salt: $0.01\,m\,s^{-1} \pm$ a factor of 3,
- giant sea salt particles and materials carried by them: $0.03\,m\,s^{-1} \pm$ a factor of 2.

Despite the widespread use of these values, it should be emphasized that they are only estimates and that the error range is, if anything, probably optimistic. For example, wind speed has a great influence on deposition velocities: for PM ~ 0.1–1.0 μm in diameter, the deposition velocity ranges from $c.$ $0.005\,cm\,s^{-1}$ at $5\,m\,s^{-1}$ to $c.$ $0.1\,cm\,s^{-1}$ at $15\,m\,s^{-1}$.

Deposition of Aerosols to the Oceans

In this section, we present estimates of the deposition of a number of PM species that are potentially important for biogeochemical processes in the oceans. Estimates of deposition to specific locations can be made using the above relationships, assuming that the necessary aerosol concentration data are available. Larger-scale estimates of deposition are best obtained with atmospheric chemical transport models as discussed below. These models are subject to large uncertainties because they generally rely on estimates of aerosol properties and concentrations over the oceans and they incorporate highly parametrized removal schemes. For illustrative purposes, we present results for mineral dust, selected trace elements, and a group of nitrogen-containing species. A wide range of natural and anthropogenic organic species are also transported to the oceans and deposited there. Of these, certain persistent organic pollutants are known to have a harmful impact on marine biological systems. There are, however, relatively little data on the large-scale transport of organics that would enable us to address this issue on a global scale. Consequently, we do not include organic species in this report.

Deposition of Mineral Dust and Eolian Iron

In many ocean regions, primary (photosynthetic) biological production is limited by the classical nutrients such as nitrate and phosphate. In nutrient-rich surface waters, biological activity is usually high which results in high chlorophyll concentrations. But in large areas of the world's oceans, nutrient concentrations are high, yet chlorophyll remains low which suggests low productivity. These are termed high-nutrient, low-chlorophyll (HNLC) waters; prominent examples include the equatorial Pacific and much of the high-latitude southern oceans. In the 1980s, it was found that primary production in some HNLC regions was limited by the availability of iron, an essential micronutrient in certain enzymes involved in photosynthesis.

Remote ocean regions are largely dependent on atmospheric dust for the input of iron. The deposition of this eolian iron and its impact on productivity has important implications for the global CO_2 budget and, hence, climate. Increased iron fluxes could conceivably fertilize the oceans, thereby increasing productivity and drawing down atmospheric CO_2. In addition, certain nitrogen fixers (e.g., *Trichodesmium* sp.) have a high iron requirement; an increased eolian iron flux could stimulate the growth of nitrogen fixers, thereby increasing nitrate levels and further contributing to the CO_2 drawdown.

Much effort has gone into estimating the temporal and spatial patterns of dust deposition to the oceans. Some studies have used satellite aerosol measurements coupled with network measurements of dust to calculate wet and dry deposition fluxes using scavenging ratios and deposition velocities.

Recently, regional- and global-scale models have been developed to provide estimates of dust emissions, transport, and deposition. Dust is generally included as a passive tracer, and its removal is highly parametrized. Dry deposition is calculated using the model's dust size distribution and size-dependent deposition (see section 'Dry deposition' above). Wet removal is also modeled, but the interaction of dust with clouds is not well constrained, partly because aerosol–cloud interactions in general are not well understood, and also because there are few measurements of cloud microphysical measurements in dusty regions. Mineral dust is not readily soluble in water; so some models assume that mineral aerosols do not interact with clouds directly, but rather are scavenged via subcloud removal – hence, simple scavenging ratios are used (see section 'Wet deposition' above). As a consequence of these uncertainties, current models show large differences in dust wet deposition lifetimes, ranging from 10 to 56 days.

A typical model estimate of dust deposition rates to the oceans is shown in **Figure 2**. In the tropical North Atlantic, rates typically range from 2 to $10 \, \text{g m}^{-2} \text{yr}^{-1}$; over the Arabian Sea, they are as high as $20 \, \text{g m}^{-2} \text{yr}^{-1}$. Over much of the North Pacific, rates are in the range $0.5–1 \, \text{g m}^{-2} \text{yr}^{-1}$, increasing to $1–2 \, \text{g m}^{-2} \text{yr}^{-1}$ closer to the coast of Asia. Dust deposition rates in **Figure 2** tend to mirror the PM distribution shown in **Figure 1**, which, as previously stated, is largely linked to the presence of dust and, in some regions, smoke from biomass burning.

Table 2 shows estimates of deposition rates to the major ocean basins produced by eight commonly used models. There is considerable agreement for the North Atlantic which is heavily impacted by African dust. In contrast, there is a considerable spread in the estimates for other regions, especially the Indian Ocean and South Pacific. Despite these differences, current models yield a reasonable, albeit broad, match with sediment trap measurements in the oceans.

These various studies show that North Africa is clearly the world's most active dust source followed by the Middle East and Central Asia. In effect, these combine to form a global dust belt that dominates transport to the oceans. These sources account for the much greater deposition rates to the northern oceans compared with southern oceans. Nonetheless, there are some substantial and important dust sources in the Southern Hemisphere in Australia, southern Africa, and southern South America. Within these continental regions, certain specific environments are particularly active dust sources, and they are sensitive to changes in climate, especially rainfall and wind speed. The presence of large, deep, alluvial deposits, usually deposited in the Pleistocene or

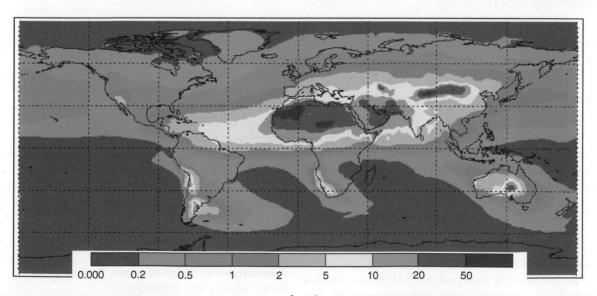

Figure 2 Model estimates of dust deposition rates (units: $\text{g m}^{-2} \text{yr}^{-1}$) to the continents and the oceans. Reproduced by permission of American Geophysical Union from Mahowald NM, Baker AR, Bergametti G, *et al.*, Atmospheric global dust cycle and iron inputs to the ocean. *Global Biogeochemical Cycles*, vol. 19, GB4025, 2005. Copyright 2005 American Geophysical Union.

Table 2   Estimate of mean annual dust deposition to the global ocean and to various ocean basins

Reference	Annual dust deposition rate[a] (10^{12} g yr^{-1})						
	GO	NAO	SAO	NPO	SPO	NIO	SIO
Duce et al. (1991)	910	220	24	480	39	100	44
Prospero (1996)	358	220	5	96	8	20	9
Ginoux et al. (2001)	478	184	20	92	28	154	
Zender et al. (2003)	314	178	29	31	8	36	12
Luo et al. (2003)	428	230	30	35	20	113	
Ginoux et al. (2004)	505	161	20	117	28	164	15
Tegen et al. (2004)	422	259	35	56	11	61	
Kaufman et al. (2005)	140						

[a]GO, Global Oceans; NAO, North Atlantic Ocean; SAO, South Atlantic Ocean; NPO, North Pacific Ocean; SPO, South Pacific Ocean; NIO, North Indian Ocean; SIO, South Indian Ocean.
Adapted from Engelstaedter S, Tegen I, and Washington R (2006) North African dust emissions and transport. *Earth-Science Reviews* 79(1–2): 73–100.

Table 3   Atmospheric and riverine fluxes of dissolved and particulate trace metals to the ocean[a]

Metal	Atmospheric transports			Riverine transports			Dissolved transports
	AtmDiss	AtmPart	Ratio Diss/Part	RivDiss	RivPart	Ratio Diss/Part	Ratio AtmDiss/RivDiss
Iron	3200	28 000	0.1	1100	11 000	0.100	2.9
Copper	30	5	6.6	10	1500	0.007	3.0
Nickel	10	16	0.6	11	1400	0.008	0.9
Zinc	102	33	3.1	6	3900	0.002	16.9
Arsenic	4	2	1.7	10	80	0.125	0.4
Cadium	3	1	4.7	0.3	15	0.020	8.7
Lead	75	9	8.3	2	1600	0.001	37.5

[a]Units: 10^9 g yr^{-1}.
Atm, atmospheric; Riv, riverine; Diss, Dissolved; Part, particulate.
Adapted from Duce RA, Liss PS, Merrill JT, et al. (1991) The atmospheric input of trace species to the world ocean. *Global Biogeochemical Cycles* 5: 193–259.

Holocene, is a common prerequisite for strong dust sources.

Trace Element Deposition

Some oceanographers study the biogeochemical cycling of trace elements and seek to quantify the elements' oceanic sources and sinks. In regions dominated by mineral dust, the ratios of many trace elements (e.g., Al, Ba, Ca, Cs, Fe, Hf, Mn, Rb, Sc, Ta, Th, and Yb) are similar to those in geological materials such as soils, thus implicating mineral dust as their main source. Several elements (Co, Cr, Eu, Mg, and Na) show slight enrichments over crustal values while others such as As, Cd, Cu, Ni, Pb, Sb, Se, V, and Zn are strongly enriched. Such large enrichments are typically associated with pollution impacts, but emissions from natural sources such as volcanoes can also be responsible.

The impact of trace element deposition on ocean biogeochemistry depends to a great extent on the degree to which the elements are soluble in seawater. Extensive studies of trace element solubilities in various natural aqueous media or in aqueous solutions of similar composition yield a wide range of solubilities; these depend on the types of aerosols used, the exposure times, and other experimental variables, especially pH. Thus it is difficult to convert the estimated air/sea exchange rates into an effective or bioavailable flux of trace elements, and therefore it is difficult to accurately assess the impact of PM deposition on ocean processes.

Comparison of Trace Element Transport by Rivers and by the Atmosphere

Rivers carry large quantities of dissolved and suspended materials to the oceans. **Table 3** compares the amounts of selected trace elements carried by rivers with that carried by winds and also the relative amounts of particulate and dissolved or soluble phases in the transported material.

Riverine transports of trace elements are overwhelmingly in the particulate phases and the ratio of dissolved to particulate phases ranges from about 0.001 to 0.13. In comparison, the corresponding ratio for atmospheric transport is much larger, ranging from 0.1 to 8.3. **Table 3** also shows that the dissolved or soluble inputs to the ocean from the atmosphere exceed those from rivers; in almost all cases the ratio is greater than unity, in some cases much larger.

The comparison of river versus atmospheric inputs is based on the measured concentrations in rivers before they reach the oceans. However, much of the material carried by rivers is rapidly deposited when the rivers reach the sea; therefore, the impact of air/sea exchange on ocean systems is in reality much greater than suggested by **Table 3**. We emphasize, however, that the data in **Table 3** are rather old. Recent work shows that the concentrations of some trace elements have changed significantly over time. For example, during the mid-1900s, aerosol lead greatly increased in response to increasing pollution emissions but in later years concentrations decreased as controls were implemented. In recent decades, other trace elements have increased due to growing industrialization in developing nations. Also, recent research suggests that there is considerably more uncertainty in PM trace metal solubility than shown in **Table 3**. Nonetheless, one would still expect that the impact of atmospheric transport is much greater than river transport, especially for the open ocean.

Nitrogen Deposition

Anthropogenic activities have greatly increased the amounts of nitrogenous materials that enter the atmosphere and find their way into the rivers (**Table 4**). There is interest in the possible impacts of these materials on the marine environment, especially about chemicals such as nitrate that can affect primary production. This issue is of particular importance in regions where nitrogen is the limiting nutrient, for example, the oligotrophic central oceanic gyres where an enhancement in productivity would increase the drawdown of CO_2 and hence affect climate. In coastal waters, atmospheric inputs could contribute to eutrophication although one would expect the inputs from rivers to dominate.

There are two broad classes of nitrogen compounds of interest: oxidized and reduced. The most important oxidized species are NO and NO_2

Table 4 Atmospheric emissions of fixed nitrogen, 1993[a]

Sources	NO_x	NH_3
Anthropogenic		
Biomass burning	6.4	4.6
Agricultural activity	2.6	39.7
Fossil fuel combustion	20.9	0.1
Industry	6.4	2.8
Total anthropogenic	36.3	47.2
Natural		
Soils, vegetation, and animals	2.9	4.6
Lightning	5.4	0.0
Natural fires	0.8	0.8
Stratosphere exchange	0.6	0.0
Ocean exchange	0.0	5.6
Total natural	6.8	11.0
Grand total	43.1	58.2
Ratio: anthropogenic/natural	5.3	4.3

[a]Units Tg/g(N) yr^{-1}.
Adapted from Jickells TD (2006) The role of air–sea exchange in the marine nitrogen cycle. *Biogeosciences* 3: 271–280.

(collectively referred to as $NO_{x)}$ and NO_y (termed reactive odd nitrogen) which is comprised of NO_x plus the compounds produced from its oxidation, including HNO_3 and other compounds. NO_x is rapidly oxidized in the atmosphere to a wide range of compounds, many of which are ultimately converted to HNO_3 and aerosol NO_3^-. In the marine boundary layer, HNO_3 reacts rapidly with sea salt particles and promptly deposits on the sea surface. Indeed, NO_3^- is the N-containing compound of greatest interest in terms of impact on the oceans, and it is the N compound most commonly measured and modeled.

Table 4 lists the major sources of oxidized and reduced N emitted to the atmosphere. The primary natural sources of NO_x are biological fixation and lightning, the latter being rather minor. In modern times, fossil-fuel combustion along with industry and biomass burning dominate the oxidized N cycle. The ratio of anthropogenic NO_x emissions to that of natural sources is 2.5 and continues to increase. While most research has focused on inorganic N (IN), there is evidence that organic nitrogen (ON) also may play an important role in marine biogeochemical cycling. However, there are relatively few data on ON compounds and most focus only on dissolved ON.

The major reduced nitrogen species (NH_x) are aerosol NH_4^+ and NH_3, the latter being the only gas-phase species that significantly titrates atmospheric acidity. The major natural sources of NH_3 (**Table 4**) include soils, vegetation, and excreta from wild animals. However, the emissions of NH_3 to the atmosphere are now dominated by fertilizers and the excreta from dairy and beef cattle. Indeed, the ratio of anthropogenic and natural NH_3 emissions is 9. The oceans can also be a source of NH_3 under certain conditions in some regions, but the continental sources clearly dominate.

Models provide estimates of the present-day atmospheric N fluxes to the oceans and their spatial distribution. **Figure 3** presents the deposition rate of reactive nitrogen $NO_y + NH_x$ for the year 2000. As was the case for dust, emissions and deposition in the Northern Hemisphere are much greater than those of the Southern Hemisphere. Deposition rates are extremely high adjacent to the continental coastlines; thus one would expect that the adjacent water bodies would be most heavily impacted. The total N flux to the ocean (NO_y and NH_x but not including ON) in

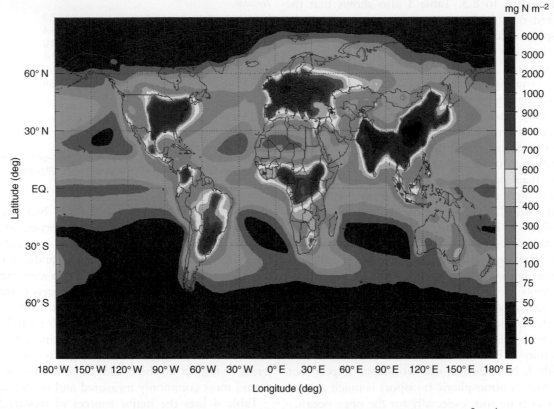

Figure 3 Model estimates of the deposition rate of total reactive nitrogen ($NO_y + NH_x$) (units: $mg\,N\,m^{-2}\,yr^{-1}$) in the year 2000. Reprinted with permission from Dentener F, Stevenson D, Ellingsen K, *et al.* (2006) The global atmospheric environment for the next generation. *Environmental Science and Technology* 40(11): 3586–3594 (doi:10.1021/es0523845). Copyright (2006) American Chemical Society.

Table 5 NO_y and NH_x deposition for the year 2000

Deposition region	NO_y		NH_x		$NO_y + NH_x$
	Total $(Tg(N)\,yr^{-1})$	Mean rate $(mg(N)\,m^{2}\,yr^{-1})$	Total $(Tg(N)\,yr^{-1})$	Mean rate $(mg(N)\,m^{2}\,yr^{-1})$	Total $(Tg(N)\,yr^{-1})$
Ocean	23	61	24	63	47
Coastal ocean	4	192	4	206	8
NH	38	150	48		87
SH	14	54	16		30
World	52	102	65	126	117
Ratio: ocean/ world	0.43		0.36		0.40

Adapted from Dentener F, Drevet J, Lamarque JF, *et al.* (2006) Nitrogen and sulfur deposition on regional and global scales: A multimodel evaluation. *Global Biogeochemical Cycles* 20: GB4003 (doi:10.1029/2005GB002672).

2000 was $46\,\mathrm{Tg\,N\,yr}^{-1}$ of which $8\,\mathrm{Tg\,N\,yr}^{-1}$ is deposited to the coastal ocean (**Table 5**). The deposition of oxidized forms (NO_y) is essentially equal to that of reduced forms (NH_x). The ocean N deposition amounts to c. 40% of global emissions.

The total reactive N transport by rivers is about $48\,\mathrm{Tg\,N\,yr}^{-1}$, essentially the same as the atmospheric source. However, there is evidence that fluvial nitrogen inputs to the oceans are denitrified on the shelf and that the shelf region is a sink rather than a source of nitrogen for the open oceans. Thus it appears that air/sea transfer is the major source of N transported to the open ocean.

ON compounds could also be playing a significant role in total N fluxes. Studies from many different environments suggest that ON constitutes about a third of the total atmospheric reactive nitrogen. Thus, ON could add significantly to the total global flux to the oceans, conceivably raising the total to about $69\,\mathrm{Tg\,N\,yr}^{-1}$.

Conclusions

It is now recognized that atmospheric transport plays a central role in ocean biogeochemical processes. There is increased interest in the chemically coupled ocean/atmosphere system, how this system has changed over time, and how it might respond to global change. Although many models are currently focusing on this question, the development of these models is handicapped by the dearth of measurements over many ocean regions. It remains a formidable challenge to the community to carry out the necessary measurements over such large ocean areas.

See also

Atmospheric Input of Pollutants. Iron Fertilization. Nitrogen Cycle.

Further Reading

Arimoto R, Kim YJ, Kim YP, et al. (2006) Characterization of Asian dust during ACE-Asia, global and planetary change. *Monitoring and Modelling of Asian Dust Storms* 52(1–4): 23–56.

Arimoto R, Ray BJ, Lewis NF, Tomza U, and Duce RA (1991) Mass-particle size distributions of atmospheric

dust and the dry deposition of dust to the remote ocean. *Journal of Geophysical Research – Atmospheres* 102(D13): 15867–15874.

Boyd PW, Watson A, Law CS, et al. (2000) A mesoscale phytoplankton bloom in the polar Southern Ocean stimulated by iron fertilization of waters. *Nature* 407: 695–702.

Dentener F, Drevet J, Lamarque JF, et al. (2006) Nitrogen and sulfur deposition on regional and global scales: A multimodel evaluation. *Global Biogeochemical Cycles* 20: GB4003 (doi:10.1029/2005GB002672).

Dentener F, Stevenson D, Ellingsen K, et al. (2006) The global atmospheric environment for the next generation. *Environmental Science and Technology* 40(11): 3586–3594 (doi:10.1021/es0523845).

Duce RA, Liss PS, Merrill JT, et al. (1991) The atmospheric input of trace species to the world ocean. *Global Biogeochemical Cycles* 5: 193–259.

Engelstaedter S, Tegen I, and Washington R (2006) North African dust emissions and transport. *Earth-Science Reviews* 79(1–2): 73–100.

Harrison SP, Kohfeld KE, Roeland C, and Claquin T (2001) The role of dust in climate today, at the Last Glacial Maximum and in the future. *Earth-Science Reviews* 54: 43–80.

Husar RB, Prospero JM, and Stowe LL (1997) Characterization of tropospheric aerosols over the oceans with the NOAA advanced very high resolution radiometer optical thickness operational product. *Journal of Geophysical Research* 102(D14): 16889–16909.

Jickells TD (2006) The role of air–sea exchange in the marine nitrogen cycle. *Biogeosciences* 3: 271–280.

Jurado E, Jaward F, Lohmann R, et al. (2005) Wet deposition of persistent organic pollutants to the global oceans. *Environmental Science and Technology* 39(8): 2426–2435 (doi:10.1021/es048599g).

Mahowald NM, Baker AR, Bergametti G, et al. (2005) Atmospheric global dust cycle and iron inputs to the ocean. *Global Biogeochemical Cycles* 19: GB4025 (doi:10.1029/2004GB002402).

Parekh P, Follows MJ, and Boyle EA (2005) Decoupling of iron and phosphate in the global ocean. *Global Biogeochemical Cycles* 19: GB2020 (doi:10.1029/2004GB002280).

Prospero JM (1996) The atmospheric transport of particles to the ocean. In: Ittekkot V, Schäfer P, Honjo S and Depetris PJ (eds.) *Particle Flux in the Ocean. SCOPE Report 57*, pp. 19–52. Chichester: Wiley.

Wesely ML and Hicks BB (2000) A review of the current status of knowledge on dry deposition. *Atmospheric Environment* 34(12–14): 2261–2282.

ATMOSPHERIC INPUT OF POLLUTANTS

R. A. Duce, Texas A&M University, College Station, TX, USA

Introduction

For about a century oceanographers have tried to understand the budgets and processes associated with both natural and human-derived substances entering the ocean. Much of the early work focused on the most obvious inputs – those carried by rivers and streams. Later studies investigated sewage outfalls, dumping, and other direct input pathways for pollutants. Over the past decade or two, however, it has become apparent that the atmosphere is also not only a significant, but in some cases dominant, pathway by which both natural materials and contaminants are transported from the continents to both the coastal and open oceans. These substances include mineral dust and plant residues, metals, nitrogen compounds from combustion processes and fertilizers, and pesticides and a wide range of other synthetic organic compounds from industrial and domestic sources. Some of these substances carried into the ocean by the atmosphere, such as lead and some chlorinated hydrocarbons, are potentially harmful to marine biological systems. Other substances, such as nitrogen compounds, phosphorus, and iron, are nutrients and may enhance marine productivity. For some substances, such as aluminum and some rare earth elements, the atmospheric input has an important impact on their natural chemical cycle in the sea.

In subsequent sections there will be discussions of the input of specific chemicals via the atmosphere to estuarine and coastal waters. This will be followed by considerations of the atmospheric input to open ocean regions and its potential importance. The atmospheric estimates will be compared with the input via other pathways when possible. Note that there are still very large uncertainties in all of the fluxes presented, both those from the atmosphere and those from other sources. Unless otherwise indicated, it should be assumed that the atmospheric input rates have uncertainties ranging from a factor of 2 to 4, sometimes even larger.

Estimating Atmospheric Contaminant Deposition

Contaminants present as gases in the atmosphere can exchange directly across the air/sea boundary or they may be scavenged by rain and snow. Pollutants present on particles (aerosols) may deposit on the ocean either by direct (dry) deposition or they may also be scavenged by precipitation. The removal of gases and/or particles by rain and snow is termed wet deposition.

Direct Deposition of Gases

Actual measurement of the fluxes of gases to a water surface is possible for only a very few chemicals at the present time, although extensive research is underway in this area, and analytical capabilities for fast response measurements of some trace gases are becoming available. Modeling the flux of gaseous compounds to the sea surface or to rain droplets requires a knowledge of the Henry's law constants and air/sea exchange coefficients as well as atmospheric and oceanic concentrations of the chemicals of interest. For many chemicals this information is not available. Discussions of the details of these processes of air/sea gas exchange can be found in other articles in this volume.

Particle Dry Deposition

Reliable methods do not currently exist to measure directly the dry deposition of the full size range of aerosol particles to a water surface. Thus, dry deposition of aerosols is often estimated using the dry deposition velocity, v_d. For dry deposition, the flux is then given by:

$$F_d = v_d \cdot C_a \qquad [1]$$

where F_d is the dry deposition flux (e.g., in $g\,m^{-2}\,s^{-1}$), v_d is the dry deposition velocity (e.g., in $m\,s^{-1}$), and C_a is the concentration of the substance on the aerosol particles in the atmosphere (e.g., in $g\,m^{-3}$). In this formulation v_d incorporates all the processes of dry deposition, including diffusion, impaction, and gravitational settling of the particles to a water surface. It is very difficult to parameterize accurately the dry deposition velocity since each of these processes is acting on a particle population, and they are each dependent upon a number of factors, including wind speed, particle size, relative humidity, etc. The following are dry deposition velocities that have been used in some studies of atmospheric deposition of particles to the ocean:

- Submicrometer aerosol particles, $0.001\,m\,s^{-1} \pm$ a factor of three

- Supermicrometer crustal particles not associated with sea salt, $0.01\,m\,s^{-1} \pm$ a factor of three
- Giant sea-salt particles and materials carried by them, $0.03\,m\,s^{-1} \pm$ a factor of two

Proper use of eqn [1] requires that information be available on the size distribution of the aerosol particles and the material present in them.

Particle and Gas Wet Deposition

The direct measurement of contaminants in precipitation samples is certainly the best approach for determining wet deposition, but problems with rain sampling, contamination, and the natural variability of the concentration of trace substances in precipitation often make representative flux estimates difficult using this approach. Studies have shown that the concentration of a substance in rain is related to the concentration of that substance in the atmosphere. This relationship can be expressed in terms of a scavenging ratio, S:

$$S = C_r \cdot \rho \cdot C_{a/g}^{-1} \qquad [2]$$

where C_r is the concentration of the substance in rain (e.g., in $g\,kg^{-1}$), ρ is the density of air ($\sim 1.2\,kg\,m^{-3}$), $C_{a/g}$ is the aerosol or gas phase concentration in the atmosphere (e.g., in $g\,m^{-3}$), and S is dimensionless. Values of S for substances present in aerosol particles range from a few hundred to a few thousand, which roughly means that $1\,g$ (or $1\,ml$) of rain scavenges $\subseteq 1\,m^3$ of air. For aerosols, S is dependent upon such factors as particle size and chemical composition. For gases, S can vary over many orders of magnitude depending on the specific gas, its Henry's law constant, and its gas/water exchange coefficient. For both aerosols and gases, S is also dependent upon the vertical concentration distribution and vertical extent of the precipitating cloud, so the use of scavenging ratios requires great care, and the results have significant uncertainties. However, if the concentration of an atmospheric substance and its scavenging ratio are known, the scavenging ratio approach can be used to estimate wet deposition fluxes as follows:

$$F_r = P \cdot C_r = P \cdot S \cdot C_{a/g} \cdot \rho^{-1} \qquad [3]$$

where F_r is the wet deposition flux (e.g., in $g\,m^{-2}\,year^{-1}$) and P is the precipitation rate (e.g., in $m\,year^{-1}$), with appropriate conversion factors to translate rainfall depth to mass of water per unit area. Note that $P \cdot S \cdot \rho^{-1}$ is equivalent to a wet deposition velocity.

Atmospheric Deposition to Estuaries and the Coastal Ocean

Metals

The atmospheric deposition of certain metals to coastal and estuarine regions has been studied more than that for any other chemicals. These metals are generally present on particles in the atmosphere. Chesapeake Bay is among the most thoroughly studied regions in North America in this regard. Table 1 provides a comparison of the atmospheric and riverine deposition of a number of metals to Chesapeake Bay. The atmospheric numbers represent a combination of wet plus dry deposition directly onto the Bay surface. Note that the atmospheric input ranges from as low as 1% of the total input for manganese to as high as 82% for aluminum. With the exception of Al and Fe, which are largely derived from natural weathering processes (e.g., mineral matter or soil), most of the input of the other metals is from human-derived sources. For metals with anthropogenic sources the atmosphere is most important for lead (32%).

There have also been a number of investigations of the input of metals to the North Sea, Baltic Sea, and Mediterranean Sea. Some modeling studies of the North Sea considered not only the direct input pathway represented by the figures in Table 1, but also considered Baltic Sea inflow, Atlantic Ocean inflow and outflow, and exchange of metals with the sediments, as well as the atmospheric contribution to all of these inputs. Figure 1 shows schematically

Table 1 Estimates of the riverine and atmospheric input of some metals to Chesapeake Bay

Metal	Riverine input ($10^6\,g\,year^{-1}$)	Atmospheric input ($10^6\,g\,year^{-1}$)	% Atmospheric input
Aluminum	160	700	81
Iron	600	400	40
Manganese	1300	13	1
Zinc	50	18	26
Copper	59	3.5	6
Nickel	100	4	4
Lead	15	7	32
Chromium	15	1.5	10
Arsenic	5	0.8	14
Cadmium	2.6	0.4	13

Data reproduced with permission from Scudlark JR, Conko KM and Church TM (1994) Atmospheric wet desposition of trace elements to Chesapezke Bay: (CBAD) study year 1 results. *Atmospheric Environment* 28: 1487–1498.

some modeling results for lead, copper, and cadmium. Note that for copper, atmospheric input is relatively unimportant in this larger context, while atmospheric input is somewhat more important for cadmium, and it is quite important for lead, being approximately equal to the inflow from the Atlantic Ocean, although still less than that entering the North Sea from dumping. As regards lead, note that approximately 20% of the inflow from the Atlantic to the North Sea is also derived from the atmosphere. This type of approach gives perhaps the most accurate and in-depth analysis of the importance of atmospheric input relative to all other sources of a chemical in a water mass.

Nitrogen Species

The input of nitrogen species from the atmosphere is of particular interest because nitrogen is a necessary nutrient for biological production and growth in the ocean. There has been an increasing number of studies of the atmospheric input of nitrogen to estuaries and the coastal ocean. Perhaps the area most intensively studied is once again Chesapeake Bay. **Table 2** shows that approximately 40% of all the nitrogen contributed by human activity to Chesapeake Bay enters via precipitation falling directly on the Bay or its watershed. These studies were different from most earlier studies because the atmospheric contributions were

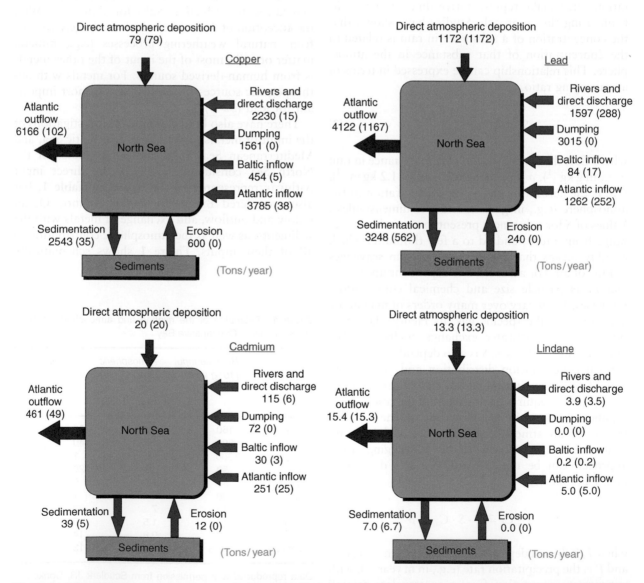

Figure 1 Input of copper, lead, cadmium, and lindane to the North Sea. Values in parentheses denote atmospheric contribution. For example, for copper the atmospheric contribution to rivers and direct discharges is 15 tons per year. (Figure reproduced with permission from Duce, 1998. Data adapted with permission from van den Hout, 1994.)

Table 2 Estimates of the input of nitrogen to Chesapeake Bay

Source	Total input (10^9g year^{-1})	Areal input rate (g m^{-2}year^{-1})	% of the total
Animal waste	5	0.4	3
Fertilizers	48	4.2	34
Point sources	33	2.9	24
Atmospheric precipitation			
nitrate	35	3.1	25
ammonium	19	1.7	14
Total	140	12.3	100

Data reproduced with permission from Fisher D, Ceroso T, Mathew T and Oppenheimer M (1988) *Polluted Coastal Waters: The Role of Acid Rain.* New York: Environmental Defense Fund.

considered not only to be direct deposition on the water surface, but also to include that coming in via the atmosphere but falling on the watershed and then entering the Bay. Note from **Table 2** that atmospheric input of nitrogen exceeded that from animal waste, fertilizers, and point sources. In the case of nitrate, about 23% falls directly on the Bay, with the remaining 77% falling on the watershed. These results suggest that studies that consider only the direct deposition on a water surface (e.g., the results shown in **Table 1**) may significantly underestimate the true contribution of atmospheric input. The total nitrogen fertilizer applied to croplands in the Chesapeake Bay region is ~ 5.4 g m^{-2} year^{-1}, while the atmospheric nitrate and ammonium nitrogen entering the Bay is ~ 4.8 g m^{-2} year^{-1}. Chesapeake Bay is almost as heavily fertilized from atmospheric nitrogen, largely anthropogenic, as the croplands are by fertilizer in that watershed!

Results from studies investigating nitrogen input to some other estuarine and coastal regions are summarized in **Table 3**. In this table atmospheric sources

for nitrogen are compared with all other sources, where possible. The atmospheric input ranges from 10% to almost 70% of the total. Note that some estimates compare only direct atmospheric deposition with all other sources and some include as part of the atmospheric input the portion of the deposition to the watershed that reaches the estuary or coast.

Synthetic Organic Compounds

Concern is growing about the input of a wide range of synthetic organic compounds to the coastal ocean. To date there have been relatively few estimates of the atmospheric fluxes of synthetic organic compounds to the ocean, and these estimates have significant uncertainties. These compounds are often both persistent and toxic pollutants, and many have relatively high molecular weights. The calculation of the atmospheric input of these compounds to the coastal ocean is complicated by the fact that many of them are found primarily in the gas phase in the atmosphere, and most of the deposition is related to

Table 3 Estimates of the input of nitrogen to some coastal areas[a]

Region	Total atmospheric input[b] (10^9g year^{-1})	Total input all sources (10^9g year^{-1})	% Atmospheric input
North Sea	400[c]	1500	27[c]
Western Mediterranean Sea	400[c]	577[d]	69[c]
Baltic Sea	500	$\sim$1200	42
Chesapeake Bay	54	140	39
New York Bight	–	–	13[c]
Long Island Sound	11	49	22
Neuse River Estuary, NC	1.7	7.5	23

[a]Data from several sources in the literature.
[b]Total from direct atmospheric deposition and runoff of atmospheric material from the watershed.
[c]Direct atmospheric deposition to the water only.
[d]Total from atmospheric and riverine input only.

Table 4 Estimates of the input of synthetic organic compounds to the North Sea

Organic compound	Atmospheric input (10^6 g year^{-1})	Input from other sources (10^6 g year^{-1})	% Atmospheric input
PCB	40	3	93
Lindane	36	3	92
Polycyclic aromatic hydrocarbons	80	90	47
Benzene	400	500	44
Trichloroethene	300	80	80
Trichloroethane	90	60	94
Tetrachloroethene	100	10	91
Carbon tetrachloride	6	40	13

Data reproduced with permission from Warmerhoven JP, Duiser JA, de Leu LT and Veldt C (1989) *The Contribution of the Input from the Atmosphere to the Contamination of the North Sea and the Dutch Wadden Sea.* Delft, The Netherlands: TNO Institute of Environmental Sciences.

the wet and dry removal of that phase. The atmospheric residence times of most of these compounds are long compared with those of metals and nitrogen species. Thus the potential source regions for these compounds entering coastal waters can be distant and widely dispersed.

Figure 1 shows the input of the pesticide lindane to the North Sea. Note that the atmospheric input of lindane dominates that from all other sources. **Table 4** compares the atmospheric input to the North Sea with that of other transport paths for a number of other synthetic organic compounds. In almost every case atmospheric input dominates the other sources combined.

Atmospheric Deposition to the Open Ocean

Studies of the atmospheric input of chemicals to the open ocean have also been increasing lately. For many substances a relatively small fraction of the material delivered to estuaries and the coastal zone by rivers and streams makes its way through the near shore environment to open ocean regions. Most of this material is lost via flocculation and sedimentation to the sediments as it passes from the freshwater environment to open sea water. Since aerosol particles in the size range of a few micrometers or less have atmospheric residence times of one to several days, depending upon their size distribution and local precipitation patterns, and most substances of interest in the gas phase have similar or even longer atmospheric residence times, there is ample opportunity for these atmospheric materials to be carried hundreds to thousands of kilometers before being deposited on the ocean surface.

Metals

Table 5 presents estimates of the natural and anthropogenic emission of several metals to the global atmosphere. Note that ranges of estimates and the best estimate are given. It appears from **Table 5** that anthropogenic sources dominate for lead, cadmium, and zinc, with essentially equal contributions for copper, nickel, and arsenic. Clearly a significant fraction of the input of these metals from the atmosphere to the ocean could be derived largely from anthropogenic sources.

Table 6 provides an estimate of the global input of several metals from the atmosphere to the ocean and compares these fluxes with those from rivers. Estimates are given for both the dissolved and particulate forms of the metals. These estimates suggest that rivers are generally the primary source of particulate metals in the ocean, although again a significant fraction of this material may not get past the coastal zone. For the dissolved phase atmospheric and riverine inputs are roughly equal for metals such as

Table 5 Emissions of some metals to the global atmosphere

Metal	Anthropogenic emissions (10^9 g year^{-1})		Natural emissions (10^9 g year^{-1})	
	Range	Best estimate	Range	Best estimate
Lead	289–376	332	1–23	12
Cadmium	3.1–12	7.6	0.15–2.6	1.3
Zinc	70–194	132	4–86	45
Copper	20–51	35	2.3–54	28
Arsenic	12–26	18	0.9–23	12
Nickel	24–87	56	3–57	30

Data reproduced with permission from Duce *et al.*, 1991.

Table 6 Estimates of the input of some metals to the global ocean

Metal	Atmospheric input		Riverine input	
	Dissolved $(10^9 \, g \, year^{-1})$	Particulate $(10^9 \, g \, year^{-1})$	Dissolved $(10^9 \, g \, year^{-1})$	Particulate $(10^9 \, g \, year^{-1})$
Iron	1600–4800	14 000–42 000	1100	110 000
Copper	14–45	2–7	10	1 500
Nickel	8–11	14–17	11	1 400
Zinc	33–170	11–55	6	3 900
Arsenic	2.3–5	1.3–3	10	80
Cadmium	1.9–3.3	0.4–0.7	0.3	15
Lead	50–100	6–12	2	1 600

Data reproduced with permission from Duce *et al.*, 1991.

iron, copper, and nickel; while for zinc, cadmium, and particularly lead atmospheric inputs appear to dominate. These estimates were made based on data collected in the mid-1980s. Extensive efforts to control the release of atmospheric lead, which has been primarily from the combustion of leaded gasoline, are now resulting in considerably lower concentrations of lead in many areas of the open ocean. For example, **Figure 2** shows that the concentration of dissolved lead in surface sea water near Bermuda has been decreasing regularly over the past 15–20 years, as has the atmospheric lead concentration in that region. This indicates clearly that at least for very particle-reactive metals such as lead, which has a short lifetime in the ocean (several years), even the

open ocean can recover rather rapidly when the anthropogenic input of such metals is reduced or ended. Unfortunately, many of the other metals of most concern have much longer residence times in the ocean (thousands to tens of thousands of years).

Figure 3 presents the calculated fluxes of several metals from the atmosphere to the ocean surface and from the ocean to the seafloor in the 1980s in the tropical central North Pacific. Note that for most metals the two fluxes are quite similar, suggesting the potential importance of atmospheric input to the marine sedimentation of these metals in this region. Lead and selenium are exceptions, however, as the atmospheric flux is much greater than the flux to the seafloor. The fluxes to the seafloor represent average

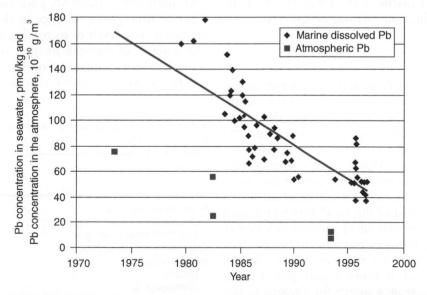

Figure 2 Changes in concentration of atmospheric lead at Bermuda and dissolved surface oceanic lead near Bermuda from the mid-1970s to the mid-1990s. (Data reproduced with permission from Wu J and Boyle EA (1997) Lead in the western North Atlantic Ocean: Completed response to leaded gasoline phaseout. *Geochimica et Cosmochimica Acta* 61: 3279–3283; and from Huang S, Arimoto R and Rahn KA (1996) Changes in atmospheric lead and other pollution elements at Bermuda. *Journal of Geophysical Research* 101: 21 033–21 040.)

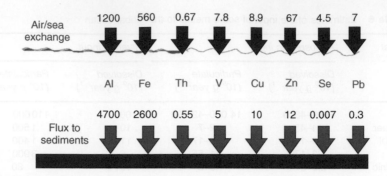

Figure 3 A comparison of the calculated fluxes of aluminum (Al), iron (Fe), thorium (Th), vanadium (V), copper (Cu), zinc (Zn), selenium (Se), and lead (Pb) (in 10^{-9} g cm^{-2} year^{-1}) from the atmosphere to the ocean and from the ocean to the sediments in the central tropical North Pacific. For each metal note the relative similarity in the two fluxes, except for lead and selenium. (Reproduced with permission from Duce, 1998.)

fluxes over the past several thousand years, whereas the atmospheric fluxes are roughly for the present time. The atmospheric lead flux is apparently much larger than the flux of lead to the sediments, primarily because of the high flux of anthropogenic lead from the atmosphere to the ocean since the introduction of tetraethyllead in gasoline in the 1920s. (The atmospheric flux is much lower now than in the 1980s, as discussed above.) However, in the case of selenium the apparently higher atmospheric flux is an artifact, because most of the flux of selenium from the atmosphere to the ocean is simply marine-derived selenium that has been emitted from the ocean to the atmosphere as gases, such as dimethyl selenide (DMSe). DMSe is oxidized in the atmosphere and returned to the ocean, i.e., the selenium input is simply a recycled marine flux. Thus, care must be taken when making comparisons of this type.

Nitrogen Species

There is growing concern about the input of anthropogenic nitrogen species to the global ocean. This issue is of particular importance in regions where nitrogen is the limiting nutrient, e.g., the oligotrophic waters of the central oceanic gyres. Estimates to date suggest that in such regions atmospheric nitrogen will in general account for only a few percent of the total 'new' nitrogen delivered to the photic zone, with most of the 'new' nutrient nitrogen derived from the upwelling of nutrient-rich deeper waters and from nitrogen fixation in the sea. It is recognized, however, that the atmospheric input is highly episodic, and at times it may play a much more important role as a source for nitrogen in surface waters. **Table 7** presents a recent estimate of the current input of fixed nitrogen to the global ocean from rivers, the atmosphere, and nitrogen fixation. From the numbers given it is apparent that all three

sources are likely important, and within the uncertainties of the estimates they are roughly equal. In the case of rivers, about half of the nitrogen input is anthropogenic for atmospheric input perhaps the most important information in **Table 7** is that the organic nitrogen flux appears to be equal to or perhaps significantly greater than the inorganic (i.e., ammonium and nitrate) nitrogen flux. The source of the organic nitrogen is not known, but there are indications that a large fraction of it is anthropogenic in origin. This is a form of atmospheric nitrogen input to the ocean that had not been considered until very recently, as there had been few measurements of organic nitrogen input to the ocean before the mid-1990s. The chemical forms of this organic nitrogen are still largely unknown.

Of particular concern are potential changes to the input of atmospheric nitrogen to the open ocean in the future as a result of increasing human activities. The amount of nitrogen fixation (formation of reactive nitrogen) produced from energy sources (primarily as NO_x, nitrogen oxides), fertilizers, and

Table 7 Estimates of the current input of reactive nitrogen to the global ocean

Source	Nitrogen input (10^{12} g year^{-1})
From the atmosphere	
Dissolved inorganic nitrogen	28–70
Dissolved organic nitrogen	28–84
From rivers (dissolved inorganic + organic nitrogen)	
Natural	14–35
Anthropogenic	7–35
From nitrogen fixation within the ocean	14–42

Data reproduced with permission from Cornell S, Rendell A and Jickells T (1995) Atmospheric inputs of dissolved organic nitrogen to the oceans. *Nature* 376: 243–246.

Table 8 Estimates of anthropogenic reactive nitrogen production, 1990 and 2020

Region	Energy (NO_x)					Fertilizer				
	1990 (10^{12}g N year^{-1})	2020	Δ	Factor	% of total increase	1990 (10^{12}g N year^{-1})	2020	Δ	Factor	% of total increase
USA/Canada	7.6	10.1	2.5	1.3	10	13.3	14.2	0.9	1.1	1.6
Europe	4.9	5.2	0.3	1.1	1	15.4	15.4	0	1.0	0
Australia	0.3	0.4	0.1	1.3	0.4	—	—	—	—	—
Japan	0.8	0.8	0	1.0	0	—	—	—	—	—
Asia	3.5	13.2	9.7	3.8	39	36	85	49	2.4	88
Central/South America	1.5	5.9	4.4	3.9	18	1.8	4.5	2.7	2.5	5
Africa	0.7	4.2	3.5	6.0	15	2.1	5.2	3.1	2.5	6
Former Soviet Union	2.2	5.7	3.5	2.5	15	10	10	0	1.0	0
Total	21	45	24	2.1	100	79	134	55	1.7	100

Data adapted with permission from Galloway et al., 1995.

legumes in 1990 and in 2020 as a result of human activities as well as the current and predicted future geographic distribution of the atmospheric deposition of reactive nitrogen to the continents and ocean have been evaluated recently. **Table 8** presents estimates of the formation of fixed nitrogen from energy use and production and from fertilizers, the two processes which would lead to the most important fluxes of reactive nitrogen to the atmosphere. Note that the most highly developed regions in the world, represented by the first four regions in the table, are predicted to show relatively little increase in the formation of fixed nitrogen, with none of these areas having a predicted increase by 2020 of more than a factor of 1.3 nor a contribution to the overall global increase in reactive nitrogen exceeding 10%. However, the regions in the lower part of **Table 8** will probably contribute very significantly to increased anthropogenic reactive nitrogen formation in 2020. For example, it is predicted that the production of reactive nitrogen in Asia from energy sources will increase ≤ fourfold, and that Asia will account for almost 40% of the global increase, while Africa will have a sixfold increase and will account for 15% of the global increase in energy-derived fixed nitrogen. It is predicted that production of reactive nitrogen from the use of fertilizers in Asia will increase by a factor of 2.4, and Asia will account for ~88% of the global increase from this source. Since both energy sources (NO_x, and ultimately nitrate) and fertilizer (ammonia and nitrate) result in the extensive release of reactive nitrogen to the atmosphere, the predictions above indicate that there should be very significant increases in the atmospheric deposition to the ocean of nutrient nitrogen species downwind of such regions as Asia, Central and South America, Africa, and the former Soviet Union.

This prediction has been supported by numerical modeling studies. These studies have resulted in the generation of maps of the 1980 and expected 2020 annual deposition of reactive nitrogen to the global ocean. **Figure 4** shows the expected significant increase in reactive nitrogen deposition from fossil fuel combustion to the ocean to the east of all of Asia, from Southeast Asia to the Asian portion of the former Soviet Union; to the east of South Africa, northeast Africa and the Mideast and Central America and southern South America; and to the west of northwest Africa. This increased reactive nitrogen transport and deposition to the ocean will provide new sources of nutrient nitrogen to some regions of the ocean where biological production is currently nitrogen-limited. There is thus the possibility of significant impacts on regional biological primary production, at least episodically, in these regions of the open ocean.

Synthetic Organic Compounds

The atmospheric residence times of many synthetic organic compounds are relatively long compared with those of the metals and nitrogen species, as mentioned previously. Many of these substances are found primarily in the gas phase in the atmosphere, and they are thus very effectively mobilized into the atmosphere during their production and use. Their long atmospheric residence times of weeks to months leads to atmospheric transport that can often be hemispheric or near hemispheric in scale. Thus atmospheric transport and deposition in general dominates all other sources for these chemicals in sea water in open ocean regions.

Table 9 compares the atmospheric and riverine inputs to the world ocean for a number of synthetic

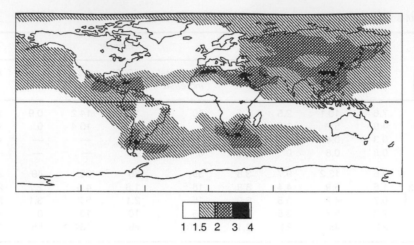

1 1.5 2 3 4

Figure 4 The ratio of the estimated deposition of reactive nitrogen to ocean and land surfaces in 2020 relative to 1980. (Figure reproduced with permission from Watson AJ (1997) Surface Ocean–Lower Atmosphere Study (SOLAS). *Global Change Newsletter, IGBP* no. 31, 9–12; data in figure adapted with permission from Galloway JN, Levy H and Kasibhatla PS (1994) Year 2020: Consequences of population growth and development on deposition of oxidized nitrogen. *Ambio* 23: 120–123.).

Table 9 Estimates of the atmospheric input of organochlorine compounds to the global ocean

Ocean	ΣHCH (10^6g year^{-1})	ΣDDT (10^6g year^{-1})	ΣPCB (10^6g year^{-1})	HCB (10^6g year^{-1})	Dieldrin (10^6g year^{-1})	Chlordane (10^6g year^{-1})
North Atlantic	850	16	100	17	17	8.7
South Atlantic	97	14	14	10	2.0	1.0
North Pacific	2600	66	36	20	8.9	8.3
South Pacific	470	26	29	19	9.5	1.9
Indian	700	43	52	11	6.0	2.4
Global input via the atmosphere	∼4700	∼170	∼230	∼80	∼40	∼22
Global input via rivers	∼60	∼4	∼60	∼4	∼4	∼4
% Atmospheric input	∼99%	∼98%	∼80%	∼95%	∼91%	∼85%

(Data reproduced with permission from Duce *et al.*, 1991.)

organochlorine compounds. Note that the atmosphere in most cases accounts for 90% or more of the input of these compounds to the ocean. **Table 9** also presents estimates of the input of these same organochlorine compounds to the major ocean basins. Since most of these synthetic organic compounds are produced and used in the northern hemisphere, it is not surprising that the flux into the northern hemisphere ocean is greater than that to southern hemisphere marine regions. There are some differences for specific compounds in different ocean basins. For example, HCH (hexachlorocyclohexane) and DDT have a higher input rate to the North Pacific than the North Atlantic, largely because of the greater use of these compounds in Asia than in North America or Europe. On the other hand, the input of PCBs (polychlorinated biphenyls) and dieldrin is higher to the North Atlantic than the North Pacific, primarily because of their greater use in the continental regions adjacent to the North Atlantic.

Conclusions

The atmosphere transports materials to the ocean that are both harmful to marine life and that are essential for marine biological productivity. It is now apparent that atmospheric transport and deposition of some metals, nitrogen species, and synthetic organic compounds can be a significant and in some cases dominant pathway for these substances entering both estuarine and coastal waters as well as some open ocean regions. Atmospheric input clearly must be considered in any evaluation of material fluxes to marine ecosystems. However, the uncertainties in the atmospheric fluxes of these materials to the ocean are large. The primary reasons for these large uncertainties are:

● The lack of atmospheric concentration data over vast regions of the coastal and open ocean, particularly over extended periods of time and under varying meteorological conditions;

- The episodic nature of the atmospheric deposition to the ocean;
- The lack of accurate models of air/sea exchange, particularly for gases;
- The inability to measure accurately the dry deposition of particles; and
- The inability to measure accurately the air/sea exchange of gases.

Further Reading

Duce RA (1998) *Atmospheric Input of Pollution to the Oceans*, pp. 9–26. Proceedings of the Commission for Marine Meteorology Technical Conference on Marine Pollution, World Meteorological Organization TD-No. 890, Geneva, Switzerland.

Duce RA, Liss PS, Merrill JT, *et al.* (1991) The atmospheric input of trace species to the world ocean. *Global Biogeochemical Cycles* 5: 193–259.

Galloway JN, Schlesinger WH, Levy H, Michaels A, and Schnoor JL (1995) Nitrogen fixation: anthropogenic enhancement – environmental reponse. *Global Biogeochemical Cycles* 9: 235–252.

Jickells TD (1995) Atmospheric inputs of metals and nutrients to the oceans: their magnitude and effects. *Marine Chemistry* 48: 199–214.

Liss PS and Duce RA (1997) *The Sea Surface and Global Change*. Cambridge: Cambridge University Press.

Paerl HW and Whitall DR (1999) Anthropogenically-derived atmospheric nitrogen deposition, marine eutrophication and harmful algal bloom expansion: Is there a link? *Ambio* 28: 307–311.

Prospero JM, Barrett K, Church T, *et al.* (1996) Nitrogen dynamics of the North Atlantic Ocean – Atmospheric deposition of nutrients to the North Atlantic Ocean. *Biogeochemistry* 35: 27–73.

van den Hout KD (ed.) (1994) *The Impact of Atmospheric Deposition of Non-Acidifying Pollutants on the Quality of European Forest Soils and the North Sea*. Report of the ESQUAD Project, IMW-TNO Report No. R 93/329.

AIR-SEA CHEMICAL EXCHANGES AND CYCLES

AIR–SEA GAS EXCHANGE

B. Jähne, University of Heidelberg, Heidelberg, Germany

wavy water surface are still not known. A number of new imaging techniques are described which give direct insight into the transfer processes and promise to trigger substantial theoretical progress in the near future.

Introduction

The exchange of inert and sparingly soluble gases, including carbon dioxide, methane, and oxygen, between the atmosphere and oceans is controlled by a 20–200-μm-thick boundary layer at the top of the ocean. The hydrodynamics in this layer is significantly different from boundary layers at rigid walls since the orbital motion of the waves is of the same order as the velocities in the viscous boundary layer. Laboratory and field measurements show that wind waves and surfactants significantly influence the gas-transfer process. Because of limited experimental techniques, the details of the mechanisms and the structure of the turbulence in the boundary layer at a

Theory

Mass Boundary Layers

The transfer of gases and volatile chemical species between the atmosphere and oceans is driven by a concentration difference and the transport by molecular and turbulent motion. Both types of transport processes can be characterized by 'diffusion coefficients', denoted by D and K_c, respectively (**Table 1**). The resulting flux density j_c is proportional to the diffusion coefficient and the concentration gradient. Thus,

$$j_c = (D + K_c(z))\nabla c \qquad [1]$$

Table 1 Diffusion coefficients for various gases and volatile chemical species in deionized water and in some cases in seawater

Species	Molecular mass	A $(10^{-5}\,cm^2\,s^{-1})$	E_a $(kJ\,mol^{-1})$	σ(Fit) (%)	Diffusion coefficient $(10^{-5}\,cm^2\,s^{-1})$			
					5 °C	15 °C	25 °C	35 °C
Heat		379.2	2.375		135.80	140.72	145.48	150.08
^{3}He[b,c]	3.02	941	11.70	2.1	5.97	7.12	8.39	9.77
^{4}He	4.00	818	11.70	2.1	5.10	6.30	7.22	8.48
^{4}He[a]		886	12.02	1.8	4.86	5.88	7.02	8.03
Ne	20.18	1608	14.84	3.5	2.61	3.28	4.16	4.82
Kr	83.80	6393	20.20	1.6	1.02	1.41	1.84	2.40
Xe	131.30	9007	21.61	3.5	0.77	1.12	1.47	1.94
^{222}Rn[b]	222.00	15877	23.26	11	0.68	0.96	1.34	1.81
H_2	2.02	3338	16.06	1.6	3.17	4.10	5.13	6.23
H_2[a]		1981	14.93	4.3	3.05	3.97	4.91	5.70
CH_4	16.04	3047	18.36	2.7	1.12	1.48	1.84	2.43
CO_2	44.01	5019	19.51	1.3	1.07	1.45	1.91	2.43
DMS[b]	62.13	2000	18.10		0.80	1.05	1.35	1.71
CH_3Br[b]	94.94	3800	19.10		0.98	1.31	1.71	2.20
F12[b] (CCl_2F_2)	120.91	4100	20.50		0.58	0.79	1.05	1.37
F11[b] (CCl_3F)	137.37	3400	20.00		0.60	0.81	1.07	1.38
SF_6[b]	146.05	2900	19.30		0.69	0.92	1.20	1.55

[a] In seawater.
[b] Values of diffusion coefficients from fit, not measured values.
[c] Set 15% higher than ^{4}He.
Columns 3 and 4 contain the parameters for the fit of the diffusion coefficient: $D = A\,\exp[-E_a/(RT)]$, the last four columns the diffusion coefficients for 5, 15, 25, and 35 °C.
Data collected from Jähne B, Heinz G, and Dietrich W (1987) Measurement of the diffusion coefficients of sparingly soluble gases in water. *Journal of Geophysical Research* 92: 10767–10776; and King DB, De Bruyn WJ, Zheng M, and Saltzman ES (1995) Uncertainties in the molecular diffusion coefficient of gases in water for use in the estimation of air–sea exchange. In: Jähne B and Monahan E (eds.) *Air–Water Gas Transfer*, pp. 13–22. Hanau: Aeon.

In a stationary homogeneous case and without sinks and sources by chemical reactions, the flux density j is in vertical direction and constant. Then integration of [1] yields vertical concentration profiles

$$C(z_r) - C(0) = j_c \int_0^{z_r} \frac{1}{D + K_c(z)} dz \qquad [2]$$

The molecular diffusion coefficient is proportional to the velocity of the molecules and the free length between collisions. The same concept can be applied to turbulent diffusion coefficients. Far away from the interface, the free length (called 'mixing length') is set proportional to the distance from the interface and the turbulent diffusion coefficient K_c for mass transfer is

$$K_c = \frac{\kappa}{Sc_t} u_* z \qquad [3]$$

where $\kappa = 0.41$ is the von Kármán constant, u_*, the friction velocity, a measure for the velocity fluctuations in a turbulent flow, and $Sc_t = K_m/K_c$ the turbulent Schmidt number. Closer to the interface, the turbulent diffusion coefficients are decreasing even faster. Once a critical length scale l is reached, the Reynolds number $Re = u_* l/\nu$ (ν is the kinematic viscosity, the molecular diffusion coefficient for momentum) becomes small enough so that turbulent motion is attenuated by viscosity. The degree of attenuation depends on the properties of the interface. At a smooth solid wall, $K_c \propto z^3$, at a free water interface it could be in the range between $K_c \propto z^3$ and $K_c \propto z^2$ depending on surface conditions.

Boundary layers are formed on both sides of the interface (**Figure 1**). When the turbulent diffusivity becomes equal to the kinematic viscosity, the edge of the 'viscous boundary layer' is reached. As the name implies, this layer is dominated by viscous dissipation and the velocity profile becomes linear because of a constant diffusivity. The edge of the 'mass boundary layer' is reached when the turbulent diffusivity becomes equal to the molecular diffusivity. The relative thickness of both boundary layers depends on the dimensionless ratio $Sc = \nu/D$ (Schmidt number).

The viscous and mass boundary layers are of about the same thickness in the air, because values of D for various gaseous species and momentum are about the same (Sc_{air} is 0.56 for H_2O, 0.63 for heat, and 0.83 for CO_2). In the liquid phase the situation is completely different. With Schmidt numbers in the range from 100 to 3000 (**Figure 2** and **Table 2**), molecular diffusion for a dissolved volatile chemical species is two to three orders of magnitude slower than diffusion of momentum. Thus the mass boundary layer

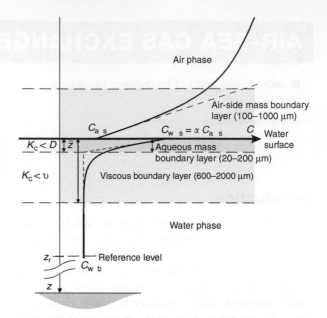

Figure 1 Schematic graph of the mass boundary layers at a gas–liquid interface for a tracer with a solubility $\alpha = 3$.

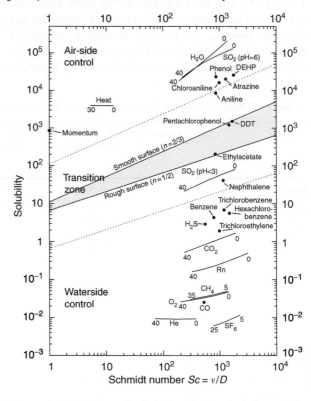

Figure 2 Schmidt number/solubility diagram including various volatile tracers, momentum, and heat for a temperature range (°C) as indicated. Filled circles refer to only a temperature of 20 °C. The regions for air-side, mixed, and waterside control of the transfer process between gas and liquid phase are marked. At the solid lines the transfer resistance is equal in both phases. The following dimensional transfer resistances were used: $r_a = 31$, $r_w = 12 Sc^{2/3}$ (smooth), $r_w = 6.5 Sc^{1/2}$ (wavy surface) with $r_a = R_a u_{*a}$ and $r_w = R_w u_{*w}$. Adapted from Jähne B and Haußecker H (1998) Air–water gas exchange. *Annual Review of Fluid Mechanics* 30: 443–468.

Table 2 Schmidt numbers of various gases and volatile species in the temperature range from 0 to 40 °C

Species	Schmidt number								
	0 °C	5 °C	10 °C	15 °C	20 °C	25 °C	30 °C	35 °C	40 °C
Heat	13.45	11.19	9.46	8.10	7.02	6.14	5.42	4.82	4.32
^{3}He	329	254	200	160	130	107	88	74	63
^{4}He	379	293	230	184	149	122	102	85	72
$^{4}He^{a}$	402	310	244	194	157	129	107	89	75
Ne	768	579	445	347	276	221	180	148	122
Kr	2045	1478	1090	819	625	483	379	301	241
Xe	2701	1930	1409	1047	791	606	471	370	294
^{222}Rn	3168	2235	1611	1182	883	669	514	400	314
H_2	633	473	360	278	219	174	140	114	94
$H_2{}^{a}$	648	488	375	293	232	186	151	124	103
CH_4	1908	1400	1047	797	616	483	383	308	250
CO_2	1922	1397	1036	782	600	466	367	293	236
DMS	2593	1905	1428	1089	844	662	527	423	344
CH_3Br	2120	1545	1150	870	669	522	412	329	266
F12 (CCl_2F_2)	3639	2624	1931	1447	1102	851	666	527	422
F11 (CCl_3F)	3521	2549	1883	1416	1082	839	658	523	420
SF_6	3033	2208	1640	1239	952	741	585	467	376

^{a}In seawater.

is significantly thinner than the viscous boundary layer in the liquid phase. This means that the transfer of gases is much slower and almost entirely controlled by the tiny residual turbulence in the small top fraction of the viscous boundary layer. This basic fact makes it difficult to investigate the mechanism of air–water gas transfer both theoretically and experimentally. In addition, the transfer process depends strongly on the water temperature because the Schmidt number decreases by about a factor of 6 from 0 to 35 °C (**Table 2**).

Transfer velocity and transfer resistance The amount of species exchanged between the air and water across the interface can be described by a quantity with the units of a velocity. It represents the velocity with which a tracer is pushed by an imaginary piston across the surface. This quantity is known as the 'transfer velocity' k (also known as the 'piston velocity', 'gas exchange rate', or 'transfer coefficient'). It is defined as the flux density divided by the concentration difference between the surface and the bulk at some reference level z_r:

$$k = \frac{j_c}{C_s - C_b} \qquad [4]$$

The inverse of the transfer velocity is known as the 'transfer resistance' R:

$$R = \frac{C_s - C_b}{j_c} \qquad [5]$$

The indices s and b denote the surface and bulk. Both quantities can directly be related to vertical concentration profiles by introducing [4] and [5] into [2]:

$$R = \frac{1}{k} = \int_0^{z_r} \frac{1}{D + K_c(z)} dz \qquad [6]$$

Thus the transfer resistances over several height intervals can be added in the same way as electrical resistances that are connected in series. Typical values of the transfer velocity across the waterside mass boundary layer are 10^{-6}–10^{-5} m s^{-1} (1–10 m d^{-1}). With respect to typical mixed layer depths in the ocean of about 100 m, gas transfer is a very slow process. It takes a time constant $\tau = h/k = 10$–100 days for the concentration of dissolved gases in the mixed layer to come into equilibrium with the atmosphere.

Boundary layer thickness The 'boundary layer thickness' $\tilde{z}$ is defined as the thickness of a fictional layer in which the flux is maintained only by molecular transport: $j = D(C_s - C_b)/\tilde{z}$. Then with [4] the boundary layer thickness is given by

$$\tilde{z} = \frac{D}{k} \qquad [7]$$

Geometrically, $\tilde{z}$ is given as the intercept of the tangent to the concentration profile at the surface and the bulk (**Figure 1**). With thicknesses between 20 and 200 µm, the mass boundary layer is extremely thin.

Boundary layer time constant The 'time constant' $\tilde{t}$ for the transport across the mass boundary layer is given by

$$\tilde{t} = \frac{\tilde{z}}{k} = \frac{D}{k^2} \qquad [8]$$

Typical values for $\tilde{t}$ are 0.04–4 s. Any chemical reaction with a time constant larger than $\tilde{t}$ does not significantly affect the transfer process. Therefore, CO_2 can be regarded as an inert gas and not fast hydrating acid gases such as SO_2.

The definitions of the three parameters k, $\tilde{z}$, and $\tilde{t}$, are generally valid and do not depend on any models of the boundary layer turbulence. According to [7] and [8] they are coupled via the molecular diffusion coefficient. Therefore only one of them needs to be measured to get knowledge of all three parameters, provided the molecular diffusion coefficient of the species is known.

Partitioning of Transfer between Air and Water

Because a mass boundary layer exists on both sides, it is important to determine which one controls the transfer, that is, exhibits the largest transfer resistance (or lowest transfer velocity). At the surface itself the thermodynamic solubility equilibrium is assumed to be established between the tracer concentrations c_a in the gas phase and c_w in the liquid phase

$$c_{ws} = \alpha c_{as} \qquad [9]$$

where α is the dimensionless solubility (partition coefficient). A solubility $\alpha \neq 1$ causes a concentration jump at the surface (**Figure 1**). Thus, the resulting total transfer velocity k_t or transfer resistance R_t can be viewed from either the gas phase or the liquid phase. Adding them up, the factor α must be considered to conserve the continuity of the concentration profile:

$$\text{air side:} \quad \frac{1}{k_{at}} = \frac{1}{k_a} + \frac{1}{\alpha k_w}, \quad R_{at} = R_a + R_w/\alpha$$

$$\text{waterside:} \quad \frac{1}{k_{wt}} = \frac{\alpha}{k_a} + \frac{1}{k_w}, \quad R_{wt} = \alpha R_a + R_w$$

$$[10]$$

The total transfer velocities in air and water differ by the factor α: $k_{at} = \alpha k_{wt}$. The ratio $\alpha k_w/k_a$ determines which boundary layer controls the transfer process. A high solubility shifts control of the transfer process to the gas-phase boundary layer, and a low solubility to the aqueous layer. The solubility value for a transition from air-sided to watersided control depends on the ratio of the transfer velocities. Typically, k_w is about

100–1000 times smaller than k_a. Thus, the transfer of even moderately soluble volatile chemical species with solubilities up to 30 is controlled by the waterside. Some environmentally important species lie in a transition zone where it is required to consider both transport processes (**Figure 2**). The transfer of highly soluble volatile and/or chemically reactive gas is controlled by the air-side transfer process and thus analogous to the transfer of water vapor. The following considerations concentrate on the waterside transfer process.

Gas Exchange at Smooth Water Surfaces

At smooth water surfaces, the theory of mass transfer is well established because it is equivalent to mass transfer to a smooth solid wall. The turbulent diffusivity can be described by the classic approach of Reichhardt with an initial z^3 increase that smoothly changes to a linear increase in the turbulent layer as in [3]. Then integration of [6] yields the following approximation for Schmidt numbers higher than 60:

$$k_w = u_{*w} \frac{1}{12.2} Sc^{-2/3}, \quad Sc > 60 \qquad [11]$$

This equation establishes the basic analogy between momentum transfer and gas exchange. The transfer coefficient is proportional to the friction velocity in water, which describes the shear stress (tangential force per unit area) $\tau = \rho_w u_{*w}^2$ applied by the wind field at the water surface. Assuming stress continuity at the water surface, the friction velocity in water is related to the friction velocity in air by

$$u_{*w} = u_{*a} \left(\frac{\rho_a}{\rho_w} \right)^{1/2} \qquad [12]$$

The friction velocity in air, u_{*a}, can further be linked via the drag coefficient to the wind speed U_R at a reference height: $c_D = (u_{*a}/U_R)^2$. Depending on the roughness of the sea surface, the drag coefficient has values between 0.8 and 2.4×10^{-3}. In this way the gas exchange rate is directly linked to the wind speed. The gas exchange further depends on the chemical species and the water temperature via the Schmidt number.

Gas Exchange at Rough and Wavy Water Surfaces

A free water surface is neither solid nor is it smooth as soon as short wind waves are generated. On a free water surface velocity fluctuations are possible. Thus, there can be convergence or divergence zone at the surface; surface elements may be dilated or contracted. At a clean water surface dilation or contraction of a surface element does not cause restoring forces, because surface tension only tries to minimize the total free surface area, which is not changed by this process. As a consequence of this hydrodynamic boundary condition, the turbulent diffusivity normal to the interface can now increase with the distance squared from the interface, $K_c \propto z^2$. Then

$$k_w = u_{*w} \frac{1}{\beta} Sc^{-1/2} \qquad [13]$$

where β is a dimensionless constant.

In comparison to the smooth case in [11], the exponent n of the Schmidt number drops from $-2/3$ to $-1/2$. This increases the transfer velocity for a Schmidt number of 600 by about a factor of 3. The total enhancement depends on the value of the constant β.

Influence of Surface Films

A film on the water surface creates pressure that works against the contraction of surface elements. This is the point at which the physicochemical structure of the surface influences the structure of the near-surface turbulence as well as the generation of waves. As at a rigid wall, a strong film pressure at the surface maintains a two-dimensional continuity at the interface just as at a rigid wall. Therefore, [11] should be valid for a smooth film-covered water surface and has indeed been verified in wind/wave tunnel studies as the lower limit for the transfer velocity. As a consequence, both [11] and [13] can only be regarded as limiting cases. A more general approach is required that has not yet been established. One possibility is a generalization of [11] and [13] to

$$k_w = u_{*w} \frac{1}{\beta(s)} Sc^{-n(s)} \qquad [14]$$

where both β and n depend on dimensionless parameters describing the surface conditions s. Even films with low film pressure may easily decrease the gas transfer rate to half of its value at clean water surface conditions. But still too few measurements at sea are available to establish the influence of surfactants on gas transfer for oceanic conditions more quantitatively.

Influence of Waves

Wind waves cannot be regarded as static roughness elements for the liquid flow because their characteristic particle velocity is of the same order of magnitude as the velocity in the shear layer at the surface.

This fact causes a basic asymmetry between the turbulent processes on the air and on the water sides of the interface. Therefore, the wave effect on the turbulent transfer in the water is much stronger and of quite different character than in the air. This basic asymmetry can be seen if the transfer velocity for CO_2 is plotted against the transfer velocity for water vapor (**Figure 3(a)**). At a smooth water surface the points fall well on the theoretical curve predicted by the theory for a smooth rigid wall. However, as soon as waves occur at the water surface, the transfer velocity of CO_2 increases significantly beyond the predictions.

Even at high wind speeds, the observed surface increase is well below 20%. When waves are generated by wind, energy is not only transferred via shear stress into the water but a second energy cycle is established. The energy put by the turbulent wind into the wave field is transferred to other wave numbers by nonlinear wave–wave interaction and finally dissipated by wave breaking, viscous dissipation, and turbulence. The turbulent wave dissipation term is the least-known term and of most importance for enhanced near-surface turbulence. Evidence for enhanced turbulence levels below wind waves has been reported from field and laboratory measurements. Experimental results also suggest that the gas transfer rate is better correlated with the 'mean square slope' of the waves as an integral measure for the nonlinearity of the wind wave field than with the wind speed.

It is not yet clear, however, to what extent 'microscale wave breaking' can account for the observed enhanced gas transfer rates. A gravity wave becomes unstable and generates a steep train of capillary waves at its leeward face and has a turbulent wake. This phenomenon can be observed even at low wind speeds, as soon as wind waves are generated. At higher wind speeds, the frequency of microscale wave breaking increases.

Influence of Breaking Waves and Bubbles

At high wind speeds, wave breaking with the entrainment of bubbles may enhance gas transfer further. This phenomenon complicates the gas exchange between atmosphere and the oceans considerably. First, bubbles constitute an additional exchange surface. This surface is, however, only effective for gases with low solubility. For gases with high solubility, the gas bubbles quickly come into equilibrium so that a bubble takes place in the exchange only for a fraction of its lifetime. Thus, bubble-mediated gas exchange depends – in contrast to the exchange at the free surface – on the solubility of the gas tracer.

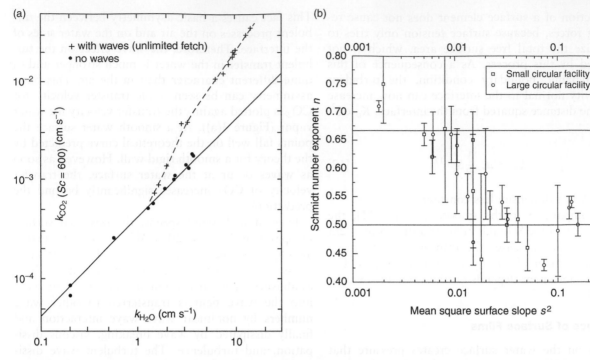

Figure 3   (a) Transfer velocity of CO_2 plotted against the transfer velocity of water vapor as measured in a small circular wind wave facility. (b) Schmidt number exponent n as a function of the mean square slope. (a) From Jähne B (1980) *Zur Parameterisierung des Gasaustausches mit Hilfe von Laborexperimenten.* Dissertation, University of Heidelberg. (b) From Jähne B and Haußecker H (1998) Air–water gas exchange. *Annual Review of Fluid Mechanics* 30: 443–468.

Second, bubble-mediated gas transfer shifts the equilibrium value to slight supersaturation due to the enhanced pressure in the bubbles by surface tension and hydrostatic pressure. Third, breaking waves also enhance near-surface turbulence during the breaking event and the resurfacing of submerged bubbles.

Experimental data are still too sparse for the size and depth distribution of bubbles and the flux of the bubbles through the interface under various sea states for a sufficiently accurate modeling of bubble-mediated air–sea gas transfer and thus a reliable estimate of the contribution of bubbles to the total gas transfer rate. Some experiments from wind/wave tunnels and the field suggest that significant enhancements can occur, other experiments could not observe a significant influence of bubbles.

Empiric Parametrization

Given the lack of knowledge all theories about the enhancement of gas transfer by waves are rather speculative and are not yet useful for practical application. Thus, it is still state of the art to use semi-empiric or empiric parametrizations of the gas exchange rate with the wind speed. Most widely used is the parametrization of Liss and Merlivat. It identifies three physically well-defined regimes (smooth,

wave-influenced, and bubble-influenced) and proposes a piecewise linear relation between the wind speed U and the transfer velocity k:

$$k = 10^{-6}$$

$$\times \begin{cases} 0.472U(Sc/600)^{-2/3}, & U \leq 3.6 \text{ m s}^{-1} \\ 7.917(U - 3.39)(Sc/600)^{-1/2}, & U > 3.6 \text{ m s}^{-1} \text{ and } U \leq 13 \text{ m s}^{-1} \\ 16.39(U - 8.36)(Sc/600)^{-1/2}, & U > 13 \text{ m s}^{-1} \end{cases}$$

$$[15]$$

At the transition between the smooth and wavy regime, a sudden artificial jump in the Schmidt number exponent n from 2/3 to 1/2 occurs. This actually causes a discontinuity in the transfer rate for Schmidt number unequal to 600.

The empiric parametrization of Wanninkhof simply assumes a quadratic increase of the gas transfer rate with the wind speed:

$$k = 0.861 \times 10^{-6}(\text{s m}^{-1})U^2(Sc/600)^{-1/2} \quad [16]$$

Thus, this model has a constant Schmidt number exponent $n = 1/2$. The two parametrizations differ significantly (see **Figure 4**). The Wanninkhof parametrization predicts significantly higher values. The discrepancy between the two parametrizations

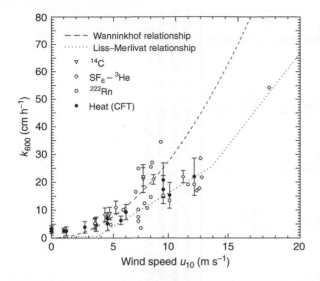

Figure 4 Summary of gas exchange field data normalized to a Schmidt number of 600 and plotted vs. wind speed together with the empirical relationships of Liss and Merlivat and Wanninkhof. Adapted from Jähne B and Haußecker H (1998) Air–water gas exchange. *Annual Review of Fluid Mechanics* 30: 443–468.

(up to a factor of 2) mirrors the current uncertainty in estimating the air–sea gas transfer rate.

Experimental Techniques and Results

Laboratory Facilities

Laboratory facilities play an important role in the investigation of air–sea gas transfer. Only laboratory studies allow a systematic study of the mechanisms and are thus an indispensible complement to field experiments. Almost all basic knowledge about gas transfer has been gained by laboratory experiments in the past. Among other things this includes the discovery of the influence of waves on air–water gas exchange (**Figure 3(a)**) and the change in the Schmidt number exponent (**Figure 3(b)**). Many excellent facilities are available worldwide (**Table 3**). Some of the early facilities are no longer operational or were demolished. However, some new facilities have also been built recently which offer new experimental opportunities for air–water gas transfer studies.

Geochemical Tracer Techniques

The first oceanic gas exchange measurements were performed using geochemical tracer methods such as the ^{14}C, ^{3}He/T, or $^{222}Rn/^{226}Ra$ methods. The volume and time-average flux density is given by mass balance of the tracer concentration in a volume of water V_w:

$$V_w \dot{c}_w = F_w j \quad \text{or} \quad j = h_w \dot{c}_w \qquad [17]$$

where F_w and h_w are the surface area and the effective

height V_w/F_w of a well-mixed water body, respectively. The time constant $\tau_w = h_w/k$ is in the order of days to weeks. It is evident that the transfer velocities obtained in this way provide only values integrated over a large horizontal length scales and timescales in the order of τ_w. Thus, a parametrization of the transfer velocity is only possible under steady-state conditions over extended periods. Moreover, the mass balance contains many other sources and sinks besides air–sea gas exchange and thus may cause severe systematic errors in the estimation of the transfer velocity. Consequently, mass balance methods are only poorly suited for the study of the mechanisms of air–water gas transfer.

Tracer Injection

The pioneering lake studies for tracer injection used sulfur hexafluoride (SF_6). However, the tracer concentration decreases not only by gas exchange across the interface but also by horizontal dispersion of the tracer. This problem can be overcome by the 'dual tracer technique' (Watson and co-workers) simultaneously releasing two tracers with different diffusivities (e.g., SF_6 and ^{3}He). When the ratio of the gas transfer velocities of the two tracers is known, the dilution effect by tracer dispersion can be corrected, making it possible to derive gas transfer velocities. But the basic problem of mass balance techniques, that is, their low temporal resolution, remains also with artificial tracer approaches.

Eddy Correlation Flux Measurements

Eddy correlation techniques are used on a routine basis in micrometeorology, that is, for tracers controlled by the boundary layer in air (momentum, heat, and water vapor fluxes). Direct measurements of the air–sea fluxes of gas tracers are very attractive because the flux densities are measured directly and have a much better temporal resolution than the mass balance-based techniques. Unfortunately, large experimental difficulties arise when this technique is applied to gas tracers controlled by the aqueous boundary layer. The concentration difference in the air is only a small fraction of the concentration difference across the aqueous mass boundary layer. But after more than 20 years of research has this technique delivered useful results. Some successful measurements under favorable conditions have been reported and it appears that remaining problems can be overcome in the near future.

The Controlled Flux Technique

The basic idea of this technique is to determine the concentration difference across the mass boundary layer when the flux density j of the tracer across the

Table 3 Comparison of the features of some major facilities for small-scale air–sea interaction studies (operational facilities are typeset in boldface)

	HH	M	D	SIO	C	UM	W	SU	HD1	HD2	HD3	HD4
Length (mean perimeter) (m)	15	40	100	40	33	15	18.3	119	1.57	11.6	29.2	3.90
Width of water channel (m)	1.8	2.6	8.0	2.4	0.76	1.0	0.91	2.0	0.10	0.20	0.62	0.37
Outer diameter (m)								40	0.60	4.0	9.92	
Inner diameter (m)								36	0.40	3.4	8.68	
Total height (m)	1.5	2.0	3.0	2.4	0.85	1.0	1.22	5.6	0.50	0.70	2.40	0.33
Max. water depth (m)	0.3	0.8	0.8	1.5	0.25	0.5	0.76	3.0	0.08	0.25	1.20	0.10
Water surface area (m²)	27	104	800	96	24.8	15	16.7	239	0.16	3.5	18.0	1.44
Water volume (m³)	8	83	768	144	8	10	?	716	0.01	0.87	20.7	0.14
Maximum wind speed (m s⁻¹)	25	15	15	12	25	30	25	19	11	12	15	8
Suitable for sea water	N	N	N	Y	N	Y	N	Y	Y	Y	Y	Y
Wave maker	N	Y	Y	Y	Y	Y	Y	N	N	N	N	N
Water current generator (m s⁻¹)	Y	Y	Y	N	±0.6	±0.5	±0.5	N	N	N	<−0.6	<−0.1
Water temperature control (°C)	N	Y	N	N	Y	Y	Y	N	5–35	N	5–35	Y
Air temperature control (°C)	N	Y	N	N	N	N	Y	N	5–35	N	5–35	N
Air humidity control	N	Y	N	N	N	Y	N	N	Y	N	Y	N
Gastight air space	N	N	N	N	Y	Y	Y	N	Y	Y	Y	Y

HH, Bundesanstalt für Wasserbau, Hamburg; M, IMST, Univ. Marseille, France; D, Delft Hydraulics, Delft, The Netherlands (no longer operational); SIO, Hydraulic Facility, Scripps Institution of Oceanography, La Jolla, USA; C, Canada Center for Inland Waters (CCIW); UM, University of Miami; W, NASA Air–Sea Interaction Research Facility, Wallops; SU, Storm basin, Marine Hydrophysical Institute, Sevastopol, Ukraine (no longer operational), HD1, Small annular wind/wave flume, Univ. Heidelberg (no longer operational), HD2, Large annular wind/wave flume, Univ. Heidelberg (dismantled); HD3, Aeolotron, Univ. Heidelberg (HD3, in operation since June 2000); HD4, Teflon-coated small Heidelberg linear wind/wave flume (N = No, Y = Yes).

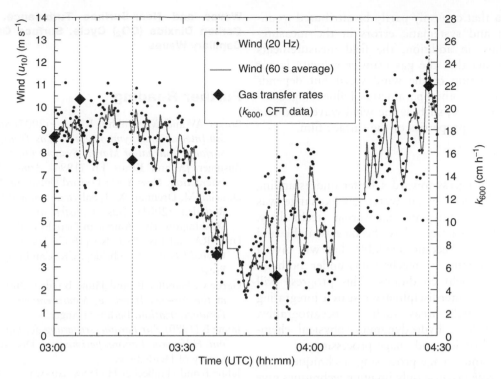

Figure 5 Wind speeds and gas transfer velocities computed with the controlled flux technique (CFT) during the 1995 MBL/CoOP West Coast experiment (JD133) for a period of 90 min. The transfer velocities are normalized to Schmidt number 600 and averaged over 4 min each. From Jähne B and Haußecker H (1998) Air–water gas exchange. *Annual Review of Fluid Mechanics* 30: 443–468.

interface is known. The local transfer velocity can be determined by simply measuring the concentration difference Δc across the aqueous boundary layer (cold surface skin temperature) according to [4] with a time constant $\tilde{t}$ for the transport across the boundary layer [8]. This technique is known as the 'controlled flux technique' (CFT).

Heat proves to be an ideal tracer for the CFT. The temperature at the water surface can then be measured with high spatial and temporal resolution using IR thermography. A known and controllable flux density can be applied by using infrared radiation. Infrared radiation is absorbed in the first few 10 μm at the water surface. Thus, a heat source is put right at top of the aqueous viscous boundary layer. Then the CFT directly measures the waterside heat transfer velocity.

A disadvantage of the CFT is that the transfer velocity of gases must be extrapolated from the transfer velocity of heat. The large difference in the Schmidt number (7 for heat, 600 for CO_2) casts some doubt whether the extrapolation to so much higher Schmidt numbers is valid.

Two variants of the technique proved to be successful. Active thermography uses a CO_2 laser to heat a spot of several centimeters in diameter on the water surface. The heat transfer rates are estimated from the temporal decay of the heated spot. Passive thermography uses the naturally occurring heat fluxes caused by latent heat flux j_l, sensible heat flux j_s, and long wave emission of radiation j_r. The net heat flux $j_n = j_l + j_s + j_r$ results according to [4] in a temperature difference across the interface of $\Delta T = j_h/(\rho c_p k_h)$. Because of the turbulent nature of the exchange process any mean temperature difference is associated with surface temperature fluctuations which can be observed in thermal images. With this technique the horizontal structure of the boundary layer turbulence can be observed. Surface renewal is directly observable in the IR image sequences, which show patches of fluid being drawn away from the surface.

With some knowledge about the statistics of the temperature fluctuations, the 'temperature difference' ΔT across the interface as well as the time constant $\tilde{t}$ of heat transfer can be computed from the temperature distribution at the surface. Results obtained with this technique are shown in **Figure 5** and also in the overview graph (**Figure 4**).

Summary of Field Data

A collection of field data is shown in **Figure 4**. Although the data show a clear increase of the transfer velocity with wind speed, there is substantial scatter in the data that can only partly be attributed to uncertainties and systematic errors in the measurements. Thus, in addition, the field measurements reflect the fact that the gas transfer velocity is not simply a function of the wind speed but depends significantly on other parameters influencing near-surface turbulence, such as the wind-wave field and the viscoelastic properties of the surface film.

Outlook

In the past, progress toward a better understanding of the mechanisms of air–water gas exchange was hindered by inadequate measuring technology. However, new techniques have become available and will continue to become available that will give a direct insight into the mechanisms under both laboratory and field conditions. This progress will be achieved by interdisciplinary research integrating different research areas such as oceanography, micrometeorology, hydrodynamics, physical chemistry, applied optics, and image processing.

Optical- and image-processing techniques will play a key role because only imaging techniques give direct insight to the processes in the viscous, heat, and mass boundary layers on both sides of the air–water interface. Eventually all key parameters including flow fields, concentration fields, and waves will be captured by imaging techniques with sufficient spatial and temporal resolution. The experimental data gained with such techniques will stimulate new theoretical and modeling approaches.

Nomenclature

D $(cm^2 s^{-1})$	molecular diffusion coefficient
j_c $(Mol\, cm^{-2} s^{-1})$	concentration flux density
k $(cm\, s^{-1})$	transfer velocity
K_c $(cm^2 s^{-1})$	turbulent diffusion coefficient
R $(cm^{-1} s)$	transfer resistance
$Sc = v/D$	Schmidt number
$\tilde{t} = \tilde{z}/k$ (s)	boundary layer time constant
u_* $(cm\, s^{-1})$	friction velocity
$\tilde{z}$ (cm)	boundary layer thickness
α	dimensionless solubility
v $(cm^2 s^{-1})$	kinematic viscosity

See also

Air–Sea Transfer: Dimethyl Sulfide, COS, CS$_2$, NH$_4$, Non-Methane Hydrocarbons, Organo-Halogens. Air–Sea Transfer: N$_2$O, NO, CH$_4$, CO. Breaking Waves and Near-Surface Turbulence. Bubbles. Carbon Dioxide (CO$_2$) Cycle. Surface Gravity and Capillary Waves.

Further Reading

Borger AV and Wanninkhof R (eds.) (2007) *Special Issue: 5th International Symposium on Gas Transfer at Water Surfaces. Journal of Marine Systems* 66: 1–308.

Businger JA and Kraus EB (1994) *Atmosphere–Ocean Interaction.* New York: Oxford University Press.

Donelan M, Drennan WM, Saltzman ES, and Wanninkhof R (eds.) (2001) *Gas Transfer at Water Surfaces.* Washington, DC: American Geophysical Union.

Duce RA and Liss PS (eds.) (1997) *The Sea Surface and Global Change.* Cambridge, UK: Cambridge University Press.

Garbe C, Handler R, and Jähne B (eds.) (2007) *Transport at the Air–Sea Interface, Measurements. Models, and Parameterization.* Berlin: Springer.

Jähne B (1980) *Zur Parameterisierung des Gasaustausches mit Hilfe von Laborexperimenten.* Dissertation, University of Heidelberg.

Jähne B and Haußecker H (1998) Air–water gas exchange. *Annual Review of Fluid Mechanics* 30: 443–468.

Jähne B, Heinz G, and Dietrich W (1987) Measurement of the diffusion coefficients of sparingly soluble gases in water. *Journal of Geophysical Research* 92: 10767–10776.

Jähne B and Monahan E (eds.) (1995) *Air–Water Gas Transfer.* Hanau: Aeon.

King DB, De Bryun WJ, Zheng M, and Saltzman ES (1995) Uncertainties in the molecular diffusion coefficient of gases in water for use in the estimation of air–sea exchange. In: Jähne B and Monahan E (eds.) *Air–Water Gas Transfer*, pp. 13–22. Hanau: Aeon.

Liss PS and Merlivat L (1986) Air–sea gas exchange rates: Introduction and synthesis. In: Buat-Menard P (ed.) *The Role of Air–Sea Exchange in Geochemical Cycles*, pp. 113–127. Dordrecht: Reidel.

McGilles WR, Asher WE, Wanninkhof R, and Jessup AT (eds.) (2004). *Special Issue: Air Sea Exchange. Journal of Geophysical Research* 109.

Wanninkhof R (1992) Relationship between wind speed and gas exchange over the ocean. *Journal of Geophysical Research* 97: 7373–7382.

Wilhelms SC and Gulliver JS (eds.) (1991) *Air–Water Mass Transfer.* New York: ASCE.

Relevant Websites

http://www.ifm.zmaw.de
– Institute of Oceanography, Universität Hamburg.
http://www.solas-int.org
– SOLAS.

CARBON CYCLE

C. A. Carlson, University of California, Santa Barbara, CA, USA

N. R. Bates, Bermuda Biological Station for Research, St George's, Bermuda, USA

D. A. Hansell, University of Miami, Miami FL, USA

D. K. Steinberg, College of William and Mary, Gloucester Pt, VA, USA

Introduction

Why is carbon an important element? Carbon has several unique properties that make it an important component of life, energy flow, and climate regulation. It is present on the Earth in many different inorganic and organic forms. Importantly, it has the ability to form complex, stable carbon compounds, such as proteins and carbohydrates, which are the fundamental building blocks of life. Photosynthesis provides marine plants (phytoplankton) with an ability to transform energy from sunlight, and inorganic carbon and nutrients dissolved in sea water, into complex organic carbon materials. All organisms, including autotrophs and heterotrophs, then catabolize these organic compounds to their inorganic constituents via respiration, yielding energy for their metabolic requirements. Production, consumption, and transformation of these organic materials provide the energy to be transferred between all the trophic states of the ocean ecosystem.

In its inorganic gaseous phases (carbon dioxide, CO_2; methane, CH_4; carbon monoxide, CO), carbon has important greenhouse properties that can influence climate. Greenhouse gases in the atmosphere act to trap long-wave radiation escaping from Earth to space. As a result, the Earth's surface warms, an effect necessary to maintain liquid water and life on Earth. Human activities have led to a rapid increase in greenhouse gas concentrations, potentially impacting the world's climate through the effects of global warming. Because of the importance of carbon for life and climate, much research effort has been focused on understanding the global carbon cycle and, in particular, the functioning of the ocean carbon cycle. Biological and chemical processes in the marine environment respond to and influence climate by helping to regulate the concentration of CO_2 in the atmosphere. We will discuss (1) the importance of the ocean to the global carbon cycle; (2) the mechanisms of carbon exchange between the ocean and atmosphere; (3) how carbon is redistributed throughout the ocean by ocean circulation; and (4) the roles of the 'solubility', 'biological,' and 'carbonate' pumps in the ocean carbon cycle.

Global Carbon Cycle

The global carbon cycle describes the complex transformations and fluxes of carbon between the major components of the Earth system. Carbon is stored in four major Earth reservoirs, including the atmosphere, lithosphere, biosphere, and hydrosphere. Each reservoir contains a variety of organic and inorganic carbon compounds ranging in amounts. In addition, the exchange and storage times for each carbon reservoir can vary from a few years to millions of years. For example, the lithosphere contains the largest amount of carbon (10^{23} g C), buried in sedimentary rocks in the form of carbonate minerals ($CaCO_3$, $CaMgCO_3$, and $FeCO_3$) and organic compounds such as oil, natural gas, and coal (fossil fuels). Carbon in the lithosphere is redistributed to other carbon reservoirs on timescales of millions of years by slow geological processes such as chemical weathering and sedimentation. Thus, the lithosphere is considered to be a relatively inactive component of the global carbon cycle (though the fossil fuels are now being added to the biologically active reservoirs at unnaturally high rates). The Earth's active carbon reservoirs contain approximately 43×10^{18} g of carbon, which is partitioned between the atmosphere (750×10^{15} g C), the terrestrial biosphere (2190×10^{15} g C), and the ocean ($39\,973 \times 10^{15}$ g C; **Figure 1**). While the absolute sum of carbon found in the active reservoirs is maintained in near steady state by slow geological processes, more rapid biogeochemical processes drive the redistribution of carbon among the active reservoirs.

Human activities, such as use of fossil fuels and deforestation, have significantly altered the amount of carbon stored in the atmosphere and perturbed the fluxes of carbon between the atmosphere, the terrestrial biosphere, and the ocean. Since the emergence of the industrial age 200 years ago, the release of CO_2 from fossil fuel use, cement manufacture, and deforestation has increased the partial pressure of atmospheric CO_2 from 280 ppm to present day values of 360 ppm; an increase of 25% in the last century (**Figure 2**). Currently, as a result of human activities, approximately 5.5×10^{15} g of

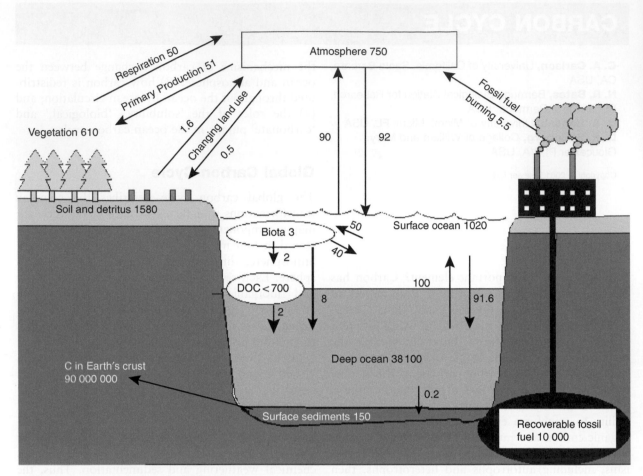

Figure 1 The global carbon cycle. Arrows indicate fluxes of carbon between the various reservoirs of the atmosphere, lithosphere, terrestrial biosphere, and the ocean. All stocks are expressed as 10^{15} g C. All fluxes are decadal means and expressed as 10^{15} g C y^{-1}. (Adapted with permission from Sigenthaler and Sarmiento, 1993, copyright 1993, Macmillan Magazines Ltd.). Data used to construct this figure came from Sigenthaler and Sarmiento (1993), Hansell and Carlson (1998), and Sarmiento and Wofsy (1999).

'anthropogenic' carbon is added to the atmosphere every year. About half of the anthropogenic CO_2 is retained in the atmosphere, while the remaining carbon is transferred to and stored in the ocean and the terrestrial biosphere. Carbon reservoirs that remove and sequester CO_2 from the atmosphere are referred to as carbon 'sinks'. The partitioning of anthropogenic carbon between oceanic and terrestrial sinks is not well known. Quantifying controls on the partitioning is necessary for understanding the dynamics of the global carbon cycle. The terrestrial biosphere may be a significant sink for anthropogenic carbon, but scientific understanding of the causative processes is hindered by the complexity of terrestrial ecosystems.

Global ocean research programs such as Geochemical Ocean Sections (GEOSEC), the Joint Global Ocean Flux Study (JGOFS), and the JGOFS/ World Ocean Circulation Experiment (WOCE) Ocean CO_2 Survey have resulted in improvements in our understanding of physical circulation and biological processes of the ocean. These studies have also allowed oceanographers to better constrain the role of the ocean in CO_2 sequestration compared to terrestrial systems. Based on numerical models of ocean circulation and ecosystem processes, oceanographers estimate that 70% (2×10^{15} g C) of the anthropogenic CO_2 is absorbed by the ocean each year. The fate of the remaining 30% (0.75×10^{15} g) of anthropogenic CO_2 is unknown. Determining the magnitude of the oceanic sink of anthropogenic CO_2 is dependent on understanding the interplay of various chemical, physical, and biological factors.

Oceanic Carbon Cycle

The ocean is the largest reservoir of the Earth's active carbon, containing $39\,973 \times 10^{15}$ g C. Oceanic carbon occurs as a variety of inorganic and organic

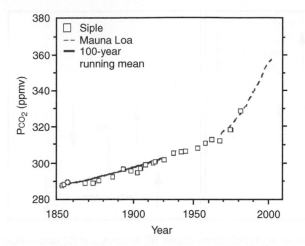

Figure 2 Atmospheric CO_2 concentrations from 1850 to 1996. These data illustrate an increase in atmospheric CO_2 concentration from pre-industrial concentration of 280 ppmv to present-day concentrations of 360 ppmv. Human activities of fossil fuel burning and deforestation have caused this observed increase in atmospheric CO_2. (Adapted from Houghton *et al.* (1996) with permission from Intergovernmental Panel on Climate Change (IPCC). The original figure was constructed from Siple ice core data and (from 1958) data collected at the Mauna Loa sampling site.)

forms, including dissolved CO_2, bicarbonate (HCO_3^-), carbonate (CO_3^{2-}) and organic compounds. CO_2 is one of the most soluble of the major gases in sea water and the ocean has an enormous capacity to buffer changes in the atmospheric CO_2 content.

The concentration of dissolved CO_2 in sea water is relatively small because CO_2 reacts with water to form a weak acid, carbonic acid (H_2CO_3), which rapidly dissociates (within milliseconds) to form HCO_3^- and CO_3^{2-} (eqn [I]).

$$CO_2(gas) + H_2O \rightleftharpoons H_2CO_3(aq) \rightleftharpoons H^+(aq) + HCO_3^-(aq) \rightleftharpoons 2H^+(aq) + CO_3^{2-}(aq) \quad [I]$$

For every 20 molecules of CO_2 absorbed by the ocean, 19 molecules are rapidly converted to HCO_3^- and CO_3^{2-}; at the typical range of pH in sea water (7.8–8.2; see below), most inorganic carbon is found in the form of HCO_3^-. These reactions (eqn [I]) provide a chemical buffer, maintain the pH of the ocean within a small range, and constrain the amount of atmospheric CO_2 that can be taken up by the ocean.

The amount of dissolved CO_2 in sea water cannot be determined analytically but can be calculated after measuring other inorganic carbon species. Dissolved inorganic carbon (DIC) refers to the total amount of CO_2, HCO_3^- plus CO_3^{2-} in sea water, while the partial pressure of CO_2 (P_{CO_2}) measures the contribution of CO_2 to total gas pressure. The alkalinity of sea water (A) is a measure of the bases present in sea

water, consisting mainly of HCO_3^- and CO_3^{2-} ($A[HCO_3^-] + 2[CO_3^{2-}]$) and minor constituents such as borate (BO_4) and hydrogen ions (H^+). Changes in DIC concentration and alkalinity affect the solubility of CO_2 in sea water (i.e., the ability of sea water to absorb CO_2) (see below).

The concentrations of inorganic carbon species in sea water are controlled not only by the chemical reactions outlined above (i.e., eqn [I]) but also by various physical and biological processes, including the exchange of CO_2 between ocean and atmosphere; the solubility of CO_2; photosynthesis and respiration; and the formation and dissolution of calcium carbonate ($CaCO_3$).

Typical surface sea water ranges from pH of 7.8 to 8.2. On addition of acid (i.e., H^+), the chemical reactions shift toward a higher concentration of CO_2 in sea water (eqn [IIa]) and pH decreases from 8.0 to 7.8 and then pH will rise from 8.0 to 8.2.

$$H^+ + HCO_3^- \rightarrow H_2CO_3 \rightarrow CO_2(aq) + H_2O \quad [IIa]$$

If base is added to sea water (eqn [IIb]), then pH will rise.

$$H_2CO_3 \rightarrow H^+ + HCO_3 \quad [IIb]$$

Solubility and Exchange of CO_2 between the Ocean and Atmosphere

The solubility of CO_2 in sea water is an important factor in controlling the exchange of carbon between the ocean and atmosphere. Henry's law (eqn [1]) describes the relationship between solubility and sea water properties, where S equals the solubility of gas in liquid, k is the solubility constant (k is a function mainly of temperature) and P is the overlying pressure of the gas in the atmosphere.

$$S = kP \quad [1]$$

Sea water properties such as temperature, salinity, and partial pressure of CO_2 determine the solubility of CO_2. For example, at $0°C$ in sea water, the solubility of CO_2 is double that in sea water at $20°C$; thus colder water will tend to absorb more CO_2 than warmer water.

Henry's law also describes the relationship between the partial pressure of CO_2 in solution (P_{CO_2}) and its concentration (i.e., $[CO_2]$). Colder waters tend to have lower P_{CO_2} than warmer waters: for every $1°C$ temperature increase, sea water P_{CO_2} increases by $\sim 4\%$. Sea water P_{CO_2} is also influenced by complicated thermodynamic relationships

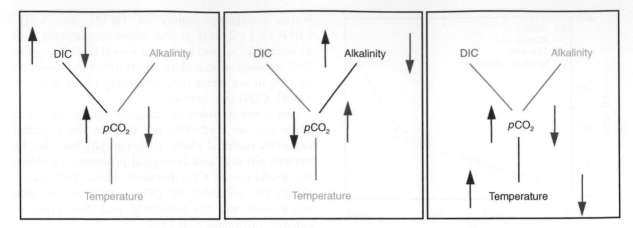

Figure 3 The response of P_{CO_2} to changes in the sea water properties of (A) DIC concentration, (B) alkalinity and (C) temperature. Each panel describes how P_{CO_2} will respond to the changes in the relevant sea water property. The blue arrows illustrate the response of P_{CO_2} to an increase in the sea water property and the red arrows illustrate the response to a decrease in the sea water property. For example, as DIC or temperature increases, P_{CO_2} increases; whereas an increase in alkalinity results in a decrease in P_{CO_2}.

between the different carbon species. For example, a decrease in sea water DIC or temperature acts to decrease P_{CO_2}, while a decrease in alkalinity acts to increase P_{CO_2} (**Figure 3**).

Carbon dioxide is transferred across the air–sea interface by molecular diffusion and turbulence at the ocean surface. The flux (F) of CO_2 between the atmosphere and ocean is driven by the concentration difference between the reservoirs (eqn [2]).

$$F = \Delta P_{CO_2} K_W \qquad [2]$$

In eqn [2] ΔP_{CO_2} is the difference in P_{CO_2} between the ocean and atmosphere and K_w is the transfer coefficient across the air–sea interface, termed the piston velocity. In cold waters, sea water P_{CO_2} tends to be lower than atmospheric P_{CO_2}, thus driving the direction of CO_2 gas exchange from atmosphere to ocean (**Figure 3**). In warmer waters, sea water P_{CO_2} is greater than atmospheric P_{CO_2}, and CO_2 gas exchange occurs in the opposite direction, from the ocean to the atmosphere. The rate at which CO_2 is transferred between the ocean and the atmosphere depends not only on the P_{CO_2} difference but on turbulence at the ocean surface. The piston velocity of CO_2 is related to solubility and the strength of the wind blowing on the sea surface. As wind speed increases, the rate of air–sea CO_2 exchange also increase. Turbulence caused by breaking waves also influences gas exchange because air bubbles may dissolve following entrainment into the ocean mixed layer.

Ocean Structure

Physically, the ocean can be thought of as two concentric spheres, the surface ocean and the deep ocean, separated by a density discontinuity called the pycnocline. The surface ocean occupies the upper few hundred meters of the water column and contains approximately 1020×10^{15} g C of DIC (**Figure 1**). The absorption of CO_2 by the ocean through gas exchange takes place in the mixed layer, the upper portion of the surface ocean that makes direct contact with the atmosphere. The surface ocean reaches equilibrium with the atmosphere within one year. The partial pressure of CO_2 in the surface ocean is slightly less than or greater than that of the atmosphere, depending on the controlling variables as described above, and varies temporally and spatially with changing environmental conditions. The deeper ocean represents the remainder of the ocean volume and is supersaturated with CO_2, with a DIC stock of $38\ 100 \times 10^{15}$ g C (**Figure 1**), or 50 times the DIC contained in the atmosphere.

CO_2 absorbed by the ocean through gas exchange has a variety of fates. Physical and biological mechanisms can return the CO_2 back to the atmosphere or transfer carbon from the surface ocean to the deep ocean and ocean sediments through several transport processes termed the 'solubility', 'biological', and 'carbonate' pumps.

The Solubility Pump, Oceanic Circulation, and Carbon Redistribution

The 'solubility pump' is defined as the exchange of carbon between the atmosphere and the ocean as mediated by physical processes such as heat flux, advection and diffusion, and ocean circulation. It assists in the transfer of atmospheric CO_2 to the deep ocean. This transfer is controlled by circulation patterns of the surface ocean (wind-driven circulation) and the deep ocean (thermohaline circulation). These circulation patterns assist in the transfer of atmospheric CO_2 to the deep ocean and help to maintain the vertical gradient of DIC found in the ocean (**Figure 4**). The ability of the ocean to take up anthropogenic CO_2 via the solubility pump is limited by the physical structure of the ocean, the distribution of oceanic DIC, ocean circulation patterns, and the exchange between the surface and deep ocean layers. To be an effective sink for anthropogenic carbon, CO_2 must be transferred to the deep ocean by mixing and biological processes (see below).

Wind-driven circulation occurs as a consequence of friction and turbulence imparted by wind blowing over the sea surface. This circulation pattern is primarily horizontal in movement and is responsible for transporting warm water from lower latitudes (warm) to higher latitudes (cold). Surface currents move water and carbon great distances within ocean

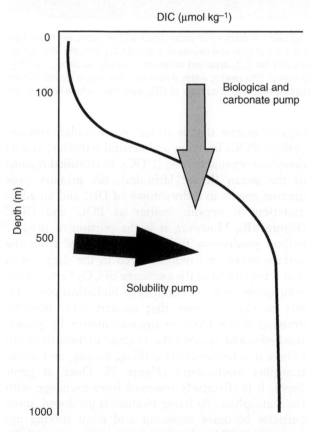

Figure 4 Illustration of the vertical gradient of DIC in the ocean. The uptake of DIC by phytoplankton and conversion into sinking organic matter ('biological pump'; gray arrow) and sinking calcium carbonate skeletal matter ('carbonate pump'; gray arrow) contributes to the maintenance of the vertical gradient. Introduction of DIC to the deep waters via the 'solubility pump' at high latitudes and subsequent deep water formation also helps maintain this vertical gradient (black arrow; see **Figure 5**).

basins on timescales of months to years. As surface sea water moves from low latitudes to high latitudes, the increasing solubility of CO_2 in the sea water (due to sea surface cooling) allows atmospheric CO_2 to invade the surface mixed layer (**Figure 5** and **6A**).

Exchange of surface waters with the deep ocean through wind-driven mixing is limited because of strong density stratification of the water column over the majority of the world's oceans. However, thermohaline (overturning) circulation at high latitudes provides a mechanism for surface waters to exchange with the deep ocean. Passage of cold and dry air masses over high-latitude regions, such as the Greenland and Labrador Seas in the North Atlantic or the Weddell Sea in the Southern Ocean, forms cold and very dense sea water ('deep water' formation). Once formed, these dense water masses sink vertically until they reach a depth at which water is of similar density (i.e., 2000–4000 m deep). Following sinking, the dense waters are transported slowly throughout all of the deep ocean basins by advection and diffusion, displacing other deep water that eventually is brought back to the surface by upwelling (**Figure 5**).

Because of the smaller volume and faster circulation, the residence time of the surface ocean is only one decade compared to 600–1000 years for the deep ocean. The process of deep water formation transfers CO_2, absorbed from the atmosphere by the solubility pump, into the deep ocean. The effect is that DIC concentration increases with depth in all ocean basins (**Figure 4, 5** and **6A,**). As a result of the long residence time of the deep ocean, carbon, once removed from the surface ocean to the deep ocean through the effects of solubility and deep water formation, is stored without contact with the atmosphere for hundreds to thousands of years. At present, deep water formed at the surface that is in equilibrium with the atmosphere (sea water P_{CO_2} of ~ 360 ppm), carries more CO_2 to depth than deep water formed prior to the industrial age (e.g., ~ 280 ppm). Furthermore, P_{CO_2} of upwelled deep water is less than that in the recently formed deep water, indicating that the deep water formation and the 'solubility pump' allow the ocean to be a net sink for anthropogenic CO_2. The vertical gradient in DIC (**Figure 4**) and the ability of the ocean to take up atmospheric CO_2 is augmented by biological processes known as the 'biological pump'.

The Biological Pump

Although the standing stock of marine biota in the ocean is relatively small (3×10^{15} g C), the activity

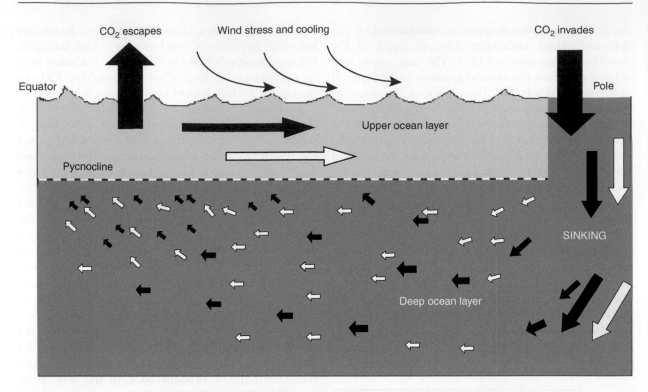

Figure 5 Conceptual model of the 'solubility pump'. White arrows represent movement of water; black arrows represent movement of CO_2 within, and into and out of, the ocean. Cooling increases the solubility of CO_2 and results in a flux of CO_2 from the atmosphere to the surface ocean. At subpolar latitudes the water density increases and the CO_2-enriched water sinks rapidly. At depth, the CO_2-enriched water moves slowly as is it is dispersed throughout the deep ocean. The sinking water displaces water that is returned to the surface ocean in upwelling regions. As the water warms, P_{CO_2} increases, resulting in escape of CO_2 from the surface water to the atmosphere.

associated with the biota is extremely important to the cycling of carbon between the atmosphere and the ocean. The largest and most rapid fluxes in the global carbon cycle are those that link atmospheric CO_2 to photosynthetic production (primary production) on the land and in the ocean. Globally, marine phytoplankton are responsible for more than one-third of the total gross photosynthetic production (50×10^{15} g C y^{-1}). In the sea, photosynthesis is limited to the euphotic zone, the upper 100–150 m of the water column where light can penetrate. Photosynthetic organisms use light energy to reduce CO_2 to high-energy organic compounds. In turn, a portion of these synthesized organic compounds are utilized by heterotrophic organisms as an energy source, being remineralized to CO_2 via respiration. Eqn [III] represents the overall reactions of photosynthesis and respiration.

$$CO_2(gas) + H_2O \xrightleftharpoons[\text{Metabolic energy (respiration)}]{\text{Light energy (photosynthesis)}} (CH_2O)_n \qquad [III]$$
$$+ O_2(gas)$$

In the sea, net primary production (primary production in excess of respiration) converts CO_2 to organic matter that is stored as particulate organic carbon (POC; in living and detrital particles) and as dissolved organic carbon (DOC). In stratified regions of the ocean (lower latitudes), net primary production results in a drawdown of DIC and an accumulation of organic matter as POC and DOC (**Figure 6B**). However, it is the portion of organic carbon production that can be exported from the surface ocean and remineralized in the deep ocean that is important in the exchange of CO_2 between the atmosphere and the ocean. The biological pump refers to the processes that convert CO_2 (thereby drawing down DIC) to organic matter by photosynthesis, and remove the organic carbon to depth (where it is respired) via sinking, mixing, and active transport mechanisms (**Figure 7**). Once at great depth, it is effectively removed from exchange with the atmosphere. As living biomass is produced, some particles becomes senescent and form sinking aggregates, while other particles are consumed by herbivores and sinking fecal pellets (POC) are formed. These sinking aggregates and pellets remove carbon from the surface to be remineralized at depth via decomposition by bacteria or consumption by zooplankton and fish (**Figure 7**). In addition, DOC produced by phytoplankton or by animal excretion

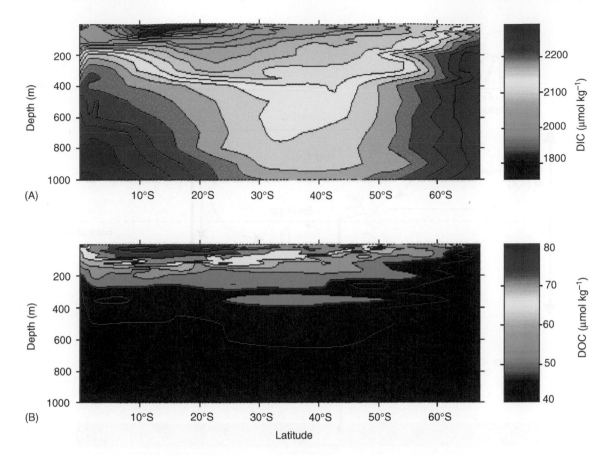

Figure 6 Contour plot of (A) DIC and (B) DOC along a transect line in the South Pacific between the equator (0°; 170° W) and the Antarctic Polar Front (66° S; 170° W). Note that in the low-latitude stratified waters DIC concentrations are depleted in surface water relative to deep water, as a result of net primary production and air–sea exchange. DOC concentrations are elevated relative to deep water. In high-latitude regions, DIC concentration are elevated in the surface water as a result of increased solubility of cooler surface waters.

in surface waters can also be transported downward by subduction or convective mixing of surface waters (**Figure 7**). Finally, vertically migrating zooplankton that feed in the surface waters at night and return to deep waters during the day actively transport dissolved and particulate material to depth, where a portion is metabolized (**Figure 7**).

Production via photosynthesis can occur only in the surface ocean, whereas remineralization can occur throughout the water column. The biological pump serves to spatially separate the net photosynthetic from net respiratory processes. Thus, the conversion of DIC to exportable organic matter acts to reduce the DIC concentration in the surface water and its subsequent remineralization increases DIC concentration in the deep ocean (**Figure 6**). The biological pump is important to the maintenance of a vertical DIC profile of undersaturation in the surface and supersaturation at depth (**Figure 4** and **5A**). Undersaturation of DIC in the surface mixed layer, created by the biological pump, allows for the influx

of CO_2 from the atmosphere (see Henry's law above; **Figure 7**).

Gross export of organic matter out of the surface waters is approximately 10×10^{15} g C y^{-1} (**Figure 1**). Less than 1% of the organic matter exported from the surface waters is stored in the abyssal sediment. In fact, most of the exported organic matter is remineralized to DIC in the upper 500 m of the water column. It is released back to the atmosphere on timescales of months to years via upwelling, mixing, or ventilation of high-density water at high latitudes. It is that fraction of exported organic matter that actually reaches the deep ocean (>1000 m) that is important for long-term atmospheric CO_2 regulation. Once in the deep ocean, the organic matter either remains as long-lived DOC or is remineralized to DIC and is removed from interaction with the atmosphere on timescales of centuries to millennia. Thus, even though less than 1% of the exported carbon is stored in marine sediments, the activities of the biological pump are very important in mediating

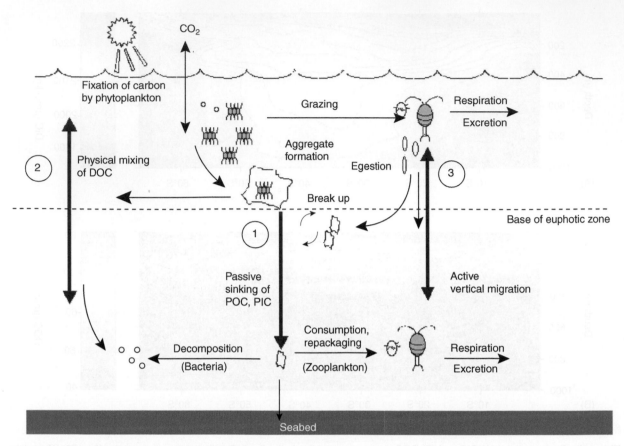

Figure 7 Conceptual diagram depicting components of the 'biological pump'. CO_2 is taken up by phytoplankton and organic matter is produced. As this organic matter is processed through the marine food web, fecal pellets or aggregates are produced, a portion of which sink from the surface waters to depth (1). As organic matter is processed through the food web, DOC is also produced. DOC is removed from the surface waters to depth via physical mixing of the water by convective overturn (2). DOC and DIC are also actively transported to depth by vertically migrating organisms such as copepods that feed in surface waters and excrete and respire the consumed organic carbon at depth (3).

the air–sea transfer of CO_2. Without this pump in action, atmospheric CO_2 concentration might be as high as 500 to 1000 ppm versus the 360 ppm observed today.

Contribution of POC Versus DOC in the Biological Pump

Historically, sinking particles were thought of as the dominant export mechanism of the biological pump and the primary driver of respiration in the ocean's interior. However, downward mixing of surface water can also transport large quantities of DOC trapped within the sinking water mass. In order for DOC to be an important contributor to the biological pump two sets of conditions must exist. First, the producer–consumer dynamics in the surface waters must yield DOC of a quality that is resistant to rapid remineralization by bacteria and lead to net DOC production. Second, the physical system must undergo periods of deep convective mixing or

subduction in order to remove surface waters and DOC to depth. Although approximately 80% of the globally exported carbon is in the form of POC, DOC can represent 30–50% of carbon export in the upper 500 m of the water column at specific ocean sites. The biological/physical controls on DOC export are complex and are currently being assessed for various regions of the worlds' ocean.

Factors That Affect the Efficiency of the Biological Pump

An efficient biological pump means that a large fraction of the system's net production is removed from the surface waters via export mechanisms. Factors that affect the efficiency of the biological pump are numerous and include nutrient supply and plankton community structure.

Nutrient supply Does an increased partial pressure of atmospheric CO_2 lead to a more efficient biological pump? Not necessarily, since net primary production is limited by the availability of other inorganic nutrients such as nitrogen, phosphorus, silicon and iron. Because these inorganic nutrients are continuously being removed from the surface waters with vertical export of organic particles, their concentrations are often below detection limits in highly stratified water columns. As a result, primary production becomes limited by the rate at which these nutrients can be re-supplied to the surface ocean by mixing, by atmospheric deposition, or by heterotrophic recycling. Primary production supported by the recycling of nutrients in the surface ocean is referred to as 'regenerative' production and contributes little to the biological pump. Primary production supported by the introduction of new nutrients from outside the system, via mixing from below or by atmospheric deposition (e.g. dust), is referred to as 'new' production. New nutrients enhance the amount of net production that can be exported (new production). Because CO_2 is not considered to

be a limiting nutrient in marine systems, the increase in atmospheric CO_2 is not likely to stimulate net production for most of the world's ocean unless it indirectly affects the introduction of new nutrients as well.

Community structure Food web structure also plays an important role in determining the size distribution of the organic particles produced and whether the organic carbon and associated nutrients are exported from or recycled within the surface waters. The production of large, rapidly settling cells will make a greater contribution to the biological pump than the production of small, suspended particles. Factors such as the number of trophic links and the size of the primary producers help determine the overall contribution of sinking particles. The number of trophic steps is inversely related to the magnitude of the export flux. For example, in systems where picoplankton are the dominant primary producers there may be 4–5 steps before reaching a trophic level capable of producing

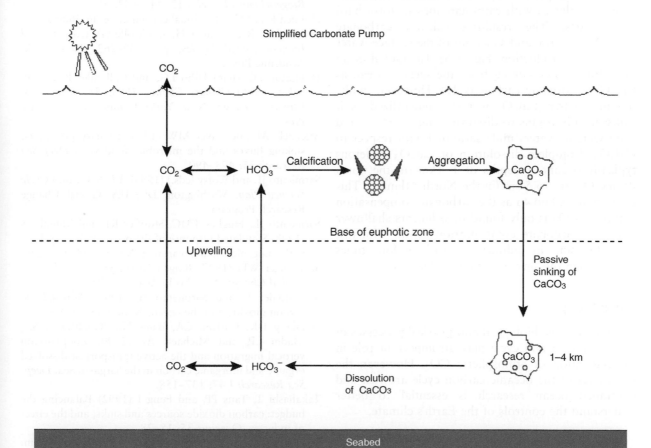

Figure 8 Conceptual diagram of a simplified 'carbonate pump'. Some marine organisms form calcareous skeletal material, a portion of which sinks as calcium carbonate aggregates. These aggregates are preserved in shallow ocean sediments or dissolve at greater depths (3000–5000 m), thus increasing DIC concentrations in the deep ocean. The calcium and bicarbonate are returned to the surface ocean through upwelling.

sinking particles. With each trophic transfer, a percentage (50–70%) of the organic carbon is respired, so only a small fraction of the original primary production forms sinking particles. Although picoplankton may dominate primary production in oceanic systems, their production is considered 'regenerative' and contributes little to the production of sinking material. Alternatively, production by larger phytoplankton such as diatoms ($>20 \mu$m in size) may represent a smaller fraction of primary production, but their contribution to the biological pump is larger because fewer trophic steps are taken to produce sinking particles.

The Carbonate Pump

A process considered part of the biological pump (depending how it is defined) is the formation and sinking of calcareous skeletal material by some marine phytoplankton (e.g., coccolithophores) and animals (e.g., pteropods and foraminifera). Calcification is the process by which marine organisms combine calcium with carbonate ions to form hard body parts. The resulting calcium carbonate ($CaCO_3$) is dense and sinks out of the surface water with export production (**Figure 8**). The global mean ratio for carbon sinking from the surface ocean as $CaCO_3$ or organic carbon is 1:4. However, unlike organic matter, $CaCO_3$ is not remineralized as it sinks; it only begins to dissolve in intermediate and deep waters, waters undersaturated with respect to $CaCO_3$. Complete dissolution of $CaCO_3$ skeletons typically occurs at depths of 1–4 km (in the north Pacific Ocean) to 5 km (in the North Atlantic). This depth zone is known as the carbonate compensation depth. $CaCO_3$ is only found in sediments shallower than the carbonate compensation depth. Globally, the CO_2 sink in sedimentary rock is four times greater than the sink in organic sediments.

Summary

In summary, the biological and physical processes of the oceanic carbon cycle play an important role in the regulation of atmospheric CO_2. However, the intricacies of the oceanic carbon cycle are vast and continued ocean research is essential to better understand the controls of the Earth's climate.

See also

Atmospheric Input of Pollutants. Carbon Dioxide (CO_2) Cycle.

Further Reading

Bates NR, Michaels AF, and Knap AH (1996) Seasonal and interannual variability of oceanic carbon dioxide species at the U.S. JGOFS Bermuda Atlantic Time-series Study (BATS) site. *Deep-Sea Research II* 43: 347–383.

Bolin B (ed.) (1983) *The Major Biogeochemical Cycles and Their Interactions: SCOPE 21.* New York: Wiley.

Carlson CA, Ducklow HW, and Michaels AF (1994) Annual flux of dissolved organic carbon from the euphotic zone in the northwestern Sargasso Sea. *Nature* 371: 405–408.

Denman K, Hofman H, and Marchant H (1996) Marine biotic responses to environmental change and feedbacks to climate. In: Houghton JT, Meira Filho LG, and Callander BA, *et al.* (eds.) *Climate Change 1995: The Science of Climate Change.* New York: Cambridge University Press.

Follows MJ, Williams RG, and Marshall JC (1996) The solubility pump of carbon in the subtropical gyre of the North Atlantic. *Journal of Marine Research* 54: 605–630.

Hansell DA and Carlson CA (1998) Net community production of dissolved organic carbon. *Global Biogeochemical Cycles* 12: 443–453.

Holmén K (1992) The global carbon cycle. In: Butcher SS, Charlson RJ, Orians GH, and Wolfe GV (eds.) *Global Biogeochemical Cycles*, pp. 239–262. New York: Academic Press.

Houghton JT, Meira Filho LG, and Callander BA, *et al.* (eds.) (1996) *Climate Change 1995: The Science of Climate Change.* New York: Cambridge University Press.

Michaels AF and Silver MW (1988) Primary producers, sinking fluxes and the microbial food web. *Deep-Sea Research* 35: 473–490.

Sarmiento JL and Wofsy (eds.) (1999) *A U.S. Carbon Cycle Science Plan.* Washington, DC: U.S. Global Change Research Program.

Sarmiento JL, Hughes TMC, Stouffer RJ, and Manabe S (1998) Simulated response of the ocean carbon cycle to anthropogenic climate warming. *Nature* 393: 245–249.

Schlesinger WH (1997) *Biogeochemistry: An Analysis of Global Change.* New York: Academic Press.

Siegenthaler U and Sarmiento JL (1993) Atmospheric carbon dioxide and the ocean. *Nature* 365: 119–125.

Steinberg DK, Carlson CA, Bates NR, Goldthwait SA, Madin LP, and Michaels AF (2000) Zooplankton vertical migration and the active transport of dissolved organic and inorganic carbon in the Sargasso Sea. *Deep-Sea Research I* 47: 137–158.

Takahashi T, Tans PP, and Fung I (1992) Balancing the budget: carbon dioxide sources and sinks, and the effect of industry. *Oceanus* 35: 18–28.

Varney M (1996) The marine carbonate system. In: Summerhayes CP and Thorpe SA (eds.) *Oceanography an Illustrated Guide*, pp. 182–194. London: Manson Publishing.

CARBON DIOXIDE (CO$_2$) CYCLE

T. Takahashi, Lamont Doherty Earth Observatory, Columbia University, Palisades, NY, USA

Introduction

The oceans, the terrestrial biosphere, and the atmosphere are the three major dynamic reservoirs for carbon on the earth. Through the exchange of CO_2 between them, the atmospheric concentration of CO_2 that affects the heat balance of the earth, and hence the climate, is regulated. Since carbon is one of the fundamental constituents of living matter, how it cycles through these natural reservoirs has been one of the fundamental questions in environmental sciences. The oceans contain about 50 times as much carbon (about 40 000 Pg-C or 10^{15} g as carbon) as the atmosphere (about 750 Pg-C). The terrestrial biosphere contains about three times as much carbon (610 Pg-C in living vegetation and 1580 Pg-C in soil organic matter) as the atmosphere. The air–sea exchange of CO_2 occurs via gas exchange processes across the sea surface; the natural air-to-sea and sea-to-air fluxes have been estimated to be about 90 Pg-C y^{-1} each. The unperturbed uptake flux of CO_2 by global terrestrial photosynthesis is roughly balanced with the release flux by respiration, and both have been estimated to be about 60 Pg-C y^{-1}. Accordingly, atmospheric CO_2 is cycled through the ocean and terrestrial biosphere with a time scale of about 7 years.

The lithosphere contains a huge amount of carbon (about 100 000 000 Pg-C) in the form of limestones ((Ca, Mg) CO$_3$), coal, petroleum, and other forms of organic matter, and exchanges carbon slowly with the other carbon reservoirs via such natural processes as chemical weathering and burial of carbonate and organic carbon. The rate of removal of atmospheric CO_2 by chemical weathering has been estimated to be of the order of 1 Pg-C y^{-1}. Since the industrial revolution in the nineteenth century, the combustion of fossil fuels and the manufacturing of cement have transferred the lithospheric carbon into the atmosphere at rates comparable to the natural CO_2 exchange fluxes between the major carbon reservoirs, and thus have perturbed the natural balance significantly (6 Pg-C y^{-1} is about an order of magnitude less than the natural exchanges with the oceans (90 Pg-C y^{-1}) and land (60 Pg-C y^{-1})). The industrial carbon emission rate has been about 6 Pg-C y^{-1} for the 1990s, and the cumulative industrial emissions since the nineteenth century to the end of the twentieth century have been estimated to be about 250 Pg-C. Presently, the atmospheric CO_2 content is increasing at a rate of about 3.5 Pg-C y^{-1} (equivalent to about 50% of the annual emission) and the remainder of the CO_2 emitted into the atmosphere is absorbed by the oceans and terrestrial biosphere in approximately equal proportions. These industrial CO_2 emissions have caused the atmospheric CO_2 concentration to increase by as much as 30% from about 280 ppm (parts per million mole fraction in dry air) in the pre-industrial year 1850 to about 362 ppm in the year 2000. The atmospheric CO_2 concentration may reach 580 ppm, double the pre-industrial value, by the mid-twenty first century. This represents a significant change that is wholly attributable to human activities on the Earth.

It is well known that the oceans play an important role in regulating our living environment by providing water vapor into the atmosphere and transporting heat from the tropics to high latitude areas. In addition to these physical influences, the oceans partially ameliorate the potential CO_2-induced climate changes by absorbing industrial CO_2 in the atmosphere.

Therefore, it is important to understand how the oceans take up CO_2 from the atmosphere and how they store CO_2 in circulating ocean water. Furthermore, in order to predict the future course of the atmospheric CO_2 changes, we need to understand how the capacity of the ocean carbon reservoir might be changed in response to the Earth's climate changes, that may, in turn, alter the circulation of ocean water. Since the capacity of the ocean carbon reservoir is governed by complex interactions of physical, biological, and chemical processes, it is presently not possible to identify and predict reliably various climate feedback mechanisms that affect the ocean CO_2 storage capacity.

Units

In scientific and technical literature, the amount of carbon has often been expressed in three different units: giga tons of carbon (Gt-C), petagrams of carbon (Pg-C) and moles of carbon or CO_2. Their relationships are: 1 Gt-C = 1 Pg-C = 1×10^{15} g of carbon = 1000 million metric tonnes of carbon = $(1/12) \times 10^{15}$ moles of carbon. The equivalent quantity as CO_2 may be obtained by multiplying the above numbers by 3.67 (= 44/12 = the molecular weight of CO_2 divided by the atomic weight of carbon).

The magnitude of CO_2 disequilibrium between the atmosphere and ocean water is expressed by the difference between the partial pressure of CO_2 of ocean water, $(pCO_2)sw$, and that in the overlying air, $(pCO_2)air$. This difference represents the thermodynamic driving potential for CO_2 gas transfer across the sea surface. The pCO_2 in the air may be estimated using the concentration of CO_2 in air, that is commonly expressed in terms of ppm (parts per million) in mole fraction of CO_2 in dry air, in the relationship:

$$p(CO_2)air = (CO_2 \text{ conc.})air \times (Pb - pH_2O) \quad [1]$$

where Pb is the barometric pressure and pH_2O is the vapor pressure of water at the sea water temperature. The partial pressure of CO_2 in sea water, $(pCO_2)sw$, may be measured by equilibration methods or computed using thermodynamic relationships. The unit of microatmospheres (μatm) or 10^{-6} atm is commonly used in the oceanographic literature.

History

The air–sea exchange of CO_2 was first investigated in the 1910s through the 1930s by a group of scientists including K. Buch, H. Wattenberg, and G.E.R. Deacon. Buch and his collaborators determined in land-based laboratories CO_2 solubility, the dissociation constants for carbonic and boric acids in sea water, and their dependence on temperature and chlorinity (the chloride ion concentration in sea water). Based upon these dissociation constants along with the shipboard measurements of pH and titration alkalinity, they computed the partial pressure of CO_2 in surface ocean waters. The Atlantic Ocean was investigated from the Arctic to Antarctic regions during the period 1917–1935, especially during the METEOR Expedition 1925–27, in the North and South Atlantic. They discovered that temperate and cold oceans had lower pCO_2 than air (hence the sea water was a sink for atmospheric CO_2), especially during spring and summer seasons, due to the assimilation of CO_2 by plants. They also observed that the upwelling areas of deep water (such as African coastal areas) had greater pCO_2 than the air (hence the sea water was a CO_2 source) due to the presence of respired CO_2 in deep waters.

With the advent of the high-precision infrared CO_2 gas analyzer, a new method for shipboard measurements of pCO_2 in sea water and in air was introduced during the International Geophysical Year, 1956–59. The precision of measurements was improved by more than an order of magnitude. The global oceans were investigated by this new method, which rapidly yielded high precision data. The equatorial Pacific was identified as a major CO_2 source area. The GEOSECS Program of the International Decade of Ocean Exploration, 1970–80, produced a global data set that began to show systematic patterns for the distribution of CO_2 sink and source areas over the global oceans.

Methods

The net flux of CO_2 across these a surface, Fs-a, may be estimated by:

$$\begin{aligned} Fs\text{-}a &= E \times [(pCO_2)sw - (pCO_2)air] \\ &= k \times \alpha \times [(pCO_2)sw - (pCO_2)air] \end{aligned} \quad [2]$$

where E is the CO_2 gas transfer coefficient expressed commonly in (moles $CO_2/m^2/y/$uatm); k is the gas transfer piston velocity (e.g. in (cmh^{-1})) and α is the solubility of CO_2 in sea water at a given temperature and salinity (e.g. (moles CO_2 kg-sw^{-1} atm^{-1})). If $(pCO_2)sw < (pCO_2)$ air, the net flux of CO_2 is from the sea to the air and the ocean is a source of CO_2; if $(pCO_2)sw < (pCO_2)$ air, the ocean water is a sink for atmospheric CO_2. The sea–air pCO_2 difference may be measured at sea and α has been determined experimentally as a function of temperature and salinity. However, the values of E and k that depend on the magnitude of turbulence near the air–water interface cannot be simply characterized over complex ocean surface conditions. Nevertheless, these two variables have been commonly parameterized in terms of wind speed over the ocean. A number of experiments have been performed to determine the wind speed dependence under various wind tunnel conditions as well as ocean and lake environments using different nonreactive tracer gases such as SF_6 and ^{222}Rn. However, the published results differ by as much as 50% over the wind speed range of oceanographic interests.

Since ^{14}C is in the form of CO_2 in the atmosphere and enters into the surface ocean water as CO_2 in a timescale of decades, its partition between the atmosphere and the oceans yields a reliable estimate for the mean CO_2 gas transfer rate over the global oceans. This yields a CO_2 gas exchange rate of 20 ± 3 mol CO_2 m^{-2} y^{-1} that corresponds to a sea–air CO_2 transfer coefficient of 0.067 mol CO_2 m^{-2} y^{-1} uatm^{-1}. Wanninkhof in 1992 presented an expression that satisfies the mean global CO_2 transfer coefficient based on ^{14}C and takes other field and wind tunnel results into consideration. His equation for variable wind speed conditions is:

$$k\left(cm\,h^{-1}\right) = 0.39 \times (u_{av})^2 \times (Sc/660)^{-0.5} \quad [3]$$

where u_{av} is the average wind speed in ms^{-1} corrected to $10\,m$ above sea surface; Sc(dimensionless) is the Schmidt number (kinematic viscosity ofwater)/ (diffusion coefficient of CO_2 gas inwater); and 660 represents the Schmidt number for CO_2 in seawater at 20°C.

In view of the difficulties in determining gas transfer coefficients accurately, direct methods for CO_2 flux measurements aboard the ship are desirable. Sea–air CO_2 flux was measured directly by means of the shipboard eddy-covariance method over the North Atlantic Ocean by Wanninkhof and McGillis in 1999. The net flux of CO_2 across the sea surface was determined by a covariance analysis of the tri-axial motion of air with CO_2 concentrations in the moving air measured in short time intervals ($\sim$ms) as a ship moved over the ocean. The results obtained over awind speed range of 2–$13.5\,m\,s^{-1}$ are consistent with eqn [3] within about $\pm 20\%$. If the data obtainedin wind speeds up to $15\,m\,s^{-1}$ are taken into consideration, they indicate that the gas transfer piston velocity tends to increase as a cubeof wind speed. However, because of a large scatter ($\pm 35\%$) of the flux values at high wind speeds, further work is needed to confirm the cubic dependence.

In addition to the uncertainties in the gas transfer coefficient (or piston velocity), the CO_2 fluxestimated with eqn [2] is subject to errors in (pCO_2)sw caused by the difference between the bulk water temperature and the temperature of the thin skin of ocean water at the sea–air interface. Ordinarily the (pCO_2)sw is obtained at the bulk seawater temperature, whereas the relevant value for the flux calculation is (pCO_2)sw at the 'skin'temperature, that depends on the rate of evaporation, the incoming solar radiation, the wind speed, and the degree of turbulence near the interface. The 'skin' temperature is often cooler than the bulk water temperatureby as much as 0.5°C if the water evaporates rapidly to a dry air mass, but is not always so if a warm humid air mass covers over the ocean. Presently, the time–space distribution of the 'skin' temperature is not well known. This, therefore, could introduce errors in (pCO_2)sw up to about $6\,\mu atm$ or 2%.

CO₂ Sink/Source Areas of the Global Ocean

The oceanic sink and source areas for atmospheric CO_2 and the magnitude of the sea–air CO_2 flux over the global ocean vary seasonally and annually as well as geographically. These changes are the manifestation of changes in the partial pressure of sea water,

(pCO_2)sw, which are caused primarily by changes in the water temperature, in the biological utilization of CO_2, and in the lateral/vertical circulation of ocean waters including the upwelling of deep water rich in CO_2. Over the global oceans, sea water temperatures change from the pole to the equator by about 32°C. Since the pCO_2 in sea water doubles with each 16°C of warming, temperature changes should cause a factor of 4 change in pCO_2. Biological utilization of CO_2 over the global oceans is about $200\,\mu mol$ $CO_2\,kg^{-1}$, which should reduce pCO_2 in sea water by a factor of 3. If this is accompanied with growths of $CaCO_3$-secreting organisms, the reduction of pCO_2 could be somewhat smaller. While these effects are similar in magnitude, they tend to counteract each other seasonally, since the biological utilization tends to be large when waters are warm. In subpolar and polar areas, winter cooling of surface waters induces deep convective mixing that brings high pCO_2 deep waters to the surface. The lowering effect on CO_2 by winter cooling is often compensated for or some times over compensated for by the increasing effect of the upwelling of high CO_2 deep waters. Thus, in high latitude oceans, surface waters may become a source for atmospheric CO_2 during the winter time when the water is coldest.

In **Figure 1**, the global distribution map of the sea–air pCO_2 differences for February and August 1995, are shown. These maps were constructed on the basis of about a half million pairs of atmospheric and seawater pCO_2 measurements made at sea over the 40-year period, 1958–98, by many investigators. Since the measurements were made in different years, during which the atmospheric pCO_2 was increasing, they were corrected to a single reference year (arbitrarily chosen to be 1995) on the basis of the following observations. Warm surface waters in subtropical gyres communicate slowly with the underlying subsurface waters due to the presence of a strong stratification at the base of the mixed layer. This allows a long time for the surface mixed-layer-waters ($\sim$75\,m thick) to exchange CO_2 with the atmosphere. Therefore, their CO_2 chemistry tends to follow the atmospheric CO_2 increase. Accordingly, the pCO_2 in the warm water follows the increasing trend of atmospheric CO_2, and the sea–air pCO_2 difference tends to be independent of the year of measurements. On the other hand, since surface waters in high latitude regions are replaced partially with subsurface waters by deep convection during the winter, the effect of increased atmospheric CO_2 is diluted to undetectable levels and their CO_2 properties tend to remain unchanged from year to year. Accordingly, the sea–air pCO_2 difference measured in a given year increases as the atmospheric CO_2

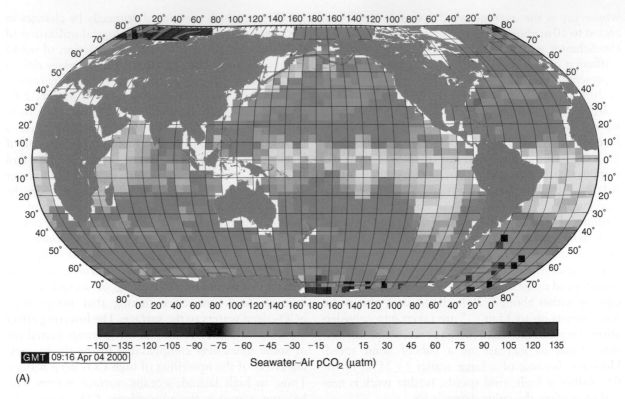

(A)

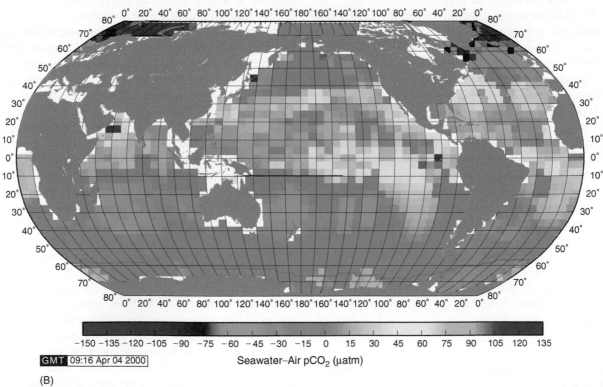

(B)

Figure 1 The sea–air pCO₂ difference in μatm (ΔpCO₂) for (A) February and (B) August for the reference year 1995. The purple-blue areas indicate that the ocean is a sink for atmospheric CO₂, and the red-yellow areas indicate that the ocean is source. The pink lines in the polar regions indicate the edges of ice fields.

concentration increases with time. This effect was corrected to the reference year using the observed increase in the atmospheric CO$_2$ concentration. During El Niño periods, sea–air pCO$_2$ differences over the equatorial belt of the Pacific Ocean, which are large in normal years, are reduced significantly and observations are scarce. Therefore, observations made between 10°N and 10°S in the equatorial Pacific for these periods were excluded from the maps. Accordingly, these maps represent the climatological means for non-El Niño period oceans for the past 40 years. The purple-blue areas indicate that the ocean is a sink for atmospheric CO$_2$, and the red-yellow areas indicate that the ocean is a source.

Strong CO$_2$ sinks (blue and purple areas) are present during the winter months in the Northern (**Figure 1A**) and Southern (**Figure 1B**) Hemispheres along the poleward edges of the subtropical gyres, where major warm currents are located. The Gulf Stream in the North Atlantic and the Kuroshio Current in the North Pacific are both major CO$_2$ sinks (**Figure 1A**) due primarily to cooling as they flow from warm tropical oceans to subpolar zones.

Similarly, in the Southern Hemisphere, CO$_2$ sink areas are formed by the cooling of poleward-flowing currents such as the Brazil Current located along eastern South America, the Agulhus Current located south of South Africa, and the East Australian Current located along south-eastern Australia. These warm water currents meet with cold currents flowing equator ward from the Antarctic zone along the northern border of the Southern (or Antarctic) Ocean. As the sub Antarctic waters rich in nutrients flow northward to more sunlit regions, CO$_2$ is drawn down by photosynthesis, thus creating strong CO$_2$ sink conditions, as exemplified by the Falkland Current in the western South Atlantic (**Figure 1A**). Confluence of subtropical waters with polar waters forms broad and strong CO$_2$ sink zones as a result of the juxta position of the lowering effects on pCO$_2$ of the cooling of warm waters and the photosynthetic drawdown of CO$_2$ in nutrient-rich subpolar waters. This feature is clearly depicted in azone between 40°S and 60°S in **Figure 1A** representing the austral summer, and between 20°S and 40°S in **Figure 1B** representing the austral winter.

During the summer months, the high latitude areas of the North Atlantic Ocean (**Figure 1A**) and the Weddell and Ross Seas, Antarctica(**Figure 1B**), are intense sink areas for CO$_2$. This is attributed to the intense biological utilization of CO$_2$ within the strongly stratified surface layer caused by solar warming and ice melting during the summer. The winter convective mixing of deep waters rich in CO$_2$ and nutrient seliminates the strong CO$_2$ sink and replenishes the depleted nutrients in the surface waters.

The Pacific equatorial belt is a strong CO$_2$ source which is caused by the warming of upwelled deep waters along the coast of South America as well as by the upward entrainment of the equatorial under current water. The source strengths are most intense in the eastern equatorial Pacific due to the strong upwelling, and decrease to the west as a result of the biological utilization of CO$_2$ and nutrients during the westward flow of the surface water.

Small but strong source areas in the north-western subArctic Pacific Ocean are due to the winter convective mixing of deep waters (**Figure 1A**). The lowering effect on pCO$_2$ of cooling in the winter is surpassed by the increasing effect of highCO$_2$ concentration in the upwelled deep waters. During the summer (**Figure 1B**), however, these source areas become a sink for atmospheric CO$_2$ due to the intense biological utilization that overwhelms the increasing effect on pCO$_2$ of warming. A similar area is found in the Arabian Sea, where upwelling of deepwaters is induced by the south-west monsoon during July–August(**Figure 1B**), causing the area tobecome a strong CO$_2$ source. This source area is eliminated by the photosynthetic utilization of CO$_2$ following the end of the upwelling period (**Figure 1A**).

As illustrated in **Figure 1A** and **B**, the distribution of oceanic sink and source areas for atmospheric CO$_2$ varies over a wide range in space and time. Surface ocean waters are out of equilibrium with respect to atmospheric CO$_2$ by as much as $\pm 200\,\mu$atm (or by$\pm 60\%$).The large magnitudes of CO$_2$ disequilibrium between the sea and theair is in contrast with the behavior of oxygen, another biologically mediated gas, that shows only up to $\pm 10\%$ sea–air disequilibrium. The large CO$_2$ disequilibrium may be attributed to the fact that the internal ocean processes that control pCO$_2$ in sea water, such as the temperature of water, the photosynthesis, and the upwelling of deep waters,occur at much faster rates than the sea–air CO$_2$ transfer rates. The slow rate of CO$_2$ transfer across the sea surface is due to the slow hydration rates of CO$_2$ as well as to the large solubility of CO$_2$ in sea water attributable to the formation of bicarbonate and carbonate ions. The latter effect does not exist at all for oxygen.

Net CO$_2$ Flux Across the Sea Surface

The net sea–air CO$_2$ flux over the global oceans may be computed using eqns [2] and [3]. **Figure 2** shows the climatological mean distribution of the annual sea–air CO$_2$ flux for the reference year 1995 using the following set of information. (1) The monthly mean ΔpCO_2 values in $4° \times 5°$ pixel areas for the reference year 1995 (**Figure 1A** and **B** for all other months); (2) the Wanninkhof formulation, eqn [3], for the effect of wind speed on the CO_2 gas transfer coefficient; and (3) the climatological mean wind speeds for each month compiled by Esbensen and Kushnir in 1981. This set yields a mean global gas transfer rate of 0.063 mole $CO_2\,m^{-2}\,\mu atm^{-1}\,y^{-1}$, that is consistent with 20 moles $CO_2\,m^{-2}\,y^{-1}$ estimated on the basis of carbon-14 distribution in the atmosphere and the oceans.

Figure 2 shows that the equatorial Pacific is a strong CO_2 source. On the other hand, the areas along the poleward edges of the temperate gyres in both hemispheres are strong sinks for atmospheric CO_2. This feature is particularly prominent in the southern Indian and Atlantic Oceans between $40°S$ and $60°S$, and is attributable to the combined effects of negative sea–air pCO_2 differences with strong winds ('the roaring 40 s') that accelerate sea–air gas transfer rates. Similarly strong sink zones are formed in the North Pacific and North Atlantic between $45°N$ and $60°N$. In the high latitude Atlantic, strong

sink areas extend into the Norwegian and Greenland Seas. Over the high latitude Southern Ocean areas, the sea–air gas transfer is impeded by the field of ice that covers the sea surface for $\geqslant 6$ months in a year.

The net sea–air CO_2 fluxes computed for each ocean basin for the reference year of 1995, representing non-El Niño conditions, are summarized in **Table 1**. The annual net CO_2 uptake by the global ocean is estimated to be about 2.0 Pg-C y^{-1}. This is consistent with estimates obtained on the basis of a number of different ocean–atmosphere models including multi-box diffusion advection models and three-dimensional general circulation models.

The uptake flux for the Northern Hemisphere ocean (north of $14°N$) is 1.2 Pg-C y^{-1}, whereas that for the Southern Hemisphere ocean (south of $14°S$) is 1.7 Pg-C y^{-1}. Thus, the Southern Hemisphere ocean is astronger CO_2 sink by about 0.5 Pg-C y^{-1}. This is due partially to the much greater oceanic areas in the Southern Hemisphere. In addition, the Southern Ocean south of $50°S$ is an efficient CO_2 sink, for it takes up about 26% of the global ocean CO_2 uptake, while it has only 10% of the global ocean area. Cold temperature and moderate photosynthesis are both

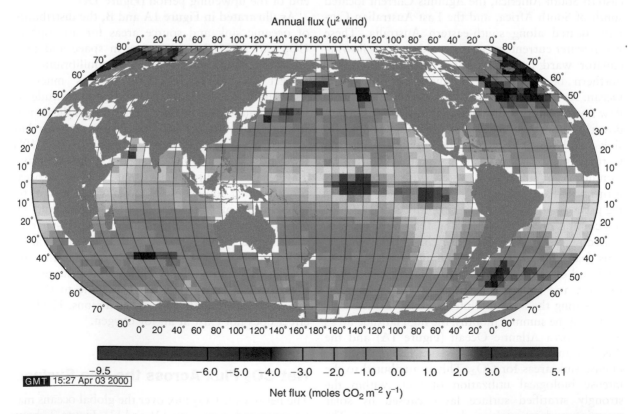

Annual flux (u² wind)

Net flux (moles $CO_2\,m^{-2}\,y^{-1}$)

Figure 2 The mean annual sea–air flux of CO_2 in the reference year 1995. The red-yellow areas indicate that the flux is from sea to air, whereas blue-purple areas indicate that the flux is from air to sea. The flux is given in moles of $CO_2\,m^{-2}\,y^{-1}$. The map gives a total annual air-to-sea flux of 2.0 Pg-C y^{-1}.

Table 1 The net sea–air flux of CO$_2$ estimated for a reference year of 1995 using the effect of wind speed on the CO$_2$ gas transfer coefficient, eqn [3], of Wanninkhof and the monthly wind field of Esbensen and Kushnir

Latitudes	Pacific Ocean	Atlantic Ocean	Indian Ocean	Southern Ocean	Global Oceans
	Sea–air flux in 10^{15}g Carbon y^{-1}				
North of 50°N	− 0.02	− 0.44	—	—	− 0.47
50°N–14°N	− 0.47	− 0.27	+ 0.03	—	− 0.73
14°N–14°S	+ 0.64	+ 0.13	+ 0.09	—	+ 0.86
14°S–50°S	− 0.37	− 0.20	− 0.60	—	− 1.17
South of 50°S	—	—	—	− 0.52	− 0.52
Total	− 0.23	− 0.78	− 0.47	− 0.52	− 2.00
%Uptake	11%	39%	24%	26%	100%
Area (10^6 km^2)	151.6	72.7	53.2	31.7	309.1
Area (%)	49.0%	23.5%	17.2%	10.2%	100%

Positive values indicate sea-to-air fluxes, and negative values indicate air-to-sea fluxes.

responsible for the large uptake by the Southern Ocean.

The Atlantic Ocean is the largest net sink for atmospheric CO$_2$ (39%); the Southern Ocean (26%) and the Indian Ocean (24%) are next; and the Pacific Ocean (11%) is the smallest. The intense biological drawdown of CO$_2$ in the high latitude areas of the North Atlantic and Arctic seasduring the summer months is responsible for the Atlantic being a major sink. This is also due to the fact that the upwelling deep waters in the North Atlantic contain low CO$_2$ concentrations, which are in turn caused primarily by the short residence time ($\sim$80y) of the North Atlantic Deep Waters. The small uptake flux of the Pacific can be attributed to the fact that the combined sink flux of the northern and southern subtropical gyres is roughly balanced by the source flux from the equatorial Pacific during non-El Niño periods. On the other hand, the equatorial Pacific CO$_2$ source flux is significantly reduced or eliminated during El Niño events. As a result the equatorial zone is covered with the eastward spreading of the warm, low pCO$_2$ western Pacific waters in response to the relaxation of the trade wind. Although the effects of El Niño and Southern Ocean Oscillation may be far reaching beyond the equatorial zone as far as to the polar areas, the El Niño effects on the equatorial Pacific alone could reduce the equatorial CO$_2$ source. Hence, this could increase the global ocean uptake flux by up to 0.6 Pg-C y^{-1} during an El Niño year.

The sea–air CO$_2$ flux estimated above is subject to three sources of error: (1) biases in sea–air ΔpCO$_2$ values interpolated from relatively sparse observations, (2) the 'skin' temperature effect, and(3) uncertainties in the gas transfer coefficient estimated on the basis of the wind speed dependence. Possible biases in ΔpCO$_2$ differences have been tested using sea surface temperatures (SST) as a proxy. The systematic error in the global sea–air CO$_2$ flux resulting from sampling and interpolation has been estimated to be about $\pm$30% or $\pm$0.6 Pg-C y^{-1}. The'skin' temperature of ocean water may affect ΔpCO$_2$ by as much as $\pm$6 µatm depending upon time and place, as discussed earlier.

Although the distribution of the 'skin' temperature over the global ocean is not known, it may be cooler than the bulk water temperature by a few tenths of a degree on the global average. This may result in an under estimation of the ocean uptake by 0.4 Pg-C y^{-1}. The estimated global sea–air flux depends on the wind speed data used. Since the gas transfer rate increases nonlinearly with wind speed, the estimated CO$_2$ fluxes tend to be smaller when mean monthly wind speeds are used instead of high frequency wind data.

Furthermore, the wind speed dependence on the CO$_2$ gas transfer coefficient in high wind speed regimes is still questionable. If the gas transfer rate is taken to be a cubic function of wind speed instead of the square dependence as shown above, the global ocean uptake would be increased by about 1 Pg-C y^{-1}. The effect is particularly significant over the high latitude oceans where the winds are strong. Considering various uncertainties discussed above, the global ocean CO$_2$ uptake presented in **Table 1** is uncertain by about 1 Pg-C y^{-1}.

See also

Air–Sea Gas Exchange. Carbon Cycle. Ocean Carbon System, Modeling of. Wind Driven Circulation.

Further Reading

Broecker WS and Peng TH (1982) *Tracers in the Sea.* Palisades, NY: Eldigio Press.

Broecker WS, Ledwell JR, Takahashi, *et al.* (1986) Isotopic versus micrometeorologic ocean CO_2 fluxes a: serious conflict. *Journal of Geophysical Research* 91: 10517–10527.

Keeling R, Piper SC, and Heinmann M (1996) Global and hemispheric CO_2 sinks deduced from changes in atmospheric O_2 concentration. *Nature* 381: 218–221.

Sarmiento JL, Murnane R, and Le Quere C (1995) Air–sea CO_2 transfer and the carbon budget of the North Atlantic. *Philosophical Transactions of the Royal Society of London, series B* 343: 211–219.

Sundquist ET (1985) Geological perspectives on carbon dioxide and carbon cycle. In: Sundquist ET and Broecker WS (eds.) *The Carbon Cycle and Atmospheric CO_2 N:atural Variations, Archean to Present, Geophysical Monograph 32*, pp. 5–59. Washington, DC: American Geophysical Union.

Takashahi T, Olafsson J, Goddard J, Chipman DW, and Sutherland SC (1993) Seasonal variation of CO_2 and nutrients in the high-latitude surface oceans a: comparative study. *Global Biogeochemical Cycles* 7: 843–878.

Takahashi T, Feely RA, Weiss R, *et al.* (1997) Global air–sea flux of CO_2 a:n estimate based on measurements of sea–air pCO_2 difference. *Proceedings of the National Academy of science USA* 94: 8292–8299.

Tans PP, Fung IY, and Takahashi T (1990) Observational constraints on the global atmospheric CO_2 budget. *Sciece* 247: 1431–1438.

Wanninkhof R (1992) Relationship between wind speed and gas exchange. *Journal of Geophysical Research* 97: 7373–7382.

Wanninkhof R and McGillis WM (1999) A cubic relationship between gas transfer and wind speed. *Geophysical Research Letters* 26: 1889–1893.

OCEAN CARBON SYSTEM, MODELING OF

S. C. Doney and D. M. Glover, Woods Hole
Oceanographic Institution, Woods Hole, MA, USA

Introduction

Chemical species such as radiocarbon, chloro-fluorocarbons, and tritium–[3]He are important tools for ocean carbon cycle research because they can be used to trace circulation pathways, estimate time-scales, and determine absolute rates. Such species, often termed chemical tracers, typically have rather simple water-column geochemistry (e.g., conservative or exponential radioactive decay), and reasonably well-known time histories in the atmosphere or surface ocean. Large-scale ocean gradients of nutrients, oxygen, and dissolved inorganic carbon reflect a combination of circulation, mixing, and the production, transport, and oxidation (or remineralization) of organic matter. Tracers provide additional, often independent, information useful in separating these biogeochemical and physical processes. Biogeochemical and tracer observations are often framed in terms of ocean circulation models, ranging from simple, idealized models to full three-dimensional (3-D) simulations. Model advection and diffusion rates are typically calibrated or evaluated against transient tracer data. Idealized models are straightforward to construct and computationally inexpensive and are thus conducive to hypothesis testing and extensive exploration of parameter space. More complete and sophisticated dynamics can be incorporated into three-dimensional models, which are also more amenable for direct comparisons with field data. Models of both classes are used commonly to examine specific biogeochemical process, quantify the uptake of anthropogenic carbon, and study the carbon cycle responses to climate change. All models have potential drawbacks, however, and part of the art of numerical modeling is deciding on the appropriate model(s) for the particular question at hand.

Carbon plays a unique role in the Earth's environment, bridging the physical and biogeochemical systems. Carbon dioxide (CO_2), a minor constituent in the atmosphere, is a so-called greenhouse gas that helps to modulate the planet's climate and temperature. Given sunlight and nutrients, plants and some microorganisms convert CO_2 via photosynthesis into organic carbon, serving as the building blocks and energy source for most of the world's biota. The concentration of CO_2 in the atmosphere is affected by the net balance of photosynthesis and the reverse reaction respiration on land and in the ocean. Changes in ocean circulation and temperature can also change CO_2 levels because carbon dioxide is quite soluble in sea water. In fact, the total amount of dissolved inorganic carbon (DIC) in the ocean is about 50 times larger than the atmospheric inventory. The air–sea exchange of carbon is governed by the gas transfer velocity and the surface water partial pressure of CO_2 (pCO_2), which increases with warmer temperatures, higher DIC, and lower alkalinity levels. The natural carbon cycle has undergone large fluctuations in the past, the most striking during glacial periods when atmospheric CO_2 levels were about 30% lower than preindustrial values. The ocean must have been involved in such a large redistribution of carbon, but the exact mechanism is still not agreed upon.

Human activities, including fossil-fuel burning and land-use practices such as deforestation and biomass burning, are altering the natural carbon cycle. Currently about $7.5\,Pg\,C\,yr^{-1}$ ($1\,Pg = 10^{15}g$) are emitted into the atmosphere, and direct measurements show that the atmospheric CO_2 concentration is indeed growing rapidly with time. Elevated atmospheric CO_2 levels are projected to heat the Earth's surface, and the evidence for climate warming is mounting. Only about 40% of the released anthropogenic carbon remains in the atmosphere, the remainder is taken up in about equal portions (or $2\,Pg\,C\,yr^{-1}$) by land and ocean sinks (**Figure 1**). The future magnitude of these sinks is not well known, however, and is one of the major uncertainties in climate simulations.

Solving this problem is complicated because human impacts appear as relatively small perturbations on a large natural background. In the ocean, the reservoir of organic carbon locked up as living organisms, mostly plankton, is only about 3 Pg C. The marine biota in the sunlit surface ocean are quite productive though, producing roughly 50 Pg of new organic carbon per year. Most of this material is recycled near the ocean surface by zooplankton grazing or microbial consumption. A small fraction, something like 10–20% on average, is exported to the deep ocean as sinking particles or as dissolved organic matter moving with the ocean circulation. Bacteria and other organisms in the deep ocean feed on this source of organic matter from above, releasing DIC and associated nutrients back into the

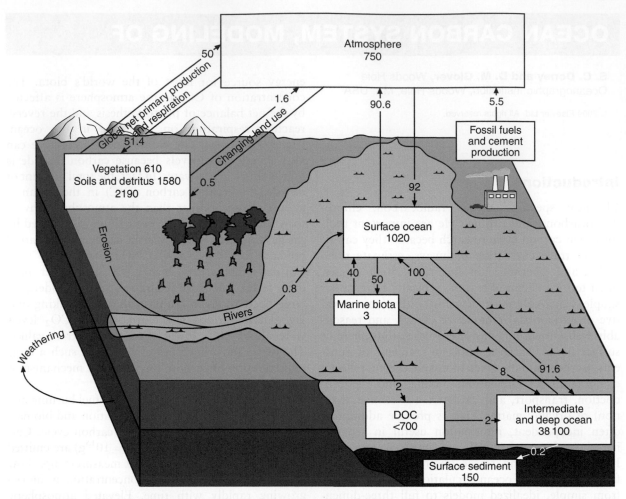

Figure 1 Schematic of global carbon cycle for the 1980s including natural background and human perturbations. Carbon inventories are in Pg C (1 Pg = 10^{15} g) and fluxes are in PgC yr^{-1}. DOC, dissolved organic carbon. Adapted with permission from Schimel D, Enting IG, Heimann M, *et al.* (1995) CO_2 and the carbon cycle. In: Houghton JT, Meira Filho LG, Bruce J, *et al.* (eds.) *Climate Change 1994, Intergovernmental Panel on Climate Change*, pp. 39–71. Cambridge, UK: Cambridge University Press.

water, a process termed respiration or remineralization. The export flux from the surface ocean is a key factor driving the marine biogeochemical cycles of carbon, oxygen, nitrogen, phosphorus, silicon, and trace metals such as iron.

The surface export and subsurface remineralization of organic matter are difficult to measure directly. Biogeochemical rates, therefore, are often inferred based on the large-scale distributions of DIC, alkalinity, inorganic nutrients (nitrate, phosphate, and silicate), and dissolved oxygen. The elemental stoichiometry of marine organic matter, referred to as the Redfield ratio, is with some interesting exceptions relatively constant in the ocean, simplifying the problem of interrelating the various biogeochemical fields. Geochemical distributions have the advantage that they integrate over much of the localized time/space variability in the ocean and can be used to extrapolate to region and basin scales. Property fields, though,

reflect a combination of the net biogeochemical uptake and release as well as physical circulation and turbulent mixing. Additional information is required to separate these signals and can come from a mix of dynamical constraints, numerical models, and ocean process tracers.

The latter two approaches are related because natural and artificial tracers are used to calibrate or evaluate ocean models. A key aspect of these tracers is that they provide independent information on timescale, either because they decay or are produced at some known rate, for example, due to radioactivity, or because they are released into the ocean with a known time history. The different chemical tracers can be roughly divided into two classes. Circulation tracers such as radiocarbon, tritium–^{3}He, and the chlorofluorocarbons are not strongly impacted by biogeochemical cycling and are used primarily to quantify physical advection and mixing rates. These tracers are the major focus here. The distribution of other tracer species is more closely governed by biology and chemistry, for example, the thorium isotope series, which is used to study export production, particle scavenging, vertical transport, and remineralization rates.

Ocean Tracers and Dynamics: A One-dimensional (1-D) Example

Natural radiocarbon (^{14}C), a radioactive isotope of carbon, is a prototypical example of a (mostly) passive ocean circulation tracer. Radiocarbon is produced by cosmic rays in the upper atmosphere and enters the surface ocean as radiolabeled carbon dioxide ($^{14}CO_2$) via air–sea gas exchange. The ^{14}C DIC concentrations in the ocean decrease away from the surface, reflecting the passage of time since the water was last exposed to the atmosphere. Some radiolabeled carbon is transported to the deep ocean in sinking particulate organic matter, which can be largely corrected for in the analysis. The ^{14}C deficits relative to the surface water can be converted into age estimates for ocean deep waters using the radioactive decay half-life (5730 years). Natural radiocarbon is most effective for describing the slow thermohaline overturning circulation of the deep ocean, which has timescales of roughly a few hundred to a thousand years.

The main thermocline of the ocean, from the surface down to about 1 km or so, has more rapid ventilation timescales, from a few years to a few decades. Tracers useful in this regard are chlorofluorocarbons, tritium and its decay product ^3He, and bomb radiocarbon, which along with tritium was released into the atmosphere in large quantities in the 1950s and 1960s by atmospheric nuclear weapons testing.

When properly formulated, the combination of ocean process tracers and numerical models provides powerful tools for studying ocean biogeochemistry. At their most basic level, models are simply a mathematical statement quantifying the rates of the essential physical and biogeochemical processes. For example, advection–diffusion models are structured around coupled sets of differential equations:

$$\frac{\partial C}{\partial t} = -\nabla \cdot (\vec{u}C) + \nabla \cdot (K\nabla C) + J \qquad [1]$$

describing the time rate of change of a generic species C. The first and second terms on the right-hand side of eqn [1] stand for the local divergence due to physical advection and turbulent mixing, respectively. All of the details of the biogeochemistry

are hidden in the net source/sink term J, which for radiocarbon would include net input from particle remineralization (R) and radioactive loss ($-\lambda^{14}C$).

One of the first applications of ocean radiocarbon data was as a constraint on the vertical diffusivity, upwelling, and oxygen consumption rates in the deep waters below the main thermocline. As illustrated in **Figure 2**, the oxygen and radiocarbon concentrations in the North Pacific show a minimum at mid-depth and then increase toward the ocean seabed. This reflects particle remineralization in the water column and the inflow and gradual upwelling of more recently ventilated bottom waters from the Southern Ocean. Mathematically, the vertical profiles for radiocarbon, oxygen (O_2), and a conservative tracer salinity (S) can be posed as steady-state, 1-D balances:

$$0 = K_z \frac{d^2 S}{dz^2} - w\frac{dS}{dz} \qquad [2]$$

$$0 = K_z \frac{d^2 O_2}{dz^2} - w\frac{dO_2}{dz} + R_{O_2} \qquad [3]$$

$$0 = K_z \frac{d^2\ ^{14}C}{dz^2} - w\frac{d^{14}C}{dz} + (^{14}C : O_2)R_{O_2} - \lambda^{14}C \qquad [4]$$

K_z and w are the vertical diffusivity and upwelling rates, and $^{14}C:O_2$ is a conversion factor.

Looking carefully at eqn [2], one sees that the solution depends on the ratio K_z/w but not K_z or w separately. Similarly the equation for oxygen gives us information on the relative rates of upwelling and remineralization. It is only by the inclusion of radiocarbon, with its independent clock due to radioactive decay, that we can solve for the absolute physical and biological rates. The solutions to eqns [2]–[4] can be derived analytically, and as shown in **Figure 2** parameter values of $w = 2.3 \times 10^{-5}\,\mathrm{cm\,s^{-1}}$, $K_z = 1.3\,\mathrm{cm^2\,s^{-2}}$, and $R_{O_2} = 0.13 \times 10^{-6}\,\mathrm{mol\,kg^{-1}\,yr^{-1}}$ fit the data reasonably well. The 1-D model-derived vertical diffusivity is about an order of magnitude larger than estimates from deliberate tracer release experiments and microscale turbulence measurements in the upper thermocline. However, they may be consistent with recent observations of enhanced deep-water vertical mixing over regions of rough bottom topography.

Ocean Circulation and Biogeochemical Models

The 1-D example shows the basic principles behind the application of tracer data to the ocean carbon

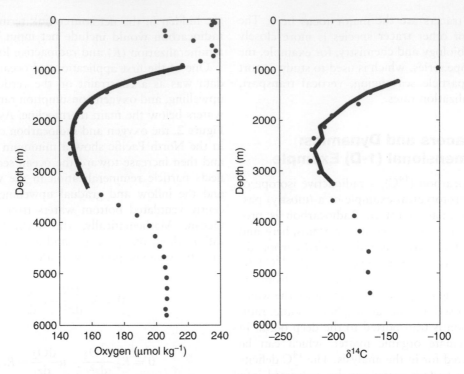

Figure 2 Observed vertical profiles of oxygen (O_2) and radiocarbon ($\delta^{14}C$) in the North Pacific. The solid curves are the model solution of a 1-D advection–diffusion equation.

cycle, but the complexity (if not always the sophistication) of the models and analysis has grown with time. Ocean carbon models can be roughly divided into idealized models (multi-box, 1-D and 2-D advection–diffusion models) and 3-D general circulation models (GCMs). Although the distinction can be blurry at times, idealized models are characterized typically by reduced dimensionality and/or kinematic rather than dynamic physics. That is, the circulation and mixing are specified rather than computed by the model and are often adjusted to best match the transient tracer data.

Global Box Models

An example of a simple, high-latitude outcrop box model is shown in **Figures 3(a)** and **3(b)**. The boxes represent the atmosphere, the high- and low-latitude surface ocean, and the deep interior. In the model, the ocean thermohaline circulation is represented by one-way flow with high-latitude sinking, low-latitude upwelling, and poleward surface return flow. Horizontal and vertical mixing is included by two-way exchange of water between each pair of boxes. The physical parameters are constrained so that model natural radiocarbon values roughly match observations. Note that the ^{14}C concentration in the deep water is significantly depleted relative to the surface

boxes and amounts to a mean deep-water ventilation age of about 1150 years.

The model circulation also transports phosphate, inorganic carbon, and alkalinity. Biological production, particle export, and remineralization are simulated by the uptake of these species in the surface boxes and release in the deep box. The model allows for air–sea fluxes of CO_2 between the surface boxes and the atmosphere. The low-latitude nutrient concentrations are set near zero as observed. The surface nutrients in the high-latitude box are allowed to vary and are never completely depleted in the simulation of modern conditions. Similar regions of 'high-nutrient, low-chlorophyll' concentrations are observed in the subpolar North Pacific and Southern Oceans and are thought to be maintained by a combination of light and iron limitation as well as zooplankton grazing. The nutrient and DIC concentrations in the deep box are higher than either of the surface boxes, reflecting the remineralization of sinking organic particles.

The three-box ocean outcrop model predicts that atmospheric CO_2 is controlled primarily by the degree of nutrient utilization in high-latitude surface regions. Marine production and remineralization occur with approximately fixed carbon-to-nutrient ratios, the elevated nutrients in the deep box are associated with an equivalent increase in DIC and pCO_2. Large adjustments in the partitioning of

(a)

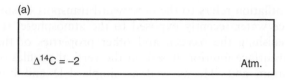

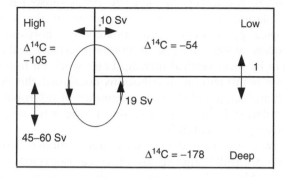

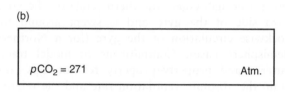

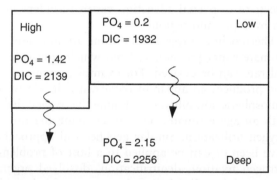

(b)

Figure 3 Results from a simple three-box ocean carbon cycle model. (a) The physical circulation and modeled radiocarbon ($\Delta^{14}C$) values. (b) The model biogeochemical fields, ocean DIC, and phosphate (PO_4) and atmospheric pCO_2. From Toggweiler JR and Sarmiento JL, Glacial to inter-glacial changes in atmospheric carbon dioxide: The critical role of ocean surface waters in high latitudes, *The Carbon Cycle and Atmospheric CO$_2$: Natural Variations Archean to Present*, Sundquist ET and Broecker WS (eds.), pp. 163–184, 1985, Copyright [1985]. American Geophysical Union. Adapted by permission of American Geophysical Union.

carbon between the ocean and atmosphere can occur only where this close coupling of the carbon and nutrient cycles breaks down. When subsurface water is brought to the surface at low latitude, production draws the nutrients down to near zero and removes to first order all of the excess seawater DIC and pCO_2. Modifications in the upward nutrient flux to the low latitudes have relatively little impact on the

model atmospheric CO_2 as long as the surface nutrient concentrations stay near zero.

At high latitudes, however, the nutrients and excess DIC are only partially utilized, resulting in higher surface water pCO_2 and, over decades to centuries, higher atmospheric CO_2 concentrations. Depending on the polar biological efficiency, the model atmosphere effectively sees more or less of the high DIC concentrations (and pCO_2 levels) of the deep ocean. Thus changes in ocean biology and physics can have a correspondingly large impact on atmospheric CO_2. On longer timescales (approaching a few millennia), these variations are damped to some extent by adjustments of the marine calcium carbonate cycle and ocean alkalinity.

The three-box outcrop model is a rather crude representation of the ocean, and a series of geographical refinements have been pursued. Additional boxes can be added to differentiate the individual ocean basins (e.g., Atlantic, Pacific, and Indian), regions (e.g., Tropics and subtropics), and depths (e.g., thermocline, intermediate, deep, and bottom waters), leading to a class of models with a half-dozen to a few dozen boxes. The larger number of unknown advective flows and turbulent exchange parameters, however, complicates the tuning procedure. Other model designs take advantage of the vertical structure in the tracer and biogeochemical profile data. The deep box (es) is discretized in the vertical, essentially creating a continuous interior akin to a 1-D advection–diffusion model. This type of model was often used in the 1970s and 1980s for the initial anthropogenic CO_2 uptake calculations, where it is important to differentiate between the decadal ventilation timescales of the thermocline and the centennial timescales of the deep water.

Intermediate Complexity and Inverse Models

In terms of global models, the next step up in sophistication from box models is intermediate complexity models. These models typically have higher resolution and/or include more physical dynamics but fall well short of being full GCMs. Perhaps the most common examples for ocean carbon cycle research are zonal average basin models. The dynamical equations are similar to a GCM but are integrated in 2-D rather than in 3-D, the third east–west dimension removed by averaging zonally across the basin. In some versions, multiple basins are connected by an east–west Southern Ocean channel. The zonal average models often have a fair representation of the shallow wind-driven Ekman and deep thermohaline overturning circulations but obviously lack western boundary currents and gyre circulations. Tracer data remain an important element in tuning some of the mixing coefficients and surface boundary conditions and in evaluating the model solutions.

Based on resolution, many inverse models can also be categorized as intermediate complexity, but their mode of operation differs considerably from the models considered so far. In an inverse model, the circulation field and biogeochemical net source/sink (the J terms in the notation above) are solved for using the observed large-scale hydrographic and tracer distributions as constraints. Additional dynamic information may also be incorporated such as the geostrophic velocity field, general water mass properties, or float and mooring velocities.

Inverse calculations are typically posed as a large set of simultaneous linear equations, which are then solved using standard linear algebra methods. The inverse techniques are most commonly applied to steady-state tracers, though some exploration of transient tracers has been carried out. The beauty of the inverse approach is that it tries to produce dynamically consistent physical and/or biogeochemical solutions that match the data within some assigned error. The solutions are often underdetermined in practice, however, which indicates the existence of a range of possible solutions. From a biogeochemical perspective, the inverse circulation models provide estimates of the net source/sink patterns, which can then be related to potential mechanisms.

Thermocline Models

Ocean process tracers and idealized models have also been used extensively to study the ventilation of the main thermocline. The main thermocline includes the upper 1 km of the tropical to subpolar ocean where the temperature and potential density vertical gradients are particularly steep. Thermocline ventilation refers to the downward transport of surface water recently exposed to the atmosphere, replenishing the oxygen and other properties of the subsurface interior. Based on the vertical profiles of tritium and ^{3}He as well as simple 1-D and box models, researchers in the early 1980s showed that ventilation of the main thermocline in the subtropical gyres occurs predominately as a horizontal process along surfaces of constant density rather than by local vertical mixing. Later work on basin-scale bomb-tritium distributions confirmed this result and suggested the total magnitude of subtropical ventilation is large, comparable to the total wind-driven gyre circulation.

Two dimensional gyre-scale tracer models have been fruitfully applied to observed isopycnal tritium–^{3}He and chlorofluorocarbon patterns. As shown in **Figure 4**, recently ventilated water (near-zero tracer age) enters the thermocline on the poleward side of the gyre and is swept around the clockwise circulation of the gyre (for a Northern Hemisphere case). Comparisons of model tracer patterns and property–property relationships constrain the absolute ventilation rate and the relative effects of isopycnal advection versus turbulent mixing by depth and region.

Thermocline tracer observations are also used to estimate water parcel age, from which biogeochemical rate can be derived. For example, remineralization produces an apparent oxygen deficit relative to atmospheric solubility. Combining the oxygen deficit with an age estimate, one can compute the rate of oxygen utilization. Similar geochemical approaches have been or can be applied to a host of problems: nitrogen fixation, denitrification, dissolved organic matter remineralization, and nutrient resupply to the upper ocean. The biogeochemical application of 2-D gyre models has not been pursued in as much detail.

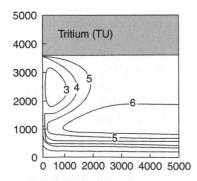
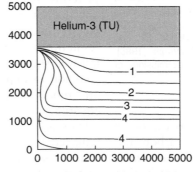
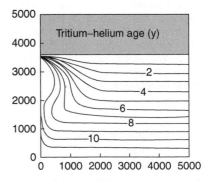

Figure 4 Tracer results from a two-dimensional gyre model. The model represents the circulation on a constant density surface (isopycnal) in the main thermocline that outcrops along the northern boundary (shaded region). Thermocline ventilation is indicated by the gradual increase of tritium–^{3}He ages around the clockwise flowing gyre circulation. TU, tritium unit (1 TU = 1 ^{3}H atom/10^{18} H atoms). Reproduced with permission from Musgrave DL (1990) Numerical studies of tritium and helium-3 in the thermocline. *Journal of Physical Oceanography* 20: 344–373.

This form of modeling, however, is particularly useful for describing regional patterns and in the areas where simple tracer age approaches breakdown.

Three-Dimensional (3-D) Biogeochemical Simulations

By their very nature, idealized models neglect many important aspects of ocean dynamics. The alternative is full 3-D ocean GCMs, which incorporate more realistic spatial and temporal geometry and a much fuller suite of physics. There are several different families of ocean physical models characterized by the underlying governing equations and the vertical discretization schemes. Within model families, individual simulations will differ in important factors of surface forcing and choice of physical parametrizations. Often these parametrizations account for complex processes, such as turbulent mixing, that occur on small space scales that are not directly resolved or computed by the model but which can have important impacts on the larger-scale ocean circulation; often an exact description of specific events is not required and a statistical representation of these subgrid scale processes is sufficient. For example, turbulent mixing is commonly treated using equations analogous to those for Fickian molecular diffusion. Ocean GCMs, particularly coarse-resolution global versions, are sensitive to the subgrid scale parametrizations used to account for unresolved processes such as mesoscale eddy mixing, surface and bottom boundary layer dynamics, and air–sea and ocean–ice interactions. Different models will be better or worse for different biogeochemical problems.

Considerable progress has been achieved over the last two decades on the incorporation of chemical tracers, biogeochemistry, and ecosystem dynamics into both regional and global 3-D models. Modeling groups now routinely simulate the more commonly measured tracers (e.g., radiocarbon, chlorofluorocarbons, and tritium–^{3}He). Most of these tracers have surface sources and, therefore, provide clear indications of the pathways and timescales of subsurface ventilation in the model simulations. The bottom panel of **Figure 5** shows from a simulation of natural radiocarbon the layered structure of the intermediate and deep water flows in the Atlantic basin, with Antarctic Intermediate and Bottom Waters penetrating from the south and southward-flowing North Atlantic Deep Water sandwiched in between.

In a similar fashion, biogeochemical modules have been incorporated to simulate the 3-D fields of nutrients, oxygen, DIC, etc. Just as with the physics, the

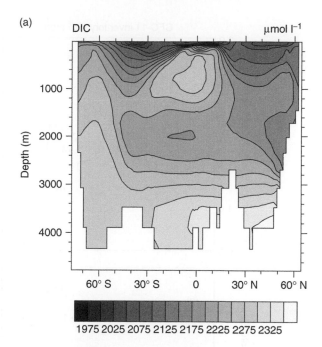

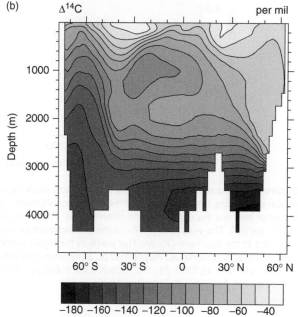

Figure 5 Simulated radiocarbon and carbon cycle results from a three-dimensional global ocean biogeochemical model. Depth–latitude sections are shown for (a) DIC concentration (μmol l^{-1}) and (b) natural (preindustrial) radiocarbon (Δ^{14}C, per mil) along the prime meridian in the Atlantic Ocean.

degree of biological complexity can vary considerably across ocean carbon models. At one end are diagnostic models where the production of organic matter in the surface ocean is prescribed based on satellite ocean color data or computed implicitly by forcing the simulated surface nutrient field to match observations (assuming that any net reduction in nutrients is driven by organic matter production). At the other

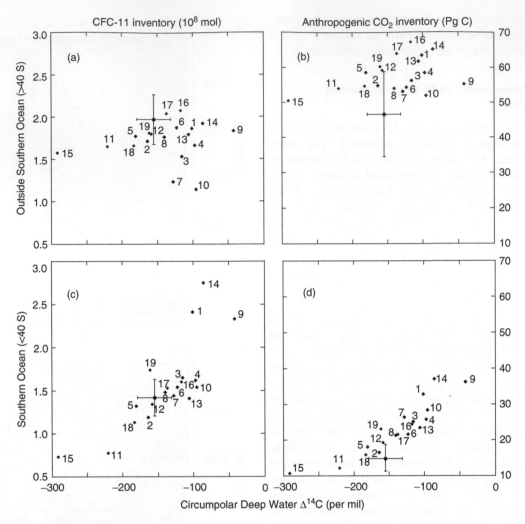

Figure 6   Transient tracer constraints on numerical model ocean carbon dynamics. Each panel shows results for the suite of global ocean simulations that participated in the international Ocean Carbon-Cycle Model Intercomparison Project (OCMIP). The points with the error bars are the corresponding observations. The *x*-axis in all panels is the average Circumpolar Deep Water radiocarbon level ($\Delta^{14}C$, per mil). The *y*-axis in the left column is the chlorofluorocarbon (CFC-11) inventory (10^8 mol) outside of the Southern Ocean (top row) and in the Southern Ocean. The *y*-axis in the right column is the corresponding anthropogenic carbon inventory (Pg C). From Matsumoto K, Sarmiento JL, Key RM, *et al.*, Evaluation of ocean carbon cycle models with data-based metrics, *Geophysical Research Letters*, Vol. 31, L07303 (doi:10.1029/2003GL018970), 2004. Copyright [2004] American Geophysical Union. Reproduced by permission of American Geophysical Union.

extreme are simulations that couple prognostic treatments of multiple phytoplankton, zooplankton, and bacteria species to the biogeochemical tracers. Transient tracers enter the picture because the biogeochemical distributions are also strongly influenced circulation. This is nicely illustrated by comparing the simulated radiocarbon and DIC sections in **Figure 5**. To first order, older waters marked by low radiocarbon exhibit high DIC levels due to the accumulation of carbon released from particle remineralization. Tracers can be used to guide the improvement of the physical circulation and highlight areas where the circulation is poor and thus the biogeochemical tracers may be suspect. They can also

be used more quantitatively to create model skill metrics; this approach can be applied across multiple model simulations to provide the best overall constraint on quantities like the ocean uptake of anthropogenic CO_2 or export production (**Figure 6**).

Models as Research Tools

Designed around a particular question or hypothesis, conceptual models attempt to capture the basic elements of the problem while remaining amenable to straightforward analysis and interpretation. They are easy to construct and computationally inexpensive, requiring only a desktop PC rather than a supercomputer. When well-formulated, idealized models provide a practical method to analyze ocean physical and biogeochemical dynamics and in some cases to quantitatively constrain specific rates. Their application is closely tied to ocean tracer observations, which are generally required for physical calibration and evaluation. Idealized models remain a valuable tool for estimating the oceanic uptake of anthropogenic carbon and the long time-scale responses (centuries to millennia) of the natural carbon cycle. Also, some of the more memorable and lasting advances in tracer oceanography are directly linked to simple conceptual models. Examples include constraints on the deep-water large-scale vertical diffusivity and demonstration of the dominance of lateral over vertical ventilation of the subtropical main thermocline.

Three-dimensional models offer more realism, at least apparently, but with the cost of greater complexity, a more limited number of simulations, and a higher probability of crucial regional errors in the base solutions, which may compromise direct, quantitative model-data comparisons. Ocean GCM solutions, however, should be exploited to address exactly those problems that are intractable for simpler conceptual and reduced dimensional models. For example, two key assumptions of the 1-D advection–diffusion model presented in **Figure 2** are that the upwelling occurs uniformly in the horizontal and vertical and that mid-depth horizontal advection is not significant. Ocean GCMs and tracer data, by contrast, show a rich three-dimensional circulation pattern in the deep Pacific.

The behavior of idealized models and GCMs can diverge, and it is not always clear that complexity necessarily leads to more accurate results. In the end, the choice of which model to use depends on the scientific problem and the judgment of the researcher. Probably the best advice is to explore solutions from a hierarchy of models and to thoroughly evaluate the skill of the models against a range of tracers and other dynamical measures. Just because a model does a good job reproducing the distribution of one tracer field does not mean that it can be applied indiscriminately to another variable, especially if the underlying dynamics or timescales differ.

Models can be quite alluring in the sense that they provide concrete answers to questions that are often difficult or nearly impossible to address from sparsely sampled field data. However, one should not forget that numerical models are simply a set of tools for doing science. They are no better than the foundations upon which they are built and should not be carried out in isolation from observations of the real

ocean. For ocean carbon cycle models, the two key elements are the ocean physical circulation and the biogeochemical processes. Even the best biogeochemical model will perform poorly in an ill-constructed physical model. Conversely, if the underlying biogeochemical mechanisms are poorly known, a model may be able to correctly reproduce the distributions of biogeochemical tracers but for the wrong reasons. Mechanistic-based models are critical in order to understand and predict natural variability and the response of ocean biogeochemistry to perturbations such as climate change.

Glossary

anthropogenic carbon The additional carbon that has been released to the environment over the last several centuries by human activities including fossil-fuel combustion, agriculture, forestry, and biomass burning.

excess ^{3}He Computed as the ^{3}He in excess of gas solubility equilibrium with the atmosphere.

export production That part of the organic matter formed in the surface layer by photosynthesis that is transported out of the surface layer and into the interior of the ocean by particle sinking, mixing, and circulation, or active transport by organisms.

radiocarbon ^{14}C Either δ^{14}C or Δ^{14}C where δ^{14}C $= [(^{14}C/^{12}C)_{sample}/(^{14}C/^{12}C)_{standard} - 1] \times 1000$ (in parts per thousand or 'per mil'); Δ^{14}C is similar but corrects the sample ^{14}C for biological fractionation using the sample $^{13}C/^{12}C$ ratio.

transient tracers Chemical tracers that contain time information either because they are radioactive or because their source, usually anthropogenic, has evolved with time.

tritium–^{3}He age An age is computed assuming that all of the excess ^{3}He in a sample is due to the radioactive decay of tritium, age $= \ln[(^3H + ^3He)/^3H]/\lambda$, where λ is the decay constant for tritium.

ventilation The physical process by which surface properties are transported into the ocean interior.

See also

Carbon Cycle. Carbon Dioxide (CO_2) Cycle. CFCs in the Ocean. Nitrogen Cycle.

Further Reading

Broecker WS and Peng T-H (1982) *Tracers in the Sea*. Palisades, NY: Lamont-Doherty Geological Observatory, Columbia University.

Charnock H, Lovelock JE, Liss P, and Whitfield M (eds.) (1988) *Tracers in the Ocean*. London: The Royal Society.

Doney SC, Lindsay K, Caldeira K, *et al.* (2004) Evaluating global ocean carbon models: The importance of realistic physics. *Global Biogeochemical Cycles* 18: GB3017 (doi:10.1029/2003GB002150).

England MH and Maier-Reimer E (2001) Using chemical tracers in ocean models. *Reviews in Geophysics* 39: 29–70.

Fasham M (ed.) (2003) *Ocean Biogeochemistry*. New York: Springer.

Jenkins WJ (1980) Tritium and ^{3}He in the Sargasso Sea. *Journal of Marine Research* 38: 533–569.

Kasibhatla P, Heimann M, and Rayner P (eds.) (2000) *Inverse Methods in Global Biogeochemical Cycles*. Washington, DC: American Geophysical Union.

Matsumoto K, Sarmiento JL, Key RM, *et al.* (2004) Evaluation of ocean carbon cycle models with data-based metrics. *Geophysical Research Letters* 31: L07303 (doi:10.1029/2003GL018970).

Munk WH (1966) Abyssal recipes. *Deep-Sea Research* 13: 707–730.

Musgrave DL (1990) Numerical studies of tritium and helium-3 in the thermocline. *Journal of Physical Oceanography* 20: 344–373.

Sarmiento JL and Gruber N (2006) *Ocean Biogeochemical Dynamics*. Princeton, NJ: Princeton University Press.

Schimel D, Enting IG, Heimann M, *et al.* (1995) CO$_2$ and the carbon cycle. In: Houghton JT, Meira Filho LG, and Bruce J, *et al.* (eds.) *Climate Change 1994, Intergovernmental Panel on Climate Change*, pp. 39–71. Cambridge, UK: Cambridge University Press.

Siedler G, Church J, and Gould J (eds.) (2001) *Ocean Circulation and Climate*. New York: Academic Press.

Siegenthaler U and Oeschger H (1978) Predicting future atmospheric carbon dioxide levels. *Science* 199: 388–395.

Toggweiler JR and Sarmiento JL (1985) Glacial to interglacial changes in atmospheric carbon dioxide: The critical role of ocean surface waters in high latitudes. In: Sundquist ET and Broecker WS (eds.) *The Carbon Cycle and Atmospheric CO$_2$: Natural Variations Archean to Present*, pp. 163–184. Washington, DC: American Geophysical Union.

Relevant Websites

http://cdiac.ornl.gov
 – Global Ocean Data Analysis Project, Carbon Dioxide Information Analysis Center.

http://w3eos.whoi.edu
 – Modeling, Data Analysis, and Numerical Techniques for Geochemistry (resource page for MIT/WHOI course number 12.747), at Woods Hole Oceanographic Institution.

http://us-osb.org
 – Ocean Carbon and Biogeochemistry (OCB) Program

http://www.ipsl.jussieu.fr
 – Ocean Carbon-Cycle Model Intercomparison Project (OCMIP), Institut Pierre Simon Laplace.

AIR–SEA TRANSFER: DIMETHYL SULFIDE, COS, CS$_2$, NH$_4$, NON-METHANE HYDROCARBONS, ORGANO-HALOGENS

J. W. Dacey, Woods Hole Oceanographic Institution, Woods Hole, MA, USA
H. J. Zemmelink, University of Groningen, Haren, The Netherlands

Copyright © 2001 Elsevier Ltd.

The oceans, which cover 70% of Earth's surface to an average depth of 4000 m, have an immense impact on the atmosphere's dynamics. Exchanges of heat and momentum, water and gases across the sea surface play major roles in global climate and biogeochemical cycling. The ocean can be thought of as a vast biological soup with myriad processes influencing the concentrations of gases dissolved in the surface waters. The quantities of mass flux across the surface interface, though perhaps small on a unit area basis, can be very important because of the extent of the ocean surface and the properties of the gases or their decomposition products in the atmosphere.

Gas exchange across the sea–air surface depends, in part, on differences in partial pressures of the gases between the ocean surface and the atmosphere. The partial pressure of a gas in the gas phase can be understood in terms of its contribution to the pressure in the gas mixture. So the partial pressure of O$_2$, for example, at 0.21 atm means that at 1 atmosphere total pressure, O$_2$ is present as 21% of the gas, or mixing, volume. Trace gases are present in the atmosphere at much lower levels, usually expressed as parts per million (10^{-6} atm), parts per billion (10^{-9} atm) or parts per trillion (10^{-12} atm, pptv). Dimethylsulfide (DMS), when present at 100 pptv, accounts for about 100 molecules per 10^{12} molecules of mixed gas phase, or about 10^{-10} of the gas volume.

In solution, a dissolved trace gas in equilibrium with the atmosphere would have the same partial pressure as the gas in the air. Its absolute concentration in terms of molecules or mass per unit volume of water depends on its solubility. Gas solubility varies over many orders of magnitude depending on the affinity of water for the gas molecules and the volatility of the gas. Gases range widely in their solubility in sea water, from the permanent gases like nitrogen (N$_2$), oxygen (O$_2$), nitrous oxide (N$_2$O) and methane (CH$_4$) that have a low solubility in sea

water to the moderately soluble carbon dioxide (CO$_2$) and dimethylsulfide (CH$_3$)$_2$S, to highly soluble ammonia (NH$_3$ and its ionized form NH$_4^+$) and sulfur dioxide (SO$_2$). Sulfur dioxide is more than 10^6 times more soluble than O$_2$ or CH$_4$. Using the example above of an atmospheric DMS concentration of 100 pptv, the equilibrium concentration of DMS in surface water would be about 0.07 nmol l^{-1}. Generally the solubility of any individual gas increases at cooler water temperatures, and solubility of gases in sea water is somewhat less than for fresh water because of the so-called 'salting out' effect of dissolved species in sea water.

At any moment the partial pressure difference between surface water and the atmosphere depends on an array of variables. The gases in this article are biogenic, meaning that their mode of formation is the result of one or more immediate or proximate biological processes. These dissolved gases may also be consumed biologically, or removed by chemical processes in sea water, or they may flux across the sea surface to the atmosphere. The rates at which the source and sink processes occur determines the concentration of the dissolved gas in solution as well as the turnover, or residence time, of each compound. Similarly, there can be several source and sink processes for the gases in the atmosphere. Long-lived compounds in the atmosphere will tend to integrate more global processes, whereas short-lived compounds are concentrated near their source and reflect relatively short-term influences of source and sink. In this sense, carbonyl sulfide is a global gas. At it has a residence time of several years in the atmosphere, its concentration does not vary in the troposphere to any appreciable degree. On the other hand, the concentration of DMS varies on a diel basis and with elevation, with higher concentrations at night when atmospheric oxidants (most notable hydroxyl) are relatively depleted.

The extent of disequilibrium between the partial pressures of a gas in the surface water and in the atmosphere determines the thermodynamic gradient which drives gas flux. The kinetics of flux ultimately depend on molecular diffusion and larger-scale mixing processes. Molecular diffusivity is generally captured in a dimensionless parameter, the Schmidt number (ratio of viscosity of water to molecular diffusivity of gas in water), and varies widely between gases depending primarily on the molecular cross-section. From moment to moment, the flux of any particular gas is dependent on interfacial turbulence which is generated by shear between the wind and the sea surface whereby higher wind speed causes increasing turbulence and thus stimulating the onset of waves and eventually the production of bubbles and sea spray. There are considerable uncertainties relating gas exchange to wind speed. These arise due to the various sea-state factors (wave height, swell, breaking waves, bubble entrainment, surfactants, and others) whose individual dependencies on actual wind speed and wind history are not well quantified. The fluxes of gases across the air–sea interface are usually calculated using a wind-speed parameterization. These estimates are considered to be accurate to within a factor of 2 or so.

This article summarizes the characteristics of several important trace gases – dimethylsulfide, carbonyl sulfide, carbon disulfide, nonmethane hydrocarbons, ammonia and methylhalides – focusing on their production and fate as it is determined by biological and chemical processes.

Dimethylsulfide

Natural and anthropogenic sulfur aerosols play a major role in atmospheric chemistry and potentially in modulating global climate. One theory holds that a negative feedback links the emission of volatile organic sulfur (mostly as DMS) from the ocean with the formation of cloud condensation nuclei, thereby regulating, in a sense, the albedo and radiation balance of the earth. The direct (backscattering and reflection of solar radiation by sulfate aerosols) and indirect (cloud albedo) effects of sulfate aerosols may reduce the climatic forcing of trace greenhouse gases like CO_2, N_2O and CH_4. The oxidation products of DMS which also contribute to the acidity of rain, particularly in marine areas, result from industrialized and/or well-populated land.

Dimethylsulfide (DMS) is the most abundant volatile sulfur compound in sea water and constitutes about half of the global biogenic sulfur flux to the atmosphere. Studies of the concentration of DMS in the ocean have shown that average surface water concentrations may vary by up to a factor of 50 between summer and winter in mid and high latitudes. Furthermore, there are large-scale variations in DMS concentration associated with phytoplankton biomass, although there are generally poor correlations between local oceanic DMS concentrations and the biomass and productivity of phytoplankton (due to differences between plankton species in ability to produce DMS).

The nature and rates of the processes involved in the production and consumption of DMS in sea water are important in determining the surface concentrations and the concomitant flux to the atmosphere. The biogeochemical cycle of DMS (**Figure 1**) begins with its precursor, β-dimethylsulfoniopropionate (DMSP). DMSP is a cellular component in certain species of phytoplankton, notably some prymnesiophytes and dinoflagellates. The function of DMSP is unclear, although there is evidence for an

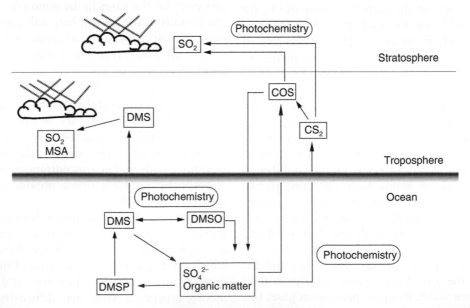

Figure 1 Fate and production of dimethylsulfide (DMS), carbonylsulfide (COS) and carbon disulfide. DMSO, dimethylsulfoxide; DMSP, dimethylsulfoniopropionate; MSA, methane sulfonic acid.

osmoregulatory role as its cellular concentrations have been found to vary with salinity. It is generally thought that healthy algal cells do not leak either DMSP or DMS, although mechanical release into the surrounding sea water can lead to DMS production during cell senescence and grazing by zooplankton or as a consequence of viral attack on phytoplankton cells. Oceanic regions dominated by prolific DMSP-producing phytoplankton tend to have high DMS and DMSP concentrations.

Breakdown of DMSP, presumably after transfer from the particulate algal (pDMSP) form to a dissolved (dDMSP) form in sea water, can proceed in different ways, mostly depending on microbiological conditions. One major pathway involves cleavage of DMSP to DMS and acrylic acid. Bacterial metabolism of dDMSP may be a major mechanism for DMS production in sea water, with acrylic acid residue acting as a carbon source for heterotrophic growth. Sulfonium compounds are vulnerable to attack by hydroxide ion; the resulting chemical elimination reaction occurs rapidly and quantitatively in strong base but only slowly at the pH of sea water.

DMS in sea water has many potential fates. The volatility of DMS and the concentration gradient across the sea–air interface lead to the ocean being the major source of DMS to the atmosphere. Estimates of the annual sulfur release (as DMS) vary from 13–37 Tg S y^{-1} (Kettle and Andreae, 1999). However, whereas the absolute flux of DMS from sea to air may be large on a global scale, sea–air exchange may represent only a minor sink for seawater DMS. It has been estimated that DMS loss to the atmosphere is only a very small percentage of the DMS sink, but this undoubtedly depends on the biogeochemical conditions in the water column at the time. Photochemical oxidation of DMS, either to dimethylsulfoxide (DMSO) or to other products, occurs via photosensitized reactions. The amount of photochemical decomposition depends on the amount of light of appropriate wavelengths and the concentration of colored organic compounds in solution to convert light energy into reactive radicals. Light declines exponentially with depth; the distribution of colored dissolved organic materials exhibits depth and seasonal variability. Microbial consumption of DMS, although extremely variable in both time and space in the ocean, appears to be a significant sink for oceanic DMS. The residence time of DMS is probably of the order of a day or two in most seawater systems.

Since the atmospheric residence time of DMS is about a day or two, the atmospheric consequences of DMS flux are mostly confined to the troposphere. In the troposphere, DMS is oxidized primarily by hydroxyl radical. The main atmospheric oxidation products are methane sulphonic acid, SO_2 and DMSO.

Carbonyl Sulfide

Carbonyl sulfide (COS, OCS) is the major sulfur gas in the atmosphere, present throughout the troposphere at 500 pptv. COS has a long atmospheric residence time (~ 4 years). Because of its relative inertness COS diffuses into the stratosphere where it oxidizes to sulfate particles and contributes in reactions involving stratospheric ozone chemistry. Unlike DMS which is photochemically oxidized in the troposphere, the major sink for COS is terrestrial vegetation and soils. COS is taken up by plants by passing through the stomata and subsequently hydrolyzing to CO_2 and H_2S through the action of carbonic anhydrase inside plant cells. There is no apparent physiological significance to the process; it appears to just occur accidentally to the normal physiology of plants.

COS is produced in the ocean by photochemical oxidation of organic sulfur compounds whereby dissolved organic matter acts as a photosensitizer. The aqueous concentration of COS manifests a strong diel cycle, with the highest concentrations in daytime (concentration range on the order of 0.03–0.1 $nmol\,l^{-1}$). COS hydrolyzes in water to H_2S at rates dependent on water temperature and pH. The flux of oceanic COS to the atmosphere may represent about one-third of the global COS flux.

Carbon Disulfide

Concentrations in surface water are around 10^{-11} mol l^{-1}. Although a number of studies have indicated that the ocean forms an important source for atmospheric CS_2, the underlying biochemical cycles still remain poorly understood. CS_2 is formed by photochemical reactions (possibly involving precursors such as DMS, DMSP and isothiocyanates). CS_2 formation has been observed to occur in bacteria in anoxic aquatic environments and in cultures of some marine algae species.

The residence time of CS_2 in the atmosphere is relatively short (about one week). Although CS_2 might contribute directly to SO_2 in the troposphere, its main significance is in the formation of COS via photochemical oxidation which results in the production of one molecule each of SO_2 and COS per molecule of CS_2 oxidized. The resulting COS may contribute to the stratospheric aerosol formation.

Concentrations around 14 $pmol\,l^{-1}$ of carbon disulfide in the mid-Atlantic Ocean were first observed

(1974); higher concentrations have been found in coastal waters. More than a decade later CS_2 concentrations in the North Atlantic were found to be comparable to the earlier observations. However, in coastal waters CS_2 concentrations were found to be a factor 10 lower, respectively 33 and 300 pmol l^{-1}. The global CS_2 flux has been estimated on 6.7 Gmol S y^{-1}, and it has been concluded that the marine emission of CS_2 provides a significant indirect source of COS, but it forms an insignificant source of tropospheric SO_2.

Nonmethane Hydrocarbons

Nonmethane hydrocarbons (NMHCs) are important reactive gases in the atmosphere since they provide a sink for hydroxyl radicals and play key roles in the production and destruction of ozone in the troposphere. NMHCs generally refer to the C_2–C_4 series, notably ethane, ethene, acetylene, propane, propene, and n-butane, but also the five-carbon compound isoprene. Of these, ethene is generally the most abundant contributing 40% to the total NMHC pool in sea water. Published data of concentrations of NMHCs in sea water vary widely sometimes exceeding a factor 100. For example, in one extensive study, ethene and propane were found to be the most abundant species in the intertropical South Pacific, with mixing ratios of 2.7 to 58 and 6 to 75 pptv, respectively; whereas in the equatorial Atlantic these species showed mixing ratios of 20 pptv and 10 pptv, respectively.

The water-column dynamics of NMHCs are poorly understood. NMHCs have been detected in the surface sea and with maxima in the euphotic zone and tend to be present at concentrations in sea water at around 10^{-10} mol l^{-1}. Evidence suggests that photochemical oxidation of dissolved organic matter results in the formation of NMHCs. There can be very little doubt that the physiology of planktonic organisms is also involved in NMHC formation. Ethene and isoprene are freely produced by terrestrial plants where the former is a powerful plant hormone but the function of the latter less well understood. It is likely that similar processes occur in planktonic algae. NMHC production tends to correlate with light intensity, dissolved organic carbon and biological production. A simplified scheme of marine NMHC production is shown in **Figure 2**.

The flux of NMHCs to the atmosphere (with estimates ranging from <10 Mt y^{-1} to 50 Mt y^{-1}) is minor on a global scale, but has a potential significance in local atmospheric chemistry. Although oceans are known to act as sources of NMHCs, the

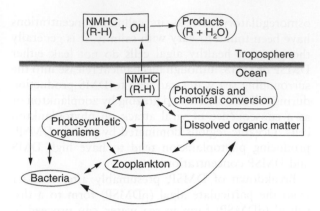

Figure 2 Simplified scheme of marine nonmethane hydrocarbon (NMHC) production. In the marine troposphere NMHC acts as a sink for hydroxyl (OH) radicals and thereby plays a key role in ozone chemistry.

sources of individual NMHCs in the marine boundary layer are not always clear. Those NMHCs with a life time of more than a week (e.g., ethane, ethyne, propane, cyclopropane) show latitudinal gradients consistent with a continental source, whereas variations of NMHCs with life times shorter than a week (all alkenes and pentane) are more consistent with a marine source.

Ammonia

Ammonia is an extremely soluble gas, reacting with water and dissociating into an ammonium ion at ambient pH. At pH 8.2, about one-tenth of dissolved ammonia is present as NH_3. Ammonium is also a rapidly cycling biological nutrient; it is taken up by bacteria and phytoplankton as a source of fixed nitrogen, and released by sundry physiological and decompositional processes in the food web. Anthropogenic loading of ammonium (and other nutrients) into the coastal marine environment results in increased phytoplankton growth in a phenomenon called eutrophication. Ammonium is oxidized to nitrate by bacteria in a process known as nitrification (**Figure 3**). Conversely, in anoxic environments, ammonium can be formed by nitrate-reducing bacteria.

Ammonia plays an important role in the acid–base chemistry in the troposphere where the unionized ammonia (NH_3) is converted into ionized ammonia (NH_4^+) via a reaction that neutralizes atmospheric acids as HNO_3 and H_2SO_4. This leads to the formation of ammonium aerosols such as the stable ammonium sulfate. Eventually the ammonia returns to the surface by dry or wet deposition.

Few data exist on the fluxes of NH_3 over marine environments. Evidence suggests that most of the ocean surface serves as a source of NH_3 to the

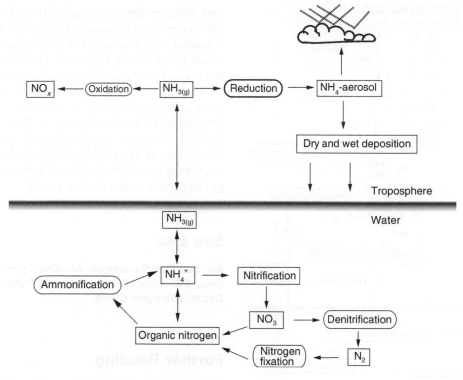

Figure 3 Simplified scheme of marine NH_x chemistry. In the marine boundary layer NO_3 acts as an initiator for the degradation of many organic compounds, in particular dimethylsulfide (DMS).

atmosphere, even in regions of very low nutrients. In the North Sea, an area situated in the middle of densely populated and industrialized countries of Western Europe, air from nearby terrestrial sources may act as a source of NH_3 into surface waters. It has been estimated that the annual biogenic emission of ammonia from European seas is around $30 \, kt \, N \, y^{-1}$, which is comparable to the emissions of smaller North European countries, leading to the conclusion (amongst others) that seas are among the largest sources of imported ammonium for maritime countries. The net emission of ammonia from coastal waters of the north-east Pacific Ocean to the atmosphere has been shown to be in the order of $10 \, \mu mol \, m^{-2} \, d^{-1}$.

Organohalogens

Halogenated compounds, such as methyl chloride (CH_3Cl), methyl bromide (CH_3Br) and methyl iodide (CH_3I) are a major source of halogens in the atmosphere, and subsequently form sources of reactive species capable of catalytically destroying ozone. Among these CH_3I is likely to play an important role in the budget of tropospheric ozone, through production of iodine atoms by photolysis. Due to their higher photochemical stability methyl chloride and

methyl bromide are more important in stratospheric chemistry; it has been suggested that BrO species are responsible for losses of tropospheric ozone in the Arctic (**Figure 4**).

Atmospheric methyl halides, measured over the ocean by several cruise surveys, have been shown to have average atmospheric mixing ratios of: CH_3Cl, 550–600 pptv; CH_3Br, 10–12 pptv; CH_3I, 0.5–1 pptv. Their temporal and spatial variations are not well understood, neither is their production mechanism in the ocean known. Measurements of atmospheric and seawater concentrations of CH_3Cl and CH_3I have indicated that the oceans form natural sources of these methyl halides. In contrast, CH_3Br appears to be undersaturated in the open ocean and exhibits moderate to 100% supersaturation in coastal and upwelling regions, leading to a global atmosphere to ocean flux of $13 \, Gg \, y^{-1}$. Coastal salt marshes, although they constitute a minor area of the global marine environment, may produce roughly 10% of the total fluxes of atmospheric CH_3Br and CH_3Cl and thus contribute significantly to the global budgets.

Macrophytic and phytoplanktonic algae produce a wide range of volatile organohalogens including di- and tri-halomethanes and mixed organohalogens. There is evidence for the involvement of enzymatic synthesis of methyl halides, but the metabolic

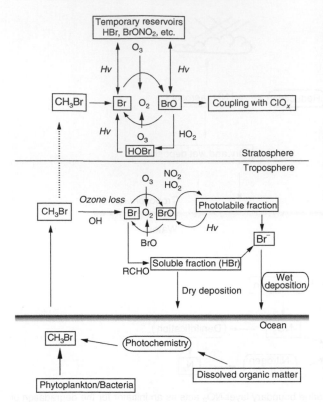

Figure 4 Schematic illustration of the circulation of methyl bromide.

production pathways are not well known. In free sea water, photochemical processes, ion substitution and, possibly the alkylation of halide ions (during the oxidation of organic matter by an electron acceptor such as Fe(III)) are also potential formation mechanisms. Sunlight or microbial mediation are not required for these reactions. In the ocean, chemical degradation of CH_3Br occurs by nucleophilic substitution by chloride and hydrolysis. Microbial consumption is also a likely sink for halogenated compounds.

Conclusions

The biogenic trace gases are influenced by the complete range of biological processes – from the biochemical and physiological to the ecological level of food web dynamics. The gases that are influenced directly by plant physiology (probably the light NMHCs and isoprene, for example) tend to be most closely related to phytoplankton biomass or primary productivity. Other gases produced during grazing and decomposition (e.g., DMS, NH_3), or gases formed by photochemical reactions in dissolved organic material show differing temporal dynamics

and different dynamics spatially and with depth in the ocean. As conditions change in an apparently warming world, changes in the dynamics of surface ocean gases can be expected. The behavior of these trace gases or even the dynamics of the planktonic community are not understood sufficiently to allow good quantitative predictions about changes in trace gas flux to be made. Changes in flux of some gases could lead to an acceleration of warming, while changes in others could lead to cooling. It is, thus, important to understand the factors controlling trace gas dynamics in the surface ocean.

See also

Air–Sea Gas Exchange. Air–Sea Transfer: N_2O, NO, CH_4, CO. Atmospheric Input of Pollutants. Carbon Cycle. Nitrogen Cycle.

Further Reading

Andreae MO (1990) Ocean–atmosphere interactions in the global biogeochemical sulfur cycle. *Marine Chemistry* 10: 1–29.

Andreae MO and Crutzen PJ (1997) Atmospheric aerosols: biogeochemical sources and role in atmospheric chemistry. *Science* 276: 1052–1058.

Barrett K (1998) Oceanic ammonia emissions in Europe and their transboundary fluxes. *Atmospheric Environment* 32(3): 381–391.

Chin M and Davis DD (1993) Global sources and sinks of OCS and CS_2 and their distributions. *Global Biogeochemical Cycles* 7: 321–337.

Cox RA, Rattigana OV, and Jones RL (1995) Laboratory studies of BrO reactions of interest for the atmospheric ozone balance. In: Bandy RA (ed.) *The Chemistry of the Atmosphere; Oxidants and Oxidation in the Earth's Atmosphere.* Cambridge: The Royal Society of Chemistry.

Crutzen PJ (1976) The possible importance of COS for the sulfate layer of the stratosphere. *Geophysical Research Letters* 3: 73–76.

Graedel TE (1995) Tropospheric budget of reactive chlorine. *Global Biogeochemical Cycles* 9: 47–77.

Kettle AJ and Andreae MO (1999) Flux of dimethylsulfide from the oceans: a comparison of updated datasets and flux models. *Journal of Geophysical Research* 105: 26793–26808.

Lovelock JE (1974) CS_2 and the natural sulfur cycle. *Nature* 248: 625–626.

Turner SM and Liss PS (1985) Measurements of various sulfur gases in a coastal marine environment. *Journal of Atmospheric Chemistry* 2(3): 223–232.

AIR–SEA TRANSFER: N₂O, NO, CH₄, CO

C. S. Law, Plymouth Marine Laboratory, The Hoe, Plymouth, UK

Introduction

The atmospheric composition is maintained by abiotic and biotic processes in the terrestrial and marine ecosystems. The biogenic trace gases nitrous oxide (N_2O), nitric oxide (NO), methane (CH_4) and carbon monoxide (CO) are present in the surface mixed layer over most of the ocean, at concentrations which exceed those expected from equilibration with the atmosphere. As the oceans occupy 70% of the global surface area, exchange of these trace gases across the air–sea interface represents a source/sink for global atmospheric budgets and oceanic biogeochemical budgets, although marine emissions of NO are poorly characterized. These trace gases contribute to global change directly and indirectly, by influencing the atmospheric oxidation and radiative capacity (the 'greenhouse effect') and, together with their reaction products, impact stratospheric ozone chemistry (**Table 1**). The resultant changes in atmospheric forcing subsequently influence ocean circulation and biogeochemistry via feedback processes on a range of timescales. This article describes the marine sources, sinks, and spatial distribution of each trace gas and identifies the marine contribution to total atmosphere

budgets. There is also a brief examination of the approaches used for determination of marine trace gas fluxes and the variability in current estimates.

Nitrous Oxide (N₂O)

The N_2O molecule is effective at retaining long-wave radiation with a relative radiative forcing 280 times that of a CO_2 molecule. Despite this the relatively low atmospheric N_2O concentration results in a contribution of only 5–6% of the present day 'greenhouse effect' with a direct radiative forcing of about 0.1 Wm^{-2}. In the stratosphere N_2O reacts with oxygen to produce NO radicals, which contribute to ozone depletion.

N_2O is a reduced gas which is produced in the ocean primarily by microbial nitrification and denitrification. N_2O is released during ammonium (NH_4^+) oxidation to nitrite (NO_2^-) (**Figure 1**), although the exact mechanism has yet to be confirmed. N_2O may be an intermediate of nitrification, or a by-product of the decomposition of other intermediates, such as nitrite or hydroxylamine. Nitrification is an aerobic process, and the N_2O yield under oxic conditions is low. However, as the nitrification rate decreases under low oxygen, the relative yield of N_2O to nitrate production increases and reaches a maximum at 10–20 μmol dm^3 oxygen (μmol $= 1 \times 10^{-6}$ mol). Conversely, denitrification is an anaerobic process in which soluble oxidized nitrogen

Table 1 The oceanic contribution and atmospheric increase and impact for methane, nitrous oxide, nitric oxide, and carbon monoxide[a]

Trace gas	Atmospheric concentration (ppbv)	Atmospheric lifetime (years)	Major impact in atmosphere	Increase in atmosphere (1980–90)	Oceanic emission as % of total global emissions
Nitrous oxide (N_2O)	315	110–180	Infrared active Ozone sink/source	0.25% (0.8 ppbv y^{-1})	7–34%
Nitric oxide (NO)	0.01	<0.2	Ozone sink/source OH sink/oxidation capacity	Not known	Not known
Methane (CH_4)	1760	10	Infrared active OH sink/oxidation capacity Ozone sink/source	0.8% (0.6 ppbv y^{-1})	1–10%
Carbon monoxide (CO)	120	0.2–0.8	OH sink/oxidation capacity Ozone sink/source Infrared active	−13 to 0.6%	0.9–9%

[a]ppbv, parts per billion by volume. (Adapted from *Houghton et al.*, 1995.)

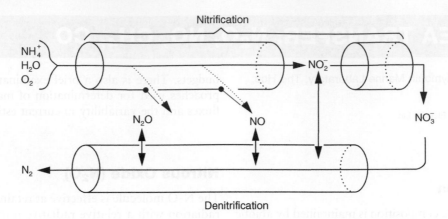

Figure 1 'Leaky Pipe' flow diagram of nitrification and denitrification indicating the potential exchange and intermediate role of NO and N₂O (Reprinted by permission from *Nature* copyright (1990), Macmillan Magazines Ltd.)

compounds, such as nitrate and nitrite, are converted to volatile reduced compounds (N_2O and N_2) in the absence of oxygen. Oxygen availability inhibits denitrification at ambient levels, and also determines the products of denitrification. An enzymatic gradient of sensitivity to oxygen results in the accumulation of N_2O under sub-oxia (3–$10 \, \mu mol \, dm^3$) due to the inhibition of the enzyme nitrous oxide reductase. At lower oxygen ($<3 \, \mu mol \, dm^3$) the reaction continues through to N_2 and so anoxic environments are sinks for N_2O. N_2O yields from nitrification are 0.2–0.5%, whereas denitrification yields may be as high as 5% at optimal levels of sub-oxia.

An inverse correlation between N_2O and oxygen, and associated linear relationship between nitrate and N_2O, suggest that N_2O in the ocean originates primarily from nitrification. This may not be the case for sediments, in which denitrification is the dominant source of N_2O under variable oxygen tension, with nitrification only contributing in a narrow suboxic band. Attribution of source is difficult as nitrification and denitrification may occur simultaneously and interact, with exchange of products and intermediates (**Figure 1**). This is further complicated, as denitrification will be limited to some extent by nitrate supply from nitrification. Isotopic data from the surface ocean in oligotrophic regions imply that N_2O originates primarily from nitrification. However, recent evidence from waters overlying oxygen-deficient intermediate layers suggests that the elevated surface mixed-layer N_2O arises from coupling between the two processes, as the observed isotope signatures cannot be explained by nitrification or denitrification alone. An additional N_2O source from the dissimilatory reduction of nitrate to ammonium is restricted to highly anoxic environments such as sediments.

The oceanic N_2O distribution is determined primarily by the oxygen and nutrient status of the water column. Estuaries and coastal waters show elevated supersaturation in response to high carbon and nitrogen loading, and the proximity of sub-oxic zones in sediment and the water column. As a result the total marine N_2O source tends to be dominated by the coastal region. The N_2O flux from shelf sea sediments is generally an order of magnitude lower than estuarine sediments, although the former have a greater spatial extent. A N_2O maximum at the base of the euphotic zone is apparent in shelf seas and the open ocean, and is attributed to production in suboxic microzones within detrital material. Oceanic surface waters generally exhibit low supersaturations ($<105\%$), although N_2O supersaturations may exceed 300% in surface waters overlying low oxygen intermediate waters and upwelling regions, such as the Arabian Sea and eastern tropical North Pacific. These 'natural chimney' regions dominate the open ocean N_2O source, despite their limited surface area (**Table 2**). The surface N_2O in upwelling regions such as the Arabian Sea originates in part from the underlying low-oxygen water column at 100–$1000 \, m$, where favorable conditions result in the accumulation of N_2O to supersaturations exceeding 1200%. N_2O transfer into the surface mixed layer will be limited by vertical transport processes and a significant proportion of N_2O produced at these depths will be further reduced to N_2.

The oceans account for 1–$5 \, Tg \, N$-N_2O per annum ($Tg = 1 \times 10^{12} \, g$) or 6–30% of total global N_2O emissions, although there is considerable uncertainty attached to this estimate (**Figure 2**). A recent estimate with greater representation of coastal sources has resulted in upward revision of the marine N_2O source to 7–$10.8 \, Tg \, N$-N_2O per annum; although this may represent an upper limit due to some bias from inclusion of estuaries with high N_2O supersaturation. However, this estimate is in agreement with a total

Table 2 N_2O and CH_4 regional surface water supersaturations (from *Bange et al.*, 1996; 1998) (supersaturation is >100%, undersaturation is <100% with equilibrium between atmosphere and water at 100%)

	Surface % N_2O saturation mean (range)	Surface % CH_4 saturation mean (range)
Estuaries	607 (101–2500)	1230 (146–29 000)
Coastal/shelf	109 (102–118)	395 (85–42 000)
Oligotrophic/transitional ocean	102.5 (102–104)	120 (80–200)
Upwelling ocean	176 (108–442)	200 (86–440)

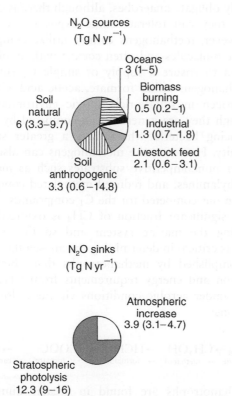

Figure 2 Atmospheric nitrous oxide sources and sinks (adapted from *Houghton et al.*, 1995). Units: $Tg = 1 \times 10^{12}\,g$.

oceanic production rate of 11 Tg N-N_2O per annum calculated from new production and nitrification.

Nitric Oxide (NO)

Nitric oxide (NO) plays a central role in atmospheric chemistry, influencing both ozone cycling and the tropospheric oxidation capacity through reactions with hydroperoxy- and organic peroxy-radicals. When the NO concentration exceeds ∼40 pptv (pptv = parts per trillion by volume) it catalyzes the production of ozone (O_3):

$$CO + OH^* + O_2 \rightarrow HO_2^* + CO_2 \qquad [1]$$

$$HO_2^* + NO \rightarrow OH^* + NO_2 \qquad [2]$$

$$NO_2 + h\nu \rightarrow NO + O(^3P) \qquad [3]$$

$$O + O_2 + M \rightarrow O_3 + M \qquad [4]$$

At high concentrations (>50 ppbv; ppbv = parts per billion by volume), O_3 in the atmospheric boundary layer becomes a toxic pollutant that also has important radiative transfer properties. The production of nitric acid from NO influences atmospheric pH, and contributes to acid rain formation. In addition, the oxidation of NO to the nitrate (NO_3) radical at night influences the oxidizing capacity of the lower troposphere. Determination of the magnitude and location of NO sources is critical to modeling boundary layer and free tropospheric chemistry.

NO cycling in the ocean has received limited attention, as a result of its thermodynamic instability and high reactivity. Photolysis of nitrite in surface waters occurs via the formation of a nitrite radical with the production of NO:

$$NO_2^- + h\nu \rightarrow NO_2^{-*} + HOH \rightarrow NO + OH^- + OH$$

This reaction may account for 10% of nitrite loss in surface waters of the Central Equatorial Pacific, resulting in a 1000-fold increase in dissolved NO at a steady-state surface concentration of 5 pmol dm^{-3} during light periods (pmol $= 1 \times 10^{-12}$ mol). This photolytic production is balanced by a sink reaction with the superoxide radical (O_2^-) to produce peroxynitrite:

$$O_2^- + NO \rightarrow {}^-OONO$$

This reaction will be dependent upon steady-state concentration of the superoxide radical; however, as the reaction has a high rate constant, NO is rapidly turned over with a half-life on the order of 10–100 seconds.

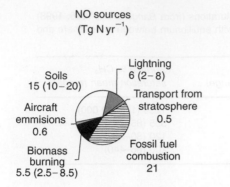

Figure 3 Atmospheric nitric oxide sources (from Graedel and Crutzen, 1992). Units: $Tg = 1 \times 10^{12}$ g.

As with N_2O, NO may also be produced as a by-product or intermediate of denitrification and nitrification (**Figure 1**). NO production by soils is better characterized than in marine systems, and is significant both in terms of nitrogen loss and the global NO budget (**Figure 3**). The greater oxygen availability in soils limits reduction of NO via denitrification and so enhances NO efflux. Sediment pore water NO maxima have been attributed to denitrification, although, as this process also represents a sink for NO (**Figure 1**), this may reflect poising at an optimal redox potential for NO production. Conversely, the NO maximum in low oxygen intermediate waters in the east tropical North Pacific derives from nitrification.

Current understanding of the oceanic NO distribution is that it is limited to the surface ocean and intermediate low oxygen water column. There is potential for higher NO concentrations in coastal and estuarine waters from sediment and photolytic sources, and nitrite photolysis to NO may also be significant in upwelling regions. Despite the short half-life of NO in surface waters, the maintenance of steady-state NO concentration suggests that photolytic production may support an, as yet unquantified, source of atmospheric NO. Surface concentrations in the Central Equatorial Pacific suggest that the oceanic NO source would not exceed 0.5 Tg N per annum, which is relatively insignificant when compared with other sources (**Figure 3**).

Methane (CH₄)

CH_4 is the most abundant organic volatile in the atmosphere and, next to CO_2, is responsible for 15% of the current greenhouse radiative forcing, with a direct radiative forcing of 0. 5 Wm^{-2}. CH_4 reacts with OH and so limits the tropospheric oxidation capacity and influences ozone and other greenhouse

gases. The reaction with OH generates a feedback that leads to a reduction in the rate of CH_4 removal.

CH_4 is a reduced gas which, paradoxically, is supersaturated in the oxidized surface waters of the ocean (see **Table 2**). CH_4 is produced biotically and abiotically, although its oceanic distribution is controlled primarily by biological processes. Methanogenesis is classically defined as the formation of CH_4 from the fermentation and remineralization of organic carbon under anoxic conditions. Methanogens require a very low reducing potential and are generally obligate anaerobes, although there is evidence that they can tolerate some exposure to oxygen. However, methanogens cannot utilize complex organic molecules and often coexist with aerobic consortia to ensure a supply of simple C_1 substrates. Methanogens utilize formate, acetic acid, CO_2, and hydrogen in sulfate-rich anoxic environments, although they are generally out-competed by sulfate-reducing bacteria which have a greater substrate affinity. However, the methanogens can also utilize other noncompetitive substrates such as methanol, methylamines, and reduced methylated compounds, when out-competed for the C_1 compounds.

A significant fraction of CH_4 is oxidized before exiting the marine system and so the oxidation rate is critical in determining the air–sea flux. This is accomplished by methanotrophs that obtain their carbon and energy requirements from CH_4 oxidation under aerobic conditions via the following reactions:

$$CH_4 \rightarrow CH_3OH \rightarrow HCHO \rightarrow HCOOC \rightarrow CO_2$$
methane → methanol → formaldehyde → formate → carbon dioxide

Methanotrophs are found in greater numbers in sediments than in oxic sea water, and consequently the oceanic water column CH_4 oxidation is an order of magnitude lower than in sediments. Methanotrophs have a high inorganic nitrogen requirement and so methanotrophy is highest at the oxic–anoxic interface where ammonium is available. Anaerobic CH_4 oxidation also occurs but is less well characterized. It is generally restricted to anaerobic marine sediments, utilizing sulfate as the only oxidant available, and is absent from anaerobic freshwater sediments which lack sulfate. A significant proportion of CH_4 produced in anaerobic subsurface layers in sediments is oxidized during diffusive transport through the sulfate-CH_4 transition zone by anaerobic oxidation and subsequently by aerobic oxidation in the overlying oxic layers. Anaerobic oxidation represents the main sink for CH_4 in marine sediments, where it may account for 97% of CH_4 production.

CH_4 production is characteristic of regions with high input of labile organic carbon such as wetlands and sediments, but is usually restricted to below the zone of sulfate depletion. The oceanic CH_4 source is dominated by coastal regions, which exhibit high CH_4 fluxes as a result of bubble ebullition from anoxic carbon-rich sediments, and also riverine and estuarine input. Some seasonality may result in temperate regions due to increased methanogenesis at higher temperatures. The predominant water column source in shelf seas and the open ocean is CH_4 production at the base of the euphotic zone. This may arise from lateral advection from sedimentary sources, and *in situ* CH_4 production. The latter is accomplished by oxygen-tolerant methanogens that utilize methylamines or methylated sulfur compounds in anoxic microsites within detrital particles and the guts of zooplankton and fish. Lateral advection and *in situ* production may be greater in upwelling regions, as suggested by the increased CH_4 supersaturation in surface waters in these regions. Oceanic CH_4 concentration profiles generally exhibit a decrease below 250 m due to oxidation. Methanogenesis is elevated in anoxic water columns, although these are not significant sources of atmospheric CH_4 due to limited ventilation and high oxidation rates. Other sources include CH_4 seeps in shelf regions from which CH_4 is transferred directly to the atmosphere by bubble ebullition, although their contribution is difficult to quantify. Abiotic CH_4 originating from high-temperature fluids at hydrothermal vents also elevates CH_4 in the deep and intermediate waters in the locality of oceanic ridges. A significant proportion is oxidized and although the contribution to the atmospheric CH_4 pool may be significant in localized regions this has yet to be constrained. Hydrates are crystalline solids in which methane gas is trapped within a cage of water molecules. These form at high pressures and low temperatures in seafloor sediments generally at depths below 500 m. Although CH_4 release from hydrates is only considered from anthropogenic activities in current budgets, there is evidence of catastrophic releases in the geological past due to temperature-induced hydrate dissociation. Although oceanic hydrate reservoirs contain 14 000 Gt CH_4, there is currently no evidence of significant warming of deep waters which would preempt release.

Other aquatic systems such as rivers and wetlands are more important sources than the marine environment. Shelf regions are the dominant source of CH_4 from the ocean (14(11–18) Tg CH_4 per annum), accounting for 75% of the ocean flux (**Table 2**). The ocean is not a major contributor to the atmospheric

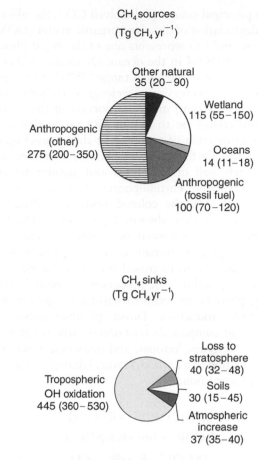

Figure 4 Atmospheric methane sources and sinks (adapted from Houghton *et al.* (1995)). Units: $Tg = 1 \times 10^{12}$ g.

CH_4 budget, as confirmed by estimates of the oceanic CH_4 source (**Figure 4**).

Carbon Monoxide (CO)

The oxidation of CO provides the major control of hydroxyl radical content in the troposphere and limits the atmospheric oxidation capacity. This results in an increase in the atmospheric lifetime of species such as CH_4, N_2O, and halocarbons, and enhances their transfer to the stratosphere and the potential for subsequent ozone destruction. It has been suggested that decreasing stratospheric ozone and the resultant increase in incident ultraviolet (UV) radiation may increase marine production and efflux of CO, thereby generating a positive feedback loop. However, this may be compensated by a negative feedback in which increased UV reduces biological production and dissolved organic matter, so reducing the CO source. CO also influences tropospheric ozone by its interaction with NOx, and is a minor greenhouse gas with a radiative forcing of 0.06 Wm^{-2} at current atmospheric concentrations.

The principal source of dissolved CO is the abiotic photodegradation of dissolved organic matter (DOM) by UV-R, and CO represents one of the major photoproducts of DOM in the ocean. Quantum yields for CO are highest in the UV-B range (280–315 nm) and decrease with increasing wavelengths. However, the UV-A (315–390 nm) and blue portion of the visible spectrum contribute to marine CO production as a greater proportion of radiation at these wavelengths reaches the Earth's surface. Humics represent approximately half of the DOM and account for the majority of the chromophoric dissolved organic matter (CDOM), the colored portion of dissolved organic matter that absorbs light energy. The CO photoproduction potential of humics is dependent upon the degree of aromaticity. Terrestrial humics are characterized by an increased prevalence of phenolic groups, and addition of precursor compounds containing phenolic moieties to natural samples stimulates CO production. Direct photo-oxidation of humics and compounds containing carbonyl groups, such as aldehydes, ketones, and quinones, occurs via the production of a carbonyl radical during α-cleavage of an adjacent bond:-

$$R + (COR')^* \rightarrow R' + CO$$
$$RCOR' + h\upsilon \rightarrow hskip200pt$$
$$(RCO)^* + R' \rightarrow R' + CO$$

CO production may also occur indirectly by a photosensitized reaction in which light energy is transferred via an excited oxygen atom to a carbonyl compound. This may occur with ketonic groups via the photosensitized production of an acetyl radical. Whereas light and CDOM are the primary factors controlling CO production there may also be additional influence from secondary factors. For example, organo-metal complexes have increased light absorption coefficients and their photo-decomposition will enhance radical formation and CO production at higher levels of dissolved metals such as iron. There is also minor biotic production of CO by methanogens, but this does not appear to be significant.

CO can be oxidized to carbon dioxide by selected microbial groups including ammonia oxidizers and methylotrophs that have a broad substrate specificity and high affinity for CO. However, only the carboxidotrophs obtain energy from this reaction, and these may be unable to assimilate CO efficiently at *in situ* concentrations. CO turnover times of 4 hours are typical for coastal waters, whereas this varies between 1 and 17 days in the open ocean. The lower oxidation rate in the open ocean may be due to light inhibition of CO oxidation. Extrapolation from laboratory measurements suggests that only 10% of photochemically produced CO is microbially oxidized.

Dissolved CO exhibits diurnal variability in the surface ocean in response to its photolytic source, although this is also indicative of a strong sink term. The decline in the surface mixed-layer CO concentration in the dark results from a combination of CO oxidation, vertical mixing, and air–sea exchange. As the equilibration time between atmosphere and oceanic surface mixed layer is on the order of a month, this suggests that the former two processes dominate. Superimposed upon the diurnal cycle of CO in the surface ocean are spatial and seasonal gradients that result from the interaction of photoproduction and the sink processes. Below the euphotic zone CO is uniformly low throughout the intermediate water column.

CO production potential is highest in wetland regions, which are characterized by high CDOM and enhanced light attenuation. Photochemical production of CO represents a potential sink for terrestrial dissolved organic carbon (DOC) in estuaries and coastal waters. This pathway may account for some of the discrepancy between the total terrestrial DOC exported and the low proportion of terrestrial DOC observed in the marine pool. Although a strong lateral gradient in CDOM exists between rivers and the open ocean, estuarine CO production may be limited by reduced UV light penetration. CO photoproduction may occur down to 80 m in the open ocean, and 20 m in the coastal zone, but is restricted to the upper 1 m in wetlands and estuaries. In addition, estuarine and coastal CO flux may also be restricted by the higher CO oxidation rates. There is evidence that upwelling regions may support enhanced CO production, in response to upwelled CDOM that is biologically refractory but photolabile.

The presence of a CO gradient in the 10 m overlying the surface ocean suggests that the photolytic source of CO may influence the marine boundary layer. The marine source of CO is poorly constrained, with estimates varying from 10 to 220 Tg CO per annum. A flux of 1200 Tg CO per annum was estimated on the assumption that low rates of oceanic CO oxidation would only remove a small proportion of photoproduced CO, and that the residual would be ventilated to the atmosphere. The discrepancy between this and other flux estimates implies that a significant CO sink has been overlooked, although this may reflect shortcomings of different techniques. The oceanic contribution to the global source is between 1 and 20%, although Extrapolation of photochemical production rates from wetlands, estuaries, and coasts suggests that these

Table 3 Atmospheric CO sources and sinks[a] (adapted from *Zuo et al.*, 1998)

CO sources (Tg CO y^{-1})		CO sinks (Tg CO y^{-1})	
Industrial/fuel combustion	400–1000	Tropospheric hydroxyl oxidation	1400–2600
Biomass burning	300–2200	Soils	250–530
Vegetation and soils	50–200	Flux to stratosphere	80–140
Methane oxidation	300–1300		
NMHC oxidation	200–1800		
Ocean	10–220		
(Coast/Shelf	300–400)		
Total sources	1260–6720	Total sinks	1730–3270

[a]Note that a separate estimate of the coastal/shelf CO source is shown for comparison, but does not contribute to the total source. Tg $= 1 \times 10^{12}$ g.

alone may account for 20% of the total global CO flux. Although the marine source is responsible for <10% of the total global flux (**Table 3**), it may still dominate atmospheric oxidation conditions in remote regions at distance from land.

Air–Sea Exchange of Trace Gases

The flux of these trace gases across the air–sea interface is driven by physical transfer processes and the surface concentration anomaly, which represents the difference between the partial pressure observed in surface water and that expected from equilibrium with the atmosphere. Direct determination of the oceanic emission of a trace gas is difficult under field conditions. Atmospheric gradient measurements above the ocean surface require enhanced analytical resolution, whereas more advanced micrometeorological techniques have yet to be applied to these trace gases. Determination of the accumulation rate in a floating surface flux chamber is a simpler approach, but may generate artefactual results from the damping of wave- and wind-driven exchange, and enhanced transfer on the inner chamber surfaces. Consequently the majority of flux estimates are calculated indirectly rather than measured. The surface anomaly is derived from the difference between the measured surface concentration (Cw), and an equilibrium concentration calculated from the measured atmospheric concentration (Cg) and solubility coefficient ($\propto$) at ambient temperature and salinity. This is then converted to a flux by the application of a dynamic term, the gas transfer velocity, k:

$$F = k(Cw - \alpha Cg)$$

The transfer velocity k is the net result of a variety of molecular and turbulent processes that operate at different time and space scales. Wind is the primary driving force for most of these turbulent processes,

and it is also relatively straightforward to obtain accurate measurements of wind speed. Consequently, k is generally parameterized in terms of wind speed, with the favored approaches assuming tri-linear and quadratic relationships between the two. These relationships are defined for CO$_2$ at 20°C in fresh water and sea water and referenced to other gases by a Schmidt number (Sc) relationship:

$$k\,gas = k\,ref\,(Sc\,gas/Sc\,ref)^n$$

where n is considered to be $-1/2$ at most wind speeds. This dependency of k is a function of the molecular diffusivity (D) of the gas and the kinematic viscosity of the water (μ), and is expressed in terms of the Schmidt number (Sc $= \mu/D$).

Determination of marine trace gas fluxes using different wind speed–transfer velocity relationships introduces uncertainty, which increases at medium-high wind speeds to a factor of two. Furthermore, additional uncertainty is introduced by the extrapolation of surface concentration gradient measurements to long-term climatological wind speeds. Current estimates of oceanic fluxes are also subject to significant spatial and temporal bias resulting from the fact that most studies focus on more productive regions and seasons. This uncertainty is compounded by the extrapolation of observational data sets to unchartered regions. With the exception of N$_2$O, the ocean does not represent a major source for these atmospheric trace gases, although spatial variability in oceanic source strength may result in localized impact, particularly in remote regions. In the near future, advances in micrometeorological techniques, improved transfer velocity parameterizations and the development of algorithms for prediction of surface ocean concentrations by remote sensing should provide further constraint in determination of the oceanic source of N$_2$O, NO, CH$_4$, and CO.

See also

Air–Sea Gas Exchange. Air–Sea Transfer: Dimethyl Sulfide, COS, CS$_2$, NH$_4$, Non-Methane Hydrocarbons, Organo-Halogens. Carbon Dioxide (CO$_2$) Cycle. Photochemical Processes. Plankton and Climate. Surface Films.

Further Reading

Bange HW, Bartel UH, *et al.* (1994) Methane in the Baltic and North Seas and a reassessment of the marine emissions of methane. *Global Biogeochemical Cycles* 8: 465–480.

Bange HW, Rapsomanikis S, and Andreae MO (1996) Nitrous oxide in coastal waters. *Global Biogeochemical Cycles* 10: 197–207.

Carpenter EJ and Capone DG (eds.) (1983) *Nitrogen in the Marine Environment*. London: Academic Press.

Graedel TE and Crutzen PJ (eds.) (1992) *Atmospheric Change: An Earth System Perspective*. London: W. H. Freeman and Co.

Houghton JT, Meira Filho M, Bruce J, *et al.* (1995) *Climate Change 1994. Radiative Forcing of Climate Change and an Evaluation of the IPCC IS92 Emission Scenarios*, Intergovernmental Panel on Climate Change. Cambridge: Cambridge University Press.

Liss PS and Duce RA (eds.) (1997) *The Sea Surface and Global Change*. Cambridge: Cambridge University Press.

Zuo Y, Guerrero MA, and Jones RD (1998) Reassessment of the ocean to atmosphere flux of carbon monoxide. *Chemistry and Ecology* 14: 241–257.

NITROGEN CYCLE

D. M. Karl, University of Hawaii at Manoa,
Honolulu, HI, USA
A. F. Michaels, University of Southern California,
Los Angeles, CA, USA

Introduction

The continued production of organic matter in the sea requires the availability of the many building blocks of life, including essential major elements such as carbon (C), nitrogen (N), and phosphorus (P); essential minor elements such as iron, zinc, and cobalt; and, for many marine organisms, essential trace organic nutrients that they cannot manufacture themselves (e.g., amino acids and vitamins). These required nutrients have diverse structural and metabolic function and, by definition, marine organisms cannot survive in their absence.

The marine nitrogen cycle is part of the much larger and interconnected hydrosphere–lithosphere–atmosphere–biosphere nitrogen cycle of the Earth. Furthermore, the oceanic cycles of carbon, nitrogen, and phosphorus are inextricably linked together through the production and remineralization of organic matter, especially near surface ocean phytoplankton production. This coordinated web of major bioelements can be viewed as the nutrient 'super-cycle.'

The dominant form of nitrogen in the sea is dissolved gaseous dinitrogen (N_2) which accounts for more than 95% of the total nitrogen inventory. However, the relative stability of the triple bond of N_2 renders this form nearly inert. Although N_2 can serve as a biologically-available nitrogen source for specialized N_2–fixing microorganisms, these organisms are relatively rare in most marine ecosystems. Consequently, chemically 'fixed' or 'reactive' nitrogen compounds such as nitrate (NO_3^-), nitrate (NO_2^-), ammonium (NH_4^+), and dissolved and particulate organic nitrogen (DON/PON) serve as the principal sources of nitrogen to sustain biological processes.

For more than a century, oceanographers have been concerned with the identification of growth-and production-rate limiting factors. This has stimulated investigations of the marine nitrogen cycle including both inventory determinations and pathways and controls of nitrogen transformations from one form to another. Contemporaneous ocean investigations have documented an inextricable link between nitrogen and phosphorus cycles, as well as the importance of trace inorganic nutrients. It now appears almost certain that nitrogen is only one of several key elements for life in the sea, neither more nor less important than the others. Although the basic features of the marine nitrogen cycle were established nearly 50 years ago, new pathways and novel microorganisms continue to be discovered. Consequently, our conceptual view of the nitrogen cycle is a flexible framework, always poised for readjustment.

Methods and Units

The analytical determinations of the various dissolved and particulate forms of nitrogen in the sea rely largely on methods that have been in routine use for several decades. Determinations of NO_3^-, NO_2^-, and NH_4^+ generally employ automated shipboard, colorimetric assays, although surface waters of open ocean ecosystems demand the use of modern high-sensitivity chemiluminescence and fluorometric detection systems. PON is measured by high-temperature combustion followed by chromatographic detection of the by-product (N_2), usually with a commercial C–N analyzer. Total dissolved nitrogen (TDN) determination employs sample oxidation, by chemical or photolytic means, followed by measurement of NO_3^-. DON is calculated as the difference between TDN and the measured dissolved, reactive inorganic forms of N (NO_3^-, NO_2^-, NH_4^+) present in the original sample. Gaseous forms of nitrogen, including N_2, nitrous oxide (N_2O), and nitric oxide (NO) are generally measured by gas chromatography.

Nitrogen exists naturally as two stable isotopes, ^{14}N (99.6% by atoms) and ^{15}N (0.4% by atoms). These isotopes can be used to study the marine nitrogen cycle by examination of natural variations in the $^{14}N/^{15}N$ ratio, or by the addition of specific tracers that are artificially enriched in ^{15}N.

Most studies of oceanic nitrogen inventories or transformations use either molar or mass units; conversion between the two is straightforward (1 mole N = 14 g N, keeping in mind that the molecular weight of N_2 gas is 28).

Components of the Marine Nitrogen Cycle

The systematic transformation of one form of nitrogen to another is referred to as the nitrogen cycle (**Figure 1**). In the sea, the nitrogen cycle revolves around the metabolic activities of selected micro-organisms and it is reasonable to refer to it as the microbial nitrogen cycle because it depends on bacteria (**Table 1**). During most of these nitrogen transformations there is a gain or loss of electrons and, therefore, a change in the oxidation state of nitrogen from the most oxidized form, NO_3^- ($+5$), to the most reduced form, NH_4^+ (-3). Transformations in the nitrogen cycle are generally either energy-requiring (reductions) or energy-yielding (oxidations). The gaseous forms of nitrogen in the surface ocean can freely exchange with the atmosphere, so there is a constant flux of nitrogen between these two pools.

The natural, stepwise process for the regeneration of NO_3^- from PON can be reproduced in a simple 'decomposition experiment' in an enclosed bottle of sea water (**Figure 2**). During a 3-month incubation period, the nitrogen contained in particulate matter is first released as NH_4^+ (the process of ammonification), then transformed to NO_2^- (first step of nitrification), and finally, and quantitatively, to NO_3^- (the second step of nitrification). These transformations are almost exclusively a result of the metabolic

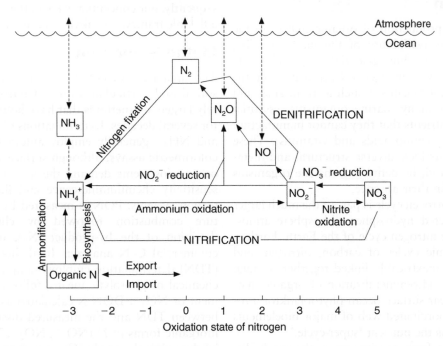

Figure 1 Schematic representation of the various transformations from one form of nitrogen to another that compose the marine nitrogen cycle. Shown at the bottom is the oxidation state of nitrogen for each of the components. Most transformations are microbiological and most involve nitrogen reduction or oxidation. (Adapted from Capone, ch. 14 of Rogers and Whitman (1991).)

Table 1 Marine nitrogen cycle

Process	Credits		
	Bacteria	Phytoplankton[a]	Zooplankton/Fish[b]
NH_4^+ production from DON/PON (ammonification)	+	+	+
NH_4^+/NO_2^-/NO_3^-/DON assimilation	+	+	−
PON ingestion	−	+	+
$NH_4^+ \rightarrow NO_2^-$ (nitrification, step 1)	+	−	−
$NO_2^- \rightarrow NO_3^-$/$N_2O$ (nitrification, step 2)	+	−	−
NO_3^-/$NO_2^- \rightarrow N_2$/N_2O (denitrification)	+	+	−
$N_2 \rightarrow NH_4^+$/organic N (N_2 fixation)	+	−	−

[a]Phytoplankton – eukaryotic phytoplankton.
[b]Zooplankton – including protozoans and metazoans.
Abbreviations: NH_4^+, ammonium; NO_2^-, nitrite; NO_3^-, nitrate; N_2O, nitrous oxide; N_2, dinitrogen; DON, dissolved organic N; PON, particulate organic N; organic N, DON and PON.

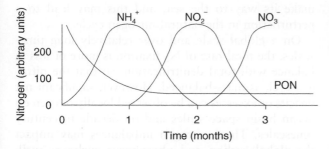

Figure 2 Stepwise decomposition of particulate organic nitrogen (PON) in arbitrary units versus time during a dark incubation. Nitrogen is transformed, first to NH_4^+ by bacterial ammonification and finally to NO_2^- and NO_3^- by the two-step process of bacterial nitrification. These same processes are responsible for the global ocean formation of NO_3^- in the deep sea. These data are the idealized results of pioneering nitrogen cycle investigators, T. von Brand and N. Rakestraw, who unraveled these processes more than 50 years ago.

activities of bacteria. This set of regeneration reactions is vital to the nitrogen cycle, and since most deep water nitrogen (excluding N_2) is in the form of NO_3^-, bacterial nitrification must be a very important process (see Nitrogen Distributions in the Sea, below).

Nitrogen Assimilation

Several forms of nitrogen can be directly transported across cell membranes and assimilated into new cellular materials as required for biosynthesis and growth. Most microorganisms readily transport NH_4^+, NO_2^-, NO_3^-, and selected DON compounds such as amino acids, urea, and nucleic acid bases. By comparison, the ability to utilize N_2 as a nitrogen source for biosynthesis is restricted to a very few species of specialized microbes. Many protozoans, including both photosynthetic and heterotrophic species, and all metazoans obtain nitrogen primarily by ingestion of PON.

Once inside the cell or organism, nitrogen is digested and, if necessary, reduced to NH_4^+. If oxidized compounds such as NO_3^- or NO_2^- are utilized, cellular energy must be invested to reduce these substrates to ammonium for incorporation into organic matter. The process of reduction of NO_3^- (or NO_2^-) for the purpose of cell growth is referred to as assimilatory nitrogen (NO_3^-/NO_2^-) reduction and most microorganisms, both bacteria and phytoplankton, possess this metabolic capability (**Table 1**). In theory, there should be a metabolic preference for NH_4^+ over either NO_3^- or NO_2^-, based strictly on energetic considerations. However, it should be emphasized that preferential utilization of NH_4^+ does not always occur. For example, two closely related and abundant planktonic cyanobacteria that coexist

in tropical and subtropical marine habitats have devised alternate metabolic strategies: *Synechococcus* prefers NO_3^- and *Prochlorococcus* prefers NH_4^+. In fact, *Prochlorococcus* cannot reduce NO_3^- to NH_4^+, presumably because the critical enzyme systems are absent.

Nitrification

As nitrogen is oxidized from NH_4^+ through NO_2^- to NO_3^-, energy is released (**Figure 1**), a portion of which can be coupled to the reduction of carbon dioxide (CO_2) to organic matter (CH_2O) by nitrifying bacteria. These specialized bacteria, one group capable only of the oxidation of NH_4^+ to NO_2^- and the second capable only of the oxidation of NO_2^- to NO_3^-, are termed 'chemolithoautotrophic' because they can fix CO_2 in the dark at the expense of chemical energy. Other related chemolithoautotrophs can oxidize reduced sulfur compounds, and this pathway of organic matter production has been hypothesized as the basis for life at deep-sea hydrothermal vents.

It is essential to emphasize an important ecological aspect of NH_4^+/NO_2^- chemolithoautotrophy. First, the oxidation of NH_4^+ to NO_2^- and of NO_2^- to NO_3^- usually requires oxygen and these processes are ultimately coupled to the photosynthetic production of oxygen in the surface water. Second, the continued formation of reduced nitrogen, in the form of NH_4^+ or organic nitrogen, is also dependent, ultimately, on photosynthesis. In this regard the CO_2 reduced via this 'autotrophic' pathway must be considered secondary, not primary, production from an ecological energetics perspective.

Marine nitrifying bacteria, especially the NO_2^- oxidizers are ubiquitous in the world ocean and key to the regeneration of NO_3^-, which dominates waters below the well-illuminated, euphotic zone. However they are never very abundant and, at least for those species in culture, grow very slowly. Certain heterotrophic bacteria can also oxidize NH_4^+ to both NO_2^- and NO_3^- during metabolism of preformed organic matter. However, very little is known about the potential for 'heterotrophic nitrification' in the sea.

Denitrification

Under conditions of reduced oxygen (O_2) availability, selected species of marine bacteria can use NO_3^- as a terminal acceptor for electrons during metabolism, a process termed NO_3^- respiration or dissimilatory NO_3^- reduction. This process allows microorganisms to utilize organic matter in low-O_2 or anoxic habitats with only a slight loss of efficiency relative to O_2-based metabolism. A majority of marine bacteria have the ability for NO_3^- respiration under the appropriate environmental conditions (**Table 1**). Potential by-products of NO_3^- respiration are NO_2^-, N_2, and N_2O; if a gas is formed (N_2/N_2O) then the process is termed denitrification because the net effect is to remove bioavailable nitrogen from the local environment. The total rate of denitrification is generally limited by the availability of NO_3^-, and a continued supply of NO_3^- via nitrification is dependent upon the availability of NH_4^+ and free O_2. Consequently denitrification typically occurs at boundaries between low-O_2 and anoxic conditions where the supply of NH_4^+ from the anoxic zone sustains a high rate of NO_3^- production via nitrification to fuel- sustained NO_3^- respiration and denitrification. Recently a new group of microorganisms has been isolated that are capable of simultaneously using both O_2 and NO_3^-/NO_2^- as terminal electron acceptors. This process is termed 'aerobic denitrification.' Likewise, there are exceptional microorganisms that are able to carry out anaerobic nitrification (oxidation of NH_4^+ in the absence of O_2). It appears difficult to establish any hard-and-fast rules regarding marine nitrogen cycle processes.

N_2 Fixation

The ability to use N_2 as a growth substrate is restricted to a relatively small group of microorganisms. Open ocean ecosystems that are chronically depleted in fixed nitrogen would appear to be ideal habitats for the proliferation of N_2-fixing microorganisms. However, the enzyme that is required for reduction of N_2 to NH_4^+ is also inhibited by O_2, so specialized structural, molecular, and behavioral adaptations have evolved to promote oceanic N_2 fixation.

Fixation of molecular nitrogen in the open ocean may also be limited by the availability of iron, which is an essential cofactor for the N_2 reduction enzyme system. Changes in iron loading are caused by climate variations, in particular the areal extent of global deserts, by the intensity of atmospheric circulation, and more recently by changes in land use practices. Conversion of deserts into irrigated croplands may cause a change in the pattern and intensity of dust production and, therefore, of iron transport to the sea. Humanity is also altering the global nitrogen cycle by enhancing the fixation of N_2 by the manufacturing of fertilizer. At the present time, the industrial fixation of N_2 is approximately equivalent to the pre-industrial, natural N_2 fixation rate. Eventually some of this artificially fixed N_2 will make its way to the sea, and this may lead to a perturbation in the natural nitrogen cycle.

On a global scale and over relatively long time-scales, the total rate of N_2 fixation is more or less in balance with total denitrification, so that the nitrogen cycle is mass-balanced. However, significant net deficits or excesses can be observed locally or even on ocean basin space scales and on decade to century timescales. These nitrogen imbalances may impact the global carbon and phosphorus cycles as well, including the net balance of CO_2 between the ocean and the atmosphere.

Nitrogen Distributions in the Sea

Required growth nutrients, like nitrogen, typically have uneven distributions in the open sea, with deficits in areas where net organic matter is produced and exported, and excesses in areas where organic matter is decomposed. For example, surface ocean NO_3^- distributions in the Pacific basin reveal a coherent pattern with excess NO_3^- in high latitudes, especially in the Southern Ocean (south of $60°$ S), and along the Equator (especially east of the dateline), and generally depleted NO_3^- concentrations in the middle latitudes of both hemispheres (**Figure 3**). These distributions are a result of the balance between NO_3^- supply mostly by ocean mixing and NO_3^- demand or net photosynthesis. The very large NO_3^- inventory in the surface waters of the Southern Ocean implies that factors other than fixed nitrogen availability control photosynthesis in these regions. It has been hypothesized that the availability of iron is key in this and perhaps other regions of the open ocean. The much smaller but very distinctive band of elevated NO_3^- along the Equator is the result of upwelling of NO_3^--enriched waters from depth to the surface. This process has a large seasonal and, especially, interannual variability, and it is almost absent during El Niño conditions.

Excluding these high-latitude and equatorial regions, the remainder of the surface waters of the North and South Pacific Oceans from about $40°N$ to $40°S$ are relatively depleted in NO_3^-. In fact surface ($0–50$ m) NO_3^- concentrations in the North Pacific subtropical gyre near Hawaii are typically below $0.01 \, \mu mol \, l^{-1}$ (**Figure 4**). Within the upper 200 m, the major pools of fixed nitrogen (e.g., NO_3^-, DON, and PON) have different depth distributions. In the sunlit surface zone, NO_3^- is removed to sustain organic matter production and export. Beneath 100 m, there is a steep concentration versus depth gradient (referred to as the nutricline), which reaches a maximum of about $40–45$ it $\mu mol \, l^{-1}$ at about 1000 m in

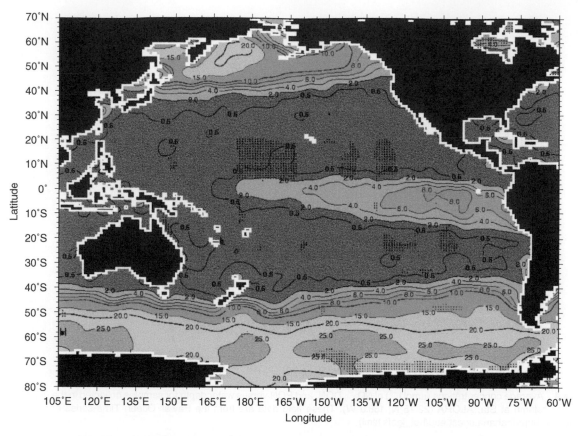

Figure 3 Mean annual NO_3^- concentration ($\mu mol\ l^{-1}$) at the sea surface for samples collected in the Pacific Ocean basin and Pacific sector of the Southern Ocean. (From Conkright et al. 1998.)

the North Pacific Ocean. PON concentration is greatest in the near-surface waters where the production of organic matter via photosynthesis is highest (**Figure 4**). PON includes both living (biomass) and nonliving (detrital) components; usually biomass nitrogen is less than 50% of the total PON in near-surface waters, and less than 10% beneath the euphotic zone ($> 150\,m$). DON concentration is also highest in the euphotic zone (~ 5–$6\,\mu mol\ l^{-1}$) and decreases systematically with depth to a minimum of 2–3 $\mu mol\ l^{-1}$ at 800–1000 m. The main sources for DON in the surface ocean are the combined processes of excretion, grazing, death, and cell lysis. Consequently, DON is a complex mixture of cell-derived biochemicals; at present, less than 20% of the total DON has been chemically characterized. Dissolved N_2 (not shown) is always high ($\sim 800\,\mu mol\ l^{-1}$) and increases systematically with depth. The major controls of N_2 concentration are temperature and salinity, which together determine gas solubility. Marine life has little impact on N_2 distributions in the open sea even though some microorganisms can utilize N_2 as a growth substrate and others can produce N_2 as a metabolic by-product. These transformations are simply too small to

significantly impact the large N_2 inventories in most regions of the world ocean.

Another important feature of the global distribution of NO_3^- is the regional variability in the deep water inventory (**Figure 5**). Deep ocean circulation can be viewed as a conveyor-belt-like flow, with the youngest waters in the North Atlantic and the oldest in the North Pacific. The transit time is in excess of 1000 y, during which time NO_3^- is continuously regenerated from exported particulate and dissolved organic matter via coupled ammonification and nitrification (**Figure 2**). Consequently, the deep Pacific Ocean has nearly twice as much NO_3^- as comparable depths in the North Atlantic (**Figure 5**).

Nitrous Oxide Production

Nitrous oxide (N_2O) is a potent greenhouse gas that has also been implicated in stratospheric ozone depletion. The atmospheric inventory of N_2O is presently increasing, so there is a renewed interest in the marine ecosystem as a potential source of N_2O. Nitrous oxide is a trace gas in sea water, with typical concentrations ranging from 5 to 50 $nmol\ l^{-1}$. Concentrations of N_2O in oceanic surface waters are

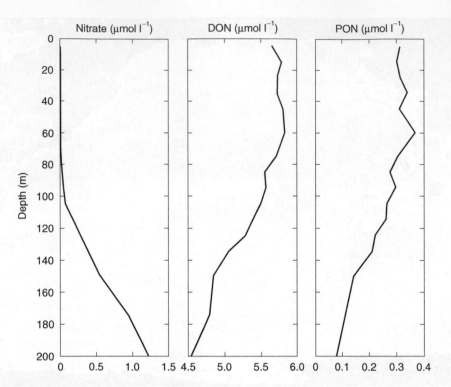

Figure 4 Average concentrations (μmol l^{-1}) of NO$_3^-$, DON, and PON versus water depth for samples collected in the upper 200 m of the water column at Sta. ALOHA (22.75°N, 158.0°W). These field data are from the Hawaii Ocean Time-series program and are available at http://hahana.soest.edu/hot_jgofs.html).

generally in slight excess of air saturation, implying both a local source and a sustained ocean-to-atmosphere flux. Typically there is a mid-water (500–1000 m) peak in N$_2$O concentration that coincides with the dissolved oxygen minimum. At these intermediate water depths, N$_2$O can exceed 300% saturation relative to atmospheric equilibrium. The two most probable sources of N$_2$O in the ocean are bacterial nitrification and bacterial denitrification, although to date it has been difficult to quantify the relative contribution of each pathway for a given habitat. Isotopic measurements of nitrogen and oxygen could prove invaluable in this regard. Because the various nitrogen cycle reactions are interconnected, changes in the rate of any one process will likely have an impact on the others. For example, selection for N$_2$-fixing organisms as a consequence of dust deposition or deliberate iron fertilization would increase the local NH$_4^+$ inventory and lead to accelerated rates of nitrification and, hence, enhanced N$_2$O production in the surface ocean and flux to the atmosphere.

Primary Nitrite (NO$_2^-$) Maximum

An interesting, almost cosmopolitan feature of the world ocean is the existence of a primary NO$_2^-$ maximum (PNM) near the base of the euphotic zone

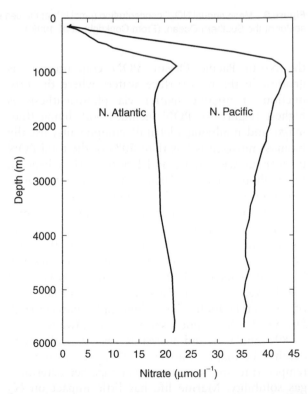

Figure 5 Nitrate concentrations (μmol l^{-1}) versus water depth at two contrasting stations located in the North Atlantic (31.8°N, 50.9°W) and North Pacific (30°N, 160.3°W) Oceans. These data were collected in the 1970s during the worldwide GEOSECS expedition, stations #119 and #212, respectively.

($\sim$100–150 m; **Figure 6**). Nitrite is a key intermediate between NO_3^- and NH_4^+, so there are several potential pathways, both oxidative and reductive, that might lead to its accumulation in sea water. First, phototrophic organisms growing on NO_3^- may partially reduce the substrate to NO_2^- as the first, and least energy-consuming, step in the assimilatory NO_3^- reduction pathway. However, the next step, reduction of NO_2^- to NH_4^+, requires a substantial amount of energy, so when energy is scarce (e.g., light limitation) NO_2^- accumulates inside the cells. Because NO_2^- is the salt of a weak acid, nitrous acid (HNO_2) forms in the slightly acidic intracellular environment, diffuses out of the cell and ionizes to form NO_2^- in the alkaline sea water. This $NO_3^- \rightarrow NO_2^-$ phytoplankton pump, under the control of light intensity, could provide a source of NO_2^- necessary to create and maintain the PNM. Alternatively, local regeneration of dissolved and particulate organic matter could produce NH_4^+ (via ammonification) that is partially oxidized in place to produce a relative excess of NO_2^- (the first step of nitrification). Kinetic controls on this process would

be rates of NH_4^+ production and NO_2^- oxidation to NO_3^- (the second and final step in nitrification). Sunlight, even at very low levels, appears to disrupt the normal coupling between NO_2^- production and NO_2^- oxidation, in favor of NO_2^- accumulation. Finally, it is possible, though perhaps less likely, that NO_3^- respiration (terminating at NO_2^-), followed by excretion of NO_2^- (into the surrounding sea water might also contribute to the accumulation of NO_2^-) near the base of the euphotic zone. Because the global ocean at the depth of the PNM is characteristically well-oxygenated, one would need to invoke microenvironments like animal guts or large particles as the habitats for this nitrogen cycle pathway. The use of ^{15}N-labeled substrates, selective metabolic inhibitors, and other experimental manipulations provides an opportunity for direct assessment of the role of each of these potential processes. In all likelihood, more than one of these processes contributes to the observed PNM. Whatever the cause, light appears to be an important determinant that might explain the relative position, with regard to depth, of this global feature.

Nitrogen Cycle and Ocean Productivity

Because nitrogen transformations include both the formation and decomposition of organic matter, much of the nitrogen used in photosynthesis is locally recycled back to NH_4^+ or NO_3^- to support another pass through the cycle. The net removal of nitrogen in particulate, dissolved, or gaseous form can cause the cycle to slow down or even terminate unless new nitrogen is imported from an external source. A unifying concept in the study of nutrient dynamics in the sea is the 'new' versus 'regenerated' nitrogen dichotomy (**Figure 7**). New nitrogen is imported from surrounding regions (e.g., NO_3^- injection from below) or locally created (e.g., NH_4^+/organic N from N_2 fixation). Regenerated nitrogen is locally recycled (e.g., NH_4^+ from ammonification, NO_2^-/NO_3^- from nitrification, or DON from grazing or cell lysis). Under steady-state conditions, the amount of new nitrogen entering an ecosystem will determine the total amount that can be exported without the system running down.

In shallow, coastal regions runoff from land or movement upward from the sediments are potentially major sources of NH_4^+, NO_3^- and DON for water column processes. In certain regions, atmospheric deposition (both wet and dry) may also supply bioavailable nitrogen to the system. However, in most open ocean environments, new sources of

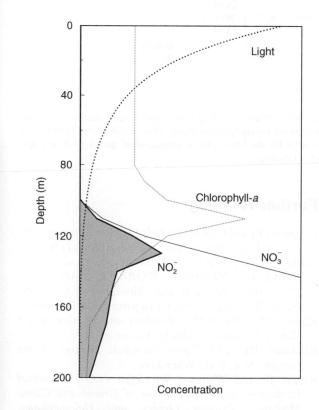

Figure 6 Schematic representation of the depth distributions of sunlight, chlorophyll-*a*, NO_2^-, and NO_3^- for a representative station in the subtropical North Pacific Ocean showing the relationship of the primary NO_2^- maximum (PNM) zone (shaded) to the other environmental variables. (Modified from J. E. Dore, Microbial nitrification in the marine euphotic zone, Ph.D. Dissertation, University of Hawaii, redrawn with permission of the author.)

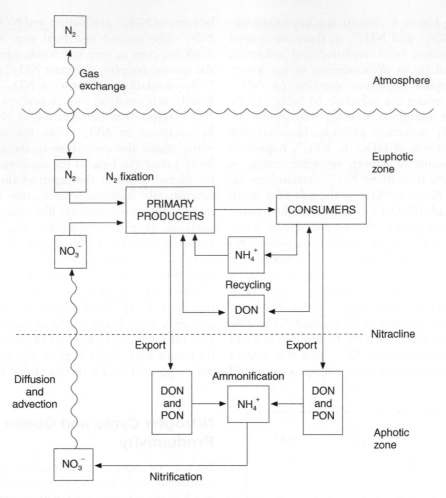

Figure 7   Schematic representation of the major pools and transformations/fluxes of nitrogen in a typical open ocean ecosystem. New sources of bioavailable N (NO_3^- and N_2 in this presentation) continuously resupply nitrogen that is lost via DON and PON export. These interactions and ocean processes form the conceptual framework for the 'new' versus 'regenerated' paradigm of nitrogen dynamics in the sea that was originally proposed by R. Dugdale and J. Goering.

nitrogen required to balance the net losses from the euphotic zone are restricted to upward diffusion or mixing of NO_3^- from deep water and to local fixation of N_2 gas. In a balanced steady state, the importation rate of new sources of bioavailable nitrogen will constrain the export of nitrogen (including fisheries production and harvesting). If all other required nutrients are available, export-rich ecosystems are those characterized by high bioavailable nitrogen loading such as coastal and open ocean upwelling regions. These are also the major regions of fish production in the sea.

See also

Atmospheric Input of Pollutants. Phosphorus Cycle. Primary Production Processes.

Further Reading

Carpenter EJ and Capone DG (eds.) (1983) *Nitrogen in the Marine Environment.* New York: Academic Press.

Conkright ME, O'Brien TD, Levitus S, *et al.* (1998) *NOAA Atlas NESDIS 37, WORLD OCEAN ATLAS 1998,* vol. II: Nutrients and Chlorophyll of the Pacific Ocean. Washington, DC: US Department of Commerce.

Harvey HW (1966) *The Chemistry and Fertility of Sea Waters.* London: Cambridge University Press.

Kirchman DL (ed.) (2000) *Microbial Ecology of the Oceans.* New York: Wiley-Liss.

Rogers JE and Whitman WB (eds.) (1991) *Microbial Production and Consumption of Greenhouse Gases: Methane, Nitrogen Oxides, and Halomethanes.* Washington, DC: American Society for Microbiology.

Schlesinger WH (1997) *Biogeochemistry: An Analysis of Global Change.* San Diego: Academic Press.

Wada E and Hattori A (1991) *Nitrogen in the Sea: Forms, Abundances, and Rate Processes.* Boca Raton, FL: CRC Press.

SINGLE COMPOUND RADIOCARBON MEASUREMENTS

T. I. Eglinton and A. Pearson, Woods Hole
Oceanographic Institution, Woods Hole, MA,
USA

Introduction

Many areas of scientific research use radiocarbon
(carbon-14, ^{14}C) measurements to determine the
age of carbon-containing materials. Radiocarbon's
~5700-year half-life means that this naturally occurring radioisotope can provide information over
decadal to millennial timescales. Radiocarbon is
uniquely suited to biogeochemical studies, where
much research is focused on carbon cycling at various spatial and temporal scales. In oceanography,
investigators use the ^{14}C concentration of dissolved
inorganic carbon (DIC) to monitor the movement of
water masses throughout the global ocean. In marine

sediment geochemistry, a major application is
the dating of total organic carbon (TOC) in order
to calculate sediment accumulation rates. Such
chronologies frequently rely on the premise that most
of the TOC derives from marine biomass production
in the overlying water column.

However, the ^{14}C content of TOC in sediments,
as well as other organic pools in the ocean (dissolved and particulate organic matter in the water
column) often does not reflect a single input source.
Multiple components with different respective ages
can contribute to these pools and can be deposited
concurrently in marine sediments (**Figure 1**). This is
particularly true on the continental margins, where
fresh vascular plant debris, soil organic matter,
and fossil carbon eroded from sedimentary rocks
can contribute a significant or even the dominant
fraction of the TOC. This material dilutes the marine input and obscures the true age of the sediment.
Although such contributions from multiple organic
carbon sources can complicate the development of

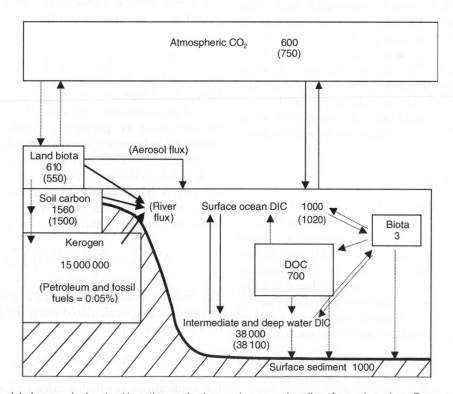

Figure 1 Major global reservoirs involved in active production, exchange and cycling of organic carbon. Reservoir sizes are shown in Gt carbon (1 GtC = 10^{15} g C). Numbers in parentheses are based on 1980s values; numbers without parentheses are estimates of the pre-anthropogenic values. Fluxes primarily mediated by biological reactions are shown with dashed arrows; physical transport processes are shown with solid arrows. (Modified after Siegenthaler and Sarmiento (1993) and Hedges and Oades (1997).)

TOC-based sediment chronologies, these sediment records hold much important information concerning the cycling of organic carbon both within and between terrestrial and marine systems. The challenge, then, is to decipher these different inputs by resolving them into their individual parts.

Most of the allochthonous, or foreign, sources represent carbon with lower ^{14}C concentrations ('older' $\Delta^{14}C$; radiocarbon ages) than the fraction of TOC originating from phytoplanktonic production. The only exception is the rapid transport and sedimentation of recently synthesized terrestrial plant material, which is in equilibrium with the ^{14}C concentration of atmospheric CO_2. Other sources of nonmarine carbon typically are of intermediate (10^3-10^4 years) or 'infinite' $\Delta^{14}C$; (beyond the detection limit of 50–60 000 years) radiocarbon age, depending on the amount of time spent in other reservoirs such as soils, fluvial deposits, or carbon-rich rocks.

It is only at the molecular level that the full extent of this isotopic heterogeneity resulting from these diverse organic carbon inputs is expressed. Isotopic analysis of individual biomarker compounds was employed originally to study the stable carbon isotope (^{13}C) distribution in lipids of geological samples. It proved to be a useful tool to describe the diversity of carbon sources and metabolic pathways as well as to link specific compounds with their biological origins. Recently, this approach was expanded into a second isotopic dimension by the development of a practical method to achieve compound-specific ^{14}C analysis. Not only do these new ^{14}C analyses of individual biomarker molecules provide a tool for dating sediments, but they are another source of fundamental information about biogeochemical processes in the marine environment.

Carbon Isotopes

Carbon in the geosphere is composed of the stable isotopes ^{12}C (98.9%) and ^{13}C (1.1%), and the cosmogenic radionucleotide, ^{14}C (radiocarbon). Upon production, ^{14}C is incorporated quickly into atmospheric CO_2, where it occurs as approximately $10^{-10}\%$ of the total atmospheric abundance of CO_2. The distribution of the minor isotopes relative to ^{12}C is governed by thermodynamic and kinetic fractionation processes[1], in addition to the radioactive decay associated with ^{14}C.

---

[1] This article assumes the reader is familiar with the conventions used for reporting stable carbon isotopic ratios, i.e., $\delta^{13}C(ppt) = 1000[R/R_{PDB}] - 1]$ where $R \equiv {}^{13}C/{}^{12}C$. For further explanation, see the additional readings listed at the end of this article.

^{14}C Systematics

Today, most radiocarbon data are obtained through the use of accelerator mass spectrometry (AMS) rather than by counting individual decay events. In particular, the advantage of AMS is its small carbon requirement (micrograms to milligrams); this ability to analyze small samples is critical to the compound-specific ^{14}C approach, where sample sizes typically range from tens to hundreds of micrograms. These sample sizes are dictated by natural concentrations of the analytes in geochemical samples (often <1 μg g^{-1} dry sediment), and by the capacity of the techniques used to isolate the individual compounds in high purity.

Raw AMS data are reported initially as fraction modern (f_m) carbon (eqn [1]).

$$f_m = \frac{R_{sn}^{14/12}}{R_{std}^{14/12}} \qquad [1]$$

$R^{14/12} \equiv {}^{14}C/{}^{12}C$ (some laboratories use $R = {}^{14}C/{}^{13}C$), sn indicates the sample has been normalized to a constant ^{13}C fractionation equivalent to $\delta^{13}C = -25$ ppt, and std is the oxalic acid I (HOxI) or II (HOxII) modern-age standard, again normalized with respect to ^{13}C.

For geochemical applications, data often are reported as $\Delta^{14}C$ values (eqn [2]).

$$\Delta^{14}C = \left[f_m \left(\frac{e^{\lambda(y-x)}}{e^{\lambda(y-1950)}} \right) - 1 \right] \times 1000 \qquad [2]$$

Here, $\lambda = 1/8267(y^{-1})(= t_{1/2}/\ln 2)$, y equals the year of measurement, and x equals the year of sample formation or deposition (applied only when known by independent dating methods, for example, by the use of ^{210}Pb). This equation standardizes all $\Delta^{14}C$ values relative to the year AD 1950. In oceanography, $\Delta^{14}C$ is a convenient parameter because it is linear and can be used in isotopic mass balance calculations of the type shown in eqn [3].

$$\Delta^{14}C_{total} = \sum_i (\chi_i \Delta^{14}C_i) \sum_i \chi_i = 1 \qquad [3]$$

The 'radiocarbon age' $\Delta^{14}C$; of a sample is defined strictly as the age calculated using the Libby half-life of 5568 years (eqn [4]).

$$Age = -8033 \, ln(f_m) \qquad [4]$$

For applications in which a calendar date is required, the calculated ages subsequently are converted are using calibration curves that account for past natural variations in the rate of formation of ^{14}C. However, the true half-life of ^{14}C is 5730 years, and this true value should be used when making decay-related corrections in geochemical systems.

^{14}C Distribution in the Geosphere

Natural Processes

Atmospheric $^{14}CO_2$ is distributed rapidly throughout the terrestrial biosphere, and living plants and their heterotrophic consumers (animals) are in equilibrium with the $\Delta^{14}C$ value of the atmosphere. Thus, in radiocarbon dating, the ^{14}C concentration of a sample is strictly an indicator of the amount of time that has passed since the death of the terrestrial primary producer. When an organism assimilates a fraction of pre-aged carbon, an appropriate 'reservoir age' $\Delta^{14}C$; must be subtracted to correct for the deviation of this material from the age of the atmosphere. Therefore, reservoir time must be considered when interpreting the ^{14}C 'ages' $\Delta^{14}C$; of all of the global organic carbon pools other than the land biota.

For example, continuous vertical mixing of the ocean provides the surface waters with some abyssal DIC that has been removed from contact with atmospheric CO_2 for up to 1500 years. This process gives the ocean an average surface water reservoir age of about 400 years ($\Delta^{14}C = -50$ ppt). A constant correction factor of 400 years often is subtracted from the radiocarbon dates of marine materials (both organic and inorganic). There are regional differences, however, and in upwelling areas the true deviation can approach 1300 years.

An example of the actual range of $\Delta^{14}C$ values found in the natural environment is shown in **Figure 2A**. This figure shows the distribution of ^{14}C in and around Santa Monica Basin, California, USA, prior to significant human influence. The basin sediments are the final burial location for organic matter derived from many of these sources, and the TOC $\Delta^{14}C$ value of -160 ppt represents a weighted average of the total organic carbon flux to the sediment surface.

Anthropogenic Perturbation

In addition to natural variations in atmospheric levels of ^{14}C, anthropogenic activity has resulted in significant fluctuations in ^{14}C content. The utilization of fossil fuels since the late nineteenth century has introduced ^{14}C- (and ^{13}C-) depleted CO_2 into the atmosphere (the 'Suess effect' $\Delta^{14}C$;). In sharp contrast to this gradual change, nuclear weapons testing in the 1950s and 1960s resulted in the rapid injection of an additional source of ^{14}C into the environment. The amount of ^{14}C in the atmosphere

nearly doubled, and the $\Delta^{14}C$ of tropospheric CO_2 increased to greater than $+900$ ppt in the early 1960s. Following the above-ground weapons test ban treaty of 1962, this value has been decreasing as the excess ^{14}C is taken up by oceanic and terrestrial sinks for CO_2. This anthropogenically derived $^{14}CO_2$ 'spike' $\Delta^{14}C$; serves as a useful tracer for the rate at which carbon moves through its global cycle. Any carbon reservoir currently having a $\Delta^{14}C$ value >0 ppt has taken up some of this 'bomb-^{14}C' $\Delta^{14}C$;. Carbon pools that exhibit no increase in $\Delta^{14}C$ over their 'pre-bomb' $\Delta^{14}C$; values have therefore been isolated from exchange with atmospheric CO_2 during the last 50 years. This contrast between 'pre-bomb' $\Delta^{14}C$; and 'post-bomb' $\Delta^{14}C$; $\Delta^{14}C$ values can serve as an excellent tracer of biogeochemical processes over short timescales. These changes in ^{14}C concentrations can be seen in the updated picture of the Santa Monica Basin regional environment shown in **Figure 2B**, where bomb-^{14}C has invaded

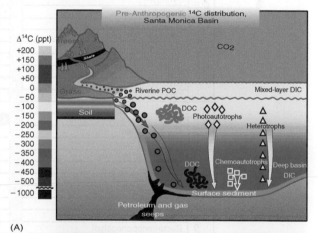

(A)

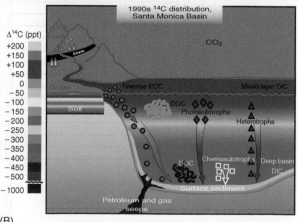

(B)

Figure 2 $\Delta^{14}C$ values for bulk carbon reservoirs in the region of Santa Monica Basin, California, USA: (A) prior to human influence ('pre-bomb' $\Delta^{14}C$;), and (B) contemporary values ('post-bomb' $\Delta^{14}C$;). Modified after Pearson (2000).)

everywhere except for the deep basin waters and older sedimentary deposits.

In general, the global distribution of organic ^{14}C is complicated by these interreservoir mixing and exchange processes. The more end-member sources contributing organic carbon to a sample, the more complicated it is to interpret a measured $\Delta^{14}C$ value, especially when trying to translate that value to chronological time. Source-specific ^{14}C dating is needed, and this requires isotopic measurements at the molecular level.

Compound-specific ^{14}C Analysis: Methods

The ability to perform natural-abundance ^{14}C measurements on individual compounds has only recently been achieved. This capability arose from refinements in the measurement of ^{14}C by AMS that allow increasingly small samples to be measured, and from methods that resolve the complex mixtures encountered in geochemical samples into their individual components. Here we describe the methods that are currently used for this purpose.

Selection of Compounds for ^{14}C Analysis

The organic matter in marine sediments consists of recognizable biochemical constituents of organisms (carbohydrates, proteins, lipids, and nucleic acids) as well as of more complex polymeric materials and nonextractable components (humic substances, kerogen). Among the recognizable biochemicals, the lipids have a diversity of structures, are comparatively easy to analyze by gas chromatographic and mass spectrometric techniques, and are resistant to degradation over time. These characteristics have resulted in a long history of organic geochemical studies aimed at identifying and understanding the

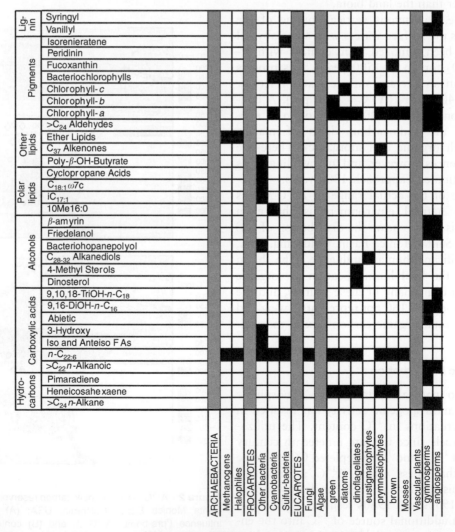

Figure 3 Common source assignments of lipid biomarkers (Modified from Hedges and Oades (1997).)

origins of 'source-specific' $\Delta^{14}C$; lipid 'biomarker' $\Delta^{14}C$; compounds. Frequently, lipids from several organic compound classes are studied within the same sample (**Figures 3** and **4**).

Although many of the most diagnostic compounds are polar lipids that are susceptible to modification during sediment diagenesis (e.g., removal of functional groups, saturation of double bonds), several retain their marker properties through the preservation of the carbon skeleton (**Figure 5**). Thus sterols (e.g., cholesterol) are transformed to sterenes and ultimately steranes. The isotopic integrity of the compound is also preserved in this way.

It is sometimes the case that families of compounds can also be characteristic of a particular source. For example, plant waxes comprise homologous series of *n*-alkanes, *n*-alkanols, and *n*-alkanoic acids (**Figure 3**). As a result, ^{14}C measurements of a compound class can yield information with similar specificity to single compound ^{14}C analysis, with the benefits of greater total analyte abundance and, potentially, simpler isolation schemes.

Compound Separation and Isolation

Procedures for single-compound ^{14}C analysis are quite involved, requiring extraction, purification, modification and isolation of the target analytes (**Figure 5**). For lipid analyses, the samples are processed by extracting whole sediment with solvents such as methylene chloride, chloroform, or methanol to obtain a total lipid extract (TLE). The TLE is then separated into compound classes using solid–liquid chromatography. The compound classes elute on the basis of polarity differences, from least polar (hydrocarbons) to most polar (free fatty acids) under normal-phase chromatographic conditions. Individual compounds for ^{14}C analysis are then isolated from these polarity fractions. Additional chromatographic steps or chemical manipulations may be included to reduce the number of components in each fraction prior to single compound isolation, or to render the compounds amenable to isolation by the method chosen. These steps may include silver nitrate-impregnated silica gel chromatography (separation of saturated from unsaturated compounds), 'molecular sieving' $\Delta^{14}C$; (e.g., urea adduction, for separation of branched/cyclic compounds from straight-chain compounds), and derivatization (for protection of functional groups, such as carboxyl or hydroxyl groups, prior to gas chromatographic separation).

For ^{14}C analysis by AMS, tens to hundreds of micrograms of each individual compound must be isolated from the sample of interest. Isolation of individual biomarkers from geochemical samples such as marine sediments and water column particulate matter requires separation techniques with high resolving power. To date, this has been most effectively achieved through the use of automated preparative capillary gas chromatography (PCGC; **Figure 6**).

A PCGC system consists of a commercial capillary gas chromatograph that is modified for work on a semipreparative, rather than analytical, scale. Modifications include a large-volume injection system; high-capacity, low-bleed 'megabore' $\Delta^{14}C$; (e.g., 60 m length × 0.53 mm inner diameter × 0.5 μm stationary phase film thickness) capillary columns; an effluent splitter; and a preparative trapping device in which isolated compounds are collected in a series of cooled U-tube traps. Approximately 1% of the effluent passes

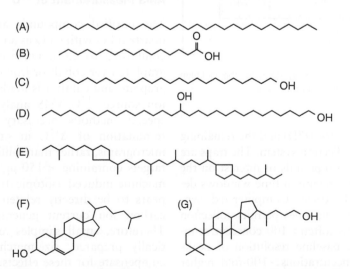

Figure 4 Selected example structures (carbon skeletons and functional groups) for biomarkers shown in Table 1: (A) *n*-C_{29} alkane; (B) *n*-$C_{16:0}$ alkanoic acid; (C) *n*-C_{24} alkanol; (D) C_{30} alkanediol; (E) $C_{40:2cy}$ isoprenoid; (F) C_{27} Δ^5-sterol (cholesterol); (G) C_{32} hopanol.

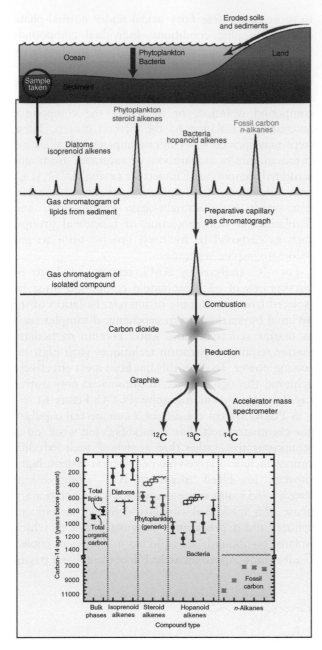

Figure 5 Schematic diagram showing steps for the isolation and ^{14}C analysis of individual sedimentary lipids.

injection, to be separated (to achieve greater resolution, typical loadings are usually about 1 µg of carbon per peak). An example of a typical PCGC separation is shown in **Figure 7**, where $\sim 40–130$ µg of individual sterols (as their acetate derivatives) were resolved and isolated from a total sterol fraction obtained from Santa Monica Basin surface sediment.

Another practical means of isolating individual components from compound mixtures is high-performance liquid chromatography (HPLC). While the resolving power of HPLC is lower, this technique is particularly suited to polar, nonvolatile, or thermally unstable analytes that are difficult to separate by GC. It also offers higher loading capacity than capillary GC.

In addition to chromatographic resolution and capacity, two additional aspects that require consideration are the potential for contamination of the analytes during the isolation procedure, and corrections for carbon associated with any derivative groups that have been appended to the molecule of interest. Regarding the former, entrainment 'bleed' $\Delta^{14}C$; of chromatographic stationary phase can result in significant carbon contamination of the isolated compound, unless steps are taken to avoid this problem (e.g., use of ultra-low bleed GC columns, removal of contaminants after the chromatographic isolation). This problem is likely to be most acute in HPLC when reversed-phase chromatographic phases are used. Comparison of yields and the $\Delta^{13}C$ compositions of the isolated compound and the CO_2 resulting from its combustion are effective means of assessing potential contamination problems.

AMS Measurement of ^{14}C

The purified compounds are sealed in evacuated quartz tubes with CuO as an oxidant. The material is combusted to CO_2, purified, and then reduced to graphite over cobalt or iron catalyst. The mixture of graphite and catalyst is loaded into a cesium sputter ion source. ^{14}C-AMS analysis is performed using special methods necessary for the accurate determination of $\Delta^{14}C$ in samples containing only micrograms, rather than milligrams, of carbon. AMS targets containing <150 µg of carbon are prone to machine-induced isotopic fractionation, which appears to be directly related to the lower levels of carbon ion current generated by these samples. Therefore, small samples are analyzed with identically prepared, size-matched small standards to compensate for these effects. The f_m values that are calculated relative to these standards no longer show a size-dependent fractionation.

to a flame ionization detector (FID) and the remaining 99% is diverted to the collection system. The traps are programmed to receive compounds of interest on the basis of chromatographic retention time windows determined from the FID trace. Computerized synchronization of the trapping times permits collection of multiple identical runs (often > 100 consecutive injections). Using PCGC, baseline resolution of peaks can be achieved at concentrations > 100-fold higher than typical analytical GC conditions, allowing up to 5 µg of carbon per chromatographic peak, per

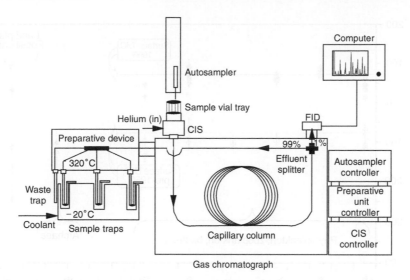

Figure 6 Diagrammatic representation of a preparative capillary gas chromatograph (PCGC) system.

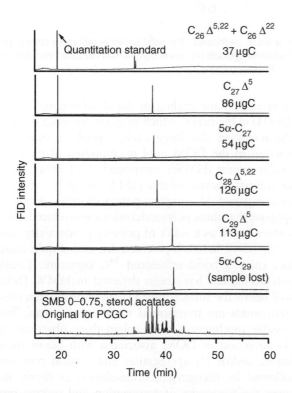

Figure 7 An example PCGC series, showing the total original mixture and the six individual, trapped compounds. In this case the analytes are sterols (as their acetate derivatives). (From Pearson (2000).)

Examples of Applications

Lipid Biomarkers in Santa Monica Basin Sediments

As one example of the application of single-compound radiocarbon analysis, we show a detailed data set for a range of lipid biomarkers extracted from marine sediments. This work focused on the upper few centimeters of a core from Santa Monica Basin. The basin has a high sedimentation rate, and its suboxic bottom waters inhibit bioturbation. As a result, laminated cores recovered from the basin depocenter allow decadal resolution of recent changes in the ^{14}C record. On the timescale of radiocarbon decay, these samples are contemporary and have no in situ ^{14}C decay. However, the $\Delta^{14}C$ values of the end-member carbon sources have changed (**Figure 2A, B**). 'Bomb-^{14}C' $\Delta^{14}C$; has invaded the modern surface ocean phytoplankton and the terrestrial biota, and through subsequent sedimentation of their organic detritus, this bomb-^{14}C is carried to the underlying sediments. The contrast between 'pre-bomb' $\Delta^{14}C$; and 'post-bomb' $\Delta^{14}C$; $\Delta^{14}C$ values, or the relative rate of bomb-^{14}C uptake, therefore is a useful tracer property. It can help distinguish biogeochemical processes that transfer carbon within years or decades (source-specific lipids that now contain bomb-^{14}C) from biogeochemical processes that do not exchange with atmospheric CO_2 on a short timescale (lipids that remain free from bomb-^{14}C).

Compound-specific $\Delta^{14}C$ values for 31 different lipid biomarker molecules are shown in **Figure 8** for sedimentary horizons corresponding to pre-bomb (before AD 1950) and post-bomb (1950–1996) eras. These organic compounds represent phytoplanktonic, zooplanktonic, bacterial, archaeal, terrestrial higher plant, and fossil carbon sources. The lipid classes include long-chain *n*-alkanes, alkanoic (fatty) acids, *n*-alcohols, C_{30} mid-chain ketols and diols, sterols, hopanols, and C_{40} isoprenoid side chains of the ether-linked glycerols of the *Archaea*.

The data show that the carbon source for the majority of the analyzed biomarkers is marine euphotic

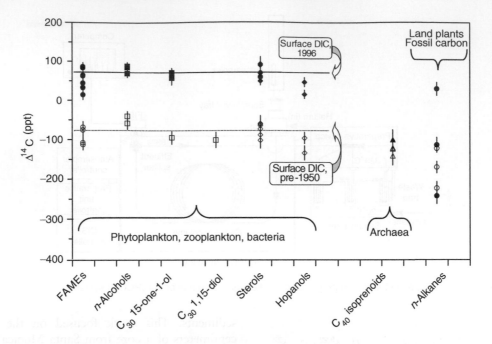

Figure 8 $\Delta^{14}C$ data for individual lipids extracted from Santa Monica Basin sediments. The solid symbols represent compounds extracted from the post-bomb sedimentary horizon (AD 1950–1996). The hollow symbols represent compounds extracted from the pre-bomb sedimentary horizon (deposited prior to AD 1950). (Modified after Pearson (2000).)

zone production. Most of the lipids from 'pre-bomb' $\Delta^{14}C$; sediments have $\Delta^{14}C$ values equal to the $\Delta^{14}C$ of surface water DIC at this time (dotted line), while most of the lipids from 'post-bomb' $\Delta^{14}C$; sediments have $\Delta^{14}C$ values equal to the $\Delta^{14}C$ of present-day surface water DIC (solid line).

However, it is clear that two of the lipid classes do not reflect carbon originally fixed by marine photoautotrophs. These are the n-alkanes, for which the $\Delta^{14}C$ data are consistent with mixed fossil and contemporary terrestrial higher plant sources, and the archaeal isoprenoids, for which the $\Delta^{14}C$ data are consistent with chemoautotrophic growth below the euphotic zone. This is just one example of the way in which compound-specific ^{14}C analysis can distinguish carbon sources and biogeochemical processes simultaneously. The large number of compounds that appear to record the $\Delta^{14}C$ of surface water DIC, and therefore marine primary production, points to the potential for numerous tracers of marine biomass; these are the target compounds of interest when developing refined sediment chronologies. In particular, the sterols appear to be particularly effective tracers of surface ocean DIC, and hence suitable for this purpose.

Monosaccharides in Oceanic High-molecular-weight Dissolved Organic Matter

The second example illustrates the utility of single compound, as well as compound class, ^{14}C measurements as ocean process tracers. In this case, the process

of interest is the cycling of dissolved organic matter (DOM) in the ocean. Much progress has been made in characterizing this large carbon pool. A significant fraction of the DOM pool is composed of high-molecular-weight (HMW) compounds (>1 kDa), and a substantial fraction of this HMW DOM is known to be comprised of complex polysaccharides. Evidence suggests that these polysaccharides are produced in the surface ocean as a result of primary productivity, and/or attendant heterotrophic activity, and should therefore carry a bomb-influenced ^{14}C signature. Similar polysaccharides have been detected in HMW DOM well below the surface mixed layer, implying that these compounds are transported to the deep ocean. Two possible mechanisms can explain these observations: (1) advection of DOM associated with ocean circulation, and/or (2) aggregation and vertical transport followed by disaggregation/dissolution at depth. Because the timescales of aggregation and sinking processes are short relative to deep water formation and advective transport, ^{14}C measurements on polysaccharides in HMW DOM provide means of determining which mechanism is dominant.

Figure 9 shows vertical ^{14}C profiles for DIC and DOM as well as ^{14}C results for selected samples of sinking and suspended particulate organic matter (POM), HMW DOM, and monosaccharides isolated from selected depths at a station in the North-east Pacific Ocean. Individual monosaccharides were obtained by hydrolysis of HMW DOM, and purified and

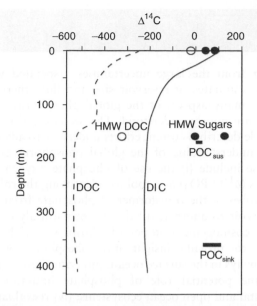

Figure 9 $\Delta^{14}C$ values (ppt) for different fractions of carbon in the North-East Pacific: solid bars labeled POC_{sus} and POC_{sink} correspond to suspended and sinking POC, respectively; solid and dashed lines show depth profiles for DIC and DOC, respectively; open circles are total HMW DOC and closed circles are individual sugars (monosaccharides) isolated from the HMW DOC fractions. (Modified from Aluwihare (1999).)

isolated by HPLC. The similarity of $\Delta^{14}C$ values of individual monosaccharides implies that they derive from a common polysaccharide source. Such similarities in ^{14}C lend support to the utility of ^{14}C measurements at the compound class level. Furthermore, the similarity between $\Delta^{14}C$ values of these compounds and surface ocean DIC indicates that they are either directly or indirectly the products of marine photoautotrophy. The deep ocean (1600 m) data shows the presence of bomb-radiocarbon in the monosaccharides. Their enrichment in ^{14}C relative to DIC, and similarity to suspended POC at the same depth, suggests that this component of HMW DOM is injected into the deep ocean by vertical transport as particles.

Summary

The ability to perform single-compound ^{14}C measurements has only recently been realized, and as a consequence its application as a tracer in ocean sciences remains in its infancy. The above examples highlight potential applications of single-compound ^{14}C measurements as tools for understanding the biogeochemical cycling of organic matter in the ocean. There are several other areas of study where this approach holds great promise. For example, ^{14}C measurements of vascular plant biomarkers (e.g., plant

waxes, lignin-derived phenols) in continental shelf sediments provide constraints on the timescales over which terrestrial organic matter is delivered to the ocean. The 'infinite ^{14}C age' $\Delta^{14}C$; signature that polycyclic aromatic hydrocarbons and other fossil fuel-derived contaminants carry provides an effective means of tracing their inputs to the coastal ocean relative to contributions from natural processes (e.g., biomass burning). As methods are streamlined, it is anticipated that single compound ^{14}C measurements will find increasing application in marine biogeochemistry.

See also

Ocean Carbon System, Modeling of.

Further Reading

Aluwihare LI (1999) *High Molecular Weight (HMW) Dissolved Organic Matter (DOM) in Seawater: Chemical Structure, Sources and Cycling*. PhD thesis, Massachusetts Institute of Technology/Woods Hole Oceanographic Institution.

Eglinton TI, Aluwihare LI, Bauer JE, Druffel ERM, and McNichol AP (1996) Gas chromatographic isolation of individual compounds from complex matrices for radiocarbon dating. *Analytical Chemistry* 68: 904–912.

Eglinton TI, Benitez-Nelson BC, Pearson A, *et al.* (1997) Variability in radiocarbon ages of individual organic compounds from marine sediments. *Science* 277: 796–799.

Faure G (1986) *Principles of Isotope Geology*. New York: Wiley.

Hayes JM (1993) Factors controlling the ^{13}C content of sedimentary organic compounds: principles and evidence. *Marine Geology* 113: 111–125.

Hedges JI (1992) Global biogeochemical cycles: progress and problems. *Marine Chemistry* 39: 67–93.

Hedges JI and Oades JM (1997) Comparative organic geochemistries of soils and marine sediments. *Organic Geochemistry* 27: 319–361.

Hoefs J (1980) *Stable Isotope Geochemistry*. New York: Springer-Verlag.

Pearson A (2000) *Biogeochemical Applications of Compound-Specific Radiocarbon Analysis*. PhD thesis, Massachusetts Institute of Technology/Woods Hole Oceanographic Institution.

Siegenthaler U and Sarmiento JL (1993) Atmospheric carbon dioxide and the ocean. *Nature* 365: 119–125.

Tuniz C, Bird JR, Fink D, and Herzog GF (1998) *Accelerator Mass Spectrometry: Ultrasensitive Analysis for Global Science*. Boca Raton, FL: CRC Press.

Volkman JK, Barrett SM, Blackburn SI et al. Microalgal biomarkers: a review of recent research developments. Organic Geochemistry 29: 1163–1179.

PHOSPHORUS CYCLE

K. C. Ruttenberg, Woods Hole Oceanographic
Institution, Woods Hole, MA, USA

Introduction

The global phosphorus cycle has four major components: (i) tectonic uplift and exposure of phosphorus-bearing rocks to the forces of weathering; (ii) physical erosion and chemical weathering of rocks producing soils and providing dissolved and particulate phosphorus to rivers; (iii) riverine transport of phosphorus to lakes and the ocean; and (iv) sedimentation of phosphorus associated with organic and mineral matter and burial in sediments (**Figure 1**). The cycle begins anew with uplift of sediments into the weathering regime.

Phosphorus is an essential nutrient for all life forms. It is a key player in fundamental biochemical reactions involving genetic material (DNA, RNA) and energy transfer (adenosine triphosphate, ATP), and in structural support of organisms provided by membranes (phospholipids) and bone (the biomineral hydroxyapatite). Photosynthetic organisms utilize dissolved phosphorus, carbon, and other essential nutrients to build their tissues using energy from the sun. Biological productivity is contingent upon the availability of phosphorus to these organisms, which constitute the base of the food chain in both terrestrial and aquatic systems.

Phosphorus locked up in bedrock, soils, and sediments is not directly available to organisms. Conversion of unavailable forms to dissolved orthophosphate, which can be directly assimilated, occurs through geochemical and biochemical reactions at various stages in the global phosphorus cycle. Production of biomass fueled by phosphorus bioavailability results in the deposition of organic matter in soil and sediments, where it acts as a source of fuel and nutrients to microbial communities. Microbial activity in soils and sediments, in turn, strongly influences the concentration and chemical form of phosphorus incorporated into the geological record.

This article begins with a brief overview of the various components of the global phosphorus cycle. Estimates of the mass of important phosphorus reservoirs, transport rates (fluxes) between reservoirs, and residence times are given in **Tables 1** and **2**. As is

clear from the large uncertainties associated with these estimates of reservoir size and flux, there remain many aspects of the global phosphorus cycle that are poorly understood. The second half of the article describes current efforts underway to advance our understanding of the global phosphorus cycle. These include (i) the use of phosphate oxygen isotopes (δ^{18}O-PO$_4$) as a tool for identifying the role of microbes in the transformer of phosphate from one reservoir to another; (ii) the use of naturally occurring cosmogenic isotopes of phosphorus (^{32}P and ^{33}P) to provide insight into phosphorus-cycling pathways in the surface ocean; (iii) critical evaluation of the potential role of phosphate limitation in coastal and open ocean ecosystems; (iv) reevaluation of the oceanic residence time of phosphorus; and (v) rethinking the global phosphorus-cycle on geological timescales, with implications for atmospheric oxygen and phosphorus limitation primary productivity in the ocean.

The Global Phosphorus Cycle: Overview

The Terrestrial Phosphorus Cycle

In terrestrial systems, phosphorus resides in three pools: bedrock, soil, and living organisms (biomass) (**Table 1**). Weathering of continental bedrock is the principal source of phosphorus to the soils that support continental vegetation (F_{12}); atmospheric deposition is relatively unimportant (F_{82}). Phosphorus is weathered from bedrock by dissolution of phosphorus-bearing minerals such as apatite (Ca$_{10}$(PO$_4$)$_6$(OH, F, Cl)$_2$), the most abundant primary phosphorus mineral in crustal rocks. Weathering reactions are driven by exposure of minerals to naturally occurring acids derived mainly from microbial activity. Phosphate solubilized during weathering is available for uptake by terrestrial plants, and is returned to the soil by decay of litterfall (**Figure 1**).

Soil solution phosphate concentrations are maintained at low levels as a result of absorption of phosphorus by various soil constituents, particularly ferric iron and aluminum oxyhydroxides. Sorption is considered the most important process controlling terrestrial phosphorus bioavailability. Plants have different physiological strategies for obtaining phosphorus despite low soil solution concentrations. For example, some plants can increase root volume

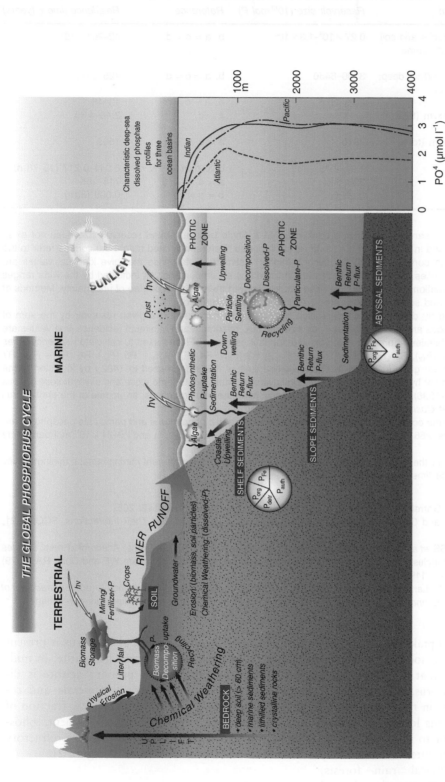

Figure 1 Cartoon illustrating the major reservoirs and fluxes of phosphorus described in the text and summarized in Tables **Table 1** and **2**. The oceanic photic zone, idealized in the cartoon, is typically thinner in coastal environments owing to turbidity from continental terrigenous input, and deepens as the water column clarifies with distance away from the continental margins. The distribution of phosphorus among different chemical/mineral forms in marine sediments is given in the pie diagrams, where the abbreviations used are: P_{org}, organic phosphorus; P_{Fe}, iron-bound phosphorus; P_{detr}, detrital apatite; P_{auth}, authigenic/biogenic apatite. The P_{org}, P_{Fe}, and P_{auth} reservoirs represent potentially reactive phosphorus pools (see text and **Tables 2** and **5** for discussion), whereas the P_{detr} pool reflects mainly detrital apatite weathered off the continents and passively deposited in marine sediments (note that P_{detr} is not an important sedimentary phosphorus component in abyssal sediments, far from continents). Continental margin phosphorus speciation data were compiled from Louchouarn P, Lucotte M, Duchemin E and de Vernal A (1997) Early diagenetic processes in recent sediments of the Gulf of St-Lawrence: Phosphorus, carbon and iron burial rates. *Marine Geology* 139(1/4): 181–200, and Ruttenberg KC and Berner RA (1993) Authigenic apatite formation and burial in sediments from non-upwelling continental margin environments. *Geochimica et Cosmochimica Acta* 57: 991–1007. Abyssal sediment phosphorus speciation data were compiled from Filippelli GM and Delaney ML (1996) Phosphorus geochemistry of equatorial Pacific sediments. *Geochimica et Cosmochimica Acta* 60: 1479)1495, and Ruttenberg KC (1990) *Diagenesis and burial of phosphorus in marine sediments: implications for the marine phosphorus budget.* PhD thesis, Yale University. The global phosphorus cycle cartoon is from Ruttenberg (2000). The global phosphorus cycle cartoon is shown in the panel to the right of the global phosphorus cycle cartoon, and are from Sverdrup HV, Johnson MW and Fleming RH (1942) *The Oceans, Their Physics, Chemistry and General Biology.* New York: Prentice Hall; used with permission. The vertical water column distributions of phosphate typically observed in the three ocean basins are shown in the panel to the right of the global phosphorus cycle cartoon, and are from Sverdrup HV, Johnson MW and Fleming RH (1942) *The Oceans, Their Physics, Chemistry and General Biology.* New York: Prentice Hall, © 1942 Prentice Hall; used with permission.

Table 1 Major reservoirs active in the global phosphorus cycle and associated residence times

Reservoir no.	Reservoir description	Reservoir size (10^{12} mol P)	Reference	Residence time τ (years)
R1	Sediments (crustal rocks and soil > 60 cm deep and marine sediments)	0.27×10^8–1.3×10^8	b, a = c = d	42–201×10^6
R2	Land ($\approx$ total soil < 60 cm deep: organic + inorganic)	3100–6450	b, a = c = d	425–2311
R3	Land biota	83.9–96.8	b, a = c = d	13–48
R4	Surface ocean, 0–300 m (total dissolved P)	87.4	a = c	2.46–4.39
R5	Deep sea, 300–3300 m (total dissolved P)	2810	a = c & d	1502
R6	Oceanic biota	1.61–4.45	b & d, a = c & d	0.044–0.217 (16–78 d)
R7	Minable P	323–645	a = c, b & d	718–1654
R8	Atmospheric P	0.0009	b = c = d	0.009 (80 h)

[a]*Notes*

(1) Ranges are reported for those reservoirs for which a consensus on a single best estimated reservoir size does not exist. Maximum and minimum estimates found in a survey of the literature are reported. References cited before the comma refer to the first (lowest) estimate, those after the comma refer to the second (higher) estimate. References that give identical values are designated by an equality sign, references giving similar values are indicated by an ampersand. As indicated by the wide ranges reported for some reservoirs, all calculations of reservoir size have associated with them a large degree of uncertainty. Methods of calculation, underlying assumptions, and sources of error are given in the references cited.

(2) Residence times are calculated by dividing the concentration of phosphorus contained in a given reservoir by the sum of fluxes out of the reservoir. Where ranges are reported for reservoir size and flux, maximum and minimum residence time values are given; these ranges reflect the uncertainties inherent in reservoir size and flux estimates. Fluxes used to calculate residence times for each reservoir are as follows: R1 (F_{12}), R2 ($F_{23} + F_{28} + F_{24(d)} + F_{24(p)}$), R3 ($F_{32}$), R4 ($F_{45} + F_{46}$), R5 ($F_{54}$), R6 ($F_{64} + F_{65}$), R7 ($F_{72}$), R8 ($F_{82} + F_{84}$). Flux estimates are given in **Table 2**. The residence time of R5 is decreased to 1492 y by inclusion of the scavenged flux of deep-sea phosphate at hydrothermal mid-ocean ridge systems, mostly onto ferric oxide and oxyhydroxide phases (Wheat CG, Feely RA and Mottl MJ (1996). Phosphate removal by oceanic hydrothermal processes: an update of the phosphorus budget in the oceans. *Geochimica et Cosmochimica Acta* 60(19): 3593–3608).

(3) Estimates for the partitioning of the oceanic reservoir between dissolved inorganic phosphorus and particulate phosphorus are given in references b and d as follows: 2581–2600×10^{12} mol dissolved inorganic phosphorus (b, d) and 20–21×10^{12} mol particulate phosphorus (d, b).

(4) The residence times estimated for the minable phosphorus reservoir reflect estimates of current mining rates; if mining activity increases or diminishes the residence time will change accordingly.

References

(a) Lerman A, Mackenzie FT and Garrels RM (1975) *Geological Society of America Memoir* 142: 205.

(b) Richey JE (1983). In: Bolin B and Cook RB (eds) *The Major Biogeochemical Cycles and Their Interactions*, SCOPE 21, pp. 51–56. Chichester: Wiley.

(c) Jahnke RA (1992) In: Butcher SS *et al.* (eds) *Global Geochemical Cycles*, pp. 301–315. San Diego: Academic Press. (Values identical to Lerman *et al.*, except the inclusion of the atmospheric reservoir estimate taken from Graham WF and Duce RA (1979) *Geochimica et Cosmochimica Acta* 43: 1195.)

(d) Mackenzie FT, Ver LM, Sabine C, Lane M and Lerman A (1993) In: Wollast R, Mackenzie FT and Chou L (eds) *Interactions of C, N, P and S Biogeochemical Cycles and Global Change*. NATO ASI Series 1, vol. 4, pp. 1–61. Berlin: Springer-Verlag.

and surface area to optimize uptake potential. Alternatively, plant roots and/or associated fungi can produce chelating compounds that solubilize ferric iron and calcium-bound phosphorus, enzymes and/or acids that solubilize phosphate in the root vicinity. Plants also minimize phosphorus loss by resorbing much of their phosphorus prior to litterfall, and by efficient recycling from fallen litter. In extremely unfertile soils (e.g., in tropical rain forests) phosphorus recycling is so efficient that topsoil contains virtually no phosphorus; it is all tied up in biomass.

Systematic changes in the total amount and chemical form of phosphorus occur during soil development. In initial stages, phosphorus is present mainly as primary minerals such as apatite. In mid-stage soils, the reservoir of primary apatite is diminished; less-soluble secondary minerals and organic phosphorus make up an increasing fraction of soil phosphorus. Late in soil development, phosphorus is partitioned mainly between refractory minerals and organic phosphorus (**Figure 2**).

Transport of Phosphorus from Continents to the Ocean

Phosphorus is transferred from the continental to the oceanic reservoir primarily by rivers (F_{24}).

Table 2 Fluxes between the major phosphorus reservoirs[a]

Flux no.	Description of flux	Flux (10^{12} mol P y^{-1})	References and comments
Reservoir fluxes			
F_{12}	Rocks/sediments → soils (erosion/weathering, soil accumulation)	0.645	a = c & d
F_{21}	Soils → rocks/sediments (deep burial, lithification)	0.301–0.603	d, a = c
F_{23}	Soils → land biota	2.03–6.45	a = c, b & d
F_{32}	Land biota → soils	2.03–6.45	a = c, b & d
$F_{24(d)}$	Soil → surface ocean (river total dissolved P flux)	0.032–0.058	e, a = c; ∼ >50% of TDP is DOP (e)
$F_{24(p)}$	Soil → surface ocean (river particulate P flux)	0.59–0.65	d, e; ∼ 40% of RSPM-P (Riverine Suspended Particulate Matter-Phosphorus) is organic P (e); it is estimated that between 25–45% is reactive once it enters the ocean (f).
F_{46}	Surface ocean → oceanic biota	19.35–35	b, d; a = c = 33.5, b reports upper limit of 32.3; d reports lower limit of 28.2
F_{64}	Oceanic biota → surface ocean	19.35–35	b, d; a & c = 32.2, b reports upper limit of 32.3, d reports lower limit of 28.2
F_{65}	Oceanic biota → deep sea (particulate rain)	1.13–1.35	d, a = c
F_{45}	Surface ocean → deep sea (downwelling)	0.581	a = c
F_{54}	Deep sea → surface ocean (upwelling)	1.87	a = c
F_{42}	Surface ocean → land (fisheries)	0.01	d
F_{72}	Minable P → land (soil)	0.39–0.45	a = c = d, b
F_{28}	Land (soil) → atmosphere	0.14	b = c = d
F_{82}	Atmosphere → land (soil)	0.1	b = c = d
F_{48}	Surface ocean → atmosphere	0.01	b = c = d
F_{84}	Atmosphere → surface ocean	0.02–0.05	c, b; d gives 0.04; ∼ 30% of atmospheric aerosol P is soluble (g)
Subreservoir fluxes: marine sediments			
sF_{ms}	Marine sediment accumulation (total)	0.265–0.280	i, j; for higher estimate (j), use of sediment P concentration below the diagenesis zone implicitly accounts for P loss via benthic remineralization flux and yields pre-anthropogenic net burial flux. For estimates of reactive P burial see note (j).
sF_{cs}	Continental margin ocean sediments → burial	0.150–0.223	j, i; values reported reflect total P, reactive P burial constitutes from 40–75% of total P (h). These values reflect pre-agricultural fluxes, modern value estimated as 0.33 (d).
sF_{as}	Abyssal (deep sea) sediments → burial	0.042–0.130	i, j; a = c gives a value of 0.055. It is estimated that 90–100% of this flux is reactive P (h). These values reflect pre-agricultural fluxes, modern value estimates range from 0.32 (d) to 0.419 (b).

(Continued)

Table 2 *Continued*

Flux no.	Description of flux	Flux (10^{12} mol P y^{-1})	References and comments
sF_{cbf}	Coastal sediments → coastal waters (remineralization, benthic flux)	0.51–0.84	d, k; these values reflect pre-agricultural fluxes, modern value estimated as 1.21 with uncertainties ± 40% (k)
sF_{abf}	Abyssal sediments → deep sea (remineralization, benthic flux)	0.41	k; this value reflects pre-agricultural fluxes, modern value estimated as 0.52, uncertainty ± 30% (k)

a *Notes*

(1) Reservoir fluxes (*F*) represent the P-flux between reservoirs #R1–R8 defined in **Table 1**. The subreservoir fluxes (s*F*) refer to the flux of phosphorus into the marine sediment portion of reservoir #1 via sediment burial, and the flux of diagenetically mobilized phosphorus out of marine sediments via benthic return flux. These subfluxes have been calculated as described in references h–k. Note that the large magnitude of these sub-fluxes relative to those into and out of reservoir #1 as a whole, and the short oceanic-phosphorus residence time they imply (**Tables 1** and **5**), highlight the dynamic nature of the marine phosphorus cycle.

(2) Ranges are reported where consensus on a single best estimate does not exist. References cited before the comma refer to the first (lowest) estimate, those after the comma refer to the second (higher) estimate. References that give identical values are designated by an equality sign, references giving similar values are indicated by an ampersand. Maximum and minimum estimates found in a survey of the literature are reported. In some cases this range subsumes ranges reported in the primary references. As indicated by the wide ranges reported, all flux calculations have associated with them a large degree of uncertainty. Methods of calculation, underlying assumptions, and sources of error are given in the references cited.

References

(a) Lerman A, Mackenzie FT and Garrels RM (1975) *Geological Society of America Memoir* 142: 205.

(b) Richey JE (1983). In: Bolin B and Cook RB (eds), *The Major Biogeochemical Cycles and Their Interactions*, SCOPE 21, pp. 51–56. Chichester: Wiley.

(c) Jahnke RA (1992) In: Butcher SS *et al.* (eds), *Global Geochemical Cycles*, pp. 301–315. San Diego: Academic Press. (Values are identical to those found in Lerman *et al.* (1975) except for atmospheric phosphorus fluxes taken from Graham WF and Duce RA (1979) *Geochimica et Cosmochimica Acta* 43: 1195.)

(d) Mackenzie FT, Ver LM, Sabine C, Lane M and Lerman A (1993) In: Wollast R, Mackenzie FT and Chou L (eds), *Interactions of C, N, P and S Biogeochemical Cycles and Global Change*, NATO ASI Series 1, vol. 4, pp. 1–61. Berlin: Springer-Verlag.

(e) Meybeck M. (1982). *American Journal of Science* 282(4): 401.

(f) The range of riverine suspended particulate matter that may be solubilized once it enters the marine realm (e.g. so-called 'reactive phosphorus') is derived from three sources. Colman AS and Holland HD ((2000). In: Glenn, C.R., Prévôt-Lucas L and Lucas J (eds) *Marine Authigenesis: From Global to Microbial*, SEPM Special Publication No. 66. pp. 53–75) estimate that 45% may be reactive, based on RSPM-P compositional data from a number of rivers and estimated burial efficiency of this material in marine sediments. Berner RA and Rao J-L ((1994) *Geochimica et Cosmochimica Acta* 58: 2333) and Ruttenberg KC and Canfield DE ((1994) *EOS, Transactions of the American Geophysical Union* 75: 110) estimate that 35% and 31% of RSPM-P is released upon entering the ocean, based on comparison of RSPM-P and adjacent deltaic surface sediment phosphorus in the Amazon and Mississippi systems, respectively. Lower estimates have been published: 8% (Ramirez AJ and Rose AW (1992) *American Journal of Science* 292: 421); 18% (Froelich PN (1988) *Limnology and Oceanography* 33: 649); 18%: (Compton J, Mallinson D, Glenn CR *et al.* (2000) In: Glenn CR, Prévôt-Lucas L and Lucas J (eds) *Marine Authigenesis: From Global to Microbial*, SEPM Special Publication No. 66, pp. 21–33). Higher estimates have also been published: 69% (Howarth RW, Jensen HS, Marine R and Postma H (1995) In: Tiessen H (ed) *Phosphorus in the Global Environment*, SCOPE 54, pp. 323–345. Chichester: Wiley). Howarth *et al.* (1995) also estimate the total flux of riverine particulate P to the oceans at 0.23×10^{12} moles P y^{-1}, an estimate likely too low because it uses the suspended sediment flux from Milliman JD and Meade RH ((1983) Journal of Geology 91: 1), which does not include the high sediment flux rivers from tropical mountainous terranes (Milliman JD and Syvitski JPM (1992) *Journal of Geology* 100: 525).

(g) Duce RA, Liss PS, Merrill JT *et al.* (1991) *Global Biogeochemical Cycles* 5: 193.

(h) Ruttenberg KC (1993) *Chemical Geology* 107: 405.

(i) Howarth RW, Jensen HS, Marino R and Postma H (1995) In: Tiessen H (ed.) *Phosphorus in the Global Environment*, SCOPE 54, pp. 323–345, Chichester: Wiley.

(j) Phosphorus-burial flux estimates as reported in Ruttenberg ((1993) *Chemical Geology* 107: 405) modified using pre-agricultural sediment fluxes updated by Colman AS and Holland HD ((2000) In: Glenn CR, Prévôt-Lucas L and Lucas J (eds) *Marine Authigenesis: From Global to Microbial*, SEPM Special Publication No. 66, pp. 53–75). Using these total phosphorus burial fluxes and the ranges of likely reactive phosphorus given in the table, the best estimate for reactive phosphorus burial flux in the oceans lies between 0.177 and 0.242×10^{12} moles P y^{-1}. Other estimates of whole-ocean reactive phosphorus burial fluxes range from, at the low end, 0.032–0.081×10^{12} moles P y^{-1} (Compton J, Mallinson D, Glenn CR *et al.* (2000) In: Glenn CR, Prévôt-Lucas L and Lucas J (eds) *Marine Authigenesis: From Global to Microbial*, SEPM Special Publication No. 66, pp. 21–33), and 0.09×10^{12} moles P y^{-1} (Wheat CG, Feely RA and Mottl MJ (1996) *Geochimica et Cosmochimica Acta* 60: 3593); to values more comparable to those derived from the table above (0.21×10^{12} moles P y^{-1}: Filippelli GM and Delaney ML (1996) *Geochimica Cosmochimica Acta* 60: 1479)

(k) Colman AS and Holland HD (2000) In: Glenn CR, Prévôt-Lucas L and Lucas J (eds) *Marine Authigenesis: From Global to Microbial*, SEPM Special Publication No. 66, pp. 53–75.

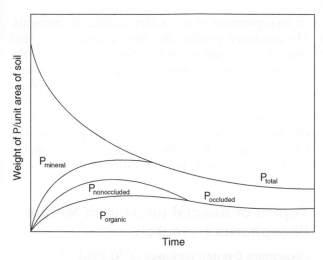

Figure 2 The fate of phosphorus during soil formation can be viewed as the progressive dissolution of primary mineral P (dominantly apatite), some of which is lost from the system by leaching (decrease in P_{total}), and some of which is reincorporated into nonoccluded, occluded and organic fractions within the soil. Nonoccluded P is defined as phosphate sorbed to surface of hydrous oxides of iron and aluminum, and calcium carbonate. Occluded P refers to P present within the mineral matrix of discrete mineral phases. The initial build-up in organic P results from organic matter return to soil from vegetation supported by the soil. The subsequent decline in $P_{organic}$ is due to progressive mineralization and soil leaching. The time-scale over which these transformations occur depends upon the initial soil composition, topographic, and climatic factors. Figure is after Walker and Syers (1976). The fate of phosphorus during pedogenesis. *Geoderma* 15: 1–19, with permission.

Deposition of atmospheric aerosols (F_{84}) is a minor flux. Groundwater seepage to the coastal ocean is a potentially important but undocumented flux.

Riverine phosphorus derives from weathered continental rocks and soils. Because phosphorus is particle-reactive, most riverine phosphorus is associated with particulate matter. By most estimates, over 90% of the phosphorus delivered by rivers to the ocean is as particulate phosphorus ($F_{24(p)}$). Dissolved phosphorus in rivers occurs in both inorganic and organic forms. The scant data on dissolved organic phosphorus suggest that it may account for 50% or more of dissolved riverine phosphorus. The chemical form of phosphorus associated with riverine particles is variable and depends upon the drainage basin geology, on the extent of weathering of the substrate, and on the nature of the river itself. Available data suggest that approximately 20–40% of phosphorus in suspended particulate matter is organic. Inorganic forms are partitioned mainly between ferric oxyhydroxides and apatite. Aluminum oxyhydroxides and clays may also be significant carriers of phosphorus.

The fate of phosphorus entering the ocean via rivers is variable. Dissolved phosphorus in estuaries at the continent–ocean interface typically displays nonconservative behavior. Both negative and positive deviations from conservative mixing can occur, sometimes changing seasonally within the same estuary. Net removal of phosphorus in estuaries is typically driven by flocculation of humic-iron complexes and biological uptake. Net phosphorus release is due to a combination of desorption from fresh water particles entering high-ionic-strength marine waters, and flux of diagenetically mobilized phosphorus from benthic sediments. Accurate estimates of bioavailable riverine phosphorus flux to the ocean must take into account, in addition to dissolved forms, the fraction of riverine particulate phosphorus released to solution upon entering the ocean.

Human impacts on the global phosphorus cycle The mining of phosphate rock (mostly from marine phosphorite deposits) for use as agricultural fertilizer (F_{72}) increased dramatically in the latter half of the twentieth century. In addition to fertilizer use, deforestation, increased cultivation, and urban and industrial waste disposal all have enhanced phosphorus transport from terrestrial to aquatic systems, often with deleterious results. For example, elevated phosphorus concentrations in rivers resulting from these activities have resulted in eutrophication in some lakes and coastal areas, stimulating nuisance algal blooms and promoting hypoxic or anoxic conditions that are harmful or lethal to natural populations.

Increased erosion due to forest clear-cutting and widespread cultivation has increased riverine suspended matter concentrations, and thus increased the riverine particulate phosphorus flux. Dams, in contrast, decrease sediment loads in rivers and therefore diminish phosphorus-flux to the oceans. However, increased erosion below dams and diagenetic mobilization of phosphorus in sediments trapped behind dams moderates this effect. The overall effect has been a 50–300% increase in riverine phosphorus flux to the oceans above pre-agricultural levels.

The Marine Phosphorus Cycle

Phosphorus in its simplest form, dissolved orthophosphate, is taken up by photosynthetic organisms at the base of the marine food web. When phosphate is exhausted, organisms may utilize more complex forms by converting them to orthophosphate via enzymatic and microbiological reactions. In the open ocean most phosphorus associated with biogenic particles is recycled within the upper water column.

Efficient stripping of phosphate from surface waters by photosynthesis combined with build-up at depth due to respiration of biogenic particles results in the classic oceanic dissolved nutrient profile. The progressive accumulation of respiration-derived phosphate at depth along the deep-water circulation trajectory results in higher phosphate concentrations in Pacific Ocean deep waters at the end of the trajectory than in the North Atlantic where deep water originates (**Figure 1**).

The sole means of phosphorus removal from the oceans is burial with marine sediments. The phosphorus flux to shelf and slope sediments is larger than the phosphorus flux to the deep sea (**Table 2**) for several reasons. Coastal waters receive continentally derived nutrients via rivers (including phosphorus, nitrogen, silicon, and iron), which stimulate high rates of primary productivity relative to the deep sea and result in a higher flux of organic matter to continental margin sediments. Organic matter is an important, perhaps primary, carrier of phosphorus to marine sediments. Owing to the shorter water column in coastal waters, less respiration occurs prior to deposition. The larger flux of marine organic phosphorus to margin sediments is accompanied by a larger direct terrigenous flux of particulate phosphorus (organic and inorganic), and higher sedimentation rates overall. These factors combine to enhance retention of sedimentary phosphorus. During high sea level stands, the sedimentary phosphorus reservoir on continental margins expands, increasing the phosphorus removal flux and therefore shortening the oceanic phosphorus residence time.

Terrigenous-dominated shelf and slope (hemipelagic) sediments and abyssal (pelagic) sediments have distinct phosphorus distributions. While both are dominated by authigenic Ca-P (mostly carbonate fluorapatite), this reservoir is more important in pelagic sediments. The remaining phosphorus in hemipelagic sediments is partitioned between ferric iron-bound phosphorus (mostly oxyhydroxides), detrital apatite, and organic phosphorus; in pelagic sediments detrital apatite is unimportant. Certain coastal environments characterized by extremely high, upwelling-driven biological productivity and low terrigenous input are enriched in authigenic apatite; these are proto-phosphorite deposits. A unique process contributing to the pelagic sedimentary Fe-P reservoir is sorptive removal of phosphate onto ferric oxyhydroxides in mid-ocean ridge hydrothermal systems.

Mobilization of sedimentary phosphorus by microbial activity during diagenesis causes dissolved phosphate build-up in sediment pore waters, promoting benthic efflux of phosphate to bottom waters or incorporation in secondary authigenic minerals. The combined benthic flux from coastal (sF_{cbf}) and abyssal (sF_{abf}) sediments is estimated to exceed the total riverine phosphorus flux ($F_{24(d+p)}$) to the ocean. Reprecipitation of diagenetically mobilized phosphorus in secondary phases significantly enhances phosphorus burial efficiency, impeding return of phosphate to the water column. Both processes impact the marine phosphorus cycle by affecting the primary productivity potential of surface waters.

Topics of Special Interest in Marine Phosphorus Research

Phosphate Oxygen Isotopes (δ^{18}O-PO$_4$)

Use of the oxygen isotopic composition of phosphate in biogenic hydroxyapatite (bones, teeth) as a paleotemperature and climate indicator was pioneered by Longinelli in the late 1960s–early 1970s, and has since been fairly widely and successfully applied. A novel application of the oxygen isotope system in phosphates is its use as a tracer of biological turnover of phosphorus during metabolic processes. Phosphorus has only one stable isotope (^{31}P) and occurs almost exclusively as orthophosphate (PO_4^{3-}) under Earth surface conditions. The phosphorus–oxygen bond in phosphate is highly resistant to nonenzymatic oxygen isotope exchange reactions, but when phosphate is metabolized by living organisms, oxygen isotopic exchange is rapid and extensive. Such exchange results in temperature-dependent fractionations between phosphate and ambient water. This property renders phosphate oxygen isotopes useful as indicators of present or past metabolic activity of organisms, and allows distinction of biotic from abiotic processes operating in the cycling of phosphorus through the environment.

Currently, the δ^{18}O-PO$_4$ system is being applied in a number of studies of marine phosphorus cycling, including (i) application to dissolved sea water inorganic phosphate as a tracer of phosphate source, water mass mixing, and biological productivity; (ii) use in phosphates associated with ferric iron oxyhydroxide precipitates in submarine ocean ridge sediments, where the δ^{18}O-PO$_4$ indicates microbial phosphate turnover at elevated temperatures. This latter observation suggests that phosphate oxygen isotopes may be useful biomarkers for fossil hydrothermal vent systems.

Reevaluating the Role of Phosphorus as a Limiting Nutrient in the Ocean

In terrestrial soils and in the euphotic zone of lakes and the ocean, the concentration of dissolved orthophosphate is typically low. When bioavailable phosphorus is exhausted prior to more abundant nutrients, it limits the amount of sustainable biological productivity. Phosphorus limitation in lakes is widely accepted, and terrestrial soils are often phosphorus-limited. In the oceans, however, phosphorus limitation is the subject of controversy and debate.

The prevailing wisdom has favored nitrogen as the limiting macronutrient in the oceans. However, a growing body of literature convincingly demonstrates that phosphate limitation of marine primary productivity can and does occur in some marine systems. In the oligotrophic gyres of both the western North Atlantic and subtropical North Pacific, evidence in the form of dissolved nitrogen : phosphorous (N:P) ratios has been used to argue convincingly that these systems are currently phosphate-limited. The N(:)P ratio of phytoplankton during nutrient-sufficient conditions is 16N:1P (the Redfield ratio) (*see*). A positive deviation from this ratio indicates probable phosphate limitation, while a negative deviation indicates probable nitrogen limitation. In the North Pacific at the Hawaiian Ocean Time Series (HOT) site, there has been a shift since the 1988 inception of the time series to N:P ratios exceeding the Redfield ratio in both particulate and surface ocean dissolved nitrogen and phosphorus (**Figure 3**). Coincident with this shift has been an increase in the prevalence of the nitrogen-fixing cyanobacterium *Trichodesmium* (**Table 3**). Currently, it appears as though the supply of new nitrogen has shifted from a limiting flux of upwelled nitrate from below the euphotic zone to an unlimited pool of atmospheric N_2 rendered bioavailable by the action of nitrogen fixers. This shift is believed to result from climatic changes that promote water column stratification, a condition that selects for N_2-fixing microorganisms, thus driving the system to phosphate limitation. A similar situation exists in the subtropical Sargasso Sea at the Bermuda Ocean Time Series (BATS) site, where currently the dissolved phosphorus concentrations (especially dissolved inorganic phosphorus (DIP)) are significantly lower than at the HOT site, indicating even more severe phosphate limitation (**Table 3**).

A number of coastal systems also display evidence of phosphate limitation, sometimes shifting seasonally from nitrogen to phosphate limitation in concert with changes in environmental features such as upwelling and river runoff. On the Louisiana Shelf in the Gulf of Mexico, the Eel River Shelf of northern California (USA), the upper Chesapeake Bay (USA), and portions of the Baltic Sea, surface water column dissolved inorganic N:P ratios indicate seasonal

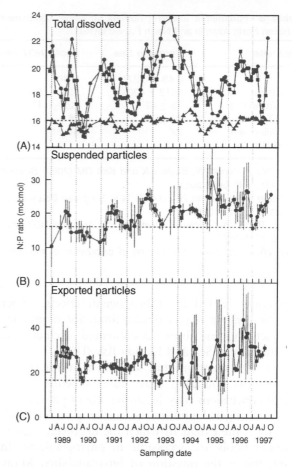

Figure 3 Time-series molar N:P ratios in (A) the dissolved pool, (B) suspended particulate matter, and (C) exported particulate matter from the HOT time series site at station ALOHA in the subtropical North Pacific near Hawaii. (A) The 3-point running mean N:P ratios for 0–100 m (circles), 100–200 m (squares), and 200–500 m (triangles). (B) The 3-point running mean ($\pm$1 SD) for the average suspended particulate matter in the upper water column (0–100 m). (C) The 3-point running mean ($\pm$1 SD) for the average N:P ratio of sediment trap-collected particulate matter at 150 m. The Redfield ratio (N:P16) is represented by a dashed line in all three panels. Particulate and upper water column dissolved pools show an increasing N:P ratio throughout the time-series, with a preponderance of values in excess of the Redfield ratio. (After Karl DM, Letelier R, Tupas L *et al.* (1997) The role of nitrogen fixation in the biogeochemical cycling in the subtropical North Pacific Ocean. *Nature* 388: 533–538, with permission.)

phosphate limitation. The suggestion of phosphate limitation is reinforced in the Louisiana Shelf and Eel River Shelf studies by the occurrence or presence of alkaline phosphatase activity, an enzyme induced only under phosphate-limiting conditions. Alkaline phosphatase has also been observed seasonally in Narragansett Bay. Although these coastal sites are recipients of anthropogenically derived nutrients (nitrogen and phosphate) that stimulate primary productivity above 'natural' levels, the processes that result in shifts in the limiting nutrient are not

Table 3 Parameters affecting nutrient limitation: comparison between North Atlantic and North Pacific Gyres

Parameter	Sargasso Sea	Pacific HOT site
DIP	0.48 ± 0.27^a	$9-40^b$
TDN (n mol l^{-1})	4512 ± 430	5680 ± 620^b
TDP (n mol l^{-1})	75 ± 42	222 ± 14^b
TDN : TDP	60 ± 7	26 ± 3^b
N$_2$-fixation rate (mmol N m^{-2} y^{-1})	72^c	$31-51^c$

After Wu J, Sunda W, Boyle EA and Karl DM (2000) *Science* 289: 759, with permission.
[a]Average DIP between 26° and 31°N in Sargasso Sea surface waters in March 1998.
[b]North Pacific near Hawaii at station ALOHA (the HOT site) during 1991–1997.
[c]See Wu *et al.* (2000) for method of calculation or measurement.

necessarily related to anthropogenic effects. Other oceanic sites where phosphorus limitation of primary productivity has been documented include the Mediterranean Sea and Florida Bay (USA).

One key question that studies of nitrogen and phosphorus limitation must address before meaningful conclusions may be drawn about phosphorus versus nitrogen limitation of marine primary productivity is the extent to which the dissolved organic nutrient pools are accessible to phytoplankton. In brief, this is the question of bioavailability. Many studies of nutrient cycling and nutrient limitation do not include measurement of these quantitatively important nutrient pools, even though there is indisputable evidence that at least some portion of the dissolved organic nitrogen (DON) and dissolved organic phosphorus (DOP) pools is bioavailable. An important direction for future research is to characterize the DON and DOP pools at the molecular level, and to evaluate what fraction of these are bioavailable. The analytical challenge of identifying the molecular composition of the DOP pool is significant. Recent advances in ^{31}P-nuclear magnetic resonance (NMR) spectroscopy have permitted a look at the high molecular weight (>1 nm) fraction of the DOP pool. However, this fraction represents only one-third of the total DOP pool; the other two-thirds of this pool, made up of smaller-molecular-weight DOP compounds, remains outside our current window of analytical accessibility.

Debate continues among oceanographers about the most probable limiting nutrient on recent and on long, geological timescales. While there is an abundant reservoir of nitrogen (gaseous N$_2$) in the atmosphere that can be rendered bioavailable by nitrogen-fixing photosynthetic organisms, phosphorus supply to the ocean is limited to that weathered off the continents and delivered by rivers, with some minor atmospheric input. As a consequence of continental weathering control on phosphorus supply to the oceans, phosphorus limitation has been considered more likely than nitrogen limitation on geological timescales. As the recent studies reviewed in this section suggest, phosphorus limitation may be an important phenomenon in the modern ocean as well. Overall, current literature indicates that there is a growing appreciation of the complexities of nutrient limitation in general, and the role of phosphorus limitation in particular, in both fresh water and marine systems.

Cosmogenic ^{32}P and ^{33}P as Tracers of Phosphorus Cycling in Surface Waters

There are two radioactive isotopes of phosphorus, ^{32}P (half-life $= 14.3$ days) and ^{33}P (half-life $= 25.3$ days). Both have been widely used in the study of biologically mediated phosphorus cycling in aquatic systems. Until very recently, these experiments have been conducted by artificially introducing radiophosphorus into laboratory incubations or, far more rarely, by direct introduction into natural waters under controlled circumstances. Such experiments necessarily involve significant perturbation of the system, which can complicate interpretation of results. Recent advances in phosphorus sampling and radioisotope measurement have made it possible to use naturally produced ^{32}P and ^{33}P as *in situ* tracers of phosphorus recycling in surface waters. This advance has permitted studies of net phosphorus recycling in the absence of experimental perturbation caused by addition of artificially introduced radiophosphorus.

^{32}P and ^{33}P are produced naturally in the atmosphere by interaction of cosmic rays with atmospheric argon nuclei. They are then quickly scavenged onto aerosol particles and delivered to the ocean surface predominantly in rain. The ratio of ^{33}P/^{32}P introduced to the oceans by rainfall remains relatively constant, despite the fact that absolute concentrations can vary from one precipitation event to another. Once the dissolved phosphorus is incorporated into a given surface water phosphorus pool (e.g., by uptake by phytoplankton or bacteria, grazing of phytoplankton or bacteria by zooplankton, or abiotic sorption), the ^{33}P/^{32}P ratio will increase in a systematic way as a given pool ages. This increase in the ^{33}P/^{32}P ratio with time results from the different half-lives of the two phosphorus radioisotopes. By measuring the ^{33}P/^{32}P ratio in rain and in different marine phosphorus pools – e.g., DIP, DOP (sometimes called soluble nonreactive phosphorus, or SNP), and particulate phosphorus of various size classes corresponding to different levels in the food chain – the net age of phosphorus in any of these reservoirs can be determined (**Table 4**). New insights into phosphorus cycling in oceanic surface waters derived from recent work using the cosmogenically produced $^{33}P/^{32}P$ ratio include the following. (1) Turnover rates of dissolved inorganic phosphorus in coastal and oligotrophic oceanic surface waters ranges from 1 to 20 days. (2) Variable turnover rates in the DOP pool range from <1 week to >100 days, suggesting differences in either the demand for DOP, or the lability of DOP toward enzymatic breakdown. (3) In the Gulf of Maine, DOP turnover times vary seasonally, increasing from 28 days in July to >100 days in August, suggesting that the DOP pool may evolve compositionally during the growing seasons. (4) Comparison of the $^{33}P/^{32}P$ ratio in different particulate size classes indicates that the age of phosphorus generally increases at successive levels in the food chain. (5) Under some circumstances, the $^{33}P/^{32}P$ ratio can reveal which dissolved pool is being ingested by a particular size class of organisms. Utilization of this new tool highlights the dynamic nature of phosphorus cycling in surface waters by revealing the rapid rates and temporal variability of phosphorus turnover. It further stands to provide new insights into ecosystem nutrient dynamics by revealing, for example, that low phosphorus concentrations can support high primary productivity through rapid turnover rates, and that there is preferential utilization of particular dissolved phosphorus pools by certain classes of organisms.

The Oceanic Residence Time of Phosphorus

As phosphorus is the most likely limiting nutrient on geological timescales, an accurate determination of its oceanic residence time is crucial to understanding how levels of primary productivity may have varied in the Earth's past. Residence time provides a means of evaluating how rapidly the oceanic phosphorus inventory may have changed in response to variations in either input (e.g., continental weathering, dust flux) or output (e.g., burial with sediments). In its role as limiting nutrient, phosphorus will dictate the amount of surface-ocean net primary productivity, and hence atmospheric CO_2 drawdown that will occur by photosynthetic biomass production. It has been suggested that this so-called 'nutrient–CO_2' connection might link the oceanic phosphorus cycle to climate change due to reductions or increases in the atmospheric greenhouse gas inventory. Oceanic phosphorus residence time and response time (the inverse of residence time) will dictate the timescales

over which such a phosphorus-induced climate effect may operate.

Over the past decade there have been several reevaluations of the marine phosphorus cycle in the literature, reflecting changes in our understanding of the identities and magnitudes of important phosphorus sources and sinks. One quantitatively important newly identified marine phosphorus sink is precipitation of disseminated authigenic carbonate fluoroapatite (CFA) in sediments in nonupwelling environments. CFA is the dominant phosphorus mineral in economic phosphorite deposits. Its presence in terrigenous-dominated continental margin environments is detectable only by indirect methods (coupled pore water/solid-phase chemical analyses), because dilution by terrigenous debris renders it below detection limits of direct methods (e.g. X-ray diffraction). Disseminated CFA has now been found in numerous continental margin environments, bearing out early proposals that this is an important marine phosphorus sink. A second class of authigenic phosphorus minerals identified as a quantitatively significant phosphorus sink in sandy continental margin sediments are aluminophosphates. Continental margins in general are quantitatively important sinks for organic and ferric iron-bound phosphorus, as well. When newly calculated phosphorus burial fluxes in continental margins, including the newly identified CFA and aluminophosphate (dominantly aluminum rare-earth phosphate) sinks are combined with older estimates of phosphorus burial fluxes in the deep sea, the overall burial flux results in a much shorter residence time than the canonical value of 100 000 years found in most text books (**Table 5**). This reduced residence time suggests that the oceanic phosphorus-cycle is subject to perturbations on shorter timescales than has previously been believed.

The revised, larger burial flux cannot be balanced by the dissolved riverine input alone. However, when the fraction of riverine particulate phosphorus that is believed to be released upon entering the marine realm is taken into account, the possibility of a balance between inputs and outputs becomes more feasible. Residence times estimated on the basis of phosphorus inputs that include this 'releasable' riverine particulate phosphorus fall within the range of residence time estimates derived from phosphorus burial fluxes (**Table 5**). Despite the large uncertainties associated with these numbers, as evidenced by the maximum and minimum values derived from both input and removal fluxes, these updated residence times are all significantly shorter than the canonical value of 100 000 years. Revised residence times on the order of 10 000–17 000 y make phosphorus-perturbations of the ocean–atmosphere CO_2

Table 4   Turnover rates of dissolved inorganic phosphorus (DIP) [a] and dissolved organic phosphorus (DOP) [b] in surface sea water

Phosphorus pool	Phosphorus turnover rate		
	Coastal	Open Ocean	References
DIP	<1 h to 10 d (>1000 d in Bedford Basin)	Weeks to months	c, d, e, f, g, h, i, j, k, l, m
Total DOP	3 to >90 d	50 to 300 d	l, m, n, o, p, q, r
Bioavailable DOP (model compounds)	2 to 30 d	1 to 4 d	k, s, t, u
Microplankton (<1 μm)	>1 to 3 d	NA	m
Phytoplankton (>1 μm)	<1 to 8 d	<1 week	m, v
Macrozooplankton (>280 μm)	14 to 40 d	30 to 80 d	m, p, q, v, w

After Benitez-Nelson CR (2000) *Earth-Science Reviews* 51: 109, with permission.
[a] DIP is equivalent to the soluble reactive phosphorus (SRP) pool, which may include some phosphate derived from hydrolysis of DOP (e.g., see Monaghan EJ and Ruttenberg KC (1998) *Limnology and Oceanography* 44(7): 1702).
[b] DOP is equivalent to the soluble nonreactive phosphorus (SNP) pool which may include dissolved inorganic polyphosphates (e.g., see Karl DM and Yanagi K (1997) *Limnology and Oceanography* 42: 1398).
[c] Pomeroy LR (1960) *Science* 131: 1731.
[d] Duerden CF (1973) PhD thesis, Dalhousie University, Halifax.
[e] Taft JL, Taylor WR and McCarthy JJ (1975) *Marine Biology* 33: 21.
[f] Harrison WG, Azam F, Renger EH and Eppley RW (1977) *Marine Biology* 40: 9.
[g] Perry MJ and Eppley RW (1981) *Deep-Sea Research* 28: 39.
[h] Smith RE, Harrison WG and Harris L (1985) *Marine Biology* 86: 75.
[i] Sorokin YI (1985) *Marine Ecology Progress Series* 27: 87.
[j] Harrison WG and Harris LR (1986) *Marine Ecology Progress Series* 27: 253.
[k] Björkman K and Karl DM (1994) *Marine Ecology Progress Series* 111: 265.
[l] Björkman K, Thomson-Bulldis AL and Karl DM (2000) *Aquatic and Microbial Ecology* 22: 185.
[m] Benitez-Nelson CR and Buesseler KO (1999) *Nature* 398: 502.
[n] Jackson GA and Williams PM (1985) *Deep-Sea Research* 32: 223.
[o] Orrett K and Karl DM (1987) *Limnology and Oceanography* 32: 383.
[p] Lal D and Lee T (1988) *Nature* 333: 752.
[q] Lee T, Barg E and Lal D (1992) *Analytica Chimica Acta* 260: 113.
[r] Karl DM and Yanagi K (1997) *Limnology and Oceanography* 42: 1398.
[s] Ammerman JW and Azam F (1985) *Limnology and Oceanography* 36: 1437.
[t] Nawrocki MP and Karl DM (1989) *Marine Ecology Progress Series* 57: 35.
[u] Björkman K and Karl DM (1999) *Marine Ecology Progress Series* (submitted).
[v] Waser NAD, Bacon MP and Michaels AF (1996) *Deep-Sea Research* 43(2–3): 421.
[w] Lee T, Barg E and Lal D (1991) *Limnology and Oceanography* 36: 1044.

reservoir on the timescale of glacial–interglacial climate change feasible.

Long Time-scale Phosphorus Cycling, and Links to Other Biogeochemical Cycles

The biogeochemical cycles of phosphorus and carbon are linked through photosynthetic uptake and release during respiration. During times of elevated marine biological productivity, enhanced uptake of surface water CO_2 by photosynthetic organisms results in increased CO_2 evasion from the atmosphere, which persists until the supply of the least abundant nutrient is exhausted. On geological timescales, phosphorus is likely to function as the limiting nutrient and thus play a role in atmospheric CO_2 regulation by limiting CO_2 drawdown by oceanic photosynthetic activity. This connection between nutrients and atmospheric CO_2 could have played a role in triggering or enhancing the global cooling that resulted in glacial episodes in the geological past. It has recently been proposed that tectonics may play the ultimate role in controlling the exogenic phosphorus mass, resulting in long-term phosphorus-limited productivity in the ocean. In this formulation, the balance between subduction of phosphorus bound up in marine sediments and underlying crust and creation of new crystalline rock sets the mass of exogenic phosphorus.

Phosphorus and oxygen cycles are linked through the redox chemistry of iron. Ferrous iron is unstable at the Earth's surface in the presence of oxygen, and oxidizes to form ferric iron oxyhydroxide precipitates, which are extremely efficient scavengers of dissolved phosphate. Resupply of phosphate to surface waters where it can fertilize biological productivity is reduced when oceanic bottom waters are well oxygenated owing to scavenging of phosphate by ferric oxyhydroxides. In contrast, during times in Earth's history when oxygen was not abundant in the atmosphere (Precambrian), and when expanses of the deep ocean were anoxic (e.g., Cretaceous), the

Table 5 Revised oceanic phosphorus input fluxes, removal fluxes, and estimated oceanic residence time

Flux description[a]		Flux (10^{12} mol P y^{-1})	Residence time (y)[e]
Input fluxes			
F_{84}	atmosphere → surface ocean	0.02–0.05	
$F_{24(d)}$	soil → surface ocean (river dissolved P flux)[b]	0.032–0.058	
$F_{24(p)}$	soil → surface ocean (river particulate P flux)[b]	0.59–0.65	
	Minimum reactive-P input flux	0.245	12 000
	Maximum reactive-P input flux	0.301	10 000
Removal fluxes			
sF_{cs}	Best estimate of total-P burial in continental margin marine sediments (**Table 2**, note j)[c]	0.150	
sF_{as}	Best estimate of total-P burial in abyssal marine sediments (**Table 2**, note j)[c]	0.130	
	Minimum estimate of reactive-P burial in marine sediments[d]	0.177	17 000
	Maximum estimate of reactive-P burial in marine sediments[d]	0.242	12 000

[a]All fluxes are from **Table 2**.

[b]As noted in **Table 2**, 30% of atmospheric aerosol phosphorus (Duce *et al.* (1991) *Global Biogeochemical Cycles* 5: 193) and 25–45% of the river particulate flux (see note (f) in **Table 2**) is believed to be mobilized upon entering the ocean. The reactive phosphorus input flux was calculated as the sum of $0.3(F_{84}) + F_{24(d)} + 0.35(F_{24(p)})$, where the mean value of the fraction of riverine particulate phosphorus flux estimated as reactive phosphorus (35%) was used. Reactive phosphorus is defined as that which passes through the dissolved oceanic phosphorus reservoir, and thus is available for biological uptake.

[c]These estimates are favored by the author, and reflect the minimum sF_{cs} and maximum sF_{as} fluxes given in **Table 2**. Because the reactive phosphorus contents of continental margin and abyssal sediments differ (see **Table 2** and note d, below), these fluxes must be listed separately in order to calculate the whole-ocean reactive phosphorus burial flux. See note (j) in **Table 2** for other published estimates of reactive-phosphorus burial flux.

[d]As noted in **Table 2**, between 40% and 75% of phosphorus buried in continental margin sediments is potentially reactive, and 90% to 100% of phosphorus buried in abyssal sediments is potentially reactive. The reactive phosphorus fraction of the total sedimentary phosphorus reservoir represents that which may have passed through the dissolved state in oceanic waters, and thus represents a true phosphorus sink from the ocean. The minimum reactive phosphorus burial flux was calculated as the sum of $0.4(sF_{cs}) + 0.9(sF_{as})$; the maximum reactive phosphorus burial flux was calculated as the sum of $0.75(sF_{cs}) + 1(sF_{as})$. Both the flux estimates and the % reactive phosphorus estimates have large uncertainties associated with them.

[e]Residence time estimates are calculated as the oceanic phosphorus inventory (reservoirs #4 and 5 (**Table 1**) = 3×10^{15} mol P) divided by the minimum and maximum input and removal fluxes.

potential for a larger oceanic dissolved phosphate inventory could have been realized due to the reduced importance of sequestering with ferric oxyhydroxides. This iron–phosphorus–oxygen coupling produces a negative feedback, which may have kept atmospheric O_2 within equable levels throughout the Phanerozoic. Thus, it is in the oceans that the role of phosphorus as limiting nutrient has the greatest repercussions for the global carbon and oxygen cycles.

Summary

The global cycle of phosphorus is truly a biogeochemical cycle, owing to the involvement of phosphorus in both biochemical and geochemical reactions and pathways. There have been marked advances in the last decade on numerous fronts of phosphorus research, resulting from application of new methods as well as rethinking of old assumptions and paradigms. An oceanic phosphorus residence time on the order of 10 000–20 000 y, a factor of 5–10 shorter than previously cited values, casts phosphorus in the role of a potential player in climate change through

the nutrient–CO_2 connection. This possibility is bolstered by findings in a number of recent studies that phosphorus does function as the limiting nutrient in some modern oceanic settings. Both oxygen isotopes in phosphate ($\delta^{18}O$-PO_4) and *in situ*-produced radiophosphorus isotopes (^{33}P and ^{32}P) are providing new insights into how phosphorus is cycled through metabolic pathways in the marine environment. Finally, new ideas about global phosphorus cycling on long, geological timescales include a possible role for phosphorus in regulating atmospheric oxygen levels via the coupled iron–phosphorus–oxygen cycles, and the potential role of tectonics in setting the exogenic mass of phosphorus. The interplay of new findings in each of these areas is providing us with a fresh look at the marine phosphorus cycle, one that is sure to evolve further as these new areas are explored in more depth by future studies.

See also

Carbon Cycle. Nitrogen Cycle.

Further Reading

Blake RE, Alt JC, and Martini AM (2001) Oxygen isotope ratios of PO$_4$: an inorganic indicator of enzymatic activity and P metabolism and a new biomarker in the search for life. *Proceedings of the National Academy of Sciences of the USA* 98: 2148–2153.

Benitez-Nelson CR (2001) The biogeochemical cycling of phosphorus in marine systems. *Earth-Science Reviews* 51: 109–135.

Clark LL, Ingall ED, and Benner R (1999) Marine organic phosphorus cycling: novel insights from nuclear magnetic resonance. *American Journal of Science* 299: 724–737.

Colman AS, Holland HD, and Mackenzie FT (1996) Redox stabilization of the atmosphere and oceans by phosphorus limited marine productivity: discussion and reply. *Science* 276: 406–408.

Colman AS, Karl DM, Fogel ML, and Blake RE (2000) A new technique for the measurement of phosphate oxygen isotopes of dissolved inorganic phosphate in natural waters. *EOS, Transactions of the American Geophysical Union* F176.

Delaney ML (1998) Phosphorus accumulation in marine sediments and the oceanic phosphorus cycle. *Global Biogeochemical Cycles* 12(4): 563–572.

Falkowski PG (1997) Evolution of the nitrogen cycle and its influence on the biological sequestration of CO$_2$ in the ocean. *Nature* 387: 272–275.

Föllmi KB (1996) The phosphorus cycle, phosphogenesis and marine phosphate-rich deposits. *Earth-Science Reviews* 40: 55–124.

Guidry MW, Mackenzie FT and Arvidson RS (2000) Role of tectonics in phosphorus distribution and cycling. In: Glenn CR, Prévt-Lucas L and Lucas J (eds) *Marine Authigenesis: From Global to Microbial*. SEPM Special Publication No. 66, pp. 35–51. (See also in this volume Compton *et al.* (pp. 21–33), Colman and Holland (pp. 53–75), Rasmussen (pp. 89–101), and others.)

Howarth RW, Jensen HS, Marino R and Postma H (1995) Transport to and processing of P in near-shore and oceanic waters. In: Tiessen H (ed.) *Phosphorus in the Global Environment*. SCOPE 54, pp. 232–345. Chichester: Wiley. (See other chapters in this volume for additional information on P-cycling.)

Karl DM (1999) A sea of change: biogeochemical variability in the North Pacific Subtropical Gyre. *Ecosystems* 2: 181–214.

Longinelli A and Nuti S (1973) Revised phosphate–water isotopic temperature scale. *Earth and Planetary Science Letters* 19: 373–376.

Palenik B and Dyhrman ST (1998) Recent progress in understanding the regulation of marine primary productivity by phosphorus. In: Lynch JP and Deikman J (eds.) *Phosphorus in Plant Biology: Regulatory Roles in Molecular, Cellular, Organismic, and Ecosystem Processes*, pp. 26–38. American Society of Plant Physiologists

Ruttenberg KC (1993) Reassessment of the oceanic residence time of phosphorus. *Chemical Geology* 107: 405–409.

Tyrell T (1999) The relative influences of nitrogen and phosphorus on oceanic productivity. *Nature* 400: 525–531.

MARINE SILICA CYCLE

D. J. DeMaster, North Carolina State University, Raleigh, NC, USA

Introduction

Silicate, or silicic acid (H_4SiO_4), is a very important nutrient in the ocean. Unlike the other major nutrients such as phosphate, nitrate, or ammonium, which are needed by almost all marine plankton, silicate is an essential chemical requirement only for certain biota such as diatoms, radiolaria, silicoflagellates, and siliceous sponges. The dissolved silicate in the ocean is converted by these various plants and animals into particulate silica (SiO_2), which serves primarily as structural material (i.e., the biota's hard parts). The reason silicate cycling has received significant scientific attention is that some researchers believe that diatoms (one of the silica-secreting biota) are one of the dominant phytoplankton responsible for export production from the surface ocean (Dugdale et al., 1995). Export production (sometimes called new production) is the transport of particulate material from the euphotic zone (where photosynthesis occurs) down into the deep ocean. The relevance of this process can be appreciated because it takes dissolved inorganic carbon from surface ocean waters, where it is exchanging with carbon dioxide in the atmosphere, turns it into particulate organic matter, and then transports it to depth, where most of it is regenerated back into the dissolved form. This process, known as the 'biological pump', along with deep-ocean circulation is responsible for the transfer of inorganic carbon into the deep ocean, where it is unable to exchange with the atmosphere for hundreds or even thousands of years. Consequently, silicate and silica play an important role in the global carbon cycle, which affects the world's climate through greenhouse feedback mechanisms. In addition, the accumulation of biogenic silica on the ocean floor can tell us where in the ocean export production has occurred on timescales ranging from hundreds to millions of years, which in turn reveals important information concerning ocean circulation and nutrient distributions.

Basic Concepts

In understanding the cycling of silicate in the oceans, the concept of mean oceanic residence time is commonly used. Mean oceanic residence time is defined as in eqn. [1].

$$\frac{\text{(amount of dissolved material in a reservoir)}}{\text{(steady-state flux into or out of the reservoir)}} \quad [1]$$

For silicate there are approximately 7×10^{16} moles of dissolved silicon in ocean water. (One mole is equal to 6×10^{23} molecules of a substance, which in the case of silicic acid has a mass of approximately 96 g.) As described later, the various sources of silicate to the ocean supply approximately 7×10^{12} mol y^{-1}, which is approximately equal to our best estimates of the removal rate. Most scientists believe that there has been a reasonably good balance between supply and removal of silicate from the oceans on thousand-year timescales because there is little evidence in the oceanic sedimentary record of massive abiological precipitation of silica (indicating enhanced silicate concentrations relative to today), nor is there any evidence in the fossil record over the past several hundred million years that siliceous biota have been absent for any extended period (indicating extremely low silicate levels). Dividing the amount of dissolved silicate in the ocean by the supply/removal rate yields a mean oceanic residence time of approximately 10 000 years. Basically, what this means is that an atom of dissolved silicon supplied to the ocean will remain on average in the water column or surface seabed (being transformed between dissolved and particulate material as part of the silicate cycle) for approximately 10 000 years before it is permanently removed from the oceanic system via long-term burial in the seabed.

Distribution of Silicate in the Marine Environment

Because of biological activity, surface waters throughout most of the marine realm are depleted in dissolved silicate, reaching values as low as a few micromoles per liter (μmol l^{-1}). When the siliceous biota die, their skeletons settle through the water column, where more than 90% of the silica is regenerated via inorganic dissolution. This process enriches the deep water in silicate, causing oceanic bottom waters to have as much as 10–100 times more silicate than surface waters in tropical and temperate regions. The magnitude of the deep-ocean silicate concentration depends on the location within the deep thermohaline circulation system. In general, deep water originates in North Atlantic and Antarctic surface waters. The deep water forming in the North Atlantic moves southward, where it joins with Antarctic water, on its way to feeding the deep Indian Ocean basin and then flowing from south to north in the Pacific basin. All along this 'conveyor belt' of deep-ocean water, siliceous biota are continually settling out from surface waters and dissolving at depth, which further increases the silicate concentration of the deep water downstream. Consequently, deep-ocean water in the Atlantic (fairly near the surface ocean source) is not very enriched in silicate (only 60 μmol l^{-1}), whereas the Indian Ocean deep water exhibits moderate enrichment (~ 100 μmol l^{-1}), and the north Pacific deep water is the most enriched (~ 180 μmol l^{-1}). This trend of increasing concentration is observed as well in the other nutrients such as nitrate and phosphate. Generalized vertical profiles of silicate are shown for the Atlantic and Pacific basins in **Figure 1**. The depth of the silicate maximum in these basins (typically 2000–3000 m depth) is deeper than the nutrient maxima for phosphate or nitrate, primarily because organic matter (the source of the phosphate and nitrate) is generally regenerated at shallower depths in the ocean than is silica. The nutrient concentrations in

oceanic deep waters can affect the chemical composition of particles settling through the water column because the vertical transport of nutrients from depth via upwelling and turbulence drives the biological production in surface waters. For example, the ratio of biogenic silica to organic carbon in particles settling between 1000 and 4000 m depth in the North Pacific Ocean (typically about 2–3) is substantially higher than that observed in the Arabian Sea (~ 0.7) and much higher than typical values in the Atlantic Ocean (< 0.3). This chemical trend in particle flux, which is caused in part by changes in planktonic species assemblage, is consistent with the systematic increase in silicate and other nutrients along the thermohaline-driven conveyor belt of deep-ocean circulation. The change in the biogenic silica to organic carbon ratio throughout the ocean basins of the world turns out to be one of the most important parameters controlling the nature of biogenic sedimentation in the world (see the global ocean sediment model of Heinze and colleagues, listed in Further Reading).

Silicate concentrations also can be used to distinguish different water masses. The most obvious example is at the Southern Ocean Polar Front (see **Figure 2**), which separates Antarctic Surface Water from the Subantarctic system. The silicate and nitrate concentration gradients across these Southern Ocean waters occur in different locations (in a manner similar to the distinct maxima in their vertical profiles). The high concentrations of silicate (50–100 μmol l^{-1}) south of the Polar Front result from wind-induced upwelling bringing silicate to the surface faster than the local biota can turn it into particulate silica. Turnover times between surface waters and

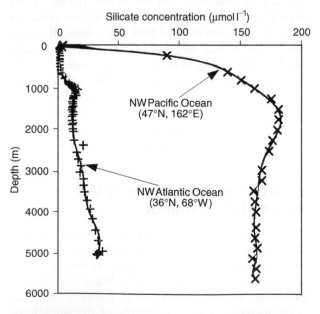

Figure 1 Vertical distribution of dissolved silicate in the Atlantic Ocean and the Pacific Ocean. The Atlantic data come from Spencer (1972), whereas the Pacific data are from Nozaki *et al.* (1997).

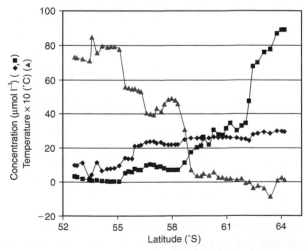

Figure 2 Distribution of silicate (■), nitrate (♦), and temperature (▲) across the Drake Passage illustrating different water masses and frontal features during November, 1999.

deep waters in the Southern Ocean are on the order of 100 years as compared to values of 500–1000 years in the Atlantic, Indian, and Pacific Oceans.

In terms of chemical equilibrium, biogenic silica is undersaturated by a factor of 10–1000-fold in surface waters and by at least a factor of 5 in deep waters. Therefore, siliceous biota must expend a good deal of energy concentrating silicate in their cells and bodies before precipitation can take place. This is quite different from the case of calcium carbonate (a material used by another type of plankton to form hard parts), which is supersaturated severalfold in most tropical and temperate surface waters. Deep waters everywhere are undersaturated with respect to biogenic silica (although to different extents). Therefore, inorganic dissolution of silica takes place in the water column as soon as the plankton's protective organic matter is removed from the biota (typically by microbial or grazing activities). It is not until the siliceous skeletal material is buried in the seabed that the water surrounding the silica even approaches saturation levels (see later discussion on sedimentary recycling), which diminishes the rate of dissolution and enhances preservation and burial.

The Marine Silica Cycle

Sources of Dissolved Silicate to the Ocean

Figure 3 shows the main features of the marine silica cycle as portrayed in a STELLA™ model. The main source of silicate to the oceans as a whole is rivers, which commonly contain ~ 150 µmol l^{-1} silicate but depending on location, climate, and local rock type, can range from 30 to 250 µmol l^{-1}. The silicate in rivers results directly from chemical weathering of rocks on land, which is most intense in areas that are warm and wet and exhibit major changes in relief (i.e., elevation). The best estimate of the riverine flux of dissolved silicate to the oceans is $\sim 6 \times 10^{12}$ mol Si y^{-1}. Other sources include hydrothermal fluxes ($\sim 0.3 \times 10^{12}$ mol Si y^{-1}), dissolution of eolian particles (0.5×10^{12} mol Si y^{-1}) submarine volcanic activity (negligible), and submarine weathering of volcanic rocks ($\sim 0.4 \times 10^{12}$ mol Si y^{-1}). Ground waters may contribute additional silicate to the marine realm, but the magnitude of this flux is difficult to quantify and it is believed to be small relative to the riverine flux. A more detailed discussion of the various sources of silicate to the marine environment can be

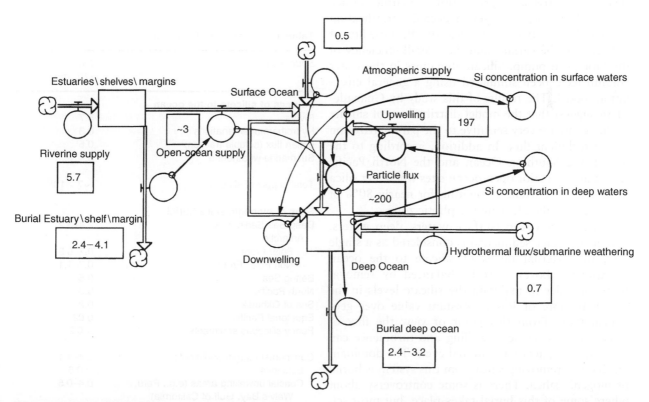

Figure 3 STELLA™ model of the global marine silica cycle showing internal and external sources of silicate to the system, internal recycling, and burial of biogenic silica in the seabed. The various reservoirs are shown as rectangles, whereas the fluxes in and out of the reservoirs are shown as arrows with regulating valves (indicating relationships and functional equations). The flux values (indicated by numbers inside the boxes) have units of 10^{12} mol y^{-1}.

found in the works by Treguer *et al.* and DeMaster (listed in Further Reading).

All of these sources of silicate to the oceanic water column are considered to be external. As shown in **Figure 3**, there are internal sources supplying silicate to oceanic surface waters and they are oceanic upwelling and turbulence. Because of the strong gradient in silicate with depth, upwelling of subsurface water (100–200 m depth) by wind-induced processes and turbulence can bring substantial amounts of this nutrient (and others) to surface waters that typically would be depleted in these valuable chemical resources as a result of biological activity. This upwelling flux ($\sim 100 - 300 \times 10^{12}$ mol Si y^{-1}), in fact, is 20–50 times greater than the riverine flux. The extraction of silicate from surface waters by siliceous biota is so efficient that nearly 100% of the nutrient reaching the surface is converted into biogenic silica. Therefore, the production of biogenic silica in oceanic surface waters is comparable to the flux from upwelling and turbulent transport. Riverine sources may be the dominant external source of silicate to the oceans, but they sustain only a negligible amount (only a few percent) of the overall marine silica production. Internal recycling, upwelling, and turbulence provide nearly all of the silicate necessary to sustain the gross silica production in marine surface waters. Therefore, changes in oceanic stratification and wind intensity may significantly affect the flux of nutrients to the surface and the overall efficiency of the biological pump. Silicate dynamics in the water column have been simulated using a general circulation model. The results of this study by Gnanadesikan suggest that the model distributions of silicate in the ocean are very sensitive to the parametrization of the turbulent flux. In addition, according to the model, the Southern Ocean and the North Pacific were the two major open-ocean sites where net silica production occurred, accounting for nearly 80% of the biogenic silica leaving the photic zone.

If the entire ocean (surface waters, deep waters, and near-surface sediments) is considered as a single box, the external fluxes of silicate to the ocean (mentioned above) must be balanced by removal terms in order to maintain the silicate levels in the ocean at more or less a constant value over geological time. From this point of view the flux of silicate from oceanic upwelling and turbulence can be treated as part of an internal cycle. The dominant mechanism removing silica from this system is burial of biogenic silica. There is some controversy about where some of this burial takes place, but most scientists believe that burial of biogenic silica (or some chemically altered by-product thereof) is the primary way that silicate is removed from the ocean.

Removal of Silica from the Ocean

The sediments with the highest rate of silica accumulation (on an areal basis) occur beneath the coastal upwelling zones, where strong winds bring extensive amounts of nutrients to the surface. Examples of these upwelling areas include the west coast of Peru, the Gulf of California, and Walvis Bay (off the west coast of South Africa). Diatom skeletons are the dominant form of biogenic silica in these deposits. The sediments in these upwelling areas accumulate at rates of 0.1–1.0 cm y^{-1} and they contain as much as 40% biogenic silica by weight. The burial rate for silica in these areas can be as high as 1.7 mol cm^{-2} y^{-1}. In calculating the total contribution of these areas to the overall marine silica budget, the accumulation rates (calculated on an areal basis) must be multiplied by the area covered by the particular sedimentary regime. Because these upwelling regimes are confined to such small areas, the overall contribution of coastal upwelling sites to the marine silica budget is quite small ($< 10\%$, see **Table 1**), despite the fact that (for a given area) they bury biogenic silica more rapidly than anywhere else in the marine realm.

The sediments containing the highest fraction of biogenic silica in the world occur in a 1000 km-wide

Table 1 The marine silica budget

Source/Sink	Flux (10^{12} mol Si y^{-1})
Sources of silicate to the ocean	
Rivers	5.7
Hydrothermal emanations	0.3
Eolian flux (soluble fraction)	0.5
Submarine weathering	0.4
Total supply of silicate	$\sim 7 \times 10^{12}$
Sites of biogenic silica burial	
Deep-Sea Sediments	2.4–3.2
Antarctic	
Polar Front	0.7–0.9
Non Polar Front	0.7–1.1
Bering Sea	0.5
North Pacific	0.3
Sea of Okhotsk	0.2
Equatorial Pacific	0.02
Poorly siliceous sediments	< 0.2
Continental margin sediments	2.4–4.1
Estuaries	< 0.6
Coastal upwelling areas (e.g., Peru, Walvis Bay, Gulf of California)	0.4–0.5
Antarctic margin	0.2
Other continental margins	1.8–2.8
Total burial of biogenic silica	$5–7 \times 10^{12}$

belt surrounding Antarctica. These sediments typically contain ~60% biogenic silica by weight (the majority of which are the skeletons of diatoms). Most of these sediments, however, accumulate at rates of only a few centimeters per thousand years, so their silica burial rate is quite small (<0.008 mol Si cm^{-2} y^{-1}), accounting for 1×10^{12} mol y^{-1} of silica burial or less than 20% of the total burial in the marine environment. Beneath the Polar Front (corresponding to the northern 200–300 km of the Southern Ocean siliceous belt), however, the sediment accumulation rates increase dramatically to values as high as 50×10^{-3} cm y^{-1}. Regional averages can be as high as 19×10^{-3} cm y^{-1}, yielding a silica burial rate of 0.08 mol Si cm^{-2} y^{-1}. Unfortunately, many of the siliceous deposits beneath the Polar Front occur in areas of very rugged bottom topography (because of the submarine Antarctic Ridge), where sediments are focused into the deeper basins from the flanks of the oceanic ridge crests. This distribution of accumulation rates would not create a bias if all of the sedimentary environments are sampled equally. However, it is more likely to collect sediment cores in the deep basins, where the deposits are thicker and accumulating more rapidly, than it is on the flanks where the sediment coverage is thinner. The effects of this sediment focusing can be assessed by measuring the amount in the seabed of a naturally occurring, particle-reactive radioisotope, thorium-230 (^{230}Th). If there were no sediment focusing, the amount of excess ^{230}Th in the sediments would equal the production from its parent, uranium-234, in the overlying water column. In some Polar Front Antarctic cores there is 12 times more excess ^{230}Th in the sediment column than produced in the waters above, indicating that sediment focusing is active. Initial estimates of the biogenic silica accumulation beneath the Polar Front were as high as 3×10^{12} mol Si y^{-1}, but tracer-corrected values are on the order of 1×10^{12} mol Si y^{-1}.

There are other high-latitude areas accumulating substantial amounts of biogenic silica, including the Bering Sea, the Sea of Okhotsk, and much of the North Pacific Ocean; however, the accumulation rates are not as high as in the Southern Ocean (see **Table 1**). The high rate of silica burial in the high-latitude sediments may be attributed in part to the facts that cold waters occurring at the surface and at depth retard the rate of silica dissolution and that many of the diatom species in high latitudes have more robust skeletons than do their counterparts in lower latitudes. Moderately high silica production rates and elevated silica preservation efficiencies (approximately double the world average) combine to yield high-latitude siliceous deposits accounting

for approximately one-third of the world's biogenic silica burial.

If the focusing-corrected biogenic silica accumulation rates are correct for the Polar Front, then a large sink for biogenic silica (~ 1–2×10^{12} mol Si y^{-1}) needs to be identified in order to maintain agreement between the sources and sinks in the marine silica budget. Continental margin sediments are a likely regime because these environments have fairly high surface productivity (much of which is diatomaceous), a relatively shallow water column (resulting in reduced water column regeneration as compared to the deep sea), rapid sediment accumulation rates (10–100×10^{-3} cm y^{-1}) and abundant aluminosilicate minerals (see Biogenic Silica Preservation below). The amount of marine organic matter buried in shelf and upper slope deposits is on the order of 3×10^{12} mol C y^{-1}. When this flux is multiplied by the silica/organic carbon mole ratio (Si/C$_{org}$) of sediments in productive continental margin settings (Si/C$_{org} = 0.6$), the result suggests that these nearshore depositional environments can account for sufficient biogenic silica burial (1.8–2.8×10^{12} mol Si y^{-1}) to bring the silica budget into near balance (i.e., within the errors of calculation).

Biogenic Silica Preservation

As mentioned earlier, all ocean waters are undersaturated with respect to biogenic silica. Surface waters may be more than two orders of magnitude undersaturated, whereas bottom waters are 5–15-fold undersaturated. The solubility of biogenic silica is greater in warm surface waters than in colder deep waters, which, coupled with the increasing silicate concentration with depth in most ocean basins, diminishes the silicate/silica disequilibrium (or corrosiveness of the water) as particles sink into the deep sea. This disequilibrium drives silica regeneration in oceanic waters along with other factors and processes such as particle residence time in the water column, organic and inorganic surface coatings, particle chemistry, particle aggregation, fecal pellet formation, as well as particle surface area. Recycling of biogenic silica occurs via inorganic dissolution; however, the organic coating that siliceous biota use to cover their skeletons (inhibiting dissolution) must be removed by microbial or zooplankton grazing prior to dissolution. This association is highlighted by the fact that bacterial assemblages can accelerate the dissolution of biogenic silica in the water column.

An important aspect of biogenic silica dissolution pertains to surface chemistry and clay-mineral formation on the surface of siliceous tests. Incorporation of aluminum in the initial skeleton as well as aluminosilicate formation on skeletal surfaces during settling and burial greatly decrease the solubility of biogenic silica, in some cases by as much as a factor of 5–10. It appears that some of this 'armoring' of siliceous skeletons occurs up in the water column (possibly in aggregates or fecal pellets), although some aluminosilicate formation may occur in flocs just above the seabed as well as deeper in the sediment column. The occurrence of clay minerals on skeletal surfaces has been documented using a variety of instruments (e.g., the scanning electron microscope). The nature of the settling particles also affects dissolution rates in the water column. If siliceous skeletons settle individually, they settle so slowly (a timescale of years to decades) that most particles dissolve before reaching the seabed. However, if the siliceous skeletons aggregate or are packaged into a fecal pellet by zooplankton, sinking velocities can be enhanced by several orders of magnitude, favoring preservation during passage through the water column. Siliceous tests that have high surface areas (lots of protruding spines and ornate surface structures) also are prone to high dissolution rates and low preservation in the water column relative to species that have more robust skeletons and more compact structures.

Very few studies have documented silica production rates in surface waters, established the vertical fluxes of silica in the water column, and then also examined regeneration and burial rates in the seabed. One place that all of these measurements have been made is in the Ross Sea, Antarctica. In this high-latitude environment, approximately one-third of the biogenic silica produced in surface waters is exported from the euphotic zone, with most of this material (27% of production) making it to the seabed some 500–900 m below. Seabed preservation efficiencies (silica burial rate divided by silica rain rate to the seafloor) vary from 1% to 86%, depending primarily on sediment accumulation rate, but average 22% for the shelf as a whole. Consequently, the overall preservation rate (water column and seabed) is estimated to be ∼6% in the Ross Sea. On a global basis, approximately 3% of the biogenic silica produced in surface waters is buried in the seabed. The total preservation efficiencies for different ocean basins vary, with the Atlantic and Indian Oceans having values on the order of 0.4–0.8% and the Pacific and Southern Oceans having values of approximately 5–10%.

Sediment accumulation rate can make a large difference in seabed preservation efficiency. In the Ross Sea, for example, increasing the sediment accumulation rate from 1–2 to 16×10^{-3} cm y^{-1}, increases the seabed preservation efficiency from 1–5% up to 50–60%. In most slowly accumulating deep-sea sediments (rates of 2×10^{-3} cm y^{-1} or less), nearly all of the biogenic silica deposited on the seafloor dissolves prior to long-term burial. Increasing the sedimentation rate decreases the time that siliceous particles are exposed to the corrosive oceanic bottom waters, by burying them in the seabed where silicate, aluminum, and cation concentrations are high, favoring aluminosilicate formation and preservation. Consequently, continental margin sediments with accumulation rates of $10–100 \times 10^{-3}$ cm y^{-1} are deposits expected to have high preservation efficiencies for biogenic silica and are believed to be an important burial site for this biogenic phase.

Estuaries extend across the river–ocean boundary and are generally regions of high nutrient flux and rapid sediment burial (0.1–10 cm y^{-1}). They commonly exhibit extensive diatom production in surface waters, but may not account for substantial biogenic silica burial because of extensive dissolution in the water column. For example, on the Amazon shelf approximately 20% of the world's river water mixes with ocean water and silicate dynamics have been studied in detail. Although nutrient concentrations are highest in the low-salinity regions of the Amazon mixing zone, biological nutrient uptake is limited because light cannot penetrate more than a few centimeters into the water column as a result of the high turbidity in the river (primarily from natural weathering of the Andes Mountains). After the terrigenous particles have flocculated in the river–ocean mixing zone, light is able to penetrate the warm surface waters, leading to some of the highest biogenic silica production rates in the world. However, resuspension on the shelf, zooplankton grazing, and high water temperatures lead to fairly efficient recycling in the water column and nearly all of the dissolved silicate coming down the river makes it out to the open ocean. The Amazon shelf seabed does appear to exhibit clay-mineral formation (primarily through replacement of dissolving diatoms), but the burial fluxes are expected to be small relative to the offshore transport of silicate and biogenic silica.

Biogenic Silica in Marine Sediments

As mentioned above, the primary biota that construct siliceous skeletons are diatoms, radiolaria, silicoflagellates, and siliceous sponges. Diatoms are marine algae. These phytoplankton account for 20–40% of the primary production in the ocean and an even greater percentage of the export production from the photic zone. Diatom skeletons are the primary form of biogenic silica in deposits associated with coastal upwelling areas, high-latitude oceans (predominantly in the Pacific and the Southern Oceans), and the continental margins (**Figure 4**). In equatorial upwelling areas radiolarian skeletons commonly occur in marine sediments along with the diatom frustules. Radiolaria are zooplankton that live in the upper few hundred meters of the water column. Their skeletons are larger and more robust than many diatoms; consequently their preservation in marine sediments is greater than that of most diatoms. Silicoflagellates account for a very small fraction of the biogenic silica in marine sediments because most of them dissolve up in the water column or in surface sediments. They have been used in some continental margin sediments as a paleo-indicator of upwelling intensity. Siliceous sponge spicules can make up a significant fraction of the near-interface sediments in areas in which the sediment accumulation rate is low ($<5 \times 10^{-3}$ cm y^{-1}). For example, on the Ross Sea continental shelf, fine sediments accumulate in the basins, whereas the topographic highs (<400 m water depth) have minimal fine-grained material (because of strong currents and turbulence). As a result, mats of siliceous sponge spicules occur in high abundance on some of these banks.

To measure the biogenic silica content of marine sediments, hot (85°C) alkaline solutions are used to dissolve biogenic silica over a period of 5–6 hours. The silicate concentration in the leaching solution is measured colorimetrically on a spectrophotometer and related to the dry weight of the original sedimentary material. In many sediments, coexisting clay minerals also may yield silicate during this leaching process; however, this contribution to the leaching solution can be assessed by measuring the silicate concentration in the leaching solution hourly over the course of the dissolution. Most biogenic silica dissolves within 2 hours, whereas clay minerals release silicate at a fairly constant rate over the entire leaching period. Consequently, the contributions of biogenic silica and clay-mineral silica can be resolved using a graphical approach (see **Figure 5**).

Measuring Rates of Processes in the Marine Silica Cycle

There are several useful chemical tracers for assessing rates of silicate uptake, silica dissolution in the water column, and particle transport in the seabed. Most of these techniques are based on various isotopes of silicon, some of which are stable and some of which are radioactive. Most of the stable silicon occurring naturally in the ocean and crust is ^{28}Si (92.2%) with minor amounts of ^{29}Si (4.7%) and ^{30}Si (3.1%). By adding known quantities of dissolved ^{29}Si or ^{30}Si to surface ocean waters, the natural abundance ratios of Si can be altered, allowing resolution of existing biogenic silica from silica produced after spiking in incubation studies. Similarly, if the silicate content of ocean water is spiked with either dissolved ^{29}Si or ^{30}Si, then, as biogenic silica dissolves, the ratio of the silicon isotopes will change in proportion

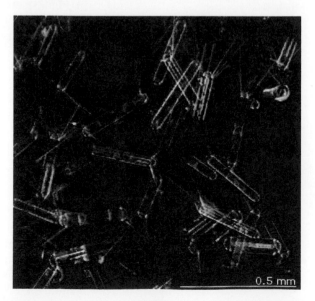

Figure 4 Micrograph of diatoms (genus *Corethron*) collected from an Antarctic plankton tow near Palmer Station.

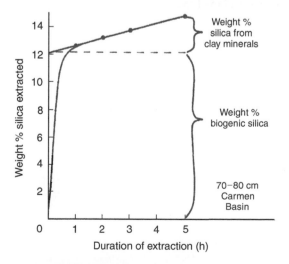

Figure 5 Graphical approach to resolving silicate originating via biogenic silica dissolution from that generated via clay-mineral dissolution during the alkaline leach technique used to quantify biogenic silica abundance. This sample was from the Gulf of California, Carmen Basin.

to the amount of silica dissolved (enabling characterization of dissolution rates). In addition, the measurement of natural silicon isotopes in sea water and in siliceous sediments has been suggested as a means of assessing the extent of silicate utilization in surface waters on timescales ranging from years to millennia. Addition of radioactive ^{32}Si (half-life 160 y) to incubation solutions recently has been used to simplify the measurement of silica production rates in surface ocean waters. In the past, ^{32}Si has been difficult to obtain, but recent advances in production and isolation protocols have made it possible to produce this radioisotope for oceanographic studies. Distributions of naturally occurring ^{32}Si in the water column and seabed can be used to determine deep-ocean upwelling rates as well as the intensity of eddy diffusion (or turbulence) in the deep ocean. This same radioactive isotope can be used to evaluate rates of bioturbation (biological particle mixing) in the seabed on timescales of hundreds of years.

See also

Carbon Cycle.

Further Reading

Craig H, Somayajulu BLK, and Turekian KK (2000) Paradox lost, silicon-32 and the global ocean silica cycle. *Earth and Planetary Science Letters* 175: 297–308.

DeMaster DJ (1981) The supply and removal of silica from the marine environment. *Geochimica et Cosmochimica Acta* 45: 1715–1732.

Dugdale RC, Wilkerson FP, and Minas HJ (1995) The role of a silicate pump in driving new production. *Deep-Sea Research* 42: 697–719.

Gnanadesikan A (1999) A global model of silicon cycling: sensitivity to eddy parameterization and dissolution. *Global Biogeochemical Cycles* 13: 199–220.

Heinze C, Maier-Reimer E, Winguth AME, and Archer D (1999) A global oceanic sediment model for long-term climate studies. *Global Biogeochemical Cycles* 13: 221–250.

Nelson DM, DeMaster DJ, Dunbar RB, and Smith WO Jr (1996) Cycling of organic carbon and biogenic silica in the southern Ocean: estimates of water-column and sedimentary fluxes on the Ross Sea continental shelf. *Journal of Geophysical Research* 101: 18519–18532.

Nelson DM, Treguer P, Brzezinski MA, Leynaert A, and Queguiner B (1995) Production and dissolution of biogenic silica in the ocean: revised global estimates, comparison with regional data and relationship to biogenic sedimentation. *Global Biogeochemical Cycles* 9: 359–372.

Nozaki Y, Zhang J, and Takeda A (1997) ^{210}Pb and ^{210}Po in the equatorial Pacific and the Bering Sea: the effects of biological productivity and boundary scavenging. *Deep-Sea Research II* 44: 2203–2220.

Ragueneau O, Treguer P, Leynaert A, *et al.* (2000) A review of the Si cycle in the modern ocean: recent progress and missing gaps in the application of biogenic opal as a paleoproductivity proxy. *Global and Planetary Change* 26: 317–365.

Spencer D (1972) GEOSECS II. The 1970 North Atlantic Station: Hydrographic features, oxygen, and nutrients. *Earth and Planetary Science Letters* 16: 91–102.

Treguer P, Nelson DM, Van Bennekom AJ, *et al.* (1995) The silica balance in the world ocean: A re-estimate. *Science* 268: 375–379.

CFCS IN THE OCEAN

R. A. Fine, University of Miami, Miami, FL, USA

Introduction

The oceans, atmosphere, continents, and cryosphere are part of the tightly connected climate system. The ocean's role in the climate system involves the transport, sequestration, and exchange of heat, fresh water, and carbon dioxide (CO_2) between the other components of the climate system. When waters descend below the ocean surface they carry with them atmospheric constituents. Some of these are gases such as carbon dioxide and chlorofluorocarbons (CFCs). The CFCs can serve as a physical analog for CO_2 because they are biologically and chemically inert in oceans. In the oceans the distribution of CFCs provides information on which waters have been in contact with the atmosphere in the past few decades. The CFCs also give information on the ocean's circulation and its variability on timescales of months to decades. The timescale information is needed to understand and to assess the ocean's role in climate change, and its capacity to take up anthropogenic constituents from the atmosphere. Thus, the advantage of using tracers like CFCs for ocean circulation studies is the added dimension of time; their time history is fairly well known, they are an integrating quantity and an analog for oceanic anthropogenic CO_2 uptake, and they provide an independent test for time integration of models.

Tracers serve as a 'dye' with which to follow the circulation of ocean waters. There are conventional ocean tracers such as temperature, salinity, oxygen, and nutrients. There are stable isotope tracers such as oxygen-18, carbon-13, and there are radioactive tracers both naturally occurring (such as the uranium/thorium series, and radium), and those produced both naturally and by the bomb tests (such as tritium and carbon-14). The bomb contributions from the latter two are called transient tracers, as are the CFCs, because they have been in the atmosphere for a short time. This implies an anthropogenic source and a nonsteady input function.

Atmospheric Source

The chlorofluorocarbons, CFCs, are synthetic halogenated methanes. Their chemical structures are as follows: CFC-11 is CCl_3F, CFC-12 is CCl_2F_2, and CFC-113 is CCl_2FCClF_2. For completeness the compound carbon tetrachloride, CCl_4, is also included in this article as its atmospheric source, measurement, and oceanic distribution are similar to those of the CFCs. The CFCs have received considerable attention because they are a double-edged environmental sword. They are a threat to the ozone layer, and a greenhouse gas. The CFCs are used as coolants in refrigerators and air conditioners, as propellants in aerosol spray cans, and as foaming agents. These chemicals were developed over 50 years ago when no one realized that they might cause environmental problems. When released CFCs are gases that have two sinks, the predominant one being the atmosphere, and to a lesser extent the oceans. Most of the CFCs go up into the troposphere, where they remain for decades. In the oceans and in the troposphere the CFCs pose no problem. However, some escape into the stratosphere where they are a threat to the ozone layer. Due to their role in UV absorption they have been correlated with the increased incidence of skin cancers. Since the recognition of the CFCs as an environmental problem in the 1970s and the signing of the Montreal Protocol in 1987, the use of CFCs has been phased out. The atmospheric concentrations have just started to decrease. This is an important international step toward correcting the dangerous trend of stratospheric ozone depletion.

The atmospheric CFC concentrations became significant after the 1940s. The concentrations increased exponentially until the mid-1970s, and then increased linearly until the 1990s at a rate of about 5% per year. The production and release data for CFCs tabulated by the Chemical Manufacturers Association (CMA) were used (**Figure 1**) to reconstruct the atmospheric time histories for the Northern and Southern Hemispheres. Since 1979 the atmospheric concentrations have been based on actual measurements at various sampling stations around the globe, and these are checked against the CMA production and release estimates. The curves in **Figure 1** show all CFCs including CCl_4 increasing with time, with CFC-11 leveling off and actually decreasing in the late 1990s. The atmospheric increase of all the CFCs slowed markedly after the Montreal Protocol. The

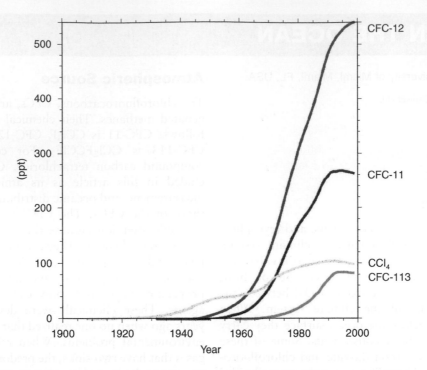

Figure 1   Northern Hemisphere atmospheric time histories. (Atmospheric data from Walker *et al.*, 2000.)

uncertainties in the reconstructed pre-1979 atmospheric histories depend on the atmospheric lifetimes of the compounds. The range of lifetimes for the compounds are 29–76 years for CFC-11, 77–185 years for CFC-12, 54–143 years for CFC-113, and 21–43 years for CCl$_4$. The uncertainties are a few percent, and they are highest for the early period. The continuous direct atmospheric measurements, which began in the late 1970s, are uncertain to within 1–2%. It is important to evaluate the uncertainties in the atmospheric source function, because they translate into uncertainties when used to put timescales on ocean processes. Because of their long atmospheric residence times, the CFCs are homogeneously distributed in the Northern and Southern Hemisphere, with the Northern Hemisphere about 8% higher than the Southern Hemisphere.

CFCs in the Oceans

Analytical Techniques

Water samples collected from the ocean are measured for CFCs and CCl$_4$ using an electron capture detection gas chromatography system. Analysis of water samples is done onboard ship, usually within hours of collection. The unit of measure is pmol kg^{-1} or 10^{-12} moles kg^{-1}. These are extremely low level concentrations that are easily susceptible to contamination from shipboard refrigerants, solvents, lubricants, etc.

Chemical Stability

Under oxygenated oceanic conditions both CFC-11 and CFC-12 are believed to be chemically stable. CFC-11 has been shown to be unstable in anoxic marine waters; both CFC-11 and CFC-12 have been shown to be unstable in anoxic sediments. The compound CCl$_4$ undergoes temperature-dependent hydrolysis, which limits its usefulness in the ocean when sea surface temperatures exceed $\sim 18°$C. CFC-113 also has some stability problems at higher temperatures.

Gas Flux and Solubility

The CFCs are gases, and like other gases they get into the ocean via air–sea exchange. There is a direct correlation between gas exchange rate and wind speed, and the direction of the gas flux between the air and ocean is from high to low concentration. For CFCs the atmospheric concentrations generally exceed those in the ocean. The concentration of CFCs dissolved in the surface layer of the oceans is dependent upon the solubility, atmospheric concentration, and other physical factors affecting the gas saturation including upwelling, entrainment due to mixing, ice cover, etc. The solubility of CFCs and CCl$_4$ has been measured in the laboratory. The accuracy of the measurements is about 1.5% and precision is about 0.7%. The solubility increases with decreasing temperature, at a rate of about 4% for 1°C. Therefore, the colder the water the higher the CFC concentration. At a constant salinity the temperature effect is about two times greater for CFCs than for oxygen. The solubility is only slightly dependent on the salinity, and it decreases with increasing salinity.

Surface Saturation

The approach to equilibrium condition or the saturation state is dependent on the mixed layer depth and air–sea transfer rate. It takes from days up to a few weeks after a change in temperature or salinity for 'normal' (not very deep) oceanic surface layers to come to equilibrium with the present atmosphere. While the surface waters of the world's oceans are close to equilibrium with the present day atmospheric concentration of CFCs, there are exceptions. At times of rapid warming, such as in the spring, the surface waters will tend to be a few percent supersaturated with the gas due to lack of time to equilibrate with the atmosphere. Likewise at times of rapid cooling the surface waters will be a few percent undersaturated with the gas. Typically there are undersaturations within a few degrees of the equator due to upwelling of deeper less saturated waters. In high latitudes, where there are deep convective mixed layers that do not readily equilibrate with the atmosphere, there are likely to be undersaturations of as much as 60%. These have been observed in the Labrador Sea. The undersaturations in the high latitude water mass source regions need to be taken into account when using the CFCs to put timescales on oceanic processes.

Oceanic Distribution

The compounds CFC-11 and CFC-12 were first measured in the oceans in the late 1970s. The first systematic and intensive survey was carried out in the tropical North and South Atlantic oceans starting in the early 1980s. Since then CFCs have been part of the measurements made during physical oceanography field work. A global survey was conducted as part of the World Ocean Circulation Experiment during the 1990s. Typical vertical profiles versus pressure for stations in the North Atlantic and North Pacific oceans are presented in **Figure 2** along with other properties. Although CFC-12 has higher concentrations in the atmosphere, CFC-11 is more soluble in sea water, so its concentrations are about twice that of CFC-12. Note that there are measurable concentrations of CFCs in the western North

Atlantic that reach to the ocean bottom, while they reach to only 1000 m in the North Pacific. The difference between the CFC concentrations of the North Atlantic as compared with the North Pacific, reflects the formation of deep waters in the North Atlantic and the absence in the North Pacific. Concentrations generally decrease as the ocean depth increases. However, there may be subsurface concentration maxima due to the lateral intrusion of water that has been in more recent contact with the atmosphere (see applications below). The concentrations of CFCs and oxygen should behave similarly except where the biological effects on the oxygen distribution cause the differences, for example, the oxygen minimum at mid-depth.

Combining a series of vertical profiles, as in **Figure 2**, will give a slice or section through the ocean. Sections through the eastern Pacific and Atlantic are shown in **Figure 3**. The absence of CFCs in the deep waters of the Pacific Ocean shows the relative isolation of the deep Pacific from contact with the atmosphere on timescales of decades. In contrast, the North Atlantic north of 35°N has CFCs in deep and bottom waters, because these waters form in the high latitudes of the North Atlantic and easily spread equatorward on timescales of 10–20 years. As part of the density-driven, thermohaline circulation some of these waters will eventually be transported into the Pacific, but it will take hundreds of years. The upper waters of both oceans are in contact with the atmosphere on much shorter timescales. These upper waters are part of the wind-driven circulation.

CFC Ages in the Ocean

Age Calculations

One of the main advantages of using CFCs as tracers of ocean circulation is that the time-dependent source function permits the calculation of timescales for these processes. A tracer age is the elapsed time since a water parcel was last exposed to the atmosphere. The tracer-derived age is the elapsed time since a subsurface water mass was last in contact with the atmosphere. Two estimates of 'age' can be calculated, one from the CFC-11/CFC-12 ratio and one from the partial pressure of either dissolved CFC. In both cases, the atmospheric value of either the ratio or partial pressure with which the water had equilibrated is compared to the atmospheric source function to determine the corresponding date.

To normalize the concentrations for the effects on the solubility of temperature and salinity CFCs are expressed in terms of their partial pressures, pCFC, where the pCFC is the concentration divided by the

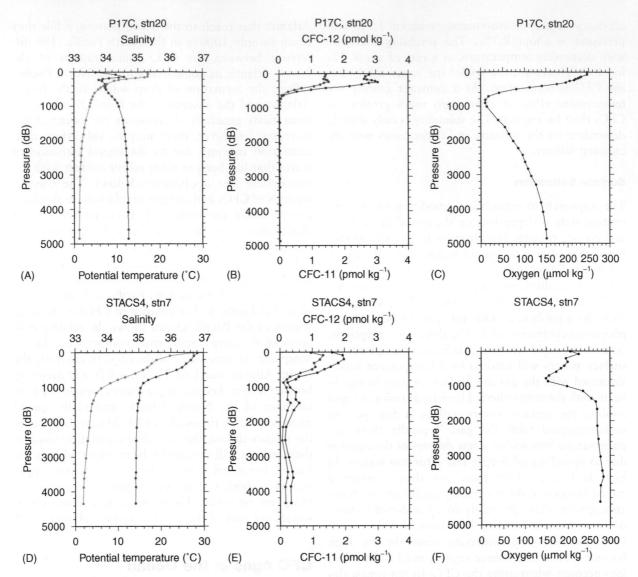

Figure 2 Vertical profiles of oceanographic data. (A) North Pacific salinity and potential temperature, (B) North Pacific CFC-11 and CFC-12, (C) North Pacific oxygen, (D) North Atlantic salinity and potential temperature, (E) North Atlantic CFC-11 and CFC-12, (F) North Atlantic oxygen. North Pacific World Ocean Circulation Experiment cruise P17C station 20, 33°N, 135°W, June 1991; North Atlantic Subtropical Atlantic Climate Studies cruise station 7, 26.5°N, 76°W, June 1990. (North Atlantic data from Johns *et al.* (1997) *Journal of Physical Oceanography* 27: 2187–2208; Pacific data from Fine *et al.* (2001) *Journal of Geophysical Research*.)

solubility of the gas. This value is then adjusted for what the percent surface saturation is thought to be based on the measured temperature and salinity, then matched to the atmospheric time histories, and a corresponding year is assigned to the water mass. This age is an average of the water parcel. The pCFC is used to calculate the age of upper ocean waters, because at low concentrations the effects of dilution will bias the age toward the older components of a mixture.

The age can also be calculated using the ratio of two CFCs; instead of using one pCFC the ratio of two pCFCs are used. In this case no assumptions are needed about surface equilibrium saturation at the

time of water mass formation. Since the atmospheric changes in the ratio of CFC-11/CFC-12 have remained unchanged since the mid-1970s, this restricts the application of the ratio age for CFC-11 and CFC-12 to waters dating back further than 1975. However, either CFC-11 or CFC-12 can be combined with CFC-113 to extend age estimates to the present. Similarly they can be combined with CCl_4 to extend age estimates further into the past. Unlike the pCFC age, the ratio ages are actually the ages of the CFC-bearing components. **Figure 4** shows sections of CFC ratio ages from the eastern North Atlantic and North Pacific oceans. Note that the intermediate and deep waters of the eastern North Atlantic (between 2000

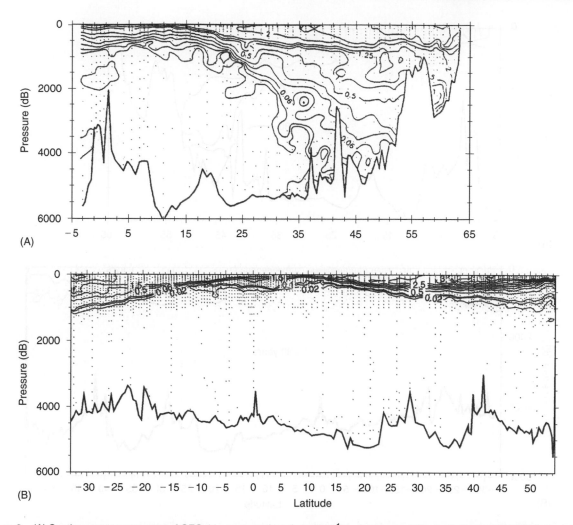

Figure 3   (A) Sections versus pressure of CFC-11 concentrations (pmol kg^{-1}) in the eastern Atlantic (latitude 65°N–5°S) along 20°W in summer 1988. (B) Sections of CFC-11 concentrations (pmol kg^{-1}) in the eastern Pacific (latitude 54°N–32°S) mostly along 135°W in summer 1991. (North Atlantic data from Doney SC and Bullister JB (1992) *Deep-Sea Research* 39: 1857–1883; Pacific data from Fine *et al.* (2001) *Journal of Geophysical Research*.)

and 4000 m) have CFCs younger than 30 years north of 45°N, because of their proximity to the formation regions, whereas this is not the case in the North Pacific. In the Pacific below 2000 m the water column has been isolated from interaction with the atmosphere on similar timescales (except for the far western South Pacific).

Caveats For Using CFC Ages

There are several caveats to the use of CFC ages. Both ages – partial pressure and ratio – may be subject to biases when there is mixing of more than one water mass component. Because of nonlinearities in the source functions and solubilities, neither age mixes linearly in multicomponent systems over the entire concentration range observed in the ocean. The atmospheric source function is nonlinear for much of

the input history; however, it can be approximated as being linear between the late 1960s and 1990. The solubilities are nonlinear functions of temperature, but they are approximately linear over ranges of a few degrees. Thus, for some regions of the ocean, these nonlinearities are not significant.

The different types of ages are appropriate for putting timescales on different processes. For thermocline ventilation, where equilibrated water is subducted and mixed isopycnally along extensively outcropping density surfaces, the water subducted within a given year mixes with water subducted in previous years. In this situation, a water parcel is a mixture of water that has left the surface over a period of several years. The average age of this water parcel can be represented by the pCFC age if the change of CFC concentration in the source region is constant with respect to time. This has been

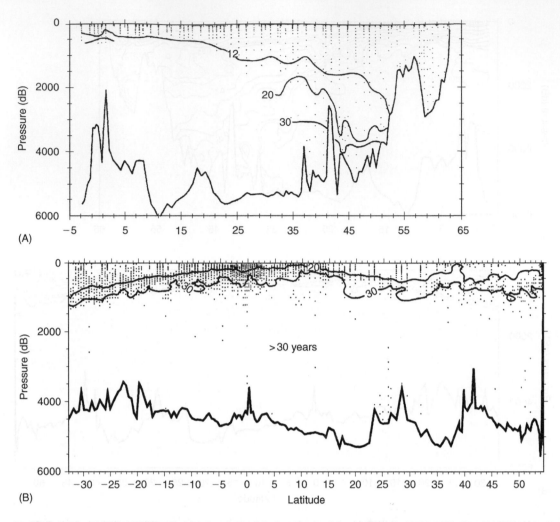

Figure 4 (A) Sections of CFC-11/CFC-12 ratio ages (years) in the eastern Atlantic (latitude 65°N–5°S) along 20°W in summer 1988. (B) Sections of CFC-11/CFC-12 ratio ages (years) in the eastern Pacific (latitude 54°N–32°S) mostly along 135°W in summer 1991. (North Atlantic data from Doney SC and Bullister JB (1992) *Deep-Sea Research* 39: 1857–1883; Pacific data from Fine *et al.* (2001) *Journal of Geophysical Research*.)

confirmed for the North Atlantic thermocline in the eastern basin by comparing pCFC ages to tritium/ He-3 ages.

In regions where surface waters are converted to deep and bottom waters which then spread into a background of low-tracer water, the high CFC concentrations of the cold surface water are diluted by entrainment and mixing. The resulting pCFC age is much too young for the average age of the mixture and much too old for the CFC-bearing component. However, a tracer ratio is conserved in this situation, and the corresponding ratio age represents that of the youngest component of the mixture, not the average age of the water parcel. Thus, there are different estimates of ages that can be derived from CFC-11 and CFC-12, and the associated timescales can be expanded in regions where CFC-113 and CCl₄ data are available.

In most high latitude intermediate and deep-water source regions the age clock is not reset to zero due to lack of time to equilibrate deep mixed layers with the atmosphere. Thus, water masses will start out with an age of a few years (rather than zero), that is, they are not completely renewed during formation. This additional age is called a relic age which can be estimated from observations of the tracers at the water mass formation regions. The relic age can then be subtracted from the tracer ages calculated downstream from the water mass formation regions.

Applications of CFCs to Ocean Processes

Examples of the application of CFCs to understanding oceanographic processes are divided into four subjects: the thermohaline circulation, upper ocean circulation, model constraints, and biogeochemical processes.

Thermohaline Circulation

There is a close coupling of the surface waters in high latitudes to the deep ocean through the density-driven thermohaline circulation. During the process of deep-water formation, atmospheric constituents such as CFCs are introduced into the newly formed water. In recent years, major advances in our knowledge of the thermohaline circulation can be attributed to information derived from transient tracer data, particularly for two reasons. First, the development of analytical techniques so that oceanographers can easily produce large quantities of high quality data. Tracer oceanographers have benefited from multi-investigator programs like the World Ocean Circulation Experiment. The following highlights some of the advances that have come about in our understanding of the thermohaline circulation to which observations of CFCs have contributed:

- Discovery of a new water mass component of North Atlantic Deep Water (NADW), called Upper Labrador Sea Water, location of its formation region and contributing processes, and timescales of eastward spreading along the equator.
- Identification of Denmark Straits Overflow Water as the primary source of bottom water of the western subpolar basin.
- Confirmation of the structure and continuity of the Deep Western Boundary Current throughout the western North Atlantic Ocean, and extension into the South Atlantic.
- Extension of the CFCs well into the interior of the western North Atlantic show the importance of deep recirculation gyres in ventilating the interior basins, and in slowing the equatorward transport to timescales of <30 years with effective spreading rates of 1–2 cm s^{-1}.
- Contribution to quantifying formation rates and decadal climate variability in the Arctic, Greenland and Labrador Seas.
- Estimates for the formation rates of Weddell Sea Deep and Bottom Waters, production rate of Antarctic Bottom Water and pathways and timescales for spreading into the North Atlantic.

Upper Ocean Circulation

The use of CFCs for upper ocean processes has involved the application of concentrations to deduce sources and circulation pathways, and application of pCFC ages. The following highlights some of the advances that have come about in our understanding of the upper ocean circulation to which observations of CFCs have contributed:

- Identification of the Sea of Okhotsk and Alaskan Gyre as important location for the ventilation of North Pacific Intermediate Water, these waters then spread into the subtropics on a timescale of <20 years.
- Quantification of the flux of water from the mixed layer into thermocline and intermediate layers of the North and South Pacific.
- Contribution to the description of sources and pathways of water masses transported from the Pacific through the Indonesian Seas into the Indian Ocean.
- Quantification of the sources of northern and southern water and the processes needed to ventilate the tropical Pacific and Atlantic, including advection, diapycnal and vertical mixing.
- Observation that pathways of the most recently ventilated Antarctic Intermediate Waters are into the eastern South Indian Ocean, while at that level there appears to be flow of older waters from the South Pacific into the western Indian Ocean.
- Quantification of subduction and formation rates for subtropical underwaters and in the North Atlantic its interannual variability that is negatively correlated with intermediate waters of the eastern subpolar gyre.

Model Constraints

In general CFC concentrations and inventories have been used in comparison with model simulated concentrations and inventories. The time-dependent nature of the CFCs provides a stringent test of a model's ability to integrate property distributions over time. The following highlights some of the advances that have come about in our ability to put constraints on models from the use of CFCs in models:

- Dilution of CFCs transported by the Deep Western Boundary Current and effect on tracer ages.
- Testing the sensitivity of a model for correct simulation of formation rates, pathways, and spreading rates.
- Testing the sensitivity of a model for correct simulation of ocean model velocity fields.
- Determining the model sensitivity to subgrid scale mixing for purposes of estimating ventilation rates.
- The importance of considering seasonal variations in the upper oceans as part of the tracer boundary

conditions when trying to simulate subduction processes.

- Demonstration in a model simulation that eddy transport is required to transport South Indian subtropical gyre waters across the equator along the western boundary.
- Use of CFCs to validate model parameterizations of gas fluxes.

Biogeochemical Processes

The tracers provide a method for calculating rates of biogeochemical fluxes that is independent of direct biological measurements. Again the age information from the CFCs is used to calculate rates for these processes. The following highlights some of the advances that have come about in our understanding of biogeochemical processes to which observations of CFCs have contributed:

- Apparent oxygen utilization rates from the central Arctic that are so high, they need to be balanced by transport of high production water from over the continental shelves.
- Quantification of moderate biological consumption and initially low oxygen concentrations in the Arabian Sea are needed to maintain the low oxygen layer.
- Calculation of denitrification rates for the Arabian Sea and Bay of Bengal.

Conclusions

The advantage of oceanic tracers like CFCs is that they can be used to provide timescale information for oceanographic processes. Direct application of the timescale information from the CFCs is used to calculate fluxes of atmospheric constituents, such as

CO_2. The oceans have taken up a considerable portion of the anthropogenic CO_2 released to the atmosphere. A large part of the uptake involves water mass formation in high latitudes. The rate at which these waters are transported into the interior will have an effect on the rate at which anthropogenic CO_2 is taken up.

See also

Air–Sea Gas Exchange. Carbon Dioxide (CO_2) Cycle. Wind Driven Circulation.

Further Reading

Broecker WS and Peng T-H (1982) *Tracers in the Sea.* Palisades, NY: Lamont-Doherty Geological Observatory, Columbia University.

Fine RA (1995) Tracers, time scales and the thermohaline circulation: the lower limb in the North Atlantic Ocean. *Reviews of Geophysics* 33: 1353–1365.

Rowland FS and Molina MJ (1994) Ozone depletion: 20 years after the alarm. *Chemical & Engineering News* 72: 8–13.

Schlosser P and Smethie WS (1994) *Transient Tracers as a Tool to Study Variability of Ocean Circulation. Natural Climate Variability on Decadal-to-Century Time Scales*, pp. 274–288. Washington, DC: National Academic Press.

Smethie WS, Fine RA, Putzka A, and Jones EP (2000) Reaching the flow of North Atlantic Deep Water using chlorofluorocarbons. *Journal of Geophysical Research* 105: 14 297–14 323.

Walker SJ, Weiss RF, and Salameth PK (2000) Reconstructed histories of the annual mean atmospheric mole fractions for the halocarbons CFC-11, CFC-12, CFC-113 and carbon tetrachloride. *Journal of Geophysical Research* 105: 14 285–14 296.

URANIUM-THORIUM SERIES ISOTOPES IN OCEAN PROFILES

S. Krishnaswami, Physical Research Laboratory, Ahmedabad, India

Natural radioactivity in the environment originates from two sources. First, primordial radionuclides which were incorporated into the Earth at the time of its formation are still present in it because of their long half-lives. ^{238}U, ^{235}U, ^{232}Th and their decay series (**Figure 1**), ^{40}K, ^{87}Rb and ^{187}Re are examples of this category. Second, cosmic ray-produced isotopes which are generated continuously in the atmosphere and earth's crust through interactions of cosmic rays with their constituents. ^{3}H, ^{14}C and ^{10}Be are some of the isotopes belonging to this group. The distribution of all these isotopes in the oceans is governed by their supply, radioactive decay, water mixing and their biogeochemical reactivity (the tendency to participate in biological and chemical processes) in sea water. Water circulation plays a dominant role in the dispersion of isotopes which are biogeochemically 'passive' (e.g. ^{3}H, Rn), whereas biological uptake and release, solute–particle interactions and chemical scavenging exert major control in the distribution of biogeochemically 'active' elements (e.g. C, Si, Th, Pb, Po). Systematic study of the isotopes of these two groups in the sea can yield important information on the physical and biogeochemical processes occurring in sea water.

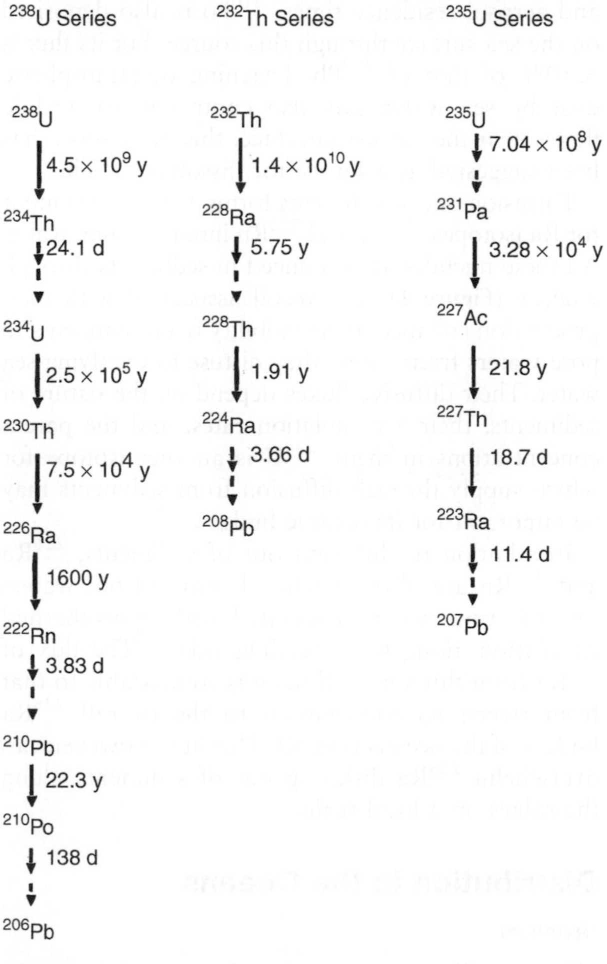

Supply of U/Th Isotopes to the Sea

These nuclides enter the oceans through three principal pathways.

Fluvial Transport

This is the main supply route for ^{238}U, ^{235}U, ^{234}U and ^{232}Th to the sea. These isotopes are transported both in soluble and suspended phases. Their dissolved concentrations in rivers depend on water chemistry and their geochemical behavior. In rivers, uranium is quite soluble and is transported mainly as uranyl carbonate, $UO_2(CO_3)_3^{-4}$, complex. The dissolved uranium concentration in rivers is generally in the range of $0.1-1.0\ \mu g\ l^{-1}$. During chemical weathering ^{235}U is also released to rivers in the same ^{235}U/^{238}U ratio as their natural abundance (1/137.8). This is unlike that of ^{234}U, a progeny of ^{238}U (**Figure 1**) which is released preferentially to solution due to α-recoil effects. As a result, the ^{234}U/^{238}U activity ratios of river waters are generally in excess of that in the host rock and the secular equilibrium value of 1.0 and often fall in the range of 1.1–1.5.

The concentration of dissolved ^{232}Th in rivers, $\sim 0.01\ \mu g\ l^{-1}$ is significantly lower than that of ^{238}U, although their abundances in the upper continental crust are comparable. This is because ^{232}Th (and other Th isotopes) is more resistant to weathering and is highly particle-reactive (the property to be associated with particles) in natural waters and hence is rapidly adsorbed from solution to particles.

Figure 1 ^{238}U, ^{232}Th and ^{235}U decay series: Only the isotopes of interest in water column process studies are shown.

It is likely that even the reported dissolved ^{232}Th concentrations are upper limits, as recent results, based on smaller volume samples and high sensitivity mass-spectrometric measurements seem to show that dissolved ^{232}Th in rivers is associated with smaller particles ($< 0.45 \mu m$ size). Similar to ^{232}Th, the bulk of ^{230}Th and ^{210}Pb is also associated with particles in rivers and hence is transported mainly in particulate form from continents.

^{226}Ra and ^{228}Ra are two other members of the U-Th series (**Figure 1**) for which dissolved concentration data are available for several rivers, these show that they are present at levels of ~ 0.1 d.p.m. l^{-1}. The available data show that there are significant differences between the abundances of U, Ra isotopes and ^{232}Th in the host rocks and in river waters. The various physicochemical processes occurring during the mobilization and transport of these nuclides contribute to these differences.

Rivers also transport U/Th series nuclides in particulate phase to the sea. These nuclides exist in two forms in the particulate phase, one as a part of their lattice structure and the other as surface coating resulting from their adsorption from solution. Analysis of suspended particulate matter from rivers shows the existence of radioactive disequilibria among the members of the same radioactive decay chain. In general, particulate phases are characterized by ^{234}U/^{238}U, ^{226}Ra/^{230}Th activity ratios < 1 and ^{230}Th/^{234}U and ^{210}Pb/^{226}Ra > 1, caused by preferential mobilization of U and Ra over Th and Pb isotopes.

Soluble and suspended materials from rivers enter the open ocean through estuaries. The interactions of sea water with the riverine materials can modify the dissolved concentrations of many nuclides and hence their fluxes to the open sea. Studies of U/Th series isotopes in estuaries show that in many cases their distribution is governed by processes in addition to simple mixing of river and sea water. For example, in the case of U there is evidence for both its addition and removal during transit through estuaries. Similarly, many estuaries have ^{226}Ra concentration higher than that expected from water mixing considerations resulting from its desorption from riverine particles and/or its diffusion from estuarine sediments. Estuaries also seem to act as a filter for riverine ^{232}Th.

The behavior of radionuclides in estuaries could be influenced by their association with colloids. Recent studies of uranium in Kalix River show that a significant part is bound to colloids which is removed in the estuaries through flocculation. Similarly, colloids seem to have a significant control on the ^{230}Th–^{232}Th distribution in estuarine waters.

In situ Production

Radioactive decay of dissolved radionuclides in the water column is an important supply mechanism for several U/Th series nuclides. This is the dominant mode of supply for ^{234}Th, ^{228}Th, ^{230}Th, ^{210}Po, ^{210}Pb, and ^{231}Pa. The supply rates of these nuclides to sea water can be precisely determined by measuring the concentrations of their parents. This is unlike the case of nuclides supplied via rivers whose fluxes are relatively more difficult to ascertain because of large spatial and temporal variations in their riverine concentrations and their modifications in estuaries.

Supply at Air–Sea and Sediment–Water Interfaces

A few of the U/Th nuclides are supplied to the sea via atmospheric deposition and diffusion through sediment pore waters. Decay of ^{222}Rn in the atmosphere to ^{210}Pb and its subsequent removal by wet and dry deposition is an important source of dissolved ^{210}Pb to the sea. As the bulk of the ^{222}Rn in the atmosphere is of continental origin, the flux of ^{210}Pb via this route depends on factors such as distance from land and aerosol residence times. ^{210}Po is also deposited on the sea surface through this source, but its flux is $< 10\%$ of that of ^{210}Pb. Leaching of atmospheric dust by sea water can also contribute to nuclide fluxes near the air–sea interface, this mechanism has been suggested as a source for dissolved ^{232}Th.

Diffusion out of sediments forms a significant input for Ra isotopes, ^{227}Ac and ^{222}Rn into overlying water. All these nuclides are produced in sediments through α-decay (**Figure 1**). The recoil associated with their production enhances their mobility from sediments to pore waters from where they diffuse to overlying sea water. Their diffusive fluxes depend on the nature of sediments, their accumulation rates, and the parent concentrations in them. ^{234}U is another isotope for which supply through diffusion from sediments may be important for its oceanic budget.

In addition to diffusion out of sediments, ^{226}Ra and ^{222}Rn are also introduced into bottom waters through vent waters associated with hydrothermal circulation along the spreading ridges. The flux of ^{226}Ra from this source though is comparable to that from rivers; its contribution to the overall ^{226}Ra budget of the oceans is small. This flux, however, can overwhelm ^{226}Ra diffusing out of sediments along the ridges on a local scale.

Distribution in the Oceans

Uranium

^{238}U and ^{235}U are progenitors of a number of particle-reactive nuclides in sea water which find applications in the study of several water column and sedimentary processes. The study of uranium distribution in the sea is therefore essential to a better understanding of the radioactive disequilibrium between $^{238}U-^{234}Th$, $^{234}U-^{230}Th$, $^{238}U-^{234}U$, and $^{235}U-^{231}Pa$ in sea water. Uranium in sea water is almost entirely in solution as $UO_2(CO_3)_3^{-4}$. Considerable data on its concentration and $^{234}U/^{238}U$ activity ratios are available in the literature, most of which are based on α-spectrometry. These results show that uranium concentration in salinity normalized open ocean sea water (35‰) are the same within experimental uncertainties, $3.3 \pm 0.2 \, \mu g \, l^{-1}$. Measurements with highly sensitive mass-spectrometric techniques also yield quite similar values, but with a much better precision ($\sim 0.2\%$) and narrower range, 3.162–3.282 ng g^{-1} 35‰ salinity water (**Figure 2**). The $\sim 3.8\%$ spread even in the recent data is intriguing and is difficult to account for as uranium is expected to be uniformly distributed in the oceans because of its long residence time, $\sim (2-4) \times 10^5$ years. More controlled sampling and analysis of uranium in sea water are needed to address this issue better. The mass-spectrometric measurements of uranium have also provided data showing that the $^{238}U/^{235}U$ atomic ratio in sea water is 137.17–138.60, identical within errors to the natural abundance ratio of 137.88.

Studies of uranium distribution in anoxic marine basins (e.g., the Black Sea and the Saanich Inlet) have been a topic of interest as sediments of such basins are known to be depositories for authigenic uranium. These measurements show that even in these basins, where H_2S is abundant, uranium exists predominantly in +6 state and its scavenging removal from the water column forms only a minor component of its depositional flux in sediments.

The preferential mobilization of ^{234}U during weathering and its supply by diffusion from deep-sea sediments causes its activity in sea water to be in excess of that of ^{238}U. The $^{234}U/^{238}U$ activity ratio of sea water, determined by α-spectrometry, indicates that it is quite homogenous in open ocean waters with a mean value of 1.14 ± 0.02. Mass-spectrometric measurements have confirmed the above observations of ^{234}U excess with a much better precision and have also led to the use of 'δ notation' to describe $^{234}U-^{238}U$ radioactive disequilibrium.

$$\delta(^{234}U) = [(R_s/R_e) - 1] \times 10^3 \qquad [1]$$

where R_s and R_e are $^{234}U/^{238}U$ atomic ratios in sample and at radioactive equilibrium respectively. The $\delta(^{234}U)$ in the major oceans (**Figure 2**) are same within analytical precision and average 144 ± 2. Coralline $CaCO_3$ and ferromanganese deposits forming from sea water incorporate $^{234}U/^{238}U$ in the ratio of 1.144, the same as that in seawater. The decay of excess ^{234}U in these deposits has been used as a chronometer to determine their ages and growth rates.

Th Isotopes

Among the U/Th series nuclides, the Th isotopes (^{232}Th, ^{230}Th, ^{228}Th, and ^{234}Th), because of their property to attach themselves to particles, are the most extensively used nuclides to investigate particle cycling and deposition in the oceans, processes which have direct relevance to carbon export, solute-particle interactions and particle dynamics. ^{232}Th, ^{230}Th and ^{228}Th are generally measured by α-spectrometry and ^{234}Th by β or γ counting. Highly sensitive mass-spectrometric techniques have now become available for precise measurements of ^{232}Th and ^{230}Th in sea water.

Dissolved ^{232}Th concentration in sea water centers around a few tens of picograms per liter. It is uncertain if the measured ^{232}Th is truly dissolved or is associated with small particles/colloids. Some ^{232}Th profiles show a surface maximum which has been attributed to its release from atmospheric dust.

^{234}Th is continuously produced in sea water from the decay of ^{238}U at a nearly uniform rate of ~ 2.4 atoms l^{-1} min^{-1}. It has been observed that ^{234}Th activity in the surface ~ 200 m is generally deficient relative to its parent ^{238}U suggesting its removal by particles, the mechanism of how this is accomplished, however, is not well understood. This result has been attested by several studies (**Figure 3**). The

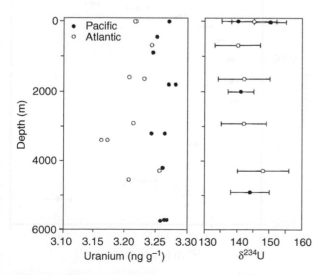

Figure 2 ^{238}U concentration (ng g^{-1} 35‰ salinity water) and $\delta(^{234}U)$ in the Pacific (●) and the Atlantic (○) waters. Data from Chen *et al.* (1986).

residence time of Th in the upper layers of the ocean is determined based on ^{234}Th–^{238}U disequilibrium and the relation;

$$\tau = \left[\frac{R}{(1-R)} \right] \tau_\lambda \qquad [2]$$

where R is the ^{234}Th/^{238}U activity ratio and τ_λ is the radioactive mean life of ^{234}Th (36.8 days). More complex models considering reversible Th exchange, particle remineralization, aggregation and breakup have also been used to treat the ^{234}Th data which allow better understanding of processes regulating both particle and Th cycling. All these studies demonstrate that Th removal by particle scavenging is ubiquitous in surface water and occurs very rapidly, on timescales of a few days to a few months. Much of this variability in the residence time of Th appears to be dictated by particle concentration, short residence times are typical of coastal and biologically productive areas where particles are generally more abundant. These observations have prompted the use of the ^{234}Th–^{238}U pair as a survey tool to determine the export fluxes of carbon from the euphotic zone. The results, though encouraging, suggest the need for a more rigorous validation of the assumptions and parameters used.

^{228}Th activity in the sea exhibits significant lateral and depth variations with higher concentration in the surface and bottom waters and low values in the ocean interior (**Figure 4**). This pattern is governed by the distribution of its parent ^{228}Ra, which determines its production (see section on Ra isotopes). Analogous to ^{234}Th, the distribution of ^{228}Th in the upper

layers of the sea is also determined by particle scavenging which causes the ^{228}Th/^{228}Ra activity ratio to be < 1, the disequilibrium being more pronounced near coasts where particles are more abundant. The residence time of Th in surface waters calculated from ^{234}Th–^{228}U and ^{228}Th–^{228}Ra pairs yields similar values. Profiles of ^{228}Th activity in bottom waters show a decreasing trend with height above the sediment–water interface. In many of these profiles ^{228}Th is in radioactive equilibrium with ^{228}Ra and in a few others it is deficient. Some of these profile data have been used as a proxy for ^{228}Ra to derive eddy diffusion rates in bottom waters.

Systematic measurements of ^{230}Th activity–depth profiles in soluble and suspended phases of sea water have become available only during the past two decades. ^{230}Th is produced from ^{234}U at a nearly uniform rate of ~ 2.7 atoms l^{-1} min^{-1}. The dissolved ^{230}Th activity in deep waters of the North Atlantic is $\sim (5$–$10) \times 10^{-4}$ d.p.m. l^{-1} and in the North Pacific it is ~ 2 times higher. In comparison, the particle ^{230}Th concentrations are about an order of magnitude lower (**Figure 5**). These values are far less than would be expected if ^{230}Th were in radioactive equilibrium with ^{234}U, ~ 2.7 d.p.m. l^{-1}, reinforcing the intense particle-reactive nature of Th isotopes and the occurrence of particle scavenging throughout the seawater column. More importantly, these studies showed that both the soluble and particulate ^{230}Th activities increase steadily with depth (**Figure 5**), an observation which led to the hypothesis of

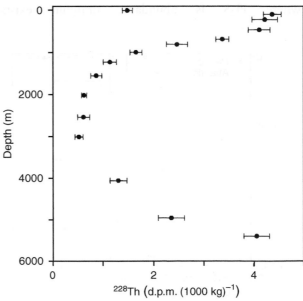

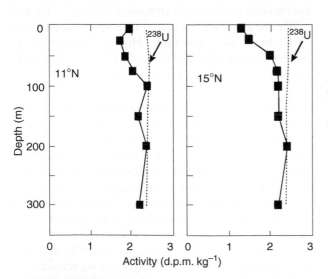

Figure 3 ^{234}Th –^{238}U profiles from the Arabian Sea. Note the clear deficiency of ^{234}Th in the upper layers relative to ^{238}U. (Modified from Sarin *et al.*, 1996.)

Figure 4 ^{228}Th distribution in the Pacific. The higher activity levels of ^{228}Th in near-surface and near-bottom waters reflect that of its parent ^{228}Ra. Data from Nozaki *et al.* (1981).

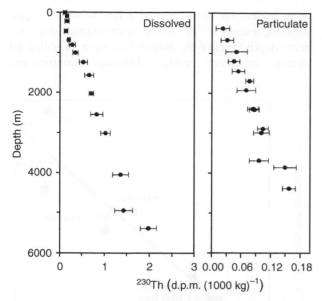

Figure 5 Water-column distributions of dissolved and particulate ^{230}Th. Dissolved ^{230}Th data from the North Pacific (Nozaki *et al.* 1981) and particulate ^{230}Th from the Indian Ocean (Krishnaswami *et al.* 1981). The steady increase in the ^{230}Th activities in both the phases is evident.

reversible exchange of Th between soluble and suspended pools to explain its distribution. In this model the equations governing the distribution of Th in the two phases are:

Suspended Th:

$$k_1 C = (\lambda + k_2)\bar{C} \qquad [3]$$

$$S\frac{d\bar{C}}{dz} + k_1 C - (\lambda + k_2)\bar{C} = 0 \qquad [4]$$

Soluble Th:

$$P + k_2\bar{C} = (\lambda + k_1)C \qquad [5]$$

where P is the production rate of ^{230}Th, C and $\bar{C}$ are the ^{230}Th concentrations in soluble and suspended phases, k_1 and k_2 are the first order adsorption and desorption rate constants, respectively, and S is the settling velocity of particles. Analysis of Th isotope data using this model suggests that adsorption of Th occurs on timescales of a year or so, whereas its release from particles to solution is much faster, i.e. a few months, and that the particles in sea are at equilibrium with Th in solution. Modified versions of the above model include processes such as particle aggregation and breakup, remineralization and release of Th to solution. The timescales of some of these processes also have been derived from the Th isotope data.

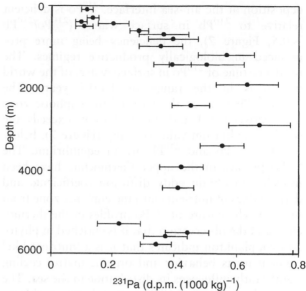

Figure 6 ^{231}Pa distribution in the north-west Pacific. Data from Nozaki and Nakanishi (1985).

^{231}Pa, ^{210}Po, and ^{210}Pb

These three isotopes share a property with Th, in that all of them are particle reactive. ^{231}Pa is a member of the ^{235}U series (**Figure 1**) and is produced in sea water at a rate of ~ 0.11 atoms l^{-1} min^{-1}. Analogous to ^{230}Th, ^{231}Pa is also removed from sea water by adsorption onto particles, causing its activity to be quite low and deficient relative to ^{235}U (**Figure 6**). The ^{231}Pa/^{235}U activity ratio in deep waters of the western Pacific is $\sim 5 \times 10^{-3}$. Measurements of ^{230}Th/^{231}Pa ratios in dissolved, suspended, and settling particles have led to a better understanding of the role of their scavenging by vertically settling particles in the open ocean in relation to their removal on continental margins. The dissolved ^{230}Th/^{231}Pa in sea water is ~ 5, less than the production ratio of ~ 10.8 and those in suspended and settling particles of ~ 20, indicating that ^{230}Th is preferentially sequestered onto settling particles. This, coupled with the longer residence time of ^{231}Pa (a few hundred years) compared to ^{230}Th (a few tens of years), has led to the suggestion that ^{231}Pa is laterally transported from open ocean areas to more intense scavenging regimes such as the continental margins, where it is removed. The measurements of settling fluxes of ^{230}Th and ^{231}Pa using sediment traps and ^{230}Th/^{231}Pa ratios in sediments from various oceanic regions support this connection.

^{210}Po is supplied to sea almost entirely through its *in situ* production from the decay of ^{210}Pb (**Figure 1**), a minor contribution comes from its atmospheric deposition at the air–sea interface. ^{210}Po is deficient relative to ^{210}Pb in surface waters (^{210}Po/^{210}Pb ~0.5, **Figure 7**), the deficiency being more pronounced in biologically productive regimes. The residence time of ^{210}Po in surface waters of the world oceans is in the range of 1 ± 0.5 years. The ^{210}Po/^{210}Pb ratio at the base of the euphotic zone falls between 1.0 and 2.0 and often exceeds the secular equilibrium value of unity (**Figure 7**), below ~200 m ^{210}Po and ^{210}Pb are in equilibrium. The ^{210}Po profiles in the upper thermocline have been modeled to obtain eddy diffusion coefficients and derive fluxes of nutrients into the euphotic zone from its base. The nature of ^{210}Po profiles in the thermocline and the observation that it is enriched in phyto- and zooplankton indicates that it is a 'nutrient like' element in its behavior and organic matter cycling significantly influences its distribution in the sea. The strong dependence of ^{210}Po removal rate on chlorophyll a abundance in various oceans (**Figure 8**) is another proof for the coupling between ^{210}Po and biological activity. In deep and bottom waters, ^{210}Po and ^{210}Pb are generally in equilibrium except in areas of hydrothermal activity where Fe/Mn oxides cause preferential removal of ^{210}Po resulting in ^{210}Po/^{210}Pb activity ratio <1.

The studies of ^{210}Pb–^{226}Ra systematics in the oceans have considerably enhanced our understanding of scavenging processes, particularly in the deep sea and the marine geochemistries of lead and its chemical homologues. ^{210}Pb occurs in excess over ^{226}Ra in surface water (**Figure 9**) resulting from its supply from the atmosphere. This excess, however, is less than that would be expected from the known supply rate of ^{210}Pb from the atmosphere if it is removed only through its radioactive decay. This led to the proposal that ^{210}Pb is scavenged from surface to

deep waters on timescales of a few years. In many profiles, excess ^{210}Pb shows exponential decrease with depth (**Figure 9**), which has been modeled to derive apparent eddy diffusion coefficients.

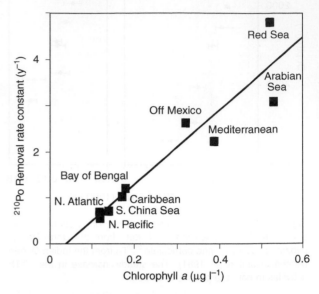

Figure 8 Interrelation between ^{210}Po scavenging rate and chlorophyll a concentrations in various oceanic regions. (Modified from Nozaki *et al.*, 1998.)

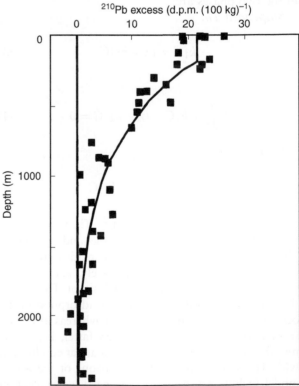

Figure 9 ^{210}Pb excess over ^{226}Ra in the upper thermocline from several stations of the Pacific. This excess results from its atmospheric deposition. (Modified from Nozaki *et al.*, 1980.)

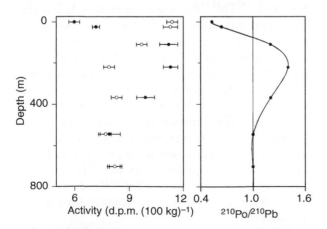

Figure 7 ^{210}Po–^{210}Pb disequilibrium in the Indian Ocean. ^{210}Po (●) is deficient relative to ^{210}Pb (○) near the surface and is in excess at 100–200 m. Data from Cochran *et al.* (1983).

Measurements of ^{210}Pb–^{226}Ra in the deep sea produced a surprise result in that ^{210}Pb was found to be deficient relative to ^{226}Ra with ^{210}Pb/^{226}Ra of ~ 0.5 (**Figure 10**). This was unexpected from the available estimates of the residence time of lead in the deep sea, i.e., a few thousands of years, orders of magnitude more than ^{210}Pb mean-life. Numerous subsequent studies have confirmed this deficiency of ^{210}Pb, though with significant variability in its extent and has led to the conclusion that ^{210}Pb is rapidly and continuously removed from the deep sea on timescales of ~ 50–200 years. The residence time is much shorter, ~ 2–5 years, in anoxic basins such as the Cariaco Trench and the Black Sea. Two other important findings of these studies are that the extent of ^{210}Pb–^{226}Ra disequilibrium increases from open ocean regimes to continental margins and topographic highs and that there is a significant concentration gradient in ^{210}Pb activity from ocean interior to ocean margins. These results coupled with ^{210}Pb data in suspended and settling particles form the basis for the proposal that ^{210}Pb is removed from deep sea both by vertically settling particles and by lateral transport to margins and subsequent uptake at the sediment–water interface. Processes contributing to enhanced uptake in continental margins are still being debated; adsorption on Fe/Mn oxides formed due to their redox cycling in sediments and the effect of higher particle fluxes, both biogenic and continental, have been suggested. It is the ^{210}Pb studies which brought to light the role of continental margins in sequestering particle-reactive species from

the sea, a sink which is now known to be important for other nuclides such as ^{231}Pa and ^{10}Be.

^{222}Rn

The decay of ^{226}Ra in water generates the noble gas ^{222}Rn; both these are in equilibrium in the water column, except near the air–sea and sea–sediment interfaces. ^{222}Rn escapes from sea water to the atmosphere near the air–sea boundary, causing it to be deficient relative to ^{226}Ra, whereas close to the sediment–water interface ^{222}Rn is in excess over ^{226}Ra due to its diffusion out of bottom sediments (**Figure 11**). These disequilibria serve as tracers for mixing rate studies in these boundary layers. In addition, the surface water data have been used to derive ^{222}Rn emanation rates and parameters pertaining to air–sea gas exchange.

^{222}Rn excess in bottom waters decreases with height above the interface, however, the ^{222}Rn activity profiles show distinct variations. Commonly the ^{222}Rn activity decreases exponentially with height above bottom (**Figure 11**) which allows the determination of eddy diffusion coefficient in these waters. In these cases the ^{222}Rn distribution is assumed to be governed by the equation:

$$K\frac{d^2C}{dz^2} - \lambda C = 0 \qquad [6]$$

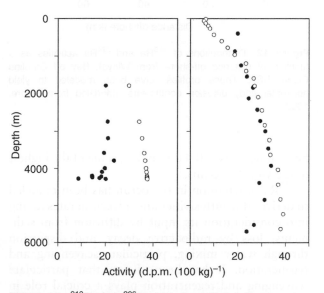

Figure 10 ^{210}Pb ($\bullet$)–^{226}Ra ($\bigcirc$) disequilibrium in sea water. The deficiency of ^{210}Pb in the ocean interior is attributed to its removal by vertically settling particles and at the ocean margins. Data from Craig et al. (1973), Chung and Craig (1980) and Nozaki et al. (1980).

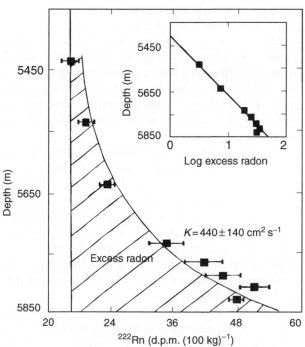

Figure 11 Example of bottom water ^{222}Rn profile in the Atlantic. The calculated vertical eddy diffusion coefficient is also given. (Modified from Sarmiento et al., 1976.)

where K is the eddy diffusion coefficient and z height above bottom with ^{222}Rn activity C. The values of K calculated from the ^{222}Rn data span about two orders of magnitude, 1–100 cm^2 s^{-1}. Other types of ^{222}Rn profiles include those with a two-layer structure and those without specific trend suggesting that its transport via advection and eddy diffusion along isopycnals and non-steady-state condition also need to be considered while describing its distribution. These studies also demonstrated a strong dependence between ^{222}Rn-based eddy diffusion and the stability of bottom water column.

Ra Isotopes

Ra isotopes, particularly, ^{226}Ra and ^{228}Ra have found extensive applications in water circulation studies. All the Ra isotopes, ^{224}Ra, ^{223}Ra, ^{228}Ra, and ^{226}Ra enter the oceans mainly through diffusion from sediments and by desorption from river particulates and are commonly measured by α and γ counting techniques. ^{224}Ra and ^{223}Ra, because of their very short half-lives (**Figure 1**), are useful for studying mixing processes occurring on timescales of a few days to a few weeks which restricts their utility to regions close to their point of injection such as coastal and estuarine waters (**Figure 12**). The half-life of ^{228}Ra is also short, 5.7 years, and hence its concentration decreases with increasing distance from its source, the sediment–water interface, e.g., from coast to open sea (**Figure 13**) surface waters to ocean interior and height above the ocean floor (**Figure 14**). These distributions have been modeled, by treating them as a balance between eddy diffusion and radioactive decay (eqn [6]), to determine the rates of lateral and vertical mixing occurring on timescales of 1–30 years in the thermocline and near bottom waters.

^{226}Ra is the longest lived among the Ra isotopes, with a half-life comparable to that of deep ocean mixing times. The potential of ^{226}Ra as a tracer to study large-scale ocean mixing was exploited using a one-dimensional vertical advection–diffusion model to describe its distribution in the water column. Subsequent studies brought to light the importance of biological uptake and cycling in influencing ^{226}Ra distribution, processes which were later included in the ^{226}Ra model.

Figure 15 shows typical profiles of ^{226}Ra in the oceans. Its concentration in surface waters falls in the range of 0.07 ± 0.01 d.p.m. l^{-1} which steadily increases with depth such that its abundance in the deep waters of the Pacific>Indian>Atlantic (**Figure 15**). ^{226}Ra concentration in the North Pacific

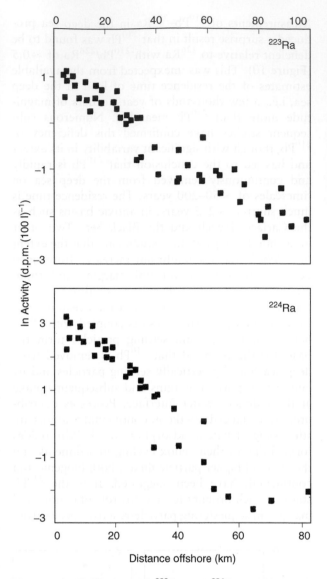

Figure 12 Distributions of ^{223}Ra and ^{224}Ra activities as a function of distance off-shore from Winyah Bay off Carolina Coast, USA. These profiles have been modeled to yield horizontal eddy diffusion coefficients. (Modified from Moore, 1999.)

bottom water is ~ 0.4 d.p.m. l^{-1}, some of the highest in the world's oceans.

^{226}Ra distribution in the ocean has been modeled to derive eddy diffusivities and advection rates taking into consideration its input by diffusion from sediments, loss by radioactive decay, and dispersion through water mixing, particulate scavenging and regeneration. It has been shown that particulate scavenging and regeneration plays a crucial role in contributing to the progressive increase in ^{226}Ra deep water concentration from the Atlantic to the Pacific. Attempts to learn more about particulate transport processes in influencing ^{226}Ra distribution

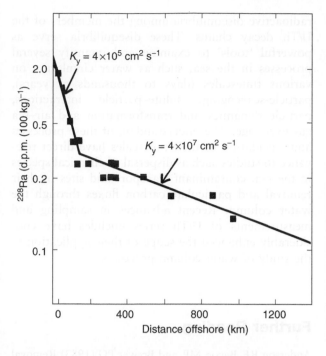

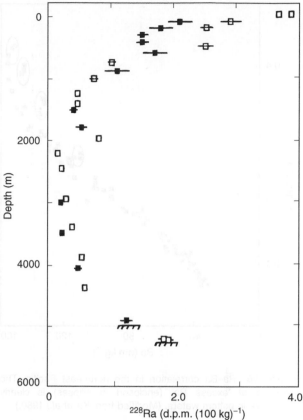

Figure 13 ^{228}Ra distribution as a function of distance from the coast off California. Values of horizontal eddy diffusion coefficient can be derived from these profiles. Note that ^{228}Ra mixes farther into the open sea than ^{223}Ra and ^{224}Ra (**Figure 12**) because of its longer half-life. (Modified from Cochran, 1992.)

Figure 14 Example of ^{228}Ra depth profile in the North Atlantic. The high concentrations near the surface and near the sediment–water interface is due to its supply by diffusion from sediments. Lateral transport also plays an important role in determining surface water concentrations. (Modified from Cochran, 1992.)

using Ba as its stable analogue and Ra–Ba and Ra–Si correlations have met with limited success and have clearly brought out the presence of more ^{226}Ra in deep waters than expected from their Ba content (**Figure 16**). This 'excess' is the nascent ^{226}Ra diffusing out of deep sea sediments and which is yet to take part in particulate scavenging and recycling. Such excesses are quite significant and are easily discernible in the bottom waters of the eastern Pacific.

^{227}Ac

The first measurement of ^{227}Ac in sea water was only reported in the mid-1980s. These results showed that its concentration increases steadily from surface to bottom water (**Figure 17**) and that its activity in ocean interior and deep waters is considerably in excess of its parent ^{231}Pa (**Figure 17**). The diffusion of ^{227}Ac out of bottom sediments is the source of its excess in bottom waters, analogous to those of Ra isotopes. Measurements of ^{227}Ac in pore waters have confirmed this hypothesis. ^{227}Ac distribution can serve as an additional tracer in studies of water mixing processes occurring on decadal timescales, thus complementing the ^{228}Ra applications.

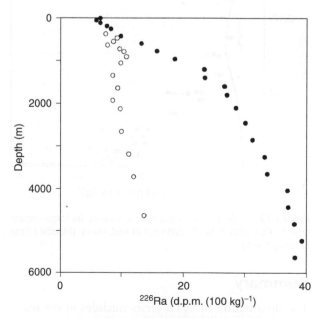

Figure 15 Typical distributions of ^{226}Ra in the water column of the Pacific (●) and Atlantic (○) oceans. Data from Broecker *et al.* (1976) and Chung and Craig (1980).

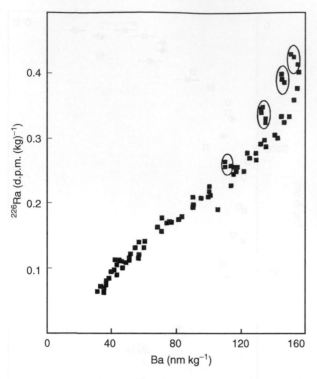

Figure 16 Ra–Ba correlation in the north-east Pacific. The presence of 'excess Ra' (enclosed in ellipses) is clearly discernible in bottom waters. (Modified from Ku *et al.*, 1980.)

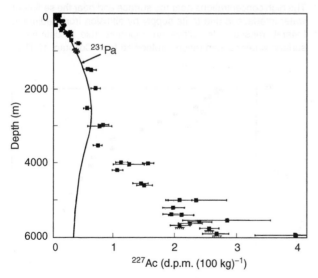

Figure 17 ^{227}Ac profile in the Pacific Ocean. Its large excess over ^{231}Pa is due to its diffusion out of sediments. (Modified from Nozaki, 1984.)

Summary

The distribution of U/Th series nuclides in the sea is regulated by physical and biogeochemical processes occurring in the water column and at the air–sea and sea–sediment interfaces. These processes often create

radioactive disequilibria among the members of the U/Th decay chains. These disequilibria serve as powerful 'tools' to examine and quantify several processes in the sea, such as water circulation on various timescales (days to thousands of years), particle-scavenging, solute–particle interactions, particle dynamics and transformation and air–sea gas exchange. The understanding of these processes and elucidation of their timescales have direct relevance to studies such as dispersal of chemical species in the sea, contaminant transport and sites of their removal and particulate carbon fluxes through the water column. Recent advances in sampling and measurements of U/Th series nuclides have considerably enhanced the scope of their application in the study of water column processes.

Further Reading

Anderson RF, Bacon MP, and Brewer PG (1983) Removal of Th-230 and Pa-231 from the open ocean. *Earth and Planetary Science Letters* 62: 7–23.

Anderson PS, Wasserburg GJ, Chen JH, Papanastassiou DA, and Ingri J (1995) ^{238}U–^{234}U and ^{232}Th–^{230}Th in the Baltic sea and in river water. *Earth and Planetary Science Letters* 130: 218–234.

Bacon MP and Anderson RF (1982) Distribution of thorium isotopes between dissolved and particulate forms in the deep sea. *Journal of Geophysical Research* 87: 2045–2056.

Bhat SG, Krishnaswami S, Lal D, and Rama and Moore WS (1969) Thorium-234/Uranium-238 ratios in the ocean. *Earth and Planetary Science Letters* 5: 483–491.

Broecker WS, Goddard J, and Sarmiento J (1976) The distribution of ^{226}Ra in the Atlantic Ocean. *Earth and Planetary Science Letters* 32: 220–235.

Broecker WS and Peng JH (1982) *Tracers in the Sea*. New York: Eldigio Press, Lamont-Doherty Geological Observatory.

Chen JH, Edwards RL, and Wesserburg GJ (1986) ^{238}U, ^{234}U and ^{232}Th in sea water. *Earth and Planetary Science Letters* 80: 241–251.

Chen JH, Edwards RL, and Wasserburg GJ (1992) Mass spectrometry and application to uranium series disequilibrium. In: Ivanovich M and Harmon RS (eds.) *Uranium Series Disequilibrium: Applications to Earth, Marine and Environmental Sciences*, 2nd edn pp. 174–206. Oxford: Clarenden Press.

Chung Y and Craig H (1980) ^{226}Ra in the Pacific Ocean. *Earth and Planetary Science Letters* 49: 267–292.

Coale KH and Bruland KW (1985) Th-234 : U-238 disequilibria within the California Current. *Limnology and Oceanography* 30: 22–33.

Cochran JK (1992) The oceanic chemistry of the uranium and thorium series nuclides. In: Ivanovich M and Harmon RS (eds.) *Uranium Series Disequilibrium Appli-*

cations to Earth, Marine and Environmental Sciences, 2nd edn, pp. 334–395. Oxford: Clarenden Press.

Cochran JK, Bacon MP, Krishnaswami S, and Turekian KK (1983) ^{210}Po and ^{210}Pb distribution in the central and eastern Indian Ocean. Earth and Planetary Science Letters 65: 433–452.

Craig H, Krishnaswami S, and Somayajulu BLK (1973) ^{210}Pb–^{226}Ra radioactive disequilibrium in the deep sea. Earth and Planetary Science Letters 17: 295.

Dunne JP, Murray JW, Young J, Balistrieri LS, and Bishop J (1997) ^{234}Th and particle cycling in the central equatorial Pacific. Deep Sea Research II 44: 2049–2083.

Krishnaswami S (1999) Thorium: element and geochemistry. In: Marshall CP and Fairbridge RW (eds.) Encyclopedia of Geochemistry, pp. 630–635. Dordrecht: Kluwer Academic.

Krishnaswami S, Sarin MM, and Somayajulu BLK (1981) Chemical and radiochemical investigations of surface and deep particles of the Indian ocean. Earth and Planetary Science Letters 54: 81–96.

Krishnaswami S and Turekian KK (1982) U-238, Ra-226 and Pb-210 in some vent waters of the Galapagos spreading center. Geophysical Research Letters 9: 827–830.

Ku TL, Huh CA, and Chen PS (1980) Meridional distribution of ^{226}Ra in the eastern Pacific along GEOSECS cruise track. Earth and Planetary Science Letters 49: 293–308.

Ku TL, Knauss KG, and Mathieu GG (1977) Uranium in open ocean: concentration and isotopic composition. Deep Sea Research 24: 1005–1017.

Moore WS (1992) Radionuclides of the uranium and thorium decay series in the estuarine environment, In: Ivanovich M and Harmon RS (eds.) Uranium Series Disequilibrium. Applications to Earth, Marine and Environmental Sciences, 2nd edn. pp. 334–395. Oxford: Clarenden Press.

Moore WS (1999) Application of ^{226}Ra, ^{228}Ra, ^{223}Ra and ^{224}Ra in coastal waters to assessing coastal mixing rates and ground water discharge to the oceans. Proceedings of the Indian Academy of Sciences (Earth and Planetary Sciences) 107: 109–116.

Nozaki Y (1984) Excess, Ac-227 in deep ocean water. Nature 310: 486–488.

Nozaki Y, Dobashi F, Kato Y, and Yamamoto Y (1998) Distribution of Ra isotopes and the ^{210}Pb and ^{210}Po balance in surface sea waters of the mid-northern hemisphere. Deep Sea Research I 45: 1263–1284.

Nozaki Y and Nakanishi T (1985) ^{231}Pa and ^{230}Th profiles in the open ocean water column. Deep Sea Research 32: 1209–1220.

Nozaki Y, Turekian KK, and Von Damm K (1980) ^{210}Pb in GEOSECS water profiles from the north Pacific. Earth and Planetary Sciences Letters 49: 393–400.

Nozaki Y, Horibe Y, and Tsubota H (1981) The water column distributions of thorium isotopes in the western north Pacific. Earth and Planetary Sciences Letters 54: 203–216.

Roy-Barman M, Chen JH, and Wasserburg GJ (1996) ^{230}Th–^{232}Th systematics in the central Pacific Ocean: the sources and fate of thorium. Earth and Planetary Science Letters 139: 351–363.

Sarin MM, Rengarajan R, and Ramaswamy V (1996) ^{234}Th scavenging and particle export fluxes from upper 100 m of the Arabian Sea. Current Science 71: 888–893.

Sarmiento JL, Feely HW, Moore WS, Bainbridge AE, and Broecker WS (1976) The relationship between vertical eddy diffusion and buoyancy gradient in the deep sea. Earth and Planetary Letters 32: 357–370.

PHOTOCHEMICAL PROCESSES

N. V. Blough, University of Maryland, College Park, MD, USA

Introduction

Life on Earth is critically dependent on the spectral quality and quantity of radiation received from the sun. The absorption of visible light (wavelengths from 400 to 700 nm) by pigments within terrestrial and marine plants initiates a series of reactions that ultimately transforms the light energy to chemical energy, which is stored as reduced forms of carbon. This complex photochemical process, known as photosynthesis, not only provides all of the chemical energy required for life on Earth's surface, but also acts to decrease the level of a major greenhouse gas, CO_2, in the atmosphere. By contrast, the absorption of ultraviolet light in the UV-B (wavelengths from 280 to 320 nm) and UV-A (wavelengths from 320 to 400 nm) by plants (as well as other organisms) can produce seriously deleterious effects (e.g. photo-inhibition), leading to a decrease in the efficiency of photosynthesis and direct DNA damage (UV-B), as well as impairing or destroying other important physiological processes. The level of UV-B radiation received at the Earth's surface depends on the concentration of ozone (O_3) in the stratosphere where it is formed photochemically. The destruction of O_3 in polar regions, leading to increased levels of surface UV-B radiation in these locales, has been enhanced by the release of man-made chlorofluorocarbons (CFCs), but may also be influenced in part by the natural production of halogenated compounds by biota.

These biotic photoprocesses have long been recognized as critical components of marine ecosystems and air–sea gas exchange, and have been studied extensively. However, only within the last decade or so has the impact of abiotic photoreactions on the chemistry and biology of marine waters and their possible coupling with atmospheric processes been fully appreciated. Light is absorbed in the oceans not only by phytoplankton and water, but also by colored dissolved organic matter (CDOM), particulate detrital matter (PDM), and other numerous trace light-absorbing species. Light absorption by these constituents, primarily the CDOM, can have a number of important chemical and biological consequences including: (1) reduction of potentially harmful UV-B and UV-A radiation within the water column; (2) photo-oxidative degradation of organic matter through the photochemical production of reactive oxygen species (ROS) such as superoxide (O_2^-), hydrogen peroxide (H_2O_2), the hydroxyl radical (OH) and peroxy radicals (RO_2); (3) changes in metal ion speciation through reactions with the ROS or through direct photochemistry, resulting in the altered biological availability of some metals; (4) photochemical production of a number of trace gases of importance in the atmosphere such as CO_2, CO, and carbonyl sulfide (COS), and the destruction of others such as dimethyl sulfide (DMS); (5) the photochemical production of biologically available low molecular weight (LMW) organic compounds and the release of available forms of nitrogen, thus potentially fueling the growth of microorganisms from a biologically resistant source material (the CDOM). These processes provide the focus of this article.

Optical Properties of the Abiotic Constituents of Sea Waters

CDOM is a chemically complex material produced by the decay of plants and algae. This material, commonly referred to as gelbstoff, yellow substance, gilvin or humic substances, can be transported from land to the oceans by rivers or be formed directly in marine waters by as yet poorly understood processes. CDOM is the principal light-absorbing component of the dissolved organic matter (DOM) pool in sea waters, far exceeding the contributions of discrete dissolved organic or inorganic light-absorbing compounds. CDOM absorption spectra are broad and unstructured, and typically increase with decreasing wavelength in an approximately exponential fashion (**Figure 1**). Spectra have thus been parameterized using the expression [1].

$$a(\lambda) = a(\lambda_0) \cdot e^{-S(\lambda - \lambda_0)} \qquad [1]$$

$a(\lambda)$ and $a(\lambda_0)$ are the absorption coefficients at wavelength λ and reference wavelength λ_0, respectively, and S defines how rapidly the absorption increases with decreasing wavelength. Absorption coefficients are calculated from relation [2], where A is the absorbance measured across pathlength, r.

$$a(\lambda) = \frac{2.303 \cdot A(\lambda)}{r} \qquad [2]$$

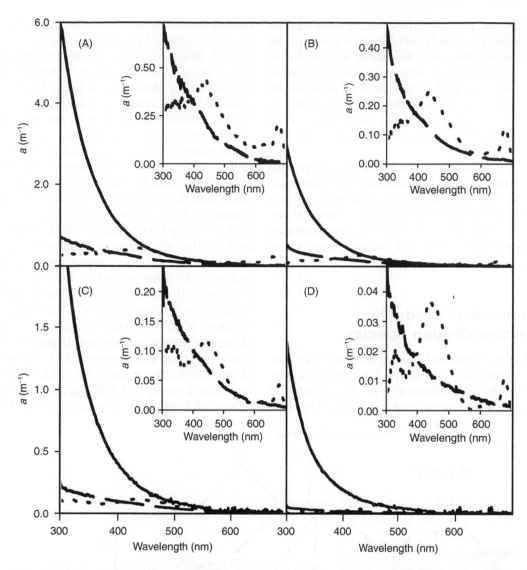

Figure 1 Absorption spectra of CDOM (—), PDM (– – –) and phytoplankton (- - -) from surface waters in the Delaware Bay and Middle Atlantic Bight off the east coast of the USA in July 1998: (A) mid-Delaware Bay at 39° 9.07′ N, 75° 14.29′ W; (B) Mouth of the Delaware Bay at 38° 48.61′ N, 75° 5.07′ W; (C) Mid-shelf at 38° 45.55′ N, 74° 46.60′ W; (D) Outer shelf at 38° 5.89′ N, 74° 9.07′ W.

Due to the exponential increase of $a(\lambda)$ with decreasing λ, CDOM absorbs light strongly in the UV-A and UV-B, and thus is usually the principal constituent within marine waters that controls the penetration depth of radiation potentially harmful to organisms (**Figure 1**). Moreover, for estuarine waters and for coastal waters strongly influenced by river inputs, light absorption by CDOM can extend well into the visible wavelength regime, often dominating the absorption by phytoplankton in the blue portion of the visible spectrum. In this situation, the amount and quality of the photosynthetically active radiation available to phytoplankton is reduced, thus decreasing primary productivity and potentially affecting ecosystem structure. High levels of absorption by CDOM in these regions can also seriously

compromise the determination of phytoplankton biomass through satellite ocean color measurements.

As described below, the absorption of sunlight by CDOM also initiates the formation of a variety of photochemical intermediates and products. The photochemical reactions producing these species ultimately lead to the degradation of the CDOM and the loss, or bleaching, of its absorption. This process can act as a feedback to alter the aquatic light field.

Particulate detrital material (PDM), operationally defined as that light-absorbing material retained on a GFF filter and not extractable with methanol, is a composite of suspended plant degradation products and sediment that also exhibits an exponentially rising absorption with decreasing wavelength (**Figure 1**); eqn [2] has thus been used to parameterize this material as well. However, the values of S acquired for this material are usually smaller than those of the CDOM. In estuarine and near-shore waters, and in shallow coastal waters subject to resuspension of bottom sediments, PDM can contribute substantially to the total water column absorption. However, in most marine waters, the PDM is a rather minor constituent. Little is known about its photochemical reactivity.

Other light-absorbing trace organic compounds such as flavins, as well as inorganic compounds such as nitrate, nitrite, and metal complexes, do not contribute significantly to the total water column absorption. However, many of these compounds are quite photoreactive and will undergo rapid transformation under appropriate light fields.

Photochemical Production of Reactive Oxygen Species

CDOM is the principal abiotic photoreactive constituent in marine waters. Available evidence suggests that the photochemistry of this material is dominated by reactions with dioxygen (O_2) in a

process known as photo-oxidation. In this process, O_2 can act to accept electrons from excited states, radicals (highly reactive species containing an unpaired electron) or radical ions generated within the CDOM by the absorption of light. This leads to the production of a variety of partially reduced oxygen species such as superoxide (O_2^-, the one-electron reduction product of O_2), hydrogen peroxide (H_2O_2, the two-electron reduction product of O_2), peroxy radicals (RO_2, formed by addition of O_2 to carbon-centered radicals, R) and organic peroxides (RO_2H), along with the concomitant oxidation of the CDOM (**Figure 2**). Many of these reduced oxygen species as well as the hydroxyl radical (OH), which is generated by other photochemical reactions, are also quite reactive. These reactive oxygen species or ROS can undergo additional secondary reactions with themselves or with other organic and inorganic seawater constituents. The net result of this complex series of reactions is the light-induced oxidative degradation of organic matter by dioxygen (**Figure 2**). This process leads to the consumption of O_2, the production of oxidized carbon gases (CO_2, CO, COS), the formation of a variety of LMW organic compounds, the release of biologically available forms of nitrogen,

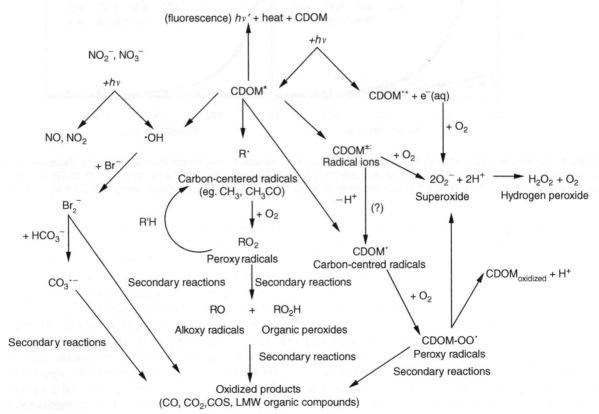

Figure 2 Schematic representation of the photochemical and secondary reactions known or thought to occur following light absorption by CDOM. For a more detailed description of these reactions see the text, Blough and Zepp (1995), and Blough (1997). Not shown in this diagram are primary and secondary reactions of metal species; for a description of these processes, see Helz *et al.* (1994) and Blough and Zepp (1995).

and the loss of CDOM absorption. Through direct photochemical reactions and reactions with the ROS, the speciation of metal ions is also affected.

These photochemical intermediates and products are produced at relatively low efficiencies. About 98–99% of the photons absorbed by CDOM are released as heat, while another ~1% are re-emitted as fluorescence. These percentages (or fractions) of absorbed photons giving rise to particular photoresponses are known as quantum yields (Φ). The Φ for the production of H_2O_2 and O_2^- (the two reduced oxygen species produced with highest efficiency), are approximately one to two orders of magnitude smaller than those for fluorescence, ranging from ~0.1% at 300 nm to ~0.01% at 400 nm. The Φ for other intermediates and products range even lower, from ~0.01% to 0.0000001% (see below). The Φ for most of the intermediates and products created from the CDOM are highest in the UV-B and UV-A, and fall off rapidly with increasing wavelength; yields at visible wavelengths are usually negligible (see for example, **Figure 3**).

The hydroxyl radical, a very powerful oxidant, can be produced by the direct photolysis of nitrate and nitrite (eqns [I]–[III]).

$$NO_3^- + hv \rightarrow O^- + NO_2 \qquad [I]$$

$$NO_2^- + hv \rightarrow O^- + NO \qquad [II]$$

$$O^- + H_2O \rightarrow OH + OH^- \qquad [III]$$

The Φ values for these reactions are relatively high, about 7% for nitrite and about 1–2% for nitrate. However, because of the relatively low concentrations of these compounds in most marine surface waters, as well as their low molar absorptivities in the ultraviolet, the fraction of light absorbed is generally small and thus fluxes of OH from these sources also tend to be small. Recent evidence suggests that OH, or a species exhibiting very similar reactivity, is produced through a direct photoreaction of the CDOM; quinoid moieties within the CDOM may be responsible for this production. Quantum yields are low, ~0.01%, and restricted primarily to the ultraviolet. In estuarine and near-shore waters containing higher levels of iron, the production of OH may also occur through the direct photolysis of iron–hydroxy complexes or through the Fenton reaction (eqn [IV]).

$$Fe^{2+} + H_2O_2 \rightarrow Fe^{3+} + OH + OH^- \qquad [IV]$$

Compounds that do not absorb light within the surface solar spectrum are also subject to photochemical modification through indirect or 'sensitized' photoreactions. In this case, the ROS or intermediates

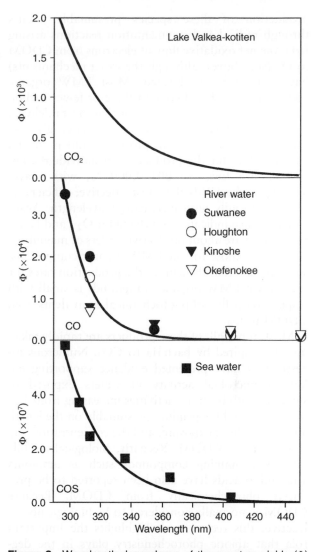

Figure 3 Wavelength dependence of the quantum yields (Φ) for the photochemical production of CO_2, CO, and COS. Data have been replotted from those dependencies originally reported in Vähätalo et al. (2000), Valentine and Zepp (1993), and Weiss et al. (1995) for CO_2, CO, and COS, respectively.

produced by direct photoreactions of a light-absorbing constituent such as CDOM can react secondarily with the nonabsorbing compounds. DMS and COS, two trace gases of some importance to the atmosphere, are thought to be destroyed and created, respectively, by sensitized photoreactions in marine surface waters.

Photochemical Production and Consumption of Low Molecular Weight Organic Compounds and Trace Gases

The photolysis of CDOM produces a suite of LMW organic compounds and a number of trace gases. The production of these species presumably occurs through radical and fragmentation reactions arising from the net oxidative flow of electrons from CDOM to O_2 (see above), although the exact mechanism(s) have yet to be established. Most LMW organic compounds produced contain three or fewer carbon atoms and include such species as acetaldehyde, acetate, acetone, formaldehyde, formate, glyoxal, glyoxalate, methylglyoxal, propanal, and pyruvate. The Φ values for the production of individual compounds are low, $\sim 0.001–0.0001\%$, with wavelengths in the UV-B the most effective; efficiencies decrease rapidly with increasing wavelength. Available evidence indicates that the Φ for O_2^- and H_2O_2 production are about one to two orders of magnitude larger than those for the LMW organic compounds, so it appears that the sum of the production rates for the known LMW organic compounds is small with respect to the flux of photochemical equivalents from CDOM to O_2.

Most, if not all, of these products are rapidly taken up and respired by bacteria to CO_2. Numerous investigators have presented evidence supporting enhanced microbial activity in waters exposed to sunlight, with bacterial activities increasing from 1.5- to almost 6-fold depending presumably on the length and type of light exposure, and the concentration and source of the CDOM. Recently, biologically labile nitrogen-containing compounds such as ammonia and amino acids have also been reported to be produced photochemically from CDOM. Because CDOM is normally considered to be biologically refractive, this recent work highlights the important role that abiotic photochemistry plays in the degradation of CDOM, not only through direct photoreactions, but also through the formation of biologically available products that can be respired to CO_2 or used as nutrients by biota. A recent estimate suggests that the utilization of biologically labile photoproducts could account for as much as 21% of the bacterial production in some near-surface waters.

Carbon dioxide and carbon monoxide are major products of the direct photolysis of CDOM (**Figure 3**). Quantum yields for CO production are about an order of magnitude smaller than those for O_2^- and H_2O_2 production, ranging from $\sim 0.01\%$ at 300 nm to $\sim 0.001\%$ at 400 nm. Available data indicate that the Φ for CO_2 range even higher, perhaps as much as 15–20-fold. The yields for CO_2 production must thus approach, if not exceed, those for O_2^- and H_2O_2. This result is somewhat surprising, since it implies that about one CO_2 is produced for each electron transferred from the CDOM to O_2, further implying a high average redox state for CDOM. Although CDOM (i.e. humic substances) is

known to contain significant numbers of carboxyl moieties that could serve as the source of the CO_2, the yield for CO_2 production, relative to O_2^- and H_2O_2, would be expected to fall rapidly as these groups were removed photochemically; available evidence suggests that this does not occur. An alternative explanation is that other species, perhaps the CDOM itself, is acting as an electron acceptor. Regardless of mechanism, existing information indicates that CO_2 is the dominant product of CDOM photolysis (**Figure 3**).

A recent estimate suggests that the annual global photoproduction of CO in the oceans could be as high as 0.82×10^{15} g C. Assuming that CO_2 photoproduction is 15–20 times higher than that for CO, values for CO_2 formation could reach from 12 to 16×10^{15} g C y^{-1}. To place these numbers in perspective, the estimated annual input of terrestrial dissolved organic carbon to the oceans (0.2×10^{15} g C y^{-1}) is only 1.3–1.7% of the calculated annual CO_2 photoproduction, which is itself about 2–3% of the oceanic dissolved organic carbon pool. These calculated CO_2 (and CO) photoproduction rates may be high due to a number of assumptions, including (1) the complete absorption of UV radiation by the CDOM throughout the oceans, (2) constant quantum yields (or action spectra) for production independent of locale or light history, and (3) neglecting mass transfer limitations associated with physical mixing. Nevertheless, these estimates clearly highlight the potential impact of abiotic photochemistry on the oceanic carbon cycle. Moreover, the products of this photochemistry are generated in near-surface waters where exchange with the atmosphere can take place readily.

Like the LMW organic compounds, bacteria can oxidize CO to CO_2; this consumption takes place in competition with the release of CO to the atmosphere. Due to its photochemical production, CO exists at supersaturated concentrations in the surface waters of most of the Earth's oceans. Recent estimates indicate that global oceanic CO emissions could range from 0.013×10^{15} g y^{-1}–1.2×10^{15} g y^{-1} (see above). The upper estimate is based on calculated photochemical fluxes (see below) and the assumption that all CO produced is emitted to the atmosphere. The lower estimate was calculated using air–sea gas exchange equations and extensive measurements of CO concentrations in the surface waters and atmosphere of the Pacific Ocean. The source of the significant discrepancy between these two estimates has yet to be resolved. Depending on the answer, CO emitted to the atmosphere from the oceans could play a significant role in controlling OH levels in the marine troposphere.

Carbonyl sulfide (COS) is produced primarily in coastal/shelf waters, apparently by the CDOM-photosensitized oxidation of organosulfur compounds. UV-B light is the most effective in its formation, with Φ decreasing rapidly from $\sim 6 \times 10^{-7}$ at 300 nm to $\sim 1 \times 10^{-8}$ by 400 nm (**Figure 3**). The principal sinks of seawater COS are release to the atmosphere and hydrolysis to CO_2 and H_2S. Accounting for perhaps as much as one-third of the total source strength, the photochemical production of COS in the oceans is probably the single largest source of COS to the atmosphere, although more recent work has revised this estimate downward. Smaller amounts of carbon disulfide (CS_2) are also generated photochemically in surface waters through CDOM sensitized reaction(s); Φ values decrease from $\sim 1 \times 10^{-7}$ at 313 nm to 5×10^{-9} at 366 nm. The CS_2 emitted to the atmosphere can react with OH to form additional COS in the troposphere. Although it was previously thought that the oxidation of COS in the stratosphere to form sulfate aerosol could be important in determining Earth's radiation budget and perhaps in regulating stratospheric ozone concentrations, more recent work suggests that other sources contribute more significantly to the background sulfate in the stratosphere.

Dimethyl sulfide (DMS), through its oxidation to sulfate in the troposphere, acts as a source of cloud condensation nuclei, thus potentially influencing the radiative balance of the atmosphere. DMS is formed in sea water through the microbial decomposition of dimethyl sulfonioproprionate (DMSP), a compound believed to act as an osmolyte in certain species of marine phytoplankton. The flux of DMS to the atmosphere is controlled by its concentration in surface sea waters, which is controlled in turn by the rate of its decomposition. Estimates indicate that 7–40% of the total turnover of DMS in the surface waters of the Pacific Ocean is due to the photosensitized destruction of this compound, illustrating the potential importance of this pathway in controlling the flux of DMS to the atmosphere.

In addition to these compounds, the photochemical production of small amounts of nonmethane hydrocarbons (NMHC) such as ethene, propene, ethane, and propane has also been reported. Production of these compounds appears to result from the photolysis of the CDOM, with Φ values of the order of 10^{-7}–10^{-9}. The overall emission rates of these compounds to the atmosphere via this source are negligible with respect to global volatile organic carbon emissions, although this production may play some role in certain restricted locales exhibiting stronger source strengths, or in the marine environment remote from the dominant terrestrial sources.

The photolysis of nitrate and nitrite in sea water produces nitrogen dioxide (NO_2) and nitric oxide (NO), respectively (eqns [I] and [II]). Previous work indicated that the photolysis of nitrite could act as a small net source of NO to the marine atmosphere under some conditions. However, this conclusion seems to be at odds with estimates of the steady-state concentrations of superoxide and the now known rate constant for the reaction of superoxide with nitric oxide ($6.7 \times 10^9 \, M^{-1} \, s^{-1}$) to form peroxynitrite in aqueous phases (eqn [V]).

$$O_2^- + NO \rightarrow {}^- OONO \qquad [V]$$

The peroxynitrite subsequently rearranges in part to form nitrate (eqn [VI]).

$$^- OONO \rightarrow NO_3^- \qquad [VI]$$

Even assuming a steady-state concentration of O_2^- (10^{-12} M) that is about two orders of magnitude lower than that expected for surface sea waters ($\sim 10^{-10}$ M), the lifetime of NO in surface sea waters would be only ~ 150 s, a timescale too short for significant exchange with the atmosphere except for a thin surface layer. Moreover, even in this situation, the atmospheric deposition of additional HO_2 radicals to this surface layer (to form O_2^-) would be expected to act as an additional sink of the NO (flux capping). It appears that most if not all water bodies exhibiting significant steady-state levels of O_2^-, produced either photochemically or thermally, should act as a net sink of atmospheric NO and probably of NO_2 as well. Further, although less is known about the steady-state levels of peroxy radicals in sea waters due largely to their unknown decomposition routes, their high rate constants for reaction with NO (1–$3 \times 10^9 \, M^{-1} \, s^{-1}$) indicate that they should also act as a sink of NO. In fact, methyl nitrate, a trace species found in sea waters, may in part be produced through the aqueous phase reactions (eqns [VII] and [VIII]) with the methylperoxy radical (CH_3OO) generated through a known photochemical reaction of CDOM (or through atmospheric deposition) and the NO arising from the photolysis of nitrite (or through atmospheric deposition).

$$CH_3OO + NO \rightarrow CH_3OONO \qquad [VII]$$

$$CH_3OONO \rightarrow CH_3ONO_2 \qquad [VIII]$$

The concentrations of NO and NO_2 in the troposphere are important because of the involvement of these gases in the formation of ozone.

The atmospheric deposition of ozone to the sea surface can cause the release of volatile iodine compounds to the atmosphere. There is also evidence that methyl iodide can be produced (as well as destroyed) by photochemical processes in surface sea waters. The release of these volatile iodine species from the sea surface or from atmospheric aqueous phases (aerosols) by these processes may act as a control on the level of ozone in the marine troposphere via iodine-catalyzed ozone destruction.

Trace Metal Photochemistry

A lack of available iron is now thought to limit primary productivity in certain ocean waters containing high nutrient, but low chlorophyll concentrations (the HNLC regions). This idea has spurred interest in the transport and photochemical reactions of iron in both seawaters and atmospheric aerosols. Very little soluble Fe(II) is expected to be available at the pH and dioxygen concentration of surface seawaters due to the high stability of the colloidal iron (hydr)oxides. The photoreductive dissolution of colloidal iron oxides by CDOM is known to occur at low pH; this process is also thought to occur in seawaters at high pH, but the reduced iron appears to be oxidized more rapidly than its detachment from the oxide surface. However, some workers have found that CDOM-driven cycles of reduction followed by oxidation increases the chemical availability, which was strongly correlated with the growth rate of phytoplankton. Significant levels of Fe(II) are also known to be produced photochemically in atmospheric aqueous phases (at lower pH) and could serve as a source of biologically available iron upon deposition to the sea surface.

Manganese oxides are also subject to reductive dissolution by light in surface seawaters. This process produces Mn(II), which is kinetically stable to oxidation in the absence of bacteria that are subject to photoinhibition. These two effects lead to the formation of a surface maximum in soluble Mn(II), in contrast to most metals which are depleted in surface waters due to biological removal processes. Other examples of the impact of photochemical reactions on trace metal chemistry are provided in Further Reading.

Photochemical Calculations

Global and regional estimates for the direct photochemical production (or consumption) of a particular photoproduct (or photoreactant) can be acquired with knowledge of the temporal and spatial variation of the solar irradiance reaching the Earth's surface combined with a simple photochemical model (eqn [3]).

$$F(\lambda, z) = E_D(\lambda, z) \cdot \Phi_i(\lambda) \cdot a_{Di}(\lambda) \qquad [3]$$

Here $F(\lambda, z)$ is the photochemical production (or consumption) rate; $E_D(\lambda, z)$ is the downwelling irradiance at wavelength, λ, and depth, z, within the water column; a_{Di} is the diffuse absorption coefficient for photoreactive constituent i; $\Phi_i(\lambda)$ is the quantum yield of this ith constituent. $E_D(\lambda, z)$ is well approximated by eqn [4].

$$E_D(\lambda, z) = E_{D0}(\lambda) \cdot e^{-K_d(\lambda) \cdot z} \qquad [4]$$

$E_{D0}(\lambda)$ is the downwelling irradiance just below the sea surface and $K_d(\lambda)$ is the vertical diffuse attenuation coefficient of downwelling irradiance. $K_d(\lambda)$ can be approximated by eqn [5].

$$K_d(\lambda) \approx \frac{\sum a_i(\lambda) + \sum b_{bi}(\lambda)}{\mu_D} \qquad [5]$$

where $\sum a_i(\lambda)$ and $\sum b_{bi}(\lambda)$ are the total absorption and backscattering coefficients, respectively, of all absorbing and scattering constituents within the water column, and μ_D is the average cosine of the angular distribution of the downwelling light. This factor accounts for the average pathlength of light in the water column, and for direct solar light is approximately equal to $\cos \theta$, where θ is the solar zenith angle (e.g. $\mu_D \sim 1$ when the sun is directly overhead). The diffuse absorption coefficient, a_{Di}, is given by eqn [6].

$$a_{Di} = \frac{a_i}{\mu_D} \qquad [6]$$

This model assumes that the water column is homogeneous, that $K_d(\lambda)$ is constant with depth, and that upwelling irradiance is negligible relative to $E_D(\lambda, z)$. Combining eqns [3], [4] and [6] gives eqn [7].

$$F(\lambda, z) = \frac{E_{D0}(\lambda) . e^{-K_d(\lambda) \cdot z} \cdot \Phi_i(\lambda) \cdot a_i(\lambda)}{\mu_D} \qquad [7]$$

This equation allows calculation of the spectral dependence of the production (consumption) rate as a function of depth in the water column, assuming knowledge of $E_{D0}(\lambda)$, $K_d(\lambda)$, $a_i(\lambda)$ and $\Phi_i(\lambda)$, all of which can be measured or estimated (**Figure 4**). Integrating over wavelength provides the total production (consumption) rate at each depth.

Integration of eqn [7] from the surface to depth z provides the spectral dependence of the photochemical flux (Y) over this interval

$$Y(\lambda, z) = \frac{E_{D0}(\lambda) . \left(1 - e^{-K_d(\lambda) \cdot z}\right) \cdot \Phi_i(\lambda) \cdot a_i(\lambda) / K_d(\lambda)}{\mu_D} \qquad [8]$$

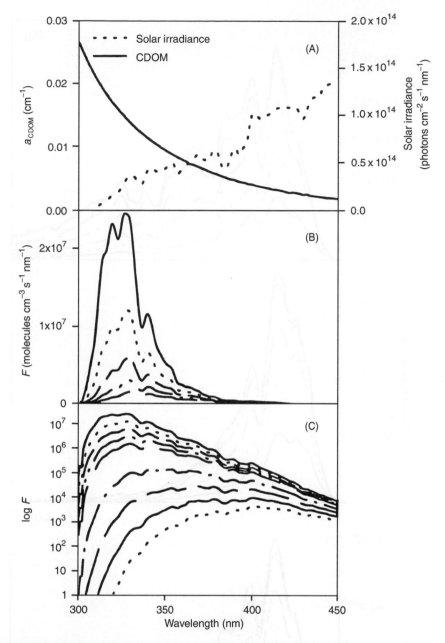

Figure 4   Spectral dependence of CO photoproduction rates with depth, plotted on a linear (B) and logarithmic (C) scale. Depths in (B) are (from top to bottom): surface, 0.5, 1, 1.5, and 2 m. Depths in (C) are (from top to bottom): surface, 0.5, 1, 1.5, 2, 4, 6, 8, and 10 m. These spectral dependencies were calculated using eqn [7], the wavelength dependence of the quantum yield for CO shown in **Figure 3**, and the CDOM absorption spectrum and surface solar irradiance shown in (A). The attenuation of irradiance down the water column in this spectral region was assumed to be only due to CDOM absorption, a reasonable assumption for coastal waters (see **Figure 1**). Note the rapid attenuation in production rates with depth in the UV-B, due to the greater light absorption by CDOM in this spectral region.

which upon substitution of eqn [4] becomes,

$$Y(\lambda, z) = E_{D0}(\lambda) \cdot \left(1 - e^{-K_d(\lambda) \cdot z}\right) \cdot \Phi_i(\lambda) \cdot \frac{a_i(\lambda)}{\sum a_i(\lambda) + \sum b_{bi}(\lambda)} \quad [9]$$

In most, but not all seawaters, the total absorption will be much greater than the total backscatter, $\sum a_i(\lambda) \gg \sum b_{bi}(\lambda)$, and thus the backscatter can be ignored; this approximation is not valid for most estuarine waters and some coastal waters, where a more sophisticated treatment would have to be applied. This approximation leads to the final

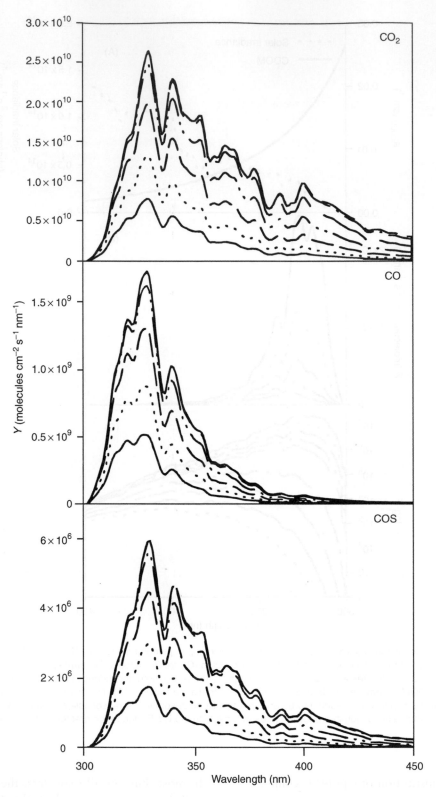

Figure 5 Spectral dependence of the photochemical flux with depth for CO_2, CO, and COS. Fluxes with depth are from the surface to 0.25, 0.5, 1.0, 2.0, and 4 m, respectively (bottom spectrum to top spectrum). Below 4 m, increases in the flux are nominal. These spectral dependencies were calculated using eqn [10], the wavelength dependence of the quantum yields for CO_2, CO and COS shown in **Figure 3**, and the surface solar irradiance shown in **Figure 4A**. CDOM is assumed to absorb all photons in this spectral region (see **Figures 1** and **4**).

expression for the variation of the spectral dependence of the flux with depth (**Figure 5**),

$$F(\lambda, z) = E_{D0}(\lambda).(1 - e^{-K_d(\lambda)\cdot z}) \cdot \Phi_i(\lambda) \cdot \frac{a_i(\lambda)}{\sum a_i(\lambda)} \quad [10]$$

The spectral dependence of the total water column flux $(z \to \infty)$ is then given by,

$$F(\lambda) = E_{D0}(\lambda) \cdot \Phi_i(\lambda) \cdot \frac{a_i(\lambda)}{\sum a_i(\lambda)} \quad [11]$$

with the total flux obtained by integrating over wavelength,

$$F \int_\lambda E_{D0}(\lambda).\Phi_i(\lambda) \cdot \frac{a_i(\lambda)}{\sum a_i(\lambda)} d\lambda \quad [12]$$

To obtain global estimates of photochemical fluxes, many investigators assume that the absorption due to CDOM, a_{CDOM}, dominates the absorption of all other seawater constituents in the ultraviolet, and thus that $a_{CDOM}(\lambda)/\sum a_i(\lambda) \approx 1$. While this approximation is reasonable for many coastal waters, it is not clear that this approximation is valid for all oligotrophic waters. This approximation leads to the final expression for flux,

$$Y \int_\lambda E_{D0}(\lambda).\Phi_i(\lambda) d\lambda \quad [13]$$

which relies only on the surface downwelling irradiance and the wavelength dependence of the quantum yield for the photoreaction of interest. Uncertainties in the use of this equation for estimating global photochemical fluxes include (1) the (usual) assumption that $\Phi(\lambda)$ acquired for a limited number of samples is representative of all ocean waters, independent of locale or light history, and (2) differences in the spatially and temporally averaged values of $E_{D0}(\lambda)$ utilized by different investigators.

Conclusions

The absorption of solar radiation by abiotic sea water constituents initiates a cascade of reactions leading to the photo-oxidative degradation of organic matter and the concomitant production (or consumption) of a variety of trace gases and LMW organic compounds (**Figure 2**), as well as affecting trace metal speciation. The magnitude and impact of these processes on upper ocean biogeochemical cycles and their coupling with atmospheric processes are just beginning to be fully quantified and understood. There remains the need to examine possible couplings between atmospheric gas phase reactions and photochemical reactions in atmospheric aqueous phases.

See also

Air–Sea Transfer: Dimethyl Sulfide, COS, CS₂, NH₄, Non-Methane Hydrocarbons, Organo-Halogens. Air–Sea Transfer: N₂O, NO, CH₄, CO.

Further Reading

Blough NV (1997) Photochemistry in the sea-surface microlayer. In: Liss PS and Duce R (eds.) *The Sea Surface and Global Change*, pp. 383–424. Cambridge: Cambrige University Press.

Blough NV and Green SA (1995) Spectroscopic characterization and remote sensing of non-living organic matter. In: Zepp RG and Sonntag C (eds.) *The role of Non-living Organic Matter in the Earth's Carbon Cycle*, pp. 23–45. New York: John Wiley.

Blough NV and Zepp RG (1995) Reactive oxygen species in natural waters. In: Foote CS, Valentine JS, Greenberg A, and Liebman JF (eds.) *Reactive Oxygen Species in Chemistry*, pp. 280–333. New York: Chapman & Hall.

de Mora S, Demers S, and Vernet M (eds.) (2000) *The Effects of UV Radiation in the Marine Environment*. Cambridge: Cambridge University Press.

Häder D-P, Kumar HD, Smith RC, and Worrest RC (1998) Effects of UV-B radiation on aquatic ecosystems. *Journal of Photochemistry and Photobiology B* 46: 53–68.

Helz GR, Zepp RG, and Crosby DG (eds.) (1994) *Aquatic and Surface Photochemistry*. Ann Arbor, MI: Lewis Publishers.

Huie RE (1995) Free radical chemistry of the atmospheric aqueous phase. In: Barker JR (ed.) *Progress and Problems in Atmospheric Chemistry*, pp. 374–419. Singapore: World Scientific Publishing Co.

Kirk JTO (1994) *Light and Photosynthesis in Aquatic Ecosystems*. Cambridge: Cambridge University Press.

Moran MA and Zepp RG (1997) Role of photoreactions in the formation of biologically labile compounds from dissolved organic matter. *Limnology and Oceanography* 42: 1307–1316.

Thompson AM and Zafiriou OC (1983) Air–sea fluxes of transient atmospheric species. *Journal of Geophysical Research* 88: 6696–6708.

Vähätalo AV, Salkinoja-Salonen M, Taalas P, and Salonen K (2000) Spectrum of the quantum yield for photochemical mineralization of dissolved organic carbon in a humic lake. *Limnology and Oceanography* 45: 664–676.

Valentine RL and Zepp RG (1993) Formation of carbon monoxide from the photodegradation of terrestrial dissolved organic carbon in natural waters. *Environmental Science Technology* 27: 409–412.

Weiss EW, Andrews SS, Johnson JE, and Zafiriou OC (1995) Photoproduction of carbonyl sulfide in south Pacific Ocean waters as a function of irradiation wavelength. *Geophysical Research Letters* 22: 215–218.

Zafiriou OC, Blough NV, Micinski E, *et al.* (1990) Molecular probe systems for reactive transients in natural waters. *Marine Chemistry* 30: 45–70.

Zepp RG, Callaghan TV, and Erickson DJ (1998) Effects of enhanced solar ultraviolet radiation on biogeochemical cycles. *Journal of Photochemistry and Photobiology B* 46: 69–82.

THE SEA SURFACE, WAVES AND UPPER OCEAN PROCESSES

SURFACE FILMS

W. Alpers, University of Hamburg, Hamburg, Germany

Introduction

Surface films floating on the sea surface are usually attributed to anthropogenic sources. Such films consist, for example, of crude oil discharged from tankers during cleaning operations or accidents. However, much more frequently surface films that are of natural origin are encountered at the sea surface. These natural surface films consist of surface-active compounds that are secreted by marine plants or animals. According to their physico-chemical characteristics the film-forming substances tend to be either enriched at the sea surface (more hydrophobic character, sometimes referred to as 'dry surfactant') or they prevail within the upper water layer (more hydrophobic character 'wet surfactants'). The first type of surface-active substances ('dry surfactants') are able to form monomolecular slicks at the air–water interface and damp the short-scale surface waves (short-gravity capillary waves) much more strongly than the second type. This implies that they have a strong effect on the mass, energy, and momentum transfer processes at the air–water interface. They also affect these transfer processes by reducing the turbulence in the subsurface layer which is instrumental in transporting water from below to the surface.

Both types of surface films are easily detectable by radars because radars are roughness sensors and surface films strongly reduce the short-scale sea surface roughness.

Orgin of Surface Films

Surface films at the sea surface can be either of anthropogenic or natural origin. Anthropogenic surface films consist, e.g., of crude or petroleum oil spilled from ships or oil platforms ('spills'), or of surface-active substances discharged from municipal or industrial plants ('slicks'). Natural surface films may also consist of crude oil which is leaking from oil seeps on the seafloor, but usually they consist of surface-active substances, which are produced by biogenic processes in the sea ('biogenic slicks'). In general, the biogenic surface slicks consisting of sufficiently hydrophobic substances ('dry surfactants') are only one molecular layer thick (approximately 3 nm). This implies that it needs only few liters of surface-active material to cover an area of $1 \, km^2$. The prime biological producers of natural surface films in the sea are algae and some bacteria. Also zooplankton and fish produce surface-active materials, but the amount is usually small in comparison with the primary biological production. Primary production depends on the quantity of light energy available to the organism and the availability of inorganic nutrients. In the higher latitudes light energy depends strongly on the season of the year which results in a seasonal variation of the primary biological production in the ocean and thus of the slick coverage by natural surface films. At times when the biological productivity is high, i.e., during plankton blooms, the probability of encountering natural biogenic surface films is strongly enhanced.

In other regions of the world's ocean where the primary production is not mainly determined by the quantity of light energy available to the organisms, but by the nutrient levels, the seasonal variation of the slick coverage is smaller, however, still observable. Surface slicks of biogenic origin are mainly encountered in sea regions where the nutrition factor favors productivity. This is the case on continental shelves, slopes, and in upwelling regions where nutrition-rich cold water is transported to the sea surface.

The areal extent, the concentration and the composition of the surface films vary strongly with time. At high wind speeds (typically above 8–$10 \, m \, s^{-1}$) breaking waves disperse the films by entrainment into the underlying water such that they disappear from the sea surface. In general, the probability of encountering surface films of biogenic origin decreases with wind speed. Furthermore, after storms, enhanced coverage of the sea surface with biogenic slicks consisting of 'dry surfactants' is often observed which is due to the fact that, firstly, the secretion of surface-active material by plankton is being increased during higher wind speed periods, and secondly, the surface-active substances are being transported to the sea surface from below by turbulence and rising air bubbles generated by breaking waves. The composition of the sea slicks varies also with time because constituents of the surface films are selectively removed by dissolution, evaporation, enzymatic degradation and photocatalytic oxidation.

Modifications of Air–Sea Interaction by Surface Films

Numerous processes that take place at the air–sea interface are affected by surface films. Among other things, surface films: (1) attenuate the surface waves; (2) reduce wave breaking; (3) reduce gas transfer; (4) increase the sea-surface temperature; (5) change the reflection of sunlight; and (6) reduce the intensity of the radar backscatter.

Attenuation of Surface Waves

Two main factors contribute to the damping of short-scale surface waves by surface films:

1. the enhanced viscous dissipation in a thin water layer below the water surface caused by strong velocity gradients induced by the presence of viscoelastic films at the water surface; and
2. the decrease in energy transfer from the wind to the waves due to the reduction of the aerodynamic roughness of the sea surface.

The enhanced viscous dissipation caused by the surface films results from the fact that, due to the different boundary condition imposed by the film at the sea surface, strong vertical velocity gradients are encountered in a thin layer below the water surface. This layer, also called shear layer, has a thickness of the order of 10^{-4} m. Here strong viscous dissipation takes place. In the case of mineral oil films floating on the sea surface, the shear layer may lie completely within the oil layer, but more often it extends also into the upper water layer since the thickness of mineral oil films is typically in the range of 10^{-3}–10^{-6} m.

In the case of biogenic monomolecular surface films accumulating at a rough sea surface, the surface films are compressed and dilated periodically, causing variations of the concentration of the molecules and thus of the surface tension. In this case not only the well-known gravity-capillary waves (surface waves) are excited, but also the so-called Marangoni waves. The Marangoni waves are predominantly longitudinal waves in the shear layer. They are heavily damped by viscous dissipation; at a distance of only one wavelength from their source their amplitude has already decreased to less than one-tenth of the original value. This is the reason why Marangoni waves escaped detection until 1968. When these gravity-capillary waves and Marangoni waves are in resonance as given by linear wave theory, the surface waves experience maximum damping. Depending on the viscoelastic properties of the surface film, maximum damping of the surface waves usually occurs in the centimeter to decimeter wavelength region.

Reduction of Wave Breaking

Since surface films reduce roughness of the sea surface, the stress exerted by the wind on the sea surface is reduced. Furthermore, the steepness of the short-scale waves is decreased which leads to less wave breaking.

Reduction of Gas Transfer

Biogenic surface films do not constitute a direct resistance for gas transfer. However, they do have a major effect on the structure of subsurface turbulence and thus on the rate at which surface water is renewed by water from below. Furthermore, surface films reduce the air turbulence above the ocean surface and thus also the surface renewal. As a consequence, the gas transfer rate across the air–sea interface is reduced in the presence of surface films.

Change of Sea Surface Temperature

In infrared images the sea surface areas covered with biogenic surface films usually appear slightly warmer than the adjacent slick-free sea areas (typical temperature increase 0.2–0.5 K). This is due to the fact that surface films reduce the mobility of the near-surface water molecules and slow down the conventional overturn of the surface layer by evaporation.

Change of Reflection of Sunlight

Slick-covered areas of the sea surface are easily visible by eye when they lie in the sun-glitter area. This is an area where facets on the rough sea surface are encountered that have orientations that reflect the sunlight to the observer. When the surface is covered with a surface film, the sea surface becomes smoother and thus the orientation of the facets is changed such that the amount of light reflected to the observer is increased. Thus surface slicks become detectable in sun-glitter areas as areas of increased brightness. Outside the sun-glitter area they are sometimes also visible, but with a much fainter contrast. In this case they appear as areas of reduced brightness relative to the surrounding.

Radar Backscattering

Surface slicks floating on the sea surface also become visible on radar images because they reduce the short-scale sea surface roughness. Since the intensity of the radar backscatter is determined by the amplitude of short-scale surface waves, slick-covered sea surfaces appear on radar images as areas of reduced radar backscattering. Since radars have their own illumination source and transmit electromagnetic waves with

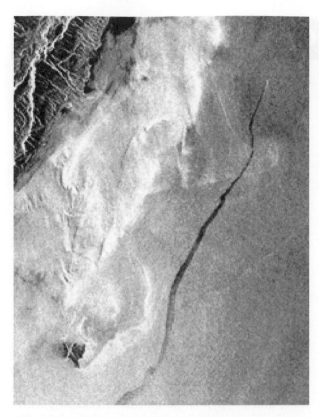

Figure 1 Radar image acquired by the synthetic aperture radar (SAR) aboard the First European Remote Sensing satellite (ERS-1) on 20 May 1994 over the coastal waters east of Taiwan. The imaged area is 70 km × 90 km. It shows a ship (the bright spot at the front of the black trail) discharging oil. The oil trail, which is approximately 80 km long, widens towards the rear because the oil disperses with time. Copyright © 2000, European Space Agency.

Figure 2 Radar image acquired by the SAR aboard the Second European Remote Sensing satellite (ERS-2) on 10 May 1998 over the Western Baltic Sea which includes the Bight of Lübeck (Germany). The imaged area is 90 km × 100 mm. Visible are the lower left coastal areas of Schleswig-Holstein with the island of Fehmarn (Germany) and in the upper right part of the Danish island of Lolland. The black areas are sea areas covered with biogenic slicks which are particularly abundant in this region during the time of the spring plankton bloom. The slicks follow the motions of the sea surface and thus render oceanic eddies visible on the radar image. Copyright © 2000, European Space Agency.

wavelengths in the centimeter to decimeter range, radar images of the sea surface can be obtained independent of the time of the day and independent of cloud conditions. This makes radar an ideal instrument for detecting oil pollution and natural surface films at the sea surface. Consequently, most oil pollution surveillance aircraft which patrol coastal waters for locating illegal discharges of oil from ships are equipped with imaging radars.

Unfortunately the reduction in backscattered radar intensity caused by mineral oil films is often of the same order (typically 5–10 decibels) as that of natural surface films. This makes it difficult by using the information contained in the reduction of the backscattered radar intensisty to differentiate whether the black patches visible on radar images of the sea surface originate from one or the other type of film. However, in many cases the shape of the black patches on the radar images can be used for discrimination. A long elongated dark patch is indicative of an oil spill originating from a travelling ship. Examples of radar images on which both types of surface films are visible are shown in **Figures 1** and **2**.

See also

Air–Sea Gas Exchange. Satellite Remote Sensing of Sea Surface Temperatures. Surface Gravity and Capillary Waves.

Further Reading

Alpers W and Hühnerfuss H (1989) The damping of ocean waves by surface films: a new look at an old problem. *Journal of Geophysical Research* 94: 6251–6265.

Levich VG (1962) *Physico-Chemical Hydrodynamics.* Englewood Cliffs, NJ: Prentice-Hall.

Lucassen J (1982) Effect of surface-active material on the damping of gravity waves: a reappraisal. *Journal of Colloid Interface Science* 85: 52–58.

Tsai WT (1996) Impact of surfactant on turbulent shear layer under the air–sea interface. *Journal of Geophysical Research* 101: 28557–28568.

SURFACE GRAVITY AND CAPILLARY WAVES

W. K. Melville, Scripps Institution of Oceanography,
La Jolla CA, USA

Introduction

Ocean surface waves are the most common oceanographic phenomena that are known to the casual observer. They can at once be the source of inspiration and primal fear. It is remarkable that the complex, random wave field of a storm-lashed sea can be studied and modeled using well-developed theoretical concepts. Many of these concepts are based on linear or weakly nonlinear approximations to the full nonlinear dynamics of ocean waves. Early contributors to these theories included such luminaries as Cauchy, Poisson, Stokes, Lagrange, Airy, Kelvin and Rayleigh. Many of the current challenges in the study of ocean surface waves are related to nonlinear processes which are not yet well understood. These include dynamical coupling between the atmosphere and the ocean, wave–wave interactions, and wave breaking.

For the purposes of this article, surface waves are considered to extend from low frequency swell from distant storms at periods of 10 s or more and wavelengths of hundreds of meters, to capillary waves with wavelengths of millimeters and frequencies of $O(10)$ Hz. In between are wind waves with lengths of $O(1–100)$ m and periods of $O(1–10)$ s. **Figure 1** shows a spectrum of surface waves measured from the Research Platform FLIP off the coast of Oregon. The spectrum, Φ, shows the distribution of energy in the wave field as a function of frequency. The wind wave peak at approximately 0.13 Hz is well separated from the swell peak at approximately 0.06 Hz.

Ocean surface waves play an important role in air–sea interaction. Momentum from the wind goes into both surface waves and currents. Ultimately the waves are dissipated either by viscosity or breaking, giving up their momentum to currents. Surface waves affect upper-ocean mixing through both wave breaking and their role in the generation of Langmuir circulations. This breaking and mixing influences the temperature of the ocean surface and thus the thermodynamics of air–sea interaction. Surface waves impose significant structural loads on ships and other structures. Remote sensing of the ocean

surface, from local to global scales, depends on the surface wave field.

Basic Formulations

The dynamics and kinematics of surface waves are described by solutions of the Navier–Stokes equations for an incompressible viscous fluid, with appropriate boundary and initial conditions. Surface waves of the scale described here are usually generated by the wind, so the complete problem would include the dynamics of both the water and the air above. However, the density of the air is approximately 800 times smaller than that of the water, so many aspects of surface wave kinematics and dynamics may be considered without invoking dynamical coupling with the air above.

The influence of viscosity is represented by the Reynolds number of the flow, $R_e = UL/\mu$, where U is a characteristic velocity, L a characteristic length scale, and $v = \mu/\rho$ is the kinematic viscosity, where μ is the viscosity and ρ the density of the fluid. The Reynolds number is the ratio of inertial forces to viscous forces in the fluid and if $R_e \gg 1$, the effects of viscosity are often confined to thin boundary layers, with the interior of the fluid remaining essentially inviscid ($v = 0$). (This assumes a homogeneous fluid. In contrast, internal waves in a continuously stratified fluid are rotational since they introduce baroclinic generation of vorticity in the interior of the fluid). Denoting the fluid velocity by $\mathbf{u} = (u, v, w)$, the vorticity of the flow is given by $\zeta = \nabla \times \mathbf{u}$. If $\zeta = 0$, the flow is said to be irrotational. From Kelvin's circulation theorem, the irrotational flow of an incompressible ($\nabla.\mathbf{u} = 0$) inviscid fluid will remain irrotational as the flow evolves. The essential features of surface waves may be considered in the context of incompressible irrotational flows.

For an irrotational flow, $\mathbf{u} = \nabla \phi$ where the scalar ϕ is a velocity potential. Then, by virtue of incompressibility, ϕ satisfies Laplace's equation

$$\nabla^2 \phi = 0 \qquad [1]$$

We denote the surface by $z = (x, y, t)$, where (x, y) are the horizontal coordinates and t is time. The kinematic condition at the impermeable bottom at $z = -h$, is one of no flow through the boundary:

$$\frac{\partial \phi}{\partial z} = 0 \quad \text{at} \quad z = -h \qquad [2]$$

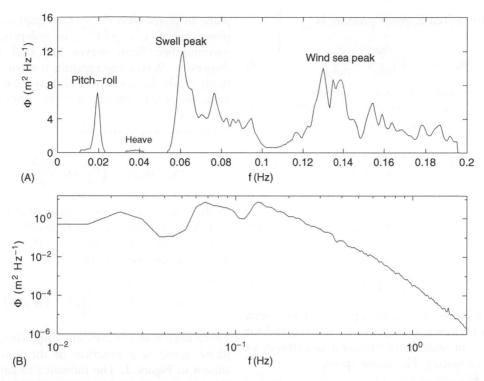

Figure 1 (A) Surface displacement spectrum measured with an electromechanical wave gauge from the Research Platform FLIP in 8 m s^{-1} winds off the coast of Oregon. Note the wind-wave peak at 0.13 Hz, the swell at 0.06 Hz and the heave and pitch and roll of FLIP at 0.04 and 0.02 Hz respectively. (B) An extension of (A) with logarithmic spectral scale, note that from the wind sea peak to approximately 1 Hz the spectrum has a slope like f^{-4}, common in wind-wave spectra. (Reproduced with permission from Felizardo FC and Melville WK (1995). Correlations between ambient noise and the ocean surface wave field. *Journal of Physical Oceanography* 25: 513–532.)

There are two boundary conditions at $z = \eta$:

$$\frac{\partial \eta}{\partial t} + u\frac{\partial \eta}{\partial x} + v\frac{\partial \eta}{\partial y} = w \qquad [3]$$

$$\frac{\partial \phi}{\partial t} + \frac{1}{2}\mathbf{u}^2 + g\eta = (p_a - p)/\rho \qquad [4]$$

The first is a kinematic condition which is equivalent to imposing the condition that elements of fluid at the surface remain at the surface. The second is a dynamical condition, a Bernoulli equation, which is equivalent to stating that the pressure p_- at $z = \eta_-$, an infinitesimal distance beneath the surface, is just a constant atmospheric pressure, p_a, plus a contribution from surface tension. The effect of gravity is to impose a restoring force tending to bring the surface back to $z = 0$. The effect of surface tension is to reduce the curvature of the surface.

Although this formulation of surface waves is considerably simplified already, there are profound difficulties in predicting the evolution of surface waves based on these equations. Although Laplace's equation is linear, the surface boundary conditions are nonlinear and apply on a surface whose specification is a part of the solution. Our ability to accurately predict the evolution of nonlinear waves is

limited and largely dependent on numerical techniques. The usual approach is to linearize the boundary conditions about $z = 0$.

Linear Waves

Simple harmonic surface waves are characterized by an amplitude a, half the distance between the crests and the troughs, and a wavenumber vector $\mathbf{k}$ with $|\mathbf{k}| = k = 2\pi/\lambda$, where λ is the wavelength. The surface displacement, (unless otherwise stated, the real part of complex expressions is taken)

$$\eta = ae^{i(\mathbf{k}\cdot\mathbf{x} - \sigma t)} \qquad [5]$$

where $\sigma = 2\pi/T$ is the radian frequency and T is the wave period. Then ak is a measure of the slope of the waves, and if $ak \ll 1$, the surface boundary conditions can be linearized about $z = 0$.

Following linearization, the boundary conditions become

$$\frac{\partial \eta}{\partial t} = w \qquad [6]$$

$$\frac{\partial \phi}{\partial t} + g\eta = \frac{\Gamma}{\rho}\left(\frac{\partial^2 \eta}{\partial x^2} + \frac{\partial^2 \eta}{\partial y^2}\right) \quad \text{at} \quad z = 0 \qquad [7]$$

where the linearized Laplace pressure is

$$p_a - p_- = \Gamma \left(\frac{\partial^2 \eta}{\partial x^2} + \frac{\partial^2 \eta}{\partial y^2} \right) \qquad [8]$$

where Γ is the surface tension coefficient.

Substituting for η and satisfying Laplace's equation and the boundary conditions at $z = 0$ and $-h$ gives

$$\phi = \frac{ig'a}{\sigma} \frac{\cosh k(z+h)}{\cosh kh} \qquad [9]$$

where

$$\sigma^2 = g'k \tanh kh \qquad [10]$$

and

$$g' = g(1 + \Gamma k^2/\rho) \qquad [11]$$

Equations relating the frequency and wavenumber, $\sigma = \sigma(k)$, are known as dispersion relations, and for linear waves provide a fundamental description of the wave kinematics. The phase speed,

$$c = \sigma/k = \left(\frac{g'}{k} \tanh kh \right)^{1/2} \qquad [12]$$

is the speed at which lines of constant phase (e.g., wave crests) move.

For waves propagating in the x-direction, the velocity field is

$$u = \frac{g'ak}{\sigma} \frac{\cosh (z+h)}{\cosh kh} e^{i(kx-\sigma t)} \qquad [13]$$

$$v = 0 \qquad [14]$$

$$w = -\frac{ig'ak}{\sigma} \frac{\sinh (z+h)}{\cosh kh} e^{i(kx-\sigma t)} \qquad [15]$$

and the pressure

$$p = \rho g' \eta \frac{\cosh (z+h)}{\cosh kh} \qquad [16]$$

The velocity decays with depth away from the surface, and, to leading order, elements of fluid execute elliptical orbits as the waves propagate.

For shallow water, $kh \ll 1$,

$$(u, v, w, p) = \left(\frac{g'k}{\sigma}, 0, 0, \rho g' \right) \eta \qquad [17]$$

so that there is no vertical motion, just a uniform sloshing backwards and forwards in the horizontal

plane in phase with the surface displacement η. The phase speed $c = (g'h)^{1/2}$, is independent of the wavenumber. Such waves are said to be non-dispersive. Waves propagating towards shore eventually attain this condition, and, as the depth tends to zero, nonlinear effects become important as ak increases.

For very deep water, $kh \gg 1$,

$$(u, v, w, p) = \left(\frac{g'k}{\sigma}, 0, -\frac{ig'k}{\sigma}, \rho g' \right) e^{kz} \eta \qquad [18]$$

so that the water particles execute circular motions that decay exponentially with depth. The horizontal motion is in phase with the surface displacement, and the phase speed of the waves

$$c = (g'/k)^{1/2} = \left[\frac{g}{k} (1 + \Gamma k^2/\rho g) \right]^{1/2} \qquad [19]$$

These deep-water waves are dispersive; that is, the phase speed is a function of the wavenumber as shown in **Figure 2**. The influence of surface tension relative to gravity is determined by the value of the dimensionless parameter $\Sigma = \Gamma k^2/\rho g$. When $\Sigma = 1$, the wavelength $\lambda = 1.7$cm and the phase speed is a minimum at $c = 23$ cm s^{-1}. When $\Sigma \gg 1$, surface tension is the dominant restoring force, the wavelength is less than 1.7 cm, and the phase speed increases as the wavelength decreases. When $\Sigma \ll 1$, gravity is the dominant restoring force, the wavelength is greater than 1.7 cm, and the phase speed increases as the wavelength increases.

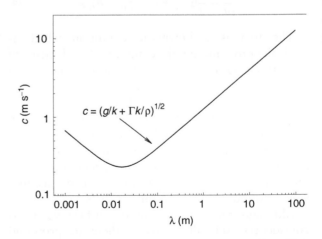

Figure 2 The phase speed of surface gravity-capillary waves as a function of wavelength λ. A minimum phase speed of 23 cm s^{-1} occurs for $\lambda = 0.017$ m. Shorter waves approach pure capillary waves, whereas longer waves become pure gravity waves. Note that there are both capillary and gravity waves for a given phase speed. This is the basis of the generation of parasitic capillary waves on the forward face of steep gravity waves.

The Group Velocity

Using the superposition principle over a continuum of wavenumbers a general disturbance (in two spatial dimensions) can be represented by

$$\eta(x,t) = \int_{-\infty}^{\infty} a(k)e^{i(kx-\sigma t)}dk \qquad [20]$$

where, as above, only the real part of the integral is taken. Assuming the disturbance is confined to wavenumbers in the neighbourhood of k_o, and expanding $\sigma(k)$ about k_o gives

$$\sigma(k) = \sigma(k_o) + (k - k_o)\frac{d\sigma}{dk}\Big|_{k=k_0} + \dots \qquad [21]$$

whence

$$\eta(x,t) \doteq e^{i(k_o x - \sigma(k_o)t)} \int_{-\infty}^{\infty} a(k)e^{i(k-k_o)(x-c_g t)}dk + \dots \qquad [22]$$

where

$$c_g = \left|\frac{d\sigma}{dk}\right|_{k=k_o} \qquad [23]$$

is the group velocity. Eqn [22] demonstrates that the modulation of the pure harmonic wave propagates at the group velocity. This implies that an isolated packet of waves centered around the wavenumber k_o will propagate at the speed c_g, so that an observer wishing to follow waves of the same length must travel at the group velocity. Since the energy density is proportional to a^2 (see below), it is also the speed at which the energy propagates. These properties of the group velocity apply to linear waves, and more subtle effects may become important at large slopes. In general, $c_g \neq c$.

For deep-water gravity waves,

$$c_g = \frac{1}{2}c = \frac{1}{2}\left(\frac{g}{k}\right)^{1/2} \qquad [24]$$

so the wave group travels at half the phase speed, with waves appearing at the rear of a group propagating forward and disappearing at the front of the group.

For deep-water capillary waves,

$$\sigma^2 = \Gamma k^3/\rho, \quad c = (\Gamma k/\rho)^{1/2}, \quad c_g = \frac{3}{2}c \qquad [25]$$

so waves appear at the front of the group and disappear at the rear of the group as it propagates.

For shallow water gravity waves, $kh \ll 1$, $c_g = c$.

Second Order Quantities

The energy density (per horizontal surface area) of surface waves is

$$E = \frac{1}{2}\rho g'a^2 \qquad [26]$$

being the sum of the kinetic and potential energies. In the case of gravity waves, the potential energy results from the displacement of the surface about its equilibrium horizontal position. For capillary waves, the potential energy arises from the stretching of the surface against the restoring force of surface tension.

The mean momentum density $\mathbf{M}$ is given by

$$\mathbf{M} = \frac{1}{2}\rho\sigma a^2 \coth kh e = \frac{E}{c}\mathbf{e} \qquad [27]$$

where the unit vector $e = \mathbf{k}/k$.

To leading order, linear gravity waves transfer energy without transporting mass; however, there is a second order mass transport associated with surface waves. In a Lagrangian description of the flow it can be shown that for irrotational inviscid wave motion the mean horizontal Lagrangian velocity (Stokes drift) of a particle of fluid originally at $z = z_o$ is

$$\mathbf{u}_l = \sigma k a^2 \frac{\cosh 2k(z+h)}{2\sinh^2 kh}\mathbf{e} \qquad [28]$$

which reduces to $(ak)^2 ce^{2kz_o}\mathbf{e}$ when $kh \gg 1$. This second order velocity arises from the fact that the orbits of the particles of fluid are not closed. Integrating eqn [28] over the depth it can be shown that this mean Lagrangian velocity accounts for the wave momentum $\mathbf{M}$ in the Eulerian description. The Stokes drift is important for representing scalar transport near the ocean surface, but this transport is likely to be significantly enhanced by the intermittent larger velocities associated with wave breaking.

Longer waves, or swell, from distant storms can travel great distances. An extreme example is the propagation of swell along great circle routes from storms in the Southern Ocean to the coast of California. For waves to travel so far, the effects of dissipation must be small. In deep water, where the wave motions have decayed away to negligible levels at depth, the contributions to the dissipation come from the thin surface boundary layer and the rate of strain of the irrotational motions in the bulk of the fluid. It can be shown that the integral is dominated by the latter contributions, and the timescale for the decay of the wave energy is just

$$\tau_e = -\left(\frac{1}{E}\frac{dE}{dt}\right)^{-1} = (4\nu k^2)^{-1} \qquad [29]$$

or $\sigma/8\pi v k^2$ wave periods. This gives negligible dissipation for long-period swell in deep water over scales of the ocean basins. More realistic models of wave dissipation must take into account breaking and near surface turbulence which is sometimes parameterized as a 'super viscosity' or eddy viscosity, several orders of magnitude greater than the molecular value. When waves propagate into shallow water, the dominant dissipation may occur in the bottom boundary layer.

Eqn [27] shows that dissipation of wave energy is concomitant with a reduction in wave momentum, but since momentum is conserved, the reduction of wave momentum is accompanied by a transfer of momentum from waves to currents. That is, net dissipative processes in the wave field lead to the generation of currents.

Waves on Currents: Action Conservation

Waves propagating in varying currents may exchange energy with the current, thus modifying the waves. Perhaps the most dramatic examples of this effect come when waves propagating against a current become larger and steeper. Examples occur off the east coast of South Africa as waves from the Southern Ocean meet the Aghulas Current; as North Atlantic storms meet the northward flowing Gulf Stream, or at the mouths of estuaries as shoreward propagating waves meet the ebb tide.

For currents $\mathbf{U} = (U, V)$ that only change slowly on the scale of the wavelength, and a surface displacement of the form

$$\eta = a(x, y, t)e^{i\theta(x,y,t)} \qquad [30]$$

where a is the slowly varying amplitude and θ is the phase. The absolute local frequency $\omega = -\partial\theta/\partial t$, and the x- and y-components of the local wavenumber are given by $k = \partial\theta/\partial x, l = \partial\theta/\partial y$. The frequency seen by an observer moving with the current $\mathbf{U}$ is

$$-\left(\frac{\partial\theta}{\partial t} + \mathbf{U}.\nabla\theta\right) \qquad [31]$$

which is equal to the intrinsic frequency σ. Thus

$$\sigma = \omega - \mathbf{U}.\mathbf{k} \qquad [32]$$

which is just the Doppler relationship.

We also have,

$$\frac{\partial\mathbf{k}}{\partial t} + \nabla\omega = 0, \qquad [33]$$

which can be interpreted as the conservation of wave crests, where $\mathbf{k}$ is the spatial density of crests and ω the wave flux.

The velocity of a wave packet along rays is

$$\frac{dx_i}{dt} = U_i + \frac{\partial\sigma}{\partial x_i} = U_i + c_{gi} \qquad [34]$$

which is simply the vector sum of the local current and the group velocity in a fluid at rest. Furthermore, refraction is governed by

$$\frac{dk_i}{dt} = -k_j\frac{\partial U_j}{\partial x_i} - \frac{\partial\sigma}{\partial x_i} \qquad [35]$$

where the first term on the right represents refraction due to the current and the second is due to gradients in the waveguide, such as changes in the depth. It is this latter term which results in waves, propagating from deep water towards a beach, refracting so that they propagate normal to shore.

For steady currents, the absolute frequency is constant along rays but the intrinsic frequency may vary, and the dynamics lead to a remarkable and quite general result for linear waves. If E is the energy density then the quantity $\mathscr{A} = E/\sigma$, the wave action, is conserved:

$$\frac{\partial\mathscr{A}}{\partial t} + \frac{\partial}{\partial x_i}[(U_i + c_{gi})\mathscr{A}] = 0 \qquad [36]$$

In other words, the variations in the intrinsic frequency σ and the energy density E, are such as to conserve the quotient.

This theory permits the prediction of the change of wave properties as they propagate into varying currents and water depths. For example, in the case of waves approaching an increasing counter current, the waves will move to shorter wavelengths (higher k), larger amplitudes, and hence greater slopes, ak. As the speed of the adverse current approaches the group velocity, the waves will be 'blocked' and be unable to propagate further. In this simplest theory, a singularity occurs with the wave slope becoming infinite, but higher order effects lead to reflection of the waves and the same blocking effect. This theory also forms the basis of models of long-wave–short-wave interaction that are important for wind-wave generation and the interpretation of remote sensing measurements of the ocean surface, including the remote sensing of long nonlinear internal waves.

Nonlinear Effects

The nonlinearity of surface waves is represented by the wave slope, ak. For typical gravity waves at the ocean surface the average slope may be $O(10^{-2} - 10^{-1})$; small, but not negligibly so. Non-linear effects may be weak and can be described as a perturbation to the linear wave theory, using the slope as an expansion parameter. This approach, pioneered by Stokes in the mid-nineteenth century, showed that for uniform approach deep-water gravity waves,

$$\sigma^2 = gk(1 + a^2 k^2 + ...),\qquad [37]$$

and

$$\eta = a\cos\theta + \frac{1}{2}a^2 k\cos 2\theta + ...\qquad [38]$$

Weakly nonlinear gravity waves have a phase speed greater than linear waves of the same wavelength. The effect of the higher harmonics on the shape of the waves leads to a vertical asymmetry with sharper crests and flatter troughs.

The largest such uniform wave train has a slope of $ak = 0.446$ a phase speed of $1.11c$, and a discontinuity in slope at the crest containing an included angle of $120°$. This limiting form has sometimes been used as the basis for the models of wave breaking; however, uniform wave trains are unstable to side-band instabilities at significantly lower slopes, and it is unlikely that this limiting form is ever achieved in the ocean.

With the assumption of both weak nonlinearity and weak dispersion (or small bandwidth, $\delta k/k_o \ll 1$), it may be shown that if

$$\eta(x,y,t) = \mathscr{R}_e[A(x,y,t)e^{i(k_o x - \sigma_o t)}]\qquad [39]$$

where $\sigma_o = \sigma(k_o)$ and $\mathscr{R}_e$ means that the real part is taken, then the complex wave envelope $A(x,y,t)$ satisfies a nonlinear Schrödinger equation or one of its variants. Solutions of the nonlinear Schrödinger equation for initial conditions that decay sufficiently rapidly in space evolve into a series of envelope solitons and a dispersive tail. Solitons propagate as waves of permanent form and survive interactions with other solitons with just a change of phase. Attempts have been made to describe ocean surface waves as fields of interacting envelope solitons; however, instabilities of the two-dimensional soliton solutions, and the effects of higher-order non-linearities, random phase and amplitude fluctuations in real wave fields give pause to the applicability of these idealized theoretical results.

Resonant Interactions

Modeling the generation, propagation, interaction, and dissipation of wind-generated surface waves is of great importance for a variety of scientific, commercial and social reasons. A rigorous theoretical foundation for all components of this problem does not yet exist, but there is a rational theory for weakly nonlinear wave–wave interactions.

For linear waves freely propagating away from a storm, the spectral content at any later time is explicitly defined by the initial storm conditions. For a nonlinear wave field, wave–wave interactions can lead to the generation of wavenumbers different from those comprising the initial disturbance. For surface gravity waves, these nonlinear effects lead to the generation of waves of lower and higher wavenumber with time. The timescale for this evolution in a random homogeneous wave field is of the order of $(ak)^4$ times a characteristic wave period; slow, but significant over the life of a storm.

The foundation of weakly nonlinear interactions between surface waves is the resonant interaction between waves satisfying the linear dispersion relationship. It is a simple consequence of quadratic nonlinearity that pairs of interacting waves lead to the generation of waves having sum and difference frequencies relative to the original waves. Thus

$$\mathbf{k}_3 = \pm\mathbf{k}_1\pm\mathbf{k}_2, \sigma_3 = \pm\sigma_1\pm\sigma_2\qquad [40]$$

If in addition, $\sigma_i(i = 1, 2, 3)$ satisfies the dispersion relationship, then the interaction is resonant. In the case of surface waves, the nonlinearities arise from the surface boundary conditions, and resonant triads are possible for gravity capillary waves, and gravity waves in water of intermediate depth.

For deep-water gravity waves, cubic nonlinearity is required before resonance occurs between a quartet of wave components:

$$\mathbf{k}_1\pm\mathbf{k}_2\pm\mathbf{k}_3\pm\mathbf{k}_4 = 0,$$
$$\sigma_1\pm\sigma_2\pm\sigma_3\pm\sigma_4... = 0,\quad \sigma_i = (gk_i)^{1/2}\qquad [41]$$

These quartet interactions comprise the basis of nonlinear wave–wave interactions in operational models of surface gravity waves. Exact resonance is not required, since even with detuning significant energy transfer can occur across the spectrum. The formal basis of these theories may be cast as problems of multiple spatial and temporal scales, and higher-order interactions should be considered as these scales increase, and the wave slope increases.

Parasitic Capillary Waves

The longer gravity waves are the dominant waves at the ocean surface, but recent developments in air–sea interaction and remote sensing, have placed increasing importance on the shorter gravity-capillary waves. Measurements of gravity-capillary waves at sea are very difficult to make and much of the detailed knowledge is based on laboratory experiments and theoretical models.

Laboratory measurements suggest that the initial generation of waves at the sea surface occurs in the gravity-capillary wave range, initially at wavelengths of $O(1)$ cm. As the waves grow and the fetch increases, the dominant waves, those at the peak of the spectrum, move into the gravity-wave range. A simple estimate of the effects of surface tension based on the surface tension parameter Σ using the gravity wavenumber k would suggest that they are unimportant, but as the wave slope increases and the curvature at the crest increases, the contribution of the Laplace pressure near the crest increases. A consequence is that so-called parasitic capillary waves may be generated on the forward face of the gravity wave (**Figure 3**).

The source of these parasitic waves can be represented as a perturbation to the underlying gravity wave caused by the localized Laplace pressure component at the crest. This is analogous to the 'fish-line' problem of Rayleigh, who showed that due to the differences in the group velocities, capillary waves are found ahead of, and gravity waves behind, a localized source in a stream. In this context the capillary waves are considered to be steady relative to the crest. The possibility of the direct resonant generation of capillary waves by perturbations moving at or near the phase speed of longer gravity waves is implied by the form of the dispersion curve in **Figure 2**. Free surfaces of large curvature, as in parasitic capillary waves, are not irrotational and so the effects of viscosity in transporting vorticity and dissipating energy must be accounted for. Theoretical and numerical studies show that the viscous dissipation of the longer gravity waves is enhanced by one to two orders of magnitude by the presence of parasitic capillary waves. These studies also show

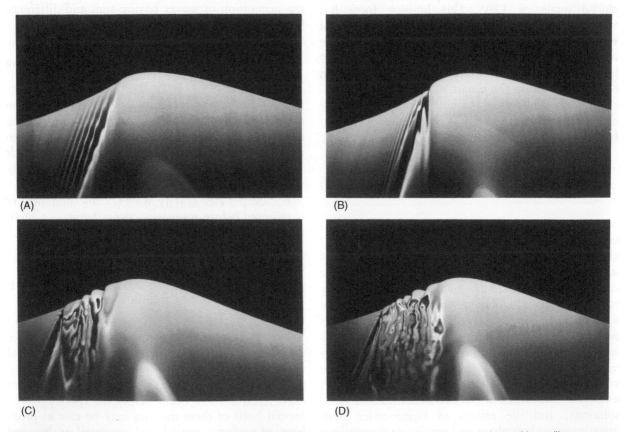

(A)

(B)

(C)

(D)

Figure 3 (A)–(D) Evolution of a gravity wave towards breaking in the laboratory. Note the generation of parasitic capillary waves on the forward face of the crest. (Reproduced with permission from Duncan JH *et al.* (1994) The formation of a spilling breaker. *Physics of Fluids* 6: S2.)

(A)

(B)

(C)

Figure 4 Waves in a storm in the North Atlantic in December 1993 in which winds were gusting up to 50–60 knots and wave heights of 12–15 m were reported. Breaking waves are (A) large, (B) intermediate and (C) small scale. (Photographs by E. Terrill and W.K. Melville; reproduced with permission from Melville, (1996).)

that the observed high wavenumber cut-off in the surface wave spectrum that has been observed at wavelengths of approximately $O(10^{-3}–10^{-2})$ m can be explained by the properties of the spectrum of parasitic capillary waves bound to short steep gravity waves.

Wave Breaking

Although weak resonant and near-resonant interactions of weakly nonlinear waves occur over slow timescales, breaking is a fast process, lasting for times comparable to the wave period. However, the turbulence and mixing due to breaking may last for a considerable time after the event. Breaking, which is a transient, two-phase, turbulent, free-surface flow, is the least understood of the surface wave processes. The energy and momentum lost from the wave field in breaking are available to generate turbulence and surface currents, respectively. The air entrained by breaking may, through the associated buoyancy force on the bubbles, be dynamically significant over times comparable to the wave period as the larger bubbles rise and escape through the surface. The sound generated with the breakup of the air into bubbles is perhaps the dominant source of high frequency sound in the ocean, and may be used diagnostically to characterize certain aspects of air–sea interaction. **Figure 4** shows examples of breaking waves in a North Atlantic storm.

Since direct measurements of breaking in the field are so difficult, much of our understanding of breaking comes from laboratory experiments and simple modeling. For example, laboratory experiments and similarity arguments suggest that the rate of energy loss per unit length of the breaking crest of a wave of phase speed c is proportional to $\rho g^{-1}c^5$, with a proportionality factor that depends on the wave slope, and perhaps other parameters. Attempts are underway to combine such simple modeling along with field measurements of the statistics of breaking fronts to give an estimate of the distribution of dissipation across the wave spectrum. Recent developments in the measurement and modeling of breaking using optical, acoustical microwave and numerical techniques hold the promise of significant progress in the next decade.

See also

Breaking Waves and Near-Surface Turbulence. Bubbles. Heat and Momentum Fluxes at the Sea Surface. Langmuir Circulation and Instability. Surface Films. Wave Generation by Wind. Whitecaps and Foam.

Further Reading

Komen GJ, Cavaleri L, Donelan M, *et al.* (1994) *Dynamics and Modelling of Ocean Waves.* Cambridge: Cambridge University Press.

Lamb H (1945) *Hydrodynamics.* New York: Dover Publications.

LeBlond PH and Mysak LA (1978) *Waves in the Ocean.* Amsterdam: Elsevier.

Lighthill J (1978) *Waves in Fluids*. Cambridge: Cambridge University Press.

Mei CC (1983) *The Applied Dynamics of Ocean Surface Waves*. New York: John Wiley.

Melville WK (1996) The role of wave breaking in air–sea interaction. *Annual Review of Fluid Mechanics* 28: 279–321.

Phillips OM (1977) *The Dynamics of the Upper Ocean*. Cambridge: Cambridge University Press.

Whitham GB (1974) *Linear and Nonlinear Waves*. New York: John Wiley.

Yuen HC and Lake BM (1980) Instability of waves on deep water. *Annual Review of Fluid Mechanics* 12: 303–334.

WAVE GENERATION BY WIND

J. A. T. Bye, The University of Melbourne, Melbourne, VIC, Australia
A. V. Babanin, Swinburne University of Technology, Melbourne, VIC, Australia

Introduction

The prime focus in this article is on ocean waves (which have always captured the scientific imagination), although results from wind-wave tank studies are also introduced wherever appropriate. Growth mechanisms fall naturally into three phases: (a) the onset of waves on a calm sea surface, (b) mature growth in the confused sea state under moderate winds, and (c) sea-spray-dominated wave environments under very high wind speeds. Of these three phases, (b) has the greatest general importance, and numerous practical formulas have been developed over the years to represent its properties. **Figure 1** illustrates the sea state which occurs at the top end of phase (b) in a strong gale (wind speed $c.$ 25 m s^{-1}, Beaufort force 9).

An important consideration is that wave generation by wind involves three main physical processes: (1) direct input from the wind, (2) nonlinear transfer between wavenumbers, and (3) wave dissipation. This article is specifically dedicated to (1); however, we briefly review (2) and (3) below.

Nonlinear interactions within the wave system can only be neglected for infinitesimal waves. To a first approximation, the wind wave can be regarded as almost sinusoidal with negligible steepness (i.e., linear), but its very weak mean nonlinearity (i.e., finite steepness and deviation of its shape from the sinusoid) is

generally believed to control the evolution of the wave field. Theoretical models of the air–sea boundary layer indicate that the input of momentum from the wind is centered in the short gravity waves. The wind pumps energy mostly into short (high-frequency) and slowly moving waves of the wave field which then transfer this energy across the continuous spectrum of waves of all scales mainly toward longer (lower-frequency) components, which may be traveling at speeds close to the wind speed, thus allowing them to grow into the dominant waves of frequencies close to the peak frequency of the wave (energy) spectrum. The transfer of energy toward shorter (higher-frequency) waves where it is dissipated occurs at a much less significant rate.

Wave breaking is the major player in the third important mechanism, which drives wave evolution – wave energy dissipation. The Southern Ocean has the greatest potential for wave growth due to the never ceasing progression of intense storm systems over vast expanses of sea surface, unimpeded by land masses. Yet, wave models (http://www.knmi.nl/waveatlas/) indicate that the significant wave height (the average crest-to-trough height of the one-third highest waves) rarely goes beyond 10 m. The process, which controls the wave growth, is the dissipation by wave breaking, and to a lesser extent radiation of wave energy away from the storm centers, and into the adjacent seas.

Theories of Wave Growth

Phase (a): The Onset of Waves

We consider firstly the initial generation of waves over a flat water surface, independently of the simultaneous generation of a surface drift current. The key theoretical result is that the initial wavelength which can be excited on the air–water interface is a wave of wavelength 17 mm, which is the capillary gravity wave of minimum phase speed 230 mm s^{-1}, controlled by gravity and surface tension. The classical Kelvin–Helmholtz analysis completed in 1871, which relies on random natural disturbances present on the water surface, shows that this wave can only be excited by a velocity shear across the sea surface exceeding 6.5 m s^{-1}.

Observations, however, show that waves are generated at much lower wind speeds, of order 1–2 m s^{-1}. In order to resolve this dilemma, another mechanism was proposed by Phillips in 1957. It takes into account the turbulent structure of wind flow. Turbulent pressure pulsations in the air create infinitesimal hollows

Figure 1 The sea state during a strong gale.

and ridges in the water surface, which, once the pressure pulsation is removed, may start propagating as free waves (similarly to the waves from a thrown stone). If the phase speed of such free waves is the same as the advection speed of the pressure pulsations by the wind, a resonant coupling can occur which will then lead these waves to grow beyond the infinitesimal stage. The first wave to be generated as the wind speed increases is likely to be the wave of minimum phase speed, propagating at an angle to the wind direction. Laboratory observations indicate that at slightly higher wind speeds, wave growth results from a shear flow instability mechanism. These two processes acting in the open ocean give rise to cat's paws, which are groups of capillary-gravity wavelets (ripples) generated by wind gusts.

These results are applicable for clean water surfaces. In the presence of surfactants (surface-active agents), which lower the surface tension, ripple growth is inhibited, and at a sufficiently high surfactant concentration it may be totally suppressed. Phytoplankton are a major source of surfactants that produce surface films, and hence slicks, which are regions of relatively smooth sea surface.

Phase (b): Mature Growth

Once the finite-height waves exist, other and much more efficient processes take over the air–sea interaction.

Jeffreys in 1924 and 1925 pioneered the analytical research of the wind input to the existing waves by employing effects of the wave-induced pressure pulsations in the air. Potential theory predicts such pressure fluctuations to be in antiphase with the waves, which results in zero average momentum/energy flux. Jeffreys hypothesised a wind-sheltering effect due to presence of the waves which causes a shift of the induced pressure maximum toward the windward wave face and brings about positive flux from the wind to the waves.

The original theory of Jeffreys was based on an assumed phenomenon of the air-flow separation over wave crests. Experiments conducted between 1930 and 1950 with wind blown over solid waves found such an effect to be small and the theory fell into a long disrepute. Jeffreys' sheltering ideas are now coming back, with both experimental and theoretical evidence lending support to his qualitative conclusions.

The period from 1957 until the beginning of the new century was dominated by the Miles theory (MT) of wave generation. This linear and quasi-laminar theory, originally suggested by Miles, was later modified by Janssen to allow for feedback changes of the airflow due to growing wind-wave seas. MT regards the air turbulence to be important only in forming the mean boundary-layer wind profile. In such a profile, a critical height exists where the wind speed equals the phase speed of the waves (**Figure 2**). Wave-induced air motion at this height leads to water-slope-coherent air-pressure perturbations at the water surface and hence to energy transfer to the waves.

MT however fails to comprehensively describe known features of the air–sea interaction. For example, for adverse winds the critical height does not exist and therefore no wind-wave energy transfer is expected, but attenuation of waves by such winds is observed. Therefore, a number of nonlinear and fully turbulent alternatives have been developed over the past 40 years.

One of the most consistent fully turbulent approaches is the two-layer theory first suggested by Townsend, and advanced by Belcher and Hunt (TBH). TBH revives the sheltering idea in a new form: by

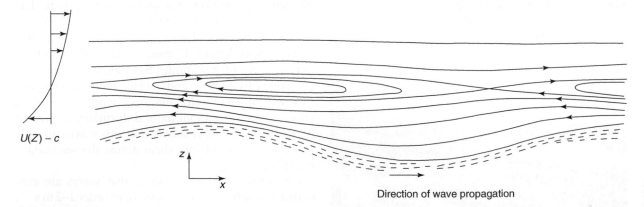

Figure 2 Mean streamlines in the turbulent flow over waves according to the MT, in a frame of reference moving with the wave. The critical layer occurs at the height (Z) where the wave speed (C) equals the wind speed ($U(Z)$). Reproduced from Phillips OM (1966) *The Dynamics of the Upper Ocean*, figure 4.3. Cambridge, UK: Cambridge University Press, with permission from Cambridge University Press.

considering perturbations of the turbulent shear stresses, which are asymmetric along the wave profile. While still in need of experimental verification, particularly for realistic non-monochromatic three-dimensional wave fields, this theory has been extensively and successfully utilized in phase-resolvent numerical simulations of the air–sea interaction by Makin and Kudryavtsev. TBH and similar theories attract serious attention because the nature of the air–sea interface is often nonlinear and always fully turbulent.

Air–sea interaction is also superimposed by a variety of physical phenomena, which alter the wave growth. Wave breaking appears to cause air-flow separation, which brings the ideas of Jeffreys back in their original form; and gustiness and non-stationarity of the wind, the presence of swell and wave groups, nonlinearity of wave shapes, modulation of surface roughness by the longer waves have all been found to cause either a reduction or an enhancement of the wind-wave input.

These processes of active wave generation give rise to the windsea in which a simple measure of the sea state, relevant to wave growth, is the wave age (c/u_*) where c is the wave speed of the dominant waves, and u_* is the friction velocity in the air (the square root of the wind stress divided by the air density). The age of the windsea increases with fetch (the distance from the coast over which the wind is blowing), and the windsea becomes 'fully developed', that is, the energy flux from the wind and the dissipation flux are in balance, at a wave age of about 35. Empirical relations for the properties of the fully developed sea in terms of the wind speed (U) at 10 m (approximately the height of the bridge on large ships) given by Toba are: $H_s = 0.30U^2/g$ and $T_s = 8.6U/g$ in which T_s ($= 2\pi c/g$) is the significant wave period and g is the acceleration of gravity. As the fetch increases, H_s and T_s both increase toward their fully developed values, and the wave spectrum spreads to lower frequencies. Older seas of wave age greater than 35 can also exist after the wind has moderated.

The observations of the velocity structure in the atmospheric boundary layer by Hristov, Miller, and Friehe have shown directly the existence of the MT critical layer mechanism for fast-moving waves of wave age about 30. It is not yet known whether it operates for younger wave age, where a quasi-laminar theory may not be appropriate.

Phase (c): Very High Wind-speed Wave Environments

The processes discussed in the previous two subsections are all grounded in two-layer fluid dynamics in which there exists a sharp interface between the two fluids. In recent times, it has been realized that this model is inadequate, especially at very high wind speeds. The link between the two phases is the breaking wave. In moderate winds (less than about $25 \, \text{m s}^{-1}$) the sea state is characterized by whitecapping due to the production of foam in a roller on the wave crests, and also foam streaks on the sea surface (**Figure 1**), whereas at very high wind speeds (greater than about $30 \, \text{m s}^{-1}$, Beaufort force 12) the air is filled with foam.

This transition arises from the structure of the breaking waves. In moderate winds, the roller remains attached to the parent wave and dissipates by the formation of foam streaks down its forward face, the trailing face of the wave remaining almost foam free. In this situation the airflow separates over the troughs and reattaches at the crests of the wave, producing Jeffreys-like phase shifts between the pressure and the underlying wave surface which enhance the energy flux to the wave. At very high wind speeds, on the other hand, the foam detaches from the wave crests, and is jetted forward into the air where it disperses vertically and horizontally before returning to the water surface. This process implies a return of momentum to the atmosphere, and hence the sea surface drag coefficient (which is an overall measure of the efficiency of momentum transfer from the atmosphere to the ocean both to waves and turbulence), which has been rising in phase (b), becomes 'capped' and possibly even reduces in phase (c). The all-pervasive presence of spray in extreme winds has prompted the anecdotal statement that "in hurricane conditions the air is too thick to breathe and too thin to swim in."

In summary, at very high wind speeds, the airflow effectively streams over the wave elements, which are reduced to acting as sources of spray. The spray then stabilizes the wind profile, and caps the sea surface drag coefficient, and interestingly, this feedback most likely allows the hurricanes to exist in the first place. This analysis has been greatly stimulated by the dropwindsonde observations of Powell, Vickery, and Reinhold in which wind profiles in hurricanes were measured for the first time, and also subsequently by experiments in high-wind-speed wind-wave tanks.

Experiments and Observations

Direct Measurements of Wave Growth Rates

The wind-to-wave energy input, which for each wave component is proportional to the time average of the product of the sea surface slope and sea surface atmospheric pressure, is the only source function, responsible for wave development, which can so far be measured directly, although this is an extremely difficult experimental task, and only a handful of attempts have been undertaken. The principal theoretical difficulty is that the sea surface atmospheric pressure must be estimated by extrapolating downward from the measurement level.

The pressure pulsations of interest are of the order of $10^{-5} - 10^{-4}$ of the mean atmospheric pressure and therefore require very sensitive probes. The surface-coherent oscillations are superposed, at the same frequencies, by random turbulent fluctuations, which are tens and hundreds of times greater in magnitude. This implies that a sophisticated data analysis is required to separate the signal buried in the noise. The wave-induced pressure decays rapidly away from the wavy surface and thus, particularly for short wave scales, it has to be sensed very close to the surface, below the wave crests of dominant waves. At the same time, the air-pressure probes have to stay dry. The last requirement leads either to measurements being conducted above the crests, which limits the estimates to the amplification of the dominant waves only, or to the use of a wave-following technique. The latter has a limited capability beyond the laboratory conditions and involves further complications due to multiple corrections needed to recover the signal contaminated by air motion in the tubes connecting the pressure probes with pressure transducers.

The first field experiment of the kind, conducted by Snyder and others in 1981, resulted in a parameterization of wind input across the wave spectrum, which has been frequently used until now. Most of these measurements, however, were taken by stationary wave probes above the wave crests, and the winds involved were very light, mostly around $4 \, \mathrm{m \, s}^{-1}$. Waves at such winds are known not to break, and this fact implies an air-sea energy balance, very different from that at moderate and strong winds. Therefore, extrapolation of these results into normal wave conditions has to be exercised with great caution.

Another field experiment was conducted by Hsiao and Shemdin in 1983. It used a wave-following technology and thus was able to obtain a spectral set of measurements somewhat beyond the dominant wave scales. This study produced a parameterization in which the growth rates were very low. Its drawback comes from the fact, that in the majority of circumstances the measured waves were 'quite old', half of the records being above the limit for the fully developed windsea. For such waves, the growth rates are expected to be very small if not zero, and

given the measurement and analysis errors, the interpretation of the low growth values becomes quite uncertain.

On the other hand, a set of precision wave-following measurements conducted by Donelan in 1999 in a wind-wave tank where the waves were very young ($c/u_* \approx 1$), produced a growth rate 2.5 times that of Hsiao and Shemdin's, and also demonstrated a very significant wave attenuation rate by the adverse wind.

The differences between these two data sets stimulated the latest campaign undertaken by Donelan and others in 2006. The Lake George experiment in Australia employed precision laboratory instruments in a field site. The site was chosen such that it provided a variety of wind-wave conditions, including very strongly wind-forced and very steep waves normally unavailable for measuring in the open ocean. The results revealed some new properties of the air–sea interaction, in which wave growth rates merged with previous results at moderate winds, but deviated significantly in strong wind conditions with continually breaking steep waves, in which full flow separation, that is, detachment of the streamlines of the airflow at the wave crest and reattachment well up the windward face of the preceding wave, occurred leading to a reduction of the wind input. This reduction means that as the winds become stronger the wind-to-wave input will keep growing, but the growth rates will be reduced compared to simple extrapolations to extreme conditions of the input measured at moderate winds. This behavior, which is consistent with that in very high wind speeds in the open ocean, did not appear to be associated with spray production.

It is worth mentioning that one of the key properties of the wind input – its directional distribution – has never been measured. It was assumed to be a cosine function by Plant, but no data on the wind input directional distribution are available. Such measurements cannot be made adequately in a wind-wave tank, and are a formidable task in the field where a spatial array of wave-following pressure probes would have to be operated. Directional wave input distribution, nevertheless, is an integral part of any wave forecast model and therefore this problem remains a major challenge for the experimentalists.

Reverse Momentum Transfer

Nonlinear interactions transfer energy to the longer, faster-propagating waves, which after leaving the region of generation are known as swell. The swell may travel at a speed greater than the local wind speed, and even propagate in the opposite direction to the wind, leading to the possibility of reverse momentum transfer from the waves to the wind.

The direct effect of waves propagating faster than the wind has been measured in a wind-wave tank by Donelan; however, when the results were applied to swell propagating in the ocean, the damping effect was found to be much too large. This is well known to surfers, who rely on the arrival of swells from distant storms: their propagation across the Pacific Ocean (over a distance of *c*. 10 000 km) was measured in a classical campaign conducted by Snodgrass and others in 1963.

Reverse momentum transfer has also been observed in wind profiles. In Lake Ontario, while a swell was running against a very light wind, the wind speed increased downward (toward the sea surface) due to the propagation of the swell, rather than the normal decrease. This is a clear example of reverse momentum transfer arising from the presence of a wave train of nonlocal origin. Reverse momentum transfer, however, is a ubiquitous process in windseas in which part of the wind input is returned to the atmosphere by the dissipation process, especially the injection of spray.

Numerical Modeling of the Wind Input

Over the past few decades, numerical modeling of ocean waves has developed into a largely independent field of study. Two different kinds of models have been used to study the wind input. Historically, spectral models based on known physics were the first. Their progress is described in great detail in the book by Komen and others. Given the uncertainties of such predictions due to simultaneous action of the multiple wave dynamics processes, the capacity of such models to scrutinize the wind input function is limited. For example, very high quality synoptic analyses of weather systems are necessary in order to discriminate between the various coupling mechanisms for wave growth by comparing observational wave data from wave buoys with the predictions of coupled wind-wave simulations. In these models the formulation of the wave energy dissipation is based on tuning the total energy balance. Csanady, in a lucid textbook on air–sea interaction, notes that in a fetch-limited windsea only about 6% of the momentum transferred from the wind to the water supports the downwind growth of the dominant waves, the remainder being accounted for locally by the dissipation stress, that is, the rate of loss of momentum from the wave field to the ocean.

The phase-resolvent models are another kind of numerical simulations of air–sea interaction, which reproduce wind input and wave evolution in physical rather than wavenumber space. Such models solve the basic fully nonlinear equations of fluid mechanics explicitly and recent advances in numerical techniques allow us to reproduce the water surface, airflow, and wave motion with potentially absolute precision and unlimited temporal and spatial resolution. Use of such models to forecast waves globally is obviously not feasible, but they now constitute a very effective tool for dedicated studies of wind–wave interaction. The interested reader is referred to recent research by Makin and Kudryavtsev, and by Chalikov and Sheinin.

Conclusions

It is clear from this article that there are still many tasks ahead to fully understand wave generation by wind. The nonlocal aspects of wave generation by wind are a particularly challenging topic. Contemporary interest lies with our climate system. The interface between the atmosphere and the ocean is vital in this regard. This holistic view calls urgently for further study, especially of extreme events in which major momentum transfers occur, affecting the land through the initiation of hurricanes, and the sea through mixing below the wave boundary layer into the deep ocean.

See also

Breaking Waves and Near-Surface Turbulence. Rogue Waves. Surface Gravity and Capillary Waves. Wind- and Buoyancy-Forced Upper Ocean.

Further Reading

Belcher SE and Hunt JCR (1998) Turbulent flow over hills and waves. *Annual Review of Fluid Mechanics* 30: 507–538.

Bye JAT and Jenkins AD (2006) Drag coefficient reduction at very high wind speeds. *Journal of Geophysical Research* 111: C03024 (doi:10.1029/2005JC003114).

Chalikov D and Sheinin D (2005) Modeling extreme waves based on equations of potential flow with a free surface. *Journal of Computational Physics* 210: 247–273.

Csanady GT (2001) *Air–Sea Interaction Laws and Mechanisms*, 239pp. Cambridge, UK: Cambridge University Press.

Donelan MA (1999) Wind-induced growth and attenuation of laboratory waves. In: Sajjadi SG, Thomas NH, and Hunt JCR (eds.) *Wind-Over-Wave Couplings: Perspectives and Prospects*, pp. 183–194. Oxford, UK: Clarendon.

Donelan MA, Babanin AV, Young IR, and Banner ML (2006) Wave follower measurements of the wind input spectral function. Part 2: Parameterization of the wind input. *Journal of Physical Oceanography* 36: 1672–1688.

Hristov T, Friehe C, and Miller S (2003) Dynamical coupling of wind and ocean waves through wave-induced air flow. *Nature* 422: 55–58.

Jones ISF and Toba Y (eds.) (2001) *Wind Stress over the Ocean*, 307pp. Cambridge, UK: Cambridge University Press.

Komen GI, Cavaleri L, Donelan M, Hasselmann K, Hasselmann S, and Janssen PAEM (1994) *Dynamics and Modelling of Ocean Waves*, 532pp. Cambridge, UK: Cambridge University Press.

Kudryavtsev VN and Makin VK (2007) Aerodynamic roughness of the sea surface at high winds. *Boundary-Layer Meteorology* 125: 289–303.

Makin VK and Kudryavtsev VN (2003) Wind-over-waves coupling. In: Sajjadi SG and Hunt LJ (eds.) *Wind Over Waves II: Forecasting and Fundamentals of Applications*, pp. 46–56. Chichester: Horwood Publishing.

Phillips OM (1966) *The Dynamics of the Upper Ocean*. Cambridge, UK: Cambridge University Press.

Powell MD, Vickery PJ, and Reinhold TA (2003) Reduced drag coefficient for high wind speeds in tropical cyclones. *Nature* 422: 279–283.

Snodgrass FE, Groves GW, Hasselmann KF, Miller GR, Munk WH, and Powers WH (1966) Propagation of ocean swell across the Pacific. *Philosophical Transactions of the Royal Society of London* 259: 431–497.

Toba Y (1972) Local balance in the air–sea boundary processes. Part I: On the growth process of wind waves. *Journal of the Oceanographical Society of Japan* 28: 15–26.

Young IR (1999) *Wind Generated Ocean Waves*, 288pp. Oxford, UK: Elsevier.

Relevant Website

http://www.knmi.nl/waveatlas
– The KNMI/ERA-40 Wave Atlas.

ROGUE WAVES

K. Dysthe, University of Bergen, Bergen, Norway
H. E. Krogstad, NTNU, Trondheim, Norway
P. Müller, University of Hawaii, Honolulu, HI, USA

Introduction

The terms 'rogue' or 'freak' waves have long been used in the maritime community for waves that are much higher than expected, given the surrounding sea conditions. For the seafarer these unexpected waves represent a frightening and often life-threatening experience. There are many accounts of such waves hitting passenger and container ships, oil tankers, fishing boats, and offshore and coastal structures, sometimes with catastrophic consequences. It is believed that more than 22 supercarriers were lost to rogue waves between 1969 and 1994 (**Figure 1**).

Rogue waves are not mariners' tales. They have been observed and documented, most succinctly from oil platforms. Two well-studied examples are the Draupner 'New Year's Wave' and the Gorm platform waves discussed below (**Figure 2**).

With their sometimes catastrophic impact the motivation for investigating rogue waves is clear, and the scientific community has studied the topic for some time, more intensely since 2000.

Despite these efforts, there is no generally accepted explanation or theory for the occurrence of rogue waves. There is even no consensus of how to define a rogue wave. Some of the inherent difficulties are related to the random nature on ocean waves: a wave recording will show waves of different sizes and shapes. In discussing rogue waves one introduces the notion of 'one wave', which is the recorded elevation over one wave period, containing one crest and one trough. One also distinguishes between the wave height H (the distance from trough to crest) and the

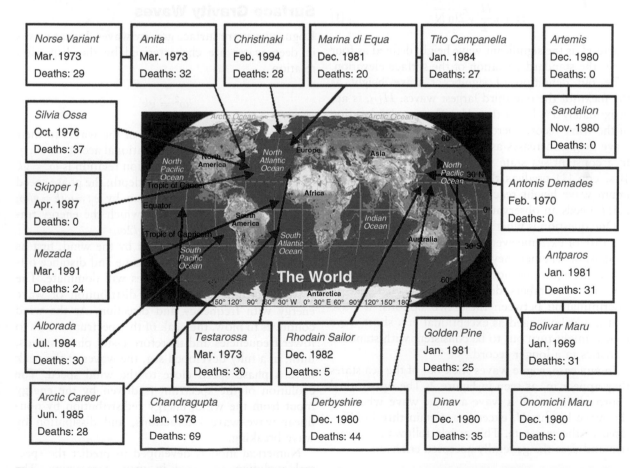

Figure 1 Locations of 22 supercarriers assumed to be lost after collisions with rogue waves between 1969 and 1994. © C. Kharif and E. Pelinovsky. Used with permission.

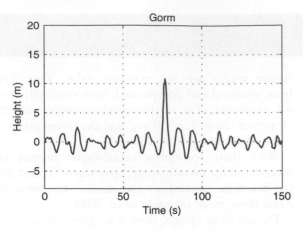
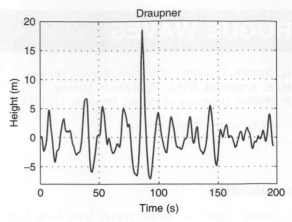

Figure 2 Two examples of rogue waves. 'Gorm' is one of the abnormal waves recorded at the Gorm field in the North Sea on 17 Nov.1984. The wave that stands out has a crest height of 11 m, which exceeds the significant wave height of 5 m by a factor of 2.2. 'Draupner' is the 'New Year Wave' recorded at the Draupner platform in the North Sea 1 Jan. 1995. The crest height is about 18.5 m and exceeds the significant wave height of 11.8 m by a factor of 1.54. Reprinted, with permission, from the *Annual Review of Fluid Mechanics*, Volume 40 © 2008 by Annual Reviews.

crest height η_{cr} (the distance from mean sea level to crest). Early wave statistics from the 1950s suggests that the most probable maximum wave height, H_M, in a wave record containing N waves is given by

$$H_M \simeq \frac{H_s}{2}\sqrt{2\ln N} \qquad [1]$$

where H_s is the significant wave height defined as four times the standard deviation of the surface elevation. (The old definition of significant wave height as the mean of the one-third largest waves, $H_{1/3}$, is approximately 5% lower than H_s.) Thus, as the duration of the wave record increases, the expected maximum wave height increases as well, although quite slowly. In a constant sea state with a mean wave period of 10 s, eqn [1] predicts that the most probable maximum wave height reaches $2H_s$ after 8.3 h, whereas $2.5H_s$ needs an observation period of c. 1 month.

One question is whether observed rogue waves are just rare or extreme events within standard statistical models, or whether they are due to some exceptional physical conditions not contained in these models. The above numbers demonstrate the challenge in discriminating between these two options: a wave that might be tagged as exceptional in a short wave record might turn out to be consistent with standard statistics in a longer record.

In any case, rogue waves stick out of the sea states they appear in, as seen in **Figure 2**. The operational approach is to call a wave a rogue wave whenever the wave height, H, exceeds a certain threshold related to the sea state. This article follows this practice and uses the generally accepted criterion

$$H/H_s > 2 \qquad [2]$$

A wave observation far beyond any reasonable statistical expectation will be called an abnormal rogue wave.

Surface Gravity Waves

Rogue waves are surface gravity waves. Linear waves in deep water are characterized by the dispersion relation

$$\omega^2 = gk \qquad [3]$$

which relates the frequency ω of the waves to their wavenumber, k, and the gravitational acceleration, g. In shallower water, the dispersion relation is $\omega^2 = gk \tanh(kd)$, where d is the water depth; the phase speed (for deep-water waves) is given by $c_p = \omega/k = g/\omega$; and the group velocity (with which the energy travels) is given by $c_g = \partial\omega/\partial k = g/2\omega$.

Surface waves are generated by the wind, first as short ripples. Given sufficient time and distance offshore, longer and longer waves will dominate. The wave spectrum describes the distribution of wave energy with frequency and direction. As the wind continues to blow, the peak of the spectrum moves to lower frequencies and therefore faster phase speeds, until, in a fully developed sea, the waves at the peak have a phase speed close to the wind speed. The evolution of the spectrum is driven by the energy input from the wind, energy redistribution by nonlinear wave–wave interactions, and dissipation by wave breaking.

Numerical models developed to predict the spectral evolution are used in wave forecasting. The spectral models predict the average partition of energy among the different waves, from which statistical quantities like H_s can be found. They do not deal with the wave phases and therefore cannot give information about individual waves, let alone rogue waves.

Physical Mechanisms

A rogue wave represents a very high concentration of wave energy. The energy of a wave of height $H = 2H_s$ is roughly a factor of 10 times larger than the average energy of the surrounding waves. The most important mechanisms capable of concentrating energy appear to be superposition and spatial, dispersive, and nonlinear focusing.

Superposition

At any given location of the ocean, waves meet with varying wavelengths and directions. Occasionally several waves add up constructively to produce a much larger wave. This is a rough interpretation of the standard linear model which considers the surface to be a superposition of independent waves (see below). So-called second-order models only slightly modify this picture, while higher-order models take into account the weak resonant interactions among the waves.

Spatial Focusing

Spatial focusing can be achieved by the refraction of waves by variable bottom topography or currents.

Topographic focusing. As waves propagate into shallower water and their wavelength becomes comparable to the water depth, the waves refract, align their crests with the topography, and steepen. Along irregular coastlines, this might lead to

focusing of wave energy in particular places and to extreme waves.

Wave–current interaction. The simplest example is a wave propagating from still water into an opposing current. A wave of phase speed c_p in still water can be completely blocked by an opposing current of only $c_p/4$. Although storm waves with $c_p \sim 15\,\mathrm{m\,s^{-1}}$ are not stopped, they will be retarded and their wavelengths reduced, when running against a current. The flux of wave action (energy divided by intrinsic frequency) is conserved implying that the current puts energy into the waves as it squeezes them.

For intense current jets like parts of the Aguhlas current off the SE coast of Africa, the situation may be more serious. When storm waves or swell are going in the opposite direction to the jet, they may be trapped where the jet is widening or at meanders, as shown by ray tracing in **Figure 3**. The trapped waves are refracted toward the jet center, causing the oscillating paths with reflection (or caustics) near the edges as seen in the figure. At reflection, the wave amplitude is significantly amplified compared to its value at the jet center. If waves from neighboring rays add up constructively around reflection, the result may be a rogue wave.

Dispersive Focusing

Gravity waves are dispersive with phase and group velocities being inversely proportional to the frequency, that is, long waves traveling faster than short waves. This fact is utilized for producing a large wave at a given position d in a wave tank by creating a short wave train where the frequency decreases with time as $\omega(t) = \omega_0 - gt/2d$, a so-called chirp (**Figure 4**). It has been shown that a chirped wave

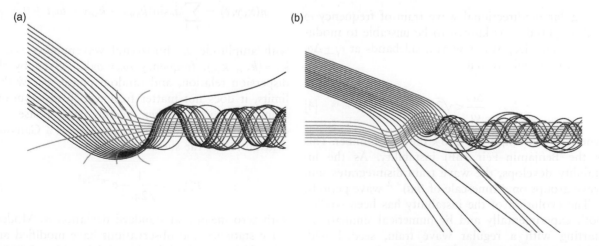

Figure 3   (a) Rays of waves getting trapped when moving upstream into a widening current jet (green streamlines). (b) Trapping of waves at a current meander.

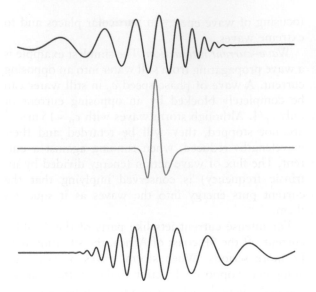

Figure 4 A chirped wave group, with the short waves in front of the long ones, contracts to a largewave. Then, the longer waves overtake the short ones giving a mirror image of the initial situation.

train that produces strong focusing in the absence of other waves will still do so when a random wave field is added. The chirp may actually be dwarfed by the random waves so that it remains invisible until it focuses.

The dispersive focusing is basically a linear effect and occurs even in a linear Gaussian sea in those rare circumstances in which waves moving in the same direction happen to have the contrived phase relations necessary to form a chirped wave train. Physical mechanisms able to produce such phase relations and chirped wave trains have not been identified for the ocean.

Nonlinear Focusing

A regular unidirectional wave train of frequency ω and amplitude a is known to be unstable to modulations. For deep-water waves, sidebands at $\omega \pm \Delta\omega$ will grow provided that

$$\frac{\Delta\omega}{\omega} < \sqrt{2}ka \qquad [4]$$

where a is the amplitude and k the wavenumber. This is the Benjamin–Feir (BF) instability. As the instability develops, the wave train disintegrates into wave groups on a timescale of $(ka)^{-2}$ wave periods.

The evolution of the instability has been studied both experimentally and by numerical simulations, starting with a regular wave train, seeded with sidebands at $\omega \pm \Delta\omega$ satisfying [4]. As the groups are formed, further focusing takes place within the

groups, producing some very large waves having a maximum surface elevation η_{max} much larger than the initial amplitude a of the wave train. Simulations show enhancement factors η_{max}/a between 3 and 4, while wave tank experiments show somewhat smaller values. Although impressive wave focusing can be achieved through this nonlinear effects, the initial states from which they develop are rather special and are not likely to occur spontaneously in a storm-generated wave field.

For narrow-band random waves, however, it has been shown, using the nonlinear Schrödinger equation (NLS), that the BF instability persists provided the relative bandwidth $\delta = \Delta\omega/\omega$ satisfies the criterion

$$\delta < s \qquad [5]$$

where s is the steepness defined as $s = k\bar{a}$ and $\bar{a}$ is the rms value of the amplitude. The ratio s/δ is called the Benjamin–Feir index (BFI). Theory and simulations in one horizontal dimension have shown that wave spectra with BFI > 1 are unstable and develop on the timescale of s^{-2} wave periods toward marginal stability. While the instability develops, there is an increase in the population of extreme waves. This effect has also been verified experimentally in a wave flume. As will be pointed out below, three-dimensional, two horizontal dimensions and time simulations and recent wave basin experiments indicate that the effect only appears for very long-crested waves.

Statistics of Large Waves

The Gaussian Sea

The Gaussian sea (or standard linear model) considers a random superposition of independent waves

$$\eta(x, y, t) = \sum_{n=1}^{\infty} a_n \sin(k_{xn}x + k_{yn}y - \omega_n t + \theta_n) \quad [6]$$

with amplitude a_n, horizontal wavenumber vector $\mathbf{k}_n = (k_{xn}, k_{yn})$, frequency $\omega_n = \omega(\mathbf{k}_n)$ given by the dispersion relation, and random phases θ_n. Within limits, it does not matter from which distributions amplitude, wavenumber vector, and phase are drawn, the central limit theorem assures a Gaussian distribution of the surface elevation:

$$P(\eta) = \frac{1}{\sqrt{2\pi\sigma^2}} e^{-\eta^2/2\sigma^2} \qquad [7]$$

with zero mean and standard deviation σ. Modern wave statistics and observations have modified and refined the standard linear model, taking into account nonlinearity to second order in the steepness s.

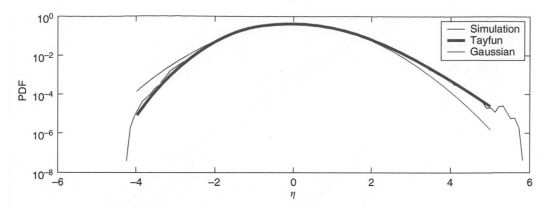

Figure 5   Typical probability density function (PDF) of the sea surface elevation η (scaled by the standard deviation σ) as simulated by a third-order numerical model (black), compared to second-order theory (red) and linear Gaussian theory (blue). Reprinted, with permission, from the *Annual Review of Fluid Mechanics*, Volume 40 © 2008 by Annual Reviews.

In **Figure 5**, a second-order modification of [7] is compared to data from 3-D nonlinear simulations. It is seen that the modified distribution breaks the symmetry of the Gaussian, having lower probability of a deep trough than a high crest.

Single Point Extremes

For realistic wave fields, state-of-the-art wave statistics expresses the wave height exceedance probabilities in terms of Weibull distributions:

$$P(H > xH_s) = \exp\left(-\frac{x^\alpha}{\beta}\right) \qquad [8]$$

where the free parameters α and β are found to vary only slightly with the sea state (characterized by the average wave steepness and the directional spread), at least for simple wind seas.

Numerical simulations and observations suggest that the probability distribution for the maximum wave height within a record of N waves can be obtained by assuming the waves to be independent. It then follows from eqn [8] that the most probable maximum wave height within a record of N waves is

$$H_M = H_s(\beta \ln N)^{1/\alpha} \qquad [9]$$

The expression [1] of the so-called narrow-band linear model is recovered for $\alpha = 2$ and $\beta = 1/2$.

The exceedance probabilities of the crest height η_{cr} are also given by Weibull distributions, where the parameters α and β are now, however, found to vary significantly with some of the sea state parameters.

Figure 6 shows the probability of exceedance of the wave and crest heights for some of the currently used expressions. Forristall's model for crest heights, based on numerical simulations that include second-order nonlinear effects, shows good agreement with

observed data. The same is true for Næss' model, N, for wave heights, based on a Gaussian sea with a typical wave spectrum. Observe that whereas the Gaussian model (G) severely underpredicts the probability of observing large crests, it gives reliable predictions for the wave height (N). Also note that the tail of Forristall's crest height distribution depends strongly on wave steepness.

In the introduction we defined a rogue wave by the criterion $H/H_s > 2$. According to Næss' model this corresponds to an exceedance probability of about 10^{-4}. The same exceedance probability is obtained for the model F1 if one chooses for the crest height the criterion

$$\eta_{cr}/H_s > 1.25 \qquad [10]$$

Both these rogue wave criteria are used interchangeably.

We call a rogue wave an abnormal rogue wave when it cannot plausibly be explained by state-of-the-art wave statistics. As mentioned above, this depends to some extent on the size of the data set. Nevertheless, if one observes a wave with a height or crest exceedance probability more than 2 orders of magnitude below the expected probability, this will indeed be exceptional and indicative of an abnormal rogue wave.

Space–Time Extremes

Whereas the extreme value theory of wave records taken at single points has been the subject of extensive research, the corresponding theory for spatial data is less developed. One problem is that the concept of neighboring maxima and minima and hence the concept of wave height is not well defined in a 2-D field. There exist, however, accurate asymptotic

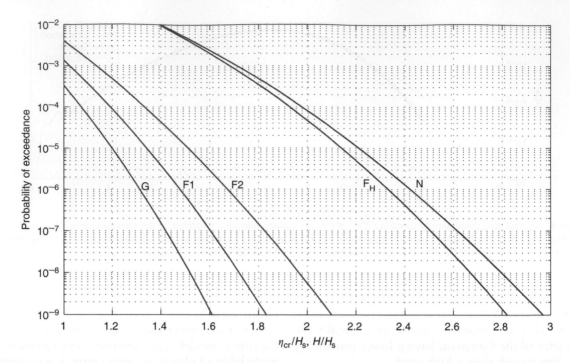

Figure 6 Probability of exceedance for crest heights (left) and wave heights (right). G, Linear Gaussian model; F1, Forristall's second-order model for medium wave steepness; F2, Forristall's second-order model for high wave steepness; N, Næss' wave height model for Gaussian seas and typical wind wave spectra; F_H, Forristall's empirical wave height model based on buoy data from the Mexican gulf. Reprinted, with permission, from the *Annual Review of Fluid Mechanics*, Volume 40 © 2008 by Annual Reviews.

expressions for the maximum crest height for multidimensional Gaussian fields. The following simple example illustrates that even Gaussian theory predicts very high crests when maxima are considered over an extended spatial area.

Consider a storm over an area $100 \, \text{km} \times 100 \, \text{km}$ and lasting for 6 h. With a mean wave period of 10 s, we expect a mean wavelength $\lambda_p \approx 200 \, \text{m}$. For a directional spread of about $20°$ one further expects a mean crest length $\lambda_c \approx 450 \, \text{m}$. If we define $\lambda_p \lambda_c$ as the characteristic area of one wave, then there are at each instant of time about 10^5 waves within the 2-D storm area. By the Gaussian theory we then have that the expected maximum crest height:

- over the storm area at a fixed time is

$$E(\eta_{\text{max}})_{\text{space}} = 1.32 H_s \qquad [11]$$

- at a fixed location over a 6-h period is

$$E(\eta_{\text{max}})_{\text{time}} = 1.02 H_s \qquad [12]$$

- and over the storm area during the 6-h period is

$$E(\eta_{\text{max}})_{\text{space+time}} = 1.69 H_s \qquad [13]$$

This last value is quite high, and far into what would be considered to be a rogue wave. This example

demonstrates that even the conservative Gaussian theory predicts waves with much larger crest heights when the spatial dimensions of the wave field are taken into account.

The Shape of Large Waves

Rogue waves have been described as walls of water or pyramidal waves, surrounded by holes in the ocean. Apart from a few research stereoimaging systems, there exist no operational instruments that directly measure the height of the surface over a sufficiently large area for extensive time periods. Thus, the 2-D shape of the surface elevation, η, of large waves has mainly to be inferred from numerical simulations and analytic methods.

A useful tool is the so-called Slepian model representation (SMR) of a stationary Gaussian stochastic surface. Consider a wave with a high maximum at $\mathbf{x} = 0$ at a fixed instant of time. According to the Slepian theory, the surface $\eta(\mathbf{x})$ around the maximum where $\nabla \eta = 0$ may be written as

$$\eta(\mathbf{x}) = \eta(0) \frac{\rho(\mathbf{x})}{\rho(0)} + \Delta(\mathbf{x}) \qquad [14]$$

where $\rho(\mathbf{x})$ is the covariance function and the residual process $\Delta(\mathbf{x})$ is Gaussian with zero mean.

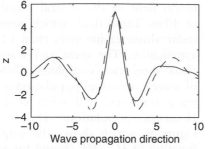
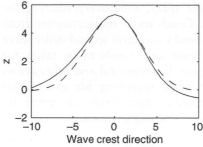

Figure 7 Averaged and scaled surface profile in the wave propagation and crest directions at an extreme wave crest obtained from large scale third-order simulations (full curve). The dashed curves is the scaled spatial covariance function. The horizontal distance is scaled by the wavelength at the spectral peak divided by 2π. Reprinted, with permission, from the *Annual Review of Fluid Mechanics*, Volume 40 © 2008 by Annual Reviews.

The approximation $\eta(\mathbf{x}) \sim \eta(0)\rho(\mathbf{x})/\rho(0)$ is only reasonable in a region where $\Delta(\mathbf{x})$ is small, which typically surrounds the maximum out to about one wave/crest length. Thus, for a Gaussian surface the average wave profile around a very high crest is that of the scaled covariance function of the wave field.

Due to the symmetry of the Gaussian [7], the probability distribution of the crest height η_{cr} is identical to that of the trough depth η_{tr} and the average shape of a deep trough is the mirror image of the one of a high crest. The inclusion of nonlinearity breaks this symmetry, as has already been demonstrated in **Figure 5**. The average shape of an extreme wave can also be expected to change. A comparison of the Slepian model with data from large-scale third-order simulations is shown in **Figure 7**. It is seen that the simulated crest becomes more narrow and the neighboring troughs less deep. The ratio R between the extreme crest height and the nearest trough depth is 1.5 and 2.3, respectively. Observations of extreme waves indicate that R is scattered around a mean value of 2.2.

It follows from the above discussion that on an average large wave events typically occur in short groups. As the waves pass through a group envelope like in **Figure 8**, they exhibit various shapes.

Experiments and Observations

Wave Tank Experiments

Controlled experiments in wave tanks have long been used to study the effect of waves on vessels and structures. Most of this work deals with unidirectional waves which are forced to violent breaking or extreme crest heights through dispersive focusing. Three-dimensional wave basins can carry out similar experiments using spatial focusing.

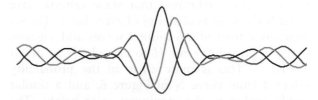

Figure 8 During half a wave period, as the waves move trough the group envelope, a large wave event may be seen as a large wave crest (black), a large wave height (green), or a deep trough (red) (here seen in the group velocity frame).

Wave tanks are essential for testing vessels and structures in extreme and violent conditions. Modern wave tank facilities are able to reconstruct accurately rogue wave profiles from field observations and to record the response of ships and structures to these waves. However, most of the field wave data are point observations of 3-D waves, in contrast to the unidirectional reconstructions in the wave tank.

Field Measurements

Instrumentation able to measure individual wave properties consists of laser and radar altimeters, buoys, and subsurface instruments (pressure gauges and surface tracking acoustic devices). Although some instruments, in particular buoys and radars, give consistent results for the wave height, the data deviate considerably when measuring the wave crests. When measured from subsurface instrumentation and buoys, the crest wave statistics is below the Gaussian theory, whereas narrow beam radar and, in particular, laser altimeters invariably show crest heights above the Gaussian theory. The differences are explained by the lateral motion of buoys tending to avoid high crests, and the inherent area averaging that occurs for pressure gage measurements and radars with broad footprints. On the other hand, it is often suggested that laser recordings are sensitive to sea spray, thus overpredicting the real crest height. Good wave measurements from fixed installations tend to confirm second-order wave theory, although some care needs to be taken for platform interference. Instrumental and other errors pose a challenge when searching for exceptional waves in wave records, where erroneous spikes are prone to be mistaken for rogue waves.

Frigg data   The wave elevation measurements with a Plessey wave radar at the Frigg oil field in the northern North Sea (depth 100 m) is a long-term quality-checked data set containing 10 000 time series each of 20-min duration where $H_s > 2$ m. The criteria $H > 2H_s$ and $\eta_c > 1.25H_s$ yield 79 and 74 rogue waves, confirming that these criteria have roughly the same probability of exceedance. The set contains a total of 1.6 million waves and suggests that the probability for H to exceed $2H_s$ is about 5×10^{-5}. This is quite close to the probability inferred from curve N in **Figure 6**, and a similar result applies to the maximum crest height. The conclusion is therefore that the Frigg data, although containing some rather extreme height and crest ratios, do not really show any abnormal rogue waves deviating from the theory. Moreover, the ratio of the crest height to the trough depth is scattered around 2.1 for the rogue waves, in accordance with nonlinear numerical simulations.

Gorm data   The Frigg findings are in strong contrast to the much-referred-to data set from the Gorm field in the central North Sea at a depth of 40 m. These data, which were collected with a radar altimeter for more than 12 years, are indeed astonishing. The crest and wave height distributions show two completely different populations of waves: a normal population adhering to current wave statistics, and what could be denoted an abnormal population. This is clearly seen in **Figure 9**, which shows the empirical distribution function of H_{max}/H_s (found for each of the 20-min time series). The Næss model for typical storm spectra is included as a dashed curve. The corresponding plot for the crest height looks similar.

The data contain 24 waves where H/H_s ranges from 2.2 to 2.94 (the most extreme case is seen in **Figure 2**). Perhaps coincidentally, the deviation from the Næss' theory happens around the rogue wave criterion. It should be observed, however, that the significant wave height is only 2–4 m for the most extreme waves.

The Draupner Incident   The Draupner New Year wave occurred on January 1995 and is shown in **Figure 2**. It was recorded by a laser instrument at an unmanned satellite platform, and minor damage on a temporary deck below the main platform deck supports the reading. This wave record has been extensively discussed in the scientific literature, and independent ship observations have confirmed that the weather situation was extreme. The crest height is 18.5 m above the mean water level and the wave height is 26 m. Although the crest and wave height

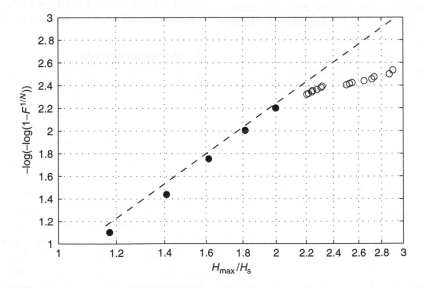

Figure 9 Empirical distribution function of H_{max}/H_s for ~5000 20-min records from the Gorm field. The filled circles are representative points, whereas the open circles represent individual records. The dashed straight line is Næss' wave height model. Points falling below the straight line indicate a larger frequency of occurrence. Reprinted, with permission, from the *Annual Review of Fluid Mechanics*, Volume 40 © 2008 by Annual Reviews.

values are slightly below the so-called 100-year values for the Draupner site, the wave was indeed quite unexpected for the observed $H_s \approx 12$ m.

Satellite and Radar Measurements of Rogue Waves

Space-borne synthetic aperture radar (SAR) is currently the only instrument that has the capacity of observing wave fields over large spatial areas. The satellites scan the world's oceans daily, although on sparse tracks. SAR measures the intensity of the backscattered signal (which depends on the amplitude and slope of the scattering waves) and the Doppler shift (which depends on the waves' orbital velocity). Recently, there have been attempts to infer the surface elevation from SAR data, and the possibility of globally measuring rogue waves from a satellite has caught the media and the space organizations with excitement. However, the algorithms converting the radar backscatter signal to surface elevation have not yet been validated and it is unclear whether the results obtained so far are correct and whether the method is viable. Marine radars, situated onshore, on platforms, or on ships, meet with the same problems.

Numerical Simulations

Large-scale simulations of random ocean waves have so far not been feasible without simplifications of the full nonlinear equations. Although fully nonlinear simulations are gradually becoming available, major simulation tools still apply approximate versions of the full equations.

One example of an approximate equation is the modified nonlinear Schrödinger equation (MNLS). It assumes a narrow spectral band $\Delta\omega$ around a peak frequency ω_p and describes the evolution of the wave field for small steepness s, were the cubic wave–wave interactions suffice. Solutions of MNLS compare favorably both with tank experiments and fully nonlinear 3-D simulations over a time horizon of $(\omega_p s^2)^{-1}$, provided $\Delta\omega/\omega_p = O(s^{1/2})$. **Figure 10** shows three cases initiated with a Joint North Sea Wave Observation Project (JONSWAP) spectrum and different angular distributions. The model allows for internal spectral energy transfer and the spectrum changes on the timescale $(\omega_p s^2)^{-1}$, most pronounced for the long-crested case C (BFI ≥ 1 for all cases).

The large computational domain, containing c. 10^4 waves, admits calculation of probability distributions at any time during the evolution process. **Figure 11** shows the probability of exceedance of the scaled crest height for the three cases of **Figure 10**, at times 25, 50, and 100 peak periods (T_p). The cases A and B show little change and fit the theoretical second-order statistics quite well. The long-crested case C however, has a significant increase in the occurrence of large waves during the early development ($25 T_p$) of the spectral instability. This last result has been confirmed by experiments in wave flumes and represents presently a hot topic in rogue wave research.

Conclusions

Rogue waves are surface gravity waves and operationally defined as waves which satisfy one or both of the criteria [2] and [10]. Their unexpectedness causes a special danger to ships and to offshore and coastal structures.

Current wave statistics predicts the exceedance probability of extremely high waves as functions of the sea state. Although most rogue wave observations seem to be consistent with the statistical predictions, there is a small group of rogue wave observations falling outside the predictions. No generally accepted

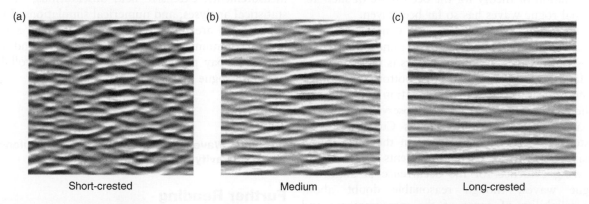

(a)   (b)   (c)

Short-crested   Medium   Long-crested

Figure 10 Simulated surfaces of a large-scale third-order simulation. Approximately 2% of the computational domain is shown at an early stage of the simulation for a short-(A), medium-(B), and long-(C) crested case. Reprinted, with permission, from the *Annual Review of Fluid Mechanics*, Volume 40 © 2008 by Annual Reviews.

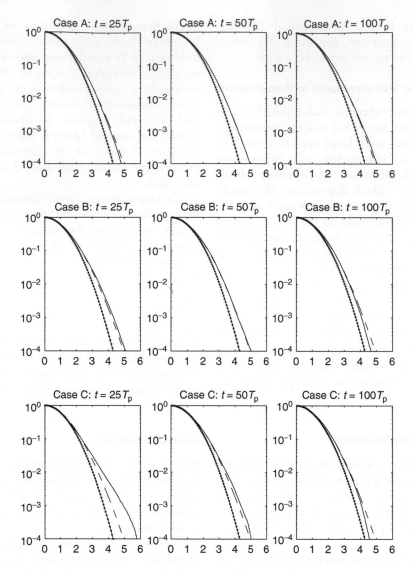

Figure 11 Simulated probability of exceedance (along vertical axes) of the crest height η_c (scaled by the standard deviation σ along horizontal axes) for the cases A, B, and C at times and 25, 50, and 100 peak periods (T_p). Solid line: simulations; dotted line: Gaussian theory; dashed line: second-order theory.

explanation or theory for the occurrence of such abnormal rogue waves has so far been given.

There seems, however, to be a consensus among researchers that occurrences of unexpected and dangerous waves in coastal waters is mostly caused by focusing due to refraction by bottom topography or current gradients. This explanation might also be valid for the extreme waves observed in intense ocean currents like the Agulhas Current off the eastern coast of South Africa. In the open ocean away from strong current gradients there is also growing evidence for the occurrence of abnormal rogue waves, though reasonable doubt about the reliability of some of the measurements and observations is still warranted. Progress is expected to come from a combination of more reliable

measurements, extensive field observations, careful statistical analyses, and numerical simulations. Tank experiments are of vital importance for the engineering community but their controlled condition might not pay proper tribute to the uncontrollable nature of rogue waves in the world's oceans.

See also

Breaking Waves and Near-Surface Turbulence. Surface Gravity and Capillary Waves.

Further Reading

Dysthe K, Krogstad H, and Müller P (2008) Oceanic rogue waves. *Annual Review of Fluid Mechanics* 40: 287–310.

Kharif C and Pelinovski E (2003) Physical mechanisms of the rogue wave phenomenon. *European Journal of Mechanics B/Fluids* 22: 603–634.

Lavrenov I (2003) *Wind-Waves in Oceans. Dynamics and Numerical Simulation.* New York: Springer.

Müller P and Henderson D (eds.) (2005) Rogue Waves. *Proceedings, 14th 'Aha Huliko'a Hawaiian Winter Workshop, Special Publication, 193pp.* University of Hawaii at Manoa, School of Ocean and Earth Science and Technology.

Olagnon M and Athanassoulis G (eds.) (2000) *Rogue Waves 2000: Proceedings of a Workshop in Brest, France, 29–30 Nov. 2000.* Plouzané, France: Editions IFREMER.

Olagnon M and Prevosto M (eds.) (2004) *Rogue Waves 2004: Proceedings of a Workshop in Brest, France, 20–22 Oct. 2004.* Plouzané, France: Editions IFREMER.

Tucker MJ and Pitt EG (2001) *Ocean Engineering Book Series, Vol. 5: Waves in Ocean Engineering*, 521pp. Amsterdam: Elsevier.

BREAKING WAVES AND NEAR-SURFACE TURBULENCE

J. Gemmrich, University of Victoria, Victoria, BC, Canada

Introduction

Most readers will associate wave breaking with breaking surf at shallow beaches. This article, however, deals with breaking wind waves in deep water where wave and turbulence fields are not affected by the presence of the seafloor.

Deep-water surface waves are sometimes compared to a gearbox linking the atmosphere to the oceans. In this analogy, breaking waves would indicate a high gear. They play a dominant role in many upper ocean processes, such as momentum transfer from wind to ocean currents, dissipation of wave energy, entrainment of air bubbles, disruption of surface films, and the generation of sea spray and aerosols, besides being a source of ambient noise. Wave breaking causes enhanced turbulent kinetic energy levels in the near-surface layer and thus governs turbulent transport of heat, gases, and particles in the near-surface zone. The forces exerted by breaking waves on ships and offshore structures are up to 10 times larger than for nonbreakers, and especially larger breaking waves pose significant danger to seafarers.

Turbulence generated by breaking waves is very intermittent and coexists with turbulence generated by other sources such as shear stress, convection, internal waves, and Langmuir circulation. Presently, it is not understood how wave-induced turbulence interacts with this background turbulence, and these complex interactions are not discussed in any detail in this article.

Breaking Waves

Wave breaking occurs on a wide range of scales and strength. At low wind speeds, the breaking of short wind waves of centimeter to meter wave lengths commonly does not generate any visible air entrainment and is called microbreaking. These small breakers disrupt the molecular boundary layer and play an important role in air–sea gas exchange. Microbreakers break up the cool surface skin of the ocean, a process best detected with infrared sensors. At moderate to high wind speed, breaking waves start to generate small air bubbles which can be observed as whitecaps. The onset of whitecap generation is associated with a minimum wind speed of order $5 \, \mathrm{m \, s^{-1}}$, although no absolute threshold wind speed applies universally. Depending on the whitecap generation mechanism, the waves are labeled spilling breakers or plunging breakers. The most dramatic form of breaking is found in plunging breakers, where the waves overturn and a sheet of moving water plunges down at some distance forward from the wave crest, creating a large air cavity as well as smaller bubbles surrounding the intruding jet. This type of breaker is most common in shallow water and on beaches, but has also been observed in open ocean conditions of wind waves propagating against swell. In deep water nearly all breakers are spilling breakers in which a turbulent current at the wave crest entrains air, leading to a whitecap at the forward edge of the crest.

Initial air fractions in breaking waves can be as high as 70–80%, but very rapidly decrease to <10%. In spilling breakers these high air fractions are limited to a shallow depth of $O(0.1 \, \mathrm{m})$. At depths below $1 \, \mathrm{m}$, air fractions of order 10^{-3}, decaying within one wave period, to order 10^{-5} are observed.

It is apparent to the casual observer that wave breaking activity and whitecap coverage increase with increasing wind speed. However, this is only an indirect dependence and vastly different breaking rates are observed on different occasions with similar wind speeds but varied other factors such as fetch, wave age, or underlying currents. It has been long known that waves on the ocean often form groups consisting of roughly four to eight individual waves with wave crests near the center of the group being much higher than waves at the beginning or end. This group structure also affects the breaking occurrence. Observing the coastal ocean from a high vantage point or a low-altitude airplane, one finds that successive dominant whitecaps are often separated by one wave length but the period between the onsets of successive breakers is equal to two wave periods. This is because waves tend to break near the center of the group where their steepness is at a maximum and in deep water, the wave group propagates at half the phase speed of the individual waves. Therefore, one wave period after the breaking onset, the breaking crest has propagated forward of the group center (its speed relative to the group being half its phase speed), so that its amplitude and thus its steepness are reduced. However, after two wave periods, when the group has propagated by one wave length, the next crest has reached the center of the group and is likely to break, leading to the observed regularity (see **Figure 1**). This process is most prevalent in narrow-banded, nearly unidirectional wave fields, most commonly observed in coastal waters or on larger lakes. The larger directional spreading and broader spectral shape of open-ocean wave fields often obscure the idealistic breaker regularity described above.

Recent theoretical and observational studies suggest that the onset of breaking is determined by the redistribution of wave energy due to nonlinear wave–wave interactions rather than by direct wind forcing. These nonlinear wave hydrodynamics are reflected in the shape of the wave energy spectrum $S(\omega)$, where ω is the wave frequency. In particular, the wave saturation $\sigma(\omega) = 2g^{-2}\omega^5 S(\omega)$, where g is the gravitational acceleration, provides a good indicator of breaking activity. It also illustrates that wave breaking cannot be ascribed to individual wave properties but rather to the complex interplay of many wave components. Observations suggest that a certain wave saturation must be exceeded for breaking to set in, and the breaking rate increases with increasing wave saturation. The exact form of this threshold behavior is not yet well established.

It seems impossible to define precise threshold conditions for the onset of breaking of individual waves within a random wave spectrum. Theory predicts that waves break when the Lagrangian downward acceleration $-\partial^2\eta/\partial t^2$ of the fluid at the crest exceeds a portion of the gravitational acceleration, $-\partial^2\eta/\partial t^2 > \alpha g$, where $\alpha = 0.5$ for the limiting Stokes wave but $\alpha < 0.39$ for the so-called almost steepest waves, and $\eta(t)$ is the surface elevation record at a fixed point. This threshold mechanism has been the basis for several theoretical studies of wave breaking; however, it is not conclusively supported by observations, where wave breaking has been observed at accelerations below the threshold value. Similarly, the theoretical maximum steepness of the limiting Stokes wave, $ak = \pi/7$, where a and k are wave amplitude and wave number, respectively, is hardly ever observed in the field, and extensive observations reveal that breaking and nonbreaking waves cannot be separated on the basis of local steepness alone.

Wave breaking may occur at all wave scales. However, the spectral distribution of breakers depends on the wave development and the most common breakers are associated with waves of higher frequencies than ω_p, the frequency at the peak of the energy spectrum. In fact, only in young wave fields,

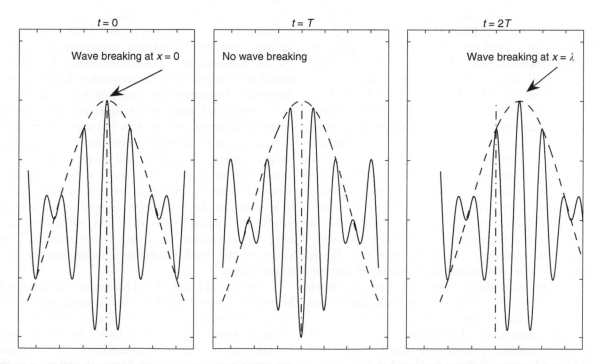

Figure 1 Idealized wave breaking periodicity. Waves tend to break in the center of wave groups. Thus, breakers are separated by one wave length λ, and the period between successive breakers equals two wave periods T.

where the wave saturation at the peak is sufficiently large, are breaking waves observed at all scales, including the dominant waves. As the wave field develops the distribution of breaking scales narrows and its peak shifts further away from the dominant wave frequency toward higher frequencies.

The total breaking rate, defined as the number of breaking waves of all frequencies passing a fixed location, depends on the distribution of breaking scales and thus on the shape of the wave saturation $\sigma(\omega)$. Limiting the breaking spectrum to breaking waves that result in visible whitecaps, one typically finds about 50–100 breakers per hour for open-ocean conditions and 12-m s^{-1} wind speed.

Turbulence beneath Breaking Waves

Direct measurement of the fine-scale velocity field in the ocean and especially in the near-surface layer is extremely challenging. Surface waves are a source of enhanced turbulence. However, they also create the major difficulties in near-surface velocity measurements. Typical turbulent velocity fluctuations are O(10^{-2}–10^{-1} m s^{-1}) and thus are 10–100 times smaller than the wave-related velocities. In terms of kinetic energy, the wave motion contains 2–4 orders of magnitude higher energy levels than the turbulent motion. Furthermore, the nonlinear advection associated with the wave orbital motion modulates the turbulent flow observed at a fixed mooring and affects the dissipation estimates obtained from single point velocity records that rely on Taylor's hypothesis of frozen turbulence. Nevertheless, significant progress in measuring wave-induced turbulence under natural conditions has been made, starting around the mid-1980s. However, most detailed information stems from controlled laboratory experiments, although the breaking characteristics in these studies are often very different to those of natural breaking waves. Focused superposition of dispersive mechanically generated waves leads to well-defined wave breaking even in the absence of wind forcing. This setup allows repeatable turbulence measurements. Turbulence beneath these breaking waves is seen to spread downward; within the first two wave periods this spreading is approximately a linear function of time and occurs more slowly thereafter. The final spreading depth of roughly twice the wave height is reached after four wave periods and by then about 90% of the energy lost by the breaking wave has been dissipated. Thereafter, the remaining decaying turbulence spreads only slightly further and may persist for tens of wave periods but can only be detected in an otherwise rather calm environment.

Behind the breaking crest, vortices of size comparable to the wave height are generated. These rotors may play an important role in mixing gases and pollutants such as small oil droplets. Nearly half of the energy lost from the breaking wave is associated with the entrainment of air bubbles, although part of it will be converted into turbulence kinetic energy (TKE) as larger bubbles rise through the water column.

Under natural conditions, turbulence is commonly characterized by the dissipation rate of TKE ε, which may be inferred from the turbulence velocity shear $\partial u/\partial z$ or rate of strain $\partial u/\partial x$, or from wavenumber velocity spectra $S(k)$. Thus, two fundamentally different approaches exist in oceanic turbulence measurements: (1) observation of the velocity shear or the rate of velocity strain, and (2) observation of the velocity field in space or time.

In isotropic turbulence, the rate of dissipation is related to the rate of strain or the turbulence shear by

$$\varepsilon = 15\nu \overline{\left(\frac{\partial u}{\partial x}\right)^2} = \frac{15}{2}\nu \overline{\left(\frac{\partial u}{\partial z}\right)^2}$$

where ν is the kinematic viscosity of the fluid, u the horizontal velocity component, and x and z are the horizontal and vertical coordinates, respectively.

These relations are the basis for the pioneering studies of near-surface turbulence measurements made with towed hot-film anemometers, electromagnetic current meters, and the common microstructure profilers utilizing airfoil shear probes. The second class of turbulence measurements makes use of Kolmogorov's inertial subrange hypothesis; within a subrange of the wavenumber band the one-dimensional wavenumber spectrum $S(k) = A\varepsilon^{2/3}k^{-5/3}$ has a universal form which depends only on the energy dissipation, where k is the wavenumber and A is a universal constant. This simple relationship between energy dissipation and wavenumber spectra allows the estimation of ε from velocity measurements. Due to recent advances in sonar technology it is now possible to resolve instantaneous velocity profiles at spatial scales of a few millimeters and temporal resolution of a tenth of a second. These scales are suitable for turbulence measurements in the upper ocean.

A common reference level for turbulence studies in boundary layers is the flow along a rigid wall. In this classic reference case, often labeled wall layer or constant stress layer, the velocity profile is logarithmic and the turbulent stress in the inner boundary layer $\tau = \rho u_*^2$ is nearly constant, where ρ is the fluid density and u_* the friction velocity. The TKE dissipation per unit mass is given by $\varepsilon = u_*^3(\kappa z)^{-1}$, with $\kappa = 0.4$ being the von Kármán constant.

Many studies show that turbulence in the ocean surface layer is enhanced compared to turbulence in a constant stress layer and there is strong evidence that the turbulence enhancement is due to breaking waves. The magnitude as well as the depth dependence of the time-averaged TKE dissipation in the near-surface layer of a wind-driven ocean departs significantly from the classic constant-stress-layer form. Observations indicate that the surface layer may be divided into three regimes. In the top layer wave breaking directly injects TKE down to a depth z_b. In this injection layer, dissipation is highest and most likely depth-independent. Below this layer, the wave-induced turbulence diffuses downward and dissipates, as has been also demonstrated in the laboratory experiments. In this diffusive region, the decay of turbulence with depth is stronger than the wall-layer dependence $\varepsilon \propto z^{-1}$. However, the exact depth dependence of the wave-induced turbulence is not well established and profiles consistent with $\varepsilon \propto z^n$, with n in the range -4 to -2, as well as exponential profiles, $\varepsilon \propto e^{-z}$, have been observed. Some open ocean observations under strong wind forcing and significant swell revealed enhanced dissipation values but depth dependence consistent with wall-layer scaling. Further down in the water column at a depth z_t, sufficiently far from the air–sea interface, the contribution of waves becomes small compared to local shear production, and turbulence properties are well described by the constant stress layer scaling. There is, as yet, no conclusive observational evidence for the vertical extension of the different regimes. In particular, the depth of direct TKE injection dominates the total dissipation. Thus, z_b is a crucial parameter in turbulence closure models, where it is implemented as the surface mixing length, which will also affect the mean profiles of tracers such as salt or heat.

Mean dissipation profiles are commonly referenced to the mean water surface, and the oscillation of the sea surface poses a challenge for observation and interpretation of near-surface turbulence. Mooring or tower-based observations are limited to observations below the troughs, and depth is referenced to the mean still water line. Surface-following measurements from floats or ships are, in principle, also suitable to monitor the region above the troughs and depth reference is made with respect to the instantaneous surface. For example, in waves of 0.5-m wave height, a nominal 1-m depth observations from a tower is equivalent to a surface-referenced depth varying from 0.75 m at the location of the wave trough to 1.25 m in the crest region. The same

observation from a float would be converted to depth values ranging from 1.25 m at the trough to 0.75 m at the crest, if referenced to the still water line (see **Figure 2**). Due to the strong depth dependence of dissipation, the choice of coordinate systems affects the resultant dissipation profile.

Microstructure profilers operated in a rising mode are capable of observing turbulence up to the sea surface. However, enhanced dissipation levels associated with wave breaking are very intermittent, and the profiling frequency is too low to adequately resolve these events.

Despite the observational challenges, quality turbulence data in the aquatic near-surface layer in the presence of breaking waves have been collected starting in the mid-1980s. (Pioneering studies started in the 1960s.) The general consensus is that the enhancement of average dissipation $\varepsilon_n = \varepsilon/(u_*^3(\kappa z)^{-1})$ in the diffusive layer is of order 10–100, where ε is taken as the mean over a few minutes, and the depth of the wave enhanced layer (z_t) is confined to a depth corresponding to a few times the significant wave height, that is 2–6 m typically. However, under severe storm conditions, the extent of the wave enhanced layer could be more than 10 m. Bubble clouds generated by wave breaking have been observed to such depths, but it is not known to what extents wave turbulence or coherent structures such as Langmuir circulation are the responsible bubble transport mechanisms. The depth of the injection layer z_b is not well established. Estimates range from $z_b = O(0.1\,H_s)$ to $z_b \gtrsim H_s$, or about 0.2–1 m.

Instantaneous dissipation levels can be much higher than seen in the mean profiles. Beneath an active breaking wave, turbulence enhancements $\varepsilon_n = O(10^4)$ have been observed. These high values persist only for a few seconds. Turbulence beneath individual breaking waves decays as $\varepsilon \propto t^m$, where observations indicate $m \approx -4$ in the diffusion layer and $m \approx -7.6$ in the injection layer. Approximately five wave periods after the onset of breaking, the turbulence levels have decayed to the background level of wall-layer flows. Thus, sufficiently fast sampled dissipation measurements reveal the coexistence of two distinct contributions, a wide distribution centered on constant-stress-layer turbulence levels $(\log(\varepsilon_n) \approx 0)$ and a smaller and narrower distribution representing breaking waves and centered on $\log(\varepsilon_n) \geq 2$. The broader distribution of lower enhancement rates is associated with periods between breaking events and is broadly consistent with a wall-layer flow. The largest turbulence levels, occurring beneath the actively breaking crest, play an important role in the breakup of air cavities and thus determine the initial bubble size distribution.

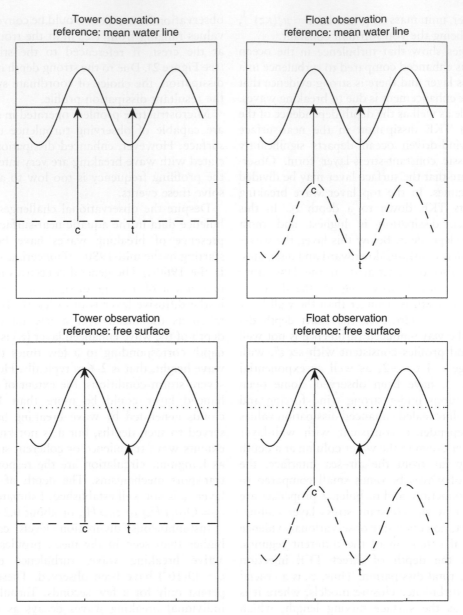

Figure 2   Perceived depth of a measurement, depending on observation platform and choice of surface reference.

The balance between surface tension γ_w and turbulent pressure forces leads to the so-called Hinze scale a_H, which describes the resulting bubble radius: $a_H = A(\gamma_w/\rho)^{3/5}\varepsilon^{-2/5}$, where A is a constant in the range 0.36–0.5. Field observations yielded $a_H \simeq 10^{-3}m$, but more observations are required to establish a possible range of these initial bubble sizes.

In spilling breakers, wave breaking occurs on the wave crest and turbulence levels have decayed significantly by the time the succeeding trough is reached. Therefore, turbulence levels in the crest region are larger than in the trough region and more than half of the energy is dissipated above the mean water line.

Wave breaking is a very intermittent phenomenon and the resulting turbulence fields are very patchy. Therefore, long time series or a suite of several sensors are required to obtain reliable statistics and sound estimates of the contribution of wave breaking to upper ocean processes. Alternatively, it was suggested in 1985 that the scale and strength of breaking may be characterized by the length of the breaking crest and its propagation speed. This opens up the possibility of remote sensing of the integral contribution of processes associated with wave breaking. The central quantity in this concept is the spectral density function $\Lambda(c)$. It is defined in a way that $\Lambda(c)dc$ describes the average total length of breaking wave crests (perpendicular to the wave propagation), per unit area, that have speeds in the range c to $c + dc$. Within a given surface area there might be several breaking crests at any given time, many of them that are only breaking along a fraction of the total crest lengths. To determine $\Lambda(c)$, breaking crests that propagate at a similar speed are combined, the total lengths of these breaking crests are added up, and then the sum is divided by the area of the observed surface patch.

The passage rate of breaking crests propagating at speed c past a fixed point is $c\Lambda(c)$. The fractional surface turnover rate per unit time is $R = \int c\Lambda(c)dc$, which can also be interpreted as the breaking frequency at a fixed point, that is the number of breakers passing a fixed location per unit time.

Furthermore, the fourth and fifth moment of this spectral density function may be related to the dynamics of wave breaking. These relationships are based on similarity scaling of breakers and were confirmed in wave tank experiments. Therefore a further challenge is the proper scaling of the laboratory experiments to wave scales observed under natural conditions. Quasi-steady breakers can be generated by towing a submerged hydrofoil along a test channel. These experiments established that the rate of energy loss per unit length of breaking crest is proportional to c^5. Therefore, the wave energy dissipation due to the breaking of waves of scale corresponding to phase speed c is $\varepsilon(c)dc = b\rho g^{-1} c^5\Lambda(c)dc$, where b is an unknown, nondimensional proportionality factor, originally assumed to be constant. However, b might depend on nondimensional expressions of, for example, the wave-scale or the wave-field nonlinearity. The total energy dissipation associated with whitecaps is $E = b\rho g^{-1} \int c^5\Lambda(c)dc$. Momentum and energy are related by $M = Ec^{-1}$ and the spectrally resolved momentum flux from breaking waves to currents is $m(c)dc = b\rho g^{-1}c^4\Lambda(c)dc$. The total momentum flux from the wave field to currents is given as $M = b\rho g^{-1} \int c^4\Lambda(c)dc$. Evaluation of this integral over all scales (phase speeds c) of breaking waves, including microbreakers, and assuming no wave growth in space yields $M = \tau_w$, where τ_w is the atmospheric momentum flux supported by the form drag of the waves. This momentum flux balance might prove to be a key relation in estimating the proportionality factor b. However, so far observations of the breaking crest density $\Lambda(c)$ in the ocean do not adequately resolve the small-scale waves and microbreakers of the breaking spectrum.

In a wind-driven sea, breaking waves provide the strongest contribution to near-surface turbulence. Other effects of surface waves are increased dissipation levels prior to the onset of air entrainment. This prebreaking turbulence is consistent with wave–turbulence interaction in a rotational wave field. Increased near-surface Reynolds stresses due to the Stokes drift of nonbreaking waves may increase horizontal transports. Interaction of the Stokes drift with the wind-driven current may trigger Langmuir circulation, conceptually described as counter-rotating cells aligned in the wind direction. Langmuir cells are large eddies and may be an important part of the turbulence field as well as influencing the turbulence generated by other processes such as breaking waves. Little is known how small-scale wave-enhanced turbulence affects lateral dispersion; however, it is likely that this process is dominated by Langmuir turbulence.

Conclusion

Especially in mid- to high latitudes breaking waves are a ubiquitous feature of the open ocean. The last two to three decades have seen increased study of breaking deep water waves and new insight has been gained from laboratory experiments and field observations as well as through theoretical wave modeling. Earlier attempts to relate breaking to geometrical or kinematic features of individual wave crests are slowly being replaced by the concept of nonlinear hydrodynamics of the wave field leading to wave breaking.

Wave breaking plays an important role in many processes of air–sea interaction, and the wave-induced turbulence is a relevant quantity in assessing its contributions. Near-surface turbulence observations show a mean dissipation enhancement of 1–2 orders of magnitude due to the effect of wave breaking. However, the detailed structure of the turbulence field and the length scales involved are still not resolved conclusively. Instantaneous dissipation levels are up to 4 orders larger than dissipation levels in a wall-layer flow. These initial high turbulence levels decay rapidly, but are likely to play a defining role in the breakup of air bubbles and thus in the setup of the bubble size distribution. Turbulence levels are highest beneath the wave crest and there is a need for more observations with adequate sampling of the region above the trough line.

The concept of relating breaking wave kinematics and dynamics to whitecap properties that can be observed remotely, say with aerial video imagery or subsurface acoustical tracking, is intriguing and holds the promise of new observational insight in wave breaking processes. However, any quantitative assessments of energy dissipation and momentum fluxes based on this concept directly depend on the proportionality factor b. Currently, only very limited data exist and estimates of b are inconclusive; in fact, it is not even established that b is constant.

Nomenclature

a	wave amplitude
a_H	Hinze scale
c	wave phase speed
E	total energy dissipation associated with wave breaking
g	gravitational acceleration
k	wave number
$m(c)$	spectrally resolved momentum flux from breaking waves to currents
M	total momentum flux associated with wave breaking
R	fractional surface turnover rate per unit time; equivalent to breaking frequency at a fixed location
$S(\omega)$	wave energy spectrum
u	horizontal velocity component
u_*	friction velocity
x	horizontal coordinate
z	vertical coordinate
z_b	injection layer depth
z_t	depth of enhanced wave-induced turbulence
γ_w	surface tension
ε	dissipation rate of turbulent kinetic energy
ε_n	enhancement of average dissipation
$\varepsilon(c)$	spectrally resolved energy dissipation by breaking waves
κ	von Kármán constant
$\Lambda(c)$	spectral density function of breaking crest lengths
ν	kinematic viscosity
ρ	fluid density
$\sigma(\omega)$	wave saturation
τ	turbulent stress
τ_w	atmospheric momentum flux supported by the form drag of the waves
ω	wave frequency
ω_p	frequency of the peak of the wave energy spectrum

See also

Air–Sea Gas Exchange. Bubbles. Langmuir Circulation and Instability. Rogue Waves. Surface Gravity and Capillary Waves. Wave Generation by Wind.

Further Reading

Banner ML (2005) Rougue waves and wave breaking – how are these phenomena related? In: Müller P and Henderson D (eds.) *Proceedings 'Aha Huliko' a Hawaiian Winter Workshop, Jan. 2005.* http://www.soest.hawaii.edu/PubServices/2005pdfs/Banner.pdf (accessed Feb. 2008).

Banner ML and Peregrine DH (1993) Wave breaking in deep water. *Annual Review of Fluid Mechanics* 25: 373–397.

Baschek B (2005) Wave-current action in tidal fronts. In: Müller P and Henderson D (eds.) *Proceedings 'Aha Huliko' a Hawaiian Winter Workshop, Jan. 2005.* http://www.soest.hawaii.edu/PubServices/2005pdfs/Baschek.pdf (accessed Feb. 2008).

Colbo K and Li M (1999) Parameterizing particle dispersion in Langmuir circulation. *Journal of Geophysical Research* 104: 26059–26068.

Donelan MA and Magnusson AK (2005) The role of focusing in generating rogue wave conditions. In: Müller P and Henderson D (eds.) *Proceedings 'Aha Huliko' a Hawaiian Winter Workshop, Jan. 2005.* http://www.soest.hawaii.edu/PubServices/2005pdfs/donelan.pdf (accessed Feb. 2008).

Garrett C, Li M, and Farmer DM (2000) The connection between bubble size spectra and energy dissipation rates in the upper ocean. *Journal of Physical Oceanography* 30: 2163–2171.

Gemmrich J (2005) A practical look at wave-breaking criteria. In: Müller P and Henderson D (eds.) *Proceedings 'Aha Huliko' a Hawaiian Winter Workshop, Jan. 2005.* http://www.soest.hawaii.edu/PubServices/2005pdfs/Gemmrich.pdf (accessed Feb. 2008).

Gemmrich JR and Farmer DM (1999) Observations of the scale and occurrence of breaking surface waves. *Journal of Physical Oceanography* 29: 2595–2606.

Gemmrich JR and Farmer DM (2004) Near surface turbulence in the presence of breaking waves. *Journal of Physical Oceanography* 34: 1067–1086.

Holthuijsen LH and Herbers THC (1986) Statistics of breaking waves observed as whitecaps in the open sea. *Journal of Physical Oceanography* 16: 290–297.

Melville WK (1996) The role of surface-wave breaking in air–sea interaction. *Annual Review of Fluid Mechanics* 26: 279–321.

Melville WK and Matusov P (2002) Distribution of breaking waves at the ocean surface. *Nature* 417: 58–62.

Müller P and Henderson D (eds.) (2005) *Proceedings 'Aha Huliko' a Hawaiian Winter Workshop, Jan. 2005.* http://www.soest.hawaii.edu/PubServices/2005pdfs/TOC2005.html (accessed Feb. 2008).

Phillips OM (1985) Spectral and statistical properties of the equilibrium range in wind-generated gravity waves. *Journal of Fluid Mechanics* 156: 505–531.

Rapp R and Melville WK (1990) Laboratory measurements of deep water breaking waves. *Philosophical Transactions of the Royal Society of London A* 331: 735–780.

Song J-B and Banner ML (2002) On determining the onset and strength of breaking for deep water waves.

Part 1: Unforced irrotational wave groups. *Journal of Physical Oceanography* 32: 2541–2558.

Sullivan PP, McWilliams JC and Melville WK (2005) Surface waves and ocean mixing: Insights from numerical simulations. In: Müller P and Henderson D (eds.) *Proceedings 'Aha Huliko' a Hawaiian Winter Workshop, Jan. 2005.* http://www.soest.hawaii.edu/PubServices/2005pdfs/Sullivan.pdf (accessed Feb. 2008).

Terray EA, Donelan MA, Agrawal YC, *et al.* (1996) Estimates of kinetic energy dissipation under breaking waves. *Journal of Physical Oceanography* 26: 792–807.

Thorpe SA (1995) Dynamical processes of transfer at the sea surface. *Progress in Oceanography* 35: 315–352.

Thorpe SA (2005) *The Turbulent Ocean.* Cambridge, UK: Cambridge University Press.

WHITECAPS AND FOAM

E. C. Monahan, University of Connecticut at Avery
Point, Groton, CT, USA

Introduction

Oceanic whitecaps and sea foam are, respectively, the transient and semipermanent bubble aggregates that are found on the surface of the ocean when certain meteorological conditions prevail. These features are of sufficient size to be detectable by eye, and an individual whitecap or foam patch can readily be recorded using standard low-resolution photographic or video systems. When they are present in sufficient number on the sea surface they alter the general visual albedo, and microwave emissivity, of that surface, thus rendering their collective presence detectable by various satellite-borne instruments. Almost all the bubbles that make up these structures were initially produced at the sea surface by breaking waves, and to understand the presence and distribution of whitecaps and foam patches it is necessary to first consider the genesis, and fate within the oceanic surface layer, of these bubbles. It will become apparent from the discussions contained in the following sections that the bubbles whose presence in great numbers is signaled by the appearance of whitecaps play a major role in the air–sea exchange of gases that are important in establishing our climate, and in the production of the sea-salt aerosol that contributes to the pool of cloud condensation nuclei in the atmosphere over the ocean. These same bubbles facilitate the sea-to-air transfer of heat and moisture, and scavenge from the bulk sea water and carry to the ocean surface various surfactant organic, and adhering inorganic, materials.

Spilling Wave Crests: Stage A Whitecaps

When a wave breaks in the more typical spilling mode, and even more so when a wave collapses in a plunging fashion, great numbers of bubbles are formed and constrained initially to a relatively small volume of water, typically extending beneath the surface a distance no greater than the height of the source wave and having lateral dimensions of only a few meters at most. Although these intense bubble clouds, often called alpha plumes, are individually often of convoluted shape, the concentration of bubbles in these alpha-plumes tends to decrease exponentially with depth, with an e-folding, or scale, depth that increases modestly from less than a meter to several meters, as the sea state increases in response to increasing wind speeds. The concentration of bubbles within an alpha-plume that has just been formed can be so great that the aggregate fraction of the water volume occupied by these bubbles, the void fraction in the terminology of the underwater acoustician, reaches 20% or even 30%. The size spectrum of the bubbles within such a plume is very broad, the bubbles present having radii ranging from several micrometers up to almost 10 mm (see **Figure 1**). Although there is no clear consensus on where the peak in the alpha-plume bubble number density spectrum lies, many authors would contend that it falls at a bubble radius of $50 \, \mu m$ or less. It has been suggested that the amplitude of this spectrum then falls off with increasing bubble radius in such a fashion that for over perhaps a decade of radius the total volume of the bubbles falling within a unit increment of size remains almost constant. It is thought that at even larger bubble radii this spectrum 'rolls off' even more rapidly, with less and less air being contained in those bubbles that fall in the larger and larger size 'bins', but there are as yet insufficient unambiguous observations to verify this contention.

The manifestation on the sea surface of an alpha-plume, the stage A whitecap, is the most readily detected category of whitecap or foam patch. Although bubbles on the surface in a stage A whitecap typically burst within a second of having arrived at the air–water interface, there is often a certain momentary 'packing' of bubbles, both vertically and laterally, on this surface, which results in this category of whitecap being truly white, with an albedo of about 0.5 which does not vary significantly over the entire visible portion of the electromagnetic spectrum. Since the visible albedo of the sea surface away from whitecaps is often 0.03–0.08, the average albedo of this surface will be noticeably increased when even a small fraction of the ocean surface is covered by spilling wave crests. Many of the satellite-borne passive microwave radiometers detect the electromagnetic emissions from the sea surface at wavelengths on the order of 10 mm. At such wavelengths a stage A whitecap is an almost perfect emitter, what in optics would be deemed a 'black body', while the rest of the sea surface at these wavelengths has a

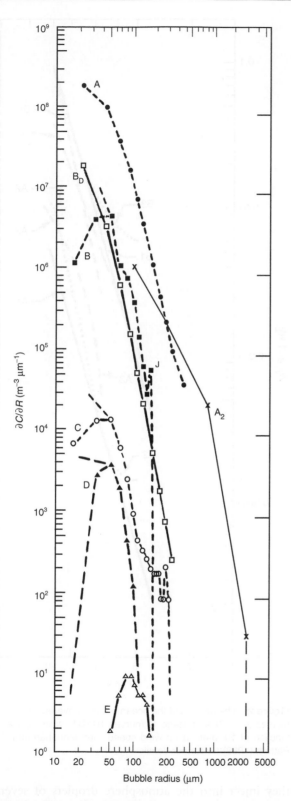

Figure 1 The number of bubbles per cubic meter of sea water, per micrometer bubble radius increment, as a function of bubble radius, as to be expected in (A) the alpha plume beneath a stage A whitecap, (B) the beta plume beneath a stage B whitecap, (C and D) in various portions of a gamma plume, and (E) the background, near-surface bubble layer. See Monahan and Van Pattern (1989) for further details.

microwave emissivity on only 30% or 40%. Thus it only requires a small fraction of the ocean surface to be covered by stage A whitecaps for there to be a measurable increase in the apparent microwave brightness temperature of this surface.

An observer located within an alpha-plume would observe, once the downward movement associated with the spilling event had been dissipated, a high level of small scale turbulence, and, superimposed on top of the rapid random motions caused by this turbulence, a clear upward movement of the larger bubbles. The reduction of gravitational potential energy associated with the upward motion of these big bubbles frees energy that then contributes to the mixing and turbulence within the plume, and this enhanced mixing, which extends to the very surface of the stage A whitecap, greatly increases the effective air–sea gas transfer coefficient, or 'piston velocity', associated with this whitecap, as compared to the gas transfer coefficient associated with the wind ruffled but whitecap-less adjacent portions of the ocean surface. These upward moving bubbles drag water along with them, and the resulting upward, buoyant, flow often induces two-dimensional, horizontal divergence at the surface; factors which also enhance the air–sea exchange of gases. Further, for gases that diffuse slowly through water, the fact that each bubble is a gas vacuole traveling from the body of the water to the air–sea interface can be an important consideration. These large bubbles, with their large cross-sectional areas and rapid rise velocities, are also important in the scavenging and transport to the sea surface of the various surface-active materials that are often present in high concentrations in the oceanic mixed layer.

Decaying Foam Patches: Stage B Whitecaps

Within seconds of a wave ceasing to break, the associated stage A whitecap has been transformed in a decaying foam patch, a stage B whitecap. As a consequence of the intense turbulence present in the alpha plume that had been present beneath the stage A whitecap, the initial lateral extent of the stage B whitecap (and of the top of the beta-plume which is located beneath it) is typically considerably greater than that of a stage A whitecap, some would contend upwards of ten times greater. The greatest discrepancies in size between parent stage A whitecaps and the initial daughter stage B whitecaps occur in those cases where the wave crest spills persistently, or episodically, as it moves along over the sea surface leaving in its wake a long stage B whitecap, or a series of decaying foam patches with short distances between them. As was the case with the stage A whitecap, most of the bubbles that come to the surface in a stage B whitecap burst within a second of their arrival at the interface, and thus a stage B whitecap owes its existence, as did its precursor stage A whitecap, to the continuing arrival at the surface of new bubbles from the dependent bubble plume or cloud. The concentration of bubbles within the associated beta-plume is much smaller than it was in the alpha-plume that preceded it, for three reasons: (1) the plume has been diffused over a greater volume of sea water; (2) many of the large bubbles that were present in the precursor alpha plume have by now reached the sea surface and burst; and (3) most of the very smallest bubbles, those with radii of only a few micrometers, have gone into solution (see **Figure 1**). (The very smallest bubbles can dissolve even when the oceanic surface layer is saturated with respect to nitrogen and oxygen, because at a depth of even a meter they are subjected to significant additional hydrostatic pressure, and because with their small radii they experience a marked increase in internal pressure due to the influence of surface tension.) The stage B whitecap decays by being torn into tattered foam patches by the turbulence of the surface layer, and by having these ever and ever smaller patches fading as the supply of bubbles from the associated portions of the beta-plume becomes exhausted. The cumulative effect of these factors is that the visually resolvable macroscopic area of a stage B whitecap decreases exponentially with time, with a characteristic e-folding time of 3–4 s. A stage B whitecap appears to the eye as a group of irregularly shaped pale blue, or green, areas clustered on the ocean surface. The visible albedo of a stage B whitecap is initially intermediate between that of a stage A whitecap and that of the ruffled sea surface, but within a few seconds its albedo approaches the low value associated with the wave-roughened sea surface. As a consequence of the relatively larger initial area of stage B whitecaps as compared to stage A whitecaps, and on account of the fact that the characteristic lifetime of a stage B whitecap is considerably greater than that of a stage A one, at any instant the fraction of the ocean surface covered by stage B whitecaps is typically at least an order of magnitude greater than the fraction covered by stage A whitecaps (see **Figure 2**).

The beta plume beneath each stage B whitecap is relatively rich in bubbles of intermediate size, with some investigators suggesting that the bubble number density spectrum for this plume has a peak at a bubble radius of about 50 μm (see **Figure 1**). When bubbles of this size burst at the surface in a whitecap

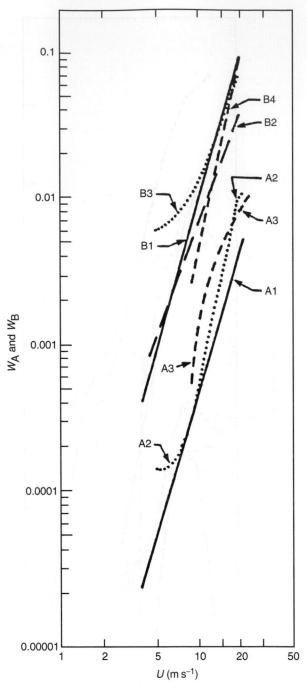

Figure 2 The fraction of the ocean surface covered by stage A (curves A1–A3) and stage B (curves B1–B4) whitecaps as a function of 10 m-elevation wind speed. See Monahan and Van Pattern (1989) for further details.

they inject into the atmosphere droplets of several micrometers radius, called jet droplets, which contribute to the sea-to-air transfer of moisture and latent, and often sensible, heat. The rupture of the upper, exposed, hemisphere of these bubbles when they burst on the sea surface also produces smaller droplets, called film droplets, which constitute a significant fraction of the cloud condensation nuclei in the maritime troposphere. The largest bubbles produced by a breaking wave, most of which reach the surface in the stage A whitecap, are even more effective at generating these film droplets when they burst.

Because bubbles are relatively scarce within stage B whitecaps, these sea surface manifestations of beta plumes have visual albedos and microwave emissivities not greatly different from those of the adjacent, wind ruffled, surface, and are thus much more difficult to detect by remote sensing than are stage A whitecaps.

Wind-dependence of Oceanic Whitecap Coverage

The frequency of wave breaking, and the average intensity of the individual breaking wave, both increase with increasing wind speed. The combined effect of these two factors is that the fraction of the sea surface covered at any moment by spilling wave crests, i.e. by stage A whitecaps, increases rapidly with strengthening wind speed. This can be seen from **Figure 2**, where the curves labeled A1, A2, etc. are summary descriptions of the dependence of stage A whitecap coverage on 10 m-elevation wind speed. Curve A1, describing the most comprehensive set of stage A whitecap observations (actually a combination of four such sets), is described by eqn [1].

$$W_A = 3.16 \times 10^{-7}U^{3.2} \qquad [1]$$

where W_A is the fraction of the sea surface covered at any instant by spilling wave crests and U is the 10 m-elevation wind speed expressed in meters per second. Understandably, the fraction of the sea surface covered instantaneously by decaying foam patches, i.e. by stage B whitecaps, shows a similar strong dependence on wind speed. This can be seen from the steep slopes of the curves B1, B2, etc., on the log–log plot in **Figure 2**. Eqn [2] defines curve B1, which is a summary description of extensive observations, from both the Atlantic and Pacific Oceans, of stage B whitecap coverage made by several investigators.

$$W_B = 3.84 \times 10^{-6}U^{3.41} \qquad [2]$$

Here W_B represents the instantaneous fraction of the sea surface covered by decaying foam patches, and U is again the 10 m-elevation wind speed. The fact that for both categories of whitecap the fraction of the ocean surface occupied by these features varies with the wind speed raised to something slightly more than the third power, is consistent with the contention that whitecap coverage varies with the friction velocity (*see* Heat and Momentum Fluxes at the Sea Surface and Wave Generation by Wind) raised to the third power. It should be stressed that although whitecap coverage, both stage A and stage B, is most sensitive to wind speed, it also varies with the thermal stability of the lower marine atmospheric boundary layer, and with wind duration and fetch. Any factor that influences sea state will also affect whitecap coverage. For near-neutral atmospheric stability, oceanic whitecap coverage begins to be noticed when the 10 m-elevation wind speed reaches 3 or 4 m s^{-1}. (There is not a distinct threshold for the onset of whitecapping at a wind speed of 7 m s^{-1} as was contended in some of the early literature on this subject.)

Since whitecap coverage, particularly stage A coverage, is readily detectable from space, and given that whitecap coverage is very sensitive to wind speed, it is apparent that satellite observations of whitecap coverage can be routinely used to infer over-water wind speeds.

Stabilized Sea Foam

Many of the first bubbles to rise to the sea surface after a breaking wave has entrained air, not only scavenge organic material from the upper meter or so of the sea but also, as they reach the air–sea interface, accrue some of the organic material that is often found on that surface (not necessarily in the form of coherent slicks). As a consequence of accreting on their surface considerable dissolved, and other, organic material, such 'early rising' bubbles may become stabilized, and hence they may not break immediately, but rather persist on the ocean surface for protracted periods. If such a bubble has managed to coat its entire upper hemisphere with such surfactant material, the markedly reduced surface tension of its film 'cap' that results from this circumstance may enable this bubble to persist indefinitely at the air–water interface. Such bubbles are certainly present at the sea surface long enough to be winnowed into windrows, those distinctive, essentially downwind, foam and seaweed streaks that appears on the sea surface when a strong wind has been blowing consistently. Often organized convective motions are present in the upper layer of the ocean. Such Langmuir cells have associated with them lines of horizontal, two-dimensional, surface convergence and divergence, oriented for the most part downwind. When such Langmuir cells are present, stabilized bubbles will be drawn into the convergence zones, and since they are buoyant, they will remain to form fairly uniformly spaced foam lines on the sea surface marking the locations of such zones (**Figure 3**). It should be noted that the 'late arriving' bubbles, representing the vast majority of the bubbles rising within any alpha-plume, do not persist on the air–sea interface for more than a second or so, even when the surface waters are quite organically rich.

The ability of bubbles to effectively scavenge surfactant organic matter from the bulk sea water and transport this material to the sea surface provides what has been described as an 'organic memory' to the upper mixed layer of the ocean. The more bubbles that have been injected into the upper layer of

the ocean by breaking waves in the recent past, the more organic material has been brought to the sea surface and remains there. Although wave action is 'a two way street', in that the same waves which upon breaking produce the bubbles that carry organic material to the air–water interface also stir and mix the surface layer, none-the-less the net effect of high sea states is to alter the partition of organic matter between the bulk fluid and the interface in favor of the interface. This can be inferred from the observation that as a high wind event persists, more and more foam lines, containing more and more stabilized bubbles, appear on the ocean surface. In stormy conditions, such foam, or spume, can be

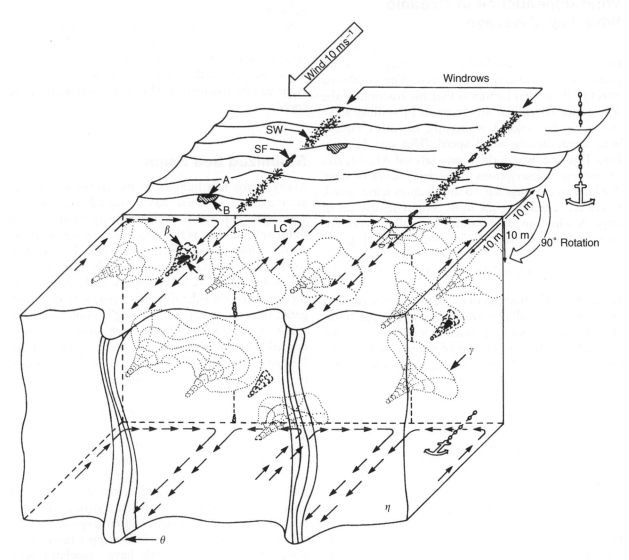

Figure 3 A view looking obliquely down at the sea surface showing stage A and stage B whitecaps, foam and spume lines, and simultaneously a view looking obliquely up toward the same sea surface showing the alpha- and beta-plumes associated with these whitecaps, the gamma-plumes, and the near-surface bubble layer. The influence on these features of a classical Langmuir circulation, which is indicated by arrows, is depicted. A, Stage A whitecap; B, Stage B whitecap; SF, stabilized foam; SW, seaweed; LC, Langmuir circulation; α, plume of stage A whitecap; β, plume of stage B whitecap; γ, old (microbubble) plume; η, background bubble layer; θ bubble curtain. From Monahan and Lu, 1990.

blown off the crests of waves, along with quite large drops of water, called 'spume drops', adding further to the indeterminacy that often prevails in such circumstances regarding the actual location of the air–water interface.

Not only do the above-mentioned Langmuir cells advent buoyant stabilized bubbles into the surface convergence zones, these same cells are believed to move the residual, long-lasting, gamma bubble plumes (those left after the dissipation of the beta plumes) into these same zones. (Alpha bubble plumes have readily detectable stage A whitecaps as their sea surface signatures, and the location of the beta-plumes into which these alpha-plumes decay can be determined from the position on the sea surface of their associated stage B whitecaps, but the large, diffuse, bubble-poor gamma-plumes into which the beta-plumes decay, have no apparent surface manifestation.) The influence of Langmuir cells on stabilized sea surface foam, on gamma-plumes, and on the near surface layer that contains an ever sparser concentration of small bubbles, is depicted in **Figure 3**.

Global Implications

As can be seen from the curves in **Figure 2**, even at quite high wind speeds such as $15 \, \mathrm{m \, s^{-1}}$ (33.5 miles $\mathrm{h^{-1}}$), only a small fraction of the sea surface is covered by stage B whitecaps (0.04 or 4%), and an even smaller fraction of that surface is covered by stage A whitecaps (0.002 or 0.2%). Yet the total area of all the world's oceans is very great ($3.61 \times 10^{14} \, \mathrm{m^2}$), and as a consequence the total area of the global ocean covered by whitecaps at any instant is considerable. If a wind speed of $7 \, \mathrm{m \, s^{-1}}$ is taken as a representative value, then at any instant some $7.0 \times 10^{10} \, \mathrm{m^2}$, i.e. some $70 \, 000 \, \mathrm{km^2}$, of stage A whitecap area is present on the surface of the global ocean. Following from this, and including such additional information as the terminal rise velocity of bubbles, it can be deduced that some $7.2 \times 10^{11} \mathrm{m^2}$, i.e. some $720 \, 000 \, \mathrm{km^2}$ of individual bubble surface area are destroyed each second in all the stage A whitecaps present on the surface of all the oceans, and an equal area of bubble surface is being generated in the same interval. The vast amount of bubble surface area destroyed each second on the surface of all the world's oceans, and the great volume of water (some $2.5 \times 10^{11} \, \mathrm{m^3}$) swept by all the bubbles that burst on the sea surface each second, have profound implications for the global rate of air–sea exchange of moisture, heat and gases. An additional preliminary calculation following along these lines, suggests that all the bubbles breaking on the sea surface each year collect some 2 Gt of carbon during their rise to the ocean surface.

See also

Heat and Momentum Fluxes at the Sea Surface. Wave Generation by Wind.

Further Reading

Andreas EL, Edson JB, Monahan EC, Rouault MP, and Smith SD (1995) The spray contribution to net evaporation from the sea review of recent progress. *Boundary-Layer Meteorology* 72: 3–52.

Blanchard DC (1963) The electrification of the atmosphere by particles from bubbles in the sea. *Progress in Oceanography* 1: 73–202.

Bortkovskii RS (1987) *Air–Sea Exchange of Heat and Moisture During Storms*, revised English edition. Dordrecht: D. Reidel [Kluwer].

Liss PS and Duce RA (eds.) (1997) *The Sea Surface and Global Change*. Cambridge: Cambridge University Press.

Monahan EC and Lu M (1990) Acoustically relevant bubble assemblages and their dependence on meteorological parameters. *IEEE Journal of Oceanic Engineering* 15: 340–349.

Monahan EC and MacNiocaill G (eds.) (1986) *Oceanic Whitecaps, and Their Role in Air–Sea Exchange Processes*. Dordrecht: D. Reidel [Kluwer].

Monahan EC and O'Muircheartiaigh IG (1980) Optimal power-law description of oceanic whitecap coverage dependence on wind speed. *Journal of Physical Oceanography* 10: 2094–2099.

Monahan EC and O'Muircheartiaigh IG (1986) Whitecaps and the passive remote sensing of the ocean surface. *International Journal of Remote Sensing* 7: 627–642.

Monahan EC and Van Patten MA (eds.) (1989) *Climate and Health Implications of Bubble-Mediated Sea–Air Exchange*. Groton: Connecticut Sea Grant College Program.

Thorpe SA (1982) On the clouds of bubbles formed by breaking wind waves in deep water, and their role in air–sea gas transfer. *philosophical Transactions of the Royal Society* [London] A304: 155–210.

BUBBLES

D. K. Woolf, Southampton Oceanography Centre, Southampton, UK

Introduction

Air–sea interaction does not solely occur directly across the sea surface, but also occurs across the surface of bubbles suspended in the upper ocean, and across the surface of droplets in the lower atmosphere. This article describes the role of bubbles in air–sea interaction.

There are three quite different types of bubbles in the oceans that can be distinguished by their sources (atmospheric, benthic, and cavitation). Benthic sources of bubbles include vents and seeps and consist of gases escaping from the seafloor. Common gases from benthic sources include methane and carbon-dioxide. Cavitation is largely an unintentional by-product of man's activities; typically occurring in the wake of ship propellors. It consists of the rapid growth and then collapse of small bubbles composed almost entirely of water vapor. Cavitation may be thought of as localized boiling, where the pressure of the water falls briefly below the local vapor pressure. Cavitation is important in ocean engineering due to the damage inflicted on man-made structures by collapsing bubbles. Both cavitation bubbles and bubbles rising from the seafloor are encountered in the upper ocean, but are peripheral to air–sea interaction. Atmospheric sources of bubbles are a product of air–sea interaction and, once generated, the bubbles are themselves a peculiar feature of air–sea interaction. The major sources of bubbles in the upper ocean are the entrapment of air within the flow associated with breaking waves and with rain impacting on the sea surface.

Once air is entrapped at the sea surface, there is a rapid development stage resulting in a cloud of bubbles. Some bubbles will be several millimeters in diameter, but the majority will be <0.1 mm in size. Each bubble is buoyant and will tend to rise towards the sea surface, but the upper ocean is highly turbulent and bubbles may be dispersed to depths of several meters. Small particles and dissolved organic compounds very often collect on the surface of a bubble while it is submerged. Gas will also be slowly exchanged across the surface of bubbles, resulting in a continual evolution of the size and composition of each bubble. The additional pressure at depth in the ocean will compress bubbles and will tend to force the enclosed gases into solution. Some bubbles will be forced entirely into solution, but generally the majority of the bubbles will eventually surface carrying their coating and altered contents. At the surface, a bubble will burst, generating droplets that form most of the sea salt aerosol suspended in the lower marine atmosphere.

The measurement of bubbles in the upper ocean depends largely on their acoustical and optical properties. At the same time, the effect of bubbles on ocean acoustics has long been a major motivation for bubble studies. The generation of noise at bubble inception may be exploited. For example, acoustic measurements of rainfall depend on bubble phenomena. Fully formed bubble clouds attenuate and scatter both sound and light in the upper ocean.

Climatologies of the distribution of bubbles in the upper ocean are based on both acoustical and optical measurements of bubbles. The global distribution of bubbles reflects the dominance of wave breaking as a source of bubbles, and the high sensitivity of wave breaking to wind speed. Bubbles are an important component of global geochemical cycling through their transport of material in the upper ocean and surface microlayer (*see* Surface Films), and especially their role in the air–sea exchange of gases and particles.

Sources of Bubbles

As described already, bubbles may originate in a variety of ways, but this section will concentrate on the major natural processes of air bubble formation. The atmosphere is clearly a potential source of air bubbles, and generation involves the 'pinching off' of part of the atmosphere, or the 'condensation' of gases dissolved from the atmosphere within a body of water.

Generation of bubbles within the body of water, when the surface water is sufficiently supersaturated with air, is similar to 'vapor' cavitation, but involves the major constituents of the atmosphere (nitrogen, oxygen, etc.) rather than water vapor alone. In the absence of hydrodynamic pressure effects associated with flow, the radial pressure into a cavity, P_b, is the sum of the atmospheric pressure, P_a, the hydrostatic pressure at a depth, z, and a component associated with the surface tension, γ, and the curvature of the cavity (or radius 'r'):

$$P_b = P_a + \rho gz + 2\gamma/r$$

For a bubble to grow, the pressure within a bubble (equal to the sum of partial pressures of the gases diffusing into the bubble), must exceed atmospheric pressure by a margin that increases both with water depth and the curvature of the cavity. A sufficiently large initial cavity is necessary for inception. The explosive dynamics of 'true' cavitation are associated with the rapid transport of water vapor across the surface of the cavity. However, the conditions for water vapor cavitation can only be achieved at normal temperatures where pressure is very low. A sufficient pressure anomaly may occur in an intense acoustic pulse, or in the wake of a fast moving solid object, but is not a common natural phenomenon. The conditions for growth of a bubble by diffusion of atmospheric gases are possible within the normal range of natural variability. For gases other than water vapor, molecular transport of the dissolved gases near the surface of the bubble is sufficiently slow that a virtual equilibrium between the internal and external pressures on the bubble must exist. We might observe bubble generation at home within a bucket of water, or in a soda bottle, where warming induces supersaturation (the solubility of most gases decreases with increasing temperature) and defects in the container walls provide the initial cavity. Warming, mixing, or bubble injection may occasionally force supersaturations of several percent at sea, in which case growth of bubbles on natural particles and microbubbles may release the excess pressure.

Entrapment of air at the sea surface is more common than inception within the body of the water. Most of us are familiar with plumes of bubbles generated by paddling and by boats, but the entrapment of air in the absence of a solid boundary is less intuitive. In general, air is rarely entrapped by enclosure of a large air volume, but is usually drawn into the interior ('entrained') where there is intense and convergent flow of water at the sea surface. Sufficiently energetic convergence occurs where precipitation impacts on the sea surface, and where waves break.

Bubble formation is associated with all common forms of precipitation (rain, hail, and snow), but the details of bubble formation are highly specific to the details of the precipitation. In particular, bubble formation by rain is known to be sensitive to the size, impact velocity, and incidence angle of the rain drops. Large drops, exceeding 2.2 mm in diameter, entrain most air. In heavy tropical downfalls, the volume of air entrained can be fairly significant ($\sim 10^{-6}\, \mathrm{m^3\, m^{-2}\, s^{-1}}$), although much lower than rates associated with wave breaking in high winds. Bubbles up to 1.8 mm in radius are entrained by large rain drops, but smaller drops (0.8–1.1 mm in diameter) generate bubbles of only 0.2 mm in radius.

When waves break at the seashore, the large 'dominant' waves dissipate their energy partly in entraining and submerging quite large volumes of air. On the open ocean, some of the largest and longest waves break, but wave breaking also occurs at much smaller scales. Some very small breaking events may be too weak to entrain air; however, small but numerous breaking events entraining small volumes of air occur on steep waves as short as 0.3 m in wavelength. The energy dissipated in wave breaking is derived from wind forcing of surface waves, and the amount of wave breaking and air entrainment is very sensitive to wind speed. The stage of development of the wave field also has some influence on air entrainment – the size of the largest breaking event is limited to the largest wave that has developed.

An important feature of bubble generation at the sea surface is that a myriad of very small (<0.1 mm radius) bubbles is produced. Very large cavities several millimeters in diameter are likely to be torn apart by large shear forces at the sea surface, but it is difficult to explain how bubbles of <1 mm might be fragmented. Also, the same processes in fresh water (e.g. a lake or a waterfall) do not produce many small bubbles. The explanation can be found in the influence of dissolved salts on surface forces. In sea water, a surface deformation will tend to grow more and more contorted, so that when a large bubble is fragmented it will often shatter into numerous much smaller bubbles. The same factors will usually prevent the coalescence of bubbles in sea water.

Dispersion and Development

Bubbles entrained by a breaking wave may be carried rapidly to a depth of the order of the height of the breaking wave by its energetic turbulent plume. For some wave breaking and other forms of bubble production the initial injection will be much shallower ($\sim$ 1–100 mm). Most of the bubbles are very small, but the majority of the volume of air is comprised of fairly large ($\sim$ 1 mm) bubbles entrained by breaking waves. Most of these larger bubbles will soon rise to the surface (typically in $\sim$ 1 s) in a highly dynamic plume close behind the breaking wave. The less buoyant, smaller bubbles are generally carried to a greater depth and are easily dispersed by mixing processes in the upper ocean.

Bubbles are mixed into the ocean by small-scale turbulence associated with the 'wind-driven upper ocean boundary layer', but also by relatively large and coherent turbulent structures, especially Langmuir circulation (*see* Langmuir Circulation and Instability). Langmuir circulation comprises sets of paired vortices (cells) aligned to the wind. Bubbles will be drawn to the downwelling portions of the Langmuir cells, producing lines of enhanced bubble concentration, parallel to the wind. Langmuir cells can be up to tens of meters deep and wide, and downwelling speeds may exceed $0.1\,\mathrm{m\,s^{-1}}$. In principle, even quite large bubbles may be forced downwards, but generally bubbles of only $\sim 20\,\mu\mathrm{m}$

radius are most common at depths of $\geq 1\,\mathrm{m}$. Concentrations fall off rapidly with increasing radius, at radii exceeding the modal radius.

The development of a bubble cloud does not solely concern the movement of bubbles, but also concerns the development of each and every bubble. Material will be transferred between the bubble and the surrounding water as a result of the flow of water around the bubble (largely induced by the buoyant rise of bubbles relative to their surroundings) and molecular diffusion close to the surface of the bubble. This transport plays a large part in the role of bubbles in geochemical cycling, which is illustrated schematically in **Figure 1**. The transport of both

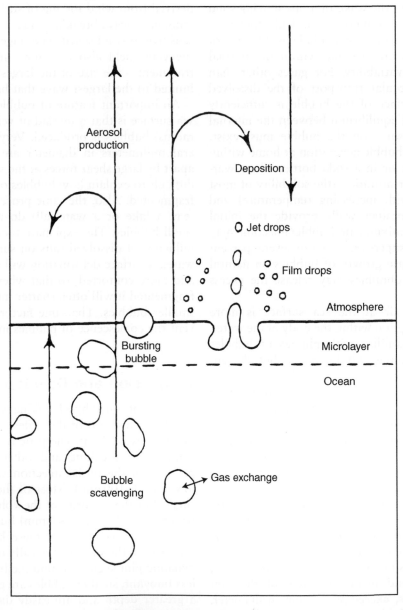

Figure 1   A schematic illustration of the role of bubbles in geochemical cycling.

volatile (i.e. gases) and nonvolatile substances is of interest.

Nonvolatile substances will not penetrate the bubble itself, but may be transported between the surface of the bubble and the surrounding water. Many substances are 'surface-active', that is, they tend to stick to the surface and alter the dynamic properties of the surface. Some substances will already be adsorbed on the surface at the point of formation at the sea surface. During the lifetime of a bubble, further material (both dissolved and small particles) will accumulate on the surface of a bubble. One consequence of the 'bubble scavenging' process is the cycling of surface-active substances. Also, the surface-active material will alter the dynamic properties of the bubble, critically affecting the rise velocity of the bubble and transport across the surface of the bubble. A pure water surface is 'mobile', but it may be immobilized by surface-active material. The flow near a mobile (or free) surface and a rigid surface is quite different. Generally, a 'dirty' bubble with a contaminated, rigid surface will rise more slowly and will exchange gas at a much slower rate compared with a 'clean' bubble of the same size. The surface of small bubbles is immobilized by only a small amount of contamination, and bubbles <100 mm radius are likely to behave as dirty bubbles for most or all of their life. Larger bubbles will also be contaminated, but their dynamic behavior may remain close to that of a 'clean' bubble for several seconds (depending on bubble radius and the contamination level of the water).

The transfer of gases across the surface of bubbles is important to the evolution of each bubble, and to the atmosphere–ocean transport of gases. Gases will diffuse across the surface of a bubble. The net transport of each gas across the surface of a single bubble depends on its concentration in the two media and the mechanics of transport:

$$\text{bubble–water flux} = -j4\pi r^2[C_w - Sp_b]$$

As explained in the previous section, the gases within a bubble are compressed so that the pressure of gases in the bubble generally exceeds those in the atmosphere. This excess leads to a tendency for bubbles to force supersaturation of gases in the upper ocean. Many bubbles may be forced entirely into solution (possibly leaving a fragment enclosed in a shell of organics and small particles – a microbubble). The total (integral) effect of bubble clouds on air–sea gas exchange can be described by the following formula (*see* Air–Sea Gas Exchange):

$$\text{air–sea flux} = -K_T[C_w - Sp_a(1 + \Delta)]$$

(per unit area of sea surface)

$$= K_b[(1 + \delta)C_a/H - C_w] + K_o[C_a/H - C_w]$$

The effect of bubbles on air–sea exchange is described by two coefficients: the contribution to the transfer coefficient, K_b, and a 'saturation anomaly', Δ. Both of these coefficients depend greatly on the solubility of the gas and the bubble statistics. For relatively soluble gases, such as carbon dioxide, the saturation anomaly due to bubble injection is generally negligible, but for less soluble gases, including oxygen the anomaly is usually significant, particularly at high wind speeds. The contribution to the transfer coefficient is again greater for less soluble gases, but is likely to be significant for most gases, at least for high wind speeds.

We have focused on unstable bubbles that will either surface or dissolve within a few minutes of their creation. When a bubble totally dissolves it may leave a conglomeration of the particles and the organic material it accumulated. Some of the bubbles may not entirely dissolve, but may be stabilized at a radius of a few micrometers by their collapsed coating. (The mechanism of stabilization is rather mysterious, external pressures will be high and the coating can not entirely prevent the diffusion of gas, therefore total collapse must be resisted by the structural integrity of the coating – perhaps like a traditional stone wall.) Stable microbubbles might also be generated by a biological mechanism. Microbubble populations are denser in coastal waters where biological productivity and organic loading are generally higher. Microbubbles influence the acoustic properties of natural waters and are a common nucleus for cavitation.

Surfacing and Bursting

Many small bubbles dissolve in the upper ocean, but generally the majority of the bubbles (and almost all the large bubbles) eventually surface. Phenomena that occur when a bubble surfaces are again significant to geochemical cycling (**Figure 1**). The release of gas from a bubble to the atmosphere completes the process of air–sea gas exchange mediated by the bubble. The approach of a bubble, or more especially a plume of bubbles, can disrupt the surface microlayer, enhancing turbulent transport directly across the sea surface. The bubble carries material to the sea surface accumulated by scavenging within the upper ocean. Most important are the energetic processes that occur when a bubble bursts on the sea surface. Bubble bursting is responsible for ejecting droplets into the atmosphere, creating the sea salt aerosol.

Droplets can also be torn directly from wave crests, but bubbles generate almost all of the very small droplets that are easily suspended in the lower atmosphere and that will be dispersed over large distances. When a bubble surfaces its upper surface will project beyond the sea surface. This 'film cap' will drain and shatter. The shattering of the film cap produces 'film drops'. In some cases, the film cap can shatter into many remarkably small ($<1\,\mu m$) droplets, while in other cases a few large $\sim 10\,\mu m$ radius droplets will be produced. The open cavity left after the film cap shatters will collapse inwards, leading to the upward ejection of a 'Worthington jet'. This jet will pinch off into a few 'jet drops'. The drop radii will typically be one-tenth of the radius of the parent bubble, producing drops from $\sim 2\,\mu m$ to tenths of a millimeter in radius from a typical bubble population. The droplets generated by bursting bubbles will be enriched by material brought to the sea surface by the bubble and drawn from the sea surface. The sea salt aerosol will include organic material, metals, viruses, and bacteria.

Acoustical and Optical Properties

Our knowledge of bubble distributions in the upper ocean is based on acoustical and optical measurements. Bubbles also have a significant impact on the acoustical and optical properties of the upper ocean. The acoustic properties of bubbles have attracted a great deal of attention. The generation of bubbles, both by breaking waves and rain, is an important source of noise in the upper ocean. Bubbles also absorb and scatter sound. The scattering of sound by an individual bubble is frequency-dependent with three primary regimes: close to, above, and below the 'breathing frequency' of the bubble. The breathing frequency of a bubble is the natural frequency at which a bubble will oscillate radially ('breathe') and is determined by its radius, surface tension, and the external pressure. The breathing frequency is inversely related to bubble radius, and in the upper ocean, bubbles of different radii will respond in resonance to acoustic frequencies from 10 kHz to a few hundred kHz. Scattering cross-sections close to resonance are very high. When the acoustic frequency is much higher than the breathing frequency of the bubble, the scattering by the bubble is related simply to its physical size ('geometric scattering'). At low acoustic frequencies the acoustic cross-section of an individual bubble is much lower than its geometric cross-section (Rayleigh scattering). The scattering by a bubble is equal in every direction (isotropic) at most practical frequencies, but

becomes more anisotropic at very low frequencies. At low acoustic frequencies ($<10\,kHz$), scattering is largely a communal response of clouds rather than of individual bubbles.

Many measurements of bubbles have taken advantage of the resonant response of bubbles to sound. In particular, measurements at a number of acoustic frequencies can be inverted to calculate the size distribution of bubbles. A pair of transmitting and receiving 'transducers' can measure backscatter remotely along a profile. This technique has been used to infer the concentration and size of bubbles as a function of depth. The very high scattering by the concentrated plumes near breaking waves defy remote measurement. Instead bubbles near the surface may be studied by measuring absorption or scattering along a short path length. Other techniques include applying the influence of air void on the conductivity of the water, and optical measurements.

Casual observation of the milky water marking a developing bubble cloud is enough to understand that bubbles in the upper ocean can alter the optical properties (e.g. color and brightness) of the sea, but among the numerous and complicated influences on ocean optics, bubbles have received relatively little attention. Bubble populations have been measured photographically, but for the sparse populations a meter or so beneath the sea surface this method is tedious if ultimately effective. Video footage of wave breaking and the early development of bubble plumes can be used to understand the many processes involved.

Summary of Bubble Distribution

Measurements of bubbles in the ocean are still fairly sparse, and the relationship of wave breaking and bubble injection to environmental conditions is only partly understood, but we can at least summarize the general relationship of unstable bubble populations to wind forcing. Away from the immediate plume of a breaking wave, the mean concentration of bubbles of radius, r, at a depth z, typically follows a distribution of the form,

$$N \propto r^{-4} \exp(-z/L)$$

for radii as small as $30\,\mu m$, but there is a maximum in N typically at $25\,\mu m$ radius. A typical attenuation depth, L, is 1 m. Some studies have suggested only a weak, approximately linear relationship between the attenuation depth and wind speed, but recent extensive studies imply attenuation depths proportional to the square of wind speed. There are fewer measurements of bubbles in the upper ocean within half a meter of the sea surface, but it is clear that concentrations are much higher near a breaking wave, and larger bubbles are far more common. The injection rate of bubbles is expected to increase with the third or fourth power of the wind speed. As vertical dispersion of the bubbles (and the attenuation depth) also increase with wind speed, the concentration of bubbles below the sea surface is extremely sensitive to wind speed. Air–sea gas exchange, scavenging, and other geochemical transport processes associated with bubbles will share this sensitivity to wind speed, suggesting that a large fraction of activity may occur in fairly rare storm conditions.

See also

Air–Sea Gas Exchange. Air–Sea Transfer: Dimethyl Sulfide, COS, CS$_2$, NH$_4$, Non-Methane Hydrocarbons, Organo-Halogens. Air–Sea Transfer: N$_2$O, NO, CH$_4$, CO. Breaking Waves and Near-Surface Turbulence. Carbon Dioxide (CO$_2$) Cycle. Evaporation and Humidity. Heat and Momentum Fluxes at the Sea Surface. Langmuir Circulation and Instability. Photochemical Processes. Surface Films. Upper Ocean Mixing Processes. Wave Generation by Wind. Whitecaps and Foam.

Further Reading

Blanchard DC (1983) The production, distribution, and bacterial enrichment of the sea-salt aerosol. In: Liss PS and Slinn WGN (eds.) *The Air–Sea Exchange of Gases and Particles*, pp. 407–454. Dordrecht: Kluwer.

Leighton TG (1994) *The Acoustic Bubble*. San Diego: Academic Press.

Medwin H and Clay CS (1998) *Fundamentals of Acoustical Oceanography*. San Diego: Academic Press.

Monahan EC (1986) The ocean as a source for atmospheric particles. In: Buat-Ménard P (ed.) *The Role of Air–Sea Exchange in Geochemical Cycling*, pp. 129–163. Dordrecht: Kluwer.

Woolf DK (1997) Bubbles and their role in gas exchange. In: Liss PS and Duce RA (eds.) *The Sea Surface and Global Change*, pp. 173–205. Cambridge: Cambridge University Press.

UPPER OCEAN CIRCULATION
AND STRUCTURE

UPPER OCEAN MIXING PROCESSES

J. N. Moum and W. D. Smyth, Oregon State
University, Corvallis, OR, USA

Introduction

The ocean's effect on weather and climate is governed largely by processes occurring in the few tens of meters of water bordering the ocean surface. For example, water warmed at the surface ona sunny afternoon may remain available to warm the atmosphere that evening, or it may be mixed deeper into the ocean not to emerge for many years, depending on near-surface mixing processes. Local mixing of the upper ocean is predominantly forced from the state of the atmosphere directly above it. The daily cycle of heating and cooling, wind, rain, and changes in temperature and humidity associated with mesoscale weather features produce a hierarchy of physical processes that act and interact to stir the upper ocean. Some of these are well understood, whereas others have defied both observational description and theoretical understanding.

This article begins with an example of *in situ* measurements of upper ocean properties. These observations illustrate the tremendous complexity of the physics, and at the same time reveal some intriguing regularities. We then describe a set of idealized model processes that appear relevant to the observations and in which the underlying physics is understood, at least at a rudimentary level. These idealized processes are first summarized, then discussed individually in greater detail. The article closes with a brief survey of methods for representing upper ocean mixing processes in large-scale ocean models.

Over the past 20 years it has become possible to make intensive turbulence profiling observations that reveal the structure and evolution of upper ocean mixing. An example is shown in **Figure 1**, which illustrates mixed-layer[1] evolution, temperature

structure and small-scale turbulence. The small white dotsin **Figure 1** indicate the depth above which stratification is neutral or unstable and mixing is intense, and below which stratification is stable and mixing is suppressed. This represents a means of determining the vertical extent of the mixed layer directly forced by local atmospheric conditions. (We will call the mixed-layer depth D.) Following the change in sign from negative (surface heating) to positive (surface cooling) of the surface buoyancy flux, J_b^0, the mixed layer deepens. (J_b^0 represents the flux of density (mass per unit volume) across the sea surface due to the combination of heating/cooling and evaporation/precipitation.) The mixed layer shown in **Figure 1** deepens each night, butthe rate of deepening and final depth vary. Each day, following the onset of daytime heating, the mixed layer becomes shallower.

Significant vertical structure is evident within the nocturnal mixed layer. The maximum potential temperature (θ) is found at mid-depth. Above this, θ is smaller and decreases toward the surface at the rate of about 2 mK in 10 m. The adiabatic change in temperature, that due to compression of fluid parcels with increasing depth, is 1 mKin 10 m. The region above the temperature maximum is superadiabatic, and hence prone to convective instability. Below this super-adiabatic surface layer is a layer of depth 10–30 m in which the temperature change is less than 1 mK. Within this mixed layer, the intensity of turbulence, as quantified by the turbulent kinetic energy dissipation rate, ε, is relatively uniform and approximately equal to J_b^0. (ε represents the rate at which turbulent motions in a fluid are dissipated to heat. It is an important term in the evolution equation for turbulent kinetic energy, signifying the tendency for turbulence to decay inthe absence of forcing.) Below the mixed layer, ε generally (but not always) decreases, whereas above, ε increases by 1–2 factors of 10.

Below the mixed layer is a region of stable stratification that partially insulates the upper ocean from the ocean interior. Heat, momentum, and chemical species exchanged between the atmosphere and the ocean interior must traverse the centimeters thick cool

---

[1]Strictly, a mixed layer refers to a layer of fluid which is not stratified (vertical gradients of potential temperature, salinity and potential density, averaged horizontally or in time, are zero. The terminology is most precise in the case of a convectively forced boundary layer. Elsewhere, oceanographers use the term loosely to describe the region of the ocean that responds most directly to surface processes. Late in the day, following periods of strong

heating, the mixed layer may be quite shallow (a few meters or less), extending to the diurnal thermocline. In winter and following series of storms, the mixed layer may extend vertically to hundreds of meters, marking the depth of the seasonal thermocline at midlatitudes.

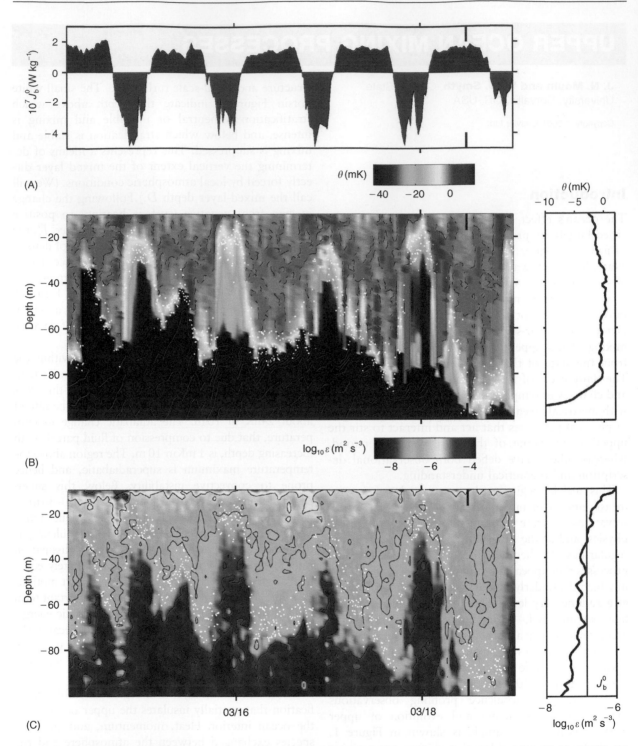

Figure 1 Observations of mixing in the upper ocean over a five-dayperiod. These observations were made in March 1987 in the North Pacific using a vertical turbulence profiler and shipboard meteorological sensors. (A) The variation in the surface buoyancy flux, J_b^0, which is dominated by surface heating and cooling. The red (blue) areas represent daytime heating (nighttime cooling). Variations in the intensity of nighttime cooling are primarily due to variations in winds. (B) Potential temperature referenced to the individual profile mean in order to emphasize vertical rather than horizontal structure (θ; K). To the right is an averaged vertical profile from the time period indicated by the vertical bars at top and bottom of each of the left-hand panels. (C) The intensity of turbulence as indicated by the viscous dissipation rate of turbulence kinetic energy, ε. To the right is an averaged profile with the mean value of J_b^0 indicated by the vertical blue line. The dots in (B) and (C) represent the depth of the mixed layer as determined from individual profiles.

skin at thevery surface, the surface layer, and the mixed layer to modify the stable layer below. These vertical transports are governed by a combination of processes, including those that affect only the surface itself (rainfall, breaking surface gravity waves), those that communicate directly from the surface throughout the entire mixed layer (convective plumes) or a good portion of it (Langmuir circulations) and also those processes that are forced at the surface but have effects concentrated at themixed-layer base (inertial shear, Kelvin–Helmholtz instability, propagating internal gravity waves). Several of these processes are represented in schematic form in **Figure 2**. Whereas **Figure 1** represents the observed time evolution of the upper ocean at a single location, **Figure 2** represents an idealized three-dimensional snapshot of some of the processes that contribute to this time evolution.

Heating of the ocean's surface, primarily by solar (short-wave) radiation, acts to stabilize the water column, thereby reducing upper ocean mixing. Solar radiation, which peaks at noon and is zero at night, penetrates the air–sea interface (limited by absorption and scattering to a few tens of meters), but

heat is lost at the surface by long-wave radiation, evaporative cooling and conduction throughout both day and night. The ability of the atmosphere to modify the upper ocean is limited by the rate at which heat and momentum can be transported across the air–sea interface. The limiting factor here is theviscous boundary layer at the surface, which permits only molecular diffusion through to the upper ocean. This layer is evidenced by the ocean's coolskin, a thin thermal boundary layer (a few millimeters thick), across which a temperature difference of typically 0.1 K is maintained. Disruption ofthe cool skin permits direct transport by turbulent processes across theair–sea interface. Once disrupted, the cool skin reforms over a period of some tens of seconds. A clear understanding of processes that disrupt the coolskin is crucial to understanding how the upper ocean is mixed.

Convection

Cooling at the sea surface creates parcels of cool, dense fluid, which later sink to a depth determined

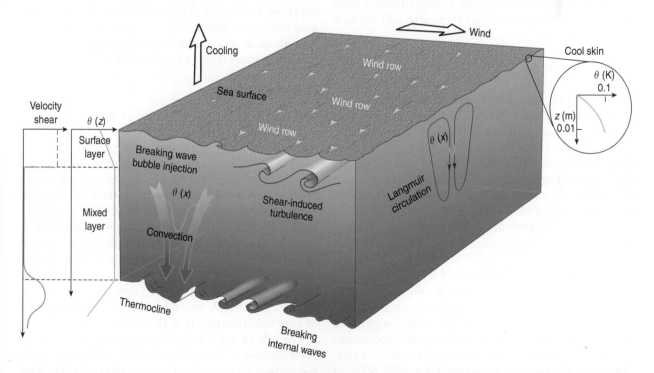

Figure 2 Diagram showing processes that have been identified by a widerange of observational techniques as important contributors to mixing the upperocean in association with surface cooling and winds. The temperature (θ) profiles shown here have the adiabatic temperature (that due to compression of fluid parcels with depth) removed; thisis termed potential temperature. The profile of velocity shear (vertical gradient of horizontal velocity) indicates no shear in the mixed layer and nonzero shear above. The form of the shear in the surface layer is a current area of research. Shear-induced turbulence near the surface may be responsible for temperature ramps observed from highly resolved horizontal measurements. Convective plumes and Langmuir circulations both act to redistribute fluid parcels vertically; during convection, they tend to movecool fluid downward. Wind-driven shear concentrated at the mixed-layer base (thermocline) may be sufficient to allow instabilities to grow, from which internal gravity waves propagate and turbulence is generated. At the surface, breaking waves inject bubbles and highly energetic turbulence beneath the sea surface and disrupt the ocean's cool skin, clearing a pathway for more rapid heat transfer into the ocean.

by the local stratification in a process known as convection. Cooling occurs almost every night and also sometimes in daytime in association with weather systems such as cold air outbreaks from continental landmasses. Convection may also be causedby an excess of evaporation over precipitation, which increases salinity, and hence density, at the surface. Winds aid convection by a variety of mechanisms that agitate the sea surface, thereby disrupting the viscous sublayer and permitting rapid transfer of heat through the surface (see below). Convection in the ocean is analogous to that found in the daytime atmospheric boundary layer, which is heated from below, and which has been studied in great detail. Recourse to atmospheric studies of convection has helped in understanding the ocean's behavior.

Surface tension and viscous forces initially prevent dense, surface fluid parcels from sinking. Once the fluid becomes sufficiently dense, however, these forces are overcome and fluid parcels sinkin the form of convective plumes. The relative motion of the plumes helps to generate small-scale turbulence, resulting in a turbulent field encompassing a range of scales from the depth of the mixed layer (typically 100 m) to a few millimeters.

A clear feature of convection created bysurface cooling is the temperature profile of the upper ocean (**Figure 1**). Below the cool skin is an unstable surface layer that is the signature of plume formation. Below that is a well-mixed layer in which density (as well as temperature and salinity) is relatively uniform. The depth of convection is limited bythe local thermocline. Mixing due to penetrative convection into the thermocline represents another source of cooling of the mixed layer above. Within the convecting layer, there is an approximate balance between buoyant production of turbulent kinetic energy and viscous dissipation, as demonstrated by the observation $\varepsilon \approx J_b^0$.

The means by which the mixed layer is restratified following nighttime convection are not clear. Whereas someone-dimensional models yield realistic time series of sea surface temperature, suggesting that restratification is a one-dimensional process (see below), other studies of this issue have shown one-dimensional processes to account for only 60% of the stratification gained during the day. It has been suggested that lateral variations in temperature, due to lateral variations in surface fluxes, or perhaps lateral variations in salinity due to rainfall variability, may be converted by buoyant forces into vertical stratification. These indicate the potential importance of three-dimensional processes to restratification.

Wind Forcing

Convection is aided by wind forcing, in part because winds help to disrupt the viscous sublayer at the sea surface, permitting more rapid transport of heat through the surface. In the simplest situation, winds produce a surface stress and a sheared current profile, yielding a classic wall-layer scaling of turbulence and fluxes in the surface layer, similar to the surface layer of the atmosphere. (Theory, supported by experimental observations, predicts a logarithmic velocity profile and constant stress layer in the turbulent layer adjacent to a solid boundary. This is typically found in the atmosphere during neutral stratification and is termed wall-layer scaling.) This simple case, however, seems to berare. The reason for the difference in behaviors of oceanic and atmospheric surface layers is the difference in the boundaries. The lower boundary of the atmosphere is solid (at least over land, where convection is well-understood), but the ocean's upper boundary is free to support waves, ranging from centimeter-scale capillary waves, through wind waves (10s of meters) to swell (100s of meters). Thesmaller wind waves lose coherence rapidly, and are therefore governed by local forcing conditions. Swell is considerably more persistent, and may therefore reflect conditions at a location remote in space and time from the observation, e.g., a distant storm.

Breaking Waves

Large scale breaking of waves is evidenced at the surface by whitecapping and surface foam, allowing visual detection from above. This process, which is not at all well understood, disrupts the ocean's cool skin, a fact highlighted by acoustic detection of bubbles injected beneath the sea surface by breaking waves. Small-scale breaking, which has no visible signature (and is even less well understood but is thought to be due to instabilities formed in concert with the superposition of smaller-scale waves) also disrupts the ocean's cool skin. An important challenge for oceanographers is to determine the prevalence of small-scale wave breaking and the statistics of cool skin disruption at the sea surface.

The role of wave breaking in mixing is an issue of great interest at present. Turbulence observations in the surface layer under a variety of conditions have indicated that at times (generally lower winds and simpler wave states) the turbulence dissipation rate (and presumably other turbulence quantities including fluxes) behaves in accordance with simple wall-layer scaling and is in this way similar to the atmospheric surface layer. However, under higher winds, and perhaps more complicated wave states, turbulence dissipation rates greatly exceed those predicted by wall-layer scaling. This condition has been observed to depths of 30 m, well below a significant wave height from the surface, and constituting a significant fraction of the ocean's mixed layer. (The significant wave height is defined as the average height of the highest third of surface displacement maxima. A few meters is generally regarded as a large value.) Evidently, an alternative to wall-layer scaling is needed for these cases. This is a problem of great importance in determining both transfer rates across the air–sea interface to the mixed layer below and the evolution of the mixed layer itself. It is at times when turbulence is most intense that most of the air–sea transfers and most of the mixed layer modification occurs.

Langmuir Circulation

Langmuir circulations are coherent structures within the mixed layer that produce counter rotating vortices with axes aligned parallel to the wind. Their surface signature is familiar as windrows: lines of bubbles and surface debris aligned with the wind that mark the convergence zones between the vortices. These convergence zones are sites of downwind jets in the surface current. They concentrate bubble clouds produced by breaking waves, or bubbles produced by rain, which are then carried downward, enhancing gas-exchange rates with the atmosphere. Acoustical detection of bubbles provides an important method for examining the structure and evolution of Langmuir circulations.

Langmuir circulations appear to be intimately related to the Stokes drift, a small net current parallel to the direction of wave propagation, generated by wave motions. Stokes drift is concentrated at the surface and is thus vertically sheared. Small perturbations in the wind-driven surface current generate vertical vorticity, which is tilted toward the horizontal (downwind) direction by the shear of the Stokes drift. The result of this tilting is a field of counterrotating vortices adjacent to the ocean surface, i.e., Langmuir cells. It is the convergence associated with these vortices that concentrates the wind-driven surface current into jets. Langmuir cells thus grow by a process of positive feedback. Ongoing acceleration of the surface current by the wind, together with convergence of the surface current by the Langmuir cells, provides a continuous source of coherent vertical vorticity (i.e., the jets), which is tilted by the mean shear to reinforce the cells.

Downwelling speeds below the surface convergence have been observed to reach more than 0.2 m s^{-1}, comparable to peak downwind horizontal flow speeds. By comparison, the vertical velocity scale associated with convection, $w^* = (J_b^0 D)^{1/3}$ is closer to 0.01 m s^{-1}. Upward velocities representing the return flow to the surface appear to be smaller and spread over greater area. Maximum observed velocities are located well below the sea surface but also well above the mixed-layer base. Langmuir circulations are capable of rapidly moving fluid vertically, thereby enhancing and advecting the turbulence necessary to mix the weak near-surface stratification which forms in response to daytime heating. However, this mechanism does not seem to contribute significantly to mixing the base of the deeper mixed layer, which is influenced more by storms and strong cooling events.

In contrast, penetration of the deep mixed layer base during convection (driven by the conversion of potential energy of dense fluid plumes created by surface cooling/evaporation into kinetic energy and turbulence) is believed to be an important means of deepening the mixed layer. So also is inertial shear, as explained next.

Wind-Driven Shear

Wind-driven shear erodes the thermocline at the mixed-layer base. Wind-driven currents often veer with depth due to planetary rotation (cf. the Ekman spiral). Fluctuations in wind speed and direction result in persistent oscillations at near-inertial frequencies. Such oscillations are observed almost everywhere in the upper ocean, and dominate the horizontal velocity component of the internal wave field. Because near-inertial waves dominate the vertical shear, they are believed to be especially important sources of mixing at the base of the mixed layer. In the upper ocean, near-inertial waves are generally assumed to be the result of wind forcing. Rapid diffusion of momentum through the mixed layer tends to concentrate shear at the mixed layer base. This concentration increases the probability of small-scale instabilities. The tendency toward instability is quantified by the Richardson number, $Ri = N^2/S^2$, where $N^2 = -(g/\rho) \cdot d\rho/dz$, represents the stability of the water column, and shear, S, represents an energy source for instability. Small values of Ri ($< 1/4$) are associated with Kelvin–Helmholtz instability. Through this instability, the inertial shear is concentrated into discrete vortices (Kelvin–Helmholtz billows) with axes aligned horizontally and perpendicular to the current. Ultimately, the billows overturn and generate small-scale turbulence and mixing. Some of the energy released by the instabilities propagates along the stratified layer as high frequency internal gravity waves. These processes are depicted in **Figure 2**. The mixing of fluid from below the mixed layer by inertial shear contributes to increasing the density of the mixed-layer and to mixed-layer deepening.

Temperature Ramps

Another form of coherent structure in the upper ocean has been observed in both stable and unstable conditions. In the upper few meters temperature ramps, aligned with the wind and marked by horizontal temperature changes of 0.1 K in 0.1 m, indicate the upward transport of cool/warm fluid during stable/unstable conditions. This transport is driven by an instability triggered by the wind and perhaps similar to the Kelvin–Helmholtz instability discussed above. It is not yet clearly understood. Because it brings water of different temperature into close contact with the surface, and also because it causes large lateral gradients, this mechanism appears to be a potentially important factor in near-surface mixing.

Effects of Precipitation

Rainfall on the sea surface can catalyze several important processes that act to both accentuate and reduce upper ocean mixing. Drops falling on the surface disrupt the viscous boundary layer, and may carry air into the water by forming bubbles.

Rain is commonly said to 'knock down the seas.' The evidence for this is the reduction in breaking wave intensity and whitecapping at the sea surface. Smaller waves (<20 cm wavelength) may be damped by subsurface turbulence as heavy rainfall acts to transport momentum vertically, causing drag on the waves. The reduced roughness of the small-scale waves reduces the probability of the waves exciting flow separation on the crests of the long waves, and hence reduces the tendency of the long waves to break.

While storm winds generate intense turbulence near the surface, associated rainfall can confine this turbulence to the upper few meters, effectively insulating the water below from surface forcing. This is due to the low density of fresh rainwater relative to the saltier ocean water. Turbulence must work against gravity to mix the surface water downward, and turbulent mixing is therefore suppressed. So long as vertical mixing is inhibited, fluid heated during the day will be trapped near the sea surface. Preexisting turbulence below the surface will continue to mix fluid in the absence of direct surface forcing, until it decays due to viscous dissipation plus mixing, typically over the time scale of a buoyancy period, N^{-1}.

Deposition of pools of fresh water on the sea surface, such as occurs during small-scale squalls, raises some interesting prospects for both lateral spreading and vertical mixing of the fresh water. In the warm pool area of the western equatorial Pacific, intense squalls are common. Fresh light puddles at the surface cause the surface density field to be

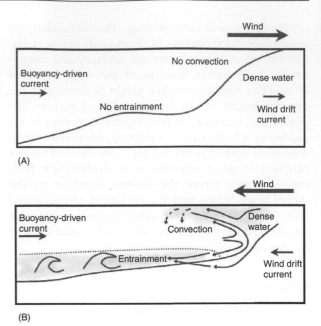

Figure 3 Two ways in which the frontal interface of a fresh surface pool may interact with ambient winds and currents. (A) The case in which the buoyancy-driven current, wind and ambient current are all in the same direction. In this case, the buoyancy-driven current spreads and thins unabated. In (B), the buoyancy-driven current is opposed by wind and ambient current. In this case, the frontal interface of the buoyancy-driven current may plunge below the ambient dense water, so that convection near the surface intensifies mixing at the frontal interface. Simultaneously, shear-forced mixing at the base of the fresh puddle may increase entrainment of dense water from below.

heterogeneous. Release of the density gradient may then occur as an internal bore forming on the surface density anomaly, causing a lateral spreading of the fresh puddle. Highly resolved horizontal profiles of temperature, salinity and density reveal sharp frontal interfaces, the features of which depend on the direction of the winds relative to the buoyancy-driven current. These are portrayed in **Figure 3**. When the wind opposes the buoyancy current, the density anomaly at the surface is reduced, possibly as a result of vertical mixing in the manner suggested in **Figure 3B**(B). This mechanism results in a rapid vertical redistribution of fresh water from the surface pool and a brake on the propagation of the buoyancy front. Similarly, an opposing ambient current results in shear at the base of the fresh layer, which may lead to instability and consequent mixing. The nature of these features has yet to be clearly established, as has the net effect on upper ocean mixing.

Ice on the Upper Ocean

At high latitudes, the presence of an ice layer (up to a few meters thick) partially insulates the ocean a gainst surface forcing. This attenuates the effects of wind forcing on the upper ocean except at the lowest frequencies. The absence of surface waves prevents turbulence due to wave breaking and Langmuir circulation. However, a turbulence source is provided by the various topographic features found on the underside of the ice layer. These range in size from millimeter-scaledendritic structures to 10 m keels, and can generate significant mixing nearthe surface when the wind moves the ice relative to the water below or currents flow beneath the ice.

Latent heat transfer associated with melting and freezing exerts a strong effect on the thermal structure of the upper ocean. Strong convection can occur under ice-free regions, in which the water surface is fully exposed to cooling and evaporative salinity increase. Such regions include leads (formed by diverging ice flow) and polynyas (where wind or currents remove ice as rapidly as it freezes). Convection can also be caused by the rejection of salt by newly formed ice, leaving dense, salty water near the surface.

Parameterizations of Upper Ocean Mixing

Large-scale ocean and climate models are incapable of explicitly resolving the complex physics of the upper ocean,and will remain so for the foreseeable future. Since upper ocean processes are crucial in determining atmosphere–ocean fluxes, methods for their representation in large-scale models, i.e., parameterizations, are needed. The development of upper-ocean mixing parameterizations has drawn on extensive experience in the more general problem of turbulence modeling. Some parameterizations emphasize generality, working from first principles as much as possible, whereas others sacrifice generality to focus on properties specific to the upper ocean. An assumption common to all parameterizations presently in use is that the upper ocean is horizontally homogeneous, i.e., the goal is to represent vertical fluxes in terms of vertical variations in ocean structure, leaving horizontal fluxes to be handled by other methods. Such parameterizations are referred to as'one-dimensional' or 'column' models.

Column modeling methods may be classified aslocal or nonlocal. In a local method, turbulent fluxes at a given depth are represented as functions of water column properties at that depth. For example, entrainment at the mixed-layer base may be determined solely by the local shear and stratification. Nonlocal methods allow fluxes to be influenced directly by remote events. For example, during nighttime

convection, entrainment at the mixed-layer base may be influenced directly by changes in the surface cooling rate. In this case, the fact that large convection rolls cannot be represented explicitly in a column model necessitates the nonlocal approach. Nonlocal methods include 'slab' models, in which currents and water properties do not vary at all across the depth of the mixed layer. Local representations may often be derived systematically from the equations of motion, whereas nonlocal methods tend tobe *ad hoc* expressions of empirical knowledge. The most successful models combine local and nonlocal approaches.

Many processes are now reasonably well-represented in upper ocean models. For example, entrainment via shear instability is parameterized using the local gradient Richardson number and/or a nonlocal (bulk) Richardson number pertaining to the whole mixed layer. Other modeling issues are subjects of intensive research. Nonlocal representations of heat fluxes have resulted in improved handling of nighttime convection, but the corresponding momentum fluxes have not yet been represented. Perhaps the most important problem at present is there presentation of surface wave effects. Local methods are able to describe the transmission of turbulent kinetic energy generated at the surface into the ocean interior. However, the dependence of that energy flux on surface forcingis complex and remains poorly understood. Current research into the physics of wave breaking, Langmuir circulation, wave-precipitation interactions, and other surface wave phenomena will lead to improved understanding, and ultimately to useful parameterizations.

See also

Breaking Waves and Near-Surface Turbulence. Bubbles. Heat and Momentum Fluxes at the Sea Surface. Langmuir Circulation and Instability. Penetrating Shortwave Radiation. Surface Gravity and Capillary Waves. Under-Ice Boundary Layer. Upper Ocean Vertical Structure. Wave Generation by Wind. Whitecaps and Foam.

Further Reading

Garrett C (1996) Processes in the surface mixed layer of the ocean. *Dynamics of Atmospheres and Oceans* 23: 19–34.

Thorpe SA (1995) Dynamical processes at the sea surface. *Progress in Oceanography* 35: 315–352.

LANGMUIR CIRCULATION AND INSTABILITY

S. Leibovich, Cornell University, Ithaca, NY, USA

Introduction

The surface of a wind-driven sea often is marked by streaks roughly aligned with the wind direction. These streaks, or windrows, are visible manifestations of coherent subsurface motions extending throughout the bulk of the ocean surface mixed layer, extending from the surface down to the seasonal thermocline. These may be regarded as the large scales of the turbulence in the mixed layer. Windrows and their subsurface origins were first systematically studied and described by Irving Langmuir in 1938, and the phenomenon since has become known as Langmuir circulation. The existence of a simple deterministic description making these large scales theoretically accessible distinguishes this problem from coherent structures in other turbulent flows. The theory traces these patterns to a convective instability mechanically driven by the wind waves and currents. Recent advances in instrumentation and computational data analysis have led to field observations of Langmuir circulation of unprecedented detail. Although the body of observational data obtained since Langmuir's own work is mainly qualitative, ocean experiments now can yield quantitative measurements of velocity fields in the near-surface region. New measurement methods are capable of producing data comprehensive enough to characterize the phenomenon, and its effect on the stirring and maintenance of the mixed layer, although the labor and difficulties involved and the shear complexity of the processes occurring in the surface layer leave much work to be done before this can be said to be accomplished. Nevertheless, the combination of new experimental techniques and a simple and testable theoretical mechanism has stimulated rapid progress in the exploration of the stirring of the ocean surface mixed layer.

Description of Langmuir Circulation

Langmuir circulation takes the ideal form of vortices with axes aligned the wind, as in the schematic drawing in **Figure 1**. The appearance resembles convective rolls driven by thermal convection, but all evidence indicates that the motions are due to mechanical processes through the action of the wind, as Langmuir originally indicated. At the surface, rolls act to sweep surface water from regions of surface divergence overlying upwelling water into convergence zones overlying downwelling water. Floating material is collected into lines of surface convergence visible as windrows.

In confined bodies of water, such as lakes and ponds, windrows are very nearly parallel to the wind, and can have a nearly uniform spacing as shown in **Figure 2**. In the open ocean, evidence indicates windrows tend to be oriented at small angles to the wind (typically to the right in the Northern Hemisphere), spacing is more variable, and individual windrows can be traced only for a modest multiple of the mean spacing. A windrow may either terminate, perhaps due to local absence of surface tracers, coalesce with an adjacent windrow, or split into two daughter windrows. Thus in the ocean, the general surface appearance is of a network of lines, occasionally interecting, yet roughly aligned with the wind.

Windrows are visible in nature only when both Langmuir circulation and surface tracers are present. In the ocean, bubbles from breaking waves are the most readily available tracers, and Langmuir

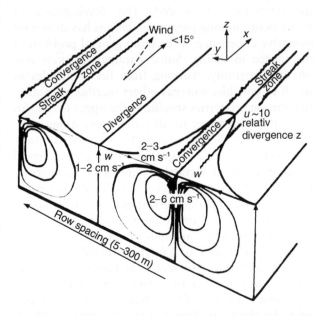

Figure 1 Sketch of Langmuir circulation. From Pollard RT (1977) Observations and theories of Langmuir circulations and their role in near surface mixing. In: Angel M (ed.) *A Voyage of Discovery: G. Deacon 70th Anniversary Volume, Re*, pp. 235–251. Oxford, UK: Pergamon.

A hierarchy of horizontal scales is observed, with windrow spacing ranging in the ocean from a few meters up to approximately 3 times the depth of the surface mixed layer. The largest scales are the most energetic and the most persistent, and extend to depths comparable to that of the mixed layer, so a cell extending from surface divergence to surface convergence, and from the surface to maximum depth of penetration, is approximately square. The maximum penetration depth in the open ocean is comparable to the depth of the seasonal thermocline. Langmuir reported the penetration depth in Lake George to be comparable to the epilimnion when the lake was stratified, and believed the epilimnion to be created by the mixing caused by the wind-driven convective motion. Whether the seasonal thermocline location is fixed by the scale of Langmuir circulation, or whether the penetration depth of the circulation is limited by the strong buoyancy at the thermocline acting like a bottom is not yet clear, although the latter seems more likely. In shallow water, the seabed or lake floor of course fixes the maximum penetration depth. While horizontal scales of up to 3 times the mixed layer depth are reported, it is not clear that even larger scales exist that have not been detected in experiments.

The smaller scales are advected and presumably eventually swept up by the largest scales. If a fixed number of permanent surface markers were used, as in some experiments where computer cards were released to serve as markers, the larger scales ultimately would be more prevalent. Most observations do depend on Lagrangian tracers, in particular bubbles with definite lifetimes that are continually but episodically created. The regeneration of bubbles on the surface between the large-scale windrows permits the smaller scales to be seen.

The largest observed windrow scales consolidate in about 20 min after a shift in wind direction, or in cases of sudden wind onset. This appears to establish the formation time, at least that required to sweep surface material into windrows. The largest windrow scales in the Pacific have been observed to persist for hours.

Downwelling speeds below windrows are substantially higher than upwelling speeds occurring below surface divergences. Furthermore, the downwind surface speed is larger in windrows than between them. A simple explanation for the speed excess is given in the section on instability. Although the observed speed increase is commonly reported, it has not often been quantified. It appears, however, that the speed increases in surface jets seem to be comparable to the maximum downwelling speeds.

Figure 2 Photograph of windrows in Rodeo Lagoon, California. From Szeri AJ (1996) Langmuir circulations on Rodeo Lagoon. *Monthly Weather Review* 124(2): 341–342.

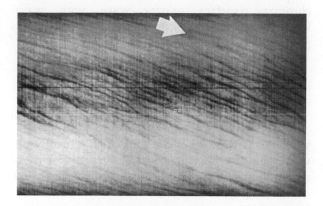

Figure 3 Infrared aerial photograph of thermal streaks in the Pacific. The arrow shows the wind direction. The wind speed is $4\,\mathrm{m\,s}^{-1}$. The image is about 750 m wide. From McLeish W (1968) On the mechanisms of wind-slick generation. *Deep-Sea Research* 15: 461–469.

circulation and bubbles both appear to exist when speeds exceed some threshold. Threshold wind levels are not absolute, since swell, wind duration, fetch, and currents existing before the onset of wind forcing play a role, but windrows are commonly reported in winds of $3\,\mathrm{m\,s}^{-1}$ or more. Tracers other than bubbles may produce windrows revealing underlying Langmuir circulation – all forms of flotsam serve. Langmuir first noticed windrows from the deck of a ship in the Sargasso Sea, which contained windrows of *Sargassum*. Organic films on the water surface are compressed in windrows, causing capillary waves to be preferentially damped; in light winds, windrows thereby are made visible as bands of smoother water. Observations in the infrared (see **Figure 3**) reveal windrows due to variation of surface temperature created by Langmuir circulation.

Theory

Theory promulgated in the 1970s has influenced experiments addressing Langmuir circulation. The theory most commonly utilized, and now commonly referred to as the Craik–Leibovich theory after its originators, begins with Langmuir's conclusion that the cellular motion bearing his name derives from the wind. Wind blowing over a water surface has two simultaneous consequences: currents are generated as horizontal momentum is transmitted from wind to water; and waves are generated on the water surface due to an instability of the air–sea interface under wind shear.

A detailed treatment of the shear flow in the presence of wind waves is not feasible. The eddy turnover timescale (time required for a fluid particle to traverse the convective cell) in Langmuir circulation is on the order of tens of minutes. Surface waves have a substantially shorter timescale. Wind-driven water waves can be thought of as comprised of the superposition of wavelets with a continuous range of wavelengths and frequencies, and the amplitudes of waves in a given band of wavelength or frequency can be characterized by an energy spectrum. For the Pierson–Moskowitz empirical wind wave spectrum, the waves at the peak of the spectrum of a wind-driven sea under wind speed U have period approximately $7g/U$. This is a typical value for the energetic part of the wind wave spectrum. For wind speeds leading to observable Langmuir circulation, a characteristic peak wave period is of the order of 10 s. Averaging over the waves is therefore useful. Surface gravity waves are approximately irrotational, and orbital speeds near the surface are an order of magnitude larger than the mean speeds in the current system. Although a wind-driven wave field is complex, and generally must be treated probabilistically, the theory depends only on averaged effects due to the net mass drift caused by the waves, and this often can be computed.

Wave Effects: The Stokes Drift

Nonbreaking surface gravity waves have small slopes, and produce a small (second order in wave slope) mean mass motion in the direction of wave propagation. This net water motion is known as the Stokes drift. If $\mathbf{u}_w$ is the Eulerian velocity due to the water waves, the Stokes drift velocity $\mathbf{u}_s$ is approximately given by

$$\mathbf{u}_s = \left\langle \int^t \mathbf{u}_w \mathrm{d}t \cdot \nabla \mathbf{u}_w \right\rangle \qquad [1]$$

where the angle brackets represent an average over time. An acceptable averaging time to calculate Stokes drift is a modest multiple of the wave period associated with the peak frequency of the energy spectrum. Provided the currents are small compared to the wave orbital speeds, as is usually assumed in the ocean, the net mass drift due to waves can be approximated by this formula. If the surface wave characteristics can be determined with specified wind fields – for example, if the wave spectrum is measured, or computed by a wave forecasting model, or if it is assumed to follow a standard empirical wave spectrum – then the Stokes drift can be calculated using [1]. The form of the vertical variation will depend on the spectrum, and the resulting functional form will depend on two parameters (the surface value of the drift, U_s, and a length scale, ℓ), representing the characteristic decay depth of the drift. For example, for a small-amplitude monochromatic surface gravity wave with wavelength λ and amplitude, a, the Stokes drift velocity is given by

$$|\mathbf{u}_s| = U_s \exp(x_3/\ell), \quad U_s = \frac{2\pi a^2}{\lambda} \sqrt{\frac{2\pi g}{\lambda}}, \quad \ell = \frac{\lambda}{4\pi} \tag*{[2]}$$

where x_3 is the coordinate measured vertically upward from the mean free surface. In the absence of swell, the direction of the drift is parallel to the wind.

Langmuir Force

Assuming the surface waves can be approximated as irrotational and have orbital speeds large compared to current speeds, averaging yields an apparent extra force on the averaged motion equal to $\mathbf{u}_s \times \omega_a$, where $\mathbf{u}_s$ is the Stokes drift velocity associated with the wave motion, and ω_a is the averaged absolute vorticity of the water body. The apparent extra force captures mean effects of the wave–current interaction. It has been called the vortex force in the literature, although it might be more appropriate to call it the Langmuir force, which is the term adopted here. The mean Lagrangian velocity, or mass drift, of a fluid particle at position $\mathbf{x}$ at time t exceeds the mean Eulerian velocity of the water by $\mathbf{u}_s$. If the system is referred to a frame rotating with constant angular velocity vector $\mathbf{\Omega}$, then the Langmuir force $\mathbf{F}_L$ is

$$\mathbf{F}_L(\mathbf{x}, t) = \mathbf{u}_s(\mathbf{x}, t) \times (2\mathbf{\Omega} + \omega(\mathbf{x}, t)) \qquad [3]$$

where $\omega = \mathrm{curl}\,\mathbf{u}$. Note that this quantity has dimensions of a force per unit mass. Here $\mathbf{u}$ is the

Eulerian mean velocity as seen in the rotating frame, and so ω is the relative mean vorticity. The full instantaneous velocity field includes fluctuations due to turbulence as well as waves, and so the wave-averaged mean velocity filters out not only the waves, but also turbulence with timescales comparable to or less than the averaging time. The Boussinesq equations for momentum and temperature, when averaged over a timescale of several wave periods, are

$$\frac{\partial \mathbf{u}}{\partial t} + \mathbf{u} \cdot \nabla \mathbf{u} + 2\mathbf{\Omega} \times \mathbf{u} = -\frac{1}{\rho}\nabla \pi + \beta g(T - \bar{T})e_3$$
$$+ \mathbf{u}_s \times (2\mathbf{\Omega} + \boldsymbol{\omega}) + \mathscr{F} \qquad [4]$$

$$\frac{\partial T}{\partial t} + (\mathbf{u} + \mathbf{u}_s) \cdot \nabla T = \mathscr{H} \qquad [5]$$

$$\nabla \cdot \mathbf{u} = 0 \qquad [6]$$

In these equations, π is a modified pressure including mean wave effects, centripetal force 'potential', and hydrostatic pressure variations; T is the water temperature; $\bar{T}$ is a reference temperature field that may depend on time and depth; e_3 is a unit vector pointing vertically upward; while $\mathscr{F}$ and $\mathscr{H}$ represent the divergence of momentum and heat fluxes, respectively, due to the unresolved turbulent scales. The rectified effects of the waves are represented by the Stokes drift in the Langmuir force and in the advection of the scala T. If other scalar quantities, like salt concentration, were to be included, they would be governed by equations of the same form as that for T. With suitable subgrid models for $\mathscr{F}$ and $\mathscr{H}$, these equations can be used to compute the motions, including those with turbulent fluctuations having timescales large compared to a wave period. The principal theoretical development is the incorporation of residual wave effects the Langmuir force and scalar advection augmented by the Stokes drift velocity. Without these additions, the governing equations are the same as conventional models not accounting for surface waves.

The simplest closure models are the assumptions of constant eddy viscosity, ν_T, and eddy diffusity of heat, κ_T, yielding $\mathscr{F} = \nu_T \nabla^2 \mathbf{u}$ and $\mathscr{H} = \kappa_T \nabla^2 T$. All analytical work done addressing the stability of the mean motion assumes this form, as does much of the computational work on Langmuir circulation. Large eddy simulations (LESs) of turbulent Langmuir circulation using standard subgrid closure models for $\mathscr{F}$ and $\mathscr{H}$ have been carried out by Skyllingstad and Denbo and by McWilliams *et al.*, among others. Calculations of this sort, while involving a great deal of computational effort, have the important

advantage of dispensing with empirical eddy coefficients ν_T, and κ_T. The relative effects of the Langmuir force and ordinary shear turbulence can be educed from LES using the 'turbulent Langmuir number' defined by McWilliams *et al.* to be $\sqrt{u_*/U_s}$. Here u_* is the friction velocity defined in the next section.

The discussion so far has been predicated on the existence of wind stress to produce currents and surface waves. Currents due to other causes may exist for a time in the absence of wind. The same is true of surface waves, which may have been generated in distant locations and propagate to the region of interest. The Langmuir force continues to exist, and under suitable conditions – especially when the angle between the Stokes drift and current is small – Langmuir circulation may result. Thus, the details of the current and wave fields must be known, and scalings based solely on wind or Stokes drift maybe misleading.

Scaling

Despite the disclaimer of the previous section, reported Langmuir circulation appears to be confined to situations driven by local wind stress. The motion will depend on the magnitude and (if Coriolis effects apply) direction of the applied wind stress, τ sea state as prescribed by $\mathbf{u}_s$; the prevailing density stratification; surface thermal conditions; water depth, or specification of conditions of current, temperature, and salinity below the mixed layer; the Coriolis acceleration, which depends on latitude; and initial conditions. The wind stress clearly is a primary factor; it provides a unit for speed, the friction velocity $u_* = \sqrt{|\tau|/\rho}$, ρ being the water density. Several length scales appear, and the choice of unit of length, d, will depend on circumstances. Since observation in the ocean and in stratified lakes indicates the importance of the depth of the seasonal thermocline, it provides a natural choice when it can be defined. Other choices replace this when a strong thermocline does not exist.

Clearly, the range of questions that may be encountered is extensive, and the parameter space encompassing them is too large for a general discussion. The most elementary situation producing Langmuir circulation, however, is that of a nonrotating layer of water of uniform density with finite depth (either to a solid bottom or to strong thermocline that strongly inhibits vertical motion) under the action of a wind of unlimited fetch and duration and with constant speed and direction producing a sea state with known Stokes drift. With u_* as unit for speed, d unit for length, the unit for time is d/u_*, and the problem then depends on the dimensionless parameters u_*/U_s and ℓ/d, and a parameter characterizing $\mathscr{F}d/u_*^2$. The latter depends on how $\mathscr{F}$ is modeled. If a constant eddy viscosity is used, then the appropriate parameter is Re_*^{-1}, where $Re_* = u_*d/\nu_T$ is the Reynolds number based on eddy viscosity. The designation Re_* can serve as a placeholder for the parameter appropriate for the choice of model. Thus, the simplest problem depends on three dimensionless parameters, although if the motion is restricted to be invariant in the wind direction, the parameter space can be reduced by one. For each additional physical effect, one more parameter is added to the list. For example, if rotation effects are added, the parameter $|\mathbf{\Omega}|d/u_*$, an inverse Rossby number, is needed.

For the most elementary case, the velocity vector in the water column is given by $\mathbf{u} = u_*\boldsymbol{\phi}(\mathbf{x}/d, t\, d/u_*, U_s/u_*, \ell/d, Re_*)$, where $\boldsymbol{\phi}$ is a dimensionless vector-valued function of its arguments. For fixed position and time, the motion depends on the three fixed dimensionless parameters. If the dependence on Re_* and ℓ/d were weak, hypotheses about how $\boldsymbol{\phi}$ depends on U_s/u_* suggest ways to scale experimental data. For example, if the dependence is linear, then $\mathbf{u}$ is directly proportional to U_s. If the dependence is on the square root, then $\mathbf{u}$ is directly proportional to $\sqrt{u_*U_s}$. Both of these scalings have been tried for field data, without conclusive results. In fact, constant eddy viscosity simulations using [4] and [6] fail to indicate a simple functional dependence of $\boldsymbol{\phi}$ on its parameters.

Instability

If all motions are due to wind and perhaps buoyancy forcing and the wind stress, sea state, and thermal boundary conditions are independent of horizontal position, the theoretical model has exact solutions independent of horizontal position. The natural mechanical condition to impose in such cases is that of a constant surface stress in a given direction, and, if waves are present, a Stokes drift velocity parallel to the applied stress. These solutions describe flows with no vertical motion and no patterns on the surface. Such flows therefore may be described as featureless.

These exact solutions can be unstable, however, and patterns can develop as a consequence. In the absence of wave effects $\mathbf{u}_s$ and $\mathbf{F}_L$ both vanish, and roll instabilities leading to parallel lines of surface convergence can arise in two physically distinct ways. If the Coriolis acceleration is retained and thermal effects are ignored, the featureless flow is the well-known stress-driven Ekman layer, and it is unstable for sufficiently large wind stress. (The same problem

without Coriolis effects yields plane Couette flow, which is stable.) If thermal effects are accounted for and the water is cooled from above, the motion is unstable for sufficiently large cooling rates.

In the absence of both Coriolis and thermal effects, the motion without surface waves is stable for any value of the surface stress. If waves are present, the action of the Langmuir force causes the flow to be unstable when the applied stress exceeds a threshold value. The preferred instability mode consists of rolls parallel to the wind direction.

If Coriolis accelerations and Langmuir force are simultaneously considered for typical oceanic conditions, the flow instability s close to that found for nonrotating case, and therefore is dominated by the Langmuir circulation mode. The effect of Coriolis acceleration in this case is mainly to shape the underlying featureless flow, and thereby to cause the rolls to have axes oriented to the right of the wind (in the Northern Hemisphere). Similarly, under typical wind conditions in thermally unstable conditions, the Langmuir circulation instability mode dominates thermal instability.

Langmuir circulation requires the generation of coherent vorticity in the wind (streamwise) direction. The mechanism of the Langmuir force can be seen by a simple geometric argument. Ignore Coriolis acceleration and density stratification. This would be appropriate, for example, when considering small bodies of water, such as New York's Lake George in which Langmuir conducted his extensive experiments, after seasonal overturning of the epilimnion. The equation for mean vorticity ω is then (since $\nabla \cdot \mathbf{u}_s = 0$)

$$\frac{\partial \omega}{\partial t} + (\mathbf{u} + \mathbf{u}_s) \cdot \nabla \omega = \omega \cdot \nabla(\mathbf{u} + \mathbf{u}_s) + \nu_T \nabla^2 \omega \quad [7]$$

The long downwind extent of windrows in these circumstances suggests the motions be idealized as independent of wind direction, and the Stokes drift is parallel to the wind. In this case, the wind supplies vorticity to the water column in the crosswind direction. Now suppose this flow is perturbed by a slight local increase in the windward component of velocity, as indicated in **Figure 4**. Such a perturbation can be created in a number of ways, for example, by breaking waves or by fluctuations in the applied wind stress. The disturbance generates vertical vorticity of opposite signs on either side of the velocity maximum. This vorticity is rotated by the Stokes drift to produce streamwise vorticity components as shown. The sense of the streamwise vorticity is to produce convergence near the surface toward the plane $(x_2 = 0)$ through the velocity maximum and

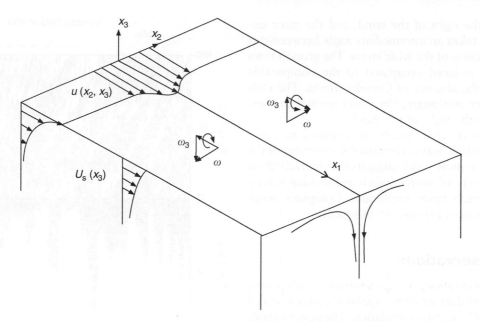

Figure 4 Sketch illustrating instability mechanism.

downwelling motion below this line. This could be a transient feature. However, the convergence transports streamwise momentum toward the plane of symmetry, accelerating the fluid at the convergence line and amplifying the original perturbation. In fact, at the plane of symmetry,

$$\frac{\partial u_1}{\partial t} = -u_3(0, x_3)\frac{\partial u_1(0, x_3)}{\partial x_3}$$

If the streamwise velocity component decreases with depth, then at a downwelling plane, $u_3(0, x_3) < 0$ and the right-hand side above is positive. The streamwise acceleration amplifies the initial perturbation, and the feedback provides a mechanism for instability. This picture is confirmed by detailed computation. Furthermore, the transport of streamwise momentum toward convergence zones provides a mechanism for the observed convergence zone jets.

The assumption of streamwise invariance implies that no streamwise vorticity is produced by the Eulerian shear – with streamwise invariance of the mean flow, generation of mean streamwise vorticity requires Stokes drift. This may be seen directly from the streamwise component of [7]. If the wind stress is in the x_1-direction, x_3 oriented vertically upward from the mean free surface, then x_2 will be crosswind in a right-handed (x_1, x_2, x_3) coordinate system with unit vectors (e_1, e_2, e_3). Then $\mathbf{u}_s = U_s(x_3)e_1$, $\partial/\partial x_1 = 0$, and the streamwise component of vorticity, ω_1, satisfies

$$\frac{\partial\omega_1}{\partial t} + \frac{\partial u_2\omega_1}{\partial x_2} + \frac{\partial u_3\omega_1}{\partial x_3} = \omega_3\frac{\partial U_s}{\partial x_3} + \nu_T\left(\frac{\partial^2\omega_1}{\partial x_2^2} + \frac{\partial^2\omega_1}{\partial x_3^2}\right)$$

Suppose the motion is periodic in the x_2-direction, or that it decays as $|x_2| \to \infty$. Multiplying by ω_1 and integrating over the water depth and over a period in the x_2-direction in the first case or all x_2 in the second, implies

$$\frac{\partial}{\partial t}\int\int\frac{1}{2}\omega_1^2\ dx_2\ dx_3 = \int\int\omega_1\omega_3\frac{\partial U_s}{\partial x_3}dx_2\ dx_3$$
$$-\nu_T\int\int\sum_{j=2}^{3}\left(\frac{\partial\omega_1}{\partial x_j}\right)^2 dx_2\ dx_3$$

The second term on the right-hand side is the dissipation of the streamwise component of enstrophy, and the first term is its production rate. Production vanishes and streamwise vorticity decays in the absence of Stokes drift. This consequence of streamwise invariance is an exact result for the instantaneous motion; the presence of waves in the instantaneous motion introduces streamwise variations, and the Stokes drift represents the residual effect of these variations in the wave-averaged equations permitting the development of averaged streamwise vorticity.

It is not possible to survey details of instability characteristics for the various cases relevant to the ocean. Qualitatively, however, it can be said that growth rates are consistent with the observed formation rates of Langmuir circulation. In the absence of Coriolis acceleration and stratification, the most unstable mode is in the form of rolls parallel to the wind, and growth is monotonic in time, so the rolls are stationary. When Coriolis acceleration is significant, the underlying featureless flow has a surface velocity to the right of the wind, and the most unstable mode takes an intermediate angle between this and the direction of the wind stress. The growth rates are slightly reduced compared to the comparable problem in the absence of Coriolis effects. The rolls are no longer stationary, but travel normal to their axes, and to the right of the wind.

Only modest wind speeds are required to cause instability in these ways. The unstable character of the ocean under typical winds suggests that perturbations due to a variety of sources, such as breaking waves, grow and reach finite amplitudes. Computer simulations using eqns [4] support this picture.

Field Observations

New instrumentation, new measurement techniques, and improved data analysis capabilities have clarified the nature of Langmuir circulation. The new methods include greatly improved current meters and sonar methods. Sonars image microbubbles that have a virtually ubiquitous presence in the upper few meters of the ocean. These bubbles are organized by Langmuir circulation, and bubble plumes collected in, and carried down by the downwelling beneath, windows produce strong sonar returns. The intensity of sonar returns permit Langmuir circulation to be visualized in extended regions of the ocean. Doppler sonar can be used to measure speeds in the current system.

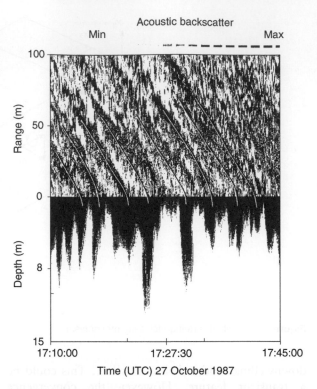

Figure 5 Simultaneous sonar images of windrows and bubble plumes. The upper picture shows bands of scatterers, and the lower picture shows bubble plumes below the bands, over a period of 35 min. Reproduced by permission of American Geophysical Union. Zedel L and Farmer DM (1991) Organized structures in subsurface bubble clouds: Langmuir circulation in the upper ocean. *Journal of Geophysical Research* 96(5): 8895. Copyright (1991) American Geophysical Union.

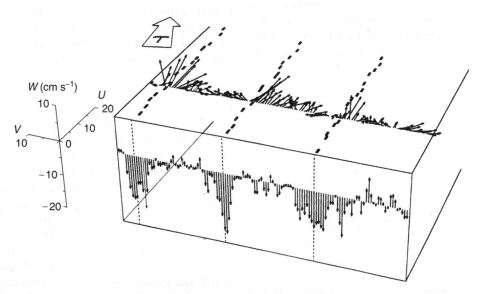

Figure 6 Data from a 50-m deep mixed layer in the Pacific. See text for for description. From Weller RA and Price JF (1988) Langmuir circulation within the oceanic mixed layer. *Deep-Sea Research* 35: 711–747.

Bubble plumes are found to extend to depths of several meters, often up to 10–12 m, below windrows. Since the micron-sized air bubbles providing the best sonar return signal go into solution at depths of this order, substantially deeper plumes do not exist. The depth of these plumes therefore provides a lower bound estimate on the strength of the downwelling motion. **Figure 5** shows an example in which bubbles are organized into bands by Langmuir circulation, and plumes are formed beneath these bands.

A quite remarkable set of measurements taken in 1982 is summarized in **Figure 6**. The data were taken by a string of current meters in a mixed layer about 50 m deep. Velocity measurements are shown near the surface and at 23-m depth, about halfway down from the surface to the base of the mixed layer. Tracks on the surface schematically indicate the disposition of computer cards distributed on the surface to provide visual makers of windrows, which lie about 15° to the right of the wind. The presence of surface jets is indicated, with maximum values of about 20 cm s^{-1}. Surprisingly, the downwelling speeds at 23 m are also about 20 cm s^{-1}. Data taken at other times also produced downwelling speeds from 20 to 30 cm s^{-1} at mid-depth in the mixed layer. These large events occurred intermittently, with less impressive activity, on the order of half as intense, occurring at other times. During these observations, the wind speeds were on the order of 8–18 m s^{-1}. Downwelling was observed to be as large as 3 cm s^{-1} under winds as light as 1.5 m s^{-1}.

The process of flotsam collection by Langmuir circulation is shown in **Figure 7**. Sulfur powder was

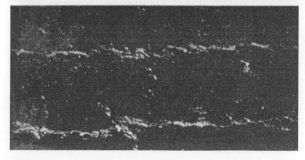

Figure 7 Sulfur powder released from an aircraft and drawn into bands of Langmuir circulation at time of (top to bottom) deployment, 1.5 min after, and 20 min after. Photograph widths are about 40 m and wind speed is 6 m s^{-1}. From McLeish W (1968) On the mechanisms of wind-slick generation. *Deep-Sea Research* 5: 461–469.

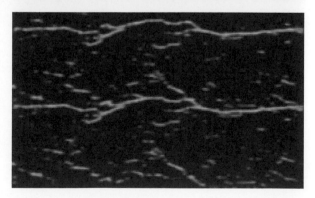

Figure 8 Computer simulation of surface tracers for a mixed layer depth of 50 m and wind speed of 9 m s^{-1}. Top panel: Initial uniform distribution of tracers. Middle: distribution after 5 min. Bottom: distribution after 20 min, separation of windrows is 150 m. From Yang G and Leibovich S (1994) Large-eddy simulation of Langmuir circulation. *Bulletin of the American Physical Society* 39(9): 1885.

spread along a line across the wind, and appears in an aerial photograph as the light blob in the upper panel immediately afterward. The second and third panels show the disposition of the dust after 1.5 min and 20 min, respectively. Surface tracers advected by the flow produced by a computer simulation of [4], [6], with a Smagorinsky turbulence subgrid scale closure model are shown in **Figure 8**. Note the similarity to **Figure 7**. Simulations carried out exactly the same way but omitting the Langmuir force fail to show these features.

Langmuir circulation is now well-established as an important mechanism, affecting mixing and dispersion in the upper ocean and providing a dynamical link between the wind-wave field and the development of the mixed layer.

See also

Breaking Waves and Near-Surface Turbulence. Bubbles. Upper Ocean Mixing Processes. Upper Ocean Time and Space Variability. Wind- and Buoyancy-Forced Upper Ocean.

Further Reading

Langmuir I (1938) Surface motion of water induced by wind. *Science* 87: 119–123.

Leibovich S (1983) The form and dynamics of Langmuir circulation. *Annual Review of Fluid Mechanics* 15: 391–427.

McLeish W (1968) On the mechanisms of wind-slick generation. *Deep-Sea Research* 15: 461–469.

McWilliams JC, Sullivan PP, and Moerig C-H (1997) Langmuir turbulence in the ocean. *Journal of Fluid Mechanics* 334: 1–30.

Phillips OM (1977) *The Dynamics of the Upper Ocean*, 2nd edn. Cambridge, UK: Cambridge University Press.

Pollard RT (1977) Observations and theories of Langmuir circulations and their role in near surface mixing. In: Angel M (ed.) *A Voyage of Discovery: G. Deacon 70th Anniversary Volume, Re*, pp. 235–251. Oxford, UK: Pergamon.

Simecek-Beatty D, Lehr WJ, Lai R, and Overstreet R (2000) Langmuir circulation and oil spill modelling. *Spill Science and Technology Bulletin* 6(3–4): 207–279.

Skyllingstad ED and Denbo DW (1995) An ocean large-eddy simulation of Langmuir circulation and convection in the surface mixed layer. *Journal of Geophysical Research* 100: 8501–8521.

Szeri AJ (1996) Langmuir Circulations on Rodeo Lagoon. *Monthly Weather Review* 124(2): 341–342.

Thorpe SA (1985) Small-scale processes in the upper ocean boundary layer. *Nature* 318: 519–522.

Thorpe SA (2004) Langmuir circulation. *Annual Review of Fluid Mechanics* 36: 55–79.

Weller RA and Price JF (1988) Langmuir circulation within the oceanic mixed layer. *Deep-Sea Research* 35: 711–747.

Yang G and Leibovich S (1994) Large-eddy simulation of Langmuir circulation. *Bulletin of the American Physical Society* 39(9): 1885.

Zedel L and Farmer DM (1991) Organized structures in subsurface bubble clouds: Langmuir circulation in the upper ocean. *Journal of Geophysical Research* 96(5): 8895.

UPPER OCEAN TIME AND SPACE VARIABILITY

D. L. Rudnick, University of California, San Diego, CA, USA

Introduction

The upper ocean is the region of the ocean in direct contact with the atmosphere. Air–sea fluxes of momentum, heat, and fresh water are the primary external forces acting upon the upper ocean (*see* Heat and Momentum Fluxes at the Sea Surface; Evaporation and Humidity; Wind- and Buoyancy-Forced Upper Ocean). These fluxes impose the temporal and spatial scales of the overlying atmosphere. The internal dynamics of the ocean cause variability at scales distinct from the forcing. This combination of forcing and dynamics creates the tapestry of oceanic phenomena at timescales ranging from minutes to decades and length scales from centimeters to thousands of kilometers.

This article is concerned primarily with the physical processes causing time and space variability in the upper ocean. The physical balances to be considered are the conservation of mass, heat, salt, and momentum. Thus, physical phenomena are discussed with special reference to their effects on the temporal and spatial variability of temperature, salinity, density, and velocity. While many other biological, chemical, and optical properties of the ocean are affected by the phenomena outlined below, their discussion is covered by other articles in this volume.

The most striking feature often seen in vertical profiles of the upper ocean is the surface mixed layer, a layer that is vertically uniform in temperature, salinity, and horizontal velocity (*see* Upper Ocean Vertical Structure and Upper Ocean Mean Horizontal Structure). The turbulence that mixes this layer derives its energy from wind and surface cooling. The region immediately below the mixed layer tends to be stratified, and is often called the seasonal thermocline because its stratification varies with the seasons. The seasonal thermocline extends down a few hundred meters to roughly 1000 m. Beneath the seasonal thermocline is the permanent thermocline whose stratification is constant on timescales of at least decades. Here the discussion is concerned with variability of the mixed layer and seasonal thermocline.

The processes discussed below are ordered roughly by increasing time and space scales (**Figure 1**). Most of the processes are covered in greater detail elsewhere in this volume. It is hoped that this section will provide a convenient introduction to the variability of the upper ocean, and that the reader can proceed to the more in-depth articles as needed.

Turbulence and Mixing

The upper ocean is distinguished from the interior of the ocean partly because of the very high levels of turbulence present (*see* Breaking Waves and Near-Surface Turbulence and Upper Ocean Mixing Processes). The smallest scale of motion worthy of note in the ocean is the Kolmogoroff scale, on the order of 1 cm, where energy is dissipated by molecular viscosity. At this scale, the ocean can be considered isotropic; that is, properties vary in the same way regardless of the direction in which they are measured. At much larger scales than the Kolmogoroff scale, the vertical stratification of the ocean becomes important.

In the seasonal thermocline, a dominant mechanism for mixing is the Kelvin-Helmholtz instability, in which a vertical shear of horizontal velocity causes the overturn of stratified water. The resulting 'billows' are observed to be on the order of 1 m thick and to decay on the order of an hour. A great deal of observational and theoretical work in the last 20 years has been devoted to relating the strength of this mixing to larger (in the order of 10 m) and more easily measurable quantities such as shear and stratification. The resulting Henyey-Gregg parameterization is one of the most fundamentally important achievements of modern oceanography.

Langmuir Circulation and Convection

Turbulence in the mixed layer is fundamentally different from that in the seasonal thermocline. Because the mixed layer is nearly unstratified, the largest eddies can be as large as the layer is thick, often about 100 m. These large eddies have come to be called Langmuir cells in honor of Irving Langmuir, the Nobel laureate in chemistry who first described them. Langmuir cells are elongated vortices whose axes are horizontal and oriented nearly parallel to the wind. The cells have radii comparable in size to the mixed layer depth, and can be as long as 1–2 km. Langmuir cells often appear in pairs with opposite senses of rotation. The cells thus create alternating regions of surface convergence and divergence. The

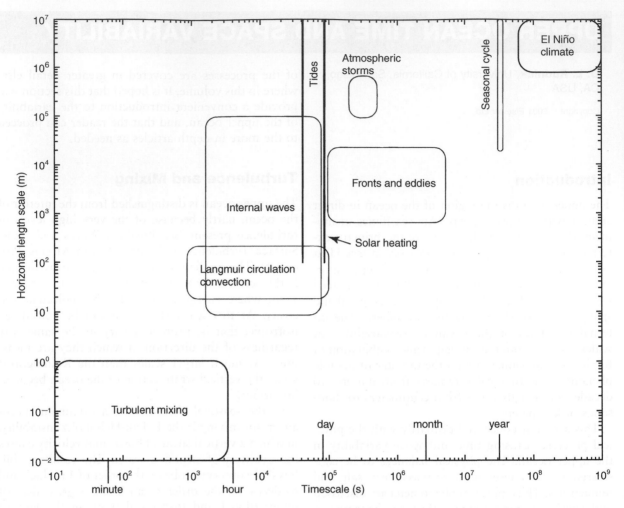

Figure 1 A schematic diagram of the distribution in time and space of upper ocean variability. The temporal and spatial limits of the phenomena should be considered approximate.

regions of convergence collect material floating on the surface such as oil and seaweed. Langmuir first became aware of these cells after noticing lines of floating seaweed during a crossing of the Atlantic. Langmuir cells are forced by a combination of wind and surface waves, and are established typically within an hour after the wind starts blowing. Langmuir cells disappear quickly after the wind stops. Recent research indicates that Langmuir cells often vacillate in strength on the timescale of roughly 15 minutes.

Convection cells forced by surface cooling also cause the mixed layer to be homogenized and to deepen. A typical feature in the mixed layer is the daily cycle of stratification, with daytime heating causing near-surface stratification and nighttime cooling causing convection that destroys this stratification and deepens the mixed layer. The vertical extent of convection cells corresponds to the depth of the mixed layer (of order 100 m); the cells have an

aspect ratio of one so their horizontal and vertical scales are equal. Because solar heating has a large, essentially global, scale the daily heating and cooling of the upper ocean is coherent and predictable over large scales. Horizontal velocity in the mixed layer also varies strongly at a 24 h period, as the daily cycle of stratification affects the depth to which the wind forces currents. The deepest mixed layers in the oceans, at high latitudes, are convectively mixed. Convection cells are thus more effective at deepening the mixed layer than are Langmuir cells.

Internal Waves

Just as there are gravity waves on the surface of the ocean, there are gravity waves in the thermocline. These thermocline gravity waves, modified by the Earth's rotation, are known as internal waves. They exist in a range of frequencies bounded at the lower end by the inertial frequency f and at the upper end by the buoyancy frequency N. A parcel of water given an initial velocity will travel in a circle under the influence of the Coriolis force. The inertial frequency f, twice the local vertical component of the Earth's rotation vector, is the frequency of rotation around such a circle. The resulting horizontal current is known as an inertial oscillation. The inertial period is 12 h at the poles, 24 h at 30° latitude, and infinite at the equator because local vertical is normal to the Earth's axis of rotation. The buoyancy frequency N, proportional to the square root of the vertical density gradient, is the frequency of oscillation of a water parcel given a displacement in the vertical. The resulting vertical motion has a frequency of less than one to several cycles per hour in typical ocean stratification. Internal waves oscillate in planes tilted from the horizontal as a function of the frequency between f and N. Internal waves have amplitudes on the order of tens of meters. They may be coherent over vertical scales that approach the depth of the ocean, particularly at high frequencies near N. Lower frequency internal waves, approaching f, have shorter vertical wavelengths often of order 100 m or less. The horizontal wavelength of an internal wave is related to its frequency and vertical wavelength through the internal wave dispersion relation. For a given vertical wavelength, a high frequency internal wave will have shorter horizontal wavelength than a low frequency wave.

At the low frequency end of the internal wave spectrum, the near-inertial waves are especially important in the upper ocean. Near-inertial waves are quite ubiquitous because they are so readily excited by wind forcing on the ocean's surface. In measurements of horizontal current, inertial oscillations are often the most obvious variability because horizontal currents 'ring' at the resonant inertial frequency. Just as a bell has a distinctive tone when struck, the ocean has inertial currents when hit, for example, by a storm. Strong inertial currents are one of the indications in the ocean of the recent passage of a hurricane. The radius of an inertial current circle is its speed divided by its rotation rate, U/f. If the current speed is $0.1 \, ms^{-1}$, then for a midlatitude inertial frequency of $10^{-4} s^{-1}$, the radius is 1 km. In the aftermath of a storm, the inertial currents and radii may be nearly an order of magnitude larger. Near-inertial waves are a dominant mechanism for transporting wind-driven momentum downward from the mixed layer to the seasonal thermocline and into the interior. Because near-inertial motions have short vertical scales, they dominate the shear spectrum in the ocean. This shear eventually leads to enhanced turbulence and mixing the penetration of inertial

shear into the ocean and the geography of shear and mixing are active topics of research.

Tides are well known to anyone who has spent at least a day at the beach. The dominant tidal periods are near one day and one-half day. Tides are most obvious to the casual observer of the sea surface, and they are easily seen in records of horizontal current in the open ocean. Internal tides exist as well, for example forced by tidal flow over bumps on the ocean bottom. These internal tides, seen as variability in density and velocity at a location, are a form of internal wave and are governed by the same dynamics. Isolated pulses of tidal internal waves, known as 'solitons,' are prevalent in certain regions of rough bottom topography, and are a field of current research.

Fronts and Eddies

While vertically uniform, the mixed layer can vary in the horizontal on a wide range of scales. We have already discussed Langmuir circulation and convection cells on scales of order 100 m, but there may be horizontal variability on longer scales. Just as there are fronts in the atmosphere, visible for example in the satellite pictures of clouds shown on the evening television news, there are fronts in the ocean. Fronts in the ocean separate regions of warm and cool water, or fresh and salty water. The most obvious fronts in the mixed layer have widths on the order of 10–100 km, and typically persist for weeks. Fronts of this size have currents directed along the front as a result of the geostrophic momentum balance. That is, the Coriolis force balances the pressure gradient due to having water of varying density across the front. The less dense (usually warmer) water is on the right side of the current in the Northern Hemisphere (the sense of the current is the opposite in the Southern Hemisphere). Fronts in the mixed layer are sites of enhanced vertical circulation on the order of tens of meters per day. Strong biological productivity at fronts is attributed to this vertical circulation which brings deeper water rich in nutrients to the surface.

Fronts at scales shorter than 10 km also exist in the mixed layer. At these shorter scales, the geostrophic balance may not be expected to hold. Typical fronts at these scales are observed to be warm and salty on one side and cold and fresh on the other such that the density contrast across the front vanishes. Such a front is often said to be compensated, since temperature and salinity gradients compensate in their effect on density. The presence of compensated fronts in the mixed layer is consistent with a horizontal mixing that is an increasing function of the horizontal density gradient. That is, small-scale horizontal density fronts do not persist as long as compensated fronts. Because of their small scale, fronts of order 1 km are poorly observed in the ocean, and are a topic of current research.

Observed fronts are usually not observed to be perfectly straight, rather they wiggle. The wiggles, or perturbations, often grow to be large in comparison with the width of the front. When the perturbations grow large enough, the front may turn back on itself and a detached eddy is formed. The eddies often have sizes on the order of 10 km, when they are confined in depth to the mixed layer. This length scale is related to the Rossby radius of deformation; at scales larger than the Rossby radius flows tend to be geostrophic. The Rossby radius for the mixed layer is given by:

$$\frac{\sqrt{gH\Delta\rho/\rho}}{f}$$

where g is acceleration due to gravity, H is the depth of the mixed layer, ρ is the density of the water, and $\Delta\rho$ is the change in density across the mixed layer base. For a typical mixed layer, H is 100 m and $\Delta\rho$ is 0.2 kg m^{-3}, g is 9.8 m s^{-2}, and ρ is 1025 kg m^{-3}, so the Rossby radius is about 6 km. Eddies that extend deeper have larger radii, as can be inferred from the formula for the Rossby radius. Large eddies can persist for as long as several months, while smaller eddies are shorter lived. The small-scale mixed layer eddies, a prominent feature in satellite photos of the sea surface, are typically observed to rotate in the counterclockwise direction in the Northern Hemisphere, and clockwise south of the equator. Again, because of their small size, they have been inadequately observed and are a topic of current research.

Wind-forced Currents (*see* Wind Driven Circulation)

One of the oldest theories of ocean circulation is due to V.W. Ekman, who in 1905 suggested a balance between the Coriolis force and the stress due to wind blowing over the ocean surface. The prediction of this theory for a steady wind is a current that spirals to the right (in the Northern Hemisphere) and decays with depth. This spiral structure was not clearly observed in the ocean until the 1980s with the advent of moorings with modern current meters. Although the details of the stress parameterization used by Ekman were found to be inadequate to describe

observations, the general picture of a spiral remains valid to this day.

An alternative theoretical construct to explain upper ocean structure is the bulk mixed layer model. Oceanic properties, such as temperature, salinity, and velocity, are assumed to be vertically uniform in the mixed layer, with a region of very strong vertical gradients at the mixed layer base. The mixed layer is then forced by air–sea fluxes of heat, fresh water, and momentum at the surface, and by turbulent fluxes at the base. The bulk mixed layer model has proven remarkably successful at predicting some basic features of the upper ocean, particularly the vertical temperature structure.

Interestingly, the disparate conceptual models of the Ekman spiral and the bulk mixed layer can be rationalized. The upper ocean velocity structure is often, but certainly not always, observed to be vertically uniform near the surface with a region of high shear beneath, in accordance with the bulk mixed layer model. On the other hand, long time averages of ocean current tend to have a spiral structure, in qualitative agreement with the Ekman spiral. This is so if the averages are long enough to span many cycles of mixed layer shoaling and deepening, as due to the daily cycle of surface heating. Thus the time-average current spiral may be very different from a typical snapshot of a nearly vertically uniform current.

The averaged wind-driven spiral extends downward to a depth comparable to, but slightly deeper than, the mixed layer. The shape of the spiral is strongly influenced by higher frequency variability in the stratification, such as the daily cycle in mixed layer depth discussed above. A spiral is observed in response to temporally variable winds, as well as to steady winds. The temporally variable spiral may have a different vertical structure to the steady spiral. In particular, the current spirals to the left with depth in response to a wind that rotates more rapidly than f in a clockwise direction, in contrast to the steady spiral to the right.

Regardless of the detailed velocity structure in the upper ocean, the net transport caused by a steady wind is 90° to the right of the wind in the Northern Hemisphere (and to the left in the Southern Hemisphere). This transport (the vertical integral of velocity) is called the Ekman transport. The Ekman transport is proportional to the wind stress and inversely proportional to the inertial frequency. Thus wind of a given strength will cause more transport near the equator than it would closer to the poles.

The spectrum of wind over the midlatitude ocean peaks at periods of a few to several days. These periods correspond to the time required for a typical storm to pass. The wind-driven current and transport is thus prominent at these periods. Atmospheric storms have typical horizontal sizes of a few to several hundred kilometers, and the direct oceanic response to these storms has similar horizontal scales. The prominent large-scale features of the wind field such as the westerlies in midlatitudes and the trade winds in the tropics directly force currents in the upper ocean. These currents have large horizontal length scales that reflect the winds.

Seasonal Cycles

Just as the seasons cause well-known changes in weather, the annual cycle is one of the most robust signals in the ocean. Summer brings greater heat flux from the atmosphere to the ocean, and warmer ocean temperatures. As the ocean warms up at the surface, stratification increases and the mixed layer becomes shallower. The heat flux reverses in many locations during the winter and the ocean cools at the surface. The resulting convection causes the mixed layer to deepen; at some high latitude locations the mixed layer can deepen to several hundred meters in the winter. Winter conditions in high and mid-latitude mixed layers are very important to the general circulation of the oceans, as it is these waters that penetrate into the thermocline and set properties that persist for decades. Along with cooler temperatures, winter brings typically stormier weather and more wind and precipitation. Wind-driven currents often peak during the winter in midlatitudes, at the same time that salinity decreases in response to the increased precipitation.

Seasonal cycles occur over the whole globe in an extremely coherent fashion, because they are driven primarily by the solar heat flux. However, the seasonal cycle can vary at different oceanic locations. For example, the seasonal cycle at the equator is smaller than that at midlatitudes because solar heat flux varies less over the year. The Arabian Sea has a pronounced semi-annual cycle. Cold northerly winds in winter cool the ocean and deepen the mixed layer as typical for midlatitudes. More unusual is a second period of relatively low ocean temperatures and deep mixed layers during the summer south-west monsoon. Wind-driven mixing causes the cooling during the south-west monsoon as cool water is mixed up to the surface. The Arabian Sea monsoon is the classic example of a seasonal wind driven by land–sea temperature differences. Monsoons also exist over the south-west USA and south-east Asia, among others. Additional local seasonal effects may be caused by river outflows and weather patterns influenced by orography.

Climatic Signals

The ocean has significant variability at periods longer than 1 year. The most well known recurrent interannual climatic phenomenon is El Niño. An El Niño occurs when trade winds reverse at the equator causing upwelling to cease off the coast of South America. The most obvious consequence of an El Niño is dramatically elevated ocean temperatures at the equator. These high temperatures progress poleward from the equator along the coast of the Americas, affecting water properties in large regions of the Pacific. El Niño has been hypothesized to start with anomalous winds in the western equatorial Pacific, eventually having an effect on the global ocean and atmosphere. El Niños occur sporadically every roughly 3–7 years, and are becoming more predictable as observations and models of the phenomenon improve. The reverse phase of El Niño, the so-called La Nina, is remarkable for exceptionally low equatorial temperatures and strong trade winds.

Oscillations with periods of a decade and longer also exist in the ocean and atmosphere. Such oscillations are apparent in the ocean as basin-scale variations in sea surface temperature, for example. Salinity and velocity are also likely variable on decadal timescales, although the observational database for these is sparse in comparison with that for temperature. Atmospheric decadal oscillations in temperature and precipitation are well established. Scientists are actively researching whether and how the ocean and atmosphere are coupled on decadal timescales. The basic idea is that the ocean absorbs heat from the atmosphere and stores it for many years because of the ocean's relatively high heat capacity. This heat may penetrate into the ocean interior and be redistributed by advective processes. The heat may resurface a decade or more later to affect the atmosphere through anomalous heat flux. The coupled ocean–atmosphere process just described is controversial, and the observations to support its existence are inadequate. A major challenge for the immediate future is to obtain the measurements needed to resolve such processes of significance to climate.

Conclusion

The upper ocean varies on a wide range of temporal and spatial scales. Processes range from mixing occurring on scales of centimeters and minutes to decadal climatic oscillations of entire ocean basins. Fundamental to the ocean is the fact that these processes can rarely be studied in isolation. That is, processes occurring on one scale affect processes on other scales. For example, decadal changes in ocean stratification are strongly affected by turbulent mixing at the smallest scales. Turbulent mixing is modulated by the internal wave field, and internal waves are focused and steered by geostrophic fronts and eddies. The interaction among processes of different scales is likely to receive increasing attention from ocean scientists in the coming years.

See also

Breaking Waves and Near-Surface Turbulence. Evaporation and Humidity. Heat and Momentum Fluxes at the Sea Surface. Upper Ocean Mean Horizontal Structure. Upper Ocean Mixing Processes. Upper Ocean Vertical Structure. Wind- and Buoyancy-Forced Upper Ocean. Wind Driven Circulation.

Further Reading

Davis RE, de Szoeke R, Halpern D, and Niiler P (1981) Variability in the upper ocean during MILE. Part I: The heat and momentum balances. *Deep-Sea Research* 28: 1427–1452.

Ekman VW (1905) On the influence of the earth's rotation on ocean currents. *Arkiv Matematik, Astronomi och Fysik* 2: 1–52.

Eriksen CC, Weller RA, Rudnick DL, Pollard RT, and Regier LA (1991) Ocean frontal variability in the Frontal Air–Sea Interaction Experiment. *Journal of Geophysical Research* 96: 8569–8591.

Gill AE (1982) *Atmosphere–Ocean Dynamics*. New York: Academic Press.

Gregg MC (1989) Scaling turbulent dissipation in the thermocline. *Journal of Geophysical Research* 94: 9686–9698.

Langmuir I (1938) Surface motion of water induced by wind. *Science* 87: 119–123.

Lighthill MJ and Pearce RP (eds.) (1981) *Monsoon Dynamics*. Cambridge: Cambridge University Press.

Munk W (1981) Internal waves and small-scale processes. In: Warren BA and Wunsch C (eds.) *Evolution of Physical Oceanography*, pp. 264–291. Cambridge, USA: MIT Press.

Philander SG (1990) *El Niño, La Niña, and the Southern Oscillation*. San Diego: Academic Press.

Roden GI (1984) Mesoscale oceanic fronts of the North Pacific. *Annals of Geophysics* 2: 399–410.

UPPER OCEAN VERTICAL STRUCTURE

J. Sprintall, University of California San Diego, La Jolla, CA, USA

M. F. Cronin, NOAA Pacific Marine Environmental Laboratory, Seattle, WA, USA

Introduction

The upper ocean connects the surface forcing from winds, heat, and fresh water, with the quiescent deeper ocean where this heat and fresh water are sequestered and released on longer time- and global scales. Classically the surface layer includes both an upper mixed layer that is subject to the direct influence of the atmosphere, and also a highly stratified zone below the mixed layer where vertical property gradients are strong. Although all water within the surface layer has been exposed to the atmosphere at some point in time, water most directly exposed lies within the mixed layer. Thus, the surface layer vertical structure reflects not only immediate changes in response to the surface forcing, but also changes associated with earlier forcing events. These forcing events may have occurred either locally in the region, or remotely at other locations and transferred by ocean currents. This article first defines the major features of the upper ocean vertical structure and discusses what causes and maintains them. We then show numerous examples of the rich variability in the shapes and forms that these vertical structures can assume through variation in the atmospheric forcing.

Major Features of the Upper Ocean Vertical Structure

The vertical structure of the upper ocean is primarily defined by the temperature and salinity, which together control the water column's density structure. Within the ocean surface layer, a number of distinct layers can be distinguished that are formed by different processes over different timescales: the upper mixed layer, the seasonal pycnocline, and the permanent pycnocline (**Figure 1**). Right at the ocean surface in the top few millimeters, a cool 'skin' exists with lowered temperature caused by the combined heat losses from long-wave radiation, sensible and latent heat fluxes. The cool skin is only a few millimeters thick, and is the actual sea surface temperature (SST) measured by airborne infrared radiometers. In contrast, *in situ* sensors generally measure the 'bulk' SST over the top few meters of the water column. The cool skin temperature is generally around 0.1–0.5 K cooler than the bulk temperature. As the air–sea fluxes are transported through the molecular layer almost instantaneously, the upper mixed layer can generally be considered to be in direct contact with the atmosphere. For this reason, when defining the depth of the surface layer, the changes in water properties are generally made relative to the bulk SST measurement.

The upper mixed layer is the site of active air–sea exchanges. Energy for the mixed layer to change its vertical structure comes from wind mixing or through a surface buoyancy flux. Wind mixing causes vertical turbulence in the upper mixed layer through waves, and by the entrainment of cooler water through the bottom of the mixed layer. Wind forcing also results in advection by upper ocean currents that can change the water properties and thus the vertical structure of the mixed layer. Surface buoyancy forcing is due to heat and fresh water fluxed across the air–sea interface. Cooling and evaporation induce convective mixing and overturning, whereas heating and rainfall cause the mixed layer to restratify in depth and display alternate levels of greater and lesser vertical

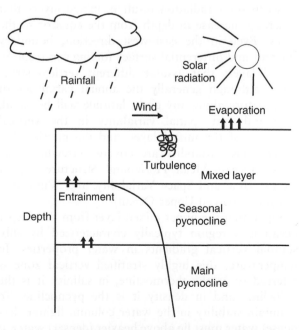

Figure 1 Conceptual diagram of the vertical structure in the surface layer, and the forcing and physics that govern its existence. The depth of the mixed layer, the seasonal pycnocline, and the main pycnocline are indicated.

property gradients. Thus, if strong enough, the wind and buoyancy fluxes can generate sufficient turbulence so that the upper portion of the surface layer has a thick, homogeneous (low vertical gradient or stratification), well-mixed layer in temperature, salinity, and density. Wind and buoyancy forcing also affect the vertical structure of the velocity or shear (vertical gradient of horizontal velocity) in the upper mixed layer. Upper ocean processes, such as inertial shear, Langmuir circulations, internal gravity waves, and Kelvin–Helmholtz instabilities, that alter the velocity profile in the surface layer are an active area of research, and are more fully discussed in Upper Ocean Mixing Processes.

Temporal and spatial variations in the strength and relative contributions of the atmospheric forcing can cause substantial variability in the water properties and thickness of the upper mixed layer. Large temporal variation can occur on daily and seasonal timescales due to changes in the solar radiation. For example, during the daily cycle the sun heats the ocean, causing the upper surface to become increasingly warm and weakly stratified. The 'classic' vertically uniform mixed layer, as depicted in **Figure 1**, may not be present in the upper ocean surface layer. As the sun sets, the surface waters are cooled and sink, generating turbulent convection that causes entrainment of water from below and mixing that produces the vertically well-mixed layer. Similarly, the mixed layer structure can exhibit significant horizontal variations. The large latitudinal differences in solar radiation result in mixed layers that generally increase in depth from the equator to the Poles. Even in the east–west direction, boundary currents and differential surface forcing can result in mixed layers that assume different vertical structures, although generally the annual variations of temperature along any given latitude will be small. Temporal and spatial variability in the vertical structure of the mixed layer, and the physics that govern this variability are covered elsewhere (*see* Upper Ocean Mean Horizontal Structure, Upper Ocean Time and Space Variability, and Wind- and Buoyancy-Forced Upper Ocean).

Separating the upper mixed layer from the deeper ocean is a region typically characterized by substantial vertical gradients in water properties. In temperature, this highly stratified vertical zone is referred to as the thermocline, in salinity it is the halocline, and in density it is the pycnocline. To maintain stability in the water column, lighter (less dense) water must lie above heavier (denser) water. It follows then, that the pycnocline is a region where density increases rapidly with depth. Although the thermocline and the halocline may not always

exactly coincide in their depth range, one or the other property will control the density structure to form the pycnocline. In mid-latitudes during summer, surface heating from the sun can cause a shallow seasonal thermocline (pycnocline) that connects the upper mixed layer to the deeper more permanent thermocline or 'main pycnocline' (see **Figure 1**). Similarly, in the subpolar regions, the seasonal summer inputs of fresh water at the surface through rainfall, rivers, or ice melt can result in a seasonal halocline (pycnocline) separating the fresh surface from the deeper saltier waters. Whereas the seasonal pycnocline disappears every winter, the permanent pycnocline is always present in these areas. The vertical density gradient in the main pycnocline is very strong, and the turbulence within the upper mixed layer induced by the air–sea exchanges of wind and heat cannot overcome the great stability of the main pycnocline to penetrate into the deeper ocean. The stability of the main pycnocline acts as a barrier against turbulent mixing processes, and beneath this depth the water has not had contact with the surface for a very long time. Therefore the main pycnocline marks the depth limit of the upper ocean surface layer.

In some polar regions, particularly in the far North and South Atlantic, no permanent thermocline exists. The presence of an isothermal water column suggests that the cold, dense waters are continuously sinking to great depths. No stable permanent pycnocline or thermocline exists as a barrier to the vertical passage of the surface water properties that extend to the bottom. In some cases, such as along the shelf in Antarctica's Weddell Sea in the South Atlantic, salinity can also play a role in dense water formation. When ice forms from the seawater in this region, it consists primarily of fresh water, and leaves behind a more saline and thus denser surface water that must also sink. The vertical flow of the dense waters in the polar regions is the source of the world's deep and bottom waters that then slowly mix and spread horizontally via the large-scale thermohaline ocean circulation to fill the deep-ocean basins.

In fact, the thermohaline circulation also plays an important role in maintaining the permanent thermocline at a relatively constant depth in the low and middle latitudes. Despite the fact that the pycnocline is extremely stable, it might be assumed that on some long-enough timescale it could be eroded away through mixing of water above and below it. Humboldt recognized early in the nineteenth century that ocean circulation must help maintain the low temperatures of the deeper oceans; the equatorward movement of the cold deep and bottom water masses are continually renewed through sinking (or 'convection') in the polar region. However, it was not until the mid-twentieth century that Stommel suggested that there was also a slow but continual upward movement of this cool water to balance the downward diffusion of heat from the surface. It is this balance, that actually occurs over very small space and timescales that sustains the permanent thermocline observed at middle and low latitudes. Thus, the vertical structure of the upper ocean helps us to understand not only the wind- and thermohaline-forced ocean circulation, but also the response between the coupled air–sea system and the deeper ocean on a global scale.

Definitions

Surface Layer Depth

There is no generally accepted definition of the surface layer depth. Conceptually the surface layer includes the mixed layer, where active air–sea exchanges are occurring, plus those waters in the seasonal thermocline that connect the mixed layer and to the permanent thermocline. Note the important detail that the surface layer includes the mixed layer, a fact that has often been blurred in the criteria used to determine their respective depth levels. A satisfactory depth criterion for the surface layer should thus include all the major features of the upper ocean surface layer described above and illustrated in **Figure 1**. Further, the surface layer depth criterion should be applicable to all geographic regimes, and include those waters that have recently been in contact with the atmosphere, at least on timescales of up to a year. Finally, the definition should preferably be based on readily measurable properties such as temperature, salinity, or density.

Ideally then, we could specify the surface layer to be the depth where, for instance, the temperature is equal to the previous winter's minimum SST. However in practice, this surface layer definition would vary temporally, making it difficult to decipher the year-to-year variability. Oceanographers therefore generally prefer a static criterion, and thus modify the definition to be the depth where the temperature is equal to the coldest SST ever observed using any historical data available at a particular geographic location. This definition is analogous to a local 'ventilation' depth: the deepest surface to which recent atmospheric influence has been felt at least over the timescale of the available historical data. The definition suggested for the surface layer is also primarily one-dimensional, involving only the temperature and salinity information from a given location. Lateral advective effects have not been included. The roles of velocity and shear, and other three-dimensional processes in the surface layer structure (e.g., Langmuir circulations, internal gravity waves, and Kelvin–Helmholtz instabilities), may on occasion be important. However, their roles are harder to quantify and have not, as yet, been adequately incorporated into a working definition for the depth of the surface layer.

Mixed Layer Depth

The mixed layer is the upper portion of the surface layer where active air–sea exchanges generate surface turbulence which causes the water to mix and become vertically uniform in temperature and salinity, and thus density. Very small vertical property gradients can still occur within the mixed layer in response to, for example, adiabatic heating or thermocline erosion. Direct measurements of the upper layer turbulence through dissipation rates provide an accurate and instant measurement of the active 'mixing' depth. However, while the technology is improving rapidly, turbulence scales are very small and difficult to detect, and their measurement is not widespread at present. Furthermore, the purpose of defining a mixed layer depth is to obtain more of an integrated measurement of the depth to which surface fluxes have penetrated in the recent past (daily and longer timescales). For this reason, as in the surface layer depth criterion, definitions of the mixed layer depth are most commonly based on temperature, salinity, or density. The mixed layer depth must define the depth of the transition from a homogeneous upper layer to the stratified layer of the pycnocline.

Several definitions of the mixed layer depth exist in the literature. One commonly used mixed layer depth criterion determines the depth where a critical temperature or density gradient corresponding to the top of the maximum property gradient (i.e., the thermocline or pycnocline) is exceeded. The critical gradient criteria range between 0.02 and 0.05 $^\circ$C m^{-1} in temperature, and 0.005 and 0.015 kg m^{-3} in density. This criterion may be sensitive to the vertical depth interval over which the gradient is calculated. Another mixed layer depth criterion determines a net temperature or density change from the surface isotherm or isopycnal. Common values used for the net change criterion are 0.2–1 $^\circ$C in temperature from the surface isotherm, or 0.03–0.125 kg m^{-3} from the surface isopycnal. Because of the different dynamical processes associated with the molecular skin SST, oceanographers generally prefer the readily determined bulk SST estimate as the surface reference temperature. Ranges of the temperature and density values used in

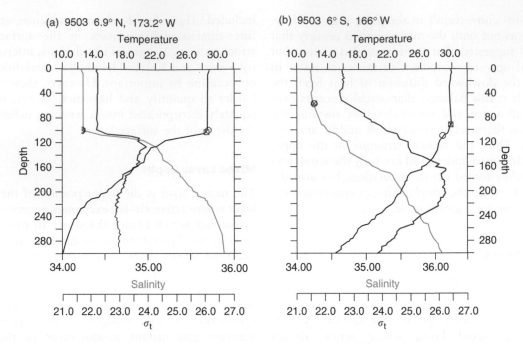

Figure 2 Temperature (black line), salinity (blue line), and density (green line) during March 1995 from expendable conductivity–temperature–depth profiles in the Pacific Ocean at (a) 6.9° N, 173.2° W and (b) 6° S, 166° W. In temperature, mixed layer depth is calculated using criteria of a net temperature change of 0.5° C (crossed box) and 1 °C (circle) from the sea surface; and a temperature gradient criteria of 0.01 °C m^{-1} (small cross). In density, mixed layer depth is determined using criteria of a net density change of 0.125σ_t units from the surface isopycnal (crossed box), a density gradient of 0.01σ_t units m^{-1} (circle), and the thermal expansion method of eqn [1] (cross). Note the barrier layer defined as the difference between the deeper isothermal layer and the shallow density-defined mixed layer in (b).

both mixed layer depth definitions will distinguish weakly stratified regions from unstratified. Another form of the net change criterion used to define the mixed layer depth (mld) takes advantage of the equivalence of temperature and density changes based upon the thermal expansion coefficient ($\alpha_0 = \delta T d\rho/dT$, where δT is the net change in temperature from the surface, e.g., 0.2–1 °C, and $d\rho/dT$ is calculated from the equation of state for seawater using surface temperature and salinity values). This criterion thus determines the depth at which density is greater than the surface density by an amount equivalent to the δT temperature change. In this way, this definition has the advantage of revealing mixed layers where salinity stratification may be important, such as in barrier layers, which are discussed further below. Criteria based on salinity changes, although inherent in the density criterion, are not evident in the literature as typically heat fluxes are large compared to freshwater fluxes, and the gravitational stability of the water column is often controlled by the temperature stratification. In addition, subsurface salinity observations are not as regularly available as temperature.

To illustrate the differences between the mixed layer depth criteria, **Figure 2(a)** shows the mixed

layer depth from an expendable conductivity–temperature–depth (XCTD: *see*) profile, using the net temperature (0.5 °C) and density (0.125 kg m^{-3}) change criteria, the gradient density criterion (0.01 kg m^{-3}), and a net change criterion based on the thermal expansion coefficient with $\delta T = 0.5$°C. In this particular case, there is little difference between the mixed layer depth determined from any method or property. However, **Figure 2(b)** shows an XCTD cast from the western Pacific Ocean, and the strong salinity halocline that defines the bottom of the upper mixed layer is only correctly identified using the density-defined criteria.

Finally, to illustrate the distinction between the surface layer and the upper mixed layer, **Figure 3(a)** shows a temperature section of the upper 300 m from Auckland to Seattle during April 1996. The corresponding temperature stratification (i.e., the vertical temperature gradient) is shown in **Figure 3(b)**. The surface layer, determined as the depth of the climatological minimum SST isotherm, and also the mixed layer depth from a 1 °C net temperature change from the surface (i.e., SST – 1 °C) are indicated on both panels. This cross-equatorial north–south section also serves to illustrate the seasonal differences expected in the mixed layer. In the early fall of the Southern

Hemisphere, the net temperature mixed layer depth criterion picks out the top of the remaining seasonal thermocline, as depicted by the increase in temperature stratification in **Figure 3(b)**. The mixed layer depth criterion therefore excludes information about the depth of the prior winter local wind stirring or heat exchange at the air–sea surface that has been successfully captured in the surface layer using the historical minimum SST criterion. In the Northern Hemisphere tropical regions where there is little seasonal cycle, the surface layer and the mixed layer criteria are nearly coincident. The depth of the mixed layer and the surface layer extend down to the main thermocline. Finally, in the early-spring northern latitudes, the mixed layer criterion again mainly picks out the upper layer of increased stratification that was likely caused through early seasonal surface heating. The surface layer definition lies deeper in the water column near the main thermocline, and below a second layer of

relatively low stratification (**Figure 3(b)**). The deeper, weakly stratified region indicates the presence of fossil layers, which are defined in the next section.

Variability in Upper Ocean Vertical Structure

Fossil Layers

Fossil layers are nearly isothermal layers that separate the upper well-mixed layer from a deeper well-stratified layer (see **Figure 3(b)**, 31–37° N). The fact that these layers are warmer than the local minimum SST defining the surface layer depth, indicates that they have at some time been subject to local surface forcing. The solar heating and reduced wind stirring of spring can cause the upper layer to become thermally restratified. The newly formed upper mixed layer of light, warm water is separated from the

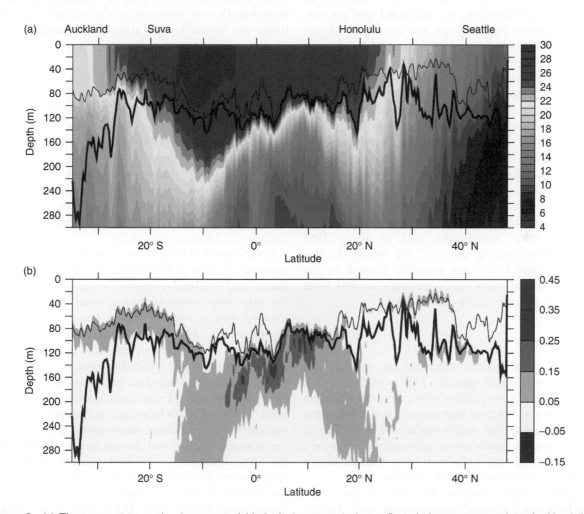

Figure 3 (a) The temperature section from expendable bathythermograph data collected along a transect from Auckland (New Zealand) to Seattle in April 1996, and (b) the corresponding temperature gradient with depth. The heavy line indicates the depth of the surface layer, according to the depth of the coldest sea surface temperature measured at each location. The light line indicates the depth of the mixed layer according to the (SST − 1 °C) criterion.

older, deeper winter mixed layer by a well-stratified thermocline. The fairly stable waters in this seasonal thermocline may isolate the lower isothermal layer and prevent further modification of its properties, so that this layer retains the water characteristics of its winter formation period and becomes 'fossilized'. Hence, fossil layers tend to form in regions with significant seasonal heating, a large annual range in wind stress, and deep winter mixed layers. These conditions can be found at the poleward edges of the subtropical gyres.

In the northeast Pacific Ocean off California and in the southwest Pacific Ocean near New Zealand, particularly deep and thick fossil layers have been associated with the formation of subtropical mode waters. As with the fossil layers, the mode waters are distinguishable by low vertical gradients in temperature and density, and thus a narrow range or 'mode' of property characteristics. The isothermal layer or thermostad of winter water trapped in the fossilized layers may be subducted into the permanent thermocline through the action of Ekman pumping, in response to a curl in the wind field. The mode waters are then transported, retaining their characteristic thermostad, with flow in the subtropical gyre.

Not all fossil layers are associated with mode water formation regions. Shallow fossil layers have also been observed where there are strong diurnal cycles, such as in the western equatorial Pacific Ocean. Here, the fossil layers are formed through the same alternating processes of heating/cooling and wind mixing as found in the mode water formation regions. Fossil layers have also been observed around areas of abrupt topography, such as along-island chains, where strong currents are found. In this case, the fossil layers are probably formed by the advection of water with properties different from those found in the upper mixed layer.

Barrier Layers

In some regions, the freshwater flux can dominate the mixed-layer thermodynamics. This is evident in the Tropics where heavy precipitation can cause a surface-trapped freshwater pool that forms a shallower mixed layer within a deeper nearly isothermal layer. The region between the shallower density-defined well-mixed layer and the deeper isothermal layer (**Figure 2(b)**) is referred to as a salinity-stratified barrier layer.

Recent evidence suggests that barrier layers can also be formed through advection of fresh surface water, especially in the equatorial region of the western Pacific. In this region, westerly wind bursts can give rise to surface-intensified freshwater jets that tilt the zonal salinity gradient into the vertical, generating a shallow halocline above the top of the thermocline. Furthermore, the vertical shear within the mixed layer may become enhanced in response to a depth-dependent pressure gradient setup by the salinity gradient and the trapping of the wind-forced momentum above the salinity barrier layer. This increased shear then leads to further surface intensified advection of freshwater and stratification that can prolong the life of the barrier layer.

The barrier layer may have important implications on the heat balance within the surface layer because, as the name suggests, it effectively limits interaction between the ocean mixed layer and the deeper permanent thermocline. Even if under light wind conditions water is entrained from below into the mixed layer, it will have the same temperature as the water in this upper layer. Thus, there is no heat flux through the bottom of the mixed layer and other sinks must come into play to balance the solar warming that is confined to the surface, or more likely, the barrier layer is transient in nature.

Inversions

Occasionally temperature stratification within the surface layer can be inverted (i.e., cool water lies above warmer water). The temperature inversion can be maintained in a stable water column since it is density-compensated by a corresponding salinity increase with depth throughout the inversion layer. Inversions are a ubiquitous feature in the vertical structure of the surface layer from the equator to subpolar latitudes, although their shape and formation mechanisms may differ.

Inversions that form in response to a change in the seasonal heating at the surface are most commonly found in the subpolar regions. They can form when the relatively warmer surface water of summer is trapped by the cooler, fresher conditions that exist during winter. The vertical structure of the surface layer has a well-mixed upper layer in temperature, salinity, and density, lying above the inversion layer that contains the halocline and subsequent pycnocline (**Figure 4(a)**). Conversely, during summer, the weak subpolar solar heating can trap the very cold surface waters of winter, sandwiching them between the warmer surface and deeper layers. In this case, the vertical structure of the surface layer consists of a temperature minimum layer below the warm stratified surface layer, and above the relatively warmer deeper layer (**Figure 4(b)**). The density-defined mixed layer occurs above the temperature minimum. With continual but slow summer heating, the cold water found in this inversion layer slowly mixes with the

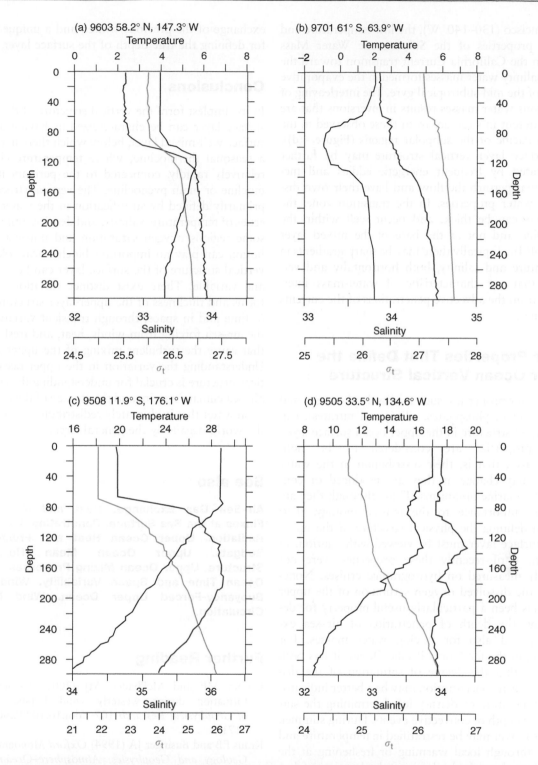

Figure 4   Temperature (black line), salinity (blue line), and density (σ_t, green line) from XCTD casts at (a) 58.2° N, 147.3° W in March 1996, (b) 61° S, 63.9° W in January 1997, (c) 11.9° S, 176.1° W in August 1998, and (d) 33.5° N, 134.6° W in May 1995. Note the presence of temperature inversions at the base of the mixed layer in all casts.

warmer water masses above and below, and erodes away.

Inversions can also form through horizontal advection of water with different properties known as water-mass interleaving. For example, in the Tropics

where there may be velocity shear between opposing currents, inversions are typically characterized as small abrupt features (often only meters thick) found at the base of a well-mixed upper layer and at the top of the halocline and pycnocline (**Figure 4(c)**). Just west of

San Francisco (130–140° W), the low temperature and salinity properties of the Subantarctic Water Mass found in the California Current transition toward the higher-salinity water masses formed in the evaporative regime of the mid-subtropical gyre. The interleaving of the various water masses results in inversions that are quite different in structure from those observed in the tropical Pacific or the subpolar regions (**Figure 4(d)**). The surface layer vertical structure may be further complicated by frequent energetic eddies and meanders that perturb the flow and have their own distinctive water properties. In the transition zone, the inversions can be thick, and occur well within the pycnocline and not at the base of the mixed layer (**Figure 4(d)**). Typically there may be sharp gradients in temperature and salinity, both horizontally and vertically, that are characteristic of water-mass interleaving from the advective penetrations of the currents and eddies.

Other Properties That Define the Upper Ocean Vertical Structure

Other water properties, such as dissolved oxygen and nutrients (e.g., phosphates, silica, and nitrates), can also vary in structure in the upper ocean surface layer. These properties are considered to be non-conservative, that is, their distribution in the water column may change as they are produced or consumed by marine organisms. Thus, although they are of great importance to the marine biology, their value in defining the physical structure of the upper ocean surface layer must be viewed with caution. In addition, until recently these properties were not routinely measured on hydrographic cruises. Nonetheless, the dissolved oxygen saturation of the upper ocean has been a particularly useful property for determining the depth of penetration of air–sea exchanges, and also for tracing water masses. For example, in the far North Pacific Ocean, it has been suggested that the degree of saturation of the dissolved oxygen concentration may be a better indicator than temperature or density for determining the surface-layer depth of convective events. During summer, the upper layer may be restratified in temperature and salinity through local warming or freshening at the surface, or through the horizontal advection of less dense waters. However, these surface processes typically do not erode the high-oxygen saturation signature of the deeper winter convection. Thus the deep high-oxygen saturation level provides a clear record of the depth of convective penetration from the air–sea

exchange of the previous winter, and a unique signal for defining the true depth of the surface layer.

Conclusions

In its simplest form the vertical structure of the upper surface layer can be characterized as having a near-surface well-mixed layer, below which there may exist a seasonal thermocline, where temperature changes relatively rapidly, connected to the permanent thermocline or main pycnocline. The vertical structure is primarily defined by stratification in the water properties of temperature, salinity, and density, although in some regions oxygen saturation and nutrient distribution can play an important biochemical role. The vertical structure of the surface layer can be complex and variable. There exist distinct variations in the forms and thickness of the upper-layer structure both in time and in space, through transient variations in the air–sea forcing from winds, heat, and fresh water that cause the turbulent mixing of the upper ocean. Understanding the variation in the upper ocean vertical structure is crucial for understanding the coupled air–sea climate system, and the storage of the heat and fresh water that is ultimately redistributed throughout the world oceans by the general circulation.

See also

Air–Sea Gas Exchange. Heat and Momentum Fluxes at the Sea Surface. Penetrating Shortwave Radiation. Upper Ocean Heat and Freshwater Budgets. Upper Ocean Mean Horizontal Structure. Upper Ocean Mixing Processes. Upper Ocean Time and Space Variability. Wind- and Buoyancy-Forced Upper Ocean. Wind Driven Circulation.

Further Reading

Cronin MF and McPhaden MJ (2002) Barrier layer formation during westerly wind bursts. *Journal of Geophysical Research* 107 (doi:10.1029/2001JC00 1171).

Kraus EB and Businger JA (1994) *Oxford Monographs on Geology and Geophysics: Atmosphere–Ocean Interaction*, 2nd edn. New York: Oxford University Press.

Philips OM (1977) *The Dynamics of the Upper Ocean*, 2nd edn. London: Cambridge University Press.

Reid JL (1982) On the use of dissolved oxygen concentration as an indicator of winter convection. *Naval Research Reviews* 3: 28–39.

UPPER OCEAN MEAN HORIZONTAL STRUCTURE

M. Tomczak, Flinders University of South Australia, Adelaide, SA, Australia

Introduction

The upper ocean is the most variable, most accessible, and dynamically most active part of the marine environment. Its structure is of interest to many science disciplines. Historically, most studies of the upper ocean focused on its impact on shipping, fisheries, and recreation, involving physical and biological oceanographers and marine chemists. Increased recognition of the ocean's role in climate variability and climate change has led to a growing interest in the upper ocean from meteorologists and climatologists.

In the context of this article the upper ocean is defined as the ocean region from the surface to a depth of 1 km and excludes the shelf regions. Although the upper ocean is small in volume when compared to the world ocean as a whole, it is of fundamental importance for life processes in the sea. It determines the framework for marine life through processes that operate on space scales from millimeters to hundreds of kilometers and on timescales from seconds to seasons. On larger space and timescales, its circulation and water mass renewal processes span typically a few thousand kilometers and several decades, which means that the upper ocean plays an important role in decadal variability of the climate system. (In comparison, circulation and water mass renewal timescales in the deeper ocean are of the order of centuries, and the water masses below the upper ocean are elements of climate change rather than climate variability.)

The upper ocean can be subdivided into two regions. The upper region is controlled by air–sea interaction processes on timescales of less than a few months. It contains the oceanic mixed layer, the seasonal thermocline and, where it exists, the barrier layer. The lower region, known as the permanent thermocline, represents the transition from the upper ocean to the deeper oceanic layers. It extends to about 1 km depth in the subtropics, is some what shallower near the equator and absent poleward of the Subtropical Front. These elements of the upper ocean will be defined and described in more detail, following an introductory overview of some elementary property fields.

Horizontal Property Fields

The annual mean sea surface temperature(SST) is determined by the heat exchange between ocean and atmosphere. If local solar heat input would be the only determinant, contours of constant SST would extend zonally around the globe, with highest values at the equator and lowest values at the poles. The actual SST field (**Figure 1**) comes close to this simple distribution. Notable departures occur for two reasons.

1. Strong meridional currents transport warm water poleward in the western boundary currents along the east coasts of continents. Examples are the Gulf Stream in the North Atlantic Ocean and the Kuroshio in the North Pacific Ocean. In contrast, cold water is transported equatorward along the west coast of continents.
2. In coastal upwelling regions, for example off the coasts of Peru and Chile or Namibia, SST is lowered as cold water is brought to the surface from several hundreds of meters depth.

The annual mean sea surface salinity(SSS) is controlled by the exchange of fresh water between ocean and atmosphere and reflects it closely (**Figure 2**), the only departures being observed as a result of seasonal ice melting in the polar regions. As a result, the subtropics with their high evaporation and low rainfall are characterized by high salinities, while the regions of the westerly wind systems with their frequent rain-bearing storms are associated with low salinities(**Figure 3**). Persistent rainfall in the intertropical convergence zone produces a regional minimum in the SSS distribution near the equator. Departures from a strict zonal distribution are again observed, for the same reasons listed for the SST distribution. In addition, extreme evaporation rates in the vicinity of large deserts are reflected in high SSS, and large river run-off produced by monsoonal rainfall over south east Asia results in low SSS in the Gulf of Bengal. As a result, the SSS distribution of the north-west Indian Ocean shows a distinct departure from the normal zonal distribution.

Seasonal variations of SST and SSS are mainly due to three factors.

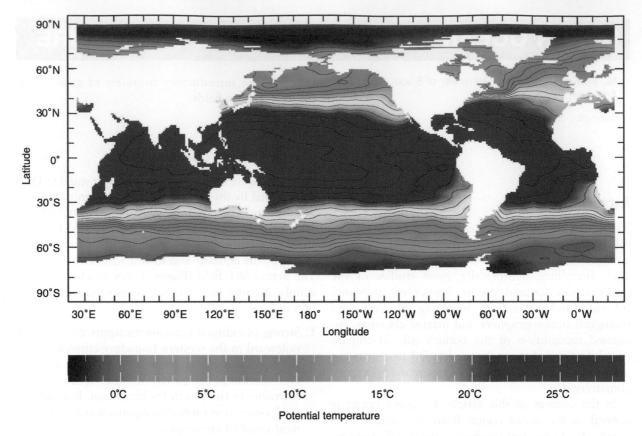

Figure 1 Annual mean sea surface temperature (°C) (the contour interval is 2°C). (Reproduced from *World Ocean Atlas 1994*.)

1. Variations in heat and freshwater exchange between ocean and atmosphere are significant for the SST distribution, which shows a drop of SST in winter and a rise in summer, but much less important for the SSS distribution, since rainfall and evaporation do not vary much over the year in most ocean regions.
2. Changes in the ocean current system, particularly in monsoonal regions where currents reverse twice a year, cause the water of some regions to be replaced by water of different SST and SSS.
3. Monsoonal variations of freshwater input from major rivers influences SSS regionally.

The temperature distribution at 500 m depth (**Figure 4**) reflects the circulation of the upper ocean. At this depth the temperature shows little horizontal variation around a mean of 8–10°C. Departures from this mean temperature are, however, observed. (1) The western basins of the subtropics have the highest temperatures in all oceans. They indicate the centres of the subtropical gyres (see below). (2) Polewardof 35° latitude temperatures fall rapidly as the polar regions are reached, an indication of the absence of the permanent thermocline (see below).

The salinity distribution at 500 m depth (**Figure 5**) shows clear similarities to the temperature distribution

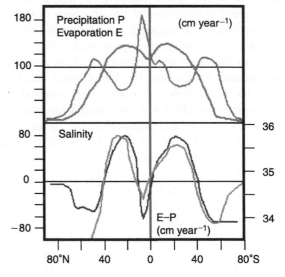

Figure 2 Mean meridional distribution of sea surface salinity and mean meridional freshwater balance (evaporation − precipitation).

and a strong correlation between high temperatures and high salinities. The salinity field displays a total-range nearly as large as the range seen at the surface (**Figure 3**). The mean salinity varies strongly between ocean basins, with the North Atlantic Ocean having

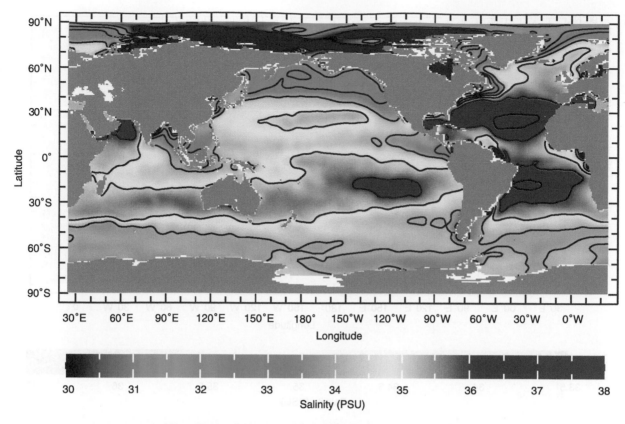

Figure 3 Annual mean sea surface salinity. (Reproduced from *World Ocean Atlas 1994*.)

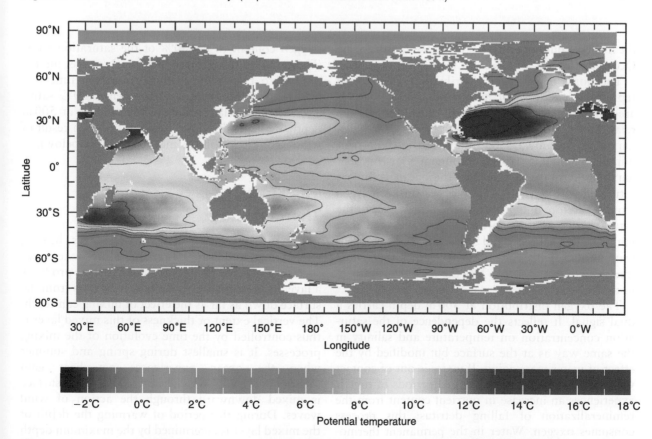

Figure 4 Annual mean potential temperature (°C) at 500 m depth. (Reproduced from *World Ocean Atlas 1994*.)

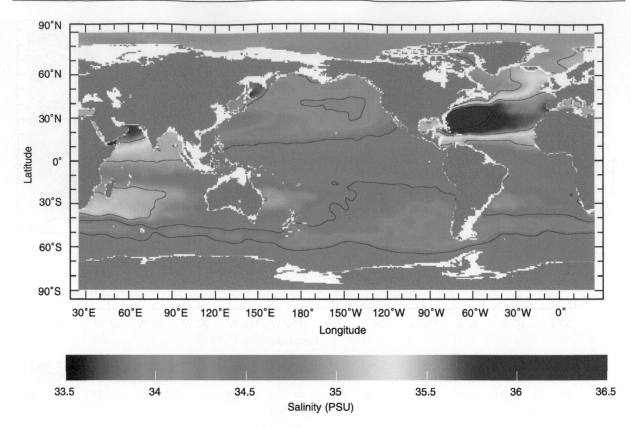

Figure 5   Annual mean salinity (PSU) at 500 m depth.(Reproduced from *World Ocean Atlas 1994.*)

the highest salinity at this depth and the North Pacific Ocean the lowest.

The horizontal oxygen distribution is chosen to represent conditions for marine life. Nutrient levels are inversely related to oxygen, and although the relationship varies between ocean basins, an oxygen maximum can always be interpreted as a nutrient minimum and an oxygen minimum as a nutrient maximum. At the sea surface the ocean is always saturated with oxygen. A map of sea surface oxygen would therefore only illustrate the dependence of the saturation concentration on temperature (and to a minor degree salinity) and show an oxygen concentration of 8 mll^{-1} or more at temperatures near freezing point and 4 mll^{-1} at the high temperatures in the equatorial region.

The oxygen distribution at 500 m depth carries a dual signal. It reflects the dependence of the saturation concentration on temperature and salinity in the same way as at the surface but modified by the effect of water mass aging. If water is out of contact with the atmosphere for extended periods of time it experiences an increase in nutrient content from the remineralization of falling detritus; this process consumes oxygen. Water in the permanent thermocline can be a few decades old, which reduces its

oxygen content to 60–80% of the saturation value (**Figure 6**). The northern Indian Ocean is an exception to this rule; its long ventilation time (see below) produces oxygen values below 20% saturation. In the polar regions oxygen values at 500 m depth are generally closer to saturation as a result of winter convection in the mixed layer (see below).

The Mixed Layer and Seasonal Thermocline

Exposed to the action of wind and waves, heating and cooling, and evaporation and rainfall, the ocean surface is a region of vigorous mixing. This produces a layer of uniform properties which extends from the surface down as far as the effect of mixing can reach. The vertical extent or thickness of this mixed layer is thus controlled by the time evolution of the mixing processes. It is smallest during spring and summer when the ocean experiences net heat gain (**Figure 7**).The heat which accumulates at the surface is mixed downward through the action of wind waves. During this period of warming the depth of the mixed layer is determined by the maximum depth which wave mixing can affect. Because winds

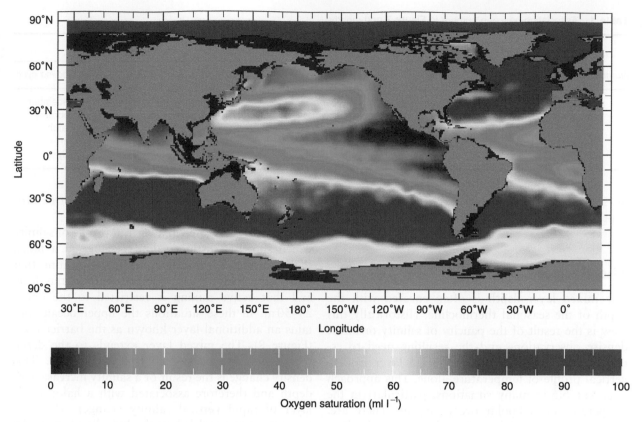

Figure 6 Annual mean oxygen saturation (%) at 500 m depth(the contour interval is 10%). (Reproduced from *World Ocean Atlas 1994*.)

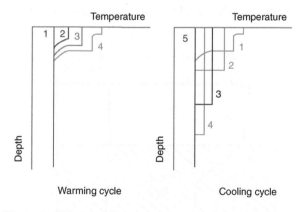

Figure 7 Time evolution of the seasonal mixed layer. Left, the warming cycle; right, the cooling cycle. Numbers can be approximately taken as successive months, with the association shown in **Table 1**.

areoften weaker during midsummer than during spring, wind mixing does not reach quite so deep during the summer months, and the mixed layer may consist of two or more layers of uniform properties (**Figure 7**, line 4 of the warming cycle).

During fall and winter the ocean loses heat. This cooling produces a density increase at the sea surface. As a result, mixing during the cooling period is no longer controlled by wave mixing but by convection. The convection depth is determined by the depth to which the layer has to be mixed until static stability is reached. The mixed layer therefore increases with time during fall and winter and reaches its greatest vertical extent just before spring.

The thin region of rapid temperature change below the mixed layer is known as the seasonal thermocline. It is strongest (i.e., is associated with the largest change in temperature) in summer and disappears in winter. In the tropics (within 20° of the equator) the heat loss during winter is not strong enough to erase the seasonal thermocline altogether, and the seasonal character of the thermocline is then only seen as a variation of the associated vertical temperature gradient.

In the subtropics the mixed layer depth varies between 20–50 m during summer and 70–120 m during winter. In subpolar regions the mixed layer depth can grow to hundreds of meters during winter. Three locations of particularly deep winter mixed layers are the North Atlantic Ocean between the Bay of Biscay and Iceland, the eastern South Indian Ocean south of the Great Australian Bight and the region to the west

Table 1 Association between number in **Figure 7** and months

| Number in **Figure 7** | Northern Hemisphere | | Southern Hemisphere | |
	Warming cycle	Cooling cycle	Warming cycle	Cooling cycle
1	February	June	August	December
2	April	August	October	February
3	May	September	November	March
4	June	October	December	April
5		January		July

of southern Chile. In these regions mixed layer depths can exceed 500 m during late winter.

The Barrier Layer

The mixed layer depth is often equated with the depth of the seasonal thermocline. Historically this view is the result of the paucity of salinity or direct density observations and the resulting need to establish information about the mixed layer from a vertical profile of temperature alone. This approach is acceptable in many situations, particularlyin the temperate and subpolar ocean regions. There are, however, situations where it can be quite misleading. A temperature profile obtained in the equatorial western Pacific Ocean, for example, can show uniform temperatures to depths of 80–100 m. Such deep homogeneity in a region where typical wind speeds

rarely exceed those of a light breeze cannot be produced by wave mixing.

The truth is revealed in a vertical profile of salinity which shows a distinct salinity change at a much shallower depth, typically 25–50 m, indicating that wave mixing does not penetrate beyond this level and that active mixing is restricted to the upper 25–50 m. In these situations the upper ocean contains an additional layer known as the barrier layer (**Figure 8**). The mixed layer extends to the depth where the first density change is observed. This density change is the result of a salinity increase with depth and therefore associated with a halocline (a layer of rapid vertical salinity change). The temperature above and below the halocline is virtually identical. The barrier layer is the layer between the halocline and thethermocline.

The barrier layer is of immense significance for the oceanic heat budget. In most ocean regions the mixed

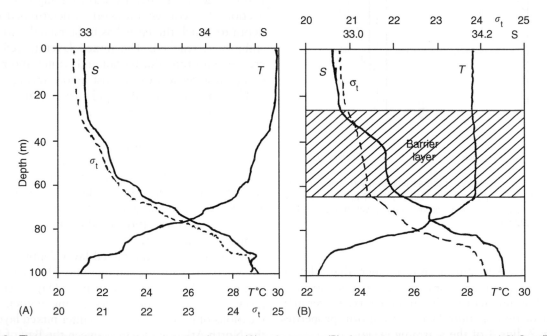

Figure 8 The structure of the upper ocean in the absence (A) and presence (B) of a barrier layer. T: temperature (°C),S: salinity, σ_t: density. Note the uniformity of temperature(T) from the surface to the bottom of the barrier layer in(B). The stations were taken in the central South China Sea during September 1994.

layer experiences a net heat gain at the surface during spring and summer and has to export heatin order to maintain its temperature in a steady state. If (as described in a previous section) the mixed layer extends down to the seasonal thermocline, this is achieved through the entrainment of colder water into the mixed layer from below. The presence of the barrier layer means that the water entrained from the region below the mixed layer is of the same temperature as the water in the mixed layer itself. The entrainment process is still active but does not achieve the necessary heat export. The barrier layer acts as a barrier to the vertical heat flux, and the heat gained by the mixed layer has to be exported through other means, mainly through horizontal advection by ocean currents and, if the mixed layer is sufficiently transparent to the incoming solar radiation, through direct downward heat transfer from the atmosphere to the barrier layer.

The existence of the barrier layer has only come to light in the last decade or two when high-quality salinity measurements became available in greater numbers. It has now been documented for all tropical ocean regions. In the Pacific Ocean the regional extent of the barrier layer is closely linked with high local rainfall in the Intertropical and South Pacific Convergence Zones of the atmosphere. This suggests that the Pacific barrier layer is formed by the lowering of the salinity in the shallow mixed layer in response to local rainfall. In contrast, the barrier layer in the Indian Ocean varies seasonally in extent, and the observed lowering of the mixed layer salinity seems to be related to the spreading of fresh water from rivers during the rainy monsoon season. In the Atlantic Ocean the barrier layer is most likely the result of subduction of high salinity water from the subtropics under the shallow tropical mixed layer. There are also observations of seasonal barrier layers in other tropical ocean regions, such as the South China Sea.

The Subtropical Gyres and the Permanent Thermocline

The permanent or oceanic the rmocline is the transition from the upper ocean to the deeper oceanic layers. It is characterized by a relatively rapid decrease of temperature with depth, with a total temperature drop of some 15°C over its vertical extent, which varies from about 800 m in the subtropics to less than 200 m near the equator. This depth range does not display the relatively strong currents experienced in the upper ocean but still forms part of the general wind driven circulation, so its water moves with the same current systems seen at the sea surface but with lesser speed.

The permanent thermocline is connected with the atmosphere through the Subtropical Convergence, broad region of the upper ocean poleward of the subtropics where the wind-driven surface currents converge, forcing water to submerge ('subduct') under the upper ocean layer and enter the permanent thermocline. This convergence is particularly intense in the subtropical front, a region of enhanced horizontal temperature change within the Subtropical Convergence found at about 35°Nand 40°S. The Subtropical Front is therefore considered the poleward limit of the permanent thermocline (**Figure 9**).

There is also a zonal variation in the vertical extent, with smallest values in the east and largest values in the west. Taken together, the permanent thermocline appears bowl shaped, being deepest in the western parts of the subtropical ocean (25–30°N and 30–35°S). The shape is the result of geostrophic adjustment in the wind-driven circulation, which produces anticyclonic water movement in the subtropics known as the subtropical gyres.

In most ocean regions the permanent thermocline is characterized by a tight temperature–salinity(TS) relationship, lower temperatures being associated with lower salinities. If temperature or salinity is plotted on a constant depth level across the permanent thermocline, the highest temperature and salinity values are found in the western subtropics (**Figures 4** and **5**). The tight TS relationship indicates the presence of a stable water mass, known as Central Water. This water mass is formed at the surface in the subtropical convergence, particularly at the downstream end of the western boundary currents, where it is subducted and from where it renews ('ventilates') the permanent thermocline by circulating in the subtropical gyres, moving equatorward in the east, westward with the equatorial current system and returning to the ventilation region in the west. As a result the age of the Central Water does not increase in a simple meridional direction from the subtropical front towards the equator but is lower in the east and higher in the west.

As the Subtropical Front is a feature of both hemispheres, each ocean, with the exception of the Indian Ocean which does not reach far enough north to have a Subtropical Front in the northern hemisphere, has Central Water of northern and southern origins (**Figure 9**). Fronts between the different varieties of Central Water are a prominent feature of the permanent thermocline. These fronts are characterized by strong horizontal temperature and salinity gradients but relatively small density change because the effect of temperature on density is partly compensated by the effect of salinity. As a result small-scale mixing processes such as double diffusion,

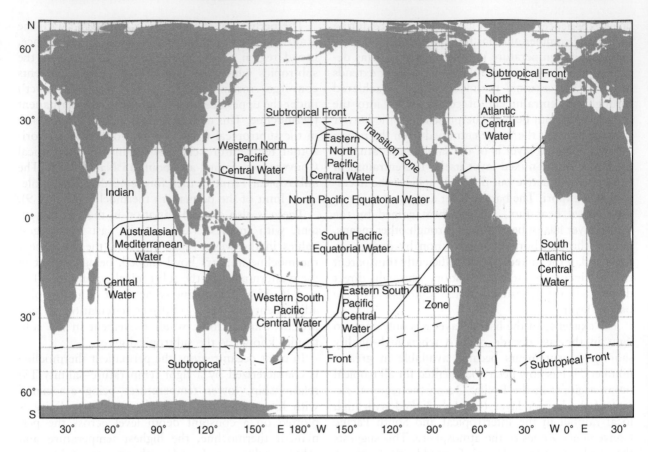

Figure 9 Regional distribution of the water masses of the permanent thermocline.

The Equatorial Region

The equatorial current system occupies the region 15°S–15°N and is thus more than 3000 km wide. Most of itis taken up by the North and South Equatorial Currents, the westward flowing equatorial elements of the subtropical gyres discussed above. Between these two currents flows the North Equatorial Countercurrent as a relatively narrow band eastward along 5°N in the Atlantic and Pacific Oceans and, during the north-east monsoon season, along 5°S in the Indian Ocean. Another eastward current, the Equatorial Undercurrent, flows submerged along the equator, where it occupies the depth range 50–250 m as a narrow band ofonly 200 km width.

Currents near the equator are generally strong, and for dynamical reasons transport across the equator is more or less restricted to the upper mixed layer and to a narrow regime of a few hundred kilometers width along the western boundary of the oceans. This restriction andthe narrow eastward

filamentation and interleaving are of particular importance in these fronts.

currents embedded in the general westward flow, shape the distribution of properties in the permanent thermocline near the equator. Insituations where subtropical gyres exist (the Atlantic and Pacific Oceans) in both hemispheres they enter the equatorial current system from the north east and from the south east, leaving a more or less stagnant region ('shadow zone') between them near the eastern coast. **Figure 10** shows the age distribution for the Atlantic Ocean. The presence of particularly old water in the east indicates a stagnant region or 'shadow zone' between the subtropical gyres.

The strong eastward flowing currents in the equatorial current system modify the age distribution in the permanent thermocline further. In **Figure 10** the Equatorial Undercurrent manifests itself as a band of relatively young water, which is carried eastward.

The Indian Ocean does not extend far enough to the north to have a subtropical convergence in the Northern Hemisphere. In the absence of a significant source of thermocline water masses north of theequator the water of the Northern Hemisphere can only be ventilated from the south. **Figure 11** shows property fields of the permanent thermocline in the

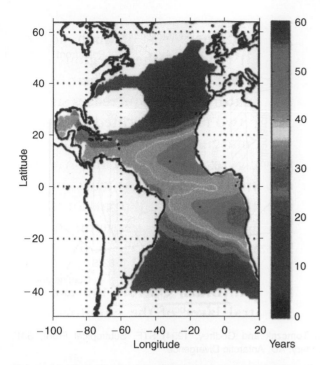

Figure 10 Pseudo age of Central Water in equatorial region of the Atlantic Ocean at 500 m depth. The quantity pseudo age expresses the time elapsed since the water had last contact with the atmosphere; it is determined by using an arbitrary but realistic oxygen consumption rate for the permanent thermocline. (Reproduced from Poole and Tomczak m (19) Optimum multiparameter analysis of the water mass structure in the Atlantic Ocean thermocline. *Deep-Sea Research* 46: 1895–1921.)

Indian Ocean and pathways of its water masses. The region between 5°S and the equator is dominated by the westward flow of Australasian Mediterranean Water (AAMW), a water mass formed in the Indonesian seas. Its mass transport is relatively modest, and it is mixed into the surrounding waters before it reaches Africa. Indian Central Water(ICW) originates near 30°S in large volume; it joins the anticyclonic circulation of the subtropical gyre and can be followed (at the depth level shown in **Figure 11** by its temperature of 11.7°C and salinity of 35.1) across the equator along the African coast and into the Northern Hemisphere. The flow into the Northern Hemisphere is thus severely restricted, and the ventilation of the northern Indian Ocean thermocline is unusually inefficient. This is reflected in the extremely low oxygen content throughout the northern Indian Ocean.

The Polar Regions

Poleward of the subtropical front the upperocean changes character. As polar latitudes are approached the distinction between upper ocean and deeper layers disappears more and more. There is no

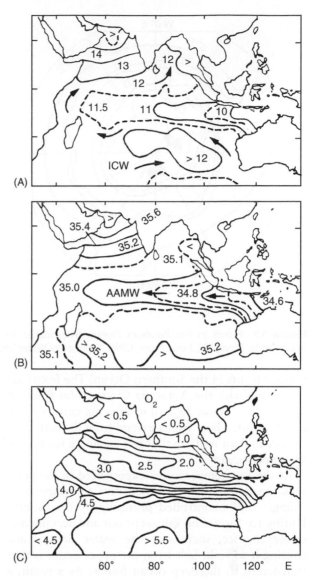

Figure 11 Climatological mean temperature (°C) (A),salinity (PSU) (B) and oxygen concentration (ml l^{-1}) (C) in the Indian Ocean for the depth range 300–450 m, with pathways for Indian Central Water (ICW)and Australasian Mediterranean Water (AAMW). (Reproduced from Tomczak and Godfrey, 1994.)

permanent thermocline; temperature, salinity and all other properties are nearly uniform with depth. The surface mixed layer is, of course, still well defined as the layer affected by wave mixing, but its significance for the heat exchange with the atmosphere is greatly reduced because frequent convection events produced by surface cooling penetrate easily into the waters below themixed layer.

Because in the polar regions the upper ocean and the deeper layers form a single dynamic unit, the horizontal structure of the upper ocean in these regions is strongly influenced by features of the deeper layers. **Figure 12** shows the arrangement of the

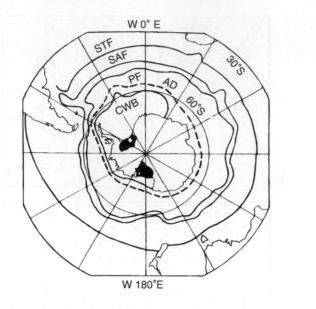

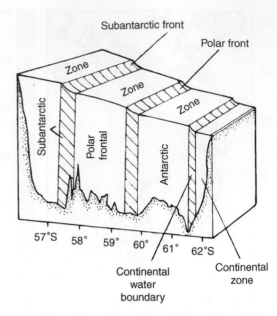

Figure 12 Fronts in the Southern Ocean. (Reproduced from Tomczak and Godfrey, 1994) STF, Subtropical Front; SAF, Subantarctic Front; PF, Polar Front; CWB, Continental Water boundary; AD, Antarctic Divergence.

various fronts in the Southern Ocean. The fronts are associated with the Antarctic Circumpolar Current. They occupy about 20% of its area but carry 75% of its transport. These fronts extend from the surface to the ocean floor and are thus not exclusive features of the upper ocean.

At the low temperatures experienced in the polar seas the density is very insensitive to temperature changes and iscontrolled primarily by the salinity. During ice formation salt seeps out and accumulates under the ice, increasing the water density and causing it to sink. Salt from the upper ocean is thus transferred to the deep ocean basins. As a result, a significant amount of fresh water is added to the upper ocean when the ice melts and floats over the oceanic water. The resulting density gradient guarantees stability of the water column even in the presence of temperature inversions. A characteristic feature of the upper ocean in the polar regions is therefore the widespread existence of shallow

temperature maxima. In the Arctic Ocean the water below the upper ocean can be as much as 4°C warmer than the mixed layer. Intermediate temperature maxima in the Antarctic Ocean are less pronounced (up to 0.5°C) but occur persistently around Antarctica.

See also

Heat Transport and Climate. Satellite Remote Sensing of Sea Surface Temperatures. Wind- and Buoyancy-Forced Upper Ocean. Wind Driven Circulation.

Further Reading

Tomczak M and Godfrey JS (1994) *Regional Oceanography: an Introduction*. Oxford: Pergamon.

UPPER OCEAN STRUCTURE: RESPONSES TO STRONG ATMOSPHERIC FORCING EVENTS

L. K. Shay, University of Miami, Miami, FL, USA

Introduction

Ocean temperature structure changes from profiler and remotely sensed data acquired during hurricane passage have been documented in the literature. These oceanic response measurements have emphasized the sea surface temperature (SST) cooling and deepening of the wind-forced ocean mixed layer (OML). The level of SST cooling and OML deepening process are associated with the oceanic current response, which has two major components (**Figure 1**). First, the momentum response is associated with the OML current divergence in the near-field with a net transport away from the storm center. This divergent flow causes upwelling of the isotherms and an upward vertical velocity. Over the next half of the cycle, currents and their transport converge toward the track, forcing downwelling of warmer water into the thermocline. This cycle of upwelling and downwelling regimes occurs over distances of an inertial wavelength and is proportional to the product of the storm translation speed and the local inertial period. Over these distances, horizontal pressure gradients couple the wind-forced OML to the thermocline as part of the three-dimensional cold wake. In the Northern Hemisphere, wind-forced currents rotate anticyclonically (clockwise) with time and depth where the period of oscillation is close to the local inertial period (referred to as near-inertial). This near-inertial current vector rotation with depth creates significant vertical current shears across the OML base and the top of the seasonal thermocline that induces vertical mixing and cooling and deepening of the layer. For these two reasons, the upper ocean current transport and vertical current shear are central to understanding the ocean's thermal response to hurricane forcing.

The SST response, and by proxy the OML temperature response, typically decreases by 1–5 °C to the right of the storm track at one to two radii of maximum winds (R_{max}) due to surface wind field asymmetries, known as the 'rightward bias'. Although warm SSTs (≥ 26 °C) are required to maintain a hurricane, maximum SST decreases and OML depth increases of 20–40 m are primarily due to entrainment mixing of the cooler thermocline water with the warmer OML water. Ocean mixing and cooling are a function of forced current shear ($\partial v/\partial z = s$) that reduce the Richardson number (defined as the ratio of buoyancy frequency (N^2) and (s^2)) to decrease below criticality. The proportion of these physical processes to the cooling of the OML heat budget are shear-driven entrainment mixing (60–85%), surface heat and moisture fluxes (Q_o) (5–15%), and horizontal advection by ocean currents (5–15%) under relatively quiescent initial ocean conditions (no background fronts or eddies). As per **Figure 1**, vertical motion (upwelling) increases the buoyancy frequency associated with more stratified water that tends to increase the Richardson number above criticality.

In strong frontal regimes (e.g., the Loop Current (LC) and warm core rings (WCRs)) with deep OML, cooling induced by these physical processes is considerably less than observed elsewhere. During hurricane Opal's passage in the Gulf of Mexico (GOM), SST cooling within a WCR was ~ 1 °C compared to ~ 3 °C on its periphery. In these regimes where the 26 °C isotherm is deep (i.e., 100 m), more turbulent mixing induced by vertical current shear is required to cool and deepen an already deep OML compared to the relatively thin OMLs. That the entrainment heat fluxes at the OML base are not significantly contributing to the SST cooling implies there is more heat for the hurricane itself via the heat and moisture surface fluxes. These regimes have less 'negative feedback' to atmosphere than typically observed over the cold wake. To accurately forecast hurricane intensity and structure change in coupled models, the ocean needs to be initialized correctly with both warm and cold fronts, rings and eddies observed in the tropical and subtropical global oceans.

The objective of this article is to build upon the article by Shay in 2001 to document recent progress in this area of oceanic response to hurricanes with a focus on the western Atlantic Ocean basin. The rationale here is that in the GOM (**Figure 1(c)**), *in situ* measurements are more comprehensive under hurricane conditions than perhaps anywhere else on the globe. Second, once a hurricane moves over this basin, it is going to make landfall along the coasts of Mexico, Cuba, and the United States. In the first

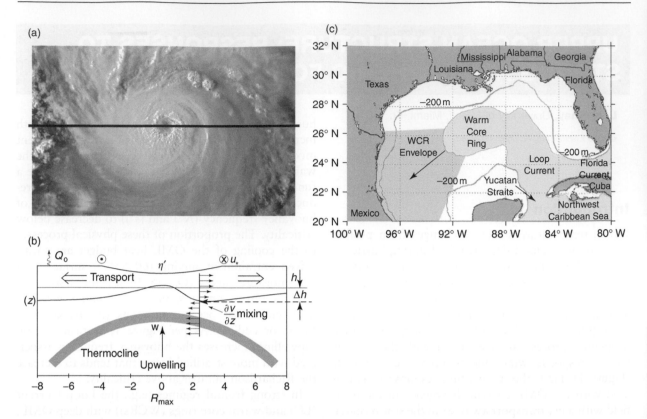

Figure 1 (a) Tropical cyclone image and (b) a cross-section schematic of the physical processes that alter the OML depth (h: light gray line) forced by hurricane winds (u_*) such as shear-induced mixing ($\partial v/\partial z =$ shear) and OML depth changes (Δh: dark gray line), upwelling (w) due to transport (arrows) by currents away from the storm center relative to the surface depression (η'), and surface heat fluxes (Q_0) from the ocean to the atmosphere, all of which may contribute to ocean cooling during TC passage. (c) States and countries surrounding the Gulf of Mexico and northwest Caribbean Sea and identification of the key oceanic features and processes and areas relative to the 200-m isobath. (a, b) Adapted from Shay LK (2001) Upper ocean structure: Response to strong forcing events. In: Weller RA, Thorpe SA, and Steele J (eds.) *Encyclopedia of Ocean Sciences*, pp. 3100–3114. London: Academic Press.

section following this, progress on understanding the wind forcing and the surface drag coefficient behavior at high winds is discussed within the context of the bulk aerodynamic formula. In the next section, the importance of temperature, current, and shear measurements with respect to model initialization are described. While cold wakes are usually observed in relatively quiescent oceans (i.e., hurricanes Gilbert (1988); Ivan, Frances (2004)), the oceanic response is not nearly as dramatic in warm features. This latter point has important consequences for coupled models to accurately simulate the atmospheric response where the sea–air transfers (e.g., surface fluxes) may not decrease to significant levels as observed over cold wakes. These physical processes for oceanic response are briefly documented here for recent hurricanes such as the LC, WCR, and cold core ring (CCR) interactions and coastal ocean response during hurricanes Lili in 2002, Ivan in 2004, and Katrina and Rita in 2005. Concluding remarks as well as suggested avenues for future research efforts are in the final section.

Atmospheric Forcing

Central to the question of storm forcing and the ocean response is the strength of the surface wind stress and the wind stress curl defined at 10 m above the surface. Within the framework of the bulk aerodynamics formula, the wind stress is given by

$$\tau = \rho_a c_d W_{10}(u_{10}i + v_{10}j)$$

where ρ_a is the air density, c_d is the surface drag coefficient, the magnitude of the 10-m wind ($W_{10} = \sqrt{(u_{10}^2 + v_{10}^2)}$, where u_{10} and v_{10} represent the surface winds at 10 m in the east (i) and north (j) directions, respectively). Momentum transfer between the two fluids is characterized by the variations of wind speed with height and a surface drag coefficient that is a function of wind speed and surface roughness.

It is difficult to acquire flux measurements for the high wind and wave conditions under the eyewall at 10 m; however, profilers have been deployed from aircraft to measure the Lagrangian wind profiles in hurricanes. These profiler data suggest a logarithmic variation of mean wind speed in the lowest 200 m of the boundary layer. Based on this variation, the surface wind stress, roughness length, and neutral stability drag coefficient determined by the profile method indicate a leveling of the surface momentum flux as winds increase above hurricane force with a slight decrease of the drag coefficient with increasing winds.

Donelan and colleagues found the characteristic behavior c_d since surface conditions change from aerodynamically smooth to aerodynamically rough (c_d increasing with wind speed) conditions. In rough flow, the drag coefficient is related to the height of the 'roughness elements' per unit distance downwind or the spatial average of the downwind slopes. In a hurricane, rapid changes in wind speed and direction occur over short distances compared to those required to approach full-wave development. The largest waves in the wind-sea move slowly compared to the wind and travel in directions differing from the surface winds. Under such circumstances, longer waves contribute to the roughness of the sea and a 'saturation' of the drag coefficient occurs after wind speeds exceed $33 \, \mathrm{m \, s^{-1}}$ (**Figure 2**). Beyond this threshold, the surface does not become any rougher. These results suggest that there may be a limiting state in the aerodynamic roughness of the sea surface.

Air–Sea Parameters

The oceanic response is usually characterized as a function of storm translation speed (U_h), radius of maximum winds ($R_{\max}$), surface wind stress at 10-m level ($\tau_{\max}$), OML depth (h), and the strength of the seasonal thermocline either by reduced gravity ($g' = g(\rho_2 - \rho_1)/\rho_2$ where ρ_1 is the density of the upper layer of depth h_1, and ρ_2 is the density in the lower layer of depth h_2 where $\rho_2 > \rho_1$) or buoyancy frequency (see **Table 1**). The latitude of the storm sets the local planetary vorticity through the local Coriolis parameter ($f = 2\Omega \sin(\varphi)$, where Ω is the angular rotation rate of the Earth ($7.29 \times 10^{-5} \, \mathrm{s^{-1}}$), and φ is the latitude). The inverse of the local Coriolis parameter (f^{-1}) is a fundamental timescale referred to as the inertial period ($\mathrm{IP} = 2\pi f^{-1}$). The local IPs decrease poleward, for example, at $10°$ N, IP ~ 70 h, at $24°$ N IP ~ 30 h, and at $35°$ N IP ~ 20 h. The relative importance of this parameter cannot be over-emphasized in that at low latitudes such as the eastern Pacific Ocean (EPAC) warm pool, the near-inertial current and shear response will require over a day to develop during hurricane passage. By contrast, at the mid-latitudes, near-inertial motions will develop significant shears across the base of the OML much more quickly. Thus, the initial SST cooling and OML deepening will be minimal at lower latitudes compared to the mid-latitudes for the same oceanic stratification and storm structure.

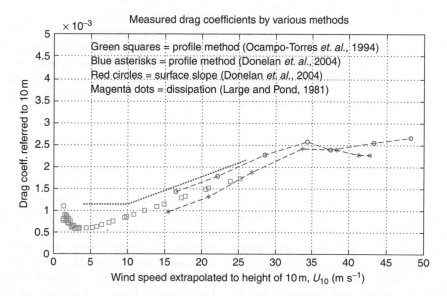

Figure 2 Laboratory measurements of the neutral stability drag coefficient ($\times 10^{-3}$) by profile, eddy correlation ('Reynolds'), and momentum budget methods. The drag coefficient refers to the wind speed measured at the standard anemometer height of 10 m. The drag coefficient formula of Large and Pond (1981) is also shown along with values from Ocampo-Torres et al. (1994) derived from field measurements. From Donelan MA, Haus BK, Reul N, et al. (2004) On the limiting aerodynamic roughness of the ocean in very strong winds. *Geophysical Research Letters* 31: L18306, **Figure 2** (doi: 1029/2004GRL019460).

Table 1   Air–sea parameters and scales for hurricanes Lili (2002) for both the LC and GOM common water, and Ivan (2004), Katrina (2005), and Rita (2005) over the GOM basin

Parameter		Lili (LC)	Lili (GOM)	Ivan	Katrina	Rita
Radius of max. winds	R_{max} (km)	25	18	32	42	19
Max. wind stress	τ_{max} (N m^{-2})	7.1	8.0	6.7	7.6	8.7
Translational speed	U_h (m^{-1} s)	6.9	7.7	5.5	6.3	4.7
Wavelength	Λ (km)	770	775	594	608	454
Mixed layer depth	h (m)	110	35	35	74	70
Inertial period	IP (d)	1.3	1.16	1.25	1.12	1.12
Thermocline thickness	b (m)	200	200	200	200	200
Barotropic phase speed	c_0 (m^{-1} s)	120	150	72	147	150
Barotropic deformation radius	α_0 (km)	2100	2400	1002	2250	2300
Baroclinic phase speed	c_1 (m^{-1} s)	1.5	2.8	2.8	2.5	1.9
Baroclinic deformation radius	α_1 (km)	26	46	40	38	29
Froude number (Fr)	U_h/c_1	2.5	2.8	2.2	2.5	2.5

Note that these parameters are based on where measurements were acquired; for example, Ivan moved over the DeSoto Canyon and over the shelf compared to Lili moving over the eastern side of the Yucatan Shelf, then into the central GOM. Katrina and Rita scales are based on the north-central GOM.

Ocean Structure

An important parameter governing the response is the wave phase speed of the first baroclinic mode due to oceanic density changes between the OML and the thermocline. In a two-layer model, both barotropic and baroclinic modes are permitted. The barotropic (i.e., depth-independent) mode is referred to as the external mode whereas the first baroclinic (depth-dependent) mode is the first internal mode associated with vertical changes in the stratification. The phase speed of the first baroclinic mode (c_1) is given by

$$c_1{}^2 = g'h_1h_2/(h_1 + h_2)$$

where the depth of the upper layer is h_1, and the depth of the lower layer is h_2. In the coastal ocean, phase speeds range from 0.1 to 0.5 m s^{-1}, whereas in the deep ocean, this phase speed ranges between 1 and 3 m s^{-1} depending on the density contrast between the two layers. The barotropic mode has a phase speed $c_0 = \sqrt{gH}$ where H represents the total depth ($h_1 + h_2$), and is typically 100 times larger than the first baroclinic mode phase speed. An important nondimensional number for estimating the expected baroclinic response depends on the Froude number (ratio of the translation speed to the first baroclinic mode phase speed $U_h c_1{}^{-1}$). If the Froude number is less than unity (i.e., stationary or slowly moving storms), geostrophically balanced currents are generated by the positive wind stress curl causing an upwelling of cooler water induced by upper ocean transport directed away from the storm track (**Figure 1**). When the hurricane moves faster than the first baroclinic mode

phase speed, the ocean response is predominantly baroclinic associated with upwelling and downwelling of the isotherms and the generation of strong near-inertial motions in a spreading three-dimensional wake.

The predominance of a geophysical process also depends upon the deformation radius of the first baroclinic mode ($\alpha_1{}^{-1}$) defined as the ratio of the first mode phase speed (c_1) and f. In the coastal regime, the deformation radius is 5–10 km, but in deeper water, it increases to 20–50 km due to larger phase speeds. For observed scales exceeding the deformation radius, Earth's rotational effects, through the variations of f, dominate the oceanic dynamics where timescales are equal or greater than IP. Thus the oceanic mixed layer response to hurricanes is characterized as rotating, stratified shear flows forced by winds and waves.

Basin-to-basin Variability

Profiles from the background GOM, LC subtropical water, and the tropical EPAC are used to illustrate differences in the buoyancy frequency profile (**Figure 3**). In an OML, the vertical density gradients (N) are essentially zero because of the vertical uniformity of temperature and salinity. Maximum buoyancy frequency (N_{max}) in the Gulf is 12–14 cycles per hour (cph) located between the mixed layer depth (40 m) and the top of the seasonal thermocline compared to c. 5–6 cph in the LC water mass distributed over the upper part of the water column. In the EPAC, however, $N_{max} \sim 20$ cph due to the sharpness of the thermocline and halocline

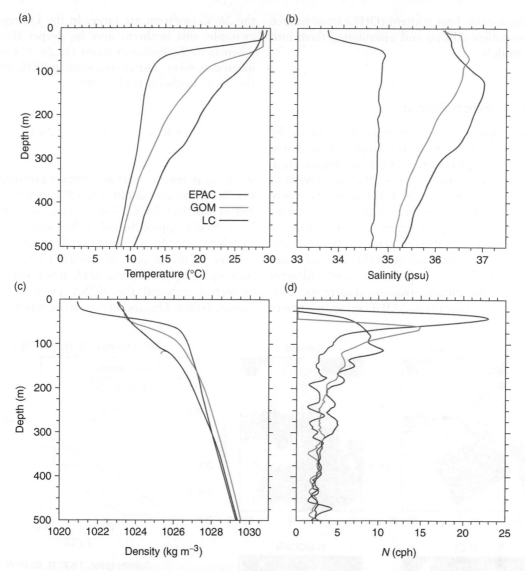

Figure 3 (a) Temperature (°C), (b) salinity (practical salinity units, psu), (c) density (kg m^{-3}), and (d) buoyancy frequency (N: cycles per hour) profiles from the eastern Pacific Ocean (red) , the GOM common water (green), and the LC water (blue) as measured from airborne expendable ocean profilers. Notice the marked difference between the gradients at the base of the OML between the three profiles.

(pycnocline) located at the base of the OML (i.e., 30 m). Beneath this maximum, buoyancy frequencies (≥ 3 cph) are concentrated in the seasonal thermocline over an approximate thermocline scale (b) of 200 m and exponentially decay with depth approaching 0.1 cph. In the LC water, N_{max} ranges from 4 to 6 cph and remains relatively constant, and below the 20 °C isotherm depth (~ 250 m), buoyancy frequency decreases exponentially.

The Richardson number increases with increases in the buoyancy frequency for a given current shear (s). This implies that a higher shear is needed in a regime like the EPAC to lower the Richardson number to below-critical values for the upper ocean to mix and cool compared to the water mass in the

GOM. Given a large N at lower latitudes (12° N) where the IP is long (~ 58 h) in the EPAC warm pool, SST cooling and OML deepening will be much less than in the GOM as observed during hurricane Juliette in September 2001 (not shown). Significant SST cooling of more than 5 °C only occurred when Juliette moved northwest where N_{max} decreased to ~ 14 cph at higher latitudes. Levels of SST cooling similar to those for the same hurricane in the GOM would be observed in the common water but not in the LC water mass since the 26 °C isotherm depth is 3–4 times deeper. These variations in the stratification represent a paradox for hurricane forecasters and are the rationale underlying the use of satellite radar altimetry in mapping isotherm depths and estimating oceanic heat content (OHC) from surface height anomalies (SHAs) and assimilating them into oceanic models.

Gulf of Mexico Basin

Warm subtropical water is transported poleward by upper-ocean currents from the tropics through the Caribbean Sea and into the GOM (see **Figure 1(c)**). This subtropical water exits the northwestern Caribbean Sea through the Yucatan Straits where the transport of $\sim 24\,\mathrm{Sv}$ ($1\,\mathrm{Sv} = 10^6\,\mathrm{m}^3\,\mathrm{s}^{-1}$) forms the LC core. Given upper ocean currents $\sim 1\,\mathrm{m\,s}^{-1}$ of the LC, horizontal density gradients between this ocean feature and surrounding GOM common water occur over smaller scales due to markedly different temperature and salinity structure (**Figure 4**). Variations in isotherm depths and OHC values relative to

the 26 °C isotherm are large. In the LC regime, for example, this isotherm may be deeper than 150 m whereas in the common water the 26 °C isotherm is located at 40 m. The corresponding OHC relative to the 26 °C isotherm is given by

$$\mathrm{OHC} = c_p \int_0^{\mathrm{D26}} \rho[T(z) - 26]\,\mathrm{d}z$$

where c_p is specific heat at constant pressure, D26 is the 26 °C isotherm depth, and OHC is zero wherever SST is less than 26 °C. Within the context of a two-layer model approach and a 'hurricane season' climatology, the 26 °C isotherm depth and its OHC relative to this depth are monitored using satellite techniques by combining SHA fields from satellite altimeters onboard the NASA Jason-1, US Navy Geosat Follow On, and European Research Satellite-2

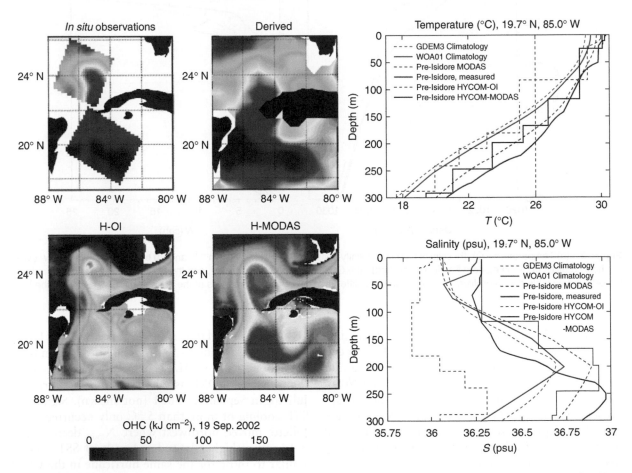

Figure 4 OHC (kJ cm^{-2}) in the northwest Caribbean Sea and southeast GOM from an objective analysis of *in situ* aircraft measurements, satellite altimetry, HYCOM NRL-CH nowcast, and HYCOM NRL-MODAS nowcast (four left panels). Temperature (right top) and salinity (right bottom) vertical profiles at a location in the northwest Caribbean Sea, where red lines are climatological profiles (GDEM3 dashed, WOA01 solid), solid blue lines are observed profiles, dashed blue lines are MODAS profiles, and black lines are model nowcasts (HYCOM-NRL dashed and HYCOM-MODAS solid). Adapted from Halliwell GR, Jr., Shay LK, Jacob SD, Smedstad OM, and Uhlhorn EW (in press) Improving ocean model initialization for coupled tropical cyclone forecast models using GODAE nowcasts. *Monthly Weather Review.*

(and Envisat) missions with observed SSTs. In the 1970s, Leipper coined the phrase 'hurricane heat potential', which represents integrated thermal structure relative to 26 °C water. In the LC regime, OHC values relative to this isotherm depth often exceed $100\,kJ\,cm^{-2}$. For oceanic response studies, the key science issue is that such deep isotherms (and OHC levels) tend to be resistive to significant storm-induced cooling by forced near-inertial current shears across the base of deep OML.

Loop Current Cycle

The LC is highly variable and when it penetrates beyond latitudes of 25° N, WCR shedding events occur at periods of 6–11 months when CCRs are located on their periphery prior to separation. By contrast, south of this latitude, WCR shedding periods increase to more than 17 months based on a series of metrics developed by Leben and colleagues. These WCRs, with diameters of approximately 200 km, then propagate west to southwest at average phase speeds of $\sim 5\,km\,day^{-1}$, and remain in the GOM for several months. At any given time, two or three WCRs may be embedded within the complex GOM circulation patterns.

Theoretical developments suggest that the LC cycle can be explained in terms of the momentum imbalance paradox theory. This theory predicts that when a northward-propagating anomalous density current (i.e., Yucatan Current) flows into an open basin (GOM) with a coast on its right (Cuba), the outflow balloons near its source forming a clockwise-rotating bulge (e.g., LC) since the outflow cannot balance the along-shelf momentum flux after turning eastward. The ballooning of the current satisfies the momentum flux balance along the northern Cuban coast. The subsequent WCR separation from the LC is due to the planetary vorticity gradients where most of the inflow forces a downstream current and the remaining inflow forms a warm ring. Subtropical water emerging from the Caribbean Sea may enter the LC bulge prior to shedding events and impact the OHC distribution, and if in phase with the height of hurricane season may spell disaster for residents along the GOM.

Model Initialization

Ocean models that assimilate data are an effective method for providing initial and boundary conditions in the oceanic component of coupled prediction models. The thermal energy available to intensify and maintain a hurricane depends on both the temperature and thickness of the upper ocean

warm layer. The ocean model must be initialized so that features associated with relatively large or small OHC are in the correct locations and $T\text{–}S$ (and density) profiles, along with the OHC, are realistic. Ocean forecast systems based on the hybrid coordinate ocean model (HYCOM) have been evaluated in the northwest Caribbean Sea and GOM for September 2002 prior to hurricanes Isidore and Lili, and in September 2004 prior to Ivan. An examination of the initial analysis prior to Isidore is from an experimental forecast system in the Atlantic basin (**Figure 4**). This model assimilates altimeter-derived SHAs and SSTs. Comparison of OHC maps by the model and observations demonstrate that the analysis (labeled NRL-CH) reproduces the LC orientation but underestimates values of the heat content. In the Caribbean Sea, the thermal structure $(T(z))$ hindcast tends to follow the September ocean climatology but does not reproduce the larger observed OHC values. The model ocean is less saline than both climatology and profiler measurements above 250 m and less saline than those between 250 and 500 m. Evaluations of model products are needed prior to coupling to a hurricane model to insure that ocean features are in the correct locations with realistic structure.

Mixing Parametrizations

One of the significant effects on the upper ocean heat budget and the heat flux to the atmosphere is the choice of entrainment mixing parametrizations at the OML base (see **Figure 1**). Sensitivity tests have been conducted using five schemes: K-profile parametrization (KPP); Goddard Institute for Space Studies level-2 closure (GISS); Mellor–Yamada level-2.5 turbulence closure scheme (MY); quasi-slab dynamical instability model (Price–Weller–Pinkel dynamical instability model, PWP); and a turbulent balance model (Kraus–Turner turbulence balance model, KT). Simulated OML temperatures for realistic initial conditions suggest similar response except that the magnitude of the cooling differs as well as its lateral extent of the cooling patterns (**Figure 5**). Three higher-order turbulent mixing schemes (KPP, MY, and GISS) seem to be in agreement with observed SST cooling patterns with a maximum of 4 °C whereas PWP (KT) over- (under-) estimate SST cooling levels after hurricane Gilbert. This case is an example of 'negative feedback' to the atmosphere given these cooling levels due primarily to shear instability at the OML base. Similar to the post-season hurricane forecast verifications, more oceanic temperature, current, and salinity measurements must be acquired to evaluate these schemes to build a larger

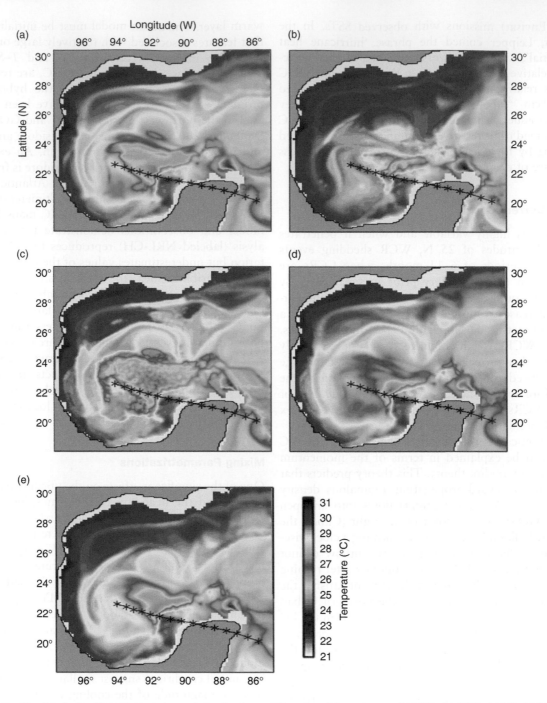

Figure 5   Simulated mixed layer temperatures during hurricane Gilbert for mixing schemes (a) KPP, (b) PWP, (c) KT, (d) MY, and (e) GISS. Differences between these five cases are visible with PWP being the coolest and KT being the warmest. Black line indicates track of the storm at 06 GMT 16 Sep. 1988.

statistical base for the oceanic response to high-wind conditions in establishing error bars for the models.

Oceanic Response

Recently observed interactions of severe hurricanes (category 3 or above) with warm ocean features such as the LC and WCR (Lili in October 2002, Katrina and Rita in August and September 2005) are contrasted with hurricanes that interact with CCR (Ivan in September 04) and cold wakes (Gilbert in September 1988) in the GOM. The levels of observed upper cooling and OML depth patterns are predicated on the amount of shear-induced mixing in the upper ocean (see **Figure 1**). The SST response is determined from an optimal interpolation scheme using the NASA TRMM microwave imager (TMI) and advanced microwave sensing radiometer (AMSR- E) where the diurnal cycle was removed from the data.

LC Interactions

Hurricanes Isidore (21 September 2002) and Lili (02 October 2002) interacted with the LC in nearly the same area spaced about 10 days apart. For negative feedback regimes, one would anticipate that after the first hurricane, there would have been a significant ocean response with little thermal energy available for the second storm as Isidore moved slowly from Cuba to the Yucatan Peninsula. The cyclonic (counterclockwise) rotating surface wind stress (in the Northern Hemisphere) should have upwelled isotherms due to divergent wind-driven transport that may have been balanced by horizontal advection due to strong northward currents through the Yucatan Straits. While observed cooling levels in the straits were less than 1 °C, the upper ocean cooled by

4.5 °C over the Yucatan Shelf. Since upwelling induced by the persistent trade wind regime maintains a seasonal thermocline close to the surface over this shelf, impulsive wind events force upwelling of colder thermocline water quickly due to transport away from the coast. Isidore remained over the Yucatan Peninsula and weakened to a tropical storm that then moved northward creating a cool wake of ~28.5 °C SSTs across the central GOM.

Lili reached hurricane status on 26 September while passing over the Caribbean Sea along a similar northwest trajectory as Isidore, making a first landfall along the Cuban coast (**Figure 6**). As Lili moved into the GOM basin, the storm intensified to a category 4 storm along the LC boundary just as R_{max} decreased to form a new eyewall (where winds are a maximum). In the common water, the SST cooling was more than 2 °C due to shear-induced mixing compared to less than 1 °C SST cooling in the LC (**Figure 6(c)**). This suggests that 'less negative feedback' (minimal ocean cooling) to hurricane Lili occurred over the LC than over the common water. Afterward, Lili began a weakening cycle to category

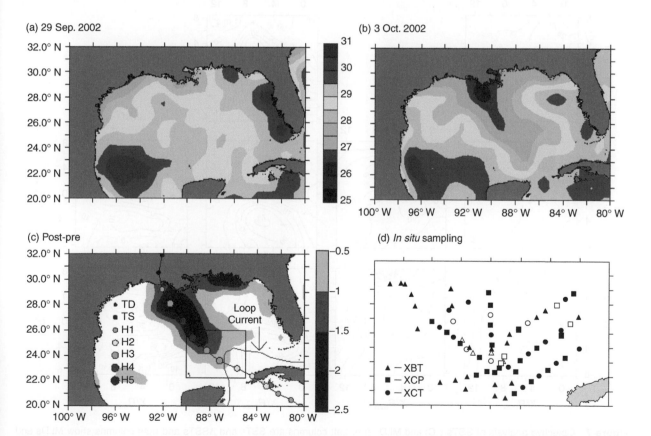

Figure 6 (a) Pre-Lili, (b) post-Lili, (c) pre–post-Lili SST (°C) field from AVHRR data, courtesy of RSMAS Remote Sensing Laboratory, and (d) measurement grid conducted by NOAA research aircraft on 2 Oct. 2002 (open symbols represent nonfunction probes). Panel (c) is relative to the track and intensity of Lili and the position of the LC. Notice the cold wake in the GOM common water compared to essentially no cold wake in the LC. More details of the response in the LC is given in **Figure 9**. Black box represents the region where *in situ* measurements from aircraft expendable were acquired during Lili's passage.

1 status due to enhanced atmospheric shear, dry-air intrusion along the western edge, and interacting with the shelf water cooled by Isidore. As shown in **Figure 6(d)**, oceanic and atmospheric profilers were deployed in the south-central part of the GOM from research aircraft. The design strategy was to measure upper-ocean response to a propagating and mature hurricane over the LC. Multiple research flights deployed profilers in the same location before, during, and after passage, which captured not only the LC response to Lili but also to Isidore as the hurricane intensified to category 3 status moving across the Yucatan Straits 10 days early. The minimal LC response highlights the importance of this current system for intensity changes.

These profiler data were objectively analyzed over a $3° × 3°$ domain in latitude and longitude with a vertical penetration to 750-m depth and aligned with the hurricane path (**Figure 7**). A day after Isidore, SSTs that remained were above 28 °C, which is suggestive of 'less negative feedback' to the storm as it crossed over the Yucatan Straits. Given the 10-day time interval between Isidore and Lili, pre-Lili SSTs warmed to over 29 °C in the experimental domain. After Lili's passage, SSTs decreased to 28.5 °C in the LC; however, along the northern extremity of the measurement domain, SSTs cooled to 27 °C in the Gulf of Mexico common water (GCW), which equates to more than 2 °C cooling. The GCW mixes quickly due to current shears across the OML base forcing the layer to deepen. In the LC itself, there was little evidence of cooling and layer deepening. Given the advective timescale (LV^{-1} where L is cross-stream scale and V is the maximum current of the LC) of about a day, heat transport from the Caribbean Sea occurs rapidly and will offset temperature decreases induced by upwelling of the isotherms and mixing as in the hurricane Isidore case. The observed current shears during the hurricane were $1.5 × 10^{-2} s^{-1}$ or about a factor of 2–3 less

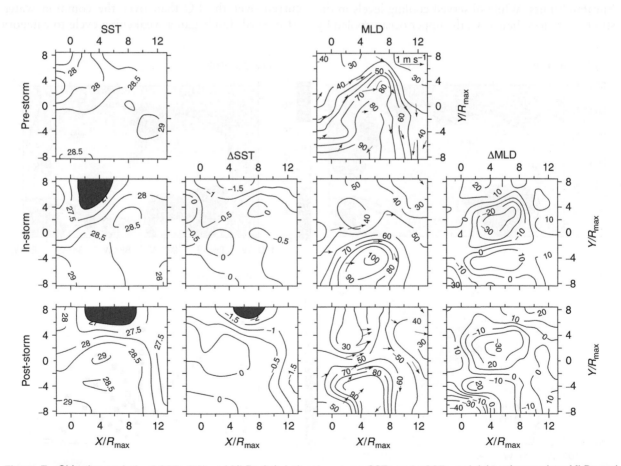

Figure 7 Objective analysis of SSTs (°C) and MLDs (m). Left columns are SSTs and ΔSSTs and right columns show MLDs and ΔMLDs for pre-storm, storm, and post-storm (Wake 1) measurements from Lili in the southeastern GOM as per **Figure 6(c)**. Panels are in storm-coordinate system for cross-track (X/R_{max}) and along-track (Y/R_{max}) based on R_{max} and the storm track orientation at 292 °T North as in **Figure 6(c)** centered at 23.2° N and 86.1° W. The ΔSST (°C) and ΔMLD (m) are estimated by subtracting the pre-storm data from the storm and post-storm data and the arrows represent current measurements from airborne expendable current profilers. Blue shaded areas are more than 2 °C consistent with satellite-derived SSTs in **Figure 6(c)**.

than observed previously in quiescent regimes due to the strength of this background upper ocean flow of the LC. This lack of shear-induced mixing has implications for hurricane intensity as they move over the deep, warm pools of the LC, which represent a reservoir of thermal energy for hurricanes to tap.

Cold and Warm Core Ring Interactions

Hurricane Ivan (Sep. 04) entered the GOM as a category 5 storm and then weakened to a category 4 storm due to a combination of lower OHC, vertical shear in the atmosphere associated with an upper-level trough, and drier air being drawn into its circulation. During its GOM trajectory, Ivan encountered two CCRs and a WCR where the surface pressure decreased by ~10 mb during a brief encounter. Shelf water, cooled by hurricane Frances (10 days earlier) along the northern GOM along with increasing atmospheric shear, acted to oppose intensification during an eyewall replacement cycle (defined as the formation of a secondary eyewall that replaces a collapsing inner eyewall).

As shown in **Figure 8**, pre- and post-SSTs to Ivan reveal the location of both WCR and CCR located

along the track of hurricane Ivan and the cold wake due to enhanced current shear instability. The SST difference field, shown in **Figure 8(c)**, indicates that both the WCR and CCR SSTs are eroded away by the strong forcing. The SSTs over the CCRs indicate cooling levels exceeding 4 °C along and to the right of Ivan's track that were embedded within the cool wake of about 3.5 °C of Ivan. The northern CCR may have been partially responsible for the observed weakening of Ivan as suggested by Walker and colleagues. Notice that just as in the case of Lili, SST cooling of less than 1 °C was observed in the LC in the southern part of the basin.

Prior to landfall, Ivan moved over 14 acoustic Doppler current profiler (ADCP) moorings that were deployed as part of the Slope to Shelf Energetics and Exchange Dynamics (SEED) project (**Figure 8(d)**), as discussed by Teague and colleagues. These profiler measurements provided the evolution of the current (and shear) structure from the deep ocean across the shelf break and over the continental shelf. The current shear response, estimated over 4-m vertical scales, is shown in **Figure 9** based on objectively analyzed data from these moorings. Over the shelf, the current shears increased due to hurricane Ivan strong

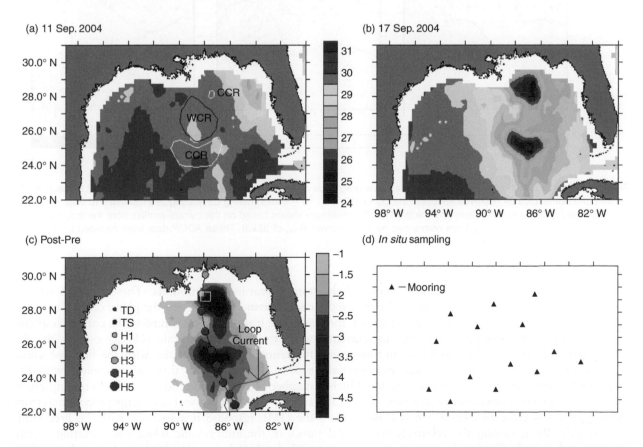

Figure 8 Same as **Figure 6** except for hurricane Ivan in Sep. 2004 and panel (d) represents ADCP mooring locations during the SEED experiment in the northern GOM in the white box in panel (c).

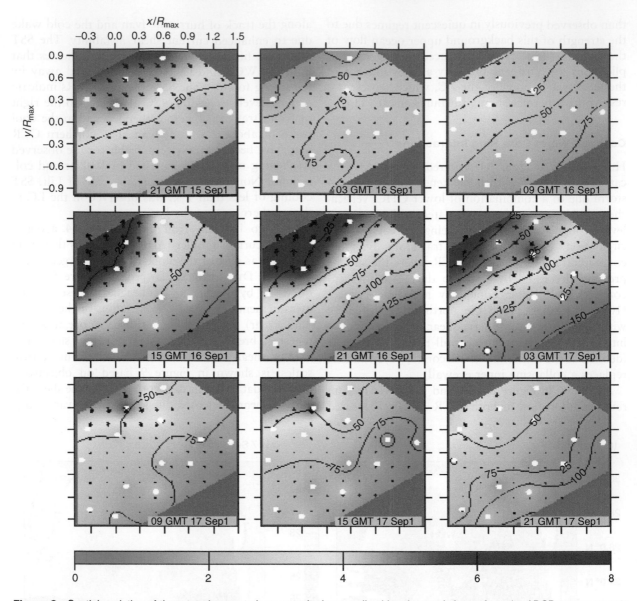

Figure 9 Spatial evolution of the rotated current shear magnitude normalized by observed shears from the ADCP measurements (white dots) normalized by observed shears in the LC of $1.5 \times 10^{-2} \, \text{s}^{-1}$ (color) during Lili starting at 2100 GMT 15 Sep. every 6 h. Black contours (25-m intervals) represent the depth of the maximum shears based on the current profiles from the moored ADCP. Cross-track (x) and along-track (y) are normalized by the observed R_{max} of 32 km. These ADCP data were provided by the Naval Research Laboratory through their SEED project.

winds. The normalized shear magnitude over the shelf (depths of 100 m) is larger by a factor of 4 compared to normalized values over the deeper part of the mooring array (500–1000 m). Notice that the current shear rotates anticyclonically (clockwise) in time over 6-h intervals associated with the forced near-inertial response (periods slightly shorter than the local inertial period). In this measurement domain, the local inertial period is close to 24 h which is close to the diurnal tide. By removing the relatively weak tidal currents and digitally filtering the records, the analysis revealed that the predominant response was due to

forced near-inertial motions. These motions have a characteristic timescale for the phase of each mode to separate from the wind-forced OML current response when the wind stress scale ($2R_{max}$) exceeds the deformation radius associated with the first baroclinic mode (~40 km). This timescale increases with the number of baroclinic modes due to decreasing phase speeds. The resultant vertical energy propagation from the OML response is associated with the predominance of the anticyclonic (clockwise) rotating energy with depth and time that is about 4 times larger than the cyclonic (counterclockwise) rotating component.

WCR Interactions

In 2005, hurricane Katrina deepened to a category 5 storm over the LC's western flank with an estimated wind stress of $\sim 7\,\mathrm{N\,m^{-2}}$. The variations of Katrina's intensity correspond well with the large OHC values in the LC and the lobe-like structure (eventually a WCR) in the northern GOM. Since SSTs exceeding 30 °C were nearly uniformly distributed in this regime, the LC structure was not apparent in the SST signals. This deeper heat reservoir of the LC provided more heat for the hurricane where satellite-inferred OHC values exceeded $120\,\mathrm{kJ\,cm^{-2}}$ or more than 5 times the threshold suggested by early studies to sustain a hurricane. Within the next 2 weeks, Rita formed and moved through the Florida Straits into the GOM basin (**Figure 10(c)**). While Rita's path did not exactly follow Katrina's trajectory in the south-central Gulf, Rita moved toward the north-northwest over

the LC and rapidly intensified to similar intensity as Katrina. After Rita interacted with the eastern tip of the WCR, the hurricane began a weakening cycle due to the cooler water associated with a CCR located on the periphery of the WCR similar to Ivan and cooler water on the shelf.

Pre- and post-SST analyses include an interval a few days prior and subsequent to hurricane passage to quantify cooling levels in the oceanic response (**Figure 11**). Prior to Katrina, SSTs exceeded 31 °C in the GOM without any clear evidence of the LC. Subsequent to Katrina, maximum cooling occurred on the right side of the track with SST decreasing to about 28 °C over the outer West Florida Shelf where OML typically lies close to the surface. Observed SSTs decreased by more than 4 °C along the LC's periphery, mainly due to shear-induced mixing and upwelling over the shelf. As Katrina moved over the LC, the SST response was less than 2 °C as expected

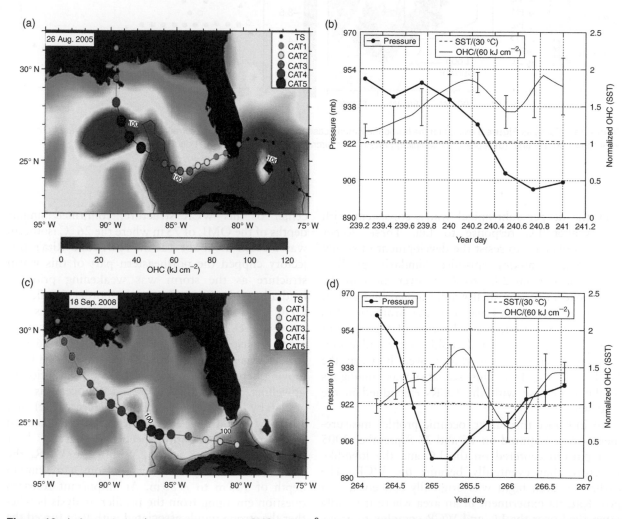

Figure 10 Left panels (a, c) represent pre-storm OHC ($\mathrm{kJ\,cm^{-2}}$: color) and 26 °C isotherm depth (m: black contour) based on a hurricane season climatology, SSTs, Jason-1, and GEOSAT Follow-on (GFO) radar altimetry measurements relative to the track and intensity of hurricane Katrina (a, b) and Rita (c, d).

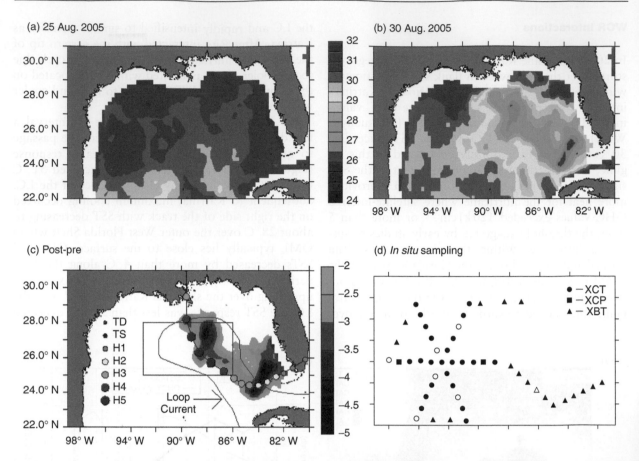

Figure 11 Same as **Figure 6** except for the hurricane Katrina case where SST were from optimally interpolated TMI data from http://www.remss.com. Panel (d) represents sampling pattern from aircraft centered on the WCR on 15 September 2005.

over the deeper subtropical water, consistent with the weaker LC response to Lili. These deeper warm pools tend to resist the development of strong shear-induced mixing episodes. Similarly, pre-Rita SSTs ranged from 28.5 to 29 °C over most of the GOM except for the shelf waters cooled by Katrina. However, after Rita's passage, the dramatic SSTs cooling of 3–4 °C occurred because of the combination of upwelling and cold water advection associated with a CCR that moved between the WCR and the LC. This scenario was analogous to the Ivan case with the CCRs embedded in the cold wake.

To illustrate this effect, oceanic profiler measurements were acquired on 15 and 26 September 2005 in a pattern centered on the LC and the lobe-like structure that eventually became the WCR. The earlier research flight was originally conceived as a post-Katrina experiment in an area where it rapidly intensified over the LC and WCR complex to assess altimeter-derived estimates of isotherm depths and OHC variations. Pairs of profilers, deployed in the

center of this WCR structure, confirmed similar depths of the OML of 75 m where the 26 °C isotherm was located at about 120 m. Hurricane Rita's trajectory clipped the northeastern part of this warm structure as the storm was weakening prior to landfall on the Texas–Louisiana border. While the OHC levels remained relatively the same in this area between pre- and post-Rita (**Figure 12**), the dramatic cooling between the LC and shed WCR on 26 September was primarily due to the advection of a CCR moving between these ocean features. In addition to upwelling, vertical mixing cooled the ocean as suggested by the vertical sections (**Figures 12(c)** and **12(d)**). Over this period, the WCR propagated westward at a translation speed of 12 km day^{-1}, or nearly double their speeds. Within the WCR, the 26 °C isotherm depth decreased from a maximum depth of 115 m to $\sim$88 m. An important research question emerging from the profiler analysis is whether the strong winds associated with Rita forced the WCR to separate prematurely and propagate faster toward the west.

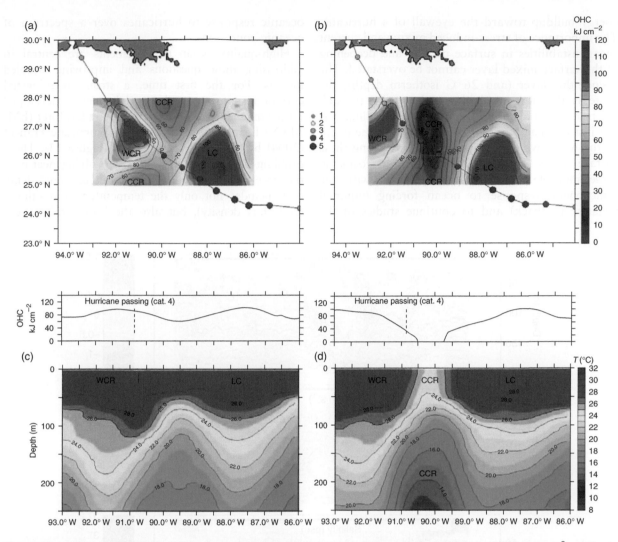

Figure 12 (a) Pre-Rita and (b) post-Rita analysis of observed (color) and satellite-inferred (contours) of OHC (kJ cm^{-2}) relative to Rita's intensity (colored circles) and track and corresponding OHC (kJ cm^{-2}: top panels) and vertical thermal structure sections (°C) along 26.5° N transect from (c) pre-Rita and (d) post-Rita.

Summary

Progress has been made in understanding the basic oceanic and atmospheric processes that occur during hurricane passage. There is a continuing need to isolate fundamental physical processes involved in these interactions through focused experimental, empirical, theoretical, and numerical approaches. The GOM is one such basin where detailed process studies can focus on the oceanic response to the hurricane forcing as well as the atmospheric response to ocean forcing. Observational evidence is mounting that the warm and cold core features and the LC system are important to the coupled response during hurricane passage. This is not unique to the GOM as this behavior has also been recently observed in other regions such as the western Pacific Ocean and the Bay of Bengal. Thus, it is a global problem that needs to be addressed.

This coupled variability occurs over the storm scales that include fundamental length scales such as the radius of maximum winds and radius to gale-force winds. The fundamental science questions are that how the ocean and atmosphere are coupled, and that what are the appropriate timescales of this interaction? These questions are not easily answered, given especially the lack of coupled measurements spanning the spectrum of hurricane parameters such as strength, radius, and speed. One school of thought is that the only important process with respect to the ocean is under the eyewall where ocean cooling occurs. However, observed cooling under the eyewall is not just due to the surface flux alone (see **Figure 1**). In this regime, the maximum winds and heat and moisture fluxes occur; however, the broad surface circulation over the ocean also has nonzero fluxes that contribute the thermal energy buildup toward the eyewall of a hurricane. The importance of stress-induced mixing and current shear instabilities in surface cooling and deepening of the surface mixed layer cannot be overstated. The deeper this layer (and 26 °C isotherm depth), the more is the heat available to the storm through the heat and moisture fluxes. Notwithstanding, it is not just the magnitude of the OHC, since the depth of the warm water is important to sustaining these surface fluxes. Future research needs to focus on these multiple scale aspects associated with the atmospheric response to ocean forcing (minimal negative feedback) and to continue studies of the oceanic response to hurricanes over a spectrum of oceanic conditions.

High-quality ocean measurements are central to addressing these questions and improving coupled models. For the first time, a strong near-inertial current response was observed by newly developed Electromagnetic Autonomous Profiling Explorer (EM-APEX) floats deployed in front of hurricane Frances (2004) by Sanford and colleagues (**Figure 13**). These profiling floats have provided the evolving near-inertial, internal wave radiation in unprecedented detail that includes not only the temperature and salinity (and thus density), but also the horizontal current

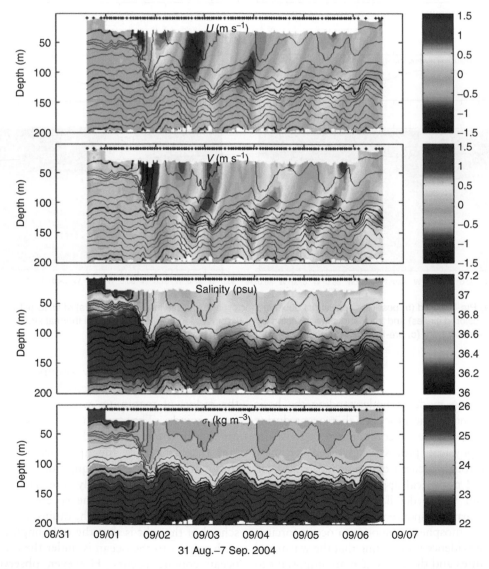

Figure 13 Current (U, V in m s^{-1}), salinity (psu), and density or σ_t (kg m^{-3}) response at R_{max} during the passage of hurricane Frances (2004) as measured by an EM-APEX float deployed from USAF WC-130 1 day ahead of the storm. Three floats were successfully deployed in the projected cross-track direction as part of the ONR Coupled Boundary Layer Air–Sea Transfer program. Reprinted from Sanford TB, Dunlap JH, Carlson JA, Webb DC, and Girton JB (2005) Autonomous velocity and density profiler: EM-APEX. In: *Proceedings of the IEEE/OES 8th Working Conference on Current Measurement Technology*, IEEE Cat No. 05CH37650, pp. 152–156 (ISBN: 0-7803-8989-1), @ 2005 IEEE.

structure. Notice that the phase propagation of the forced near-inertial currents is upward associated with downward energy propagation from the OML as current vectors rotate anticyclonically (clockwise) with time and depth (in the Northern Hemisphere). Velocity shears associated with these near-inertial currents force mixing events as manifested in a large fraction of the observed SST cooling of more than 2 °C (and layer deepening). Given these measurements of the basic state variables, the evolution of the Richardson numbers forced by a hurricane can be determined to evaluate mixing parametrization schemes used in coupled models for forecasting at the national centers.

The variability of the surface drag coefficient has received considerable attention over the last 5 years. Several treatments have concluded that there is a leveling off or a saturation value at $\sim 30 \pm 3 \mathrm{\,m\,s}^{-1}$. The ratio of the enthalpy (heat and moisture) coefficient and the drag coefficient is central to air–sea fluxes impacting the hurricane boundary layer. In this context, the relationship between the coupled processes such as wave breaking and the generation of sea spray and how this is linked to localized air–sea

fluxes remains a fertile research area. A key element of this topic is the atmospheric response to the oceanic forcing where there seem to be contrasting viewpoints. One argument is that the air–sea interactions are occurring over surface wave (wind-wave) time and space scales and cause significant intensity changes by more than a category due to very large surface drag coefficients. While, these sub-mesoscale phenomena may affect air–sea fluxes, the first-order balances are primarily between the atmospheric and oceanic mixed layers. The forced surface waves modulate the heat and momentum fluxes.

Future Research

A promising avenue of research has focused on the upper ocean's role on intensity change. Climatologically, for the western Atlantic basin, the expected number of category 5 storms is one approximately every 3 years. Over the last 4 years, there have been a total of six category 5 storms, well above this mean. Based on extensive deliberations by the international tropical cyclone community, intensity and structure changes are primarily due to environmental

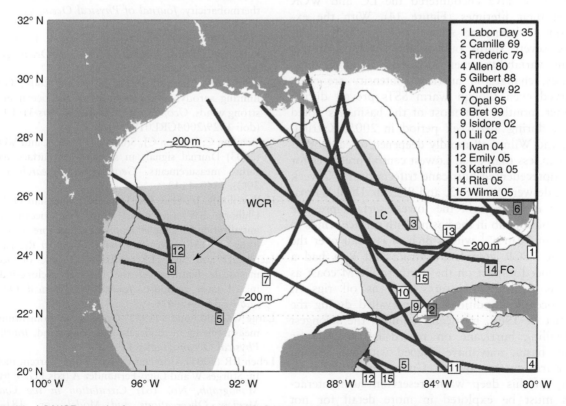

Figure 14 LC/WCR complex based on satellite-derived 26 °C isotherm depth (gray area) and generalized westward propagation of the WCR in the GOM (darker gray) based on 2005 altimeter data relative to the storm tracks (red: severe storms) over several decades (legend) based on best track data from the NHC and 200-m contour (black). FC represents the Florida current that flows through the Straits of Florida. TC best tracks were provided by the National Hurricane Center through their website http://www.nhc.noaa.gov/pastall.shtml.

conditions such as atmospheric circulation, internal dynamics, and oceanic circulation processes. Gyre-scale ocean circulation redistributes ocean heat throughout the basins primarily through poleward advection and transport along its western boundary. While there is an open scientific question whether the increased frequency of occurrence of severe hurricanes is due to global warming or natural cycle associated with geophysical processes, the severe hurricanes during the 2005 season interacted with the warm Caribbean Current and the LC. As shown here and in recently published papers, the oceanic response over these regimes differs considerably from that observed quiescent regimes. The key issue is the level of observed ocean cooling in these regimes that is considerably less (i.e., 'less negative feedback') than compared to other areas where the cooling is more dramatic. Since winds begin to mix the thin 'skin' layer of SST well in front of the storm, the surface temperature reflects the temperature of the oceanic mixed layer under high winds. This point is often overlooked in atmospheric models where SSTs are prescribed or weakly coupled to an ocean where the basic state is at rest.

As discussed above, intense hurricanes in the GOM may have encountered the LC and WCR during their lifetimes (**Figure 14**). With the exception of hurricane Allen (1980), which maintained severe status outside the envelope of this oceanic variability, when hurricanes encounter these features, changes in hurricane intensity are often observed even though warm SSTs prevail during summer months over most of the basin. As noted above, during a 7-week period in 2005, Katrina, Rita, and Wilma all rapidly deepened to catagory 5 status in less than 24 h. Lowest central pressures for this unprecedented hurricane trifecta over a 7-week timescale were 896, 892, and 882 mb. Until Wilma, Gilbert in 1988 held the lowest surface pressure record of 888 mb in the basin. With surface winds in excess of $70 \, \text{m s}^{-1}$ within 36 h of landfall over the LC and WCR complex, Katrina and Rita had a pronounced impact on the northern GOM coast as well as offshore structures such as oil rigs. If these oceanic conditions had prevailed during the summer of 1969, hurricane Camille, the strongest land-falling hurricane on record in the Atlantic Ocean basin, may have aligned with the axis of this warm current system. Given the natural variability of this deep warm reservoir, such interactions must be explored in more detail for not only the oceanic response, but also the potential feedbacks to the hurricanes where ocean cooling is minimized with respect to the next-generation forecast models.

Acknowledgments

L.K. Shay gratefully acknowledges NSF support and the support of the NOAA Aircraft Operations Center. Mr. Bill Teague provided Ivan current data; and Drs. Mark Donelan, Brian Haus, George Halliwell, S. Daniel Jacob, and Tom Sanford shared material. SSTs were provided by Remote Sensing Systems website (http://www.remss.com), courtesy of Dr. Chelle Gentemann. Benjamin Jaimes, Eric Uhlhorn, and Jodi Brewster also contributed to this article.

See also

Breaking Waves and Near-Surface Turbulence. Upper Ocean Mixing Processes. Upper Ocean Time and Space Variability. Upper Ocean Vertical Structure.

Further Reading

Chassignet EP, Smith L, Halliwell GR, and Bleck R (2003) North Atlantic simulations with the hybrid coordinate ocean model (HYCOM): Impact of the vertical coordinate choice and resolution, reference density, and thermobaricity. *Journal of Physical Oceanography* 33: 2504–2526.

D'Asaro EA (2003) The ocean boundary layer under hurricane Dennis. *Journal of Physical Oceanography* 33: 561–579.

Donelan MA, Haus BK, Reul N, et al. (2004) On the limiting aerodynamic roughness of the ocean in very strong winds. *Geophysical Research Letters* 31: L18306 (doi: 1029/2004GRL019460).

Gentemann C, Donlon CJ, Stuart-Menteth A, and Wentz F (2003) Diurnal signals in satellite sea surface temperature measurements. *Geophysical Research Letters* 30(3): 1140–1143.

Halliwell GR, Jr., Shay LK, Jacob SD, Smedstad OM, and Uhlhorn EW (in press) Improving ocean model initialization for coupled tropical cyclone forecast models using GODAE nowcasts. *Monthly Weather Review.*

Jacob SD and Shay LK (2003) The role of oceanic mesoscale features on the tropical cyclone-induced mixed layer response. *Journal of Physical Oceanography* 33: 649–676.

Large WG and Pond S (1981) Open ocean momentum flux measurements in moderate to strong wind. *Journal of Physical Oceanography* 11: 324–336.

Leben RR (2005) Altimeter derived Loop Current metrics. In: Sturges W and Lugo-Fernandez A (eds.) *Geophysical Monograph, No. 161: Circulation in the Gulf of Mexico: Observations and Models*, pp. 181–201. Washington, DC: American Geophysical Union.

Lugo-Fernandez A (2007) Is the Loop Current a chaotic oscillator? *Journal of Physical Oceanography* 37: 1455–1469.

Nof D (2005) The momentum imbalance paradox revisited. *Journal of Physical Oceanography* 35: 1928–1939.

Ocampo-Torres FJ, Donelan MA, Merzi N, and Jai F (1994) Laboratory measurements of mass transfer of carbon dioxide and water vapour for smooth and rough flow conditions. *Tellus* 46B: 16–32.

Powell MD, Vickery PJ, and Reinhold TA (2003) Reduced drag coefficient for high wind speeds in tropical cyclones. *Nature* 422: 279–283.

Sanford TB, Dunlap JH, Carlson JA, Webb DC, and Girton JB (2005) Autonomous velocity and density profiler: EM-APEX. In: *Proceedings of the IEEE/OES 8th Working Conference on Current Measurement Technology*, IEEE Cat No. 05CH37650, pp. 152–156 (ISBN: 0-7803-8989-1).

Shay LK (2001) Upper ocean structure: Response to strong forcing events. In: Weller RA, Thorpe SA, and Steele J (eds.) *Encyclopedia of Ocean Sciences*, pp. 3100–3114. London: Academic Press.

Shay LK and Uhlhorn EW (2008) Loop Current response to hurricanes Isidore and Lili. *Monthly Weather Review* 136 (doi: 10.1175/2008MWR2169).

Sturges W and Leben R (2000) Frequency of ring separations from the Loop Current in the Gulf of Mexico: A revised estimate. *Journal of Physical Oceanography* 30: 1814–1819.

Teague WJ, Jarosz E, Carnes MR, Mitchell DA, and Hogan PJ (2006) Low frequency current variability observed at the shelf break in the northern Gulf of Mexico: May–October 2004. *Continental Shelf Research* 26: 2559–2582 (doi:10.1016/j.csr.2006.08.002).

Vukovich FM (2007) Climatology of ocean features in the Gulf of Mexico using satellite remote sensing data. *Journal of Physical Oceanography* 37: 689–707.

Walker N, Leben RR, and Balasubramanian S (2005) Hurricane forced upwelling and chlorophyll a enhancement within cold core cyclones in the Gulf of Mexico. *Geophysical Research Letter* 32: L18610 (doi: 10.1029/2005GL023716).

Relevant Website

http://www.remss.com
– Remote Sensing Systems Home Page.

UPPER OCEAN HEAT AND FRESHWATER BUDGETS

P. J. Minnett, University of Miami, Miami, FL, USA

Introduction

Most of the solar energy reaching the surface of the Earth is absorbed by the upper ocean. Some of this is released locally, often within the course of the following night, but some heat is retained for longer periods and is moved around the planet by the oceanic surface currents. Subsequent heat release to the atmosphere helps determine the patterns of weather and climate around the globe.

While maps of sea surface temperature measured from satellites are now commonplace, it is the underlying reservoir of heat stored in the upper ocean that has the impact on the atmospheric circulation and weather, not only over the oceans but also over the continents downstream. Because the specific heat of water is much greater than that for air, the thermal capacity of a layer of the ocean about 3-m thick is the same as that of the entire atmosphere above. The upper ocean heat content, however, is not so accessible to measurements by satellite-borne instruments and is therefore less well described, and its properties less well understood.

The density of seawater is determined in a nonlinear fashion by temperature and salinity and, to a much lesser degree, by pressure. Warmer, fresher seawater is less dense than cooler, saltier water. The viscosity of seawater is very low and so the fluid is very sensitive to flow generation by density differences. However, as a result of the rotation of the Earth, oceanic flow is not simply a redistribution of mass so that the surfaces of constant density coincide with surfaces of constant gravitational force; deviations are supported by balancing the horizontal pressure forces, caused by the variable distribution of density, with the Coriolis force (geostrophy). Vertical exchanges between the upper ocean and the deeper layers are inhibited by layers of density gradients, called pycnoclines, some of which are permanent features of the ocean, and others, generally close to the surface, are transient, existing for a day or less. Upper ocean salinity, through its contribution to controlling the ocean density, is therefore an important variable in determining the density distribution of the upper ocean and the availability of oceanic heat to drive atmospheric processes.

The range of sea surface temperatures, and, by extension, the mixed layer temperature, extends from $-1.8\,°C$, the freezing point of seawater, to above $30\,°C$ in the equatorial regions, especially in the western Pacific Ocean and eastern Indian Ocean. In particularly favorable situations, surface temperatures in excess of $35\,°C$ may be found, such as in the southern Red Sea.

The lowest upper ocean salinities are found in the vicinity of large river outflows and are close to zero. For most of the open ocean, upper ocean salinities lie in the range of 34–37. (Ocean salinity is measured as a dimensionless ratio with a multiplier of 10^{-3}. A salinity of 35 means that 1 kg of seawater contains 35 g of dissolved salts.) Unlike elevated surface temperatures that result in a lowering of the surface density and a stable near-surface water column, increasing surface salinities by evaporation lead to increasing density and an unstable situation where the denser surface waters sink.

Governing Processes

The upper ocean heat and salt (or freshwater) distributions are determined by the fluxes of heat and moisture through the ocean surface, the horizontal divergence of heat and salinity by advection, and by fluxes through the pycnocline at the base of the upper ocean 'mixed layer'. This can be expressed for heat content per unit area, H, by

$$\frac{\Delta H}{\Delta t} = Q_{\text{surf}} + Q_{\text{horiz}} + Q_{\text{base}}$$

where Q_{surf} represents the heat fluxes through the ocean surface, Q_{horiz} the divergence of advective heat flux in the column extending from the surface to the depth of the mixed layer, and Q_{base} is the vertical heat flux through the pycnocline at the base of the mixed layer, often presumed to be small in comparison with the surface exchanges. The surface heat flux has three components: the radiative fluxes, the turbulent fluxes, and the heat transport by precipitation. The radiative fluxes are the sum of the shortwave contribution from the sun, and the net infrared flux, which is in turn the difference between the incident atmospheric emission and the emission from the sea surface. The turbulent fluxes comprise those of sensible and latent heat. A similar expression can be used for the upper ocean freshwater budget, where the fluxes are simply those of water. The surface exchanges are the difference between the mass fluxes due to precipitation and evaporation, and the horizontal advective fluxes can be best framed in terms of the divergence of salinity.

The depth of the mixed layer is often not easy to determine and there are several approaches used in the literature, including the depth at which the temperature is cooler than the surface temperature, and values of 0.1, 0.2, or 0.5 K are commonly used. Another is based on an increase in density, and a value of $0.125 \, \mathrm{kg \, m^{-3}}$ is often used. These are both proxies for the parameter that is really desired, which is the depth to which turbulent mixing occurs, thereby connecting the atmosphere to the heat stored in the upper ocean. In situations of low wind speed and high insolation, a significant shallow pycnocline can develop through temperature stratification, and this decouples the 'mixed' layer beneath from the atmosphere above. Nevertheless, in most discussions of the surface heat and salt budget, these diurnal effects are discounted and the depth of integration is to the top of the seasonal pycnocline, or in the absence of the seasonal pycnocline, to the depth of the top of the permanent pycnocline.

Surface Heat Exchanges

The heat input at the surface is primarily through the absorption of insolation. Of course this heating occurs only during daytime and is very variable in the course of a day because of the changing solar zenith angle, and by modulation of the atmospheric transparency by clouds, aerosols, and variations in water vapor. At a given location, there is also a seasonal modulation. In the Tropics, with the sun overhead on a very clear day, the instantaneous insolation can exceed $1000 \, \mathrm{W \, m^{-2}}$. The global average of insolation is about $170 \, \mathrm{W \, m^{-2}}$. The reflectivity of the sea surface in the visible part of the spectrum is low and depends on the solar zenith angle and the surface roughness, and thereby on surface wind speed. For a calm surface with the sun high in the sky, the integrated reflectance, the surface albedo, is about 0.02, with an increase to ~ 0.06 for a solar zenith angle of 60°. Having passed through the sea surface the solar irradiance, L_λ, is absorbed along the propagation path, z, according to Beer's law:

$$\mathrm{d}L_\lambda / \mathrm{d}z = -\kappa_\lambda L_\lambda$$

where the absorption coefficient, κ_λ, is dependent on the wavelength of the light (red being absorbed more quickly than blue) and on the concentration of suspended and dissolved material in the surface layer, such as phytoplankton. When the wind is low, the near-surface density stratification that results from the absorption of heat causes the temperature increase to be confined to the near-surface layers, causing the growth of a diurnal thermocline. This is usually eroded by heat loss back to the atmosphere during the following night. If the wind speed during the day is sufficiently high, greater than a few meters per second, the subsurface turbulence spreads the heat throughout the mixed layer. There are a few locations where the insolation is high, the water is very clear, and the mixed layer depth sufficiently shallow that a small fraction of the solar radiation penetrates the entire mixed layer and is absorbed in the underlying pycnocline.

Although the absorption and emission of thermal infrared radiation are confined to the ocean surface skin layer of a millimeter or less, the net infrared budget is a component of the surface heat flux that indirectly contributes to the upper ocean heat budget. The infrared budget is the difference between the emission, given by $\varepsilon \sigma T^4$, where ε is the broadband infrared surface emissivity, σ is Stefan–Boltzmann constant, and T is the absolute temperature of the sea surface. For $T = 20 \, °\mathrm{C}$, the surface emission is $\sim 410 \, \mathrm{W \, m^{-2}}$. The incident infrared radiation is the emission from greenhouse gases (such as CO_2 and H_2O), aerosols, and clouds, and as such is very variable. For a dry, cloud-free polar atmosphere, the incident atmospheric radiation can be $< 200 \, \mathrm{W \, m^{-2}}$, whereas for a cloudy tropical atmosphere, $400 \, \mathrm{W \, m^{-2}}$ can be exceeded. The net infrared flux at the surface is generally in the range of $0–100 \, \mathrm{W \, m^{-2}}$, with an average of about $50 \, \mathrm{W \, m^{-2}}$.

The turbulent heat fluxes at the ocean surface are so called because the vertical transport is accomplished by turbulence in the lower atmosphere. They can be considered as having two components: the sensible heat flux that results from a temperature difference between the sea surface and the overlying atmospheric boundary layer, and the latent heat flux that results from evaporation at the sea surface. The sensible heat flux depends on the air–sea temperature difference and the latent heat flux on the atmospheric humidity near the sea surface. Both have a strong wind speed dependence. Since the ocean is usually warmer than the atmosphere in contact with the sea surface, and since the atmosphere is rarely saturated at the surface, both components usually lead to heat being lost by the ocean. The global average of latent heat loss is about $90 \, \mathrm{W \, m^{-2}}$ but sensible heat loss is only about $10 \, \mathrm{W \, m^{-2}}$. Extreme events, such as cold air outbreaks from the eastern coasts of continents over warm western boundary currents, can lead to much higher turbulent heat fluxes, even exceeding $1 \, kW \, m^{-2}$.

The final component of the surface heat budget is the sensible heat flux associated with precipitation. Rain is nearly always cooler than the sea surface and so precipitation causes a reduction of heat content in the upper ocean. Typical values of this heat loss are about $2-3 \, W \, m^{-2}$ in the Tropics, but in cases of intense rainfall values of up to $200 \, W \, m^{-2}$ can be attained.

Surface Freshwater Exchanges

Over most of the world's ocean, the flux of fresh water through the ocean surface is the difference between evaporation and precipitation. The loss of fresh water at the sea surface through evaporation is linked to the latent heat flux through the latent heat of evaporation. Clearly, precipitation exhibits very large spatial and temporal variability, especially in the Tropics where torrential downpours associated with individual cumulonimbus clouds can be very localized and short-lived. Estimates of annual, globally averaged rainfall over the oceans is about 1 m of fresh water per year, but there are very large regional variations with higher values in areas of heavy persistent rain, such as the Intertropical Convergence Zone (ITCZ) which migrates latitudinally with the seasons. Over much of the mid-latitude oceans, drizzle is the most frequent type of precipitation according to ship weather reports.

The global distribution of evaporation exceeds that of oceanic precipitation, with the difference being made up by the freshwater influx from rivers and melting glaciers.

Advective Fluxes

The determination of the amount of heat and fresh water moved around the upper oceans is not straightforward as the currents are not steady, exhibiting much temporal and spatial variation. The upper ocean currents are driven both by the surface wind stress, including the large-scale wind patterns such as the trade winds and westerlies, and by the large-scale density differences that give rise to the thermohaline circulation that links all oceans at all depths. The strong western boundary surface currents, such as the Gulf Stream, carry much heat poleward, but have large meanders and shed eddies into the center of the ocean basins. Indeed, the ocean appears to be filled with eddies. Thus the measurements of current speed and direction, and temperature and salinity taken at one place at one time could be quite different when repeated at a later date.

Measurements

Much of what we know about the upper ocean heat and salt distribution has been gained from analysis of measurements from ships. Large databases of shipboard measurements have been compiled to produce a 'climatological' description of upper ocean heat and salt content. In some ocean areas, such as along major shipping lanes, the sampling density of the temperature measurements is sufficient to provide descriptions of seasonal signals, and the length of measurements sufficiently long to indicate long-term climate fluctuations and trends, but these interpretations are somewhat contentious. In other ocean areas, the data are barely adequate to confidently provide an estimate of the mean state of the upper ocean.

Temperature is a much simpler measurement than salinity and so there is far more information on the distribution of upper ocean heat than of salt. Historically temperatures were measured by mercury-in-glass thermometers which recorded temperatures at individual depths. Water samples could also be taken for subsequent chemical analysis for salinity. The introduction of continuously recording thermometers, such as platinum resistance thermometers and later thermistors, resulted in measurements of temperature profiles, and the use of expendable bathythermographs (XBTs) meant that temperature profiles could be taken from moving ships or aircraft. The continuous measurement of salinity was a harder problem to solve and is now accomplished by calculating salinity from measurements of the ocean electrical conductivity. The standard instruments for the combined measurements of temperature and conductivity are referred to as CTDs (conductivity–temperature–depth) and are usually deployed on a cable from a stationary research ship, although some have been installed in towed vehicles for measurements behind a moving ship in the fashion of a yo-yo. In recent years, CTDs have been mounted in autonomous underwater vehicles (AUVs) that record profiles along inclined saw-tooth paths through the upper ocean (say to 600 m) and which periodically break surface to transmit data by satellite telemetry. Similarly, autonomous measurements from deep-water (to 2000 m) floats are transmitted via satellite when they surface. In the ARGO project, begun in 2000, over 3000 floats have been deployed throughout the global ocean. The floats remain at depth for about 10 days, drifting with the currents, and then make CTD measurements as they come to surface where their positions are fixed by the Global Positioning System (GPS), and the profile data transmitted to shore. Where time series of profiles are required at a particular location, internally recording

CTDs can be programmed to run up and down wires moored to the seafloor. These have been used effectively in the Arctic Ocean, but the instruments have to be recovered to retrieve the data as the presence of ice prevents the use of a surface float for data telemetry. Additional sensors, such as transmissometers to measure turbidity, often augment the CTD measurements. Further information is supplied by a network of moored buoys that now span the tropical Pacific and Atlantic Oceans and which support sensors at fixed depths.

The spatial distribution of upper ocean temperatures can be derived from satellite measurements of the sea surface temperatures which can now be made with global accuracies of 0.4 K or better using infrared and microwave radiometers. Such data sets now extend back a couple of decades. In 2009 and 2010, two new low-frequency microwave radiometers capable of measuring open ocean salinity are planned for launch (Aquarius is a NASA instrument, and SMOS – Soil Moisture, Ocean Salinity – is an ESA mission). To convert these satellite measurements of surface temperature and surface salinity into upper ocean heat and salt contents requires knowledge of the mixed layer depth, and while this is not directly accessible from satellite measurements, it can be inferred from measurements of ocean surface topography, derived from satellite altimetry, through the use of a simple upper ocean model. Such upper ocean heat content estimates are now being routinely derived and used in an experimental mode to assist in hurricane forecasting and research.

Several lines across ocean basins have been sampled from a fleet of research ships in the framework of the World Ocean Circulation Experiment (WOCE) which took place between 1990 and 2002. Many of these sections are currently being reoccupied in the Repeat Hydrography Program to determine changes on decadal scales.

Distributions

Heat

The quantitative specification of the upper ocean heat content remains rather uncertain, not so much because of our ability to measure the sea surface temperature (**Figure 1**), which is generally a good estimate of the mixed layer temperature, especially at night, but in determining the depth of the mixed layer. If the objective is to estimate the heat potentially available to the atmosphere, then the depth of the mixed layer based on density stratification in the pycnocline is more appropriate than the depth based on a temperature gradient in the thermocline, although this is often used because of the availability of more data. On an annual basis, this can lead to significant differences in the estimates of the depth of the oceanic layer that can supply heat to the atmosphere (**Figure 2**).

The difference between the tops of the thermocline and pycnocline results from density stratification caused by vertical salinity gradients (a halocline). In the low-latitude oceans, this is sometimes called a 'barrier layer' and can be as thick as the overlying isothermal layer, that is, halving the thickness of the upper ocean layer in contact with the atmosphere compared to that which would be estimated using temperature profiles alone. The barrier layers are probably caused by the subduction of more saline waters underneath water freshened by rainfall or river runoff.

At high latitudes, the nonlinear relationship between seawater density and temperature and salinity means that density is nearly independent of temperature. The depth of the surface layer is therefore determined by the vertical salinity profile. During ice formation, brine is released from the freezing water and this destabilizes the surface layer, causing convective mixing. During ice melt, the release of fresh water stabilizes the upper ocean. In the Arctic Ocean, the depth of the mixed layer is determined by the depth of the halocline.

The annual means of course do not reveal the details of the seasonal cycle in heat content, which in turn reflect the seasonal patterns of the surface fluxes and advective transports. **Figure 3** shows the global distributions of the surface fluxes for January and July. The patterns in the insolation (short-wave heat flux) reflect the changes in the solar zenith angle, and the seasonal changes in cloud cover and properties. The seasonal patterns of the surface winds are apparent in the turbulent fluxes.

The seasonal changes in the sea surface temperatures and upper ocean heat content are the summations of small daily residuals of local heating and cooling. Under clear skies and low winds, the absorption of insolation in the upper ocean leads to a stabilization of the surface layer through the formation of a near-surface thermocline. While the surface heat budget remains positive (i.e., the insolation exceeds heat loss through turbulent heat loss and the net infrared radiation), the diurnal thermocline grows with an attendant increase in the sea surface temperature. As the insolation decreases and the surface heat budget changes sign, the surface heat loss results in a fall in surface temperature and the destabilization of the near-surface layer. The resultant convective instability erodes the thermal stratification, returning the upper layer to a state close to

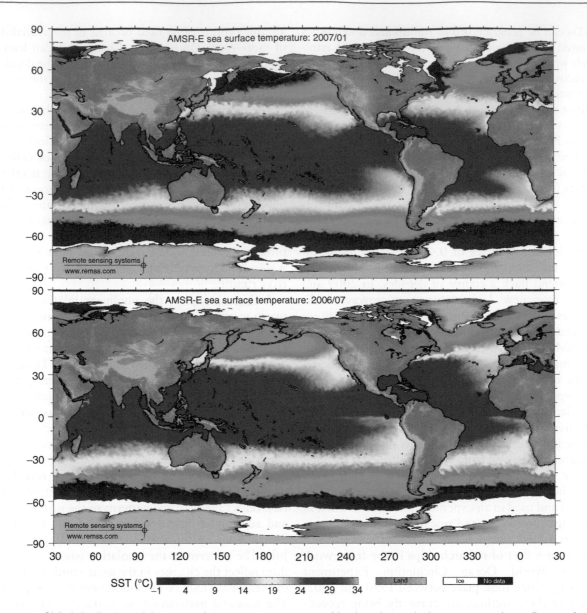

Figure 1 Global distributions of the sea surface temperature measured by the advanced microwave scanning radiometer for the Earth Observing System on the NASA satellite *Aqua*. Monthly averaged fields for Jan. 2007 and Jul. 2006 are shown. AMSR-E data are produced by Remote Sensing Systems and sponsored by the NASA Earth Science REASoN DISCOVER Project and the AMSR-E Science Team. Data are available at http://www.remss.com.

that before diurnal heating began. In the heating season, on average, there will be more heat in the upper ocean at the end of the diurnal cycle, and in the cooling season there will be less. On days when the wind speed is greater than a few meters per second, the wind-induced turbulent mixing prevents the growth of the diurnal thermocline and the heat input during the day, and removed at night, is distributed throughout the mixed layer. **Figure 4** shows measurements of the diurnal heating, expressed as a difference between the 'skin' temperature and a bulk temperature at the depth of a few meters, as a function of wind speed and time of day.

The magnitude of the surface temperature signal of diurnal warming is very strongly dependent on wind speed, and can be eroded very quickly if winds increase in the course of a day.

The upper ocean of course exhibits variability on timescales longer than a year, often with profound consequences around the globe. The best known is the El Niño–Southern Oscillation that results in a marked change in sea surface temperature, depth of the mixed layer, and consequently upper ocean heat content in the equatorial Pacific Ocean. The perturbations to the atmospheric circulation have effects on weather patterns, including rainfall, around the

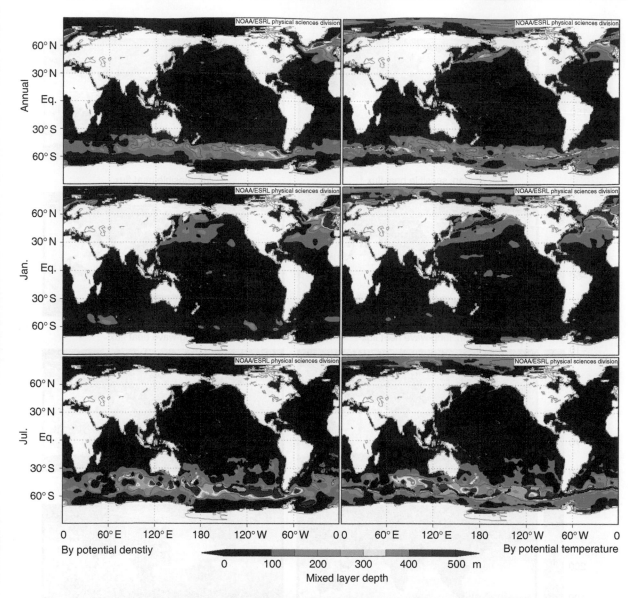

Figure 2 Maps of the mixed layer depth. Annual averages are shown along with monthly means for January and July. The left column shows mixed layer depths based on a potential density difference criterion, and the right column on a potential temperature difference. The deepest values are found at high latitudes in the winter hemisphere. A discrepancy in the estimates by a factor of 2 is seen in some regions. This translates into an equivalent uncertainty in the estimate of the heat content of the upper ocean. The figure is based on images produced at the NOAA-CIRES Climate Diagnostics Center, Boulder, Colorado, from their website at http://www.cdc.noaa.gov.

globe. Other multiyear features include the North Atlantic Oscillation, Arctic Oscillation, and Pacific Decadal Oscillation, in all of which there is a shift in both atmospheric circulation and oceanic response. In the case of the Arctic Oscillation, determined from the strength of the polar vortex relative to mid-latitude surface pressure, a negative phase results in high surface pressures in the Arctic, and a more uniform distribution of sea ice. This is considered the normal situation. The positive phase results in lower surface pressure fields over the Arctic Ocean, a

thinning of the ice cover, and intrusion of relatively warm Atlantic water into the Arctic Basin. These have consequences on the upper ocean salinity and density stratification, and on interactions with the atmosphere, although the complexities of these feedbacks are poorly understood.

The heat and salinity content advected through imaginary boundaries extending across oceans from coast to coast and from the surface to depth are a measure of the transport of heat and salt. For example, to maintain the Earth's radiative equilibrium

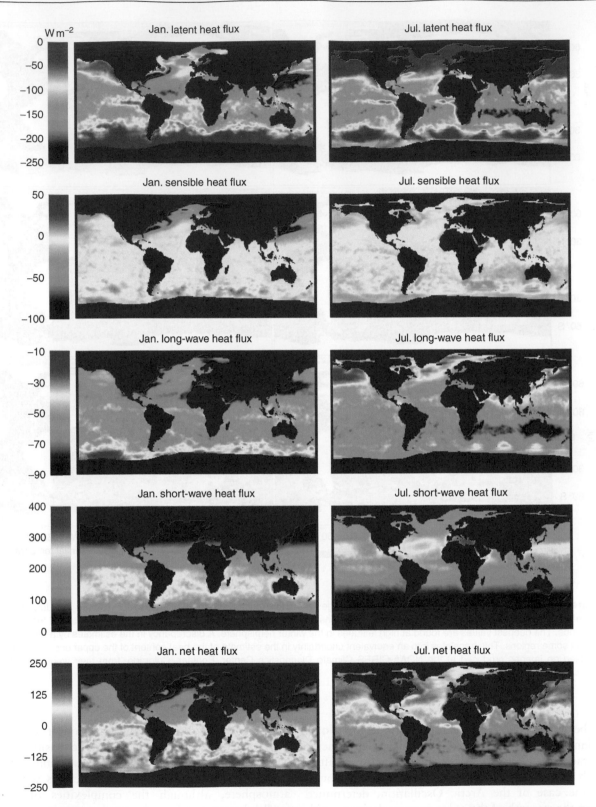

Figure 3 Distributions of the components of the surface heat fluxes for Jan. and Jul. The warm colors indicate warming or less cooling of the ocean, and the cool colors indicate cooling or less warming of the ocean. The data are from the UK National Oceanography Center surface flux climatology (v1.1) and were obtained from http://www.noc.soton.ac.uk.

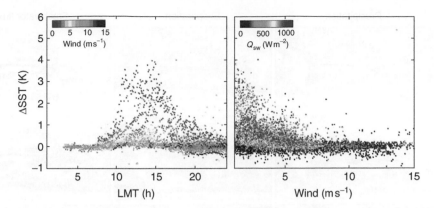

Figure 4 Signatures of diurnal heating revealed in the temperature difference between a radiometrically measured ocean skin temperature and a bulk temperature at a depth of 2 m. The measurements were taken in the Caribbean Sea from the Royal Caribbean International cruise liner *Explorer of the Seas*, which has been equipped as a research vessel. The temperature differences are plotted as a function of local mean time (LMT) and colored by wind speed (left), and as function of wind speed, colored by LMT (left). The largest temperature differences occur in early afternoon on days when the winds are low. Figure provided by Dr. C. L. Gentemann.

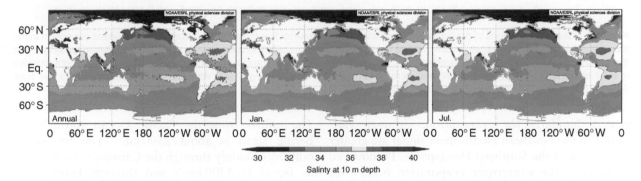

Figure 5 Global distribution of salinity at a depth of 10 m as a global average (left) and monthly averages for Jan. (center) and Jul. (right). The figure is based on images produced at the NOAA-CIRES Climate Diagnostics Center, Boulder, Colorado (http://www.cdc.noaa.gov).

with the sun and space the combined heat transport of the atmosphere and ocean from the Tropics toward the Poles is about 5.5×10^{15} W. How this is partitioned between the atmosphere and ocean is the subject of much research. Within the Atlantic Ocean, the northward transport of heat across $24°$N is about 1.3×10^{15} W. Interestingly the average heat transport in the Atlantic is northward, even south of the Equator: 0.3×10^{15} W northward at $30°$S and 0.6×10^{15} W at $11°$S, although these estimates include transport at depth. The differences in heat transport between such lines at different latitudes provide estimates of the net heat absorbed by the upper ocean or given up to the atmosphere within the surface area of the oceans enclosed by the sections. Thus 77 ± 57 W m^{-2} are estimated to be released by the Atlantic Ocean to the atmosphere between $36°$ and $48°$N, but only 8 ± 33 W m^{-2} between $22°$ and $36°$N. In the North Pacific, 39 ± 19 W m^{-2} flow to the atmosphere between $24°$ and $48°$N. Similarly the differences in the salt (or freshwater) content advected across these imaginary

boundaries indicate the imbalance between precipitation plus continental runoff and evaporation.

Fresh Water

The large-scale patterns of upper ocean salinity (**Figure 5**) mirror the distribution of the annual freshwater flux at the sea surface (**Figure 6**), which is determined by the difference between rainfall and evaporation. The patterns of the components of the freshwater flux are quite zonal in character, with a band of heavy rainfall in the ITCZ and the maxima in evaporation occurring in the regions of the trade winds. There is very little known variability in the seasonal distribution of surface salinity, with the exceptions being in coastal regions where river run off often has a seasonal modulation, especially in the Bay of Bengal where the rainfall influencing the river discharge is dominated by the monsoons.

The precipitation over the Bay of Bengal also shows a strong monsoonal influence, but over much of the oceans the seasonal variability is relatively

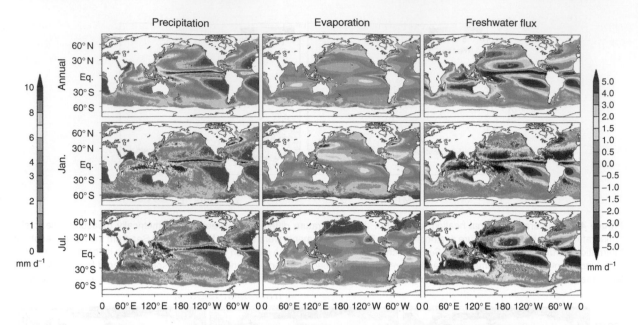

Figure 6 Global distributions of precipitation, evaporation, and freshwater flux at the ocean surface. Annual means are shown in the top row besides monthly averages for Jan. (middle row) and Jul. (bottom row). The color scale is at left for precipitation (positive mass flux into the ocean) and evaporation (positive mass flux into the atmosphere). The color scale for the freshwater flux is at right (positive mass flux into the atmosphere). The figures were generated from the Hamburg Ocean Atmosphere Parameters and Fluxes from Satellite Data Set (HOAPS) (http://www.hoaps.zmaw.de).

muted. Similarly for evaporation, although variations in the Northern Hemisphere signal are greater than those in the Southern Hemisphere. Pronounced maxima in the wintertime evaporation occur over the Gulf Stream and Kuroshio, and represent enhanced moisture fluxes from the sea surface driven by cold, dry air flowing off the continents over the warm surface waters of the north-flowing currents (**Figure 6**).

On shorter timescales, there is pronounced variability in rainfall associated with the passage of weather fronts at mid-latitudes and with individual clouds in the Tropics. These small-scale, short-duration rainfall events hinder the accurate determination of the freshwater flux into the sea surface. In the Tropics, the rainfall associated with individual cumulonimbus clouds has a diurnal signature, especially in the vicinity of islands, even small atolls, where the diurnal sea breeze can trigger convection that results in rainfall, either directly into the ocean, or as runoff from land.

The Arctic Ocean is a particularly interesting area regarding the local freshwater budget as the vertical stability is constrained by the salinity gradients in the halocline. Freshwater volumes in the Arctic Ocean are often calculated relative to a seawater salinity of 34.8. The fresher surface waters are sustained by riverine inflow, primarily from the great Siberian rivers and the Mackenzie River in Canada, that between them annually contribute about 3200 km³.

The inflow from the Pacific Ocean through the Bering Strait is about 2500 km³. The freshwater outflow is mainly through the Canadian Archipelago as liquid ($\sim$3200 km³) and through Fram Strait ($\sim$2400 km³ as liquid and $\sim$2300 km³ as ice). The contribution of precipitation minus evaporation is c. 2000 km³. The fresh water generated by brine rejection during ice formation would be $\sim$10 000 km³, which is a relatively small proportion of the riverine and Bering Strait input. The residence time of fresh water in the Arctic Ocean is about 10 years.

Severe Storms

An important consequence of variations in the upper ocean heat content is severe storm generation and intensification. The prediction of the strength and trajectory of land-falling hurricanes and cyclones benefits from knowledge of the upper ocean heat content in the path of the storm. A surface temperature of 26 °C is generally accepted as being necessary for hurricane development, but the rate of development depends on the heat in the upper ocean available to drive the storm's intensification. The passage of a severe storm leaves a wake that is identifiable as a depression of the surface temperature of several degrees and a deficit in the upper ocean heat content. These may survive for several days and can influence the development of subsequent storms should they pass over the wake. There are several well-documented cases in the Atlantic where hurricanes approaching land have suddenly lost intensity as they follow, or cross, the path of a prior storm.

The converse is also true and hurricanes can undergo sudden intensification when they pass over regions of high upper ocean heat content, as can result from the meandering of the Loop Current in the Gulf of Mexico, for example.

Monitoring the upper ocean heat content has become important for severe storm forecasting in the Tropics, especially in terms of sudden intensification. Using a combination of satellite measurements of sea surface temperature, sea surface topography, and a simple ocean model, the spatial distribution of the heat content between the surface and the estimated depth of the 26 °C isotherm is calculated on a daily basis. This is referred to as the 'tropical cyclone heat

potential' (**Figure** 7) and indicates regions where intensification of severe storms is likely.

The rate of heat transfer from ocean to atmosphere in a hurricane is very difficult to measure, and varies greatly with the size, intensity, and stage of development of the storm. Estimates range in the order of 10^{13}–10^{14} W. We have already seen that the northward heat flux in the Atlantic Ocean at 24° N is $\sim 1.3 \times 10^{15}$ W. Thus, even though severe storms grow and are sustained by large heat fluxes, the magnitudes of the associated flow are relatively small in comparison to the poleward oceanic heat transport which is ultimately released to the atmosphere.

Reactions to Climate Change

Away from polar regions, the density of seawater is a strong function of its temperature, and a consequence

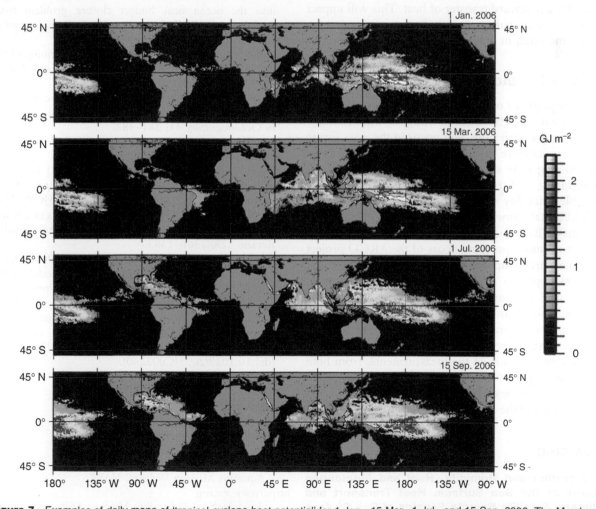

Figure 7 Examples of daily maps of 'tropical cyclone heat potential' for 1 Jan., 15 Mar., 1 Jul., and 15 Sep. 2006. The March and September dates correspond roughly to the peak of cyclone activity in each hemisphere. The maps were derived from satellite measurements of sea surface temperature, sea surface topography, and a simple ocean model. The figure was derived from images generated by the NOAA Atlantic Oceanographic and Meteorological Laboratory (AOML), Miami, Florida (http://www.aoml.noaa.gov).

of increasing temperatures as a result of global change is the expansion of the upper ocean, which will contribute to sea level rise. In fact, about half of the observed rise in global sea level during the twentieth century of 1–2 mm yr^{-1} can be attributable to expansion of the warming upper ocean.

In addition to thermal expansion, another major impact of climate change is the increase in the upper ocean freshwater budget (reduction in salinity) as the land ice (glaciers and ice caps of Greenland and Antarctica) melt and the runoff enter the high-latitude oceans. This will result in an increase in the stability of the upper ocean and a consequent likely reduction in the mixed layer depths, especially in winter (**Figure 2**). Here the atmosphere is coupled to the heat available in a very deep ocean layer. Mixing heat and fresh water from the upper ocean to depth is also a driver of the global thermohaline circulation and disruption to this will also have significant consequences on the near-surface components of the circulation and on the details of the poleward transfer of heat. This will impact global weather patterns, including rainfall over the ocean and land, in ways that are difficult to predict.

Future Developments

Improvements in our ability to determine the upper ocean heat and freshwater budgets, and monitor their changes with time, will occur in the near future with new satellite missions that will both continue the existing time series of sea surface temperature, topography and rainfall, and also introduce new variables: notably sea surface salinity.

Additional information on the subsurface distributions of heat and fresh water will be provided by the autonomous profiling floats of the ARGO project that measure temperature and salinity from about 2000-m depth to a few meters below the surface. These will be augmented by AUVs, or 'gliders', roaming the oceans taking measurements along undulating paths, transmitting the data via satellite communications when they break the surface.

The interpretation of the measurements, from both *in situ* and space-borne sensors, will be aided by increasingly complex, high-resolution models of the ocean state and the coupled ocean–atmosphere system.

See also

Evaporation and Humidity. Heat and Momentum Fluxes at the Sea Surface. Heat Transport and Climate. Satellite Remote Sensing of Sea Surface Temperatures. Satellite Remote Sensing: Salinity Measurements. Upper Ocean Mean Horizontal Structure. Upper Ocean Mixing Processes. Upper Ocean Time and Space Variability. Upper Ocean Vertical Structure. Wind-and Buoyancy-Forced Upper Ocean. Wind Driven Circulation.

Further Reading

Chen SS and Houze RA (1997) Diurnal variation and lifecycle of deep convective systems over the tropical Pacific warm pool. *Quarterly Journal of the Royal Meteorological Society* 123: 357–388.

Foltz GR, Grodsky SA, Carton JA, and McPhaden MJ (2003) Seasonal mixed layer heat budget of the tropical Atlantic Ocean. *Journal of Geophysical Research* 108: 3146 (doi:10.1029/2002JC001584).

Gill AE (1982) *Atmosphere–Ocean Dynamics.* San Diego, CA: Academic Press.

Hasegawa T and Hanawa K (2003) Decadal-scale variability of upper ocean heat content in the tropical Pacific. *Geophysical Research Letters* 30: 1272 (doi:10.1029/2002GL016843).

Josey SA, Kent EC, and Taylor PK (1999) New insights into the ocean heat budget closure problem from analysis of the SOC air–sea flux climatology. *Journal of Climate* 12: 2856–2880.

Levitus S, Antonov J, and Boyer T (2005) Warming of the world ocean, 1955–2003. *Geophysical Research Letters* 32: L02604 (doi:10.1029/2004GL021592).

Macdonald AM (1998) The global ocean circulation: A hydrographic estimate and regional analysis. *Progress in Oceanography* 41: 281–382.

Peixoto JO and Oort AH (1992) *Physics of Climate.* New York: American Institute of Physics.

Serreze MC, Barrett AP, Slater AG, et al. (2006) The large-scale freshwater cycle of the Arctic. *Journal of Geophysical Research* 111: C11010 (doi:10.1029/2005JC003424).

Shay LK, Goni GJ, and Black PG (2000) Effects of a warm oceanic feature on hurricane Opal. *Monthly Weather Review* 128: 1366–1383.

Siedler G, Church J, and Gould J (eds.) (2001) *Ocean Circulation and Climate: Observing and Modelling the Global Ocean.* San Diego, CA: Academic Press.

Willis JK, Roemmich D, and Cornuelle B (2004) Interannual variability in upper ocean heat content, temperature, and thermosteric expansion on global scales. *Journal of Geophysical Research* 109: C12036 (doi:10.1029/2003JC002260).

Relevant Websites

http://www.remss.com
– AMSR Data, Remote Sensing Systems.
http://aquarius.gsfc.nasa.gov
– Aquarius Mission Website, NASA.
http://www.esr.org
– Aquarius/SAC-D Satellite Mission, ESR.
http://www.hoaps.zmaw.de
– Hamburg Ocean Atmosphere Parameters and Fluxes from Satellite Data.

http://www.ghrsst-pp.org
   – High-Resolution SSTs from Satellites, GHRSST-PP.
http://www.noc.soton.ac.uk
   – NOC Flux Climatology, at Ocean Observing and Climate pages of the National Oceanography Centre (NOC), and The World Ocean Circulation Experiment (WOCE) 1990–2002, NOC, Southhampton.
http://ushydro.ucsd.edu
   – Repeat Hydrography Project.

http://www.cdc.noaa.gov
   – Search for Gridded Climate Data at PSD, ESRL Physical Sciences Division, NOAA.
http://www.esa.int
   – SMOS, The Living Planet Programme, ESA.
http://www.aoml.noaa.gov
   – Tropical Cyclone Heat Potential, Atlantic Oceanographic and Meteorological Laboratory (AOML).

WIND DRIVEN CIRCULATION

P. S. Bogden, Maine State Planning Office, Augusta, ME, USA

C. A. Edwards, University of Connecticut, Groton, CT, USA

Introduction

Winds represent a dominant source of energy for driving oceanic motions. At the ocean surface, such motions include surface gravity waves, which are familiar as the waves that break on beaches. Winds are also responsible for small-scale turbulent fluctuations just beneath the ocean surface. Turbulent motions can be created by breaking waves or by the nonlinear evolution of currents near the air–sea interface. Subsurface processes such as these can lead to easily observed windrows or scum lines on the sea surface. Winds also generate other complex and varied small-scale motions in the top few tens of meters of the ocean. However, the surface/wind-driven circulation described here refers instead to considerably larger-scale motions that compare in size to the ocean basins and extend as much as a kilometer or more below the surface.

The textbook notion of the surface/wind-driven circulation includes most of the well-known surface currents, such as the intense poleward-flowing Gulf Stream in the western North Atlantic (**Figure 1**). Analogues of the Gulf Stream can be found in each of the major ocean basins, including the Kuroshio in the North Pacific, the Brazil Current in the South Atlantic, the East Australian Current in the South Pacific, and the Aghulas in the Indian. These 'western boundary currents' are not isolated structures. Rather, they represent the poleward return flow for basin-scale motions that occupy middle latitudes in all major oceans. Each of the major ocean basins has an analogous set of large-scale current systems. The western boundary currents are quite intense, reaching velocities in excess of $1 \, \text{m s}^{-1}$, while the interior flow speeds are considerably smaller in magnitude.

The basin-scale patterns in the mid-latitude surface circulation are referred to as subtropical gyres. The gyres extend many hundreds of meters below the surface, reaching the bottom in some locations. Subtropical gyres rotate anticyclonically, that is, they rotate in a sense that is opposite to the sense of the earth's rotation (clockwise in the northern hemisphere and counterclockwise in the south). In the North Atlantic and North Pacific, subpolar gyres reside to the north of the subtropical gyres. They too include intense western boundary currents. However, the subpolar gyres rotate cyclonically, in the opposite sense of the subtropical gyres and in the same sense as the earth. Rotation of the wind-driven gyres is related to the rotation of the earth through a simple, though nonintuitive, physical mechanism. This mechanism is fundamental to understanding how the wind drives large-scale flows.

Our present understanding of the dynamics associated with the surface/wind-driven circulation developed largely during a 30-year period starting in the late 1940s. Before that time, oceanographers were aware of the gyre-scale features of the surface circulation. But it was not until the major theoretical advances in geophysical fluid dynamics beginning around 1947 that the surface circulation was conceptually linked to the winds.

Observations

Oceanic wind systems exhibit a large-scale pattern that is common to the major ocean basins (**Figure 2**). Near the equator, trade winds blow from east to west. Near the poles, westerly winds blow from west to east. The ocean gyres have similar distributions of east–west flow. But the reasons for this are quite subtle. Moreover, there are profound differences between oceanic and atmospheric motion. North–south flows in the ocean are much more strongly pronounced than they are in the atmosphere, and winds fail to exhibit analogues of the intense poleward western boundary currents found in the ocean.

Western boundary currents such as the Gulf Stream were evident in estimates of time-averaged surface circulation obtained over a century ago with 'ship-drift' data (**Figure 3**). Ship drift is the discrepancy between a ship's position as obtained with dead reckoning and that obtained by more accurate navigation. Thus, ship drift can be attributed at least in part to ocean currents. The patterns of the surface circulation that emerge after averaging large numbers of such measurements are qualitatively correct. In general, however, accurate measurements of currents are difficult to obtain, particularly below the surface.

Temperature and salinity measurements are relatively abundant, and they provide an alternative resource for estimating large-scale currents. Temperature and salinity determine seawater density. The density distribution provides information about the pressure

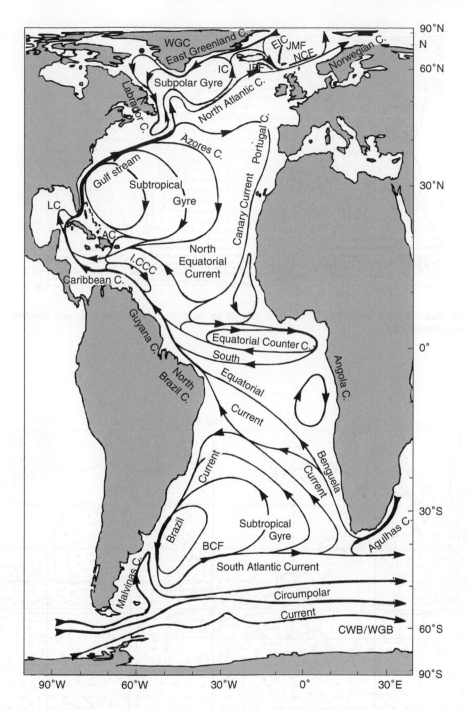

Figure 1 Schematic of large-scale surface currents in the Atlantic Ocean. (From Tomczak and Godfrey (1994) *Regional Oceanography: An Introduction*.)

field, which in turn can be used to diagnose currents. For many decades, oceanographers have been routinely measuring vertical profiles of temperature and salinity. Consequently, detailed maps of the three-dimensional density structure exist for all the ocean basins.

Such maps clearly show a region of anomalously large vertical gradients in temperature, salinity, and density known as the main thermocline. The main thermocline divides two regions of relatively less

stratified water near the surface and bottom. Thermocline depth varies substantially on the gyre scale, and can exceed 700 m depth in some regions. Lateral variations in the density field can be quite large as well, and these are associated with the currents that we refer to as the surface/wind-driven circulation.

The first step in estimating currents from density involves computation of dynamic height using the principle of isostasy. Isostasy describes, for example,

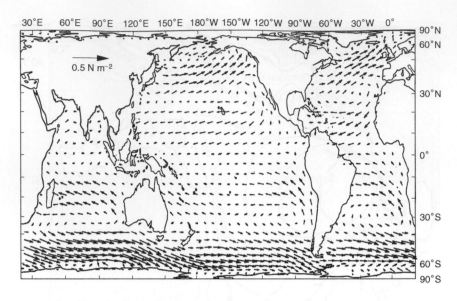

Figure 2 Global mean surface wind stress, which is related to wind [1]. (From Tomczak and Godfrey (1994) *Regional Oceanography: An Introduction.*)

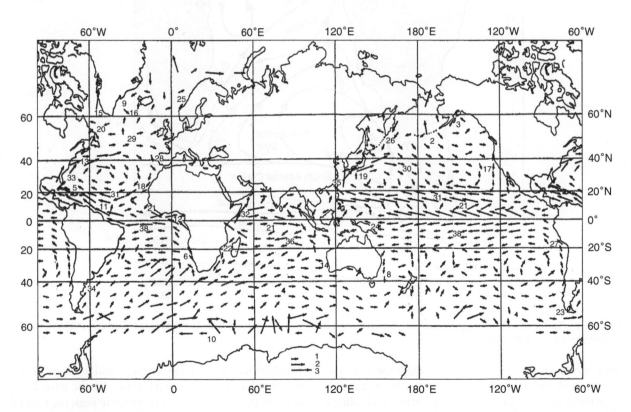

Figure 3 Surface currents inferred from ship-drift measurements. To simplify the presentation, there are three vector sizes in this figure indicated by the scale vectors at the bottom of the figure. A vector the size of vector 1 corresponds to flow speeds in the range 0–10 cm s^{-1}, vector 2 is 10–20 cm s^{-1}, vector 3 is more than 30 cm s^{-1}. While some of the values have questionable reliability, the vectors show the general patterns large-scale circulation at the ocean surface. From Stidd CK (1974) Ship Drift Components: Means and standard Deviations, SIO Reference Series 74-33 as appearing in Burkov VA 1980 *General Circulation of the World Ocean.* Gidrometeoizdat Publishers, Leningrad, published for the Division of Ocean Sciences, National Science Foundation, Washington, DC, by Amerind Publishing Co. Pvt. Ltd., New Delhi. 1993.

the pressure field in a glass of ice water. An ice cube represents a region where water is slightly less dense than its surroundings. Buoyancy forces elevate the surface of the ice cube above the surface of the surrounding fluid. Similarly, a region in the ocean with less dense water than its surroundings will have a slightly elevated sea surface. In the ocean, this result involves the tacit assumption that currents in the abyssal ocean are weak relative to those nearer the surface, as is usually the case. The ocean's surface topography implied by the density field is referred to as dynamic height. **Figure 4** shows dynamic height computed from density between 2000 and 200 m depth. The variations of a meter or more are large enough to account for the pressure gradients that force the large-scale gyres.

The connection between pressure and large-scale currents involves the principle of 'geostrophy'. Geostrophic currents arise from a balance of the forces involving pressure gradient forces and Coriolis accelerations. This balance is a consequence of the large horizontal scales of the flow combined with the rotation of the earth. If the earth were not rotating, the sea surface elevations would accelerate horizontal flows down the pressure gradient, as occurs with smaller-scale motions such as surface gravity waves. With large-scale geostrophic flows, however, the Coriolis effect gives rise to currents that flow perpendicular to the pressure gradient, as indicated by the arrows in **Figure 4**.

Geostrophic currents such as those in **Figure 4** provide evidence of the surface/wind-driven circulation.

This point requires some explanation, since the geostrophic flows are associated with large-scale pressure-gradient forces in the top kilometer of the ocean. As discussed below, winds directly drive motions in a relatively thin layer at the ocean surface known as the surface mixed layer. But these directly wind-driven flows give rise to other large-scale flows and, in turn, to the large-scale pressure gradients that can be estimated with dynamic height. Thus, it is accurate to refer to the large-scale geostrophic surface circulation as the wind-driven circulation because the pressure gradients would not exist without the wind.

Wind-driven Surface Layer

Surface Mixed Layer

The surface mixed layer is loosely defined as a part of the water column near the surface where observed temperature and salinity fields are vertically uniform. In practice this layer extends from the ocean surface to a depth where stratification in temperature or density exceeds some threshold value. Typically, the underlying water is more strongly stratified. The mixed-layer depth often undergoes large diurnal and seasonal variations, varying between 0 and 100 m. However, the surface mixed layer rarely occupies more than 1% of the total water column.

Winds provide the primary source of mechanical forcing for the motions that homogenize water properties within the mixed layer. Mixing can also result from destabilizing effects of cooling and

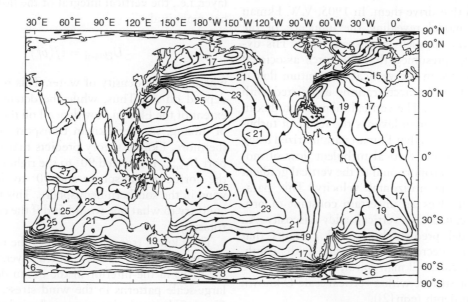

Figure 4 Dynamic height (m² s⁻²) computed using the density field between 0 m and 2000 m depth, and assuming that the pressure field at 2000 m has no horizontal variation. Dynamic height is roughly proportional to the sea-surface height, in meters, multiplied by 10. (Based on Levitus (1982).)

evaporation near the surface, as these can temporarily give rise to localized regions where the surface water overlies less dense water. Restratification of a stable water column involves solar heating from above, reduced salinity from precipitation, and other more subtle processes.

The detailed mechanisms by which wind generates small-scale motions (i.e., motions on scales smaller than the mixed-layer depth, such as breaking waves) are quite complex and incompletely understood. Nevertheless, the effect of wind on the surface mixed layer is commonly parameterized through a stress τ_w on the ocean surface. Wind stress has units of force per unit area. The standard empirical relation has the form of eqn [1].

$$\tau_w = \rho_{air} C_d u^2, \qquad [1]$$

where ρ_a is the density of air, $C_d \approx 10^{-3}$ is a drag coefficient that may depend on wind speed and atmospheric stability, and u is the wind speed 10 m above the sea surface. Ten meters is the standard height that commercial ships use to mount their anemometers, and ship reports still account for most of the direct measurements of wind over the ocean.

Ekman Dynamics

The small-scale motions that mix temperature and salinity also mix momentum. As a result, the momentum of the wind is efficiently transmitted throughout the mixed layer, thereby accelerating horizontal currents. The resulting motions have large horizontal length scales comparable to those of the wind systems that drive them. In 1905, V.W. Ekman developed a model revealing the influence of the earth's rotation on such large-scale flows. His dynamical model presumed that the force associated with a divergence in the vertical momentum flux is balanced by Coriolis accelerations associated with the horizontal flows. The vertical momentum flux in this wind-driven Ekman layer is the result of turbulent mixing processes that Ekman parametrized using Fick's law. Thus, the vertical turbulent flow of momentum is made proportional to the vertical gradient of the large-scale horizontal velocity. This parametrization involves an uncertain constant of proportionality called the vertical eddy viscosity A_v. Ekman's model predicts horizontal currents that simultaneously decrease and rotate with depth. Within this so-called Ekman spiral, currents decrease away from the surface with a vertical scale D known as the Ekman depth (eqn [2]).

$$D = (2A_v/f)^{1/2} \qquad [2]$$

This relation provides our first introduction to the Coriolis parameter $f = 2\Omega \sin \theta$, where θ is latitude, which appears here because of Coriolis accelerations in the Ekman dynamics. The angular velocity of the earth is a vector of magnitude $\Omega = 2\pi$ day that is parallel to the earth's axis of rotation. The Coriolis parameter equals twice the magnitude of the vector component that is parallel to the local vertical. The vertical component is the only component that creates horizontal Coriolis accelerations with horizontal flow. This dependence on the local vertical and the sphericity of the earth explain the $\sin \theta$ factor in the formula for f. Thus, for any given flow, Coriolis accelerations are strongest at the poles, negligible at the equator, and smoothly varying in between. As discussed below, this geometric detail has profound implications for the surface/wind-driven circulation.

Consider a typical mixed-layer depth of 30 m at mid-latitudes, where $f \approx 10^{-4} s^{-1}$. By relating these two quantities to an Ekman depth D, one deduces a vertical eddy viscosity $A_v \approx 0.05$ m^2 s^{-1}. This value is many orders of magnitude larger than the kinematic viscosity of water, $v \approx 10^{-6}$ m^2 s^{-1}. The large value of A_v indicates the efficiency of turbulent mixing compared with molecular diffusion. But A_v arises from the use of Fick's law to parametrize the turbulence, and Fick's law is an oversimplified model for turbulence. In fact, details of the wind-mixed layer that depend heavily on A_v, such as spiraling velocities, are rarely observed.

There is, nevertheless, one very important and robust conclusion from Ekman theory. The net mass transport (the Ekman transport) within the mixed layer, i.e., the vertical integral of the horizontal flow, has magnitude

$$U_{Ekman} = \tau/(\rho f) \qquad [3]$$

where ρ is the density of water. This result does not depend on A_v. Thus, while the details and vertical extent of the Ekman flow depend on the complexities of mixing, the net Ekman transport does not. Furthermore, Ekman theory predicts that the net transport U_{Ekman} is directed 90° to the right of the wind in the northern hemisphere and 90° to the left of the wind in the southern hemisphere. This result is quite contrary to what one would find if the earth were not rotating.

The Ekman transport describes the net horizontal motion in a thin surface mixed layer. Implications for flows in the interior of the ocean depend on the large-scale patterns in the wind stress, as shown in **Figure 2**. In particular, westward wind stress near the equator results in poleward Ekman mass transport and eastward wind stress at higher latitudes drives equatorward Ekman transport. This pattern results in a convergence of fluid that gives rise to an elevated sea surface at the center of the clockwise wind system. Ultimately, this horizontally convergent Ekman transport has only one direction in which to go – down. The resultant downward motion at the base of the mixed layer, called Ekman pumping, occurs in all mid-latitude ocean basins. Likewise, at higher latitudes, counterclockwise wind systems cause horizontally divergent Ekman transport, a depressed sea surface, and an upward motion known as Ekman suction.

In the classic wind-driven ocean circulation models discussed below, vertical Ekman flows drive the horizontal geostrophic flow. In fact, the net effect of all the complex motions in the mixed layer is often reduced to a simple prescription of the vertical Ekman-pumping velocity W_{Ekman} (eqn[4]).

$$W_{Ekman} = \mathrm{curl}(\tau_w/\rho f) \qquad [4]$$

where $\mathrm{curl}(\tau_w/\rho f))$ represents the curl of the surface wind-stress vector divided by ρf. Thus, it is not simply the magnitude of the wind stress that determines the Ekman pumping velocities, but its spatial distribution. The Ekman-pumping velocity is often applied as a boundary condition at the sea surface associated with a negligibly thin mixed layer.

On average, Ekman-pumping speeds rarely exceed $1\,\mu m\,s^{-1}$. Nevertheless, such minuscule vertical velocities give rise to the most massive current systems in the ocean. This remarkable fact reflects the enormous constraint that the earth's rotation plays in large-scale ocean dynamics.

Large-scale Dynamics

The directly wind-driven flows within the Ekman layer occupy only a small fraction of the total water column. In fact, the impact of the wind extends considerably deeper. The connection between the minute Ekman-pumping velocities and the tremendous horizontal flows associated with the surface/wind-driven circulation involves a balance of forces that is very different from that in the surface mixed layer.

Far from continental boundaries, and below the surface mixed layer, the basin-scale circulation varies on length scales measured in thousands of kilometers. The time-averaged horizontal velocities sometimes exceed $1\,m\,s^{-1}$, but they more generally vary between 1 and $10\,cm\,s^{-1}$. With these scales, flows are plausibly geostrophic. Furthermore, when the density is uniform, geostrophic flows exhibit no vertical variation. Rather, they behave like a horizontal continuum of vertical columns of fluid. It is reasonable to

approximate the region between the mixed layer and the thermocline as a region of constant density. In this region the geostrophic columns of fluid span many hundreds of meters. The earliest models of the ocean circulation obtained remarkable predictive skills by assuming that columnar geostrophic flow extends from the top to the bottom of the ocean.

Downward Ekman-pumping velocities, as small as they may be, effectively compress the fluid columns. Under the influence of downward Ekman pumping, fluid columns below the mixed layer compress vertically and expand horizontally, as if they were conserving their total volume. Likewise, Ekman suction causes water columns to stretch vertically and contract horizontally. Because of the earth's rotation, the effect of Ekman pumping and suction on large-scale motions is related to the principle of angular momentum conservation in classical mechanics. For example, a water column that undergoes the stretching effect of Ekman suction is not unlike a rotating figure skater who draws in her arms, thereby decreasing her moment of inertia and rotating more rapidly. (Note: The analogy is incomplete because Ekman pumping is a consequence of external forcing, whereas the spinning skater is unforced. Nevertheless, the comparison is physically relevant.) Ekman pumping is then similar to a skater extending her arms, which causes a reduction of rate of spin.

The connection between water-column stretching and horizontal flow in the ocean involves one additional subtle point: Water columns on the earth rotate by virtue of their location on the earth's surface. Vertical fluid columns that appear stationary in the earth's frame of reference are actually rotating at a rate that is proportional to the vertical component of the earth's angular velocity. The absolute rotation rate is the sum of the earth's rotation plus the rotation relative to the earth. More precisely, the fluid's absolute vorticity includes a contribution from the planetary vorticity, which has magnitude equal to the Coriolis parameter f, plus a contribution from its relative vorticity.

Relative vorticity is measured from a frame of reference that is fixed on the earth's surface. The earth itself rotates at a rate of one revolution per day. In comparison, large-scale ocean currents progress around an ocean basin at average speeds much less than $1\,m\,s^{-1}$, so the period of rotation associated with a complete circuit can be several years. Thus, the relative rotation rate of large-scale ocean currents is negligible compared with the rotation rate of the earth itself. For this reason, it is a very good approximation to neglect the relative vorticity and equate the vorticity of the large-scale circulation with f. The critically important result is that fluid columns change their vorticity by changing their latitude. Just as the skater who extends her arms starts to rotate more slowly, a water column undergoing the compression of Ekman pumping will travel toward the equator where the magnitude of f is smaller.

The physical mechanism that connects Ekman pumping and subsurface geostrophic flow was described mathematically by H.U. in 1947, and is summarized by the relation (eqn [5]).

$$\beta V = f(W_{Ekman} - W_{deep}) \quad [5]$$

V represents the meridional (positive northward) velocity integrated over the depth H of the water column, $\beta = (1/R)\partial f/\partial \theta$ is the meridional variation in the Coriolis parameter, R is the radius of the earth, W_w is the Ekman pumping velocity, and W_{deep} is a vertical velocity at depth H. In this model, V is the depth-integrated geostrophic velocity. While prescription of W_{deep} is discussed further below, the early models of the wind-driven circulation assumed that $W_{deep} = 0$ at some depth well below the main thermocline. Thus, V can be estimated using the Sverdrup relation with W_{Ekman} prescribed using eqn [1], eqn [4] and measurements of wind. V computed in this way agrees remarkably well with V computed from geostrophic flows estimated from the observed density field. The agreement is good everywhere except near the western boundaries of the ocean basins.

Westward Intensification

While the Sverdrup relation provides guidance for the ocean interior, it cannot describe the basin-wide circulation. From a mathematical viewpoint, the Sverdrup relation is a purely local relation between wind stress and meridional flow, so it does not determine east–west flow within the basin. Moreover, the Sverdrup relation predicts that V will be large only where W_{Ekman} is large. In fact, V is observed to be largest in the intense western boundary currents found in each ocean basin, such as the Gulf Stream. This is problematic because W_{Ekman} fails to exhibit the westward intensification, or a change in sign. This means that the Sverdrup relation predicts weak western boundary flow in the wrong direction! Thus, the Sverdrup balance must break down, at least in certain regions.

It is common to presume that Sverdrup theory holds everywhere in the ocean interior except near the western ocean boundary (and northern and southern boundaries if the wind-stress curl does not vanish there). Then, V can be integrated from east to west to determine the total transport required in a poleward western boundary current that returns the meridional Sverdrup transport back to its place of origin. This calculation comes close to predicting the observed transport in the poleward-traveling Gulf Stream at some locations. But models that predict the structure and location of the return flow require fundamentally different dynamics.

In 1948, H. Stommel developed a theory for the wind-driven circulation in which the ocean bottom exerts a frictional drag on the horizontal flow. In the ocean interior, the Stommel and Sverdrup dynamics are nearly indistinguishable, but bottom friction becomes important near the western boundary, allowing Stommel's model to predict a closed circulation for the entire ocean basin (**Figure 5**). The friction-dominated western boundary layer contains the intense poleward analogue of the Gulf Stream. Stommel showed the remarkable fact that westward intensification of the wind-driven gyres is fundamentally linked to the latitudinal variation of the Coriolis parameter. That is, the Gulf Stream and its western-boundary analogues in all the ocean basins exist because of the sphericity of the rotating earth (**Figure 6**).

Friction is the key to closing the circulation cell. In Stommel's model the friction parametrization was

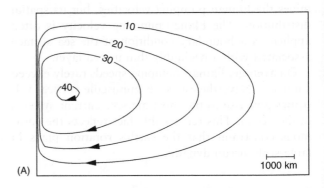

(A)

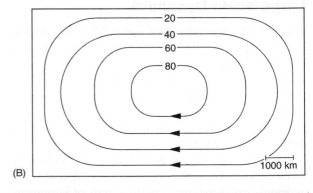

(B)

Figure 5 Streamlines from Stommel's model indicating the total flow in an idealized flat-bottom subtropical gyre. The flow is everywhere parallel to the streamlines in the direction indicated by the arrows. Flow intensity is greatest where the streamlines are closest together. (A) An idealized subtropical gyre for a rotating earth in which the Coriolis parameter varies with latitude. (B) Streamlines for a 'uniformly' rotating earth, that is, for a Coriolis parameter that does not vary with latitude. (From Stommel (1948).)

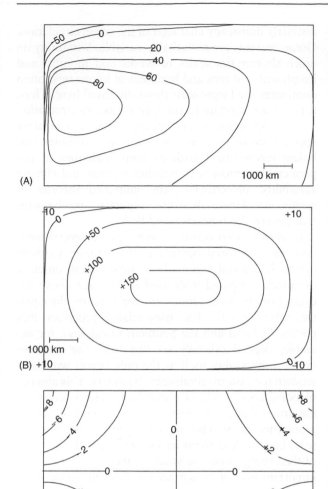

Figure 6 Contours of sea-surface height from Stommel's model. (A) Sea-surface height consistent with the stream function of **Figure 5**. Features in this figure can be directly compared with the dynamic height computed from data in **Figure 4**. (B) Sea-surface height for Stommel's model after setting the Coriolis parameter to a constant. This effectively removes the geometrical factor associated with sphericity. Stommel referred to this as the case of a 'uniformly rotation ocean'. (C) The sea-surface height for the same wind distribution as in (A) and (B), but for a nonrotating ocean. (From Stommel (1948).)

chosen for its simplicity, but it is ultimately unrealistic. In 1950, W. Munk developed a similar flat-bottom model with lateral viscosity, an entirely different form of dissipation. Nevertheless, Munk's model produces an intense western boundary current for the same reasons as does the Stommel model.

The primary results from these frictional models are robust. Both theories deduce the zonal flow within the basin, and share the central conclusion that the return flow for the interior Sverdrup transport occurs in a meridional current near the western

edge of the basin. This current is an example of a boundary layer, a narrow region governed by different physical balances from those dominating the larger domain. In the western boundary layer, fluid columns can change latitude because dissipation changes their vorticity, thereby counteracting the effects introduced by the Ekman pumping or suction. Both models show that this dissipative mechanism can only occur in an intense western boundary layer.

Dissipation in both models actually parametrizes many interesting smaller-scale phenomena. This is evident in Munk's model. The horizontal viscosity needed to produce a realistic Gulf Stream is many orders of magnitude larger than molecular viscosity, larger even than Ekman's vertical eddy viscosity, A_v. Modern theories show that these viscous parametrizations for ocean turbulence greatly oversimplify the effect of small-scale motions on the large-scale circulation. More importantly, the Stommel and Munk models neglect the fact that the ocean has variable depth and density stratification.

Topography, Stratification, and Nonlinearity

The simplified Stommel and Munk models describe the wind-driven circulation for a rectangular ocean that has uniform density, a flat bottom, and vertical side walls. It remains to put these idealized models in context for an ocean that has density stratification, mid-ocean ridges, and continental slopes and shelves. The flat-bottom constant-density models clearly oversimplify the ocean geometry. Were the mid-ocean ridges placed on land, they would stand as tall as the Rockies and the Alps. The assumption of constant density turns out to be an oversimplification of comparable proportions.

In flat-bottom models, deep currents are unimpeded by topographic obstructions. With realistic bathymetry, however, flow into regions of varying depth can lead to large vertical velocities. For rotating fluid columns, these vertical velocities affect vorticity. Computer models that add realistic bathymetry and Ekman pumping to the Stommel or Munk models show that such vertical velocities can substantially alter the horizontal flow pattern, so much so that the flows in the center of the ocean no longer resemble the observed surface circulation. Thus, in idealized constant-density models, realistic bathymetry eliminates the most remarkable similarities between the models and the ocean observations.

This conundrum can be reconciled in a model that has variable density. In a constant-density ocean, geostrophic fluid columns extend all the way to the bottom. This allows bottom topography to have an unrealistically strong influence on the flow. Density stratification reduces the vertical extent of columnar motion. Conceptually, a stratified ocean behaves almost like a series of distinct layers, each with variable thickness and constant density. For example, the main thermocline may be considered the interface between one continuum of fluid columns in a surface layer and a second continuum of fluid columns in an abyssal layer. Generalizations of the Stommel and Munk models have often treated the ocean as two distinct layers of fluid.

The main thermocline varies smoothly compared with the ocean bottom. This means that there are fewer obstructions to the columnar flow above the thermocline than below. In this sense, the thermocline effectively isolates the ocean bathymetry from the surface circulation. In fact, observed currents above the main thermocline tend to be stronger. While the Sverdrup theory applies to the top-to-bottom transport, stratification allows the flow to be surface intensified. Smaller abyssal velocities reduce the influence of bottom topography. Flat-bottom models describe a limiting case where the topographic effects are identically zero.

Without question, the vertical extent of the large-scale wind-driven circulation is linked to density stratification. Realistic models of the large-scale circulation must include thermodynamic processes that affect temperature, salinity, and density structure. For example, atmospheric processes change the heat and fresh water content of the surface mixed layer. Large-scale motions can result when the water column becomes unstable, with more dense water overlying less dense water. The resulting motion is often referred to as the thermohaline circulation, as distinct from the wind-driven circulation, but the conclusion to be drawn from the more realistic ocean-circulation models is that the thermohaline circulation and the wind-driven circulation are inextricably linked.

Additional factors come into play in the more comprehensive ocean models. For example, the persistent temperature and salinity structure of the ocean indicates that many large-scale features in the ocean have remained qualitatively unchanged for decades, perhaps even centuries. But there are no simple (linear) theories that predict the existence of the thermocline. The transport and mixing of density by ocean currents are inherently nonlinear effects. Other classes of nonlinearities inherent to fluid flow add other types of complexity. Such nonlinear effects account for Gulf Stream rings, mid-ocean eddies, and much of the

distinctly nonsteady character of the ocean circulation. Ocean currents are remarkably variable. Variability on much shorter timescales of weeks and months, and length scales of tens and hundreds of kilometers, often dominates the larger-scale flows discussed here. Thus, it is not appropriate to think of the ocean circulation as a sluggish, linear, and steady. Instead, it is more appropriate to think of the ocean as a complex turbulent environment with its own analogues of unpredictable atmospheric weather systems and climate variability. Nevertheless, the simplified theories of steady circulation illustrate important mechanisms that govern the time-averaged flows.

In closing, two ocean regions deserve special mention: the equatorial ocean and the extreme southern ocean. Equatorial regions have substantially different dynamics compared with models discussed above because Coriolis accelerations are negligible on the equator, where $f = 0$. The wind-related processes that govern El Niño and the Southern Oscillation, for example, depend critically on this fact. The southern ocean distinguishes itself as the only region without a western (or eastern) continental boundary. This absence of boundaries produces a circulation characteristic of the atmosphere, with intense zonal flows that extend around the globe. They represent some of the most intense large-scale currents in the world, and derive much of their energy from the wind. So they too represent an important part of the surface/wind-driven circulation.

See also

Surface Gravity and Capillary Waves.

Further Reading

Henderschott MC (1987) Single layer models of the general circulation. In: Abarbanel HDI and Young WR (eds.) *General Circulation of the Ocean*. New York: Springer-Verlag.

Pedlosky J (1996) *Ocean Circulation Theory*. New York: Springer-Verlag.

Salmon R (1998) *Lectures on Geophysical Fluid Dynamics*. Oxford: Oxford University Press.

Stommel H (1976) *The Gulf Stream*. Berkeley: University of California Press.

Veronis G (1981) Dynamics of large-scale ocean circulation. In: Warren BA and Wunsch C (eds.) *Evolution of Physical Oceanography*. Cambridge. MA: MIT Press.

WIND- AND BUOYANCY-FORCED UPPER OCEAN

M. F. Cronin, NOAA Pacific Marine Environmental
Laboratory, Seattle, WA, USA
J. Sprintall, University of California San Diego,
La Jolla, CA, USA

Published by Elsevier Ltd.

Introduction

Forcing from winds, heating and cooling, and rainfall and evaporation has a profound influence on the distribution of mass and momentum in the ocean. Although the effects from this wind and buoyancy forcing are ultimately felt throughout the entire ocean, the most immediate impact is on the surface mixed layer, the site of the active air–sea exchanges. The mixed layer is warmed by sunshine and cooled by radiation emitted from the surface and by latent heat loss due to evaporation (**Figure 1**). The mixed layer also tends to be cooled by sensible heat loss since the surface air temperature is generally cooler than the ocean surface. Evaporation and precipitation change the mixed layer salinity. These salinity and temperature changes define the ocean's surface buoyancy. As the surface loses buoyancy, the surface water can become denser than water below it, causing convective overturning and mixing to occur. Wind forcing can also cause near-surface overturning and mixing, as well as localized overturning at the base of the mixed layer through shear-flow instability. This wind- and buoyancy-generated turbulence causes the surface water to be well mixed and vertically uniform in temperature, salinity, and density. Furthermore, the turbulence can entrain deeper water into the surface mixed layer, causing the surface temperature and salinity to change and the layer of well-mixed, vertically uniform water to thicken. Wind forcing also sets up oceanic currents and can cause changes in the mixed layer temperature and salinity through horizontal and vertical advection.

Although the ocean is forced by the atmosphere, the atmosphere can also respond to ocean surface conditions, particularly sea surface temperature (SST). Direct thermal circulation, in which moist air rises over warm SSTs and descends over cool SSTs, is prevalent in the Tropics. The resulting atmospheric circulation cells influence the patterns of cloud, rain, and winds that combine to form the wind and buoyancy forcing for the ocean. Thus, the oceans and atmosphere form a coupled system, where it is sometimes difficult to distinguish forcing from response. Because water has a density and effective heat capacity nearly 3 orders of magnitude greater than air, the ocean has mechanical and thermal inertia relative to the atmosphere. The ocean thus acts as a memory for the coupled ocean–atmosphere system.

We begin with a discussion of air–sea interaction through surface heat fluxes, moisture fluxes, and wind forcing. The primary external force driving the ocean–atmosphere system is radiative warming from the Sun. Because of the fundamental importance of solar radiation, the surface wind and buoyancy forcing is illustrated here with two examples of the seasonal cycle. The first case describes the seasonal cycle in the North Pacific, and can be considered a classic example of a one-dimensional (involving only vertical processes) ocean response to wind and buoyancy forcing. In the second example, the seasonal cycle of the eastern tropical Pacific, the atmosphere and the ocean are coupled, so that wind and buoyancy forcing lead to a sequence of events that make cause and effect difficult to determine. The impact of wind and buoyancy forcing on the surface mixed layer and the deeper ocean is summarized in the conclusion.

Air–Sea Interaction

Surface Heat Flux

As shown in **Figure 1**, the net surface heat flux entering the ocean (Q_0) includes solar (shortwave) radiation (Q_{sw}), net infrared (long-wave) radiation (Q_{lw}), latent heat flux due to evaporation (Q_{lat}), and sensible heat flux due to air and water having different surface temperatures (Q_{sen}):

$$Q_0 = Q_{sw} + Q_{lw} + Q_{lat} + Q_{sen} \qquad [1]$$

The Earth's seasons are largely defined by the annual cycle in the net surface heat flux associated with the astronomical orientation of the Earth relative to the Sun. The Earth's tilt causes solar radiation to strike the winter hemisphere more obliquely than the summer hemisphere. As the Earth orbits the Sun, winter shifts to summer and summer shifts to winter, with the Sun directly overhead at the equator twice per year, in March and again in September. Thus, one might expect the seasonal cycle in the Tropics to be semiannual, rather than annual. However, as discussed later, in some parts of the

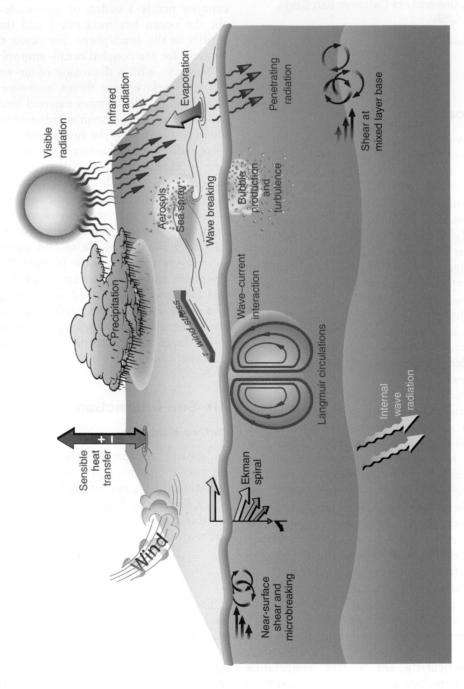

Figure 1 Schematic drawing of wind- and buoyancy-forced upper ocean processes. Courtesy Jayne Doucette, Woods Hole Oceanographic Institution.

equatorial oceans, the annual cycle dominates due to coupled ocean–atmosphere–land interactions.

Solar radiation entering the Earth's atmosphere is absorbed, scattered, and reflected by water in both its liquid and vapor forms. Consequently, the amount of solar radiation which crosses the ocean surface, Q_{sw}, also depends on the cloud structures. The amount of solar radiation absorbed by the ocean mixed layer depends on the transmission properties of light in water and can be estimated as the difference between the solar radiation entering the surface and the solar radiation penetrating through the base of the mixed layer.

The Earth's surface also radiates energy at longer wavelengths similar to a black body (i.e., proportional to the fourth power of the surface temperature in units kelvin). Infrared radiation emitted by the atmosphere and clouds can reflect against the ocean surface and become upwelling infrared radiation. Thus net long-wave radiation, Q_{lw}, is the combination of the outgoing and incoming infrared radiation and tends to cool the ocean.

The ocean and atmosphere also exchange heat via conduction ('sensible' heat flux). When the ocean and atmosphere have different surface temperatures, sensible heat flux will act to reduce this temperature difference. Thus when the ocean is warmer than the air (which is nearly always the case), sensible heat flux will tend to cool the ocean and warm the atmosphere. Likewise, the vapor pressure at the air–sea interface is saturated with water while the air just above the interface typically has relative humidity less than 100%. Thus, moisture tends to evaporate from the ocean and in doing so, the ocean loses heat at a rate of

$$Q_{lat} = -L(\rho_{fw}E) \qquad [2]$$

where Q_{lat} is the latent heat flux, L is the latent heat of evaporation, ρ_{fw} is the freshwater density, and E is the rate of evaporation. Q_{lat} has units $W\,m^{-2}$, and $(\rho_{fw}E)$ has units $kg\,s^{-1}\,m^{-2}$. The latent heat flux is nearly always larger than the sensible heat flux due to conduction. When the evaporated moisture condenses in the atmosphere to form clouds, heat is released, affecting the large-scale wind patterns.

Air–sea heat and moisture transfer occur through turbulent processes and is amplified by sea spray, bubble production, and wave breaking. Sensible and latent heat loss thus also depend on the speed of the surface wind relative to the ocean surface flow, $|\mathbf{u}_a - \mathbf{u}_s|$. Using similarity arguments, the latent (Q_{lat}) and sensible (Q_{sen}) heat fluxes can be expressed in terms of 'bulk' properties at and near the ocean surface:

$$Q_{lat} = \rho_a L C_E |\mathbf{u}_a - \mathbf{u}_s|(q_a - q_s) \qquad [4]$$

$$Q_{sen} = \rho_a c_{pa} C_H |\mathbf{u}_a - \mathbf{u}_s|(T_a - T_s) \qquad [4]$$

where ρ_a is the air density, c_{pa} is the specific heat of air, C_E and C_H are the transfer coefficients of latent and sensible heat flux, q_s is the saturated specific humidity at T_s, the SST, and q_a and T_s are, respectively, the specific humidity and temperature of the air at a few meters above the air–sea interface. The sign convention used here is that a negative flux tends to cool the ocean surface. The transfer coefficients, C_E and C_H, depend upon the wind speed and stability properties of the atmospheric boundary layer, making estimations of the heat fluxes quite difficult. Most algorithms estimate the turbulent heat fluxes iteratively, using first estimates of the heat fluxes to compute the transfer coefficients. Further, the dependence of heat flux on wind speed and SST causes the system to be coupled since the heat fluxes can change the wind speed and SST.

Figure 2 shows the climatological net surface heat flux, Q_0, and SST for the entire globe. Several patterns are evident. (Note that the spatial structure of the climatological latent heat flux can be inferred from the climatological evaporation shown in **Figure 3(a)**.) In general, the Tropics are heated more than the poles, causing warmer SST in the tropics and cooler SST at the poles. Also, there are significant zonal asymmetries in both the net surface heat flux and SST. The largest ocean surface heat losses occur over the mid-latitude western boundary currents. In these regions, latent and sensible heat loss are enhanced due to the strong winds which are cool and dry as they blow off the continent and over the warm water carried poleward by the western boundary currents. In contrast, the ocean's latent and sensible heat loss are reduced in the eastern boundary region where marine winds blow over the cool water. Consequently, the eastern boundary is a region where the ocean gains heat from the atmosphere. These spatial patterns exemplify the rich variability in the ocean–atmosphere climate system that occurs on a variety of spatial and temporal scales. In particular, seasonal conditions can often be quite different from mean climatology. The seasonal warming and cooling in the north Pacific and eastern equatorial Pacific are discussed later.

Thermal and Haline Buoyancy Fluxes

Since the density of seawater depends on temperature and salinity, air–sea heat and moisture fluxes can

(a)

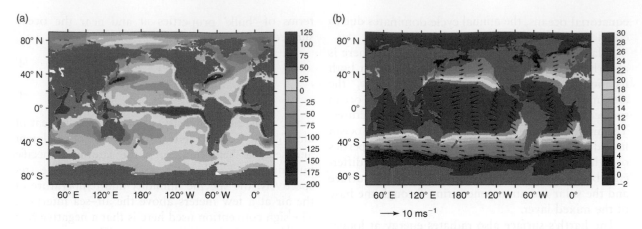

Figure 2   Mean climatologies of (a) net surface heat flux (W m⁻²) and (b) SST (°C), and surface winds (m s⁻¹). The scale vector for the winds is shown below (b). Climatologies provided by da Silva *et al.* (1994). A positive net surface heat flux acts to warm the ocean.

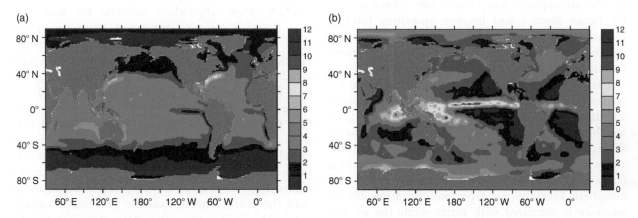

Figure 3   Mean climatologies of (a) evaporation and (b) precipitation from da Silva *et al.* (1994). Both have units of mm day⁻¹ and share the scale shown on the right.

change the surface density, making the water column more or less buoyant. Specifically, the net surface heat flux (Q_0), rate of evaporation (E), and precipitation (P) can be expressed as a buoyancy flux (B_0):

$$B_0 = -g\alpha Q_0/(\rho c_p) + g\beta(E - P)S_0 \qquad [5]$$

where g is gravity, ρ is ocean density, c_p is specific heat of water, S_0 is surface salinity, α is the effective thermal expansion coefficient ($-\rho^{-1}\partial\rho/\partial T$), and β is the effective haline contraction coefficient ($-\rho^{-1}\partial\rho/\partial S$). Q_0 has units W m⁻² and E and P have units m s⁻¹. B_0 has units m² s⁻³ and can be interpreted (when multiplied by density and integrated over a volume) as the buoyant production of turbulent kinetic energy (or destruction of available potential energy). A negative (i.e., downward) buoyancy flux, due to either surface warming or precipitation, tends to make the ocean surface more buoyant and stable. Conversely, a positive buoyancy flux, due to either

surface cooling or evaporation, tends to make the ocean surface less buoyant. As the water column loses buoyancy, it can become convectively unstable with heavy water lying over lighter water. Turbulent kinetic energy, generated by the ensuing convective overturning, can then cause deeper, generally cooler water to be entrained and mixed into the surface mixed layer (**Figure 1**). Thus entrainment mixing typically causes the SST to cool and the mixed layer to deepen. As discussed in the next section, entrainment mixing can also be generated by wind forcing, through wind stirring and shear at the base of the mixed layer.

Figure 3 shows the climatological evaporation and precipitation fields. Note that in terms of buoyancy, a 20 W m⁻² heat flux is approximately equivalent to a 5 mm day⁻¹ rain rate. Thus, in some regions of the world oceans, the freshwater flux term in eqn [5] dominates the buoyancy flux, and hence is a major factor in the mixed layer thermodynamics. For example, in the tropical regions, heavy precipitation can result in a surface-trapped freshwater pool that forms a shallower mixed layer within a deeper, nearly isothermal layer. The difference between the shallower mixed layer of uniform density and the deeper isothermal layer is referred to as a salinity-stratified barrier layer. As the name suggests, a barrier layer can effectively limit turbulent mixing of heat between the ocean surface and the deeper thermocline since the barrier layer water has nearly the same temperature as the mixed layer.

In subpolar latitudes, freshwater fluxes can also dominate the surface layer buoyancy profile. During the winter season, atmospheric cooling of the ocean, and stronger wind mixing leaves the water-column isothermal to great depths. Then, wintertime ice formation extracts fresh water from the surface layer, leaving a saltier brine that further increases the surface density, decreases the buoyancy, and enhances the deep convection. This process can lead to deep-water formation as the cold and salty dense water sinks and spreads horizontally, forcing the deep, slow thermohaline circulation. Conversely, in summer when the ice shelf and icebergs melt, fresh water is released, and the density in the surface layer is reduced so that the resultant stable halocline (pycnocline) inhibits the sinking of water.

Wind Forcing

The influence of the winds on the ocean circulation and mass field cannot be overstated. Wind blowing over the ocean surface causes a tangential stress ('wind stress') at the interface which acts as a vertical flux of horizontal momentum. Similar to the air–sea fluxes of heat and moisture, this air–sea flux of horizontal momentum, τ_0, can be expressed in terms of bulk properties as

$$\tau_0 = \rho_a C_D |\mathbf{u_a} - \mathbf{u_s}|(\mathbf{u_a} - \mathbf{u_s}) \qquad [6]$$

where ρ_a is the air density, and C_D is the drag coefficient. The direction of the stress is determined by the orientation of the surface wind, $\mathbf{u_a}$, relative to the ocean surface flow, $\mathbf{u_s}$. The units of the surface wind stress are $N\ m^{-2}$. Wind stress can also be expressed in terms of an oceanic frictional velocity, u_* (i.e., $\tau_0 = \rho u_*^2$). With frictional velocity related to the wind-generated velocity shear through the nondimensional 'von Kármán constant', κ, the shear production of turbulent kinetic energy by the wind can be expressed as: $\rho(\kappa z)^{-1} u_*^3$.

The mechanisms by which the momentum flux extends below the interface are not well understood. Some of the wind stress goes into generating ocean surface waves. However, most of the wave momentum later becomes available for generating currents through wave breaking, and wave–wave and wave–current interactions. For example, wave–current interactions associated with Langmuir circulation can set up large coherent vortices that carry momentum to near the base of the mixed layer. As with convective overturning, wind stirring can entrain cooler thermocline water into the mixed layer, producing a colder and deeper mixed layer. Likewise, current shear at the base of the mixed layer can cause 'Kelvin–Helmholtz' shear instability that further mix properties within and at the base of the mixed layer.

Variability in the depth of the well-mixed layer can be understood through consideration of the turbulent kinetic energy (TKE) budget. For example, for a stable buoyancy flux (i.e., 'forced convection'), the depth, L_{MO}, at which there is just sufficient mechanical energy available from the wind to mix the input of buoyancy uniformly is referred to as the Monin–Obukhov depth scale:

$$L_{MO} = u_*^3/(\kappa B_0) \qquad [7]$$

At depths below L_{MO}, buoyant suppression of turbulence exceeds the mechanical production and there tends to be little surface-generated turbulence. Typically, however, other terms in the TKE budget cannot be ignored. In particular, for an unstable buoyancy flux (i.e., 'free convection'), the production of potential energy through entrainment becomes important. Thus the mixed layer depth is rarely equivalent to the Monin–Obukhov depth scale.

Over timescales at and longer than roughly a day, the Earth's spinning tends to cause a rotation of the vertical flux of momentum. From the noninertial perspective of an observer on the rotating Earth, the tendency to rotate appears as a force, referred to as the Coriolis force. When the wind ceases, inertial motion tends to continue and accounts for a significant fraction of the total kinetic energy in the global ocean. Vertical shear in the currents and inertial oscillations generated by the winds can cause 'Kelvin–Helmholtz' shear instability and be a significant source of TKE.

For sustained winds beyond the inertial timescale, Coriolis turning causes the wind-forced surface layer transport ('Ekman transport') to be perpendicular to the wind stress. Because the projection of the Earth's axis onto the local vertical axis (direction in which gravity acts) changes sign at the equator, the Ekman transport is to the right of the wind stress in the Northern Hemisphere and to the left of the wind stress in the Southern Hemisphere. Convergence and divergence of this Ekman transport leads to vertical motion which can deform the thermocline and thereby generate pressure gradients that set the sub-surface waters in motion. In this way, meridional variations in the prevailing zonal wind stress drive the steady, large-scale ocean gyres.

The influence of Ekman upwelling on SST can be seen along the eastern boundary of the ocean basins and along the equator (**Figure 2(b)**). Equatorward winds along the eastern boundaries of the Pacific and Atlantic Oceans cause an offshore-directed Ekman transport. Mass conservation requires that this water be replaced with upwelled water, water that is generally cooler than the surface waters outside the upwelling zone. Likewise, in the tropics, prevailing easterly trade winds cause poleward Ekman transport. At the equator, this poleward flow results in substantial surface divergence and upwelling. As with the eastern boundary, equatorial upwelling results in relatively cold SSTs (**Figure 2(b)**). Because of the geometry of the continents, the thermal equator favors the Northern Hemisphere and is generally found several degrees of latitude north of the equator.

In the tropics, winds tend to flow from cool SSTs to warm SSTs, where deep atmospheric convection can occur. Thus, surface wind convergence in the Intertropical Convergence Zone (ITCZ) is associated with the thermal equator, north of the equator. The relationship between the SST gradient and winds accounts for an important coupling mechanism in the Tropics.

The Seasonal Cycle

The North Pacific: A One-dimensional Ocean Response to Wind and Buoyancy Forcing

From 1949 through 1981, a ship (*Ocean Weather Station Papa*) was stationed in the North Pacific at 50° N 145° W with the primary mission of taking routine ocean and atmosphere measurements. The seasonal climatology observed at this site (**Figure 4**) illustrates a classic near-one-dimensional ocean response to wind and buoyancy forcing. A one-dimensional response implies that only the vertical structure of the ocean is changed by the forcing.

During springtime, layers of warmer and lighter water are formed in the upper surface in response to the increasing solar warming. By summer, this heating has built a stable (buoyant), shallow seasonal thermocline that traps the warm surface waters. In fall, storms are more frequent and net cooling sets in. By winter, the surface layer is mixed by wind stirring and convective overturning. The summer thermocline is eroded and the mixed layer deepens to the top of the permanent thermocline.

To first approximation, horizontal advection does not seem to be important in the seasonal heat budget. The progression appears to be consistent with a surface heat budget described by

$$\partial T/\partial t = Q_0/(\rho c_p H) \qquad [8]$$

where $\partial T/\partial t$ is the local time rate of change of the mixed layer temperature, and H is the mixed layer depth. Since only vertical processes (e.g., turbulent mixing and surface forcing) affect the depth and temperature of the mixed layer, the heat budget can be considered one-dimensional.

A similar one-dimensional progression occurs in response to the diurnal cycle of buoyancy forcing associated with daytime heating and nighttime cooling. Mixed layer depths can vary from just a few meters thick during daytime to several tens of meters thick during nighttime. Daytime and nighttime SSTs can sometimes differ by $> 1\,°C$. However, not all regions of the ocean have such an idealized mixed layer seasonal cycle. Our second example shows a more complicated seasonal cycle in which the tropical atmosphere and ocean are coupled.

The Eastern Equatorial Pacific: Coupled Ocean–Atmosphere Variability

Because there is no Coriolis turning at the equator, water and air flow are particularly susceptible to horizontal convergence and divergence. Small changes in the wind patterns can cause large variations in oceanic upwelling, resulting in significant changes in SST and consequently in the atmospheric heating patterns. This ocean and atmosphere coupling thus causes initial changes to the system that perpetuate further changes.

At the equator, the Sun is overhead twice per year: in March and again in September. Therefore one might expect a semiannual cycle in the mixed layer properties. Although this is indeed found in some parts of the equatorial oceans (e.g., in the western equatorial Pacific), in the eastern equatorial Pacific the annual cycle dominates. During the warm season (February–April), the solar equinox causes a maximum in insolation, equatorial SST is warm, and the meridional SST gradient is weak. Consequently, the ITCZ is near the equator, and often a double ITCZ is observed that is symmetric about the equator. The weak winds associated with the ITCZ cause a reduction in latent heat loss, wind stirring, and upwelling, all of which lead to further warming of the equatorial SSTs. Thus the warm SST and surface heating are mutually reinforcing.

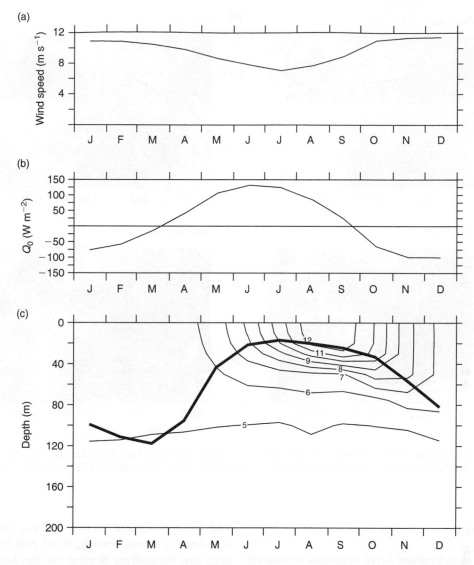

Figure 4 Seasonal climatologies at the ocean weather station *Papa* in the north Pacific: (a) Wind speed, (b) net surface heat flux, and (c) upper ocean temperature. The bold line represents the base of the ocean mixed layer defined as the depth where the temperature is 0.5 °C cooler than the surface temperature. Wind speed and net surface heat flux climatologies are from da Silva *et al.* (1994).

Beginning in about April–May, SSTs begin to cool in the far eastern equatorial Pacific, perhaps in response to southerly winds associated with the continental monsoon. The cooler SSTs on the equator cause an increased meridional SST gradient that intensifies the southerly winds and the SST cooling in the far eastern Pacific. As the meridional SST gradient increases, the ITCZ begins to migrate northward. Likewise, the cool SST anomaly in the far east sets up a zonal SST gradient along the equator that intensifies the zonal trade winds to the west of the cool anomaly. These enhanced trade winds then produce SST cooling (through increased upwelling, wind stirring, and latent heat loss) that spreads westward (**Figure 5**).

By September, the equatorial cold tongue is fully formed. Stratus clouds, which tend to form over the very cool SSTs in the tropical Pacific, cause a reduction in solar radiation, despite the equinoctial increase. The large meridional gradient in SST associated with the fully formed cold tongue causes the ITCZ to be at its northernmost latitude. After the cold tongue is fully formed, the reduced zonal SST gradient within the cold tongue causes the trade winds to weaken there, leading to reduced SST cooling along the equator. Finally, by February, the increased solar radiation associated with the approaching vernal equinox causes the equatorial SSTs to warm and the cold tongue to disappear, bringing the coupled system back to the warm season conditions.

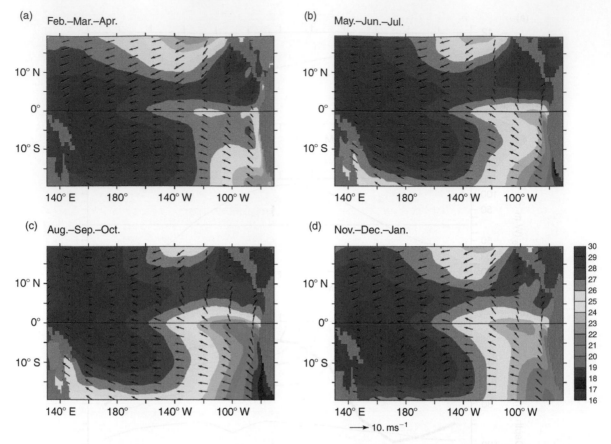

Figure 5 Seasonal climatologies of the tropical Pacific SST (°C) and wind (m s⁻¹): (a) Feb.–Mar.–Apr.; (b) May–Jun.–Jul.; (c) Aug.–Sep.–Oct.; and (d) Nov.–Dec.–Jan. Climatologies for wind are from da Silva *et al.* (1994) and for SSTs are from Reynolds and Smith (1994).

Conclusion

Because the ocean mixed layer responds so rapidly to surface-generated turbulence through wind- and buoyancy-forced processes, the surface mixed layer can often be modeled successfully using one-dimensional (vertical processes only) physics. Surface heating and cooling cause the ocean surface to warm and cool; evaporation and precipitation cause the ocean surface to become saltier and fresher. Stabilizing buoyancy forcing, whether from net surface heating or precipitation, stratifies the surface and isolates it from the deeper waters, whereas wind stirring and destabilizing buoyancy forcing generate surface turbulence that cause the surface properties to mix with deeper water. Eventually, however, one-dimensional models drift away from observations, particularly in regions with strong ocean–atmosphere coupling and oceanic current structures. The effects of horizontal advection are explicitly not included in one-dimensional models. Likewise, vertical advection depends on horizontal convergences and divergences and therefore is not truly a one-dimensional process. Finally, wind and buoyancy forcing can themselves depend on the horizontal SST patterns, blurring the distinction between forcing and response.

Although the mixed layer is the principal region of wind and buoyancy forcing, ultimately the effects are felt throughout the world's oceans. Both the wind-driven motion below the mixed layer and the thermohaline motion in the relatively more quiescent deeper ocean originate through forcing in the surface layer that causes an adjustment in the mass field (i.e., density profile). In addition, buoyancy and wind forcing in the upper ocean define the property characteristics for all the individual major water masses found in the world oceans. On a global scale, there is surprisingly little mixing between water masses once they acquire the characteristic properties at their formation region and are vertically subducted or convected from the active surface layer. As these subducted water masses circulate through the global oceans and later outcrop, they can contain the memory of their origins at the surface through their water mass properties and thus can potentially induce decadal and centennial modes of variability in the ocean–atmosphere climate system.

Nomenclature

B_0	surface buoyancy flux
C_D	drag coefficient
C_E	latent heat flux transfer coefficient
C_H	sensible heat flux transfer coefficient
c_p	specific heat capacity of water
E	rate of evaporation
g	gravity
H	mixed layer depth
L	latent heat of evaporation
L_{MO}	Monin–Obkuhov depth scale
P	precipitation
q_a	specific humidity of the air
q_s	saturated specific humidity at the sea surface temperature
Q_0	net surface heat flux entering ocean
Q_{lat}	latent heat flux due to evaporation
Q_{lw}	net infrared (long-wave) radiation
Q_{sen}	sensible heat flux
Q_{sw}	net solar (shortwave) radiation
S_0	surface salinity
T_a	air temperature
T_s	surface ocean temperature
$\mathbf{u}_a$	air velocity
$\mathbf{u}_s$	surface ocean velocity
u_*	oceanic frictional velocity
α	thermal expansion coefficient
β	haline contraction coefficient
κ	von Kármán constant
ρ	ocean density
ρ_a	air density
ρ_{fw}	density of fresh water
τ_0	wind stress

See also

Breaking Waves and Near-Surface Turbulence. Langmuir Circulation and Instability. Penetrating Shortwave Radiation. Sea Surface Exchanges of Momentum, Heat, and Fresh Water Determined by Satellite Remote Sensing. Upper Ocean Heat and Freshwater Budgets. Upper Ocean Time and Space Variability. Upper Ocean Vertical Structure. Wind Driven Circulation.

Further Reading

da Silva AM, Young CC, and Levitus S (1994) *Atlas of Surface Marine Data 1994, Vol. 1: Algorithms and Procedures, NOAA Atlas NESDIS 6*. Washington, DC: US Department of Commerce.

Fairall CF, Bradley EF, Hare JE, Grachev AA, and Edson JB (2003) Bulk parameterization of air–sea fluxes: Updates and verification for the COARE algorithm. *Journal of Climate* 16: 571–591.

Kraus EB and Businger JA (1994) *Oxford Monographs on Geology and Geophysics: Atmosphere–Ocean Interaction*, 2nd edn. New York: Oxford University Press.

Large WG (1996) An observational and numerical investigation of the climatological heat and salt balances at OWS Papa. *Journal of Climate* 9: 1856–1876.

Niiler PP and Kraus EB (1977) One-dimensional models of the upper ocean. In: Kraus EB (ed.) *Modelling and Prediction of the Upper Layers of the Ocean*, pp. 143–172. New York: Pergamon.

Philander SG (1990) *El Niño, La Niña, and the Southern Oscillation*. San Diego, CA: Academic Press.

Price JF, Weller RA, and Pinkel R (1986) Diurnal cycling: Observations and models of the upper ocean response to diurnal heating, cooling, and wind mixing. *Journal of Geophysical Research* 91: 8411–8427.

Reynolds RW and Smith TM (1994) Improved global sea surface temperature analysis using optimum interpolation. *Journal of Climate* 7: 929–948.

PLANKTON

PLANKTON

PLANKTON

M. M. Mullin[†], Scripps Institution of Oceanography, La Jolla, CA, USA

The category of marine life known as plankton represents the first step in the food web of the ocean (and of large bodies of fresh water), and components of the plankton are food for many of the fish harvested by humans and for the baleen whales. The plankton play a major role in cycling of chemical elements in the ocean, and thereby also affect the chemical composition of sea water and air (through exchange of gases between the sea and the overlying atmosphere). In the parts of the ocean where planktonic life is abundant, the mineral remains of members of the plankton are major contributors to deep-sea sediments, both affecting the chemistry of the sediments and providing a micropaleontological record of great value in reconstructing the earth's history.

'Plankton' refers to 'drifting', and describes organisms living in the water column (rather than on the bottom – the benthos) and too small and/or weak to move long distances independently of the ocean's currents. However, the distinction between plankton and nekton (powerfully swimming animals) can be difficult to make, and is often based more on the traditional method of sampling than on the organisms themselves.

Although horizontal movement of plankton at kilometer scales is passive, the metazoan zooplankton nearly all perform vertical migrations on scales of 10s to 100s of meters. This depth range can take them from the near surface lighted waters where the phytoplankton grow, to deeper, darker and usually colder environments. These migrations are generally diurnal, going deeper during the day, or seasonal, moving to deeper waters during the winter months to return to the surface around the time that phytoplankton production starts. The former pattern can serve various purposes: escaping visual predators and scanning the watercolumn for food. (It should be noted that predators such as pelagic fish also migrate diurnally.) Seasonal descent to greater depths is a common feature for several copepod species and may conserve energy at a time when food is scarce in the upper layers. However, vertical migration has another role. Because of differences in current strength and direction between surface and deeper layers in the ocean, time spent in deeper water acts as a transport mechanism relative to the near surface layers. On a daily basis this process can take plankton into different food concentrations. Seasonally, this effective 'migration' can complete a spatial life cycle.

The plankton can be subdivided along functional lines and in terms of size. The size category, picoplankton ($0.2–2.0\,\mu m$), is approximately equivalent to the functional category, bacterioplankton; most phytoplankton (single-celled plants or colonies) and protozooplankton (single-celled animals) are nano- or microplankton ($2.0–20\,\mu m$ and $20–200\,\mu m$, respectively). The metazoan zooplankton (animals, the 'insects of the sea') includes large medusae and siphonophores several meters in length. Size is more important in oceanic than in terrestrial ecosystems because most of the plants are small (the floating seaweed, *Sargassum*, being the notable exception), predators generally ingest their prey whole (there is no hard surface on which to rest prey while dismembering it), and the early life stages of many types of zooplankton are approximately the same size as the larger types of phytoplankton. Therefore, while the dependence on light for photosynthesis is characteristic of the phytoplankton, the concepts of 'herbivore' and 'carnivore' can be ambiguous when applied to zooplankton, since potential plant and animal prey overlap in size and can be equivalent sources of food. Though rabbits do not eat baby foxes on land, analogous ontogenetic role-switching is very common in the plankton.

Among the animals, holoplanktonic species are those that spend their entire life in the plankton, whereas many benthic invertebrates have meroplanktonic larvae that are temporarily part of the plankton. Larval fish are also a temporary part of the plankton, becoming part of the nekton as they grow. There are also terms or prefixes indicating special habitats, such as 'neuston' to describe zooplanktonic species whose distribution is restricted to within a few centimeters of the sea's surface, or 'abyssoplankton' to describe animals living only in the deepest waters of the ocean. Groups of such species form communities (see below).

Since the phytoplankton depend on sunlight for photosynthesis, this category of plankton occurs almost entirely from the surface to 50–200 m of the ocean – the euphotic depth (where light intensity is 0.1–1% of full surface sunlight). Nutrients such as

---

[†] Deceased.

nitrate and phosphate are incorporated into protoplasm in company with photosynthesis, and returned to dissolved form by excretion or remineralization of dead organic matter (particulate detritus). Since much of the latter process occurs after sinking of the detritus, uptake of nutrients and their regeneration are partially separated vertically. Where and when photosynthesis is proceeding actively and vertical mixing is not excessive, a near-surface layer of low nutrient concentrations is separated from a layer of abundant nutrients, some distance below the euphotic depth, by a nutricline (a layer in which nutrient concentrations increase rapidly with depth). Therefore, the spatial and temporal relations between the euphotic depth (dependent on light intensity at the surface and the turbidity of the water), the nutricline, and the pycnocline (a layer in which density increases rapidly with depth) are important determinants of the abundance and productivity of phytoplankton.

Zooplankton is typically more concentrated within the euphotic zone than in deeper waters, but because of sinking of detritus and diel vertical migration of some species into and out of the euphotic zone, organic matter is supplied and various types of zooplankton (and bacterioplankton and nekton) can be found at all depths in the ocean. An exception is anoxic zones such as the deep waters of the Black Sea, although certainly types of bacterioplankton that use molecules other than oxygen for their metabolism are in fact concentrated there.

Even though the distributions of planktonic species are dependent on currents, species are not uniformly distributed throughout the ocean. Species tend to be confined to particular large water masses, because of physiological constraints and inimical interactions with other species. Groups of species, from small invertebrates to active tuna, seem to 'recognize' the same boundaries in the oceans, in the sense that their patterns of distribution are similar. Such groups are called 'assemblages' (when emphasizing their statistical reality, occurring together more than expected by chance) or 'communities' (when emphasizing the functional relations between the members in food webs), though terms such as 'biocoenoses' can be found in older literature. Thus, one can identify 'central water mass,' 'subantarctic,' 'equatorial,' and 'boreal' assemblages associated with water masses defined by temperature and salinity; 'neritic' (i.e. nearshore) versus 'oceanic' assemblages with respect to depth of water over which they occur, and 'neustonic' (i.e. air–sea interface), 'epipelagic,' 'mesopelagic,' 'bathypelagic,' and 'abyssopelagic' for assemblages distinguished by the depth at which they occur. Within many of these there may be seasonally distinguishable assemblages of organisms, especially those with life spans of less than one year.

Regions which are boundaries between assemblages are sometimes called ecotones or transition zones; they generally contain a mixture of species from both sides, and (as in the transition zone between subpolar and central water mass assemblages) may also have an assemblage of species that occur only in the transition region.

Despite the statistical association between assemblages and water masses or depth zones, it is far from clear that the factor that actually limits distribution is the temperature/salinity or depth that physically defines the water mass or zone. It is likely that a few important species have physiological limits confining them to a zone, and the other members of the assemblage are somehow linked to those species functionally, rather than being themselves physiologically constrained. Limits can be imposed on certain life stage, such as the epipelagic larvae of meso- or bathypelagic species, creating patterns that reflect the environment of the sensitive life stage rather than the adult. Conversely, meroplanktonic larvae, such as the phyllosome of spiny lobsters, can often be found far away from the shallow waters that are a suitable habitat for the adults.

See also

Continuous Plankton Recorders.

Further Reading

Cushing DH (1995) *Population Production and Regulation in the Sea*. Cambridge: Cambridge University Press.

Longhurst A (1998) *Ecological Geography of the Sea*. New York: Academic Press.

Mullin MM (1993) *Webs and Scales*. Seattle: University of Washington Press.

PRIMARY PRODUCTION METHODS

J. J. Cullen, Department of Oceanography,
Halifax, NS, Canada

Introduction

Primary production is the synthesis of organic material from inorganic compounds, such as CO_2 and water. The synthesis of organic carbon from CO_2 is commonly called carbon fixation: CO_2 is fixed by both photosynthesis and chemosynthesis. By far, photosynthesis by phytoplankton accounts for most marine primary production. Carbon fixation by macroalgae, microphytobenthos, chemosynthetic microbes, and symbiotic associations can be locally important.

Only the measurement of marine planktonic primary production will be discussed here. These measurements have been made for many decades using a variety of approaches. It has long been recognized that different methods yield different results, yet it is equally clear that the variability of primary productivity, with depth, time of day, season, and region, has been well described by most measurement programs. However, details of these patterns can depend on methodology, so it is important to appreciate the uncertainties and built-in biases associated with different methods for measuring primary production.

Definitions

Primary production is centrally important to ecological processes and biogeochemical cycling in marine systems. It is thus surprising, if not disconcerting, that (as discussed by Williams in 1993), there is no consensus on a definition of planktonic primary productivity, or its major components, net and gross primary production. One major reason for the problem is that descriptions of ecosystems require clear conceptual definitions for processes (e.g., net daily production of organic material by phytoplankton), whereas the interpretation of measurements requires precise operational definitions, for example, net accumulation of radiolabeled CO_2 in particulate matter during a 24 h incubation. Conceptual and operational definitions can be reconciled for particular approaches, but no one set of definitions is sufficiently general, yet detailed, to serve as a framework both for measuring planktonic primary

production with a broad variety of methods and for interpreting the measurements in a range of scientific contexts. It is nonetheless useful to define three components of primary production that can be estimated from measurements in closed systems:

- **Gross primary production** (P_g) is the rate of photosynthesis, not reduced for losses to excretion or to respiration in its various forms
- **Net primary production** (P_n) is gross primary production less losses to respiration by phytoplankton
- **Net community production** (P_{nc}) is net primary production less losses to respiration by heterotrophic microorganisms and metazoans.

Other components of primary production, such as new production, regenerated production, and export production, must be characterized to describe food-web dynamics and biogeochemical cycling. As pointed out by Platt and Sathyendranath in 1993, in any such analysis, great care must be taken to reconcile the temporal and spatial scales of both the measurements and the processes they describe.

Marine primary production is commonly expressed as grams or moles of carbon fixed per unit volume, or pet unit area, of sea water per unit time. The timescale of interest is generally 1 day or 1 year. Rates are characterized for the euphotic zone, commonly defined as extending to the depth of 1% of the surface level of photosynthetically active radiation (PAR: 400–700 nm). This convenient definition of euphotic depth (sometimes simplified further to three times the depth at which a Secchi disk disappears) is a crude and often inaccurate approximation of where gross primary production over 24 h matches losses to respiration and excretion by phytoplankton. Regardless, rates of photosynthesis are generally insignificant below the depth of 0.1% surface PAR.

Photosynthesis and Growth of Phytoplankton

Primary production is generally measured by quantifying light-dependent synthesis of organic carbon from CO_2 or evolution of O_2 consistent with the simplified description of photosynthesis as the reaction:

$$CO_2 + 2H_2O \xrightarrow{\sim 8hv} (CH_2O) + H_2O + O_2 \quad [1]$$

Absorbed photons are signified by hv and the carbohydrates generated by photosynthesis are represented as CH_2O. Carbon dioxide in sea water is found in several chemical forms which exchange quickly enough to be considered in aggregate as total CO_2 (TCO_2). In principle, photosynthesis can be quantified by measuring any of three light-dependent processes: (1) the increase in organic carbon; (2) the decrease of TCO_2; or (3) the increase of O_2. However, growth of phytoplankton is not so simple: since phytoplankton are composed of proteins, lipids, nucleic acids, and other compounds besides carbohydrate, both photosynthesis and the assimilation of nutrients are required. Consequently, many chemical transformations are associated with primary production, and eqn [1] does not accurately describe the process of light-dependent growth.

It is therefore useful to describe the growth of phytoplankton (i.e., net primary production) with a more general reaction that describes how transformations of carbon and oxygen depend on the source of nutrients (particularly nitrogen) and on the chemical composition of phytoplankton. For growth on nitrate:

$$1.0NO_3^- + 5.7CO_2 + 5.4H_2O$$
$$\rightarrow (C_{5.7}H_{9.8}O_{2.3}N) + 8.25O_2 + 1.0OH^- \quad [2]$$

The idealized organic product, $C_{5.7}H_{9.8}O_{2.3}N$, represents the elemental composition of phytoplankton. Ammonium is more reduced than nitrate, so less water is required to satisfy the demand for reductant:

$$1.0NH_4^+ + 5.7CO_2 + 3.4H_2O$$
$$\rightarrow (C_{5.7}H_{9.8}O_{2.3}N) + 6.25O_2 + 1.0H^+ \quad [3]$$

The photosynthetic quotient (PQ; $mol\,mol^{-1}$) is the ratio of O_2 evolved to inorganic C assimilated. It must be specified to convert increases of oxygen to the synthesis of organic carbon. For growth on nitrate as described by eqn [2], PQ is $1.45\,mol\,mol^{-1}$; with ammonium as the source of N, PQ is 1.10. The photosynthetic quotient also reflects the end products of photosynthesis, the mixture of which varies according to environmental conditions and the species composition of phytoplankton. For example, if the synthesis of carbohydrate is favored, as can occur in high light or low nutrient conditions, PQ is lower because the reaction described in eqn [1] becomes more important. Uncertainty in PQ is often ignored. This can be justified when the synthesis of organic carbon is measured directly, but large errors can be introduced when attempts are made to infer carbon fixation from the dynamics of oxygen. Excretion of organic material would have a small influence on PQ and is not considered here.

Approaches

Primary production can be estimated from chlorophyll (from satellite color or *in situ* fluorescence) if carbon uptake per unit of chlorophyll is known. Therefore, 'global' estimates of primary production depend on direct measurements by incubation. The technical objectives are to obtain a representative sample of sea water, contain it so that no significant exchange of materials occurs, and to measure light-dependent changes in carbon or oxygen during incubations that simulate the natural environment. Methods vary widely, and each approach involves compromises between needs for logistical convenience, precision, and the simulation of natural conditions. Each program of measurement involves many decisions, each of which has consequences for the resulting measurements. Several options are listed in **Tables 1** and **2** and discussed below.

Light-dependent Change in Dissolved Oxygen

The light-dark oxygen method is a standard approach for measuring photosynthesis in aquatic systems, and it was the principal method for measuring marine primary production until it was supplanted by the ^{14}C method, which is describged below. Accumulation of oxygen in a clear container (light bottle) represents net production by the enclosed community, and the consumption of oxygen in a dark bottle is a measure of respiration. Gross primary production is estimated by subtracting the dark bottle result from that for the light bottle. It is thus assumed that respiration in the light equals that in the dark. As documented by Geider and Osborne in their 1992 monograph, this assumption does not generally hold, so errors in estimation of the respiratory component of P_g must be tolerated unless isotopically labelled oxygen is used (see below).

Methods based on the direct measurement of oxygen are less sensitive than techniques using the isotopic tracer ^{14}C. However, careful implementation of procedures using automated titration or pulsed oxygen electrodes can yield useful and reliable data, even from oligotrophic waters of the open ocean. Interpretation of results is complicated by containment effects common to all methods for direct measurement of primary production (see below). Also, a value for photosynthetic quotient must be assumed in order to infer carbon fixation from oxygen production. Abiotic consumption of oxygen through photochemical reactions with dissolved organic matter can also contribute to the measurement, primarily near the surface, where the effective ultraviolet wavelengths penetrate.

Table 1 Measurements that can be related to primary production

Measurement	Advantages	Disadvantages	Comments
Change in TCO_2	Direct measure of net inorganic C fixation	Relatively insensitive: small change relative to large background	Not generally practical for open-ocean work
Change in oxygen concentration (high precision titration)	Direct measures of O_2 dynamics can yield estimates of net and gross production	Small change relative to large background Interpretation of light-dark incubations is not simple	Very useful if applied with great care. Requires knowledge of PQ to convert to C-fixation
Incorporation of ^{14}C-bicarbonate into organic material (radioactive isotope)	Very sensitive and relatively easy. Small volumes can be used and many samples can be processed	Tracer dynamics complicate interpretations Radioactive – requires special precautions and permission	The most commonly used method in oceanography
Incorporation of ^{13}C-bicarbonate into organic material (stable isotope)	No problems with radioactivity	Less sensitive and more work than ^{14}C method Larger volumes required	A common choice when ^{14}C method is impractical
Measurement of $^{18}O_2$ production from $H_2^{18}O$	Measures photosynthesis without interference from respiration	Requires special equipment	A powerful research tool, not generally used for routine measurements

Table 2 Approaches for incubating samples for the measurement of primary production

Incubation system	Advantages	Disadvantages	Comments
Incubation in situ	Best simulation of the natural field of light and temperature	Limits mobility of the ship Vertical mixing is not simulated Artifacts possible if deployed or recovered in the light	Not perfect, but a good standard method if a station can be occupied all day
Simulated in situ	Many stations can be surveyed Easy to conduct time-courses and experimental manipulations	Special measures must be taken to stimulate spectral irradiance and temperatureo Vertical mixing not simulated	Commonly used when many stations must be sampled. Significant errors possible if incubated samples are exposed to unnatural irradiance and temperature
Photosynthesis versus irradiance (P versus E) incubator (^{14}C)	Data can be used to model photosynthesis in the water column With care, vertical mixing can be addressed	Extra expenses and precautions are required Spectral irradiance is not matched to nature Results depend on timescale of measurement Analysis can be tricky	A powerful approach when applied with caution

Light-dependent Change in Dissolved Inorganic Carbon

Changes in TCO_2 during incubations of sea water can be measured by several methods. Uncertainties related to biological effects on pH-alkalinity-TCO_2 relationships are avoided through the use of coulometric titration or infrared gas analysis after acidification. Measurement of gross primary production and net production of the enclosed community is like that for the light-dark oxygen method, but there is no need to assume a photosynthetic quotient. However, precision of the analyses is not quite as good as

for bulk oxygen methods. Extra procedures, such as filtration, would be required to assess precipitation of calcium carbonate (e.g., by coccolithophores) and photochemical production of CO_2. These processes cause changes in TCO_2 that are not due to primary production. The TCO_2 method is not used routinely for measurement of primary production in the ocean.

The ^{14}C Method

Marine primary production is most commonly measured by the ^{14}C method, which was introduced by Steemann Nielsen in 1952. Samples are collected and the dissolved inorganic carbon pool is labeled with a known amount of radioactive ^{14}C-bicarbonate. After incubation in clear containers, carbon fixation is quantified by liquid scintillation counting to detect the appearance of ^{14}C in organic form. Generally, organic carbon is collected as particles on a filter. Both dissolved and particulate organic carbon can be quantified by analyzing whole water after acidification to purge the inorganic carbon. It is prudent to correct measurements for the amount of label incorporated during incubations in the dark. The ^{14}C method can be very sensitive, and good precision can be obtained through replication and adequate time for scintillation counting. The method has drawbacks, however. Use of radioisotopes requires special procedures for handling and disposal that can greatly complicate or preclude some field operations. Also, because ^{14}C is added as dissolved inorganic carbon and gradually enters pools of particulate and dissolved matter, the dynamics of the labeled carbon cannot accurately represent all relevant transformations between organic and inorganic carbon pools. For example, respiration cannot be quantified directly. The interpretation of ^{14}C uptake (discussed below) is thus anything but straightforward.

The [13]C Method

The ^{13}C method is similar to the ^{14}C method in that a carbon tracer is used. Bicarbonate enriched with the stable isotope ^{13}C is added to sea water and the incorporation of CO_2 into particulate matter is followed by measuring changes in the [13]C:[12]C ratio of particles relative to that in the TCO_2 pool. Isotope ratios are measured by mass spectrometry or emission spectrometry. Problems associated with radioisotopes are avoided, but the method can be more cumbersome than the ^{14}C method (e.g., larger volumes are generally needed) and some sensitivity is lost.

The [18]O Method

Gross photosynthesis can be measured as the production of ^{18}O-labeled O_2 from water labeled with this heavy isotope of oxygen (see eqn [1]). Detection is carried out by mass spectrometry. Net primary production of the enclosed community is measured as the increase of oxygen in the light bottle, and respiration is estimated by difference. In principle, the difference between gross production measured with [18]O and gross production from light–dark oxygen changes is due to light-dependent changes in respiration and photochemical consumption of oxygen. Respiration can also be measured directly by tracking the production of $H_2^{18}O$ from $^{18}O_2$.

The [18]O method is sufficiently sensitive to yield useful results even in oligotrophic waters. It is not commonly used, but when the measurements have been made and compared to other measures of productivity, important insights have been developed.

Methodological Considerations

Many choices are involved in the measurement of primary production. Most influence the results, some more predictably than others. A brief review of methodological choices, with an emphasis on the ^{14}C method, reveals that the measurement of primary production is not an exact science.

Sampling

Every effort should be made to avoid contamination of samples obtained for the measurement of primary production. Concerns about toxic trace elements are especially important in oceanic waters. Trace metal-clean procedures, including the use of specially cleaned GO-FLO sampling bottles suspended from Kevlar[TM] line, prevent the toxic contamination associated with other samplers, particularly those with neoprene closure mechanisms. Frequently, facilitates for trace metal-clean sampling are unavailable. Through careful choice of materials and procedures, it is possible to minimize toxic contamination, but enrichment with trace nutrients such as iron is probably unavoidable. Such enrichment could stimulate the photosynthesis of phytoplankton, but only after several hours or longer.

Exposure of samples to turbulence during sampling can damage the phytoplankton and other microbes, altering measured rates. Also, significant inhibition of photosynthesis can occur when deep samples acclimated to low irradiance are exposed to bright light, even for brief periods, during sampling.

Method of Incubation

Samples of seawater can be incubated *in situ*, under simulated *in situ* (SIS) conditions, or in incubators illuminated by lamps. Each method has advantages and disadvantages (**Table 2**).

Incubation *in situ* ensures the best possible simulation of natural conditions at the depths of sampling. Ideally, samples are collected, prepared, and deployed before dawn in a drifting array. Samples are retrieved and processed after dusk or before the next sunrise. If deployment or retrieval occur during daylight, deep samples can be exposed to unnaturally high irradiance during transit, which can lead to artifactually high photosynthesis and perhaps to counteracting inhibitory damage. Incubation of samples *in situ* limits the number of stations that can be visited during a survey, because the ship must stay near the station in order to retrieve the samples. Specialized systems both capture and inoculate samples *in situ*, thereby avoiding some logistical problems.

Ship operations can be much more flexible if primary productivity is measured using SIS incubations. Water can be collected at any time of day and incubated for 24 h on deck in transparent incubators to measure daily rates. The incubators, or bottles in the incubators, are commonly screened with neutral density filters (mesh or perforated metal screen) to reproduce fixed percentages of PAR at the surface. Light penetration at the station must be estimated to choose the sampling depths corresponding to these light levels. Cooling comes from surface sea water. This system has many advantages, including improved security of samples compared with *in situ* deployment, convenient access to incubations for time-course measurements, and freedom of ship movement after sampling. Because the spectrally neutral attenuation of sunlight by screens does not mimic the ocean, significant errors can be introduced for samples from the lower photic zone where the percentage of surface PAR imposed by a screen will not match the percentage of photosynthetically utilizable radiation (PUR, spectrally weighted for photosynthetic absorption) at the sampling depth. Incubators can be fitted with colored filters to simulate subsurface irradiance for particular water types. Also, chillers can be used to match subsurface temperatures, avoiding artifactual warming of deep samples.

Artificial incubators are used to measure photosynthesis as a function of irradiance (*P* versus *E*). Illumination is produced by lamps, and a variety of methods are used to provide a range of light levels to as many as 24 or more subsamples. Temperature is controlled by a water bath. The duration of incubation generally ranges from about 20 min to several hours, and results are fitted statistically to a *P* versus *E* curve. If *P* versus *E* is determined for samples at two or more depths (to account for physiological differences), results can be used to describe photosynthesis in the water column as a function of irradiance. Such a calculation requires measurement of light penetration in the water and consideration of spectral differences between the incubator and natural waters. Because many samples, usually of small volume, must be processed quickly, only the ^{14}C method is appropriate for most *P* versus *E* measurements in the ocean.

Containers

Ideally, containers for the measurement of primary production should be transparent to ultraviolet and visible solar radiation, completely clean, and inert (**Table 3**). Years ago, soft glass bottles were used. Now it is recognized that they can contaminate samples with trace elements and exclude naturally occurring ultraviolet radiation. Glass scintillation vials are still used for some *P* versus *E* measurements of short duration; checks for effects of contaminants are warranted. Compared with soft glass, laboratory-grade borosilicate glass bottles (e.g., Pyrex^{TM}) have better optical properties, excluding only UV-B (320–400 nm) radiation. Also, they contaminate less. Laboratory-grade glass bottles are commonly used for oxygen measurements. Polycarbonate bottles are favored in many studies because they are relatively inexpensive, unbreakable, and can be cleaned meticulously. Polycarbonate absorbs UV-B and some UV-A 320–400 nm radiation, so near-surface inhibition of photosynthesis can be underestimated. The error can be significant very close to the surface, but not when the entire water column is considered. Teflon^{TM} bottles, more expensive than polycarbonate, are noncontaminating and they transmit both visible and UV (280–4000 nm) radiation. When the primary emphasis is an assessing effects of UV radiation, incubations are conducted in polyethylene bags or in bottles made of quartz or Teflon^{TM}.

The size of the container is an important consideration. Small containers (≤ 50 ml) are needed when many samples must be processed (e.g. for *P* versus *E*) or when not much water is available. However, small samples cannot represent the planktonic assemblage accurately when large, rare organisms or colonies are in the water. Smaller containers have greater surface-to-volume ratios, and thus small samples have greater susceptibility to contamination. If it is practical, larger samples should be used for the measurement of primary production. The problems with large samples are mostly logistical. More water, time, and materials are needed, more radioactive waste is generated, and some measurements can be compromised if handling times are too long.

Duration of Incubation

Conditions in containers differ from those in open water, and the physiological and chemical differences between samples and nature increase as the incubations proceed. Unnatural changes during incubation include: extra accumulation of phytoplankton due to exclusion of grazers; enhanced inhibition of photosynthesis in samples collected from mixed

Table 3 Containers for incubations

Container	Advantages	Disadvantages	Comments
Polycarbonate bottle	Good for minimizing trace element contamination Nearly unbreakable Affordable	Excludes UV radiation Compressible, leading to gas dissolution and filtration problems for deep samples	Many advantages for routine and specialized measurements at sea
Laboratory grade borosilicate glass (e.g., Pyrex™)	More transparent to UV Incompressible	More trace element contamination Breakable	A reasonable choice if compromises are evaluated
Borosilicate glass scintillation vials	Inexpensive Practical choice for P versus E	Contaminate samples with trace elements and Si Exclude UV radiation	Can be used with caution for short-term P versus E measurements
Polyethylene bag	Inexpensive Compact UV-transparent	More difficult to handle Requires caution with respect to contamination	Used for special projects, e.g., effect of UV
Quartz, Teflon™	UV-transparent Teflon™ does not contaminate	Relatively expensive	Used for work assessing effects of UV
Small volume (1–25 ml)	Good for P versus E Samples can be processed by acidification (no filtration)	Cannot sample large, rare phytoplankton evenly Containment effects more likely	Used for P versus E with many replicates
Large volume (1–20 l)	Some containment artifacts are minimized Potential for time-course measurements	More work Longer filtration times with possible artifacts	Required for some types of analysis, e.g., ^{13}C

layers and incubated at near-surface irradiance; stimulation of growth due to contamination with a limiting trace nutrient such as iron; and poisoning of phytoplankton with a contaminant, such as copper. When photosynthesis is measured with a tracer, the distribution of the tracer among pools changes with time, depending on the rates of photosynthesis, respiration, and grazing. All of these effects, except possibly toxicity, are minimized by restricting the time of incubation, so a succession of short incubations, or P versus E measurements, can in principle yield more accurate data than a day-long incubation. This requires much effort, however, and extrapolation of results to daily productivity is still uncertain. The routine use of dawn-to-dusk or 24 h incubations may be subject to artifacts of containment, but it has the advantage of being much easier to standardize.

Filtration or Acidification

Generally, an incubation with ^{14}C or ^{13}C is terminated by filtration. Labeled particles are collected on a filter for subsequent analysis. Residual dissolved inorganic carbon can be removed by careful rinsing with filtered sea water; exposure of the filter to acid purges both dissolved inorganic carbon and precipitated carbonate. The choice of filter can influence the result. Whatman GF/F glass-fiber filters, with nominal pore size 0.7 μm, are commonly used and widely (although not universally) considered to capture all sizes of phytoplankton quantitatively. Perforated filters with uniform pore sizes ranging from 0.2 to 5 μm or more can be used for size-fractionation. Particles larger than the pores can squeeze through, especially when vacuum is applied. The filters are also subject to clogging, leading to retention of small particles.

Labeled dissolved organic carbon, including excreted photosynthate and cell contents released through 'sloppy feeding' of grazers, is not collected on filters. These losses are generally several percent of total or less, but under some conditions, excretion can be much more. When ^{14}C samples are processed with a more cumbersome acidification and bubbling technique, both dissolved and particulate organic carbon is measured.

Interpretation of Carbon Uptake

Because the labeled carbon is initially only in the inorganic pool, short incubations with ^{14}C (≤ 1 h) characterize something close to gross production. As incubations proceed, cellular pools of organic carbon are labeled, and some ^{14}C is respired. Also, some excreted ^{14}C organic carbon is assimilated by heterotrophic microbes, and some of the phytoplankton are consumed by grazers. So, with time, the measurement comes closer to an estimate of the net primary production of the enclosed community (**Table 4**).

Table 4 Incubation times for the measurement of primary production

Incubation time	Advantages	Disadvantages	Comments
Short (≤ 1 h)	Little time for unnatural physiological changes	Usually requires artificial illumination Uncertain extrapolation to daily rates in nature	Closer to P_g
1–6 h	Convenient Appropriate for some process studies	Uncertain extrapolation to daily rates in nature	Used for P versus E, especially with larger samples
Dawn–dusk	Good for standardization of methodology	Limits the number of stations that can be sampled Containment effects Vertical mixing is not simulated, leading to artifacts	A good choice for standard method using *in situ* incubation Closer to P_{nc} near the surface; close to P_g deep in the photic zone
24 h	Good for standardization of methodology	Results may vary depending on start time. Longer time for containment effects to act	A good standard for SIS incubations. Close to P_{nc} near the surface; closer to P_g deep in the photic zone

However, many factors, including the ratio of photosynthesis to respiration, influence the degree to which ^{14}C uptake resembles gross versus net production. Consequently, critical interpretation of ^{14}C primary production measurements requires reference to models of carbon flow in the system.

Conclusions

Primary production is not like temperature, salinity or the concentration of nitrate, which can in principle be measured exactly. It is a biological process that cannot proceed unaltered when phytoplankton are removed from their natural surroundings. Artifacts are unavoidable, but many insults to the sampled plankton can be minimized through the exercise of caution and skill. Still, the observed rates will be influenced by the methods chosen for making the measurements. Interpretation is also uncertain: the ^{14}C method is the standard operational technique for measuring marine primary production, yet there are no generally applicable rules for relating ^{14}C measurements to either gross or net primary production.

Fortunately, uncertainties in the measurements and their interpretation, although significant, are not large enough to mask important patterns of primary productivity in nature. Years of data on marine primary production have yielded information that has been centrally important to our understanding of marine ecology and biogeochemical cycling. Clearly, measurements of marine primary production are useful and important for understanding the ocean. It is nonetheless prudent to recognize that the measurements themselves require circumspect interpretation.

See also

Carbon Cycle. Ocean Carbon System, Modeling of. Primary Production Processes. Tracers of Ocean Productivity.

Further Reading

Geider RJ and Osborne BA (1992) *Algal Photosynthesis: The Measurement of Algal Gas Exchange*. New York: Chapman and Hall.

Morris I (1981) Photosynthetic products, physiological state, and phytoplankton growth. In: Platt T (ed.) Physiological Bases of Phytoplankton Ecology. *Canadian Bulletin of Fisheries and Aquatic Science* 210: 83–102.

Peterson BJ (1980) Aquatic primary productivity and the ^{14}C–CO_2 method: a history of the productivity problem. *Annual Review of Ecology and Systematics* 11: 359–385.

Platt T and Sathyendranath S (1993) Fundamental issues in measurement of primary production. *ICES Marine Science Symposium* 197: 3–8.

Sakshaug E, Bricaud A, Dandonneau Y, et al. (1997) Parameters of photosynthesis: definitions, theory and interpretation of results. *Journal of Plankton Research* 19: 1637–1670.

Steemann Nielsen E (1963) Productivity, definition and measurement. In: Hill MW (ed.) *The Sea*, vol. 1, pp. 129–164. New York: John Wiley.

Williams PJL (1993a) Chemical and tracer methods of measuring plankton production. *ICES Marine Science Symposium* 197: 20–36.

Williams PJL (1993b) On the definition of plankton production terms. *ICES Marine Science Symposium* 197: 9–19.

PRIMARY PRODUCTION PROCESSES

J. A. Raven, Biological Sciences, University
of Dundee, Dundee, UK

Introduction

This article summarizes the information available on
the magnitude of and the spatial and temporal variations in, marine plankton primary productivity. The
causes of these variations are discussed in terms
of the biological processes involved, the organisms
which bring them about, and the relationships to
oceanic physics and chemistry. The discussion begins
with a definition of primary production.

Primary producers are organisms that rely on
external energy sources such as light energy (photolithotrophs) or inorganic chemical reactions (chemolithotrophs). These organisms are further characterized
by obtaining their elemental requirements from
inorganic sources, e.g. carbon from inorganic carbon
such as carbon dioxide and bicarbonate, nitrogen from
nitrate and ammonium (and, for some, dinitrogen),
and phosphate from inorganic phosphate. These
organisms form the basis of food webs, supporting all
organisms at higher trophic levels. While chemolitho-
trophy may well have had a vital role in the origin and
early evolution of life, the role of chemolithotrophs in
the present ocean is minor in energy and carbon terms
(**Table 1**), but is very important in biogeochemical
element cycling, for example in the conversion of
ammonium to nitrate.

Quantitatively a much more important process
in primary productivity on a global scale is
photolithotrophy (**Table 1**). Essentially all photolithotrophs which contribute to net inorganic carbon
removal from the atmosphere or the surface ocean
are O_2-evolvers, using water as electron donor for
carbon dioxide reduction, according to eqn [1]:

$$CO_2 + 2H_2^*O + 8 \text{ photons} \rightarrow (CH_2O) + H_2O + {}^*O_2 \qquad [1]$$

The contribution of O_2-evolving photolithotrophy
from terrestrial environments is greater than that
in the oceans, despite the sea occupying more than
two-thirds of the surface of the planet (**Table 1**).
Sunlight is attenuated by sea water to an extent
which limits primary productivity to, at most, the
top 300 m of the ocean. Since only a few percent of

the ocean floor is within 300 m of the surface, the
role of benthic primary producers (i.e. those attached
to the ocean floor) is small in terms of the total
marine primary production (**Table 1**). Despite the
relatively small area of benthic habitat for photolithotrophs in the ocean as a percentage of the total
sea area (2%), benthic primary productivity producers account for almost 10% of marine primary
productivity (**Table 1**).

Until very recently it has been assumed that the
photosynthetic primary producers in the marine
phytoplankton are the O_2-evolvers with two photochemical reactions involved in moving each electron
from water to carbon dioxide, although molecular
genetic data from around 1990 indicated the presence of erythrobacteria in surface ocean waters.
Recent work has shown that both rhodopsin-based
and bacteriochlorophyll-based phototrophy is widespread in the surface ocean. This phototrophy does
not involve O_2 evolution and, while it does not
necessarily involve net carbon dioxide fixation,
it may impact on surface ocean carbon dioxide
dynamics. Thus, growth of prokaryotes using dissolved organic carbon can occur with less carbon
dioxide produced per unit dissolved organic carbon
incorporated into organic carbon by the use of
energy from photons to replace energy that would
otherwise be transformed by oxidation of dissolved
organic carbon. It is probable that these phototrophs
which do not evolve O_2 contribute <1% to gross
carbon dioxide fixation by the surface ocean.

This article investigates the reasons for this constrained planktonic primary production in the ocean
in terms of the marine pelagic habitat and the
diversity of the organisms involved in terms of their
phylogeny and life-form.

The Habitat

The surface ocean absorbs solar radiation via the
properties of sea water, as well as of any dissolved
organic material and of particles. A very small
fraction ($\leq 1\%$ in most areas) of the 400–700 nm
component is converted to energy in organic matter
in photosynthesis, while the rest is converted to
thermal energy. In the absence of wind shear and
ocean currents, themselves ultimately caused by solar
energy input, the thermal expansion of the surface
water would cause permanent stratification, except
near the poles in winter.

Table 1 Net primary productivity of habitats, and area of the habitats, on a world basis[a]

Habitat	Total area (m^2)	Organisms	Global net primary productivity (10^{15}g C year^{-1})
Marine phytoplankton	370×10^{12}	Cyanobacteria and microalgae	46
Marine planktonic chemolithotrophs converting to and to NH_4^+ to NO_2^- and NO_2^- to NO_3^-	370×10^{12}	Bacteria	≤ 0.19
Marine benthic[b,c]	6.8×10^{12}	Cyanobacteria and microalgae	0.34
		Macroalgae	3.4
Marine benthic[b]	0.35×10^{12}	Angiosperms (salt marshes plus beds of seagrasses)	0.35
Inland waters	2×10^{12}	Phytoplankton (cyanobacteria and microalgae), benthic algae and higher plants	0.58
Terrestrial[d]	150×10^{12}	Mainly higher plants	54

[a]From Raven (1991, 1996) and Falkowski et al. (2000). All values are for photolithotrophs unless otherwise indicated.
[b]Area of the habitat the marine benthic cyanobacteria, algae, and marine benthic angiosperm is in series with that of the overlying phytoplankton habitat. The benthic habitat area is included in the habitat area for marine phytoplankton.
[c]The marine benthic cyanobacterial and algal category includes cyanobacteria and algae symbionts with protistans and invertebrates.
[d]As well as higher plants the terrestrial productivity involves cyanobacteria and microalgae, both lichenized and free-living, although there seem to be no estimates of the magnitude of nonhigher plant productivity.

Such an ocean is approximated by most parts of the tropical ocean, where ocean currents and wind are inadequate to cause breakdown of thermal stratification; the upper mixed layer shows very little seasonal variation. By contrast, at higher latitudes the varying solar energy inputs throughout the year, combined with wind shear and ocean current influences, lead to stratification with a relatively shallow upper mixed layer in the (local) summer and a much deeper one in the (local) winter, usually giving a winter mixing depth so great that net primary production is not possible as a result of the inadequate mean photon flux density (light-energy) incident on the cells.

A very important impact of stratification is the isolation of the upper mixed layer, where inorganic nutrients are taken up by phytoplankton, from the lower, dark, ocean where nutrients are regenerated by heterotrophy. The movement of organic particles from the upper to lower zones is gravitational. While there is significant recycling of inorganic nutrients in the upper mixed layer via primary productivity and activities of other parts of the food web, ultimately there is loss of particles containing nutrient elements across the thermocline. Seasonal variations in mixing depth, and upwellings, are the main processes bringing nutrient solutes back to the euphotic zone.

Global biogeochemical cycling considerations suggest that the nutrient element that limits the extent of global primary production each year is, over long time periods, phosphorus. This element has a shorter residence time than the other nutrients (such as iron) which are supplied solely from terrestrial sources. Nitrogen, by contrast, is present in the atmosphere and dissolved in the ocean as dinitrogen in such large quantities that any limitation of marine phytoplankton primary productivity by the availability of such universally available nitrogen formed as ammonium and nitrate could be offset by diazotrophy, i.e. biologically dinitrogen fixation, which can only be brought about by certain Archea and Bacteria. In the ocean the phytoplanktonic cyanobacteria are the predominant diazotrophs, as the free-living *Trichodesmium* and as symbionts such as *Richia* in such diatoms as *Hemiaulis* and *Rhizosolenia*. Diazotrophy needs energy (ultimately from solar radiation) and trace elements such as iron (always), molybdenum (usually), and vanadium (sometimes). These trace elements all have longer oceanic residence times than phosphorus, and so are less likely to limit primary productivity than is phosphorus over geologically significant time intervals. However, the balance of evidence for the present ocean suggests that nitrogen is a limiting resource for the rate and extent of primary productivity over much of the world ocean, while iron seems to be the limiting nutrient in the 'high nutrient (nitrogen, phosphorus), low chlorophyll' areas of the ocean. Even where nitrogen does appear to be limiting, this could be a result of restricted iron supply which restricts the assimilation of combined nitrogen, and especially of nitrate.

Processes at the Cell Level

The photosynthetic primary producers in the marine plankton show great variability in taxonomy, and in size and shape. The taxonomic differences reflect phylogenetic differences, including the prokaryotic bacteria (cyanobacteria, embracing the chlorophyll b-containing chloroxybacteria) and a variety of phyla (divisions) of Eukaryotes. The Eukaryotes include members of the Chlorophyta (green algae), Cryptophyta (cryptophytes), Dinophyta (dinoflagellates), Haptophyta (*Phaeocystis* and coccolithophorids), and Heterokontophyta (of which the diatoms or Bacillariophyceae are the most common marine representatives).

The phylogenetic differences determine pigmentation, with the ubiquitous chlorophyll a accompanied by phycobilins in cyanobacteria *sensu stricto*, chlorophyll b in Chloroxybacteria and green algae, chlorophyll(s) c together with significant quantities of light-harvesting carotenoids in dinoflagellates, haptophytes, and diatoms, and chlorophyll c with phycobilins in cryptophytes. These differences in pigmentation alter the capacity for a given total quantity of pigment per unit volume of cells for photon absorption in a given light field, noting that the deeper a cell lives in open-ocean water, the less longer wavelength (red-orange-yellow) light is available relative to blue-green light. This effect of different light-harvesting pigments on light absorption capacity is greatest in very small cells as a result of the package effect.

Another phylogenetic difference among phytoplankton organisms is a dependence on Si (in diatoms) and on large quantities of Ca (in coccolithophorids) in those algae which have mineralized skeletons. Furthermore, some vegetative cells move relative to their immediate aqueous environment using flagella (almost all planktonic dinoflagellates, some green algae). Movement relative to the surrounding water occurs in any organism which is denser than the surrounding water sinking (e.g. by many mineralized cells) or less dense than the surrounding water (buoyancy engendered by cyanobacterial gas vacuoles or the ionic content of vacuoles in large vacuolate cells). The variation in cell size among marine phytoplankton organisms is also partly related to taxonomy. The smallest marine phytoplankton cells are prokaryotic, with cells of *Prochlorococcus* (cyanobacteria *sensu lato*) as small as 0.5 μm diameter, while cells of the largest diatoms (*Ethmosdiscus* spp.) and the green *Halosphaera* are at least 1 mm in diameter, i.e. a range of volume from $6.25 \times 10^{-20}\,m^3$ to more than $4.19 \times 10^{-9}\,m^3$. This means a volume of the largest cells which is almost 10^{11} that of the smallest; allowing for the vacuolation of the large cells gives a ratio of almost 10^{10} for cytoplasmic volume. The size range for phytoplankton organisms is expanded by considering colonial organisms (e.g. the cyanobacterium *Trichodesmium*, and the haptophyte *Phaeocystis*) to a range of cytoplasmic volumes up to almost 10^{12}. At the level of the cell size cyanobacteria have a limited volume range, while in organisms size the range is at least 10^{12}; for haptophytes it is at least 10^{11}. Cell (or organism) size is, on physicochemical principles, very important for the effectiveness of light absorption per unit pigment, nutrient uptake as a result of surface area per unit volume, and of diffusion boundary layer thickness, and rate of vertical movement relative to the surrounding water for a given difference in density between the organisms and their environment. These physicochemical predictions are, to some extent, modified by the organisms by, for example, changed pigment per unit volume and light scattering, and modulation of density.

Determinants of Primary Productivity

Despite these variations in phylogenetic origin and in the size of the organisms, that can be related to seasonal and spatial variations over the world ocean, it is not easy to find consistent spatial and temporal variations in the 'major element' ratios (C:N:P, 106:16:1 by atoms) or Redfield ratio in space or time. This means that we should not look to differences in the phylogeny or size of phytoplankton organisms to account for differences in the requirement for major nutrients (C, N, P) in supporting primary productivity. What is less clear is the possible variations in trace element (Fe, Mn, Zn, Mo, Cu, etc.) requirements in relation to the properties of different bodies of water in the world ocean. The trace metals are essential catalysts of primary productivity through their roles in photosynthesis, respiration, nitrogen assimilation, and protection against damaging active oxygen species. Geochemical evidence for limitation by some factor other than nitrogen and phosphorus is indicated for 'high nutrient' (i.e. available nitrogen and phosphorus) 'low chlorophyll' (i.e. photosynthetic biomass and hence productivity), or HNLC, regions of the ocean (north-eastern sub-Antarctic Pacific; eastern tropical Pacific; Southern Ocean).

Before seeking other geochemical limitations on primary production to explain why these apparently available sources of nitrogen and phosphorus have not been used in primary productivity, we need to consider geophysical or 'bottom up' (mixing depth, surface photon flux density) and ecological or 'top down' factors (involvement of grazers or pathogens). While these nongeochemical 'bottom up' (control of production of biomass) and 'top down' (removal of the product of primary production) constraints on the use of nitrate and phosphate are, in principle, causes of this HNLC phenomenon, *in situ* Fe enrichments show that addition of this trace element causes drawdown of nitrate and phosphate, increases in chlorophyll and primary productivity, and increased abundance, and contribution to primary productivity of large diatoms.

These IRONEX and SOIREE experiments strongly support the notion that iron limits primary productivity in HNLC regions, as well as suggesting that Fe enrichment can impact differentially on primary producers as a function of their taxonomy and cell size. The increased importance of large diatoms as a result of Fe enrichment can be a result of the diffusion boundary layer thicknesses and surface area per unit volume, rather than of the biochemical demand for Fe to catalyze a given rate of metabolism per unit cell volume. While data are not abundant, theoretical considerations suggest that cyanobacteria should, other things being equal, have higher requirements for Fe for growth than do diatoms, haptophytes, or green algae. This prediction contrasts with observations (and production) for major nutrients such as organic C, N, and P, where cell quotas are much less variable phylogenetically than are those for micronutrients. There is, of course, much less elasticity possible for the C content of cells than for other, less abundant, nutrient elements, with the same applying to a lesser extent to N and P. It is clear that the cost of N, P, or Fe in fixing carbon dioxide is higher for growth at low (limiting) as opposed to high (saturating) photon flux densities.

To broaden the issues of limitations on primary productivity, the ultimate limitation on primary productivity in the ocean is presumably the 'geochemical' limiting element P, i.e. the nutrient element with the shortest residence time in the ocean. It has been plausibly argued that, in the short term, nitrogen has become a limiting nutrient indirectly by the short-term (geologically speaking) Fe limitation. Thus, 'new' production, depending on nitrate upwelled or eddy-diffused from the deep ocean, has a greater Fe requirement than the NH_4^+ (or organic N) assimilation in 'recycled' production in which primary production is chemically fuelled by N, P, and Fe generated by zooplankton, and more importantly, by Fe limitation (at least in the geological short-term) is seen as restricting diazotrophy. In the context of the balance of diazotrophy plus atmospheric and riverine inputs of combined N, and denitrification and sedimentation loss of combined N, Fe limitation can reduce the combined N availability relative to that of P to below the 16:1 atomic ration of the Redfield ratio. This, then, restricts the N:P ration in upwelled

sea water. Even more immediate Fe limitation is seen in the HNLC ocean, as discussed above.

Conclusions

Marine primary production accounts for almost half of the global primary production, and is carried out by a much greater phylogenetic range of organisms than is the case for terrestrial primary production. As on land, almost all marine primary production involves O_2-evolving photolithotrophs. Marine phytoplankton has a volume of cells of 6.10^{-20}–$4.10^{-9} \, m^3$. While the primary production in the oceans is, on geological grounds, ultimately limited by P, proximal (shorter-term) limitation involves N or Fe.

Glossary

HNLC high nutrient, low chlorophyll. Areas of the ocean in which combined nitrogen and phosphate are present at concentrations which might be expected to give higher rates of primary production and levels of biomass, than are observed unless some 'top down' or 'bottom up' limitation is involved.

IRONEX iron enrichment experiment. Two releases of $FeSO_4$, with SF_6 as a tracer, south of the Galapagos in the Eastern Equatorial Pacific HNLC area.

Photon flux density Units are mol photon $m^{-2} s^{-1}$. Means of expressing incident irradiance in terms of photons, i.e. the aspect of the particle/wave duality of electromagnetic radiation which is appropriate for consideration of photochemical reactions such as photosyntheses. For O_2-evolving photosynthetic organisms the appropriate wavelength range is 400–700 nm.

SOIREE southern ocean iron release experiment. A Southern Ocean analogue of IRONEX, performed between Australasia and Antarctica.

See also

Carbon Cycle. Carbon Dioxide (CO$_2$) Cycle. Nitrogen Cycle. Primary Production Methods.

Further Reading

Falkowski PG and Raven JA (1997) *Aquatic Photosynthesis*. Malden: Blackwell Science.

Falkowski PG, *et al.* (2000) The global carbon cycle: a test of our knowledge of Earth as a system. *Science* 290: 291–296.

Fuhrman JA (1999) Marine viruses and their biogeo-chemical and ecological effects. *Nature* 399: 541–548.

Martin JH (1991) Iron, Liebig's Law and the greenhouse. *Oceanography* 4: 52–55.

Platt T and Li WKW (eds.) (1986) Photosynthetic Pico-plankton. *Canadian Bulletin of Fisheries and Aquatic Sciences* 214.

Raven JA (1991) Physiology of inorganic C acquisition and implications for resource use efficiency by marine phytoplankton: relation to increased CO_2 and temperature. *Plant Cell Environment* 14: 779–794.

Raven JA (1996) The role of autotrophs in global CO_2 cycling. In: Lidstrom ME and Tabita FR (eds.)

Microbial Growth on C_1 Compounds, pp. 351–358. Dordrecht: Kluwer Academic Publishers.

Raven JA (1998) Small is beautiful: the picophytoplankton. *Funct. Ecol.* 12: 503–513.

Redfield AC (1958) The biological control of chemical factors in the environment. *American Scientist* 46: 205–221.

Stokes T (2000) The enlightened secrets across the ocean. *Trends in Plant Science* 5: 461.

Van den Hoek C, Mann DG, and Jahns HM (1995) *Algae. An Introduction to Phycology.* Cambridge: Cambridge University Press.

MARINE PLANKTON COMMUNITIES

G.-A. Paffenhöfer, Skidaway Institute of
Oceanography, Savannah, GA, USA

Introduction

By definition, a community is an interacting population of various kinds of individuals (species) in a
common location (*Webster's Collegiate Dictionary*,
1977).

The objective of this article is to provide general
information on the composition and functioning of
various marine plankton communities, which is accompanied by some characteristic details on their
dynamicism.

General Features of a Plankton Community

The expression 'plankton community' implies that
such a community is located in a water column. It
has a range of components (groups of organisms)
that can be organized according to their size. They
range in size from tiny single-celled organisms such
as bacteria (0.4–1-μm diameter) to large predators
like scyphomedusae of more than 1 m in diameter.
A common method which has been in use for
decades is to group according to size, which here is
attributed to the organism's largest dimension; thus
the organisms range from picoplankton to macro-
plankton (**Figure 1**). It is, however, the smallest
dimension of an organism which usually determines whether it is retained by a mesh, since in a
flow, elongated particles align themselves with the
flow.

A plankton community is operating/functioning
continuously, that is, physical, chemical, and biological variables are always at work. Interactions
among its components occur all the time. As one
well-known fluid dynamicist stated, "The surface of
the ocean can be flat calm but below that surface
there is always motion of the water at various
scales." Many of the particles/organisms are moving
or being moved most of the time: Those without
flagella or appendages can do so due to processes
within or due to external forcing, for example, from
water motion due to internal waves; and those with
flagella/cilia or appendages or muscles move or create motion of the water in order to exist. Oriented
motion is usually in the vertical which often results

in distinct layers of certain organisms. However,
physical variables also, such as light or density differences of water masses, can result in layering of
planktonic organisms. Such layers which are often
horizontally extended are usually referred to as
patches.

As stated in the definition, the components of a
plankton community interact. It is usually the case
that a larger organism will ingest a smaller one or a
part of it (**Figure 1**). However, there are exceptions.
The driving force for a planktonic community originates from sun energy, that is, primary productivity's
(1) direct and (2) indirect products: (1) autotrophs
(phytoplankton cells) which can range from near 2 to
more than 300-μm width/diameter, or chemotrophs;
and (2) dissolved organic matter, most of which is
released by phytoplankton cells and protozoa as
metabolic end products, and being taken up by bacteria and mixo- and heterotroph protozoa (**Figure 1**).
These two components mainly set the microbial loop
(ML) in motion; that is, unicellular organisms of different sizes and behaviors (auto-, mixo-, and heterotrophs) depend on each other – usually,
but not always, the smaller being ingested by the
larger. Most of nutrients and energy are recirculated
within this subcommunity of unicellular organisms in
all marine regions of our planet (for more details, especially the ML). These processes of the ML dominate
the transfer of energy in all plankton communities
largely because the processes (rates of ingestion,
growth, reproduction) of unicellular heterotrophs almost always outpace those of phytoplankton, and also
of metazooplankton taxa at most times.

The main question actually could be: "What is the
composition of plankton communities, and how do
they function?" **Figure 1** reveals sizes and relationships within a plankton community including the
ML. It shows the so-called 'bottom-up' and 'top-
down' effects as well as indirect effects like the
above-mentioned labile dissolved organic matter
(labile DOM), released by auto- and also by heterotrophs, which not only drives bacterial growth but
can also be taken up or used by other protozoa.
There can also be reversals, called two-way processes. At times a predator eating an adult metazoan
will be affected by the same metazoan which is able
to eat the predator's early juveniles (e.g., well-grown
ctenophores capturing adult omnivorous copepods
which have the ability to capture and ingest very
young ctenophores).

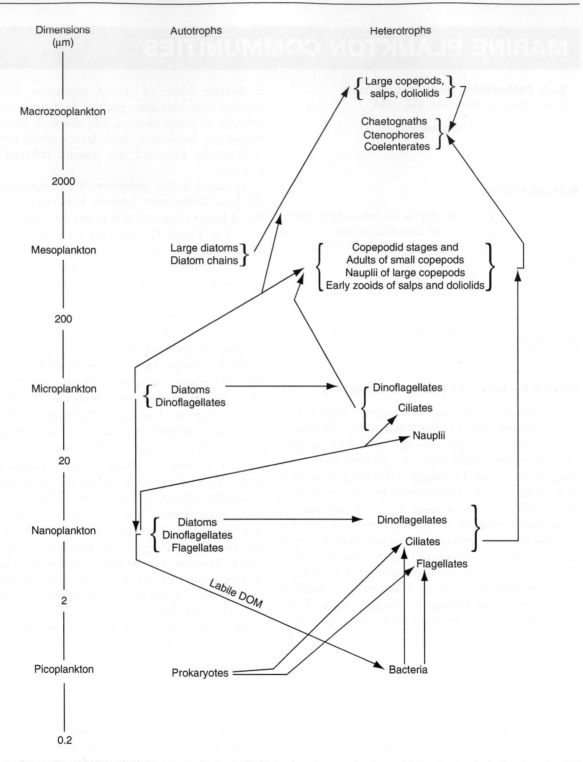

Figure 1 Interactions within a plankton community separated into size classes of auto- and heterotrophs, including the microbial loop; the arrows point to the respective grazer, or receiver of DOM; the figure is partly related to figure 9 from Landry MR and Kirchman DL (2002) Microbial community structure and variability in the tropical Pacific. *Deep-Sea Research II* 49: 2669–2693.

To comprehend the functioning of a plankton community requires a quantitative assessment of the abundances and activities of its components. First, almost all of our knowledge to date stems from *in situ* sampling, that is, making spot measurements of the abundance and distribution of organisms in the water column. The accurate determination of abundance and distribution requires using meshes or devices which quantitatively collect the respective organisms. Because of methodological difficulties and insufficient comprehension of organisms' sizes and activities, quantitative sampling/quantification of a community's main components has been often inadequate. The following serves as an example of this. Despite our knowledge that copepods consist of 11 juvenile stages aside of adults, the majority of studies of marine zooplankton hardly considered the juveniles' significance and this manifested itself in sampling with meshes which often collected merely the adults quantitatively. Second, much knowledge on rate processes comes from quantifying the respective organisms' activities under controlled conditions in the laboratory. Some *in situ* measurements (e.g., of temperature, salinity, chlorophyll concentrations, and acoustic recordings of zooplankton sizes) have been achieved 'continuously' over time, resulting in time series of increases and decreases of certain major community components. To date there are few, if any, direct *in situ* observations on the activity scales of the respective organisms, from bacteria to proto- and to metazooplankton, mainly because of methodological difficulties. In essence, our present understanding of processes within plankton communities is incomplete.

Specific Plankton Communities

We will provide several examples of plankton communities of our oceans. They will include information about the main variables affecting them, their main components, partly their functioning over time, including particular specifics characterizing each of those communities.

In this section, plankton communities are presented for three different types of marine environments: estuaries/inshore, continental shelves, and open ocean regions.

Estuaries

Estuaries and near-shore regions, being shallow, will rapidly take up and lose heat, that is, will be strongly affected by atmospheric changes in temperature, both short- and long-term, the latter showing in the seasonal extremes ranging from 2 to 32 °C in estuaries of North Carolina. Runoff of fresh water, providing continuous nutrient input for primary production, and tides contribute to rapid changes in salinity. This implies that resident planktonic taxa ought to be eurytherm as well as – therm. Only very few metazooplanktonic species are able to exist in such an environment (**Table 1**). In North Carolinian

estuaries, representative of other estuaries, they are the copepod species *Acartia tonsa*, *Oithona oculata*, and *Parvocalanus crassirostris*. In estuaries of Rhode Island, two species of the genus *Acartia* occur. During colder temperatures *Acartia hudsonica* produces dormant eggs as temperatures increase and then is replaced by *A. tonsa*, which produces dormant eggs once temperatures again decrease later in the year. Such estuaries are known for high primary productivity, which is accompanied by high abundances of heterotroph protozoa preying on phytoplankton. Such high abundances of unicellular organisms imply that food is hardly limiting the growth of the above-mentioned copepods which can graze on auto- as well as heterotrophs. However, such estuaries are often nursery grounds for juvenile fish like menhaden which prey heavily on late juveniles and adults of such copepods, especially *Acartia*, which is not only the largest of those three dominant copepod species but also moves the most, and thus can be seen most easily by those visual predators. This has resulted in diurnal migrations mostly of their adults, remaining at the seafloor during the day where they hardly eat, thus avoiding predation by such visual predators, and only entering the water column during dark hours. That then is their period of pronounced feeding. The other two species which are not heavily preyed upon by juvenile fish, however, can be affected by the co-occurring *Acartia*, because from early copepodid stages on this genus can be strongly carnivorous, readily preying on the nauplii of its own and of those other species.

Nevertheless, the usually continuous abundance of food organisms for all stages of the three copepod species results in high concentrations of nauplii which in North Carolinian estuaries can reach $100 \, l^{-1}$, as can their combined copepodid stages. The former is an underestimate, because sampling was done with a 75-μm mesh, which is passed through by most of those nauplii. By comparison, in an estuary on the west coast of Japan (Yellow Sea), dominated also by the genera *Acartia*, *Oithona*, and *Paracalanus* and sampling with 25-μm mesh, nauplius concentrations during summer surpassed $700 \, l^{-1}$, mostly from the genus *Oithona*. And copepodid stages plus adults repeatedly exceeded $100 \, l^{-1}$. Here sampling with such narrow mesh ensured that even the smallest copepods were collected quantitatively.

In essence, estuaries are known to attain among the highest concentrations of proto- and metazooplankton. The known copepod species occur during most of the year, and are observed year after year which implies persistence of those species beyond decades.

Table 1 Some characteristics of marine plankton communities

	Estuaries	Shelves	Open ocean gyres — Subarctic Pacific	Open ocean gyres — Boreal Atlantic	Epipelagic subtropical Atlantic/Pacific
Physical variables	Wide range of temperature and salinity	Intermittent and seasonal atmospheric forcing	Steady salinity, seasonal temp. variability	Major seasonal variability of temperature	Steady temperature and salinity, continuous atmospheric forcing
Nutrient supply	Continuous	Episodic	Seasonal	Seasonal	Occasional
Phytoplankton abundance	High from spring to autumn	Intermittently high	Always low	Major spring bloom	Always low
Phytoplankton composition	Flagellates, diatoms	Flagellates, diatoms, dinoflagellates	Nanoflagellates	Spring: diatoms Other: mostly nanoplankton	Mostly prokaryotes, small nano- and dinoflagellates
Primary Productivity	High at most times	Intermittently high	Maximum in spring	Max. in spring and autumn	Always low
No. of metazoan species	≤5	~10–30	>10	>20	>100
Seasonal variability of metazoan abundance	High spring and summer, low winter	Highly variable	High	High	Low
Copepod Ranges	$N^a \sim 10$–$500\,l^{-1}$ $Cop^b \sim 5$–$100\,l^{-1}$	<5–$50\,l^{-1}$ <3–$30\,l^{-1}$			3–$10\,l^{-1}$
Abundance			Up to $1000\,m^{-3}$ Neocalanus	Up to $1000\,m^{-3}$ C. finmarchicus	300–$1000\,m^{-3}$
Dominant metazooplankton taxa	Acartia Oithona Parvocalanus	Oithona Paracalanus Temora Doliolida	Neocalanus Oithona Metridia	Calanus Oithona Oncaea	Oithona Clausocalanus Oncaea

a Nauplii.
b Copepodids and adult copepods.

Continental Shelves

By definition they extend to the 200-m isobath, and range from narrow (few kilometers) to wide (more than 100-km width). The latter are of interest because the former are affected almost continuously and entirely by the nearby open ocean. Shelves are affected by freshwater runoff and seasonally changing physical variables. Water masses on continental shelves are evaluated concerning their residence time, because atmospheric events sustained for more than 1 week can replace most of the water residing on a wide shelf with water offshore but less so from near shore. This implies that plankton communities on wide continental shelves, which are often near boundary currents, usually persist for limited periods of time, from weeks to months (**Table 1**). They include shelves like the Agulhas Bank, the Campeche Banks/Yucatan Shelf, the East China Sea Shelf, the East Australian Shelf, and the US southeastern continental shelf. There can be a continuous influx year-round of new water from adjacent boundary currents as seen for the Yucatan Peninsula and Cape Canaveral (Florida). The momentum of the boundary current (here the Yucatan Current and Florida Current) passing a protruding cape will partly displace water along downstream-positioned diverging isobaths while the majority will follow the current's general direction. This implies that upstream-produced plankton organisms can serve as seed populations toward developing a plankton community on such wide continental shelves.

Whereas estuarine plankton communities receive almost continuously nutrients for primary production from runoff and pronounced benthic-pelagic coupling, those on wide continental shelves infrequently receive new nutrients. Thus they are at most times a heterotroph community unless they obtain nutrients from the benthos due to storms, or receive episodically input of cool, nutrient-rich water from greater depths of the nearby boundary current as can be seen for the US SE shelf. Passing along the outer shelf at about weekly intervals are nutrient-rich cold-core Gulf Stream eddies which contain plankton organisms from the highly productive Gulf of Mexico. Surface winds, displacing shelf surface water offshore, lead to an advance of the deep cool water onto the shelf which can be flooded entirely by it. Pronounced irradiance and high-nutrient loads in such upwellings result in phytoplankton blooms which then serve as a food source for protozoo- and metazooplankton. Bacteria concentrations in such cool water masses increase within several days by 1 order of magnitude. Within 2–3 weeks most of the smaller phytoplankton (c. <20-μm width) has been greatly reduced, usually due to grazing by protozoa and relatively slow-growing assemblages of planktonic copepods of various genera such as *Temora*, *Oithona*, *Paracalanus*, *Eucalanus*, and *Oncaea*. However, quite frequently, the Florida Current which becomes the Gulf Stream carries small numbers of Thaliacea (Tunicata), which are known for intermittent and very fast asexual reproduction. Such salps and doliolids, due to their high reproductive and growth rate, can colonize large water masses, the latter increasing from ~5 to >500 zooids per cubic meter within 2 weeks, and thus form huge patches, covering several thousands of square kilometers, as the cool bottom water is displaced over much of the shelf. The increased abundance of salps (usually in the warmer and particle-poor surface waters) and doliolids (mainly in the deeper, cooler, particle-rich waters, also observed on the outer East China shelf) can control phytoplankton growth once they achieve bloom concentrations. The development of such large and dense patches is partly due to the lack of predators.

Although the mixing processes between the initially quite cool intruding bottom (13–20 °C) and the warm, upper mixed layer water (27–28 °C) are limited, interactions across the thermocline occur, thus creating a plankton community throughout the water column of previously resident and newly arriving components. The warm upper mixed layer often has an extraordinary abundance of early copepodid stages of the poecilostomatoid copepod *Oncaea*, thanks to their ontogenetical migration after having been released by the adult females which occur exclusively in the cold intruding water. Also, early stages of the copepod *Temora turbinata* are abundant in the warm upper mixed layer; while *T. turbinata*'s late juvenile stages prefer the cool layer because of the abundance of large, readily available phytoplankton cells. As in estuaries, the copepod genus *Oithona* flourishes on warm, temperate, and polar continental shelves throughout most of the euphotic zone.

Such wide subtropical shelves will usually be well mixed during the cooler seasons, and then harbor, due to lower temperatures, fewer metazoplankton species which are often those tolerant of wider or lower temperature ranges. Such wide shelves are usually found in subtropical regions, which explains the rapidity of the development of their plankton communities. They, however, are also found in cooler climates, like the wide and productive Argentinian/Brazilian continental shelf about which our knowledge is limited. Other large shelves, like the southern North Sea, have a limited exchange of water with the open ocean but at the same time considerable influx of runoff, plus nutrient supply from the benthos due to storm events, and thus can maintain identical plankton communities over months and seasons.

In essence, continental shelf plankton communities are usually relatively short-lived, which is largely due to their water's limited residence time.

Open Ocean

The open ocean, even when not including ocean margins (up to 1000-m water column), includes by far the largest regions of the marine environment. Its deep-water columns range from the polar seas to the Tropics. All these regions are under different atmospheric and seasonal regimes, which affect plankton communities. Most of these communities are seasonally driven and have evolved along the physical conditions characterizing each region. The focus here is on gyres as they represent specific ocean communities whose physical environment can be readily presented.

Gyres represent huge water masses extending horizontally over hundreds to even thousands of kilometers in which the water moves cyclonically or anticyclonically. They are encountered in subpolar, temperate, and subtropical regions. The best-studied ones are:

- subpolar: Alaskan Gyre;
- temperate: Norwegian Sea Gyre, Labrador–Irminger Sea Gyre;
- subtropical: North Pacific Central Gyre (NPCG), North Atlantic Subtropical Gyre (NASG).

The Alaskan Gyre is part of the subarctic Pacific (**Table 1**) and is characterized physically by a shallow halocline (~110-m depth) which prevents convective mixing during storms. Biologically it is characterized by a persistent low-standing stock of phytoplankton despite high nutrient abundance, and several species of large copepods which have evolved to persist via a life cycle as shown for *Neocalanus plumchrus*. By midsummer, fifth copepodids (C5) in the upper 100 m which have accumulated large amounts of lipids begin to descend to greater depths of 250 m and beyond undergoing diapause, and eventually molt to females which soon begin to spawn. Spawning females are found in abundance from August to January. Nauplii living off their lipid reserves and moving upward begin to reach surface waters by mid- to late winter as copepodid stage 1 (C1), and start feeding on the abundant small phytoplankton cells (probably passively by using their second maxillae, but mostly by feeding actively on heterotrophic protozoa which are the main consumers of the tiny phytoplankton cells). The developing copepodid stages accumulate lipids

which in C5 can amount to as much as 50% of their body mass, which then serve as the energy source for metabolism of the females at depth, ovary development, and the nauplii's metabolism plus growth. While the genus *Neocalanus* over much of the year provides the highest amount of zooplankton biomass, the cyclopoid *Oithona* is the most abundant metazooplankter; other abundant metazooplankton taxa include *Euphausia pacifica*, and in the latter part of the year *Metridia pacifica* and *Calanus pacificus*.

In the temperate Atlantic (**Table 1**), the Norwegian Sea Gyre maintains a planktonic community which is characterized, like much of the temperate oceanic North Atlantic, by the following physical features. Pronounced winds during winter mix the water column to beyond 400-m depth, being followed by lesser winds and surface warming resulting in stratification and a spring bloom of mostly diatoms, and a weak autumn phytoplankton bloom. A major consumer of this phytoplankton bloom and characteristic of this environment is the copepod *Calanus finmarchicus*, occurring all over the cool North Atlantic. This species takes advantage of the pronounced spring bloom after emerging from diapause at >400-m depth, by moulting to adult, and grazing of females at high clearance rates on the diatoms, right away starting to reproduce and releasing up to more than 2000 fertilized eggs during their lifetime. Its nauplii start to feed as nauplius stage 3 (N3), being able to ingest diatoms of similar size as the adult females, and can reach copepodid stage 5 (C5) within about 7 weeks in the Norwegian Sea, accumulating during that period large amounts of lipids (wax ester) which serve as the main energy source for the overwintering diapause period. Part of the success of *C. finmarchicus* is found in its ability of being omnivorous. C5s either descend to greater depths and begin an extended diapause period, or could moult to adult females, thus producing another generation which then initiates diapause at mostly C5. Its early to late copepodid stages constitute the main food for juvenile herring which accumulate the copepods' lipids for subsequent overwintering and reproduction. Of the other copepods, the genus *Oithona* together with the poecilostomatoid *Oncaea* and the calanoid *Pseudocalanus* were the most abundant.

Subtropical and tropical parts of the oceans cover more than 50% of our oceans. Of these, the NPCG, positioned between *c.* 10° and 45° N and moving anticyclonically, has been frequently studied. It includes a southern and northern component, the latter being affected by the Kuroshio and westerly winds, the former by the North Equatorial Current and the trade winds. Despite this, the NPCG has been considered as an ecosystem as well as a huge plankton community. The NASG, found between *c.* 15° and 40° N and moving anticyclonically, is of similar horizontal dimensions. There are close relations between subtropical and tropical communities; for example, the Atlantic south of Bermuda is considered close to tropical conditions. Vertical mixing in both gyres is limited. Here we focus on the epipelagic community which ranges from the surface to about 150-m depth, that is, the euphotic zone. The epipelagial is physically characterized by an upper mixed layer of *c.* 15–40 m of higher temperature, below which a thermocline with steadily decreasing temperatures extends to below 150-m depth. In these two gyres, the concentrations of phytoplankton hardly change throughout the year in the epipelagic (**Table 1**) and together with the heterotrophic protozoa provide a low and quite steady food concentration (**Table 1**) for higher trophic levels. Such very low particle abundances imply that almost all metazooplankton taxa depending on them are living on the edge, that is, are severely food-limited. Despite this fact, there are more than 100 copepod species registered in the epipelagial of each of the two gyres. How can that be? Almost all these copepod species are small and rather diverse in their behavior: the four most abundant genera have different strategies to obtain food particles: the intermittently moving *Oithona* is found in the entire epipelagial and depends on moving food particles (hydrodynamic signals); *Clausocalanus* is mainly found in the upper 50 m of the epipelagial and always moves at high speed, thus encountering numerous food particles, mainly via chemosensory; *Oncaea* copepodids and females occur in the lower part of the epipelagial and feed on aggregates; and the feeding-current producing *Calocalanus* perceives particles via chemosensory. This implies that any copepod species can persist in these gyres as long as it obtains sufficient food for growth and reproduction. This is possible because protozooplankton always controls the abundance of available food particles; thus, there is no competition for food among the metazooplankton. In addition, since total copepod abundance (quantitatively collected with a 63-μm mesh by three different teams) is steady and usually $< 1000 \, m^{-3}$ including copepodid stages (pronounced patchiness of metazooplankton has not yet been observed in these oligotrophic waters), the probability of encounter (only a minority of the zooplankton is carnivorous on metazooplankton) is very low, and therefore the probability of predation low within the metazooplankton. In summary, these steady conditions make it possible that in the epipelagial more than 100 copepod species can coexist, and are in a steady state throughout much of the year.

Conclusions

All epipelagic marine plankton communities are at most times directly or indirectly controlled or affected by the activity of the ML, that is, unicellular organisms. Most of the main metazooplankton species are adapted to the physical and biological conditions of the respective community, be it polar, subpolar, temperate, subtropical, or tropical. The only metazooplankton genus found in all communities mentioned above, and also all other studied marine plankton communities, is the copepod genus *Oithona*. This copepod has the ability to persist under adverse conditions, for example, as shown for the subarctic Pacific. This genus can withstand the physical as well as biological (predation) pressures of an estuary, the persistent very low food levels in the warm open ocean, and the varying conditions of the Antarctic Ocean. Large copepods like the genus *Neocalanus* in the subarctic Pacific, and *C. finmarchicus* in the temperate to subarctic North Atlantic are adapted with respective distinct annual cycles in their respective communities. Among the abundant components of most marine plankton communities from near shore to the open ocean are appendicularia (Tunicata) and the predatory chaetognaths.

Our present knowledge of the composition and functioning of marine planktonic communities derives from (1) oceanographic sampling and time series, optimally accompanied by the quantification of physical and chemical variables; and (2) laboratory/onboard experimental observations, including some time series which provide results on small-scale interactions (microns to meters; milliseconds to hours) among components of the community. Optimally, direct *in situ* observations on small scales in conjunction with respective modeling would provide insights in the true functioning of a plankton community which operates continuously on scales of milliseconds and larger.

Our future efforts are aimed at developing instrumentation to quantify *in situ* interactions of the various components of marine plankton communities. Together with traditional oceanographic methods we would go 'from small scales to the big picture', implying the necessity of understanding the functioning on the individual scale for a comprehensive understanding as to how communities operate.

See also

Continuous Plankton Recorders. Plankton. Plankton and Climate.

Further Reading

Atkinson LP, Lee TN, Blanton JO, and Paffenhöfer G-A (1987) Summer upwelling on the southeastern continental shelf of the USA during 1981: Hydrographic observations. *Progress in Oceanography* 19: 231–266.

Fulton RS, III (1984) Distribution and community structure of estuarine copepods. *Estuaries* 7: 38–50.

Hayward TL and McGowan JA (1979) Pattern and structure in an oceanic zooplankton community. *American Zoologist* 19: 1045–1055.

Landry MR and Kirchman DL (2002) Microbial community structure and variability in the tropical Pacific. *Deep-Sea Research II* 49: 2669–2693.

Longhurst AR (1998) *Ecological Geography of the Sea*, 398pp. San Diego, CA: Academic Press.

Mackas DL and Tsuda A (1999) Mesozooplankton in the eastern and western subarctic Pacific: Community structure, seasonal life histories, and interannual variability. *Progress in Oceanography* 43: 335–363.

Marine Zooplankton Colloquium 1 (1998) Future marine zooplankton research – a perspective. *Marine Ecology Progress Series* 55: 197–206.

Menzel DW (1993) *Ocean Processes: US Southeast Continental Shelf*, 112pp. Washington, DC: US Department of Energy.

Miller CB (1993) Pelagic production processes in the subarctic Pacific. *Progress in Oceanography* 32: 1–15.

Miller CB (2004) *Biological Oceanography*, 402pp. Boston: Blackwell.

Paffenhöfer G-A and Mazzocchi MG (2003) Vertical distribution of subtropical epiplanktonic copepods. *Journal of Plankton Research* 25: 1139–1156.

Paffenhöfer G-A, Sherman BK, and Lee TN (1987) Summer upwelling on the southeastern continental shelf of the USA during 1981: Abundance, distribution and patch formation of zooplankton. *Progress in Oceanography* 19: 403–436.

Paffenhöfer G-A, Tzeng M, Hristov R, Smith CL, and Mazzocchi MG (2003) Abundance and distribution of nanoplankton in the epipelagic subtropical/tropical open Atlantic Ocean. *Journal of Plankton Research* 25: 1535–1549.

Smetacek V, DeBaar HJW, Bathmann UV, Lochte K, and Van Der Loeff MMR (1997) Ecology and biogeochemistry of the Antarctic Circumpolar Current during austral spring: A summary of Southern Ocean JGOFS cruise ANT X/6 of RV *Polarstern*. *Deep-Sea Research II* 44: 1–21 (and all articles in this issue).

Speirs DC, Gurney WSC, Heath MR, Horbelt W, Wood SN, and de Cuevas BA (2006) Ocean-scale modeling of the distribution, abundance, and seasonal dynamics of the copepod *Calanus finmarchicus*. *Marine Ecology Progress Series* 313: 173–192.

Tande KS and Miller CB (2000) Population dynamics of *Calanus* in the North Atlantic: Results from the Trans-Atlantic Study of *Calanus finmarchicus*. *ICES Journal of Marine Science* 57: 1527 (entire issue).

Webber MK and Roff JC (1995) Annual structure of the copepod community and its associated pelagic environment off Discovery Bay, Jamaica. *Marine Biology* 123: 467–479.

TRACERS OF OCEAN PRODUCTIVITY

W. J. Jenkins, University of Southampton, Southampton, UK

Introduction

Primary production is the process whereby inorganic carbon is fixed in the sunlit (euphotic) zone of the upper ocean, and forms the base of the marine food pyramid. It occurs when marine phytoplankton use sunlight energy and dissolved nutrients to convert inorganic carbon to organic material, thereby releasing oxygen. The total amount of carbon fixed during photosynthesis is called gross production, whereas the amount of carbon fixed in excess of internal metabolic costs is referred to as net production. It is understood that a significant fraction of the carbon fixed in this manner is rapidly recycled by a combination of grazing by zooplankton and *in situ* bacterial oxidation of organic material. New production is that portion of net production that is supported by the introduction of new nutrients into the euphotic zone. Traditionally, this has been regarded as production fueled by nitrate as opposed to more reduced forms of nitrogen, such as ammonia and urea. Some portion of the fixed carbon sinks out of the euphotic zone in particulate form, or is subducted or advected away as dissolved organic material from the surface layers by physical processes. This flux is regarded collectively as export production. The ratio of new (export) to net production, referred to as the f-ratio (e-ratio) can vary between 0 and 1, and is believed to be low in oligotrophic ('blue water'), low productivity regions, and higher in eutrophic, high productivity regions. Finally, net community production is the total productivity in excess of net community metabolic cost. On sufficiently long space- and time-scales, it can be argued that new, net community, and export production should be equivalent in magnitude.

Net production has been measured 'directly' by radiocarbon incubation experiments, whereby water samples are 'spiked' with radiocarbon-labeled bicarbonate, and the net rate of transfer of the radioisotope into organic matter phases determined by comparison of light versus dark incubations. Global maps of net productivity have been constructed on the basis of such measurements, and current estimates indicate a global fixation rate of order 50 GT

C a^{-1} (1 GT $= 10^{15}$ g). Rates of export, new, and net community production are more difficult to determine directly, yet are of equal importance as determinants of biogeochemically important fluxes on annual through centennial timescales.

Geochemical tracer techniques have been used to make such estimates, and offer significant advantages in that they are fundamentally nonperturbative, and integrate over relatively large space-scales and long time-scales. Conversely, such determinations must be viewed from the perspective that they are indirect measures of biogeochemical processes, and have characteristic implicit space- and time-scales, as well as boundary conditions, and sometimes ambiguities and model dependence. Further, the specific tracer or physical system used to obtain production estimates determines the type of productivity measured. Thus any treatment of geochemical tracer estimates must include a discussion of these attributes.

Measuring Oceanic Productivity with Tracers

Just a few approaches will be discussed here. Other techniques have been used with some success, particularly with relation to particle interceptor traps, but this section will concentrate on basic mass budgeting approaches using water column distributions or seasonal cycling of tracers. There are three basic, yet fundamentally independent approaches that can be used.

1. Aphotic zone oxygen consumption rates that, when vertically integrated, provide a net water column oxygen demand that can then be related stoichiometrically to a carbon export flux.
2. Seasonal timescale euphotic zone mass budgets, particularly of oxygen, carbon, and carbon isotopes, which lead to estimates of net community production.
3. Tracer flux-gauge measurements of physical mechanisms of nutrient supply to the surface ocean, which place lower bounds on rates of new production.

These techniques, summarized in **Figure 1**, yield estimates of subtly different facets of biological production. On annual timescales, however, these different modes of production should be very close to equivalent, and hence the results of these various measurement approaches should be comparable. As

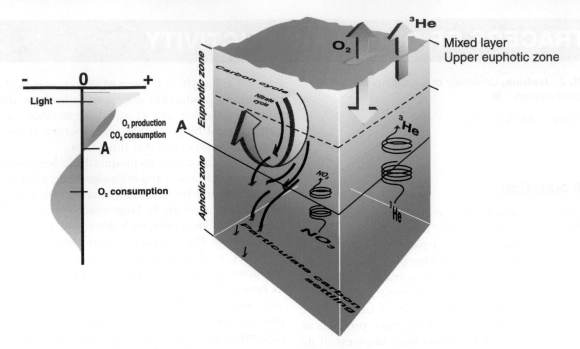

Figure 1 A schematic of the upper ocean, showing material fluxes and various tracer constraints on primary production.

shown below, their quantitative agreement coupled with their essential independence lends an inductive support to the validity of their results.

Aphotic Zone Oxygen Consumption Rates

In the surface ocean air–sea gas exchange controls the composition of dissolved gases and phytoplankton release oxygen. Below, in the aphotic (nonsunlit) zone, oxygen is generally undersaturated, because bacterially mediated oxidation of sinking organic material consumes oxygen. Credible estimates of aphotic zone oxygen consumption rates have been made since the 1950s. However, the earliest quantitative linkage to primary production was in 1982. The principle behind it is dating water masses and dividing the age of the water mass into the observed oxygen deficit. Another approach involves correlating water mass age along streamlines with oxygen concentration (older water has less oxygen). This dating can be achieved by a technique such as tritium-^{3}He dating, which uses the ingrowth of the stable, inert noble gas isotope ^{3}He from the decay of the radioactive heavy isotope of hydrogen (tritium), according to:

$$^3\text{H} \xrightarrow{12.45\text{y}} {}^3\text{He}$$

If surface waters are in good gas exchange contact with the atmosphere, then very little ^{3}He will accumulate due to tritium decay. Once isolated from the surface, this ^{3}He can accumulate. From the measurement of both isotopes in a fluid parcel, a tritium-^{3}He age can be computed according to:

$$\tau = \lambda^{-1}\ln\left(1 + \frac{[^3\text{He}]}{[^3\text{H}]}\right)$$

where λ is the decay probability for tritium, and τ is the tritium-^{3}He age (usually given in years). Under typical Northern Hemispheric conditions with current technology, times ranging from a few months to a few decades can be determined.

Although a conceptually simple approach, under normal circumstances mixing must be accounted for because it can affect the apparent tritium-^{3}He age in a nonlinear fashion. Furthermore, in regions of horizontal oxygen gradients, lateral mixing may significantly affect apparent oxygen consumption rates. For example, following a fluid parcel as it moves down a streamline, mixing of oxygen out of the parcel due to large-scale gradients will masquerade as an augmentation of oxygen consumption rates. These issues can be accounted for by determining the three-dimensional distributions of these properties, and applying the appropriate conservation equations. With additional constraints provided by geostrophic velocity calculations, these effects can be separated and absolute oxygen

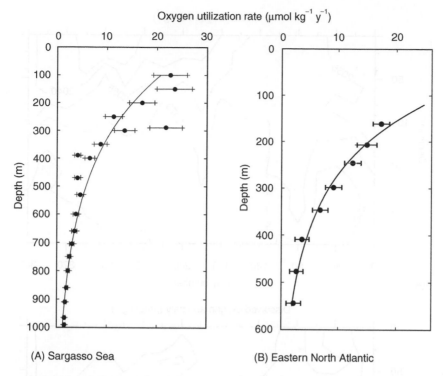

Oxygen utilization rate (μmol kg^{-1} y^{-1})

(A) Sargasso Sea

(B) Eastern North Atlantic

Figure 2 Aphotic zone oxygen consumption rates as a function of depth for two locales in the subtropical North Atlantic. These consumption rates are based on tritium-^{3}He dating and other tracer techniques.

consumption rates can be computed as a function of depth. **Figure 2** shows profiles of oxygen consumption rates as functions of depth for two locales in the subtropical North Atlantic. Integration of these curves as a function of depth gives net water column oxygen demands of 6.5 ± 1.0 mol m^{-2} a^{-1} for the Sargasso Sea and 4.7 ± 0.5 mol m^{-2} a^{-1} in the eastern subtropical North Atlantic. Using the molar ratio of oxygen consumed to carbon oxidized for organic material (170 : 117), the flux of carbon from the euphotic zone above required to support such an oxygen demand can be calculated for the two regions (4.5 ± 0.7 and 3.2 ± 0.4 mol C m^{-2} a^{-1}).

The character of these estimates bears some consideration. Firstly, according to the definitions of primary production types described earlier, this represents a determination of export productivity. Secondly, the determinations represent an average over timescales ranging from several years to a decade or more. This is the range of ages of the water masses for which the oxygen utilization rate has been determined. Thirdly, the corresponding space-scales are of order 1000 km, for this is the region over which the age gradients were determined. Fourthly, although the calculation was done assuming that the required carbon flux was particulate material, it cannot distinguish between the destruction of a particulate rain of carbon and the *in situ* degradation of dissolved organic material advected along with the

water mass from a different locale. These characteristics must be borne in mind when comparing this with other estimates.

Seasonal Euphotic Zone Mass Budgets

There have been three basically independent approaches to estimating net community production based on observation of the seasonal cycles of oxygen and carbon in the upper ocean. Photosynthesis in the euphotic zone results in the removal of inorganic carbon from the water column, and releases oxygen (**Figure 3**). Recycling of organic material via respiration and oxidation consumes oxygen and produces CO_2 in essentially the same ratios. It is only that carbon fixation that occurs in excess of these processes, i.e., processes that result in an export of organic material from the euphotic zone, or a net biomass increase, that leaves behind an oxygen or total CO_2 (ΣCO_2) signature. Estimates of productivity based on euphotic zone oxygen or carbon budgets are consequently estimates of net community production. Such productivity estimates are characterized by seasonal to annual timescales, and space-scales of order of a few hundred kilometers.

In subtropical waters, excess oxygen appears within the euphotic zone just after the onset of

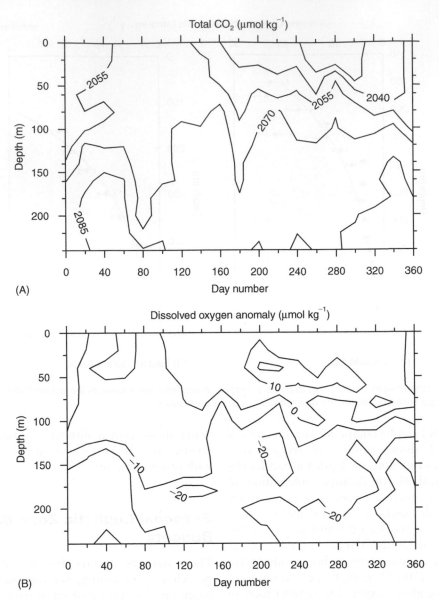

Figure 3 Euphotic zone seasonal cycles of total inorganic carbon (A) and oxygen (B) near Bermuda. Note the build-up of oxygen anomaly and reduction of total CO_2 in the euphotic zone during the summer months due to photosynthetic activity.

stratification, and continues to build up throughout summer months. Use of the seasonal accumulation of photosynthetic oxygen in the upper ocean to estimate primary production is complicated by the fact that it tends to be lost to the atmosphere by gas exchange at the surface. Furthermore, temperature changes due to seasonal heating and cooling will change the solubility of the gas, further driving fluxes of oxygen across the air–sea interface. In addition, bubble trapping by surface waves can create small supersaturations. While such processes conspire to complicate the resultant picture, it is possible to use observations of noble gases (which do not undergo biological and chemical processing) and upper ocean physical models to interpret the seasonal cycle of oxygen. These calculations have been successfully carried out at a variety of locations, including the subtropical North Atlantic and the North Pacific. In the Sargasso Sea, estimates of oxygen productivity range from 4.3 to 4.7 mol m^{-2} a^{-1}. Using the molar ratio of oxygen released to carbon fixed in photosynthesis of 1.4 : 1, the carbon fixation rate is estimated to be 3.2 ± 0.4 mol m^{-2} a^{-1}.

There is also a net seasonal decrease in ΣCO_2 attributable to photosynthesis at these locations. Such decreases are simpler to use in productivity estimates, principally because air–sea interaction has a much weaker influence on ΣCO_2. On the other hand, precise measurements are required because the photosynthetically driven changes are much smaller compared with the background ΣCO_2 levels. Because of these differences, estimates based on ΣCO_2 seasonal cycles offer an independent measure of euphotic zone mass budgets.

Finally, differences in the carbon isotopic ratio between organic and inorganic carbon, as well as atmospheric CO_2, allow the construction of yet a third mass budget for the euphotic zone. There is a clear carbon isotope signature that can be modeled as a function of primary production, air–sea exchange, and mixing with deeper waters.

Tracer Flux-gauge Determinations

The third tracer constraint that may be used to determine primary production involves the use of 'tracer flux gauges' to estimate the flux of nutrients to

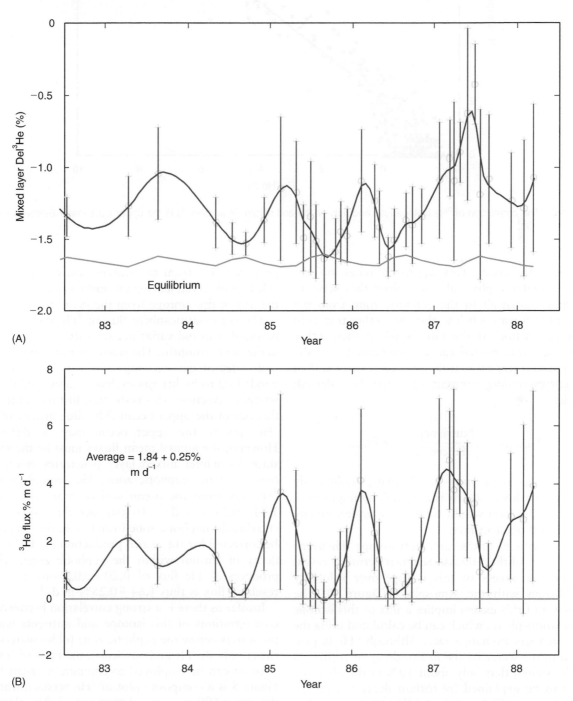

Figure 4 An approximately 6 year history of surface water ^{3}He isotope ratio anomalies (A) and computed flux to the atmosphere near Bermuda (B).

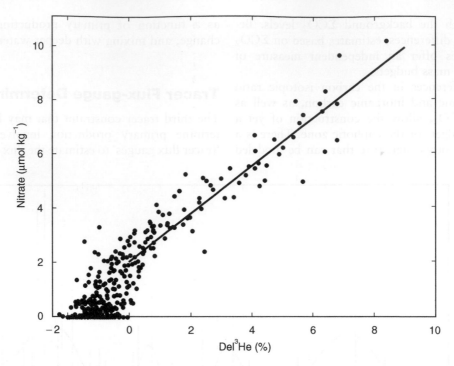

Figure 5 The correlation of ³He isotope ratio anomaly (in %) and nitrate (in $\mu mol\,kg^{-1}$) in the upper ocean near Bermuda for the period 1985–88.

the euphotic zone. This approach relies on the premise that the physical mechanisms that serve to transport nutrients to the euphotic zone from the nutrient-rich waters below also carry other tracers in fixed proportion. If the rate at which these other tracers are transported can be determined, and the nutrient to tracer ratio at the 'source' is known, then the corresponding nutrient flux may be inferred; that is:

$$F_{Nutrient} = \left[\frac{Nutrient}{Tracer} \right]_{Source} \times F_{Tracer}$$

Inasmuch as there may be alternate, biologically mediated pathways (such as zooplankton migration), such a calculation would serve as an underestimate to the total nutrient flux.

Measurements of the rare, inert isotope ³He in the mixed layer of the Sargasso Sea near Bermuda reveal a persistent excess of this isotope over solubility equilibrium with the atmosphere (**Figure 4**). The existence of this excess implies a flux of this isotope to the atmosphere, which can be calculated using the estimated gas exchange rate. Although ³He is produced in the water by the *in situ* decay of tritium, it can be shown that only about 10% of the observed flux can be explained by tritium decay within the euphotic zone. The greater portion of this ³He flux arises from the upward 'exhalation' of old tritium-produced ³He from the waters below. That is, the ³He flux observed leaving the surface ocean is largely the loss of this isotope from the main thermocline.

The ocean–atmosphere flux of ³He shows a pronounced seasonal variation, with the greatest fluxes in the winter months. The winter maximum is due to high rates of gas exchange (more vigorous winter winds lead to higher gas exchange rates) and deeper winter convection. This is the time history of the ³He flux out of the upper ocean. The time history of the ³He flux to the upper ocean may be different. However, the annual mean fluxes must be the same, since the winter mixed layer penetrates below the bottom of the euphotic zone. The annual average ³He flux from the ocean surface near Bermuda is 1.84 ± 0.25 %-m d^{-1}. To estimate the flux of ³He entering the euphotic zone from below, this flux must be corrected for the *in situ* production of ³He by the decay of tritium within the euphotic zone, which produces a ³He flux of 0.20 ± 0.02 %-m d^{-1}. The resultant flux is thus 1.64 ± 0.25 %-m d^{-1}.

Insofar as there is a strong correlation between the concentrations of this isotope and nutrients within the waters below the euphotic zone (older waters are richer in both ³He and nutrients), the ratio of ³He to nutrient can be employed to compute nutrient flux. **Figure 5** is a composite plot of ³He versus nitrate in the upper 600 m over a 3 year period. The slope of the relationship is $0.87 \pm 0.05 \, \mu mol \, kg^{-1}\text{‰}^{-1}$.

Applying the flux equation presented above, a nitrate flux of $0.56 \pm 0.16 \, \mathrm{mol \, m^{-2} \, a^{-1}}$ is computed. Using the average biological C : N ratio of 6.6, this leads to a carbon fixation rate of $3.7 \pm 1.0 \, \mathrm{mol \, m^{-2} \, a^{-1}}$. The estimate thus obtained is a local, annual-scale measure of new production.

A similar calculation can be made by observing the long-term (decade timescale) trends in thermocline ^{3}He inventories. The long-term evolution of ^{3}He inventory in the thermocline must respond to the opposing processes of production by tritium decay and 'exhalation' upward to the euphotic zone. Knowing the former gives the latter. Using nutrient-^{3}He ratios, a gyre-scale, decadal average estimate of the nutrient flux to the euphotic zone can be obtained. A detailed analysis of the long-term trends of tritium and ^{3}He in the upper 1000 m of the Sargasso Sea, coupled with the observed nitrate : ^{3}He ratios, yields an estimate of $0.70 \pm 0.20 \, \mathrm{mol \, m^{-2} \, a^{-1}}$. This leads to a somewhat higher carbon fixation rate of $4.6 \pm 1.3 \, \mathrm{mol \, m^{-2} \, a^{-1}}$. This estimate differs from the surface layer flux calculation in that it is a much longer-term average, since it depends on the very long-term evolution of isotopes in the thermocline. Moreover, it represents a very large-scale gyre-scale determination, rather than a local measure: horizons within the thermocline probably connect to regions of higher productivity further north.

Comparing Tracer-derived Estimates

Although the various techniques described here are based on differing assumptions, and measure different types of production, they should be mutually consistent on annual or greater timescales. **Table 1** is a comparison between the various estimates near Bermuda in the Sargasso Sea. A weighted average of these determinations gives a productivity of 3.6 ± 0.5 mol (C) m^{-2} a^{-1} for the Sargasso Sea near Bermuda. The determinations are within uncertainties of each other, although they utilize different tracer systems,

Table 1 Comparison of tracer-derived estimates near Bermuda in the Sargasso Sea

Type of determination	Type of production	Technique used	Carbon flux (mol m^{-2}a^{-1})
Aphotic zone oxygen consumption rates	Export production	Tritium-^{3}He dating	4.5 ± 0.7
Euphotic zone cycling	Net community	Oxygen cycling	3.2 ± 0.4
		Carbon isotopes	3.8 ± 1.3
Tracer flux-gauge	New production	Mixed layer ^{3}He	3.7 ± 1.0
		Thermocline budgets	4.6 ± 1.3

are reliant on different assumptions, and are virtually independent of each other. This agreement provides some confidence as to their accuracy.

See also

Air–Sea Transfer: N$_2$O, NO, CH$_4$, CO. Carbon Cycle. Primary Production Processes.

Further Reading

Falkowski PG and Woodhead AD (1992) *Primary Productivity and Biogeochemical Cycles in the Sea.* New York: Plenum Press.

Jenkins WJ (1995) Tracer based inferences of new and export primary productivity in the oceans. IUGG, Quadrennial Report 1263–1269.

Williams PJ and le B (1993) On the definition of plankton production terms. *ICES Marine Science Symposium* 197: 9–19.

IRON FERTILIZATION

K. H. Coale, Moss Landing Marine Laboratories, CA, USA

Introduction

The trace element iron has been shown to play a critical role in nutrient utilization and phytoplankton growth and therefore in the uptake of carbon dioxide from the surface waters of the global ocean. Carbon fixation in the surface waters, via phytoplankton growth, shifts the ocean–atmosphere exchange equilibrium for carbon dioxide. As a result, levels of atmospheric carbon dioxide (a greenhouse gas) and iron flux to the oceans have been linked to climate change (glacial to interglacial transitions). These recent findings have led some to suggest that large-scale iron fertilization of the world's oceans might therefore be a feasible strategy for controlling climate. Others speculate that such a strategy could deleteriously alter the ocean ecosystem, and still others have calculated that such a strategy would be ineffective in removing sufficient carbon dioxide to produce a sizable and rapid result. This article focuses on carbon and the major plant nutrients, nitrate, phosphate, and silicate, and describes how our recent discovery of the role of iron in the oceans has increased our understanding of phytoplankton growth, nutrient cycling, and the flux of carbon from the atmosphere to the deep sea.

Major Nutrients

Phytoplankton growth in the oceans requires many physical, chemical, and biological factors that are distributed inhomogenously in space and time. Because carbon, primarily in the form of the bicarbonate ion, and sulfur, as sulfate, are abundant throughout the water column, the major plant nutrients in the ocean commonly thought to be critical for phytoplankton growth are those that exist at the micromolar level such as nitrate, phosphate, and silicate. These, together with carbon and sulfur, form the major building blocks for biomass in the sea. As fundamental cellular constituents, they are generally thought to be taken up and remineralized in constant ratio to one another. This is known as the Redfield ratio (Redfield, 1934, 1958) and can be expressed on a molar basis relative to carbon as 106C : 16N : 1P.

Significant local variations in this uptake/regeneration relationship can be found and are a function of the phytoplankton community and growth conditions, yet this ratio can serve as a conceptual model for nutrient uptake and export.

The vertical distribution of the major nutrients typically shows surface water depletion and increasing concentrations with depth. The schematic profile in **Figure 1** reflects the processes of phytoplankton uptake within the euphotic zone and remineralization of sinking planktonic debris via microbial degradation, leading to increased concentrations in the deep sea. Given favorable growth conditions, the nutrients at the surface may be depleted to zero. The rate of phytoplankton production of new biomass, and therefore the rate of carbon uptake, is controlled by the resupply of nutrients to the surface waters, usually via the upwelling of deep waters. Upwelling occurs over the entire ocean basin at the rate of approximately 4 m per year but increases in coastal and regions of divergent surface water flow, reaching average values of 15 to 30 or greater. Thus, those

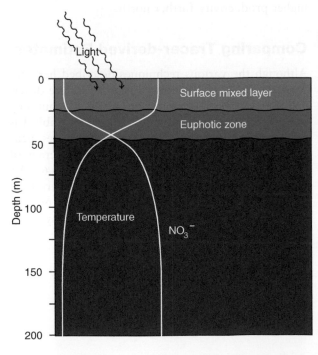

Figure 1 A schematic profile indicating the regions of the upper water column where phytoplankton grow. The surface mixed layer is that region that is actively mixed by wind and wave energy, which is typically depleted in major nutrients. Below this mixed layer temperatures decrease and nutrients increase as material sinking from the mixed layer is regenerated by microbial decomposition.

regions of high nutrient supply or persistent high nutrient concentrations are thought to be most important in terms of carbon removal.

Nitrogen versus Phosphorus Limitation

Although both nitrogen and phosphorus are required at nearly constant ratios characteristic of deep water, nitrogen has generally been thought to be the limiting nutrient in sea water rather than phosphorus. This idea has been based on two observations: selective enrichment experiments and surface water distributions. When ammonia and phosphate are added to sea water in grow-out experiments, phytoplankton growth increases with the ammonia addition and not with the phosphate addition, thus indicating that reduced nitrogen and not phosphorus is limiting. Also, when surface water concentration of nitrate and phosphate are plotted together (**Figure 2**), it appears that there is still residual phosphate after the nitrate has gone to zero.

The notion of nitrogen limitation seems counterintuitive when one considers the abundant supply of dinitrogen (N_2) in the atmosphere. Yet this nitrogen gas is kinetically unavailable to most phytoplankton because of the large amount of energy required to break the triple bond that binds the dinitrogen molecule. Only those organisms capable of nitrogen fixation can take advantage of this form of nitrogen and reduce atmospheric N_2 to biologically available nitrogen in the form of urea and ammonia. This is, energetically, a very expensive process requiring specialized enzymes (nitrogenase), an anaerobic microenvironment, and large amounts of reducing power in the form of electrons generated by photosynthesis. Although there is currently the suggestion that nitrogen fixation may have been underestimated as an important geochemical process, the major mode of nitrogen assimilation, giving rise to new plant production in surface waters, is thought to be nitrate uptake.

The uptake of nitrate and subsequent conversion to reduced nitrogen in cells requires a change of five in the oxidation state and proceeds in a stepwise fashion. The initial reduction takes place via the nitrate/nitrite reductase enzyme present in phytoplankton and requires large amounts of the reduced nicotinamide–adenine dinucleotide phosphate (NADPH) and of adenosine triphosphate (ATP) and thus of harvested light energy from photosystem II. Both the nitrogenase enzyme and the nitrate reductase enzyme require iron as a cofactor and are thus sensitive to iron availability.

Ocean Regions

From a nutrient and biotic perspective, the oceans can be generally divided into biogeochemical provinces that reflect differences in the abundance of macronutrients and the standing stocks of phytoplankton. These are the high-nitrate, high-chlorophyll (HNHC); high-nitrate, low-chlorophyll (HNLC); low-nitrate, high-chlorophyll (LNHC); and low-nitrate, low-chlorophyll (LNLC) regimes (**Table 1**). Only the HNLC and LNLC regimes are relatively stable, because the high phytoplankton growth rates in the other two systems will deplete any residual nitrate and sink out of the system. The processes that give rise to these regimes have been the

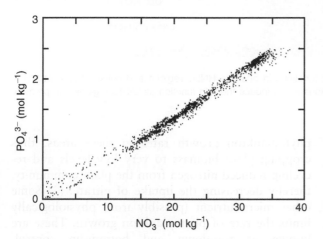

Figure 2 A plot of the global surface water concentrations of phosphate versus nitrate indicating a general positive intercept for phosphorus when nitrate has gone to zero. This is one of the imperical observations favoring the notion of nitrate limitation over phosphate limitation.

Table 1 The relationship between biomass and nitrate as a function of biogeochemical province and the approximate ocean area represented by these regimes

	High-chlorophyll	Low-chlorophyll
High-nitrate	Unstable/coastal (5%)	Stable/Subarctic/Antarctic/ equatorial Pacific (20%)
Low-nitrate	Unstable/coastal (5%)	Oligotrophic gyres (70%)

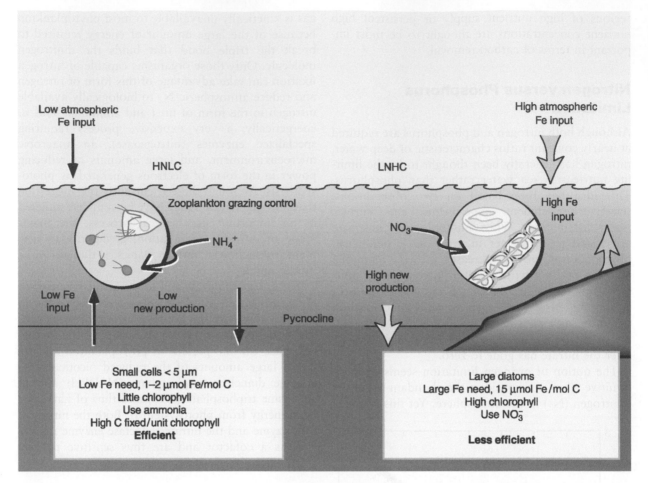

Figure 3 A schematic representation of the 'iron theory' as it functions in offshore HNLC regions and coastal transient LNHC regions. It has been suggested that iron added to the HNLC regions would induce them to function as LNHC regions and promote carbon export.

subject of some debate over the last few years and are of fundamental importance relative to carbon export (**Figure 3**).

High-nitrate, Low-chlorophyll Regions

The HNLC regions are thought to represent about 20% of the areal extent of the world's oceans. These are generally regions characterized by more than $2 \, \mu mol \, l^{-1}$ nitrate and less than $0.5 \, \mu g \, l^{-1}$ chlorophyll-*a*, a proxy for plant biomass. The major HNLC regions are shown in **Figure 4** and represent the Subarctic Pacific, large regions of the eastern equatorial Pacific and the Southern Ocean. These HNLC regions persist in areas that have high macronutrient concentrations, adequate light, and physical characteristics required for phytoplankton growth but have very low plant biomass. Two explanations have been given to describe the persistence of this condition. (1) The rates of zooplankton grazing of the phytoplankton community may balance or exceed

phytoplankton growth rates in these areas, thus cropping plant biomass to very low levels and recycling reduced nitrogen from the plant community, thereby decreasing the uptake of nitrate. (2) Some other micronutrient (possibly iron) physiologically limits the rate of phytoplankton growth. These are known as top-down and bottom-up control, respectively.

Several studies of zooplankton grazing and phytoplankton growth in these HNLC regions, particularly the Subarctic Pacific, confirm the hypothesis that grazers control production in these waters. Recent physiological studies, however, indicate that phytoplankton growth rates in these regions are suboptimal, as is the efficiency with which phytoplankton harvest light energy. These observations indicate that phytoplankton growth may be limited by something other than (or in addition to) grazing. Specifically, these studies implicate the lack of sufficient electron transport proteins and the cell's ability to transfer reducing power from the

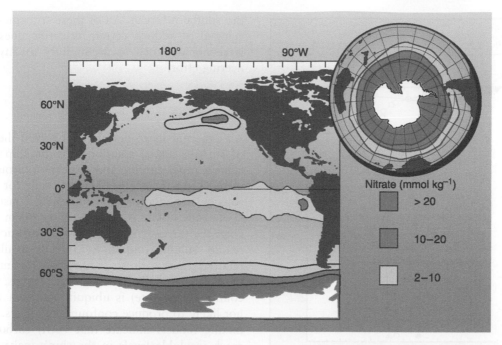

Figure 4 Current HNLC regions of the world's oceans covering an extimated 20% of the ocean surface. These regions include the Subarctic Pacific, equatorial Pacific and Southern Ocean.

photocenter. These have been shown to be symptomatic of iron deficiency.

The Role of Iron

Iron is a required micronutrient for all living systems. Because of its d-electron configuration, iron readily undergoes redox transitions between Fe(II) and Fe(III) at physiological redox potentials. For this reason, iron is particularly well suited to many enzyme and electron carrier proteins. The genetic sequences coding for many iron-containing electron carriers and enzymes are highly conserved, indicating iron and iron-containing proteins were key features of early biosynthesis. When life evolved, the atmosphere and waters of the planet were reducing and iron was abundant in the form of soluble Fe(II). Readily available and at high concentration, iron was not likely to have been limiting in the primordial biosphere. As photosynthesis evolved, oxygen was produced as a by-product. As the biosphere became more oxidizing, iron precipitated from aquatic systems in vast quantities, leaving phytoplankton and other aquatic life forms in a vastly changed and newly deficient chemical milieu. Evidence of this mass Fe(III) precipitation event is captured in the ancient banded iron formations in many parts of the world. Many primitive aquatic and terrestrial organisms have subsequently evolved the ability to sequester iron through the elaboration of specific Fe(II)-binding ligands,

known as siderophores. Evidence for siderophore production has been found in several marine dinoflagellates and bacteria and some researchers have detected similar compounds in sea water.

Today, iron exists in sea water at vanishingly small concentrations. Owing to both inorganic precipitation and biological uptake, typical surface water values are on the order of 20 pmol l^{-1}, perhaps a billion times less than during the prehistoric past. Iron concentrations in the oceans increase with depth, in much the same manner as the major plant nutrients (**Figure 5**).

The discovery that iron concentrations in surface waters is so low and shows a nutrient-like profile led some to speculate that iron availability limits plant growth in the oceans. This notion has been tested in bottle enrichment experiments throughout the major HNLC regions of the world's oceans. These experiments have demonstrated dramatic phytoplankton growth and nutrient uptake upon the addition of iron relative to control experiments in which no iron was added.

Criticism that such small-scale, enclosed experiments may not accurately reflect the response of the HNLC system at the level of the community has led to several large-scale iron fertilization experiments in the equatorial Pacific and Southern Ocean. These have been some of the most dramatic oceanographic experiments of our times and have led to a profound and new understanding of ocean systems.

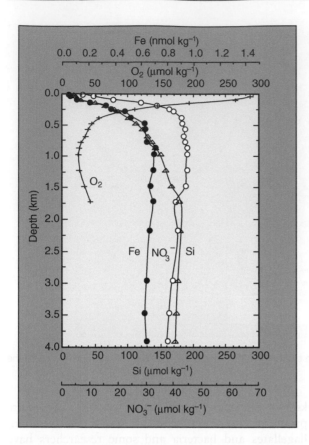

Figure 5 The vertical distributions of iron, nitrate, silicate, and oxygen in sea water. This figure shows how iron is depleted to picomolar levels in surface waters and has a profile that mimics other plant nutrients.

Open Ocean Iron Enrichment

The question of iron limitation was brought into sharp scientific focus with a series of public lectures, reports by the US National Research Council, papers, special publications, and popular articles between 1988 and 1991. What was resolved was the need to perform an open ocean enrichment experiment in order to definitively test the hypothesis that iron limits phytoplankton growth and nutrient and carbon dioxide uptake in HNLC regions. Such an experiment posed severe logistical challenges and had never been conducted.

Experimental Strategy

The mechanics of producing an iron-enriched experimental patch and following it over time was developed in four release experiments in the equatorial Pacific (IronEx I and II) and more recently in the Southern Ocean (SOIREE). At this writing, a similar strategy is being employed in the Caruso experiments now underway in the Atlantic sector of the Southern Ocean. All of these strategies were developed to address certain scientific questions and were not designed as preliminary to any geoengineering effort.

Form of Iron

All experiments to date have involved the injection of an iron sulfate solution into the ship's wake to achieve rapid dilution and dispersion throughout the mixed layer (**Figure 6**). The rationale for using ferrous sulfate involved the following considerations: (1) ferrous sulfate is the most likely form of iron to enter the oceans via atmospheric deposition; (2) it is readily soluble (initially); (3) it is available in a relatively pure form so as to reduce the introduction of other potentially bioactive trace metals; and (4) its counterion (sulfate) is ubiquitous in sea water and not likely to produce confounding effects. Although mixing models indicate that Fe(II) carbonate may reach insoluble levels in the ship's wake, rapid dilution reduces this possibility.

New forms of iron are now being considered by those who would seek to reduce the need for subsequent infusions. Such forms could include iron lignosite, which would increase the solubility and residence time of iron in the surface waters. Since this is a chelated form of iron, problems of rapid precipitation are reduced. In addition, iron lignosulfonate is about 15% Fe by weight, making it a space-efficient form of iron to transport. As yet untested is the extent to which such a compound would reduce the need for re-infusion.

Although solid forms of iron have been proposed (slow-release iron pellets; finely milled magnetite or iron ores), the ability to trace the enriched area with an inert tracer has required that the form of iron added and the tracer both be in the dissolved form.

Inert Tracer

Concurrent with the injection of iron is the injection of the inert chemical tracer sulfur hexafluoride (SF_6). By presaturating a tank of sea water with SF_6 and employing an expandable displacement bladder, a constant molar injection ratio of Fe : SF_6 can be achieved (**Figure 6**). In this way, both conservative and nonconservative removal of iron can be quantified. Sulfur hexafluoride traces the physical properties of the enriched patch; the relatively rapid shipboard detection of SF_6 can be used to track and map the enriched area. The addition of helium-3 to the injected tracer can provide useful information regarding gas transfer.

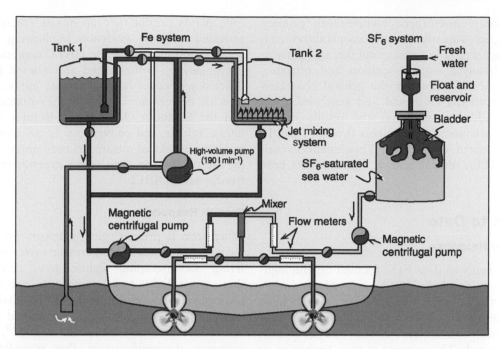

Figure 6 The iron injection system used during the IronEx experiments utilized two polyethylene tanks that could be sequentially filled with sea water and iron sulfate solution while the other was being injected behind the ship's propellers. A steel tank of sea water saturated with 40 g of sulfur hexafluoride (SF_6) was simultaneously mixed with the iron sulfate solution to provide a conservative tracer of mixing.

Fluorometry

The biophysical response of the phytoplankton is rapid and readily detectable. Thus shipboard measurement of relative fluorescence (F_v/F_m) using fast repetition rate fluorometry has been shown to be a useful tactical tool and gives nearly instantaneous mapping and tracking feedback.

Shipboard Iron Analysis

Because iron is rapidly lost from the system (at least initially), the shipboard determination of iron is necessary to determine the timing and amount of subsequent infusions. Several shipboard methods, using both chemiluminescent and catalytic colorimetric detection have proven useful in this regard.

Lagrangian Drifters

A Lagrangian point of reference has proven to be very useful in every experiment to date. Depending upon the advective regime, this is the only practical way to achieve rapid and precise navigation and mapping about the enriched area.

Remote Sensing

A variety of airborne and satellite-borne active and passive optical packages provide rapid, large-scale mapping and tracking of the enriched area. Although

SeaWiffs was not operational during IronEx I and II, AVHRR was able to detect the IronEx II bloom and airborne optical LIDAR was very useful during IronEx I. SOIREE has made very good use of the more recent SeaWiffs images, which have markedly extended the observational period and led to new hypotheses regarding iron cycling in polar systems.

Experimental Measurements

In addition to the tactical measurements and remote sensing techniques required to track and ascertain the development of the physical dynamics of the enriched patch, a number of measurements have been made to track the biogeochemical development of the experiment. These have typically involved a series of underway measurements made using the ship's flowing sea water system or towed fish. In addition, discrete measurements are made in the vertical dimension at every station occupied both inside and outside of the fertilized area. These measurements include temperature salinity, fluorescence (a measure of plant biomass), transmissivity (a measure of suspended particles), oxygen, nitrate, phosphate, silicate, carbon dioxide partial pressure, pH, alkalinity, total carbon dioxide, iron-binding ligands, $^{234}Th : ^{238}U$ radioisotopic disequilibria (a proxy for particle removal), relative fluorescence (indicator of photosynthetic competence), primary production, phytoplankton and zooplankton enumeration, grazing rates, nitrate uptake, and particulate and dissolved organic carbon and nitrogen. These parameters allow for the general characterization of both the biological and geochemical response to added iron. From the results of the equatorial enrichment experiments (IronEx I and II) and the Southern Ocean Iron Enrichment Experiment (SOIREE), several general features have been identified.

Findings to Date

Biophysical Response

The experiments to date have focused on the high-nitrate, low-chlorophyll (HNLC) areas of the world's oceans, primarily in the Subarctic, equatorial Pacific and Southern Ocean. In general, when light is abundant many researchers find that HNLC systems are iron-limited. The nature of this limitation is similar between regions but manifests itself at different levels of the trophic structure in some characteristic ways. In general, all members of the HNLC photosynthetic community are physiologically limited by iron availability. This observation is based primarily on the examination of the efficiency of photosystem II, the light-harvesting reaction centers. At ambient levels of iron, light harvesting proceeds at suboptimal rates. This has been attributed to the lack of iron-dependent electron carrier proteins at low iron concentrations. When iron concentrations are increased by subnanomolar amounts, the efficiency of light harvesting rapidly increases to maximum levels. Using fast repetition rate fluorometry and non-heme iron proteins, researchers have described these observations in detail. What is notable about these results is that iron limitation seems to affect the photosynthetic energy conversion efficiency of even the smallest of phytoplankton. This has been a unique finding that stands in contrast to the hypothesis that, because of diffusion, smaller cells are not iron limited but larger cells are.

Nitrate Uptake

As discussed above, iron is also required for the reduction (assimilation) of nitrate. In fact, a change of oxidation state of five is required between nitrate and the reduced forms of nitrogen found in amino acids and proteins. Such a large and energetically unfavorable redox process is only made possible by substantial reducing power (in the form of NADPH) made available through photosynthesis and active nitrate reductase, an iron-requiring enzyme. Without iron, plants cannot take up nitrate efficiently. This provided original evidence implicating iron deficiency as the cause of the HNLC condition. When phytoplankton communities are relieved from iron deficiency, specific rates of nitrate uptake increase. This has been observed in both the equatorial Pacific and the Southern Ocean using isotopic tracers of nitrate uptake and conversion. In addition, the accelerated uptake of nitrate has been observed in both the mesoscale iron enrichment experiments to date, IronEx and SOIREE.

Growth Response

When iron is present, phytoplankton growth rates increase dramatically. Experiments over widely differing oceanographic regimes have demonstrated that, when light and temperature are favorable, phytoplankton growth rates in HNLC environments increase to their maximum at dissolved iron concentrations generally below 0.5 nmol l^{-1}. This observation is significant in that it indicates that phytoplankton are adapted to very low levels of iron and they do not grow faster if given iron at more than 0.5 nmol l^{-1}. Given that there is still some disagreement within the scientific community about the validity of some iron measurements, this phytoplankton response provides a natural, environmental, and biogeochemical benchmark against which to compare results.

The iron-induced transient imbalance between phytoplankton growth and grazing in the equatorial Pacific during IronEx II resulted in a 30-fold increase in plant biomass (**Figure 7**). Similarly, a 6-fold increase was observed during the SOIREE experiment in the Southern Ocean. These are perhaps the most dramatic demonstrations of iron limitation of nutrient cycling, and phytoplankton growth to date and has fortified the notion that iron fertilization may be a useful strategy to sequester carbon in the oceans.

Heterotrophic Community

As the primary trophic levels increase in biomass, growth in the small microflagellate and heterotrophic bacterial communities increase in kind. It appears that these consumers of recently fixed carbon (both particulate and dissolved) respond to the food source and not necessarily the iron (although some have been found to be iron-limited). Because their division rates are fast, heterotrophic bacteria, ciliates, and flagellates can rapidly divide and respond to increasing food availability to the point where the growth rates of the smaller phytoplankton can be overwhelmed by grazing. Thus there is a much more rapid turnover of fixed carbon and

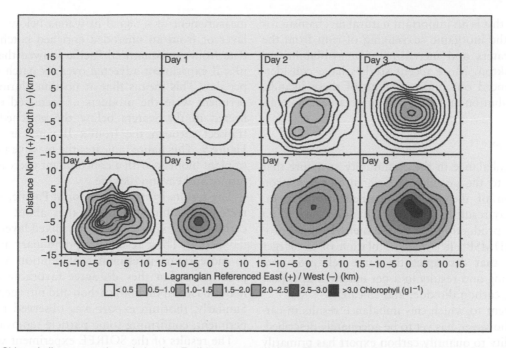

Figure 7 Chlorophyll concentrations during IronEx II were mapped daily. This figure shows the progression of the phytoplankton bloom that reached over 30 times the background concentrations.

nitrogen in iron replete systems. M. Landry and co-workers have documented this in dilution experiments conducted during IronEx II. These results appear to be consistent with the recent SOIREE experiments as well.

Nutrient Uptake Ratios

An imbalance in production and consumption, however, can arise at the larger trophic levels. Because the reproduction rates of the larger micro- and mesozooplankton are long with respect to diatom division rates, iron-replete diatoms can escape the pressures of grazing on short timescales (weeks). This is thought to be the reason why, in every iron enrichment experiment, diatoms ultimately dominate in biomass. This result is important for a variety of reasons. It suggests that transient additions of iron would be most effective in producing net carbon uptake and it implicates an important role of silicate in carbon flux. The role of iron in silicate uptake has been studied extensively by Franck and colleagues. The results, together with those of Takeda and co-workers, show that iron alters the uptake ratio of nitrate and silicate at very low levels (**Figure 8**). This is thought to be brought about by the increase in nitrate uptake rates relative to silica.

Organic Ligands

Consistent with the role of iron as a limiting nutrient in HNLC systems is the notion that organisms may

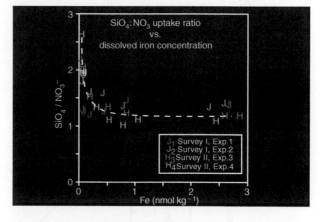

Figure 8 Bottle enrichment experiments show that the silicate : nitrate uptake ratio changes as a function of the iron added. This is thought to be due to the increased rate of iron uptake relative to silicate in these experimental treatments.

have evolved competitive mechanisms to increase iron solubility and uptake. In terrestrial systems this is accomplished using extracellularly excreted or membrane-bound siderophores. Similar compounds have been shown to exist in sea water where the competition for iron may be as fierce as it is on land. In open ocean systems where it has been measured, iron-binding ligand production increases with the addition of iron. Whether this is a competitive response to added iron or a function of phytoplankton biomass and grazing is not yet well understood.

However, this is an important natural mechanism for reducing the inorganic scavenging of iron from the surface waters and increasing iron availability to phytoplankton. More recent studies have considerably advanced our understanding of these ligands, their distribution and their role in ocean ecosystems.

Carbon Flux

It is the imbalance in the community structure that gives rise to the geochemical signal. Whereas iron stimulation of the smaller members of the community may result in chemical signatures such as an increased production of beta-dimethylsulfoniopropionate (DMSP), it is the stimulation of the larger producers that decouples the large cell producers from grazing and results in a net uptake and export of nitrate, carbon dioxide, and silicate.

The extent to which this imbalance results in carbon flux, however, has yet to be adequately described. The inability to quantify carbon export has primarily been a problem of experimental scale. Even though mesoscale experiments have, for the first time, given us the ability to address the effect of iron on communities, the products of surface water processes and the effects on the midwater column have been difficult to track. For instance, in the IronEx II experiment, a time-series of the enriched patch was diluted by 40% per day. The dilution was primarily in a lateral (horizontal/isopycnal) dimension. Although some correction for lateral dilution can be made, our ability to quantify carbon export is dependent upon the measurement of a signal in waters below the mixed layer or from an uneroded enriched patch. Current data from the equatorial Pacific showed that the IronEx II experiment advected over six patch diameters per day. This means that at no time during the experiment were the products of increased export reflected in the waters below the enriched area. A transect through the IronEx II patch is shown in **Figure 9**. This figure indicates the massive production of plant biomass with a concomitant decrease in both nitrate and carbon dioxide.

The results from the equatorial Pacific, when corrected for dilution, suggest that about 2500 t of carbon were exported from the mixed layer over a 7-day period. These results are preliminary and subject to more rigorous estimates of dilution and export production, but they do agree favorably with estimates based upon both carbon and nitrogen budgets. Similarly, thorium export was observed in this experiment, confirming some particle removal.

The results of the SOIREE experiment were similar in many ways but were not as definitive with respect to carbon flux. In this experiment biomass increased 6-fold, nitrate was depleted by $2\,\mu mol\,l^{-1}$ and carbon dioxide by 35–40 microatmospheres (3.5–4.0 Pa). This was a greatly attenuated signal relative to IronEx II. Colder water temperatures likely led to slower rates of production and bloom evolution and there was no observable carbon flux.

Original estimates of carbon export in the Southern Ocean based on the iron-induced efficient utilization of nitrate suggest that as much as 1.8×10^9 t

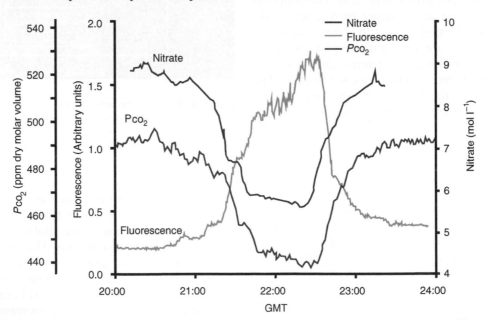

Figure 9 A transect through the IronEx II patch. The x-axis shows GMT as the ship steams from east to west through the center of the patch. Simultaneously plotted are the iron-induced production of chlorophyll, the drawdown of carbon dioxide, and the uptake of nitrate in this bloom.

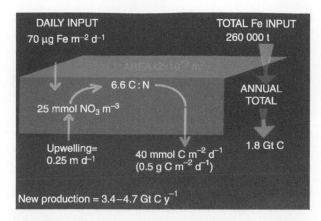

DAILY INPUT
70 µg Fe m^{-2} d^{-1}

TOTAL Fe INPUT
260 000 t

6.6 C:N

25 mmol NO$_3$ m^{-3}

ANNUAL TOTAL

Upwelling= 0.25 m d^{-1}

40 mmol C m^{-2} d^{-1}
(0.5 g C m^{-2} d^{-1})

1.8 Gt C

New production = 3.4–4.7 Gt C y^{-1}

Figure 10 Simple calculations of the potential for carbon export for the Southern Ocean. These calculations are based on the necessary amount of iron required to efficiently utilize the annual upwelled nitrate and the subsequent incorporation into sinking organic matter. An estimated 1.8×10^9 t (Gt) of carbon export could be realized in this simple model.

of carbon could be removed annually (**Figure 10**). These estimates of carbon sequestration have been challenged by some modelers yet all models lack important experimental parameters which will be measured in upcoming experiments.

Remaining Questions

A multitude of questions remain regarding the role of iron in shaping the nature of the pelagic community. The most pressing question is whether iron enrichment accelerates the downward transport of carbon from the surface waters to the deep sea? More specifically, how does iron affect the cycling of carbon in HNLC, LNLC, and coastal systems? Recent studies indicate that coastal systems may be iron-limited and the iron requirement for nitrogenase activity is quite large, suggesting that iron may limit nitrogen fixation, but there have been limited studies to test the former and none to test the latter. If iron does stimulate carbon uptake, what are the spatial scales over which this fixed carbon may be remineralized? This is crucial to predicting whether fertilization is an effective carbon sequestration mechanism.

Given these considerations, the most feasible way to understand and quantify carbon export from an enriched water mass is to increase the scale of the experiment such that both lateral dilution and sub-mixed-layer relative advection are small with respect to the size of the enriched patch. For areas such as the equatorial Pacific, this would be very large (hundreds of kilometers on a side). For other areas, it could be much smaller.

The focus of the IronEx and SOIREE experiments has been from the scientific perspective, but this focus

is shifting toward the application of iron enrichment as a carbon sequestration strategy. We have come about rapidly from the perspective of trying to understand how the world works to one of trying to make the world work for us. Several basic questions remain regarding the role of natural or anthropogenic iron fertilization on carbon export. Some of the most pressing questions are: What are the best proxies for carbon export? How can carbon export best be verified? What are the long-term ecological consequences of iron enrichment on surface water community structure, midwater processes, and benthic processes? Even with answers to these, there are others that need to be addressed prior to any serious consideration of iron fertilization as an ocean carbon sequestration option.

Simple technology is sufficient to produce a massive bloom. The technology required either for a large-scale enrichment experiment or for purposeful attempts to sequester carbon is readily available. Ships, aircraft (tankers and research platforms), tracer technology, a broad range of new Autonomous Underwater Vehicles (AUVs) and instrument packages, Lagrangian buoy tracking systems, together with aircraft and satellite remote sensing systems and a new suite of chemical sensors/*in situ* detection technologies are all available, or are being developed. Industrial bulk handling equipment is available for large-scale implementation. The big questions, however, are larger than the technology.

With a slow start, the notion of both scientific experimentation through manipulative experiments, as well as the use of iron to purposefully sequester carbon, is gaining momentum. There are now national, international, industrial, and scientific concerns willing to support larger-scale experiments. The materials required for such an experiment are inexpensive and readily available, even as industrial by-products (of paper, mining, and steel processing).

Given the concern over climate change and the rapid modernization of large developing countries such as China and India, there is a pressing need to address the increased emission of greenhouse gases. Through the implementation of the Kyoto accords or other international agreements to curb emissions (Rio), financial incentives will reach into the multi-billion dollar level annually. Certainly there will soon be an overwhelming fiscal incentive to investigate, if not implement, purposeful open ocean carbon sequestration trials.

A Societal Challenge

The question is not whether we have the capability of embarking upon such an engineering strategy but whether we have the collective wisdom to responsibly negotiate such a course of action. Posing the question another way: If we do not have the social, political and economic tools or motivation to control our own population and greenhouse gas emissions, what gives us the confidence that we have the wisdom and ability to responsibly manipulate and control large ocean ecosystems without propagating yet another massive environmental calamity? Have we as an international community first tackled the difficult but obvious problem of overpopulation and implemented alternative energy technologies for transportation, industry, and domestic use?

Other social questions arise as well. Is it appropriate to use the ocean commons for such a purpose? What individuals, companies, or countries would derive monetary compensation for such an effort and how would this be decided?

It is clear that there are major scientific investigations and findings that can only benefit from large-scale open ocean enrichment experiments, but certainly a large-scale carbon sequestration effort should not proceed without a clear understanding of both the science and the answers to the questions above.

Glossary

ATP	Adenosine triphosphate
AVHRR	Advanced Very High Resolution Radiometer
HNHC	High-nitrate high-chlorophyll
HNLC	High-nitrate low-chlorophyll
IronEx	Iron Enrichment Experiment
LIDAR	Light detection and ranging
LNHC	Low-nitrate high-chlorophyll
LNLC	Low-nitrate low-chlorophyll
NADPH	Reduced form of nicotinamide–adenine dinucleotide phosphate
SOIREE	Southern Ocean Iron Enrichment Experiment

See also

Nitrogen Cycle. Phosphorus Cycle. Primary Production Processes. Satellite Remote Sensing Sar.

Further Reading

Abraham ER, Law CS, Boyd PW, et al. (2000) Importance of stirring in the development of an iron-fertilized phytoplankton bloom. Nature 407: 727–730.

Barbeau K, Moffett JW, Caron DA, Croot PL, and Erdner DL (1996) Role of protozoan grazing in relieving iron limitation of phytoplankton. Nature 380: 61–64.

Behrenfeld MJ, Bale AJ, Kobler ZS, Aiken J, and Falkowski PG (1996) Confirmation of iron limitation of phytoplankton photosynthesis in Equatorial Pacific Ocean. Nature 383: 508–511.

Boyd PW, Watson AJ, Law CS, et al. (2000) A mesoscale phytoplankton bloom in the polar Southern Ocean stimulated by iron fertilization. Nature 407: 695–702.

Cavender-Bares KK, Mann EL, Chishom SW, Ondrusek ME, and Bidigare RR (1999) Differential response of equatorial phytoplankton to iron fertilization. Limnology and Oceanography 44: 237–246.

Coale KH, Johnson KS, Fitzwater SE, et al. (1996) A massive phytoplankton bloom induced by an ecosystem-scale iron fertilization experiment in the equatorial Pacific Ocean. Nature 383: 495–501.

Coale KH, Johnson KS, Fitzwater SE, et al. (1998) IronEx-I, an in situ iron-enrichment experiment: experimental design, implementation and results. Deep-Sea Research Part II 45: 919–945.

Elrod VA, Johnson KS, and Coale KH (1991) Determination of subnanomolar levels of iron (II) and total dissolved iron in seawater by flow injection analysis with chemiluminescence dection. Analytical Chemistry 63: 893–898.

Fitzwater SE, Coale KH, Gordon RM, Johnson KS, and Ondrusek ME (1996) Iron deficiency and phytoplankton growth in the equatorial Pacific. Deep-Sea Research Part II 43: 995–1015.

Greene RM, Geider RJ, and Falkowski PG (1991) Effect of iron lititation on photosynthesis in a marine diatom. Limnology Oceanogrography 36: 1772–1782.

Hoge EF, Wright CW, Swift RN, et al. (1998) Fluorescence signatures of an iron-enriched phytoplankton community in the eastern equatorial Pacific Ocean. Deep-Sea Research Part II 45: 1073–1082.

Johnson KS, Coale KH, Elrod VA, and Tinsdale NW (1994) Iron photochemistry in seawater from the Equatorial Pacific. Marine Chemistry 46: 319–334.

Kolber ZS, Barber RT, Coale KH, et al. (1994) Iron limitation of phytoplankton photosynthesis in the Equatorial Pacific Ocean. Nature 371: 145–149.

Landry MR, Ondrusek ME, Tanner SJ, et al. (2000) Biological response to iron fertilization in the eastern equtorial Pacific (Ironex II). I. Microplankton community abundances and biomass. Marine Ecology Progress Series 201: 27–42.

LaRoche J, Boyd PW, McKay RML, and Geider RJ (1996) Flavodoxin as an in situ marker for iron stress in phytoplankton. Nature 382: 802–805.

Law CS, Watson AJ, Liddicoat MI, and Stanton T (1998) Sulfer hexafloride as a tracer of biogeochemical and physical processes in an open-ocean iron fertilization experiment. Deep-Sea Research Part II 45: 977–994.

Martin JH, Coale KH, Johnson KS, et al. (1994) Testing the iron hypothesis in ecosystems of the equatorial Pacific Ocean. Nature 371: 123–129.

Nightingale PD, Liss PS, and Schlosser P (2000) Measurements of air–gas transfer during an open ocean algal bloom. *Geophysical Research Letters* 27: 2117–2121.

Obata H, Karatani H, and Nakayama E (1993) Automated determination of iron in seawater by chelating resin concentration and chemiluminescence detection. *Analytical Chemistry* 65: 1524–1528.

Redfield AC (1934) On the proportions of organic derivatives in sea water and their relation to the composition of plankton. *James Johnstone Memorial Volume*, pp. 177–192. Liverpool: Liverpool University Press.

Redfield AC (1958) The biological control of chemical factors in the environment. *American Journal of Science* 46: 205–221.

Rue EL and Bruland KW (1997) The role of organic complexation on ambient iron chemistry in the equatorial Pacific Ocean and the response of a mesoscale iron addition experiment. *Limnology and Oceanography* 42: 901–910.

Smith SV (1984) Phosphorus versus nitrogen limitation in the marine environment. *Limnology and Oceanography* 29: 1149–1160.

Stanton TP, Law CS, and Watson AJ (1998) Physical evolution of the IronEx I open ocean tracer patch. *Deep-Sea Research Part II* 45: 947–975.

Takeda S and Obata H (1995) Response of equatorial phytoplankton to subnanomolar Fe enrichment. *Marine Chemistry* 50: 219–227.

Trick CG and Wilhelm SW (1995) Physiological changes in coastal marine cyanobacterium *Synechococcus* sp. PCC 7002 exposed to low ferric ion levels. *Marine Chemistry* 50: 207–217.

Turner SM, Nightingale PD, Spokes LJ, Liddicoat MI, and Liss PS (1996) Increased dimethyl sulfide concentrations in seawater from *in situ* iron enrichment. *Nature* 383: 513–517.

Upstill-Goddard RC, Watson AJ, Wood J, and Liddicoat MI (1991) Sulfur hexafloride and helium-3 as sea-water tracers: deployment techniques and continuous underway analysis for sulphur hexafloride. *Analytica Chimica Acta* 249: 555–562.

Van den Berg CMG (1995) Evidence for organic complesation of iron in seawater. *Marine Chemistry* 50: 139–157.

Watson AJ, Liss PS, and Duce R (1991) Design of a small-scale *in situ* iron fertilization experiment. *Limnology and Oceanography* 36: 1960–1965.

PLANKTON AND CLIMATE

A. J. Richardson, University of Queensland, St. Lucia, QLD, Australia

Introduction: The Global Importance of Plankton

Unlike habitats on land that are dominated by massive immobile vegetation, the bulk of the ocean environment is far from the seafloor and replete with microscopic drifting primary producers. These are the phytoplankton, and they are grazed by microscopic animals known as zooplankton. The word 'plankton' derives from the Greek *planktos* meaning 'to drift' and although many of the phytoplankton (with the aid of flagella or cilia) and zooplankton swim, none can progress against currents. Most plankton are microscopic in size, but some such as jellyfish are up to 2 m in bell diameter and can weigh up to 200 kg. Plankton communities are highly diverse, containing organisms from almost all kingdoms and phyla.

Similar to terrestrial plants, phytoplankton photosynthesize in the presence of sunlight, fixing CO_2 and producing O_2. This means that phytoplankton must live in the upper sunlit layer of the ocean and obtain sufficient nutrients in the form of nitrogen and phosphorus for growth. Each and every day, phytoplankton perform nearly half of the photosynthesis on Earth, fixing more than 100 million tons of carbon in the form of CO_2 and producing half of the oxygen we breathe as a byproduct.

Photosynthesis by phytoplankton directly and indirectly supports almost all marine life. Phytoplankton are a major food source for fish larvae, some small surface-dwelling fish such as sardine, and shoreline filter-feeders such as mussels and oysters. However, the major energy pathway to higher trophic levels is through zooplankton, the major grazers in the oceans. One zooplankton group, the copepods, is so numerous that they are the most abundant multicellular animals on Earth, outnumbering even insects by possibly 3 orders of magnitude. Zooplankton support the teeming multitudes higher up the food web: fish, seabirds, penguins, marine mammals, and turtles. Carcasses and fecal pellets of zooplankton and uneaten phytoplankton slowly yet consistently rain down on the cold dark seafloor, keeping alive the benthic (bottom-dwelling) communities of sponges, anemones, crabs, and fish.

Phytoplankton impact human health. Some species may become a problem for natural ecosystems and humans when they bloom in large numbers and produce toxins. Such blooms are known as harmful algal blooms (HABs) or red tides. Many species of zooplankton and shellfish that feed by filtering seawater to ingest phytoplankton may incorporate these toxins into their tissues during red-tide events. Fish, seabirds, and whales that consume affected zooplankton and shellfish can exhibit a variety of responses detrimental to survival. These toxins can also cause amnesic, diarrhetic, or paralytic shellfish poisoning in humans and may require the closure of aquaculture operations or even wild fisheries.

Despite their generally small size, plankton even play a major role in the pace and extent of climate change itself through their contribution to the carbon cycle. The ability of the oceans to act as a sink for CO_2 relies largely on plankton functioning as a 'biological pump'. By reducing the concentration of CO_2 at the ocean surface through photosynthetic uptake, phytoplankton allow more CO_2 to diffuse into surface waters from the atmosphere. This process continually draws CO_2 into the oceans and has helped to remove half of the CO_2 produced by humans from the atmosphere and distributed it into the oceans. Plankton play a further role in the biological pump because much of the CO_2 that is fixed by phytoplankton and then eaten by zooplankton sinks to the ocean floor in the bodies of uneaten and dead phytoplankton, and zooplankton fecal pellets. This carbon may then be locked up within sediments.

Phytoplankton also help to shape climate by changing the amount of solar radiation reflected back to space (the Earth's albedo). Some phytoplankton produce dimethylsulfonium propionate, a precursor of dimethyl sulfide (DMS). DMS evaporates from the ocean, is oxidized into sulfate in the atmosphere, and then forms cloud condensation nuclei. This leads to more clouds, increasing the Earth's albedo and cooling the climate.

Without these diverse roles performed by plankton, our oceans would be desolate, polluted, virtually lifeless, and the Earth would be far less resilient to the large quantities of CO_2 produced by humans.

Beacons of Climate Change

Plankton are ideal beacons of climate change for a host of reasons. First, plankton are ecthothermic (their body temperature varies with the surroundings), so their physiological processes such as nutrient uptake, photosynthesis, respiration, and reproductive development are highly sensitive to temperature, with their speed doubling or tripling with a 10 °C temperature rise. Global warming is thus likely to directly impact the pace of life in the plankton. Second, warming of surface waters lowers its density, making the water column more stable. This increases the stratification, so that more energy is required to mix deep nutrient-rich water into surface layers. It is these nutrients that drive surface biological production in the sunlit upper layers of the ocean. Thus global warming is likely to increase the stability of the ocean and diminish nutrient enrichment and reduce primary productivity in large areas of the tropical ocean. There is no such direct link between temperature and nutrient enrichment in terrestrial systems. Third, most plankton species are short-lived, so there is tight coupling between environmental change and plankton dynamics. Phytoplankton have lifespans of days to weeks, whereas land plants have lifespans of years. Plankton systems will therefore respond rapidly, whereas it takes longer before terrestrial plants exhibit changes in abundance attributable to climate change. Fourth, plankton integrate ocean climate, the physical oceanic and atmospheric conditions that drive plankton productivity. There is a direct link between climate and plankton abundance and timing. Fifth, plankton can show dramatic changes in distribution because they are free floating and most remain so their entire life. They thus respond rapidly to changes in temperature and oceanic currents by expanding and contracting their ranges. Further, as plankton are distributed by currents and not by vectors or pollinators, their dispersal is less dependent on other species and more dependent on physical processes. By contrast, terrestrial plants are rooted to their substrate and are often dependent upon vectors or pollinators for dispersal. Sixth, unlike other marine groups such as fish and many intertidal organisms, few plankton species are commercially exploited so any long-term changes can more easily be attributed to climate change. Last, almost all marine life has a planktonic stage in their life cycle because ocean currents provide an ideal mechanism for dispersal over large distances. Evidence suggests that these mobile life stages known as meroplankton are even more sensitive to climate change than the holoplankton, their neighbors that live permanently in the plankton.

All of these attributes make plankton ideal beacons of climate change. Impacts of climate change on plankton are manifest as predictable changes in the distribution of individual species and communities, in the timing of important life cycle events or phenology, in abundance and community structure, through the impacts of ocean acidification, and through their regulation by climate indices. Because of this sensitivity and their global importance, climate impacts on plankton are felt throughout the ecosystems they support.

Changes in Distribution

Plankton have exhibited some of the fastest and largest range shifts in response to global warming of any marine or terrestrial group. The general trend, as on land, is for plants and animals to expand their ranges poleward as temperatures warm. Probably the clearest examples are from the Northeast Atlantic. Members of a warm temperate assemblage have moved more than 1000-km poleward over the last 50 years (**Figure 1**). Concurrently, species of a subarctic (cold-water) assemblage have retracted to higher latitudes. Although these translocations have been associated with warming in the region by up to 1 °C, they may also be a consequence of the stronger northward flowing currents on the European shelf edge. These shifts in distribution have had dramatic impacts on the food web of the North Sea. The cool water assemblage has high biomass and is dominated by large species such as *Calanus finmarchicus*. Because this cool water assemblage retracts north as waters warm, *C. finmarchicus* is replaced by *Calanus helgolandicus*, a dominant member of the warm-water assemblage. This assemblage typically has lower biomass and contains relatively small species. Despite these *Calanus* species being indistinguishable to all but the most trained eye, the two species contrast starkly in their seasonal cycles: *C. finmarchicus* peaks in spring whereas *C. helgolandicus* peaks in autumn. This is critical as cod, which are traditionally the most important fishery of the North Sea, spawn in spring. As cod eggs hatch into larvae and continue to grow, they require good food conditons, consisting of large copepods such as *C. finmarchicus*, otherwise mortality is high and recruitment is poor. In recent warm years, however, *C. finmarchicus* is rare, there is very low copepod biomass during spring, and cod recruitment has crashed.

Changes in Phenology

Phenology, or the timing of repeated seasonal activities such as migrations or flowering, is very sensitive to global warming. On land, many events in spring are happening earlier in the year, such as the arrival of swallows in the UK, emergence of butterflies in the US, or blossoming of cherry trees in Japan. Recent

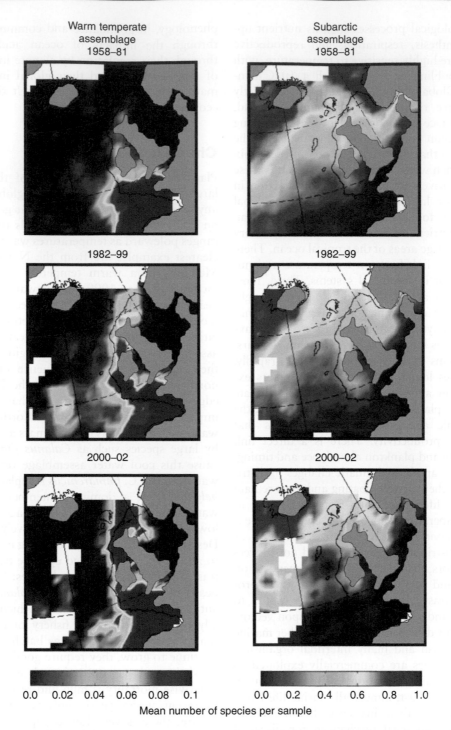

Mean number of species per sample

Figure 1 The northerly shift of the warm temperate assemblage (including *Calanus helgolandicus*) into the North Sea and retraction of the subarctic assemblage (including *Calanus finmarchicus*) to higher latitudes. Reproduced by permission from Gregory Beaugrand.

evidence suggests that phenological changes in plankton are greater than those observed on land. Larvae of benthic echinoderms in the North Sea are now appearing in the plankton 6 weeks earlier than they did 50 years ago, and this is in response to warmer temperatures of less than 1 °C. In echinoderms,

temperature stimulates physiological developments and larval release. Other meroplankton such as larvae of fish, cirrepedes, and decapods have also responded similarly to warming (**Figure 2**).

Timing of peak abundance of plankton can have effects that resonate to higher trophic levels. In the

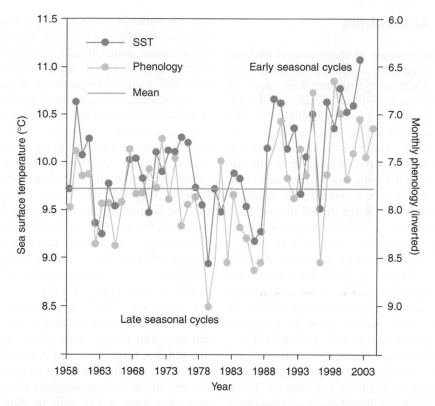

Figure 2 Monthly phenology (timing) of decapod larval abundance and sea surface temperature in the central North Sea from 1958 to 2004. Reproduced from Edwards M, Johns DG, Licandro P, John AWG, and Stevens DP (2006) Ecological status report: Results from the CPR Survey 2004/2005. *SAHFOS Technical Report* 3: 1–8.

North Sea, the timing each year of plankton blooms in summer over the last 50 years has advanced, with phytoplankton appearing 23 days earlier and copepods 10 days earlier. The different magnitude of response between phytoplankton and zooplankton may lead to a mismatch between successive trophic levels and a change in the synchrony of timing between primary and secondary production. In temperate marine systems, efficient transfer of marine primary and secondary production to higher trophic levels, such as those occupied by commercial fish species, is largely dependent on the temporal synchrony between successive trophic production peaks. This type of mismatch, where warming has disturbed the temporal synchrony between herbivores and their plant food, has been noted in other biological systems, most notably between freshwater zooplankton and diatoms, great tits and caterpillar biomass, flycatchers and caterpillar biomass, winter moth and oak bud burst, and the red admiral butterfly and stinging nettle. Such mismatches compromise herbivore survival.

Dramatic ecosystem repercussions of climate-driven changes in phenology are also evident in the subarctic North Pacific Ocean. Here a single copepod species, *Neocalanus plumchrus*, dominates the zooplankton biomass. Its vertical distribution and development are both strongly seasonal and result in an ephemeral (2-month duration) annual peak in upper ocean zooplankton biomass in late spring. The timing of this annual maximum has shifted dramatically over the last 50 years, with peak biomass about 60 days earlier in warm than cold years. The change in timing is a consequence of faster growth and enhanced survivorship of early cohorts in warm years. The timing of the zooplankton biomass peak has dramatic consequences for the growth performance of chicks of the planktivorous seabird, Cassin's auklet. Individuals from the world's largest colony of this species, off British Columbia, prey heavily on *Neocalanus*. During cold years, there is synchrony between food availability and the timing of breeding. During warm years, however, spring is early and the duration of overlap of seabird breeding and *Neocalanus* availability in surface waters is small, causing a mismatch between prey and predator populations. This compromises the reproductive performance of Cassin's auklet in warm years compared to cold years. If Cassin's auklet does not adapt to the changing food conditions, then global warming will place severe strain on its long-term survival.

Changes in Abundance

The most striking example of changes in abundance in response to long-term warming is from foraminifera in the California Current. This plankton group is valuable for long-term climate studies because it is more sensitive to hydrographic conditions than to predation from higher trophic levels. As a result, its temporal dynamics can be relatively easily linked to changes in climate. Foraminifera are also well preserved in sediments, so a consistent time series of observations can be extended back hundreds of years. Records in the California Current show increasing numbers of tropical/subtropical species throughout the twentieth century reflecting a warming trend, which is most dramatic after the 1960s (**Figure 3**). Changes in the foraminifera record echo not only increase in many other tropical and subtropical taxa in the California Current over the last few decades, but also decrease in temperate species of algae, zooplankton, fish, and seabirds.

Changes in abundance through alteration of enrichment patterns in response to enhanced stratification is often more difficult to attribute to climate change than are shifts in distribution or phenology, but may have greater ecosystem consequences. An illustration from the Northeast Atlantic highlights the role that global warming can have on stratification and thus plankton abundances. In this region, phytoplankton become more abundant when cooler regions warm, probably because warmer temperatures boost metabolic rates and enhance stratification in these often windy, cold, and well-mixed regions.

But phytoplankton become less common when already warm regions get even warmer, probably because warm water blocks nutrient-rich deep water from rising to upper layers where phytoplankton live. This regional response of phytoplankton in the North Atlantic is transmitted up the plankton food web. When phytoplankton bloom, both herbivorous and carnivorous zooplankton become more abundant, indicating that the plankton food web is controlled from the 'bottom up' by primary producers, rather than from the 'top down' by predators. This regional response to climate change suggests that the distribution of fish biomass will change in the future, as the amount of plankton in a region is likely to influence its carrying capacity of fish. Climate change will thus have regional impacts on fisheries.

There is some evidence that the frequency of HABs is increasing globally, although the causes are uncertain. The key suspect is eutrophication, particularly elevated concentrations of the nutrients nitrogen and phosphorus, which are of human origin and discharged into our oceans. However, recent evidence from the North Sea over the second half of the twentieth century suggests that global warming may also have a key role to play. Most areas of the North Sea have shown no increase in HABs, except off southern Norway where there have been more blooms. This is primarily a consequence of the enhanced stratification in the area caused by warmer temperatures and lower salinity from meltwater. In the southern North Sea, the abundance of two key HAB species over the last 45 years is positively related to warmer ocean temperatures. This work

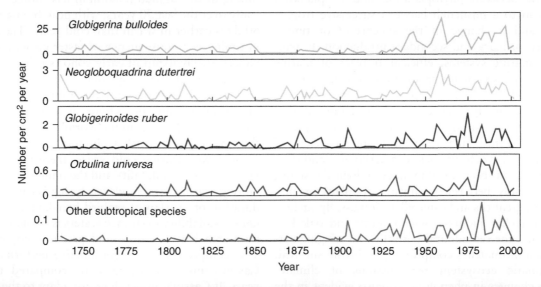

Figure 3 Fluxes of tropical/planktonic foraminifera in Santa Barbara Basin (California Current). Tropical/subtropical foraminifera showing increased abundance in the twentieth century. Reproduced from Field DB, Baumgartner TR, Charles CD, Ferreira-Bartrina V, and Ohman M (2006) Planktonic foraminifera of the California Current reflect 20th century warming. *Science* 311: 63–66.

supports the notion that the warmer temperatures and increased meltwater runoff anticipated under projected climate change scenarios are likely to increase the frequency of HABs.

Although most evidence for changes in abundance in response to climate change are from the Northern Hemisphere because this is where most (plankton) science has concentrated, there is a striking example from waters around Antarctica. Over the last 30 years, there has been a decline in the biomass of krill *Euphausia superba* in the Southern Ocean that is a consequence of warmer sea and air temperatures. In many areas, krill has been replaced by small gelatinous filter-feeding sacs known as salps, which occupy the less-productive, warmer regions of the Southern Ocean. The decline in krill is likely to be a consequence of warmer ocean temperatures impacting sea ice. It is not only that sea ice protects krill from predation, but also the algae living beneath the sea ice and photosynthesizing from the dim light seeping through are a critical food source for krill. As waters have warmed, the extent of winter sea ice and its duration have declined, and this has led to a deterioration in krill density since the 1970s. As krill are major food items for baleen whales, penguins, seabirds, fish, and seals, their declining population may have severe ramifications for the Southern Ocean food web.

Impact of Acidification

A direct consequence of enhanced CO_2 levels in the ocean is a lowering of ocean pH. This is a consequence of elevated dissolved CO_2 in seawater altering the carbonate balance in the ocean, releasing more hydrogen ions into the water and lowering pH. There has been a drop of 0.1 pH units since the Industrial Revolution, representing a 30% increase in hydrogen ions.

Impacts of ocean acidification will be greatest for plankton species with calcified (containing calcium carbonate) shells, plates, or scales. For organisms to build these structures, seawater has to be supersaturated in calcium carbonate. Acidification reduces the carbonate saturation of the seawater, making calcification by organisms more difficult and promoting dissolution of structures already formed.

Calcium carbonate structures are present in a variety of important plankton groups including coccolithophores, mollusks, echinoderms, and some crustaceans. But even among marine organisms with calcium carbonate shells, susceptibility to acidification varies depending on whether the crystalline form of their calcium carbonate is aragonite or calcite. Aragonite is more soluble under acidic conditions than calcite, making it more susceptible to dissolution. As oceans absorb more CO_2, undersaturation of aragonite and calcite in seawater will be initially most acute in the Southern Ocean and then move northward.

Winged snails known as pteropods are probably the plankton group most vulnerable to ocean acidification because of their aragonite shell. In the Southern Ocean and subarctic Pacific Ocean, pteropods are prominent components of the food web, contributing to the diet of carnivorous zooplankton, myctophids, and other fish and baleen whales, besides forming the entire diet of gymnosome mollusks. Pteropods in the Southern Ocean also account for the majority of the annual flux of both carbonate and organic carbon exported to ocean depths. Because these animals are extremely delicate and difficult to keep alive experimentally, precise pH thresholds where deleterious effects commence are not known. However, even experiments over as little as 48 h show shell deterioration in the pteropod *Clio pyrimidata* at CO_2 levels approximating those likely around 2100 under a business-as-usual emissions scenario. If pteropods cannot grow and maintain their protective shell, their populations are likely to decline and their range will contract toward lower-latitude surface waters that remain supersaturated in aragonite, if they can adapt to the warmer temperature of the waters. This would have obvious repercussions throughout the food web of the Southern Ocean.

Other plankton that produce calcite such as foraminifera (protist plankton), mollusks other than pteropods (e.g., squid and mussel larvae), coccolithophores, and some crustaceans are also vulnerable to ocean acidification, but less so than their cousins with aragonite shells. Particularly important are coccolithophorid phytoplankton, which are encased within calcite shells known as liths. Coccolithophores export substantial quantities of carbon to the seafloor when blooms decay. Calcification rates in these organisms diminish as water becomes more acidic (**Figure 4**).

A myriad of other key processes in phytoplankton are also influenced by seawater pH. For example, pH is an important determinant of phytoplankton growth, with some species being catholic in their preferences, whereas growth of other species varies considerably between pH of 7.5 and 8.5. Changes in ocean pH also affect chemical reactions within organisms that underpin their intracellular physiological processes. pH will influence nutrient uptake kinetics of phytoplankton. These effects will have repercussions for phytoplankton community

Emiliania huxleyi   *Gephyrocapsa oceanica*

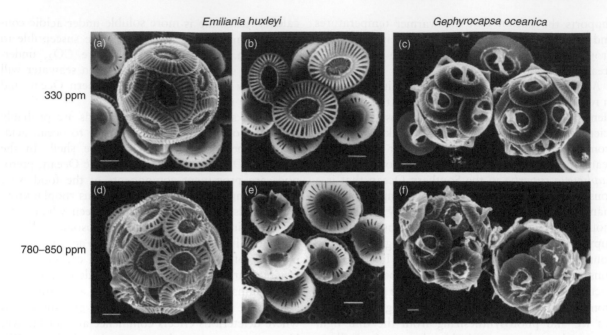

Figure 4 Scanning electron microscopy photographs of the coccolithophores *Emiliania huxleyi* and *Gephyrocapsa oceanica* collected from cultures incubated at CO_2 levels of about 300 and 780–850 ppm. Note the difference in the coccolith structure (including distinct malformations) and in the degree of calcification of cells grown at normal and elevated CO_2 levels. Scale bar = 1 mm. Reprinted by permission from Macmillan Publishers Ltd., *Nature*, Riebesell U, Zondervan I, Rost B, Tortell PD, Zeebe RE, and Morel FMM, Reduced calcification of marine plankton in response to increased atmospheric CO_2, 407: 364–376, Copyright (2000).

composition and productivity, with flow-on effects to higher trophic levels.

Climate Variability

Many impacts of climate change are likely to act through existing modes of variability in the Earth's climate system, including the well-known El Niño/Southern Oscillation (ENSO) and the North Atlantic Oscillation (NAO). Such large synoptic pressure fields alter regional winds, currents, nutrient dynamics, and water temperatures. Relationships between integrative climate indices and plankton composition, abundance, or productivity provide an insight into how climate change may affect ocean biology in the future.

ENSO is the strongest climate signal globally, and has its clearest impact on the biology of the tropical Pacific Ocean. Observations from satellite over the past decade have shown a dramatic global decline in primary productivity. This trend is caused by enhanced stratification in the low-latitude oceans in response to more frequent El Niño events. During an El Niño, upper ocean temperatures warm, thereby enhancing stratification and reducing the availability of nutrients for phytoplankton

growth. Severe El Niño events lead to alarming declines in phytoplankton, fisheries, marine birds and mammals in the tropical Pacific Ocean. Of concern is the potential transition to more frequent El Niño-like conditions predicted by some climate models. In such circumstances, enhanced stratification across vast areas of the tropical ocean may reduce primary productivity, decimating fish, mammal, and bird populations. Although it is unknown whether the recent decline in primary productivity is already a consequence of climate change, the findings and underlying understanding of climate variability are likely to provide a window to the future.

Further north in the Pacific, the Pacific Decadal Oscillation (PDO) has a strong multi-decadal signal, longer than the ENSO period of a few years. When the PDO is negative, upwelling winds strengthen over the California Current, cool ocean conditions prevail in the Northeast Pacific, copepod biomass in the region is high and is dominated by large cool-water species, and fish stocks such as coho salmon are abundant (**Figure 5**). By contrast, when the PDO is positive, upwelling diminishes and warm conditions exist, the copepod biomass declines and is dominated by small less-nutritious species, and the abundance of coho salmon plunges.

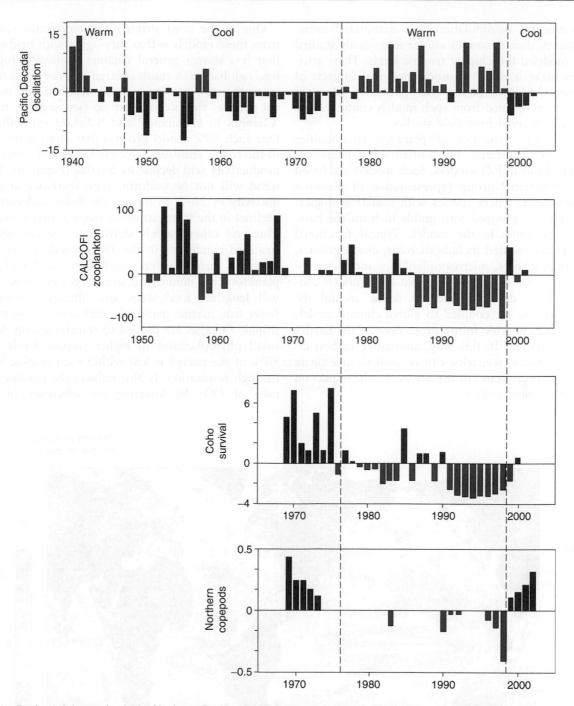

Figure 5 Annual time series in the Northeast Pacific: the PDO index from May to September; anomalies of zooplankton biomass (displacement volumes) from the California Current region (CALCOFI zooplankton); anomalies of coho salmon survival; and biomass anomalies of cold-water copepod species (northern copepods). Positive (negative) PDO index indicates warmer (cooler) than normal temperatures in coastal waters off North America. Reproduced from Peterson WT and Schwing FB (2003) A new climate regime in Northeast Pacific ecosystems. *Geophysical Research Letters* 30(17): 1896 (doi:10.1029/2003GL017528).

These transitions between alternate states have been termed regime shifts. It is possible that if climate change exceeds some critical threshold, some marine systems will switch permanently to a new state that is less favorable than present.

In Hot Water: Consequences for the Future

With plankton having relatively simple behavior, occurring in vast numbers, and amenable to experimental manipulation and automated measurements, their dynamics are far more easily studied and modeled than higher trophic levels. These attributes make it easier to model potential impacts of climate change on plankton communities. Many of our insights gained from such models confirm those already observed from field studies.

The basic dynamics of plankton communities have been captured by nutrient–phytoplankton–zooplankton (NPZ) models. Such models are based on a functional group representation of plankton communities, where species with similar ecological function are grouped into guilds to form the basic biological units in the model. Typical functional groups represented include diatoms, dinoflagellates, coccolithophores, microzooplankton, and mesozooplankton. There are many global NPZ models constructed by different research teams around the world. These are coupled to global climate models (GCMs) to provide future projections of the Earth's climate system. In this way, alternative carbon dioxide emission scenarios can be used to investigate possible future states of the ocean and the impact on plankton communities.

One of the most striking and worrisome results from these models is that they agree with fieldwork that has shown general declines in lower trophic levels globally as a result of large areas of the surface tropical ocean becoming more stratified and nutrient-poor as the oceans heat up (see sections titled 'Changes in abundance' and 'Climate variability'). One such NPZ model projects that under a middle-of-the-road emissions scenario, global primary productivity will decline by 5–10% (**Figure 6**). This trend will not be uniform, with increases in productivity by 20–30% toward the Poles, and marked declines in the warm stratified tropical ocean basins. This and other models show that warmer, more-stratified conditions in the Tropics will reduce nutrients in surface waters and lead to smaller phytoplankton cells dominating over larger diatoms. This will lengthen food webs and ultimately support fewer fish, marine mammals, and seabirds, as more trophic linkages are needed to transfer energy from small phytoplankton to higher trophic levels and 90% of the energy is lost within each trophic level through respiration. It also reduces the oceanic uptake of CO_2 by lowering the efficiency of the

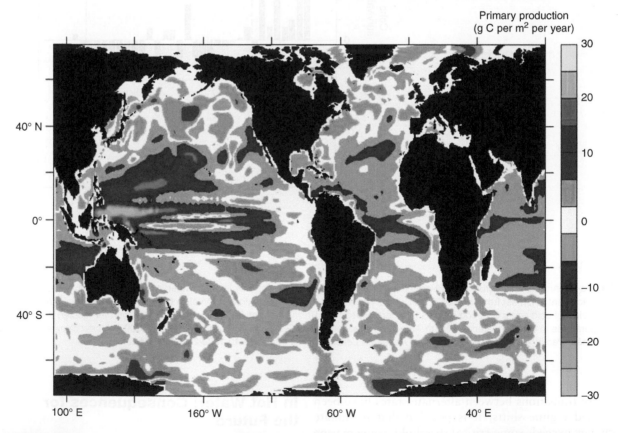

Figure 6 Change in primary productivity of phytoplankton between 2100 and 1990 estimated from an NPZ model. There is a global decline in primary productivity by 5–10%, with an increase at the Poles of 20–30%. Reproduced from Bopp L, Monfray P, Aumont O, *et al.* (2001) Potential impact of climate change on marine export production. *Global Biogeochemical Cycles* 15: 81–99.

biological pump. This could cause a positive feedback between climate change and the ocean carbon cycle: more CO_2 in the atmosphere leads to a warmer and more stratified ocean, which supports less and smaller plankton, and results in less carbon being drawn from surface ocean layers to deep waters. With less carbon removed from the surface ocean, less CO_2 would diffuse into the ocean and more CO_2 would accumulate in the atmosphere.

It is clear that plankton are beacons of climate change, being extremely sensitive barometers of physical conditions. We also know that climate impacts on plankton reverberate throughout marine ecosystems. More than any other group, they also influence the pace and extent of climate change. The impact of climate change on plankton communities will not only determine the future trajectory of marine ecosystems, but the planet.

See also

Marine Plankton Communities. Plankton.

Further Reading

Atkinson A, Siegel V, Pakhomov E, and Rothery P (2004) Long-term decline in krill stock and increase in salps within the Southern Ocean. *Nature* 432: 100–103.

Beaugrand G, Reid PC, Ibanez F, Lindley JA, and Edwards M (2002) Reorganisation of North Atlantic marine copepod biodiversity and climate. *Science* 296: 1692–1694.

Behrenfield MJ, O'Malley RT, Siegel DA, *et al.* (2006) Climate-driven trends in contemporary ocean productivity. *Nature* 444: 752–755.

Bertram DF, Mackas DL, and McKinnell SM (2001) The seasonal cycle revisited: Interannual variation and ecosystem consequences. *Progress in Oceanography* 49: 283–307.

Bopp L, Aumont O, Cadule P, Alvain S, and Gehlen M (2005) Response of diatoms distribution to global warming and potential implications: A global model study. *Geophysical Research Letters* 32: L19606 (doi:10.1029/2005GL023653).

Bopp L, Monfray P, Aumont O, *et al.* (2001) Potential impact of climate change on marine export production. *Global Biogeochemical Cycles* 15: 81–99.

Edwards M, Johns DG, Licandro P, John AWG, and Stevens DP (2006) Ecological status report: Results from the CPR Survey 2004/2005. *SAHFOS Technical Report* 3: 1–8.

Edwards M and Richardson AJ (2004) The impact of climate change on the phenology of the plankton community and trophic mismatch. *Nature* 430: 881–884.

Field DB, Baumgartner TR, Charles CD, Ferreira-Bartrina V, and Ohman M (2006) Planktonic forminifera of the California Current reflect 20th century warming. *Science* 311: 63–66.

Hays GC, Richardson AJ, and Robinson C (2005) Climate change and plankton. *Trends in Ecology and Evolution* 20: 337–344.

Peterson WT and Schwing FB (2003) A new climate regime in Northeast Pacific ecosystems. *Geophysical Research Letters* 30(17): 1896 (doi:10.1029/2003GL017528).

Raven J, Caldeira K, Elderfield H, *et al.* (2005) *Royal Society Special Report: Ocean Acidification Due to Increasing Atmospheric Carbon Dioxide*. London: The Royal Society.

Richardson AJ (2008) In hot water: Zooplankton and Climate change. *ICES Journal of Marine Science* 65: 279–295.

Richardson AJ and Schoeman DS (2004) Climate impact on plankton ecosystems in the Northeast Atlantic. *Science* 305: 1609–1612.

Riebesell U, Zondervan I, Rost B, Tortell PD, Zeebe RE, and Morel FMM (2000) Reduced calcification of marine plankton in response to increased atmospheric CO_2. *Nature* 407: 364–367.

POPULATION DYNAMICS MODELS

Francois Carlotti, C.N.R.S./Université Bordeaux 1, Arachon, France

Introduction

The general purpose of population models of plankton species is to describe and eventually to predict the changes in abundance, distribution, and production of targeted populations under forcing of the abiotic environment, food conditions, and predation. Computer-based approaches in plankton ecology were introduced during the 1970s with the application of population models to investigate large-scale population phenomena by the use of mathematical models.

Today, virtually every major scientific research project of population ecology has a modeling component. Population models are built for three main objectives: (1) to estimate the survival of individuals and the persistence of populations in their physical and biological environments, and to look at the factors and processes that regulate their variability; (2) to estimate the flow of energy and matter through a given population; and (3) to study different aspects of behavioral ecology. The study of internal properties of a population, like the various effects of individual variability, and the study of interactions between populations and successions of population are also topics related to population models. The field of biological modeling has diversified and, at present, complex mathematical approaches such as neural networks, genetic algorithms, and dynamical optimization are coming into use, along with the application of supercomputers. However, the use of models in marine research should always be accompanied by extensive field data and laboratory experiments, for initialization, verification or falsification, or continuous updating.

Approach for Modelling Plankton Populations

Population Structure and Units

A population is defined as a group of living organisms all of one species restricted to a given area and with limited exchanges of individuals from other populations. The first step in building a population model is to identify state variables (components of the population) and to describe the interactions between these state variables and external variables of the system and among the components themselves. The components of a population can be (1) the entire population (one component); (2) groups of individuals identified by a certain states: developmental stages, weight or size classes, age classes (fixed numbers of components); or (3) all individuals (varying numbers of components).

The usual unit in population dynamics models is the number of individuals per volume of water, but the population biomass can also be used (in g biomass or carbon (C) or nitrogen (N)). When all individuals or groups of individuals are represented, the individual weight can be considered as a state variable. The forcing variables influencing the population dynamics are biological factors, mainly nutrients and predators, and physical factors, mainly temperature, light, advection, and diffusion.

Individual and Demographic Processes

Population models usually work with four major processes: individual growth, development, reproduction, and mortality. Growth is computed by the rate of individual weight change. Development is represented by the change of states (phases in phytoplankton and protozoan cell division, developmental stages in zooplankton) through which each individual progresses to reach maturity. Reproduction is represented by the production of new individuals. Mortality induces loss of individuals, and can be divided in two components: natural physiological mortality (due, for instance, to starvation) and mortality due to predation. Combination of the four processes permits one to stimulate (1) increase in terms of number of individuals in the population, (2) the body growth of these individuals, and (3) by combination of the two previously simulated values, the increase in total biomass of the population (which is usually termed 'population growth').

Plankton Characteristics

Essential information to be built into models of plankton organisms are (1) the individual life duration (a few hours for bacteria; one to a few days for phytoplankton and unicellular animals; several weeks to years for zooplankton and ichthyoplankton organisms); (2) the range of change in size or weight between the beginning and the end of a life cycle; and (3) the number of individuals produced by a mother individual (from two individuals up to thousands of individuals). When developmental stages in the life cycle are identified, the stage durations are needed.

The observation time step has to be defined to adequately follow the timescale of the chosen variables, and thus should be smaller than the duration of the shortest phases.

Plankton Population Models

The most modeled component in marine planktonic ecosystems is phytoplankton production. Most of the phytoplankton models simulate the growth of phytoplankton as a whole, using only the process of photosynthesis. Few models deal with phytoplankton population growth dynamics at the species level. Existing models of other unicellular plankton organisms (bacterioplankton, species of microzooplankton) usually treat them as a single unit, except for a few models simulating phytoplankton and microbial cell cycles. In contrast, mesozooplanktonic organisms, including the planktonic larval stages of benthic species (meroplankton), and fish that have complex life cycles are extensively modeled at the population level.

Dynamics of Single Species

Population Models Described by the Total Density

When a population is observed at timescales much larger than the individual life span, and on a large number of generations, models with one variable (the total number of individuals or the total biomass in that population) are the simplest. These models postulate that the rate of change of the population number, N, is proportional to N (eqn [1], where r is the difference between birth and death rates).

$$\frac{dN}{dt} = rN \qquad [1]$$

The logistic equation ([2]) represents limitation due to the resources or space (see **Figure 1**).

$$\frac{dN}{dt} = rN\left(1 - \frac{N}{K}\right) \qquad [2]$$

where K is the carrying capacity.

Population growth of bacteria, phytoplankton, or microzooplankton can be simulated adequately by the logistic equation. With addition of a time delay term into the logistic equation, oscillations of the population can be represented.

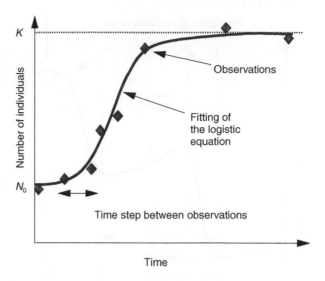

Figure 1 The growth of a plankton population with density regulation and its fitting by the logistic equation.

Population Models of Organisms with Description of the Life Cycle

If observations of a population are made with a time step shorter than the life cycle duration (**Figure 2**), the population development pattern is a succession of periods with decreasing abundance of individuals due to mortality, and increasing abundance due to recruitment of new individuals in periods of reproduction (cell division or egg production). Recruitment is defined as the input flux of individuals in a given state (stage, size class, etc.).

To represent such patterns, it is necessary to identify different phases in the life cycle, based on age, size, developmental stages, and so on. Two types of models can be developed:

- Structured population models, which consider the flux of individuals through different classes
- Individual-based models, which simulate birth, growth and development through stages and death of each individual

The major distinction between physiologically structured population models and individual-based models in a stricter sense is that individual-based models track the fate of all individuals separately over time, while physiologically structured population models follow the density of individuals of a specific type (age or size classes, stages). These models are particularly used for representing the complex life cycles of zooplankton and ichthyoplankton, but have also been useful for studying the population growth of bacteria, phytoplankton, and microzooplankton, particularly division synchrony in controlled conditions (chemostat).

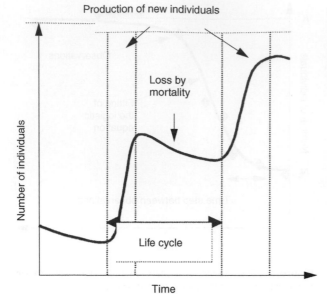

Production of new individuals

Loss by mortality

Number of individuals

Life cycle

Time

Figure 2 Total population abundance in controlled conditions. The population is initiated with newborn individuals, and decreases, first owing to mortality, until the maturation period, when a new generation is produced. At the end of the recruitment period of new individuals of second generation, the total abundance decreases, again owing to mortality. Owing to individual variability in development, loss of synchronism in population induces a broader recruitment period for the third generation. In nonlimiting food conditions, recruitment of new individuals is continued after few generations. The three generations correspond to the exponential phase presented in **Figure 1** (i.e., without density regulation).

Structured Population Models

A population can be structured with respect to age (age-structured population models), stage (stage-structured population models), or size or weight (size or weight-structured population models). Two types of equations systems are usually used: matrix models, which are discrete-time difference equation models, and continuous-time structured population models.

Matrix models constitute a class of population models that incorporate some degree of individual variability. Matrix models are powerful tools for analyzing, for example, the impact of life history characteristics on population dynamics, the influence of current population state on its growth potential, and the sensitivity of the population dynamics to quantitative changes in vital rates. Matrix models are convenient for cases where there are discrete pulses of reproduction, but not for populations with continuous reproduction. They are not suitable for studying the dynamics of populations that live in fluctuating environments.

The Leslie matrix is the simplest type of age-structured dynamic considering discrete classes.

Suppose there are m age classes numbered 1, 2, ..., m, each covering an interval τ. If $N_{j,t}$ denotes the number of individuals in age class j at time t and G_j denotes the fraction of the population in this age class that survive to enter age class $j + 1$, then eqn [3] applies.

$$N_{j+1,\,t+1} = G_j N_{j,t} \qquad [3]$$

Individuals of the first age class are produced by mature individuals from older age classes and eqn [4] applies, where F_j is the number of age class 1 individuals produced per age class i individual during the time step τ.

$$N_{1,t+1} = \sum_{j=1}^{m} F_j N_{j,t} \qquad [4]$$

The system of eqns [3] and [4] can be written in matrix form (eqn [5]).

$$\begin{bmatrix} N_1 \\ N_2 \\ N_3 \\ \vdots \\ N_m \end{bmatrix}(t+1) = \begin{bmatrix} 0 & F_2 & F_3 & \cdots & F_m \\ G_1 & 0 & 0 & \cdots & 0 \\ 0 & G_2 & 0 & \cdots & 0 \\ \vdots & \ddots & \ddots & \cdots & \vdots \\ 0 & 0 & & G_{m-1} & 0 \end{bmatrix} \begin{bmatrix} N_1 \\ N_2 \\ N_3 \\ \vdots \\ N_m \end{bmatrix}(t)$$

$$[5]$$

The Leslie matrix can easily be modified to deal with size classes, weight classes, and developmental stages as the key individual characteristics of the population. Organisms grow through a given stage or size/weight class for a given duration.

There are several variations of matrix models, differing mainly in the expression of vital rates, which can vary with time depending on external factors (e.g., temperature, food concentration, competitors, predators) or internal (e.g., density-dependent) factors.

The earlier type of continuous-time structured model is usually referred to as the McKendrick–von Foerster equation, and uses the age distribution on a continuous-time basis in partial differential equations. This type of model has been developed to the extent that it can be used to describe population dynamics in fluctuating environments. In addition, it also applies to situations in which more than one physiological trait of the individuals (e.g., age, size, weight, and energy reserves) have strong influences on individual reproduction and mortality. The movement of individuals through the different structural classes is followed over time. Age and weight are continuous variables, whereas stage is a discrete variable.

The general equation is eqn [6], where n is abundance of individuals of age a and mass m at time t.

$$\frac{\partial n(a,w,t)}{\partial t} + \frac{\partial n(a,w,t)}{\partial a} + \frac{\partial g(a,w,t)n(a,w,t)}{\partial w} = -\mu(a,w,t)n(a,w,t) \quad [6]$$

where $\mu(a,w,t)$ is the death rate of the population of age a, weight w at time t.

The von Foerster equation describes population processes in terms of continuous age and time (age-structured models) according to eqn [7].

$$\frac{\partial n(a,t)}{\partial t} + \frac{\partial n(a,t)}{\partial a} = -\mu(a,t)n(a,t) \quad [7]$$

The equation has both an initial age structure φ at $t = 0$ (eqn [8]) and a boundary condition of egg production at $a = 0$ (eqn [9]).

$$n(a,0) = \varphi_0(a) \quad [8]$$

$$n(0,t) = \int_0^\infty F(a,S_R)n(a,t)\,\mathrm{d}a \quad [9]$$

F is a fecundity function that depends on age (a) and the sex ratio of the population S_R. These kinds of equations are mathematically and computationally difficult to analyze, especially if the environment is not constant.

The same type of equation as [7] can be used where the age is replaced by the weight (weight-structured models (eqn [10]).

$$\frac{\partial n(w,t)}{\partial t} + \frac{\partial g(w,T,P)n(w,t)}{\partial w} = -\mu(w,t)n(w,t) \quad [10]$$

The weight of the individual w and the growth g are influenced by the temperature T, the food P, and by the weight itself through allometric metabolic relationships.

The equation has both an initial age structure φ at $t = 0$ (eqn [11]) and a boundary condition of egg production at $w = w_0$ (eqn [12]).

$$n(w,0) = \varphi_0(w) \quad [11]$$

$$N(0,t) = \int_0^\infty F(w,S_R)n(w,t)\,\mathrm{d}w \quad [12]$$

F is the fecundity function, which depends on weight (w) and the sex ratio of the population S_R.

The numerical realization of this equation requires a representation of the continuous distribution $n(w,t)$ by a set of discrete values $n_i(t)$ that are spaced along the weight axis at intervals $\Delta w_i = w_{i+1} - w_i$.

Using upwind difference discretization to solve the equations, and recasting the representation in terms of the number of individuals in the ith weight class, $N_i(t) \approx n_i(t)\,\Delta w_i$, the dynamic equation becomes [13], where $\mu_i(t)$ replaces $\mu(w_i,t)$

$$\frac{\mathrm{d}N_i}{\mathrm{d}t} = \left[\frac{g_{i-1}}{\Delta w_{i-1}}\right]N_{i-1} - \left[\frac{g_i}{\Delta w_i}\right]N_{i-\mu_i N_i} \quad [13]$$

This describes the dynamics of all weight classes except the first $(i = 2)$ and last $(i = Q)$. If $R(t)$ represents the total rate of recruitment of newborns to the population, and all newborns are recruited with the same weight w_1, then the dynamic of the weight class covering the range Δw_1 is described by eqn [14].

$$\frac{\mathrm{d}N_1}{\mathrm{d}t} = R - \left[\frac{g_1}{\Delta w_1}\right]N_1 - \mu_1 N_1 \quad [14]$$

If we assume that individuals in only the Qth weight class are adult, and that adult individuals expend all assimilated energy on reproduction rather than growth, the population dynamics of the adult population is given by eqn [15].

$$\frac{\mathrm{d}N_Q}{\mathrm{d}t} = \left[\frac{g_{Q-1}}{\Delta w_{Q-1}}\right]N_{Q-1} - \mu_Q N_Q \quad [15]$$

The rate of recruitment of newborns to the population is given by [16] where $\beta(t)$ represents the per capita fecundity of an average adult at time t.

$$R(t) = \beta(t)N_Q(t) \quad [16]$$

The weight intervals Δw_i increase with class number i as an allometric function. The growth rate $g(w,t)$ can be calculated by a physiological model.

Stage-structured Population Models

Plankton populations often have continuous recruitment and are followed in the field by observing stage abundances over time. A large number of zooplankton population models deal with population structures in term of developmental stage, using ordinary differential equations (ODEs).

A single ODE can be used to model each development stage or group of stages: for instance, a copepod population can be subdivided into four groups: eggs, nauplii, copepodites, and adults. The equation system is eqn [17]–[20], where R is recruitment, α is the transfer rate to next stage, and μ is the mortality rate.

$$\text{Eggs} \quad \frac{\mathrm{d}N_1}{\mathrm{d}t} = R - \alpha_1 N_1 - \mu_1 N_1 \quad [17]$$

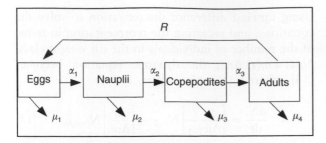

Figure 3 Schematic representation of the population dynamics mathematically represented by eqn [17]–[20]. α_i = transfer rate of stage i to stage $i + 1$; μ_i = mortality rate in stage i; R = recruitment: number of eggs produced by females per day.

$$\text{Naupli} \quad \frac{dN_2}{dt} = \alpha_1 N_1 - \alpha_2 N_2 - \mu_2 N_2 \quad [18]$$

$$\text{Copepodids} \quad \frac{dN_3}{dt} = \alpha_2 N_2 - \alpha_3 N_3 - \mu_3 N_3 \quad [19]$$

$$\text{adults} \quad \frac{dN_4}{dt} = \alpha_3 N_3 - \mu_4 N_4 \quad [20]$$

The system of ODEs is solved by Euler or Runge–Kutta numerical integration methods, usually with a short time step (approximately 1 hour).

In the model presented in **Figure 3**, the transfer rate of animals from stage to stage and the mortality at each stage are expressed as simple linear functions, which induce a rapidly stable stage distribution. To represent the delay of growth within a stage, more refined models consider age-classes within each stage, or systems of delay differential equations. They have a high degree of similarity with observed cohort development in mesocosms or closed areas (**Figure 4**).

Individual-based Models of a Population

Individual-based models (IBMs) describe population dynamics by simulating the birth, development, and eventual death of a large number of individuals in the population. IBMs have been developed for phytoplankton, zooplankton, meroplanktonic larvae, and early life history of fish populations. Object-oriented programming (OOP) and cellular automata techniques have been applied to IBMs.

As powerful computers become more accessible, numerous IBMs of plankton populations have been developed, mainly to couple them with 1D-mixed layer models (phyto- and zooplankton) and circulation models (zooplankton).

IBMs treat populations as collections of individuals, with explicit rules governing individual biology and interactions with the environment. Each

biological component can change as a function of the others. Each individual is represented by a set of variables that store its i-state (age, size, weight, nutrient or reserve pool, etc.). These variables may be grouped together in some data structure that represents a single individual, or they may be collected into arrays (an array of all the ages of the individuals, an array of all the sizes of the individuals, etc.), in which case an individual is an index number in the set of arrays. The i-state of an individual changes as a function of the current i-state, the interactions with other individuals, and the state of the local environment. The local environment can include prey and predator organisms that do not warrant explicit representation as individuals in the model. Population-level phenomena (e.g., temporal or spatial dynamics) or vital rates can then be inferred directly from the contributions of individuals in the ensemble.

The model starts with an initial population and the basic environment, then monitors the changes of each individual. At any time t, the i-state of individual j changes as eqn [21].

$$X_{i,j}(t) = X_{i,j}(t - dt) + f(X_{1,j}(t - dt), \ldots, X_{i,j}(t - dt)) \quad [21]$$

$X_{i,j}(t)$ is the value of the i-state of individual j, and f is the process modifying $X_{i,j}$, as a function of the values of different i-states of the organism and external parameters such as the temperature. When the fate of all individuals during the time-step dt has been calculated, the changes to the environment under the effects of individuals can be updated. Any stochastic process can be added to eqn [21].

This type of model can add a lot of detail in the representation of physiological functions. Individual growth can be calculated as assimilation less metabolic loss, and the interindividual variation in physiology can be represented by adding stochastic processes or parameters describing the characteristics of each individual (growth and development parameters, mortality coefficient, and parameters connected with reproduction). The end results are unique life histories, which when considered as a whole give rise to growth/size distributions that provide a measure of the state of the population.

Calibration of Parameters

Parametrization of a model can range from very simplistic to extremely complex depending upon the amount of information known about the population under consideration. Bioenergetic processes (ingestion, egestion, excretion, respiration, and egg

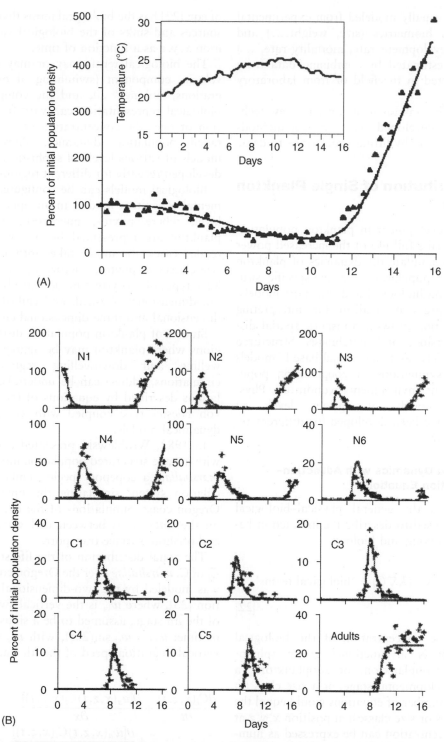

Figure 4 Simulation of the cohort development of the copepod *Euterpina acutifrons* in mesocosms with a structured stage and age-within-stage model. (A) Total population during development, with variable temperature and constant food supply (points = experimental data; line = simulation). The initial density decreases owing to mortality and then increases to newborn individuals (as the first part in **Figure 2**). (B) Naupliar stages (N1 to N6), copepodite stages (C1 to C5), and adults during development (points = experimental data; line = simulation). The simulation start with similar N1 of same age. Reproduced from Carlotti F and Sciandra A, 1989. Population dynamics model of *Euterpina acutifrons* (Copepoda: Harpacticoida) coupling individual growth and larval development. *Mar. Ecol. Prog. Ser.*, 56, 3, 225–242.

production) are usually modeled from experimental results, whereas biometrics (size, weight,...) and demographic (development rate, mortality rate, ...) parameters are estimated by combining data from life tables collected in the field or from laboratory studies.

To solve for the unknown parameters, new techniques have been developed such as inverse methods and data assimilation by fitting simulations to data.

Spatial Distribution of Single Plankton Populations

An important development in plankton population modeling is to make full use of the increased power of computers to simulate the dynamics of plankton (communities or populations) in site-specific situations by coupling biological and transport models, giving high degrees of realism for interpreting plankton population growth, transport, spatial distribution, dispersion, and patchiness. Structured population models and individual-based models allow detailed simulations of zooplankton populations in different environmental conditions. Physical–biological models of various levels of sophistication have been developed for different regions of the ocean.

Spatial Plankton Dynamics with Advection–Diffusion–Reaction Equations

Equation [22] is the general physical–biological model equation used to describe the interaction between physical mixing and biology.

$$\frac{\partial C}{\partial t} + \nabla \cdot (v_a C) - \nabla \cdot (K \nabla C) = \text{'biological teams'}$$

[22]

$C(x, y, z, t)$ is the concentration of the biological variable, which is a functional group (phytoplankton, microzooplankton, or zooplankton), a species or a developmental stage, or a size class (in which case the number of equations would equal the number of stages or size classes) at position x, y, z at time t. The concentration can be expressed as numbers of organisms or biomass of organisms per unit volume. v_a (u_a, v_a, w_a) represents the advective fluid velocities in x, y, z directions. K_x, K_y, K_z are diffusivities in x, y, z directions. $\Delta = (\partial/\partial x, \partial/\partial y, \partial/\partial z)$ is the Laplacian operator.

On the left-hand side of eqn [22], the first term is the local change of C, the second term is advection caused by water currents, and the third term is the diffusion or redistribution term. The right-hand side of eqn [22] has the biological terms that represent the sources and sinks of the biological variable at position x, y, z as a function of time.

The biological terms may or may not include a velocity component (swimming of organisms, migrations, sinking, ...), and the complexity of the biological representation can vary from the dispersion of one (the concentration of a cohort) to detailed population dynamics. Physical–biological models of various levels of sophistication have been developed recently for different regions of the ocean.

Biological models can be configured as compartmental ecosystem models in an upper-ocean mixed layer, where phyto-, microzoo-, and mesozooplankton are represented by one variable. In extended cases, the model takes into account several size classes of phyto-, microzoo-, and zooplankton. Such types of ecosystem model have been coupled to one-dimensional physical, and embedded into two-dimensional and three-dimensional circulation.

Studies of plankton population distribution in regions where plankton may be aggregated (e.g., upwelling and downwelling regions, Langmuir circulations, eddies) can be undertaken with populations described by equations of the McKendrick–von Foester type coupled with 2D or 3D hydrodynamical models.

In 1982, Wroblewski presented a clear example with a stage-structured population model of *Calanus marshallae*, a copepod species, embedded in a circulation system simulating the upwelling off the Oregon coast. Simulations of the dynamics focused on the interaction between diel vertical migration and offshore surface transport.

The zonal distribution of the life stage categories C_i of *C. marshallae* over the Oregon continental shelf was modeled by the two-dimensional (x, z, t) equation [23], where w_{bi} is the vertical swimming speed of the ith stage, assumed to be a sinusoidal function of time: $w_{bi} = w_{si} \sin(2\pi t)$, with w_{si} the maximum vertical migration speed of the ith stage.

$$\frac{\partial C_i(x, z, t)}{\partial t} + \frac{\partial [u_a(x, z, t) C_i(x, z, t)]}{\partial x}$$
$$+ \frac{\partial [w_a(x, z, t) C_i(x, z, t)]}{\partial z}$$
$$- \frac{\partial}{\partial x}\left[K(x, t)\frac{\partial C_i(x, z, t)}{\partial x}\right] - \frac{\partial}{\partial z}\left[K(z, t)\frac{\partial C_i(x, z, t)}{\partial z}\right]$$
$$= \text{population dynamics} + \frac{\partial [w_{bi}(x, z, t) C_i(x, z, t)]}{\partial z}$$

[23]

The population dynamics model was presented in eqns [17]–[20].

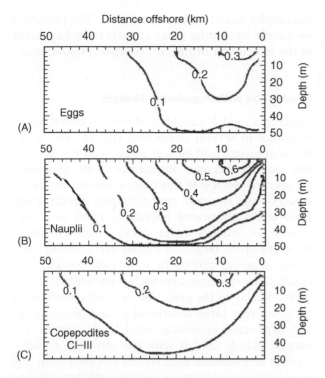

Figure 5 Model of *Calanus marshallae* in the Oregon upwelling zone. The figure shows the simulated zonal distribution of (A) eggs, (B) nauplii, and (C) early copepodites at noon on 15 August. Concentrations of each stage are expressed as a fraction of the total population (all stages m^{-3}). (Reproduced with permission from Wroblewski JS, 1982. Interaction of currents and vertical migration in maintaining *Calanus marshallae* in the Oregon upwelling zone—a simulation. *Deep Sea Research* 29: 665–686.

The upwelling zone extended 50 km from the coast down to a depth of 50 m, and was divided into a grid with spacing 2.5 m in depth and 1 km in the horizontal. The author used a finite-difference scheme with a time step of 1 h, which fell within the bounds for computational stability (**Figure 5**).

Coupling IBMs and Spatially Explicit Models

Individual-based models are more and more frequently used to assess the influence of space on the population dynamics, namely on the time course of population abundances and the pattern formation of populations in their habitats. This approach uses simulated currents from sophisticated 3D hydrodynamic models driving Lagrangian models of particle trajectories to examine dispersion processes.

The approach is relatively straightforward and is a first step in formulating spatially explicit individual based models. Given a 'properly resolved' flow field, particle (larval fish/zooplankton/meroplanktonic larvae) trajectories are computed (generally with standard Runge–Kutta integration methods of the

velocity field). Specifically, hydrodynamic models provide the velocity vector $v = (u, v, w)$ as a function of location $x = (x, y, z)$ and time t, and the particle trajectories are obtained from the integration of eqn [24].

$$\frac{dx}{dt} = v(x, y, z, t) \qquad [24]$$

The simplest model of dispersion is a random walk model in which individuals move along a line from the same starting position. These trajectories could be modified by turbulent dispersion as described in the section below. Once the larval/particle position is known, additional local physical variables can be estimated along the particle's path: e.g., temperature, turbulence, light, etc., and input to the IBM. The physical quantities are then included in biological (physiological or behavioural) formulations of IBMs (see below).

Simulations considering trajectories of plankton as passive particles are a necessary step before considering any active swimming capability of planktonic organisms. They show the importance of physical features in the aggregation or dispersion of the particles.

Plankton transport models that include biological components typically use a prescribed vertical migration strategy for all or part of an animal's life history or a vertical motion (sinking or swimming) that is determined by animal's development and growth. The simulated plankton distributions from these models tend to compare better with observed distributions than do models that use passive particles. Sensitivity studies show that behavior is an important factor in determining larval transport and/or retention.

The coupling of IBMs of zooplankton and fish populations and 3D circulation models is a recent field of study, even for fish models. Generally, models that describe the spatial heterogeneity of the habitat have been designed to answer questions about the spatial and temporal distribution of a population rather than questions about the numbers and characteristics of surviving individuals. They allow us to explore the potential effects of habitat alteration on these populations. Using this approach, biological mechanisms that are strongly dependent on habitat and that are not fully understood could be studied by examining different scenarios.

Modeling Behavioral Mechanisms, Aggregation and Schooling, and Patches

Different types of models have been built, some of them focusing on the structure and shape of aggregations depending on internal and external physical forces, others dealing with the benefits for individuals of living in groups with regard to feeding (foraging models) and to predation. The Lagrangian approach can take into account the behavior of individual organisms and the effects of the physical environment upon them. Although Eulerian approaches are mathematically tractable, the methods do not explicitly address the density dependence of aggregating individual behavior within a patch.

Dynamic optimization allows descriptions of the internal state of individuals, which may lead to both variable and fluctuating motivations among individuals over short time periods.

Interactions between Populations

Models with Plankton Populations in Interaction

Simple models of two species interactions take the form of eqns [25] and [26].

$$\frac{dN_1}{dt} = r_1 N_1 - k_1 N_1 N_2 \qquad [25]$$

$$\frac{dN_2}{dt} = r_2 N_2 - k_2 N_1 N_2 \qquad [26]$$

These population models represent some special experimental situations or typical field situations. Interactions between two species have been rarely treated by population models with description of the life cycle, although structured population models as well as IBM models can represent interactions between species such as predation, parasitism, or even cannibalism.

As an example, Gaedke and Ebenhöh presented in 1991 a study on the interaction between two estuarine species of copepods, *Acartia tonsa* and *Eurytemora affinis*. They first used a simple model based on eqns [25] and [26] including (a) predation (including self-predation of immature stages) by *Acartia* on the two, (b) a term of biomass gain of *Acartia* by this predation, and (c) a density-dependent loss term caused by predation by invertebrates or by starvation of the two species. This simple model did not result in stable coexistence between the two species with a reasonable parameter range under steady-state conditions.

These authors then used two-stage-structured population models with stage-specific interactions (with similar equations to [17]–[20]) allowing the predation of large individuals of *A. tonsa* (copepodites 4 to adults) on nauplii of both species to be represented. The results of this detailed numerical model were compared with results obtained using the simpler model with two variables. The predation on nauplii by *Acartia tonsa* appears to be key factor in the interaction of the two copepod populations.

Food Webs with Population Models

Structured models should be chosen to stimulate the dynamics of several interacting species. The stage-based approach will be acceptable with few species, but quickly become intractable with increasing numbers of species. In this case, a community model based on size structure and using prey–predator size ratio is the alternative approach. There is a continuum of models from detailed size spectrum structure up to large size classes representing functional (trophic) groups in food web models. The detailed size spectrum approach is particularly useful when simulating the predation of a fish cohort on its prey, whereas large functional groups are required for large-scale ecosystem models. Numerous examples include models with size structure of herbivorous zooplankton populations and their prey, and their interactions, in a nutrient–phytoplankton–herbivore–carnivore dynamics model. Size-based plankton model with large entities consider the size range 0.2–2000 μm, picophytoplankton, bacterioplankton, nanophytoplankton, heterotrophic flagellates, phytoplankton, microzooplankton, and mesozooplankton.

See also

Carbon Cycle. Marine Mesocosms. Nitrogen Cycle. Phosphorus Cycle. Plankton. Small-Scale Patchiness, Models of.

Further Reading

Carlotti F, Giske J, and Werner F (2000) Modelling zooplankton dynamics. In: Harris RP, Wiebe P, Lenz J, Skjoldal HR, and Huntley M (eds.) *Zooplankton Methodology Manual*, pp. 571–667. New York: Academic Press.

Caswell H (1989) *Matrix Population Models: Construction, Analysis and Interpretation*. Sunderland, MA: Sinauer Associates.

Coombs S, Harris R, Perry I and Alheit J (1998) *GLOBEC* special issue, vol. 7 (3/4).

DeAngelis DL and Gross LJ (1992) *Individual-based Models and Approaches in Ecology: Populations, Communities and Ecosystems*. New York: Chapman and Hall.

Hofmann EE and Lascara C (1998) Overview of interdisciplinary modeling for marine ecosystems.

In: Brink KH and Robinson AR (eds.) *The Sea*, vol. 10, Ch. 19, pp. 507–540. New York: Wiley.

Levin SA, Powell TM, and Steele JH (1993) *Patch Dynamics*. Berlin: Springer-Verlag. Lecture Notes in Biomathematics 96..

Mangel M and Clark CW (1988) *Dynamic Modeling in Behavioral Ecology*. Princeton, NJ: Princeton University Press.

Nisbet RM and Gurney WSC (1982) *Modelling Fluctuating Populations*. Chichester: Wiley.

Renshaw E (1991) *Modelling Biological Populations in Space and Time*. Cambridge: Cambridge University Press. Cambridge Studies in Mathematical Biology.

Tuljapurkar S and Caswell H (1997) *Structured-population Models in Marine, Terrestrial, and Freshwater Systems*. New York: Chapman and Hall. Population and Community Biology Series 18.

Wood SN and Nisbet RM (1991) *Estimation of Mortality Rates in Stage-structured Populations*. Berlin: Springer-Verlag. Lecture Notes in Biomathematics 90.

Wroblewski JS (1982) Interaction of currents and vertical migration in maintaining *Calanus marshallae* in the Oregon upwelling zone – a simulation. *Deep Sea Research* 29: 665–686.

SMALL-SCALE PATCHINESS, MODELS OF

D. J. McGillicuddy Jr., Woods Hole Oceanographic
Institution, Woods Hole, MA, USA

Introduction

Patchiness is perhaps the most salient characteristic
of plankton populations in the ocean. The scale of
this heterogeneity spans many orders of magnitude in
its spatial extent, ranging from planetary down to
microscale (**Figure 1**). It has been argued that
patchiness plays a fundamental role in the functioning of marine ecosystems, insofar as the mean
conditions may not reflect the environment to which
organisms are adapted. For example, the fact that
some abundant predators cannot thrive on the mean
concentration of their prey in the ocean implies that
they are somehow capable of exploiting small-scale
patches of prey whose concentrations are much larger
than the mean. Understanding the nature of this
patchiness is thus one of the major challenges of
oceanographic ecology.

The patchiness problem is fundamentally one of
physical–biological–chemical interactions. This inter-
connection arises from three basic sources: (1) ocean
currents continually redistribute dissolved and suspended constituents by advection; (2) space–time
fluctuations in the flows themselves impact biological
and chemical processes; and (3) organisms are capable
of directed motion through the water. This tripartite
linkage poses a difficult challenge to understanding
oceanic ecosystems: differentiation between the three
sources of variability requires accurate assessment of
property distributions in space and time, in addition to
detailed knowledge of organismal repertoires and the
processes by which ambient conditions control the
rates of biological and chemical reactions.

Various methods of observing the ocean tend to lie
parallel to the axes of the space/time domain in which
these physical–biological–chemical interactions take
place (**Figure 2**). Given that a purely observational
approach to the patchiness problem is not tractable
with finite resources, the coupling of models with
observations offers an alternative which provides a
context for synthesis of sparse data with articulations
of fundamental principles assumed to govern func-
tionality of the system. In a sense, models can be used
to fill the gaps in the space/time domain shown in
Figure 2, yielding a framework for exploring the

controls on spatially and temporally intermittent
processes.

The following discussion highlights only a few of
the multitude of models which have yielded insight
into the dynamics of plankton patchiness. Examples
have been chosen to provide a sampling of scales
which can be referred to as 'small' – that is, smaller
than the planetary scale shown in **Figure 1A**. In
addition, this particular collection of examples is
intended to furnish some exposure to the diversity of
modeling approaches which can be brought to bear
on the problem. These approaches range from abstract theoretical models intended to elucidate specific processes, to complex numerical formulations
which can be used to actually simulate observed
distributions in detail.

Formulation of the Coupled Problem

A general form of the coupled problem can be written
as a three-dimensional advection-diffusion-reaction
equation for the concentration C_i of any particular
organism of interest:

$$\underbrace{\frac{\partial C_i}{\partial t}}_{\text{local rate of change}} + \underbrace{\nabla \cdot (\mathbf{v}C_i)}_{\text{advection}} - \underbrace{\nabla \cdot (K\nabla C_i)}_{\text{diffusion}}$$

$$= \underbrace{R_i}_{\text{biological sources/sinks}} \qquad [1]$$

where the vector v represents the fluid velocity plus
any biologically induced transport through the water
(e.g., sinking, swimming), and K the turbulent diffusivity. The advection term is often written simply as
$\mathbf{v} \cdot \nabla C_i$ because the ocean is an essentially incompressible fluid (i.e., $\nabla \cdot = 0$). The 'reaction term'
R_i on the right-hand side represents the sources and
sinks due to biological activity.

In essence, this model is a quantitative statement of
the conservation of mass for a scalar variable in a fluid
medium. The advective and diffusive terms simply
represent the redistribution of material caused by
motion. In the absence of any motion, eqn [1] reduces
to an ordinary differential equation describing the
biological and/or chemical dynamics. The reader is
referred to the review by Donaghay and Osborn for a
detailed derivation of the advection-diffusion-reaction
equation, including explicit treatment of the Reynolds
decomposition for biological and chemical scalars (see
Further Reading).

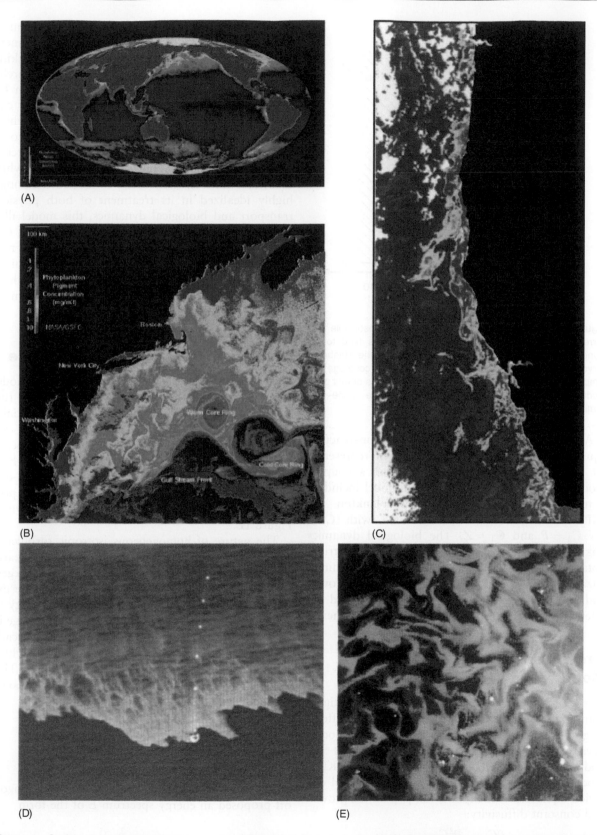

Figure 1 Scales of plankton patchiness, ranging from global down to 1 cm. (A–C) Satellite-based estimates of surface-layer chlorophyll computed from ocean color measurements. Images courtesy of the Seawifs Project and Distributed Active Archive Center at the Goddard Space Flight Center, sponsored by NASA. (D) A dense stripe of *Noctiluca scintillans*, 3 km off the coast of La Jolla. The boat in the photograph is trailing a line with floats spaced every 20 m. The stripe stretched for at least 20 km parallel to the shore (photograph courtesy of P.J.S Franks). (E) Surface view of a bloom of *Anabaena flos-aquae* in Malham Tarn, England. The area shown is approximately 1 m² (photograph courtesy of G.E. Fogg).

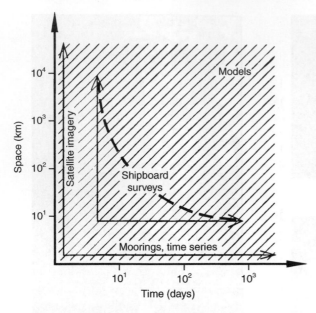

Figure 2 Space–time diagram of the scales resolvable with current observational capabilities. Measurements tend to fall along the axes; the dashed line running between the 'shipboard survey' axes reflects the trade-off between spatial coverage and temporal resolution inherent in seagoing operations of that type. Models can be used to examine portions of the space–time continuum (shaded area).

Any number of advection–diffusion–reaction equations can be posed simultaneously to represent a set of interacting state variables C_i in a coupled model. For example, an ecosystem model including nutrients, phytoplankton, and zooplankton (an 'NPZ' model) could be formulated with $C_1 = dN, C_2 = P$ and $C_3 = Z$. The biological dynamics linking these three together could include nutrient uptake, primary production, grazing, and remineralization. R_i would then represent not only growth and mortality, but also terms which depend on interactions between the several model components.

Growth and Diffusion – the 'KISS' Model

Some of the earliest models used to investigate plankton patchiness dealt with the competing effects of growth and diffusion. In the early 1950s, models developed independently by Kierstead (KI) and Slobodkin (S) and Skellam (S) – the so-called 'KISS' model – were formulated as a one-dimensional diffusion equation with exponential population growth and constant diffusivity:

$$\frac{\partial C}{\partial t} - K\frac{\partial^2 C}{\partial x^2} = \alpha C \qquad [2]$$

Note that this model is a reduced form of eqn [1]. It is a mathematical statement that the tendency for

organisms to accumulate through reproduction is counterbalanced by the tendency of the environment to disperse them through turbulent diffusion. Seeking solutions which vanish at $x = 0$ and $x = L$ (thereby defining a characteristic patch size of dimension L), with initial concentration $C(x, 0) = f(x)$, one can solve for a critical patch size $L = \pi(K/\alpha)^{\frac{1}{2}}$ in which growth and dispersal are in perfect balance. For a specified growth rate α and diffusivity K, patches smaller than L will be eliminated by diffusion, while those that are larger will result in blooms. Although highly idealized in its treatment of both physical transport and biological dynamics, this model illuminates a very important aspect of the role of diffusion in plankton patchiness. In addition, it led to a very specific theoretical prediction of the initial conditions required to start a plankton bloom, which Slobodkin subsequently applied to the problem of harmful algal blooms on the west Florida shelf.

Homogeneous Isotropic Turbulence

The physical regime to which the preceding model best applies is one in which the statistics of the turbulence responsible for diffusive transport is spatially uniform (homogeneous) and has no preferred direction (isotropic). Turbulence of this type may occur locally in parts of the ocean in circumstances where active mixing is taking place, such as in a wind-driven surface mixing layer. Such motions might produce plankton distributions such as those shown in **Figure 1E**.

The nature of homogeneous isotropic turbulence was characterized by Kolmogoroff in the early 1940s. He suggested that the scale of the largest eddies in the flow was set by the nature of the external forcing. These large eddies transfer energy to smaller eddies down through the inertial subrange in what is known as the turbulent cascade. This cascade continues to the Kolmogoroff microscale, at which viscous forces dissipate the energy into heat. This elegant physical model inspired the following poem attributed to L. F. Richardson:

Big whorls make little whorls
which feed on their velocity;
little whorls make smaller whorls,
and so on to viscosity.

Based on dimensional considerations, Kolmogoroff proposed an energy spectrum E of the form

$$E(k) = A\varepsilon^{\frac{2}{3}}k^{\frac{-5}{3}}$$

where k is the wavenumber, ε is the dissipation rate of turbulent kinetic energy, and A is a dimensionless constant. This theoretical prediction was later borne out by measurements, which confirmed the 'minus five-thirds' dependence of energy content on wavenumber.

In the early 1970s, Platt published a startling set of measurements which suggested that for scales between 10 and 10^3 m the variance spectrum of chlorophyll in the Gulf of St Lawrence showed the same $-5/3$ slope. On the basis of this similarity to the Kolmogoroff spectrum, he argued that on these scales, phytoplankton were simply passive tracers of the turbulent motions. These findings led to a burgeoning field of spectral modeling and analysis of plankton patchiness. Studies by Denman, Powell, Fasham, and others sought to formulate more unified theories of physical–biological interactions using this general approach. For example, Denman and Platt extended a model for the scalar variance spectrum to include a uniform growth rate. Their theoretical analysis suggested a breakpoint in the spectrum at a critical wavenumber k_c (**Figure 3**), which they estimated to be in the order of $1\,km^{-1}$ in the upper ocean. For wavenumbers lower than k_c, phytoplankton growth tends to dominate the effects of turbulent diffusion, resulting in a k^{-1} dependence. In the higher wavenumber region, turbulent motions overcome biological effects, leading to spectral slopes of -2 to -3. Efforts to include more biological realism in theories of this type have continued to produce interesting results, although Powell and others have cautioned that spectral characteristics may not be sufficient in and of themselves to resolve the underlying physical–biological interactions controlling plankton patchiness in the ocean.

Vertical Structure

Perhaps the most ubiquitous aspect of plankton distributions which makes them *anisotropic* is their vertical structure. Organisms stratify themselves in a multitude of ways, for any number of different purposes (e.g., to exploit a limiting resource, to avoid predation, to facilitate reproduction). For example, consider the subsurface maximum which is characteristic of the chlorophyll distribution in many parts of the world ocean (**Figure 4**). The deep chlorophyll maximum (DCM) is typically situated below the nutrient-depleted surface layer, where nutrient concentrations begin to increase with depth. Generally this is interpreted to be the result of joint resource limitation: the DCM resides where nutrients are abundant and there is sufficient light for photosynthesis. However, this maximum in chlorophyll does not necessarily imply a maximum in phytoplankton biomass. For example, in the nutrient-impoverished surface waters

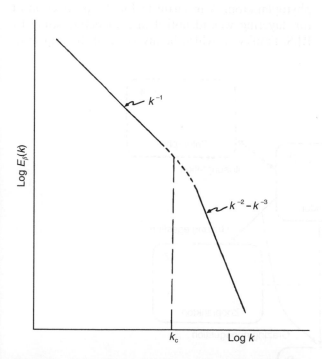

Figure 3 A theoretical spectrum for the spatial variability of phytoplankton, $E_\beta(k)$, as a function of wavenumber, k, displayed on a log-log plot. To the left of the critical wavenumber k_c, biological processes dominate, resulting in a k^{-1} dependence. The high wavenumber region to the right of k_c where turbulent motions dominate, has a dependence between k^{-2} and k^{-3}. (Reproduced with permission from Denman KL and Platt T (1976). The variance spectrum of phytoplankton in a turbulent ocean. *Journal of Marine Research* 34: 593–601.)

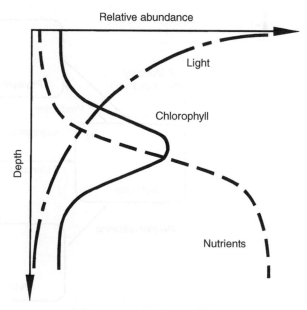

Figure 4 Schematic representation of the deep chlorophyll maximum in relation to ambient light and nutrient profiles in the euphotic zone (typically 10s to 100s of meters in vertical extent).

of the open ocean, much of the phytoplankton standing stock is sustained by nutrients which are rapidly recycled; thus relatively high biomass is maintained by low ambient nutrient concentrations. In such situations, the DCM often turns out to be a pigment maximum, but not a biomass maximum. The mechanism responsible for the DCM in this case is photoadaptation, the process by which phytoplankton alter their pigment content according to the ambient light environment. By manufacturing more chlorophyll per cell, phytoplankton populations in this type of DCM are able to capture photons more effectively in a low-light environment.

Models have been developed which can produce both aspects of the DCM. For example, consider the nutrient, phytoplankton, zooplankton, detritus (NPZD) type of model (**Figure 5**) which simulates the flows of nitrogen in a planktonic ecosystem. The various biological transformations (such as nutrient uptake, primary production, grazing, excretion, etc.) are represented mathematically by functional relationships which depend on the model state variables and parameters which must be determined empirically. Doney *et al.* coupled such a system to a one-dimensional physical model of the upper ocean (**Figure 6**). Essentially, the vertical velocity (w) and diffusivity fields from the physical model are used to drive a set of four coupled advection-diffusion-

reaction equations (one for each ecosystem state variable) which represent a subset of the full three-dimensional eqn [1]:

$$\frac{\partial C_i}{\partial t} + w\frac{\partial C_i}{\partial z} - \frac{\partial}{\partial z}\left(K\frac{\partial C_i}{\partial z}\right) = R_i \qquad [3]$$

The R_i terms represent the ecosystem interaction terms schematized in **Figure 5**. Using a diagnostic photoadaptive relationship to predict chlorophyll from phytoplankton nitrogen and the ambient light and nutrient fields, such a model captures the overall character of the DCM observed at the Bermuda Atlantic Time-series Study (BATS) site (**Figure 6**).

Broad-scale vertical patchiness (on the scale of the seasonal thermocline) such as the DCM is accompanied by much finer structure. The special volume of *Oceanography* on 'Thin layers' provides an excellent overview of this subject, documenting small-scale vertical structure in planktonic populations of many different types. One particularly striking example comes from high-resolution fluorescence measurements (**Figure 7A**). Such profiles often show strong peaks in very narrow depth intervals, which presumably result from thin layers of phytoplankton. A mechanism for the production of this layering was identified in a modeling study by P.J.S. Franks, in which he investigated the impact of

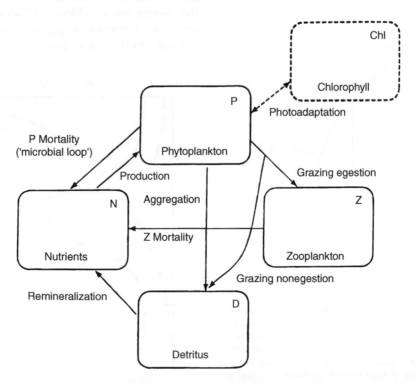

Figure 5 A four-compartment planktonic ecosystem model showing the pathways for nitrogen flow. (Reproduced with permission from Doney SC, Glover DM and Najjar RG (1996) A new coupled, one-dimensional biological–physical model for the upper ocean: applications to the JGOFS Bermuda Atlantic Time-series Study (BATS) site. *Deep-Sea Research II* 43: 591–624.)

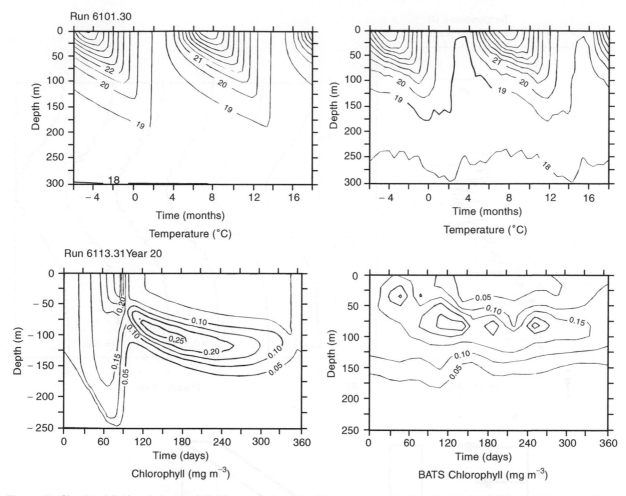

Figure 6 Simulated (left) and observed (right) seasonal cycles of temperature and chlorophyll at the BATS site. (Reproduced with permission from Doney SC, Glover DM and Najjar RG (1996) A new coupled, one-dimensional biological–physical model for the upper ocean: applications to the JGOFS Bermuda Atlantic Time-series Study (BATS) site. *Deep-Sea Research II* 43: 591–624.)

near-inertial wave motion on the ambient horizontal and vertical patchiness which exists at scales much larger than the thin layers of interest. Near-inertial waves are a particularly energetic component in the internal wave spectrum of the ocean. Their horizontal velocities can be described by:

$$u = U_0 \cos(mz - \omega t) \quad v = U_0 \sin(mz - \omega t) \quad [4]$$

where U_0 is a characteristic velocity scale, m is the vertical wavenumber, and ω the frequency of the wave. This kinematic model prescribes that the velocity vector rotates clockwise in time and counterclockwise with depth; its phase velocity is downward, and group velocity upward. In his words, 'the motion is similar to a stack of pancakes, each rotating in its own plane, and each slightly out of phase with the one below'. Franks used this velocity field to perturb an initial distribution of phytoplankton in which a Gaussian vertical distribution

(of scale σ) varied sinusoidally in both x and y directions with wavenumber K_P. Neglecting the effects of growth and mixing, and assuming that phytoplankton are advected passively with the flow, eqn [1] reduces to:

$$\frac{\partial C}{\partial t} + u\frac{\partial C}{\partial x} + v\frac{\partial C}{\partial y} = 0 \quad [5]$$

Plugging the velocity fields [4] into this equation, the initial phytoplankton distribution can be integrated forward in time. This model demonstrates the striking result that such motions can generate vertical structure which is much finer than that present in the initial condition (**Figure 7B**). Analysis of the simulations revealed that the mechanism at work here is simple and elegant: vertical shear can translate horizontal patchiness into thin layers by stretching and tilting the initial patch onto its side (**Figure 7C**).

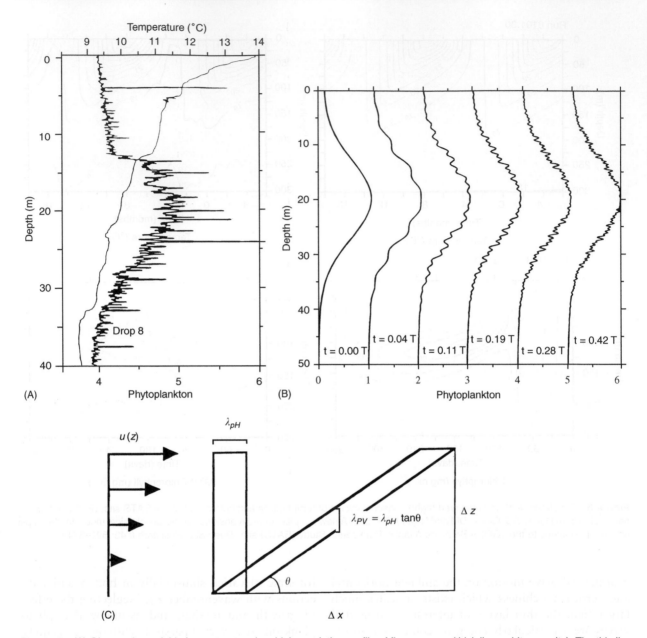

Figure 7 (A) Observations of thin layer structure in a high-resolution profile of fluorescence (thick line, arbitrary units). The thin line shows the corresponding temperature structure. (Data courtesy of Dr T. Cowles.) (B) Simulated vertical profiles of phytoplankton concentration (arbitrary units) at six sequential times. Each profile is offset from the previous by 1 phytoplankton unit. The times are given as fractions of the period of the near-inertial wave used to drive the model. (C) A schematic diagram of the layering process. Vertical shear stretches a vertical column of a property horizontally through an angle θ, creating a layer in the vertical profile. (Reproduced with permission from Franks PJS (1995) Thin layers of phytoplankton: a model of formation by near-inertial wave shear. *Deep-Sea Research I* 42: 75–91.)

Mesoscale Processes: The Internal Weather of the Sea

Just as the atmosphere has weather patterns that profoundly affect the plants and animals that live on the surface of the earth, the ocean also has its own set of environmental fluctuations which exert fundamental control over the organisms living within it.

The currents, fronts, and eddies that comprise the oceanic mesoscale, sometimes referred to as the 'internal weather of the sea', are highly energetic features of ocean circulation. Driven both directly and indirectly by wind and buoyancy forcing, their characteristic scales range from tens to hundreds of kilometers with durations of weeks to months. Their space scales are thus smaller and timescales longer than their counterparts in atmospheric weather, but the dynamics of the two systems are in many ways analogous. Impacts of these motions on surface ocean chlorophyll distributions are clearly visible in satellite imagery (**Figure 1B**).

Mesoscale phenomenologies accommodate a diverse set of physical–biological interactions which influence the distribution and variability of plankton populations in the sea. These complex yet highly organized flows continually deform and rearrange the hydrographic structure of the near-surface region in which plankton reside. In the most general terms, the impact of these motions on the biota is twofold: not only do they stir organism distributions, they can also modulate the rates of biological processes. Common manifestations of the latter are associated with vertical transports which can affect the availability of both nutrients and light to phytoplankton, and thereby the rate of primary production. The dynamics of mesoscale and submesoscale flows are replete with mechanisms that can produce vertical motions.

Some of the first investigations of these effects focused on mesoscale jets. Their internal mechanics are such that changes in curvature give rise to horizontal divergences which lead to very intense vertical velocities along the flanks of the meander systems (**Figure 8**). J. D. Woods was one of the first to suggest that these submesoscale upwellings and downwellings would have a strong impact on upper ocean plankton distributions (see his article contained in the volume edited by Rothschild; see Further Reading). Subsequent modeling studies have investigated these effects by incorporating planktonic ecosystems of the type shown in **Figure 5** into three-dimensional dynamical models of meandering jets. Results suggest that upwelling in the flank of a meander can stimulate the growth of phytoplankton (**Figure 9**). Simulated plankton fields are quite complex owing to the fact that fluid parcels are rapidly advected in between regions of upwelling and downwelling. Clearly, this complicated convolution of physical transport and biological response can generate strong heterogeneity in plankton distributions.

What are the implications of mesoscale patchiness? Do these fluctuations average out to zero, or are they important in determining the mean characteristics of the system? In the Sargasso Sea, it appears that mesoscale eddies are a primary mechanism by which nutrients are transported to the upper ocean. Numerical simulations were used to suggest that upwelling due to eddy formation and intensification causes intermittent fluxes of nitrate into the euphotic zone (**Figure 10A**). The mechanism can be conceptualized by considering a density surface with mean depth coincident with the base of the euphotic zone (**Figure 10B**). This surface is perturbed vertically by the formation, evolution, and destruction of mesoscale features. Shoaling density surfaces lift nutrients into the euphotic zone which are rapidly utilized by the biota. Deepening density surfaces serve to push nutrient-depleted water out of the well-illuminated surface layers. The asymmetric light field thus rectifies vertical displacements of both directions into a net upward transport of nutrients, which is presumably balanced by a commensurate flux of sinking particulate material. Several different lines of evidence suggest that eddy-driven nutrient flux represents a large portion of the annual nitrogen budget in the Sargasso Sea. Thus, in this instance, plankton patchiness appears to be an essential characteristic that drives the mean properties of the system.

Coastal Processes

Of course, the internal weather of the sea is not limited to the eddies and jets of the open ocean. Coastal regions contain a similar set of phenomena, in addition to a suite of processes in which the presence of a land boundary plays a key role. A canonical example of such a process is coastal upwelling, in which the surface layer is forced offshore when the wind blows in the alongshore direction with the coast to the left (right) in the northern (southern) hemisphere. This event triggers upwelling of deep water to replace the displaced surface water. The biological ramifications of this were explored in the mid-1970s by Wroblewski with one of the first coupled physical–biological models to include spatial variability explicitly. Configuring a two-dimensional advection-diffusion-reaction model in vertical plane cutting across the Oregon shelf, he studied the response of an NPZD-type ecosystem model to transient wind forcing. His 'strong upwelling' case provided a dramatic demonstration of mesoscale patch formation (**Figure 11**). Deep, nutrient-rich waters from the bottom boundary layer drawn up toward the surface stimulate a large increase in primary production which is restricted to within 10 km of the coast. The phytoplankton distribution reflects the localized enhancement of production, in addition to advective transport of the resultant biogenic material. Note that the highest concentrations of phytoplankton are displaced from the peak in primary production, owing to the offshore transport in the near-surface layers.

Although Wroblewski's model was able to capture some of the most basic elements of the biological response to coastal upwelling, its two-dimensional formulation precluded representation of alongshore variations which can sometimes be as dramatic as

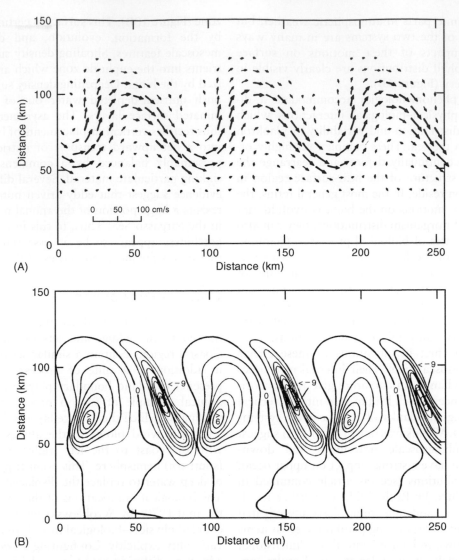

Figure 8 Simulation of a meandering mesoscale jet: (A) velocity on an isopycnal surface with a mean depth of 20 m; (B) vertical velocity (m d^{-1}) on the same isopycnal surface as in (A). Note the consistent pattern of the vertical motion with respect to the structure of the jet. (Reproduced with permission from Woods JD (1988) Mesoscale upwelling and primary production. In: Rothschild BJ (ed.) *Toward a Theory on Physical–Biological Interactions in the World Ocean*. London: Kluwer Academic.)

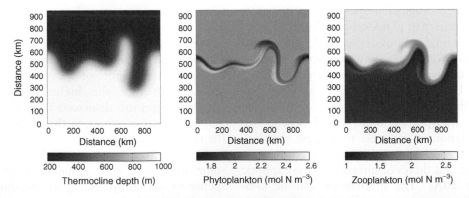

Figure 9 Results from a coupled model of the Gulf Stream: thermocline depth (left), phytoplankton concentration (middle), and zooplankton concentration (right). (Courtesy of GR Flierl, Massachusetts Institute of Technology).

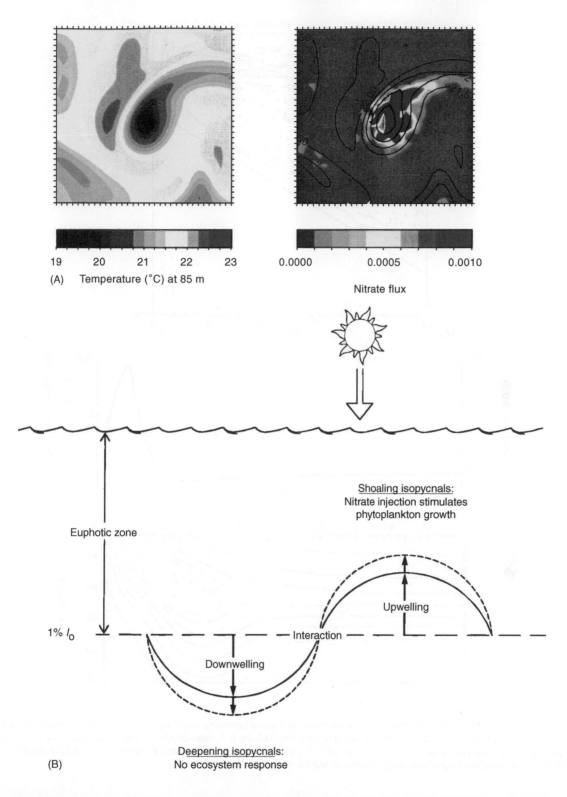

Figure 10   (A) A simulated eddy-driven nutrient injection event: snapshots of temperature at 85 m (left column, °C) and nitrate flux across the base of the euphotic zone (right column, moles of nitrogen m^{-2} d^{-1}). For convenience, temperature contours from the left-hand panels are overlayed on the nutrient flux distributions. The area shown here is a 500 km on a side domain. (The simulation is described in McGillicuddy DJ and Robinson AR (1997) Eddy induced nutrient supply and new production in the Sargasso Sea. *Deep-Sea Research I* 44(8): 1427–1450.) (B) A schematic representation of the eddy upwelling mechanism. The solid line depicts the vertical deflection of an individual isopycnal caused by the presence of two adjacent eddies of opposite sign. The dashed line indicates how the isopycnal might be subsequently perturbed by interaction of the two eddies. (Reproduced with permission from McGillicuddy DJ *et al.* (1998) Influence of mesoscale eddies on new production in the Sargasso Sea. *Nature* 394: 263–265.)

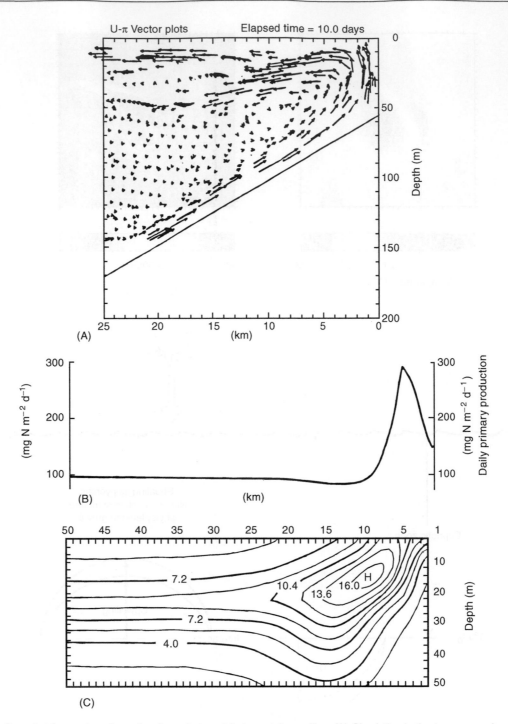

Figure 11 Snapshot from a two-dimensional coupled model of coastal upwelling. (A) Circulation in the transverse plane normal to the coast (maximum horizontal and vertical velocities are −6.1 and 0.05 cm s^{-1}, respectively); (B) daily gross primary production; (C) phytoplankton distribution (contour interval is 1.6 μg) at N l^{-1}. (Adapted with permission from Wroblewski JS (1977) A model of plume formation during variable Oregon upwelling. *Journal of Marine Research* 35(2): 357–394.)

those in the cross-shore direction. The complex set of interacting jets, eddies, and filaments characteristic of such environments (as in **Figure 1C**) have been the subject of a number of three-dimensional modeling investigations. For example, Moisan *et al.* incorporated a food web and bio-optical model into

simulations of the Coastal Transition Zone off California. This model showed how coastal filaments can produce a complex biological response through modulation of the ambient light and nutrient fields (**Figure 12**). The simulations suggested that significant cross-shelf transport of carbon can occur in episodic

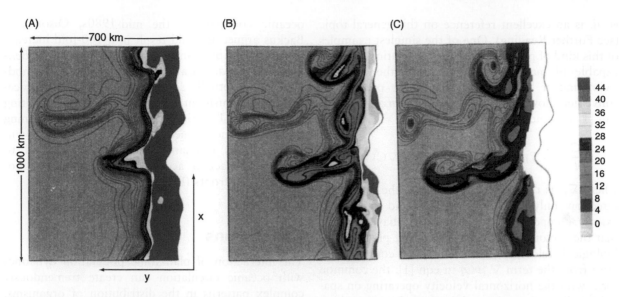

Figure 12 Modeled distributions of phytoplankton (color shading, mg nitrogen m^{-3}) in the Coastal Transition Zone off California. Instantaneous snapshots in panels (A–C) are separated by time intervals of 10 days. Contour lines indicate the depth of the euphotic zone, defined as the depth at which photosynthetically available radiation is 1% of its value at the surface. Contours range from 30 to 180 m, 40 to 180 m, and 60 to 180 m in panels (A), (B), and (C) respectively. (Reproduced with permission from Moisan et al. (1996) Modeling nutrient and plankton processes in the California coastal transition zone 2. A three-dimensional physical-bio-optical model. *Journal of Geophysical Research* 101(C10): 22677–22691.

pulses when filaments meander offshore. These dynamics illustrate the tremendous complexity of the processes which link the coastal ocean with the deep sea.

Behavior

The mechanisms for generating plankton patchiness described thus far consist of some combination of fluid transport and physiological response to the physical, biological, and chemical environment. The fact that many planktonic organisms have behavior (interpreted narrowly here as the capability for directed motion through the water) facilitates a diverse array of processes for creating heterogeneity in their distributions. Such processes pose particularly difficult challenges for modeling, in that their effects are most observable at the level of the population, whereas their dynamics are governed by interactions which occur amongst individuals. The latter aspect makes modeling patchiness of this type particularly amenable to individual-based models, in contrast to the concentration-based model described by eqn [1]. For example, many species of marine plankton are known to form dense aggregations, sometimes referred to as swarms. Okubo suggested an individual-based model for the maintenance of a swarm of the form:

$$\frac{d^2x}{dt^2} = -k\frac{dx}{dt} - \omega^2 x - \phi(x) + A(t) \qquad [6]$$

where x represents the position of an individual. This model assumes a frictional force on the organism which is proportional to its velocity (with frictional coefficient k), a random force $A(t)$ which is white noise of zero mean and variance B, and attractive forces. Acceleration resulting from the attractive forces is split between periodic (frequency ω) and static ($\phi(x)$) components. The key aspect of the attractive forces is that they depend on the distance from the center of the patch. A Fokker-Planck equation can be used to derive a probability density function:

$$p(x) = p_0 \exp\left(-\frac{\omega^2}{2B}x^2 - \int \frac{\phi(x)}{b}dx \right) \qquad [7]$$

where p_0 is the density at the center of the swarm. Thus, the macroscopic properties of the system can be related to the specific set of rules governing individual behavior. Okubo has shown that observed characteristics of insect swarms compare well with theoretical predictions from this model, both in terms of the organism velocity autocorrelation and the frequency distribution of their speeds. Analogous comparisons with plankton have proven elusive owing to the extreme difficulty in making such measurements in marine systems.

The foregoing example illustrates how swarms can arise out of purely behavioral motion. Yet another class of patchiness stems from the joint effects of behavior and fluid transport. The paper by Flierl *et al.* is an excellent reference on this general topic (see Further Reading). One of the simplest examples of this kind of process arises in a population which is capable of maintaining its depth (either through swimming or buoyancy effects) in the presence of convergent flow. With no biological sources or sinks, eqn [1] becomes:

$$\frac{\partial C}{\partial t} + \mathbf{v} \cdot \nabla_H C + C \nabla_H \cdot \mathbf{v} - \nabla \cdot (K \nabla C) = 0 \quad [8]$$

where ∇_H is the vector derivative in the horizontal direction only. Because vertical fluid motion is exactly compensated by organism behavior (recall that the vector v represents the sum of physical and biological velocities), two advective contributions arise from the term $\nabla \cdot (\mathbf{v} C)$ in eqn [1]: the common form with the horizontal velocity operating on spatial gradients in concentration, *plus* a source/sink term created by the divergence in total velocity (fluid + organism). The latter term provides a mechanism for accumulation of depth-keeping organisms in areas of fluid convergence. It has been suggested that this process is important in a variety of different oceanic contexts. In the mid-1980s, Olson and Backus argued it could result in a 100-fold increase in the local abundance of a mesopelagic fish *Benthosema glaciale* in a warm core ring. Franks modeled a conceptually similar process with a surface-seeking organism in the vicinity of a propagating front (**Figure 13**). Simply stated, upward swimming organisms tend to accumulate in areas of downwelling. This mechanism has been suggested to explain spectacular accumulations of motile dinoflagellates at fronts (**Figure 1D**).

Conclusions

The interaction of planktonic population dynamics with oceanic circulation can create tremendously complex patterns in the distribution of organisms. Even an ocean at rest could accommodate significant inhomogeneity through geographic variations in environmental variables, time-dependent forcing, and organism behavior. Fluid motions tend to amalgamate all of these effects in addition to introducing yet another source of variability: space–time fluctuations in the flows themselves which impact biological processes. Understanding the mechanisms responsible for observed variations in plankton distributions is thus an extremely difficult task.

Coupled physical–biological models offer a framework for dissection of these manifold contributions to structure in planktonic populations. Such models take many forms in the variety of approaches which have been used to study plankton patchiness. In theoretical investigations, the basic dynamics of idealized systems are worked out using techniques from applied mathematics and mathematical physics. Process-oriented numerical models offer a conceptually similar way to study systems that are too complex to be solved analytically. Simulation-oriented models are aimed at reconstructing particular data sets using realistic hydrodynamic forcing pertaining to the space/time domain of interest. Generally speaking, such models tend to be quite complex because of the multitude of processes which must be included to simulate observations made in the natural environment. Of course, this complexity makes diagnosis of the coupled system more challenging. Nevertheless, the combination of models and observations provides a unique context for the synthesis of necessarily sparse data: space–time continuous representations of the real ocean which can be diagnosed term-by-term to reveal the underlying processes. Formal union between models and observations is beginning to occur through the emergence of inverse methods and data assimilation in the field of biological oceanography. 410

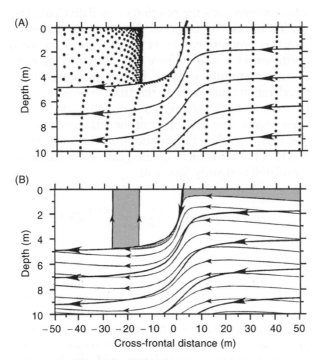

Figure 13 Surface-seeking organisms aggregating at a propagating front. Modeled particle locations (dots, panel (A)) and particle streamlines (thin lines, panel (B)) in the cross-frontal flow. The front is centered at $x = 0$, and the coordinate system translates to the right with the motion of the front. Flow streamlines are represented in both panels as bold lines; they differ from particle streamlines due to propagation of the front. The shaded area in (B) indicates the region in which cells are focused into the frontal zone, forming a dense band at $x = -20$ m. (Reproduced with permission from Franks, 1997.)

provides an up-to-date review of this very exciting and rapidly evolving aspect of coupled physical–biological modeling.

Although the field is more than a half-century old, modeling of plankton patchiness is still in its infancy. The oceanic environment is replete with phenomena of this type which are not yet understood. Fortunately, the field is perhaps better poised than ever to address such problems. Recent advances in measurement technologies (e.g., high-resolution acoustical and optical methods, miniaturized biological and chemical sensors) are beginning to provide direct observations of plankton on the scales at which the coupled processes operate. Linkage of such measurements with models is likely to yield important new insights into the mechanisms controlling plankton patchiness in the ocean.

See also

Continuous Plankton Recorders. Satellite Remote Sensing: Ocean Color. Upper Ocean Mixing Processes.

Further Reading

Denman KL and Gargett AE (1995) Biological–physical interactions in the upper ocean: the role of vertical and small scale transport processes. *Annual Reviews of Fluid Mechanics* 27: 225–255.

Donaghay PL and Osborn TR (1997) Toward a theory of biological–physical control on harmful algal bloom dynamics and impacts. *Limnology and Oceanography* 42: 1283–1296.

Flierl GR, Grunbaum D, Levin S, and Olson DB (1999) From individuals to aggregations: the interplay between behavior and physics. *Journal of Theoretical Biology* 196: 397–454.

Franks PJS (1995) Coupled physical–biological models in oceanography. *Reviews of Geophysics* supplement: 1177–1187.

Franks PJS (1997) Spatial patterns in dense algal blooms. *Limnology and Oceanography* 42: 1297–1305.

Levin S, Powell TM, and Steele JH (1993) *Patch Dynamics*. Berlin: Springer-Verlag.

Mackas DL, Denman KL, and Abbott MR (1985) Plankton patchiness: biology in the physical vernacular. *Bulletin of Marine Science* 37: 652–674.

Mann KH and Lazier JRN (1996) *Dynamics of Marine Ecosystems: Biological–Physical Interactions in the Oceans*. Oxford: Blackwell Scientific Publications.

Okubo A (1980) *Diffusion and Ecological Problems: Mathematical Models*. Berlin: Springer-Verlag.

Okubo A (1986) Dynamical aspects of animal grouping: swarms, schools, flocks and herds. *Advances in Biophysics* 22: 1–94.

Robinson AR, McCarthy JJ, and Rothschild BJ (2001) *The Sea: Biological–Physical Interactions in the Ocean*. New York: John Wiley and Sons.

Rothschild BJ (1988) *Toward a Theory on Biological–Physical Interactions in the World Ocean*. Dordrecht: D. Reidel.

Oceanography Society (1998) *Oceanography 11(1): Special Issue on Thin Layers*. Virginia Beach, VA: Oceanography Society.

Steele JH (1978) *Spatial Pattern in Plankton Communities*. New York: Plenum Press.

Wroblewski JS and Hofmann EE (1989) U.S. interdisciplinary modeling studies of coastal–offshore exchange processes: past and future. *Progress in Oceanography* 23: 65–99.

OCEAN BIOGEOCHEMISTRY AND ECOLOGY, MODELING OF

N. Gruber, Institute of Biogeochemistry and Pollutant Dynamics, ETH Zurich, Switzerland
S. C. Doney, Woods Hole Oceanographic Institution, Woods Hole, MA, USA

Introduction

Modeling has emerged in the last few decades as a central approach for the study of biogeochemical and ecological processes in the sea. While this development was facilitated by the fast development of computer power, the main driver is the need to analyze and synthesize the rapidly expanding observations, to formulate and test hypotheses, and to make predictions how ocean ecology and biogeochemistry respond to perturbations. The final aim, prediction, has gained in importance recently as scientists are increasingly asked by society to investigate and assess the impact of past, current, and future human actions on ocean ecology and biogeochemistry.

The impact of the carbon dioxide (CO_2) that humankind has emitted and will continue to emit into the atmosphere for the foreseeable future is currently, perhaps, the dominant question facing the marine biogeochemical/ecological research community. This impact is multifaceted, and includes both direct (such as ocean acidification) and indirect effects that are associated with the CO_2-induced climate change. Of particular concern is the possibility that global climate change will lead to a reduced capacity of the ocean to absorb CO_2 from the atmosphere, so that a larger fraction of the CO_2 emitted into the atmosphere remains there, further enhancing global warming. In such a positive feedback case, the expected climate change for a given CO_2 emission will be larger relative to a case without feedbacks. Marine biogeochemical/ecological models have played a crucial role in elucidating and evaluating these processes, and they are increasingly used for making quantitative predictions with direct implications for climate policy.

There are many other marine biogeochemical and/or ecological problems related to human activities, for which models play a crucial role assessing their importance and magnitude and devising possible solutions. These include, for example, coastal eutrophication, overfishing, and dispersion of invasive species. The use of marine biogeochemical/ecological models is now so pervasive that practically every field of oceanography is on this list.

The aim of this article is to provide an introduction and overview of marine biogeochemical and ecological modeling. Given the breadth of modeling approaches in use today, this overview can by design not be inclusive and authoritative. We rather focus on some basic concepts and provide a few illustrative applications. We start with a broader description of the marine biogeochemical/ecological challenge at hand, and then introduce basic concepts used for biogeochemical/ecological modeling. In the final section, we use a number of examples to illustrate some of the core modeling approaches.

The Marine Ecology and Biogeochemistry Challenge

The complexity of the ocean biogeochemical/ecological problem is daunting, as it involves a complex interplay among biology, physical variability of the oceanic environment, and the interconnected cycles of a large number of bioactive elements, particularly those of carbon, nitrogen, phosphorus, oxygen, silicon, and iron (**Figure 1**). Furthermore, the ocean is an open system that exchanges mass and many elements with the surrounding realms, such as the atmosphere, the land, and the sediments.

The engine that sets nearly all of these cycles into motion is the photosynthetic fixation of dissolved inorganic carbon and many other nutrient elements into organic matter by phytoplankton in the illuminated upper layers of the ocean (euphotic zone). The net rate of this process (i.e., net primary production) is distributed heterogeneously in the ocean, primarily as a result of the combined limitation of nutrients and light. This results in similar heterogeneity in surface chlorophyll, a direct indicator of the amount of phytoplankton biomass (**Figure 2(a)**). The large-scale surface nutrient distributions, in turn, reflect the balance between biological removal and the physical processes of upwelling and mixing that transport subsurface nutrient pools upward into the euphotic zone (**Figure 2(b)**). In addition to the traditional macronutrients (nitrate, phosphate, silicate etc.), growing evidence shows that iron limitation is

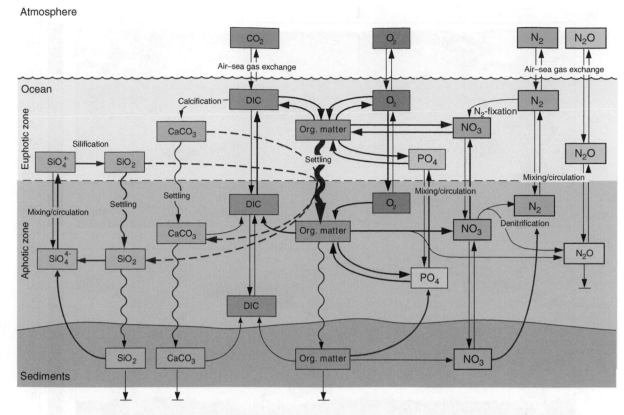

Figure 1 Schematic diagram of a number of key biogeochemical cycles in the ocean and their coupling. Shown are the cycles of carbon, oxygen, phosphorus, nitrogen, and silicon. The main engine of most biogeochemical cycles in the ocean is the biological production of organic matter in the illuminated upper ocean (euphotic zone), part of which sinks down into the ocean's interior and is then degraded back to inorganic constituents. This organic matter cycle involves not only carbon, but also nitrogen, phosphorus, and oxygen, causing a tight linkage between the cycles of these four elements. Since some phytoplankton, such as diatoms and coccolithophorids, produce shells made out of amorphous silicon and solid calcium carbonate, the silicon and $CaCO_3$ cycles also tend to be closely associated with the organic matter cycle. These ocean interior cycles are also connected to the atmosphere through the exchange of a couple of important gases, such as CO_2, oxygen, and nitrous oxide (N_2O). In fact, on timescales longer than a few decades, the ocean is the main controlling agent for the atmospheric CO_2 content.

a key factor governing net primary production in many parts of the ocean away from the main external sources for iron, such as atmospheric dust deposition and continental margin sediments. The photosynthesized organic matter is the basis for a complex food web that involves both a transfer of the organic matter toward higher trophic levels as well as a microbial loop that is responsible for most of the breakdown of this organic matter back to its inorganic constituents.

A fraction of the synthesized organic matter (about 10–20%) escapes degradation in the euphotic zone and sinks down into the dark aphotic zone, where it fuels the growth of microbes and zooplankton that eventually also remineralize most of this organic matter back to its inorganic constituents. The nutrient elements and the dissolved inorganic carbon are then eventually returned back to the surface ocean by ocean circulation and mixing, closing this 'great biogeochemical loop'. This loop is slightly leaky in that a

small fraction of the sinking organic matter fails to get remineralized in the water column and is deposited onto the sediments. Very little organic matter escapes remineralization in the sediments though, so that only a tiny fraction of the organic matter produced in the surface is permanently removed from the ocean by sediment burial. In the long-term steady state, this loss of carbon and other nutrient elements from the ocean is replaced by the input from land by rivers and through the atmosphere.

Several marine phytoplankton and zooplankton groups produce hard shells consisting of either mineral calcium carbonate ($CaCO_3$) or amorphous silica, referred to as 'opal' (SiO_2). The most important producers of $CaCO_3$ are coccolithophorids, while most marine opal stems from diatoms. Both groups are photosynthetic phytoplankton, highlighting the extraordinary importance of this trophic group for marine ecology and biogeochemistry. Upon the death of the mineral-forming organisms, these minerals

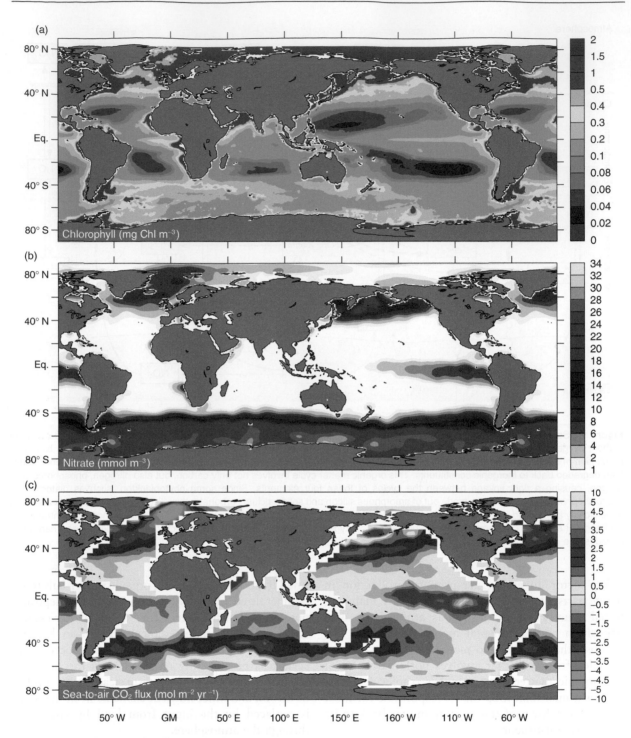

Figure 2 Global maps of the distribution of three key biogeochemical/ecological modeling targets. (a) Annual mean distribution of near-surface chlorophyll ($mg\,Chl\,m^{-3}$) as measured by the *SeaWiFS* satellite. (b) Annual mean surface distribution of nitrate ($mmol^{-3}$) compiled from *in situ* observations (from the World Ocean Atlas). (c) Annual mean air–sea CO_2 flux ($mol\,m^{-2}\,yr^{-1}$) for a nominal year of 2000 derived from a compilation of *in situ* measurements of the oceanic pCO_2 and a wind-speed gas exchange parametrization. Data provided by *T. Takahashi*, Lamont Doherty Earth Observatory of Columbia University.

also sink into the ocean's interior, thereby often acting as 'ballast' for organic matter.

This great biogeochemical loop also has a strong impact on several gases that are dissolved in seawater and exchange readily with the atmosphere, most importantly CO_2 and oxygen (O_2). Biological processes have opposite effects on these two gases: photosynthesis consumes CO_2 and liberates O_2, while respiration and bacterial degradation releases CO_2 and consumes O_2. As a result, one tends to find an inverse relationship in the oceanic distribution of these two gases. CO_2 also sets itself apart from O_2 in that it reacts readily with seawater. In fact, due to the high content of alkaline substances in the ocean, this reaction is nearly complete, so that only around 1% of the total dissolved inorganic carbon in the ocean exists in the form of CO_2.

The exchange of CO_2 across the air–sea interface is a prime question facing ocean biogeochemical/ecological research, particularly with regard to the magnitude of the oceanic sink for anthropogenic CO_2. The anthropogenic CO_2 sink occurs on top of the natural air–sea CO_2 fluxes characterized by oceanic uptake in mid-latitudes and some high latitudes, and outgassing in the low latitudes and the Southern Ocean. This distribution of the natural CO_2 flux is the result of an interaction between the exchange of heat between the ocean and the atmosphere, which affects the solubility of CO_2, and the great biogeochemical loop, which causes an uptake of CO_2 from the atmosphere in regions where the downward flux of organic carbon exceeds the upward supply of dissolved inorganic carbon, and an outgassing where the balance is the opposite. This flux has changed considerably over the last two centuries in response to the anthropogenically driven increase in atmospheric CO_2 that pushes additional CO_2 from the atmosphere into the ocean. Nevertheless, the pattern of the resulting contemporary air–sea CO_2 flux (**Figure 2(c)**) primarily still reflects the flux pattern of the natural CO_2 fluxes, albeit with a global integral flux into the ocean reflecting the oceanic uptake of anthropogenic CO_2.

Given the central role of the great biogeochemical loop, any modeling of marine biogeochemical/ecological processes invariably revolves around the modeling of all the processes that make up this loop. As this loop starts with the photosynthetic production of organic matter by phytoplankton, biogeochemical modeling is always tightly interwoven with the modeling of marine ecology, especially that of the lower trophic levels.

Core questions that challenge marine biogeochemistry and ecology are as follows:

1. What controls the mean concentration and three-dimensional (3-D) distribution of bioreactive elements in the ocean?
2. What controls the air–sea balance of climatically important gases, that is, CO_2, N_2O, and O_2?
3. What controls ocean productivity, the downward export of organic matter, and the transfer of organic matter to higher trophic levels?
4. How do ocean biogeochemistry and ecology change in time in response to climate dynamics and human perturbations?

Modeling represents a powerful approach to studying and addressing these core questions. Modeling is by no means the sole approach. In fact, integrated approaches that combine observational, experimental, and modeling approaches are often necessary to tackle this set of complex problems.

What Is a Biogeochemical/Ecological Model?

At its most fundamental level, a model is an abstract description of how some aspect of nature functions, most often consisting of a set of mathematical expressions. In the biogeochemical/ecological modeling context, a model usually consists of a number of partial differential equations, which describe the time and space evolution of a (limited) number of ecological/biogeochemical state variables. As few of these equations can be solved analytically, they are often solved numerically using a computer, which requires the discretization of these equations, that is, they are converted into difference equations on a predefined spatial and temporal grid.

The Art of Biogeochemical/Ecological Modeling

A model can never fully represent reality. Rather, it aims to represent an aspect of reality in the context of a particular problem. The art of modeling is to find the right level of abstraction, while keeping enough complexity to resolve the problem at hand. That is, marine modelers often follow the strategy of Occam's razor, which states that given two competing explanations, the one that is simpler and makes fewer assumptions is the more likely to be correct. Therefore, a typical model can be used only for a limited set of applications, and great care must be used when a model is applied to a problem for which it was not designed. This is especially true for biogeochemical/ecological models, since their underlying mathematical descriptions are for the most part not based on first principles, but often derived from empirical relationships. In fact, marine biogeochemical/ecological modeling is at present a data-limited activity because we lack data to formulate and parametrize key processes and/or to evaluate the model predictions. Further, significant simplifications are often made to make the problem more tractable. For example, rather than treating individual organisms or even species, model variables often aggregate entire functional groups into single boxes (e.g., photosynthetic organisms, grazers, and detritus decomposers), which are then simulated as bulk concentrations (e.g., $mol\, C\, m^{-3}$ of phytoplankton).

Despite these limitations, models allow us to ask questions about the ocean inaccessible from data or experiments alone. In particular, models help researchers quantify the interactions among multiple processes, synthesize diverse observations, test hypotheses, extrapolate across time – space scales, and predict past and future behavior. A well-posed model encapsulates our understanding of the ocean in a mathematically consistent form.

Biogeochemical/Ecological Modeling Equations and Approaches

In contrast with their terrestrial counterparts, models of marine biogeochemical/ecological processes must be coupled to a physical circulation model of some sort to take into consideration that nearly all relevant biological and biogeochemical processes occur either in the dissolved or suspended phase, and thus are subject to mixing and transport by ocean currents. Thus, a typical coupled physical–biogeochemical/ecological model consists of a set of time-dependent advection, diffusion, and reaction equations:

$$\frac{\partial C}{\partial t} + \mathrm{Adv}(C) + \mathrm{Diff}(C) = \mathrm{SMS}(C) \qquad [1]$$

where C is the state variable to be modeled, such as the concentration of phytoplankton, nutrients, or dissolved inorganic carbon, often in units of mass per unit volume (e.g., $mol\, m^{-3}$). $\mathrm{Adv}(C)$ and $\mathrm{Diff}(C)$ are the contributions to the temporal change in C by advection and eddy-diffusion (mixing), respectively, derived from the physical model component. The term $\mathrm{SMS}(C)$ refers to the 'sources minus sinks' of C driven by ecological/biogeochemical processes. The SMS term often involves complex interactions among a number of state variables and is provided by the biogeochemical/ecological model component.

Marine biogeochemical/ecological models are diverse, covering a wide range of complexities and applications from simple box models to globally 4-D-(space and time) coupled physical–biogeochemical simulations, and from strict research tools to climate change projections with direct societal implications. Model development and usage are strongly shaped by the motivating scientific or policy problems as well as the dynamics and time–space scales considered. The

complexity of marine biogeochemical/ecological models can be organized along two major axes (**Figure 3**): the physical complexity, which determines how the left-hand side of [1] is computed, and the biogeochemical/ecological complexity, which determines how the right-hand side of [1] is evaluated.

Due to computational and analytical limitations, there is often a trade-off between the physical and biogeochemical/ecological complexity, so that models of the highest physical complexity are often using relatively simple biogeochemical/ecological models and vice versa (**Figure 4**). Additional constraints arise from the temporal domain of the integrations. Applications of coupled physical–biogeochemical/ecological models to paleoceanographic questions require integrations of several thousand years. This can only be achieved by reducing both the physical and the biogeochemical/ecological complexity (**Figure 4**). At the same time, the continuously increasing computational power has permitted researchers to push forward along both complexity axes. Nevertheless, the fundamental tradeoff between physical and biogeochemical/ecological complexity remains.

In addition to the physical and biogeochemical/ecological complexity, models can also be categorized with regard to their interaction with observations. In the case of 'forward models', a set of equations in the form of [1] is integrated forward in time given initial and boundary conditions. A typical forward problem is the prediction of the future state of ocean biogeochemistry and ecology for a certain evolution of the Earth's climate. The solutions of such forward models are the time–space distribution of the state variables as well as the implied fluxes. The expression 'inverse models' refers to a broad palette of modeling approaches, but all of them share the goal of optimally combining observations with knowledge about the workings of a system as embodied in the model. Solutions to such inverse models can be improved estimates of the current state of the system (state estimation), improved estimates of the initial or boundary conditions, or an optimal set of parameters. A typical example of an inverse model is the optimal determination of ecological parameters, such as growth and grazing rates, given, for example, the observed distribution of phytoplankton, zooplankton, and nutrients.

Examples

Given the large diversity of marine biogeochemical/ecological models and approaches, no review can do

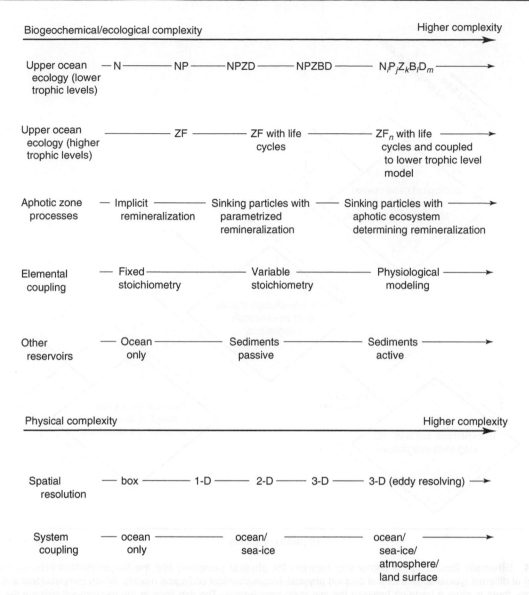

Biogeochemical/ecological complexity — Higher complexity

Upper ocean ecology (lower trophic levels) — N —— NP —— NPZD —— NPZBD —— $N_iP_jZ_kB_lD_m$

Upper ocean ecology (higher trophic levels) —— ZF —— ZF with life cycles —— ZF_n with life cycles and coupled to lower trophic level model

Aphotic zone processes — Implicit remineralization —— Sinking particles with parametrized remineralization —— Sinking particles with aphotic ecosystem determining remineralization

Elemental coupling — Fixed stoichiometry —— Variable stoichiometry —— Physiological modeling

Other reservoirs — Ocean only —— Sediments passive —— Sediments active

Physical complexity — Higher complexity

Spatial resolution — box —— 1-D —— 2-D —— 3-D —— 3-D (eddy resolving)

System coupling — ocean only —— ocean/ sea-ice —— ocean/ sea-ice/ atmosphere/ land surface

Figure 3 Schematic diagram summarizing the development of complexity in coupled physical–biogeochemical/ecological models. The two major axes of complexity are biogeochemistry/ecology and physics, but each major axis consists of many subaxes that describe the complexity of various subcomponents. Currently existing models fill a large portion of the multidimensional space opened by these axes. In addition, the evolution of models is not always necessarily straight along any given axis, but depends on the nature of the particular problem investigated. N, nutrient; P, phytoplankton; Z, zooplankton; B, bacteria; D, detritus; F, fish.

full justice. We restrict our article here to the discussion of four examples, which span the range of complexities as well as have been important milestones in the evolution of biogeochemical/ecological modeling.

Box Models or What Controls Atmospheric Carbon Dioxide?

Our aim here is to develop a model that explains how the great biogeochemical loop controls atmospheric CO_2. The key to answering this question is a quantitative prediction of the surface ocean concentration of

CO_2, as it is this surface concentration that controls the atmosphere–ocean balance of CO_2. The simplest models used for such a purpose are box models, where the spatial dimension is reduced to a very limited number of discrete boxes. Such box models have played an important role in ocean biogeochemical/ecological modeling, mostly because their solutions can be readily explored and understood. However, due to the dramatic reduction of complexity, there are also important limitations, whose consequences one must keep in mind when interpreting the results.

Box models can be formally derived from the tracer conservation eq [1] by integrating over

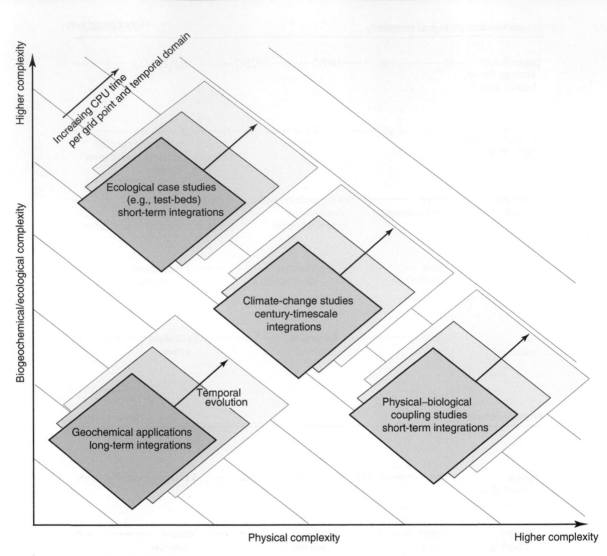

Figure 4 Schematic illustrating the relationship between the physical (abscissa) and the biogeochemical/ecological complexity (ordinate) of different typical applications of coupled physical–biogeochemical/ecological models. Given computational and analytical constraints, there is often a trade-off between the two main complexities. The thin lines in the background indicate the CPU time required for the computation of a problem over a given time period.

the volume of the box, V, and by applying Gauss' (divergence) theorem, which states that the volume integral of a flux is equal to the flux in normal direction across the boundary surfaces, S. The resulting integral form of [1] is:

$$\int \frac{\partial C}{\partial t}\, dV + \oint (\text{Adv}_n(C) + \text{Diff}_n(C))\, dS$$
$$= \int \text{SMS}(C)\, dV \qquad [2]$$

where $\text{Adv}_n(C)$ and $\text{Diff}_n(C)$ are the advective and diffusive transports in the normal direction across S. If we assume that the concentration of tracer C as well as the SMS term within the box are uniform, [2]

can be rewritten as

$$V \cdot \frac{d\overline{C}}{dt} + \sum_i \left(T_{i,\text{out}} \cdot \overline{C} - T_{i,\text{in}} \cdot C_i + v_i(\overline{C} - C_i) \right)$$
$$= V \cdot \text{SMS}(\overline{C}) \qquad [3]$$

where $\overline{C}$ is the mean concentration within the box. We represented the advective contributions as products of a mass transport, T (dimensions volume time^{-1}), with the respective upstream concentration, separately considering the mass transport into the box and out of the box across each interface i, that is, $T_{i,\text{in}}$ and $T_{i,\text{out}}$. The eddy-diffusion contributions are parametrized as products of a mixing coefficient, v_i (dimension volume time^{-1}), with the gradient across each interface $(\overline{C} - C_i)$. For simplicity, we subsequently drop the overbar, that is, use C instead of $\overline{C}$.

The simplest box-model representation of the great biogeochemical loop is a two-box model, wherein the ocean is divided into a surface and a deep box, representing the euphotic and aphotic zones, respectively (**Figure 5(a)**). Mixing and transport between the two boxes is modeled with a mixing term, v. The entire complexity of marine ecology and biogeochemistry in the euphotic zone is reduced to a single term, Φ, which represents the flux of organic matter out of the surface box and into the deep box, where the organic matter is degraded back to inorganic constituents.

Let us consider the deep-box balance for a dissolved inorganic bioreactive element, C_d (such as nitrate, phosphate, or dissolved inorganic carbon):

$$V_d \frac{dC_d}{dt} + v \cdot (C_d - C_s) = \Phi \qquad [4]$$

where V_d is the volume of the deep box and C_s is the dissolved inorganic concentration of the surface box. The steady-state solution of [4], that is, the solution when the time derivative of C_d vanishes ($dC_d/dt = 0$),

$$v \cdot (C_d - C_s) = \Phi \qquad [5]$$

represents the most fundamental balance of the great biogeochemical loop. It states that the net upward transport of an inorganic bioreactive element by physical mixing is balanced by the downward transport of this element in organic matter. This steady-state balance [5] has several important applications. For example, it permits us to estimate the downward export of organic matter by simply

analyzing the vertical gradient and by estimating the vertical exchange. The balance also states that the magnitude of the vertical gradient in bioreactive elements is proportional to the strength of the downward flux of organic matter and inversely proportional to the mixing coefficient. Since nearly all oceanic inorganic carbon and nutrients reside in the deep box, that is, $V_d \cdot C_d \gg V_s \cdot C_s$, one can assume, to first order, that C_d is largely invariant, so that the transformed balance for the surface ocean concentration

$$C_s = C_d - \frac{\Phi}{v} \qquad [6]$$

reveals that the surface concentration depends primarily on the relative magnitude of the organic matter export to the magnitude of vertical mixing. Hence, [6] provides us with a first answer to our challenge: it states that, for a given amount of ocean mixing, the surface ocean concentration of inorganic carbon, and hence atmospheric CO_2, will decrease with increasing marine export production. Relationship [6] also states that for a given magnitude of marine export production, atmospheric CO_2 will increase with increased mixing.

While very powerful, the two-box model has severe limitations. The most important one is that this model does not consider the fact that the deep ocean exchanges readily only with the high latitudes, which represent only a very small part of the surface ocean. The exchange of the deep ocean with the low latitudes is severely limited because diapcynal mixing in the ocean is small. Another limitation is that the one-box model does not take into account that the nutrient concentrations in the high latitudes tend to be much higher than in the low latitudes (**Figure 2(c)**).

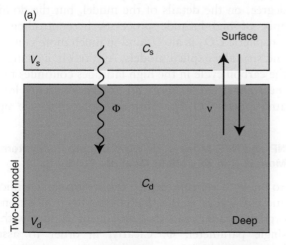
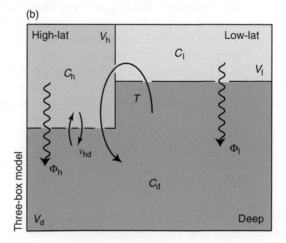

Figure 5 Schematic representation of the great biogeochemical loop in box models: (a) two-box model and (b) three-box model. C, concentrations; V, volume; v, exchange (mixing) coefficient; Φ, organic matter export fluxes; T, advective transport.

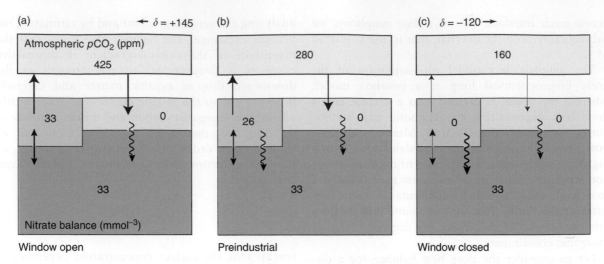

Figure 6 Schematic illustration of the impact of the high-latitude (Southern Ocean) window on atmospheric CO_2. In (a) the high-latitude window is open, high-latitude nitrate is high, and atmospheric CO_2 attains 425 ppm. Panel (b) represents the preindustrial ocean, where the window is half open, nitrate in the high latitudes is at intermediate levels, and atmospheric CO_2 is 280 ppm. In (c), the high-latitude window is closed, high-latitude nitrate is low, and atmospheric CO_2 decreases to about 160 ppm.

An elegant solution is the separation of the surface box into a high-latitude box, h, and a low-latitude box, l (**Figure 5(b)**). Intense mixing is assumed to occur only between the deep and the high-latitude boxes, v_{hd}. A large-scale transport, T, is added to mimic the ocean's global-scale overturning circulation. As before, the complex SMS terms are summarized by Φ_l and Φ_h. In steady state, the surface concentrations are given by

$$C_l = C_d - \frac{\Phi_l}{T}$$
$$C_h = C_d - \frac{\Phi_h + \Phi_l}{T + v_{hd}} \quad [7]$$

which are structurally analogous to [6], but emphasize the different dynamics setting balances in the low and high latitudes. With $v_{hd} \gg T$ and $\Phi_l \approx \Phi_h$, [7] explains immediately why nutrient concentrations tend to be higher in high latitudes than in low latitudes (**Figure 2(b)**). In fact, writing [7] out for both nitrate (NO_3^-) and dissolved inorganic carbon (DIC), and using the observation that surface NO_3^- in the low latitudes is essentially zero (**Figure 2(c)**), the DIC equations are

$$DIC_l = DIC_d - r_{C:N} \cdot [NO_3^-]_d$$
$$DIC_h = DIC_d - \frac{\Phi_h + r_{C:N} \cdot T \cdot [NO_3^-]_d}{T + v_{hd}} \quad [8]$$

where $r_{C:N}$ is the carbon-to-nitrogen ratio of organic matter. Assuming that the deep-ocean nitrate concentration is time invariant, the analysis of [8] reveals that the low-latitude DIC concentration, DIC_l,

is more or less fixed, while the high-latitude DIC concentration, DIC_h, can be readily altered by changes in either the high-latitude export flux, Φ_h, or high-latitude mixing, v_{hd}. Furthermore, if one considers the fact that the rapid communication between the high-latitude ocean and the deep ocean makes the high latitudes the primary window into the oceanic reservoir of inorganic carbon, it becomes clear that it must be the high latitudes that control atmospheric CO_2 on the millenial timescales where the steady-state approximation is justified. This high-latitude dominance in controlling atmospheric CO_2 is illustrated in **Figure 6**, which demonstrates that atmospheric CO_2 can vary between about 160 and 425 ppm by simply opening or closing this high-latitude window. The exact magnitude of the atmospheric CO_2 change depends, to a substantial degree, on the details of the model, but the dominance of high-latitude processes in controlling atmospheric CO_2 is also found in much more complex and spatially explicit models. As a result, a change in the carbon cycle in the high latitudes continues to be the leading explanation for the substantially lower atmospheric CO_2 concentrations during the ice ages.

NP and NPZ Models, or What Simple Ecosystem Models Can Say about Oceanic Productivity

So far, we have represented the ecological, chemical, and physical processes that control the production of organic matter and its subsequent export with a single parameter, Φ. Clearly, in order to assess how marine biology responds to climate change and other perturbations, it is necessary to resolve these processes in much more detail. Two fundamentally different approaches have been developed to model such ecological processes: concentration-based models and individually based models (IBMs). In the latter, the model's equations represent the growth and losses of individual organisms, and the model then simulates the evolution of a large number of these individuals through space and time. This approach is most commonly used for the modeling of organisms at higher trophic levels, where life cycles play an important role (e.g., models of zooplankton, fish, and marine mammals). In contrast, concentration-based models are almost exclusively used for the modeling of the lower trophic levels, in particular to represent the interaction of light, nutrient, and grazing in controlling the growth of phytoplankton, that is, marine primary production.

The simplest such model considers just one limiting nutrient, N, and one single phytoplankton group, P (see also **Figure 7(a)**):

$$SMS(P) = V_{max} \cdot \frac{N}{K_N + N} \cdot P - \lambda_P \cdot P$$
$$SMS(N) = -V_{max} \cdot \frac{N}{K_N + N} \cdot P + \mu_P \cdot \lambda_P \cdot P \qquad [9]$$

where the first term on the right-hand side of the phytoplankton eqn [9] is net phytoplankton growth (equal to net primary production) modeled here as a function of a nutrient-saturated growth rate V_{max}, and a hyperbolic dependence on the *in situ* nutrient concentration (Monod-type), with a single parameter, the half-saturation constant, K_N. The second term is the net loss due to senescence, viral infection, and grazing, modeled here as a linear process with a loss rate λ_P. A fraction μ_P of the phytoplankton loss is assumed to be regenerated inside the euphotic zone, while a fraction $(1 - \mu_P)$ is exported to depth (**Figure 7(a)**). These two parametrizations for phytoplankton growth and losses reflect a typical situation for marine ecosystem models in that the growth terms are substantially more elaborate, reflecting the interacting influence of various controlling parameters, whereas the loss processes are highly simplified. This situation also reflects the fact that the processes controlling the growth of marine organisms are often more amenable to experimental studies, while the loss processes are much harder to investigate with careful experiments.

When the SMS terms of [10] are inserted into the full tracer conservation eqn [1], the resulting equations form a set of coupled partial differential equations with a number of interesting consequences in steady state as shown in **Figure 8(a)**. Below a certain nutrient concentration threshold, phytoplankton growth is smaller than its loss term, so that the resulting steady-state solution is $P = 0$, that is, no phytoplankton. Above this threshold, the abundance of phytoplankton (and hence primary production) increases with increasing nutrient supply in a linear manner. Phytoplankton is successful in reducing the dissolved inorganic nutrient concentration to low levels, so that the steady state is characterized by most nutrients residing in organic form in the phytoplankton pool. This NP model reflects a situation where marine productivity is limited by bottom-up processes, that is, the supply of the essential nutrients. Comparison with the nutrient and phytoplankton distribution shown in (**Figure 2(b)**) shows that this model could explain the observations in the subtropical open ocean, where nutrients are indeed drawn down to very low levels. However, the NP model would also predict relatively high P levels to go along with the low nutrient levels, which is clearly not observed (**Figure 2(a)**). In addition, this NP model also fails clearly to explain the high-nutrient regions of the high latitudes. However, we have so far neglected the limitation by light, as well as the impact of zooplankton grazing.

The addition of a zooplankton compartment to the NP model (**Figure 7(b)**) dramatically shifts the nutrient allocation behavior (**Figure 8(b)**). In this case, as the nutrient loading increases, the phytoplankton abundance gets capped at a certain level by zooplankton grazing. Due to the reduced levels of biomass, the phytoplankton is then no longer able to consume all nutrients at high-nutrient loads, so that an increasing fraction of the supplied nutrients remains unused. This top-down limitation situation could therefore, in part, explain why macronutrients in certain regions remain untapped. Most recent research suggests that micronutrient (iron) limitation, in conjunction with zooplankton grazing, plays a more important role in causing these high-nutrient/low chlorophyll regions. Another key limitation is light, which has important consequences for the seasonal and depth evolution of phytoplankton.

Global 3-D Modeling of Ocean Biogeochemistry/Ecology

The coupling of relatively simple NPZ-type models to global coarse-resolution 3-D circulation models has proven to be a challenging task. Perhaps the most important limitation of such NPZ models is the fact that all phytoplankton in the ocean are represented by a single phytoplankton group, which is grazed upon by a single zooplankton group. This means that the tiny phytoplankton that dominate the relatively nutrient-poor central gyres of the ocean, and the

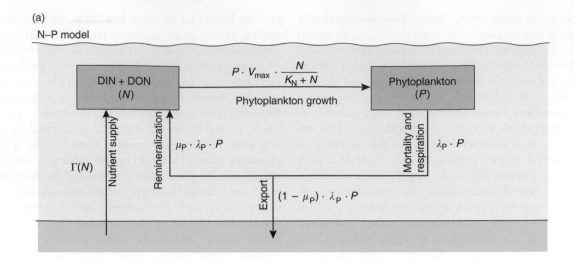

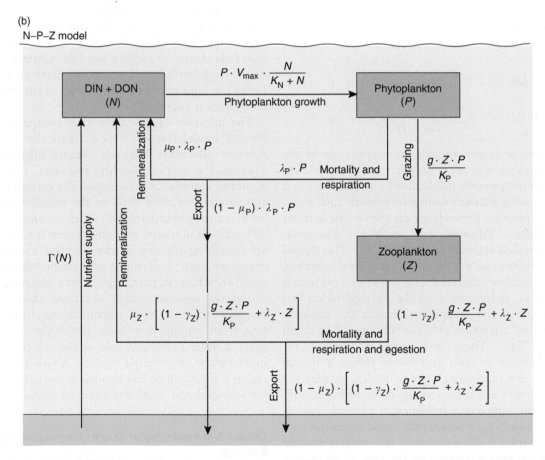

Figure 7 A schematic illustration of (a) a nitrate–phytoplankton model and (b) a nitrate–phytoplankton–zooplankton model. The mathematical expressions associated with each arrow represent typical representations for how the individual processes are parametrized in such models. V_{max}, maximum phytoplankton growth rate; K_N, nutrient half saturation concentration; g, maximum zooplankton growth rate; K_P, half saturation concentration for phytoplankton grazing; γ_Z, zooplankton assimilation efficiency; λ_P, phytoplankton mortality rate; λ_Z, zooplankton mortality rate; μ_P, fraction of dead phytoplankton nitrogen that is remineralized in the euphotic zone; μ_Z, as μ_P, but for zooplankton. Adapted from Sarmiento JL and Gruber N (2006) *Ocean Biogeochemical Dynamics*, 526pp. Princeton, NJ: Princeton University Press.

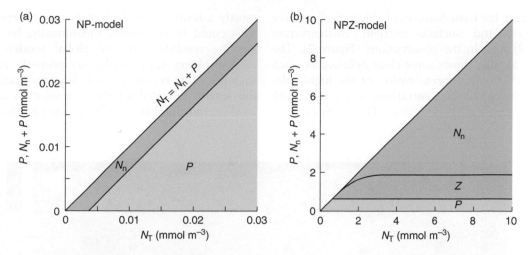

Figure 8 Nitrogen in phytoplankton (P), zooplankton (Z), and nitrate (N_n) as a function of total nitrogen. (a) Results from a two-component model that just includes N and P. (b) Results from a three-component model that is akin to that shown in **Figure 7(b)**. The vertical axis is concentration in each of the components plotted as the cumulative amount. Adapted from Sarmiento JL and Gruber N (2006) *Ocean Biogeochemical Dynamics*, 526pp. Princeton, NJ: Princeton University Press.

large phytoplankton that dominate the highly productive upwellling regions are represented with a single group, whose growth characteristics cannot simultaneously represent both. An additional challenge of the large-scale coupling problem is the interaction of oceanic circulation/mixing with the marine ecological and biogeochemical processes. Often, small deficiencies in the physical model get amplified by the ecological/biogeochemical model, so that the resulting fields of phytoplankton abundance may diverge strongly from observations.

The latter problem is addressed by improving critical aspects of the physical model, such as upper ocean mixing, atmospheric forcing, resolution, and numerical tracer transport algorithms, while the first problem is addressed by the consideration of distinct plankton functional groups. These functional groups distinguish themselves by their different size, nutrient requirement, biogeochemical role, and several other characteristics, but not necessarily by their taxa. Currently existing models typically consider the following phytoplankton functional groups: small phytoplankton (nano- and picoplankton), silicifying phytoplankton (diatoms), calcifying phytoplankton (coccolithophorids), N_2-fixing phytoplankton, large (nonsilicifying) phytoplankton (e.g., dinoflagellates and phaeocystis), with the choice of which group to include being driven by the particular question at hand. The partial differential equations for the different phytoplankton functional groups are essentially the same and follow a structure similar to [9], but are differentiated by varying growth parameters, nutrient/light requirements, and susceptibility to grazing. At present, most applications include between five and 10

phytoplankton functional groups. Some recent studies have been exploring the use of several dozen functional groups, whose parameters are generated stochastically with some simple set of rules that are then winnowed via competition for resources among the functional groups. Since the different phytoplankton functional groups are also grazed differentially, different types of zooplankton generally need to be considered as well, though fully developed models with diverse zooplankton functional groups are just emerging. An alternative is to use a single zooplankton group, but with changing grazing/growth characteristics depending on its main food source. A main advantage of models that build on the concept of plankton functional groups is their ability to switch between different phytoplankton community structures and to simulate their differential impact on marine biogeochemical processes. For example, the nutrient poor subtropical gyres tend to be dominated by nano- and picoplankton, whose biomass tends to be tightly capped by grazing by very small zooplankton. This prevents the group forming blooms, even under nutrient-rich conditions. By contrast, larger phytoplankton, such as diatoms, can often escape grazing control (at least temporarily) and form extensive blooms. Such phytoplankton community shifts are essential for controlling the export of organic matter toward the abyss, since the export ratio, that is, the fraction of net primary production that is exported, tends to increase substantially in blooms.

Results from the coupling of such a multiple functional phytoplankton ecosystem model and of a relatively simple biogeochemical model to a global 3-D circulation model show substantial success in representing the basic features of chlorophyll, air–sea CO_2 flux, and surface nutrient concentration (**Figure 9**) seen in the observations (**Figure 2**). The comparison also shows some clear deficiencies, such as the lack of the representation of the highly elevated chlorophyll concentrations associated with many continental boundaries. This deficiency is mostly a result of the lack of horizontal resolution. This could be overcome, theoretically, by increasing the resolution of the global models to the eddy-resolving scale needed to properly represent such boundary regions ($c.$ 5–10 km or so, rather than the several hundred kilometer resolution currently used for most global-scale biogeochemical/ecological

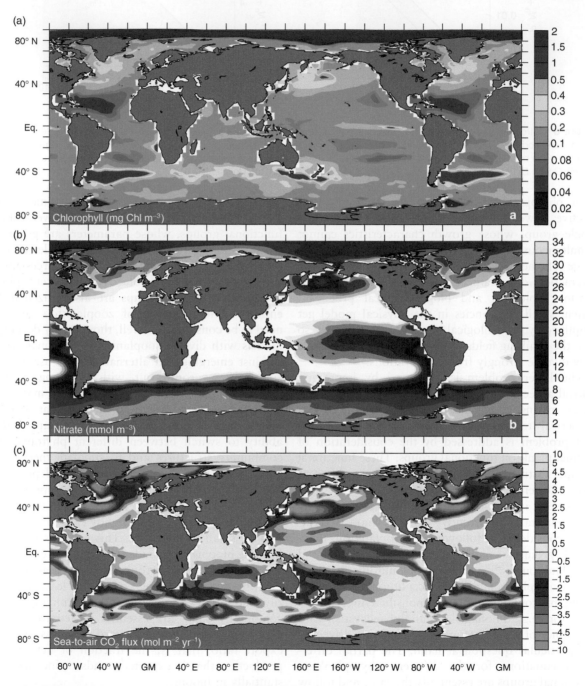

Figure 9 Global maps of model simulated key biogeochemical/ecological targets (cf. **Figure 2**) from the National Center for Atmospheric Research (NCAR) Community Climate System Model (CCSM): (a) annual mean distribution of near-surface chlorophyll (mg Chl m^{-3}); (b) annual mean surface distribution of nitrate (mmol^{-3}); and (c) annual mean air–sea CO_2 flux (mol m^{-2} yr^{-1}) for the period 1990 until 2000. All model results stem from a simulation, in which a multiple functional phytoplankton group model was coupled to a global 3-D ocean circulation model. Results provided by S.C. Doney and I. Lima, WHOI.

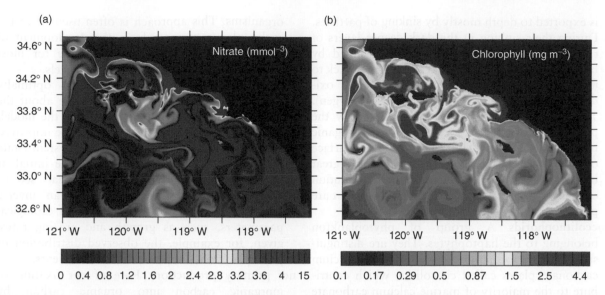

Figure 10 Snapshots of (a) surface nitrate and (b) surface chlorophyll as simulated by a 1-km, that is, eddy-resolving, regional model for the Southern California Bight. The snapshots represent typical conditions in response to a spring-time upwelling event in this region. The model consists of a multiple phytoplankton functional group model that has been coupled to the Regional Ocean Modeling System (ROMS). Results provided by H. Frenzel, UCLA.

simulations). This is currently feasible for only very short periods, so that limited domain models are often used instead.

Eddy-resolving Regional Models

Limited domain models, as shown in **Figure 10** for the Southern California Bight, that is, the southern-most region of the West Coast of the US, can be run at much higher resolutions over extended periods, permitting the investigation of the processes occurring at the kilometer scale. The most important such processes are meso- and submesoscale eddies and other manifestations of turbulence, which are ubiquitous features of the ocean. The coupling of the same ecosystem/biogeochemical model used for the global model shown in **Figure 9** to a 1-km resolution model of the Southern California Bight reveals the spatial richness and sharp gradients that are commonly observed (**Figure 10**) and that emerge from the intense interactions of the physical and biogeochemical/ecological processes at the eddy scale. Since such mesoscale processes are unlikely to be resolved at the global scale for a while, a current challenge is to find parametrizations that incorporate the net impact of these processes without actually resolving them explicitly.

Future Directions

The modeling of marine biogeochemical/ecological processes is a new and rapidly evolving field, so that predictions of future developments are uncertain. Nevertheless, it is reasonable to expect that models will continue to evolve along the major axes of complexity, including, for example, the consideration of higher trophic levels (**Figure 4**). One also envisions that marine biogeochemical/ecological models will be increasingly coupled to other systems. A good example are Earth System Models that attempt to represent the entire climate system of the Earth including the global carbon cycle. In those models, the marine biogeochemical/ecological models are just a small part, but interact with all other components, possibly leading to complex behavior including feedbacks and limit cycles. Another major anticipated development is the increasing use of inverse modeling approaches to optimally make use of the increasing flow of observations.

Glossary

advection The transport of a quantity in a vector field, such as ocean circulation.

anthropogenic carbon The additional carbon that has been released to the environment over the last several centuries by human activities including fossil-fuel combustion, agriculture, forestry, and biomass burning.

biogeochemical loop, great A set of key processes that control the distribution of bioreactive elements in the ocean. The loop starts with the photosynthetic production of organic matter in the light-illuminated upper ocean, a fraction of which is exported to depth mostly by sinking of particles. During their sinking in the dark deeper layers of the ocean, this organic matter is consumed by bacteria and zooplankton that transform it back to its inorganic constituents while consuming oxidized components in the water (mostly oxygen). The great biogeochemical loop is closed by the physical transport and mixing of these inorganic bioreactive elements back to the near-surface ocean, where the process starts again. This great biogeochemical loop also impacts the distribution of many other elements, particularly those that are particle reactive, such as thorium.

coccolithophorids A group of phytoplankton belonging to the haptophytes. They are distinguished by their formation of small mineral calcium carbonate plates called coccoliths, which contribute to the majority of marine calcium carbonate production. An example of a globally significant coccolithophore is *Emiliania huxleyi*.

diatoms A group of phytoplankton belonging to the class Bacillariophyceae. They are one of the most common types of phytoplankton and distinguish themselves through their production of a cell wall made of amorphous silica, called opal. These walls show a wide diversity in form, but usually consist of two symmetrical sides with a split between them.

diffusion The transport of a quantity by Brownian motion (molecular diffusion) or turbulence (eddy diffusion). The latter is actually some form of advection, but since its net effect is akin to diffusion, it is often represented as a diffusive process.

export production The part of the organic matter formed in the surface layer by photosynthesis that is transported out of the surface layer and into the interior of the ocean by particle sinking, mixing, and circulation, or active transport by organisms.

models, concentration based Models, in which the biological state variables are given as numbers per unit volume, that is, concentration. This approach is often used when physical dispersion plays an important role in determining the biological and biogeochemical impact of these state variables. Concentration-based models are most often used to represent lower trophic levels in the ocean.

models, forward Models, in which time is integrated forward to compute the temporal evolution of the distribution of state variables in response to a set of initial and boundary conditions.

models, individually based Models, in which the biological state variables represent individual organisms or a small group of individual organisms. This approach is often used when life cycles play a major role in the development of these organisms, such as is the case for most marine organisms at higher trophic levels.

models, inverse Models that aim to optimally combine observations with knowledge about the workings of a system as embodied in the model. Solutions to such inverse models can be improved estimates of the current state of the system (state estimation), improved estimates of the initial or boundary conditions, or an optimal set of parameters. A typical example of an inverse model is the optimal determination of ecological parameters, such as growth and grazing rates, given, for example, the observed distribution of phytoplankton, zooplankton, and nutrients.

net primary production Rate of net fixation of inorganic carbon into organic carbon by autotrophic phytoplankton. This net rate is the difference between the gross uptake of inorganic carbon during photosynthesis and autotrophic respiration.

organisms, autotrophic An autotroph (from the Greek *autos* = self and *trophe* = nutrition) is an organism that produces organic compounds from carbon dioxide as a carbon source, using either light or reactions of inorganic chemical compounds, as a source of energy. An autotroph is known as a producer in a food chain.

organisms, heterotrophic A heterotroph (Greek *heterone* = (an)other and *trophe* = nutrition) is an organism that requires organic substrates to get its carbon for growth and development. A heterotroph is known as a consumer in the food chain.

plankton Organisms whose swimming speed is smaller than the typical speed of ocean currents, so that they cannot resist currents and are hence unable to determine their horizontal position. This is in contrast to nekton organisms that can swim against the ambient flow of the water environment and control their position (e.g., squid, fish, krill, and marine mammals).

plankton, phytoplankton Phytoplankton are the (photo)autotrophic components of the plankton. Phytoplankton are pro- or eukaryotic algae that live near the water surface where there is sufficient light to support photosynthesis. Among the more important groups are the diatoms, cyanobacteria, dinoflagellates, and coccolithophorids.

plankton, zooplankton Zooplankton are heterotrophic plankton that feed on other plankton. Dominant groups include small protozoans or metazoans (e.g., crustaceans and other animals).

See also

Carbon Cycle. Carbon Dioxide (CO₂) Cycle. Nitrogen Cycle. Ocean Carbon System, Modeling of. Phosphorus Cycle.

Further Reading

DeYoung B, Heath M, Werner F, Chai F, Megrey B, and Monfray P (2004) Challenges of modeling ocean basin ecosystems. *Science* 304: 1463–1466.

Doney SC (1999) Major challenges confronting marine biogeochemical modeling. *Global Biogeochemical Cycles* 13(3): 705–714.

Fasham M (ed.) (2003) *Ocean Biogeochemistry*. New York: Springer.

Fasham MJR, Ducklow HW, and McKelvie SM (1990) A nitrogen-based model of plankton dynamics in the oceanic mixed layer. *Journal of Marine Systems* 48: 591–639.

Glover DM, Jenkins WJ, and Doney SC (2006) Course No. 12.747: Modeling, Data Analysis and Numerical Techniques for Geochemistry. http://w3eos.whoi.edu/12.747 (accessed in March 2008).

Rothstein L, Abbott M, Chassignet E, *et al.* (2006) Modeling ocean ecosystems: The PARADIGM Program. *Oceanography* 19: 16–45.

Sarmiento JL and Gruber N (2006) *Ocean Biogeochemical Dynamics*, 526pp. Princeton, NJ: Princeton University Press.

Relevant Websites

http://www.uta.edu
– Interactive models of ocean ecology/biogeochemistry.

CONTINUOUS PLANKTON RECORDERS

A. John, Sir Alister Hardy Foundation for Ocean Science, Plymouth, UK
P. C. Reid, SAHFOS, Plymouth, UK

Introduction

The Continuous Plankton Recorder (CPR) survey is a synoptic survey of upper-layer plankton covering much of the northern North Atlantic and North Sea. It is the longest running and the most geographically extensive of any routine biological survey of the oceans. Over 4 000 000 miles of towing have resulted in the analysis of nearly 200 000 samples and the routine identification of over 400 species/groups of plankton. Data from the survey have been used to study biogeography, biodiversity, seasonal and inter-annual variation, long-term trends, and exceptional events. The value of such an extensive time-series increases as each year's data are accumulated. Some recognition of the importance of the CPR survey was achieved in 1999 when it was adopted as an integral part of the Initial Observing System of the Global Ocean Observing System (GOOS).

History

The CPR prototype was designed by Alister Hardy for operation on the 1925–27 Discovery Expedition to the Antarctic, as a means of overcoming the problem of patchiness in plankton. It consisted of a hollow cylindrical body tapered at each end, weighted at the front and with a diving plane, horizontal tail fins, and a vertical tail fin with a buoyancy chamber on top (**Figure 1A**). Hardy designed a more compact version with a smaller sampling aperture for use on merchant ships and this was first deployed on a commercial ship in the North Sea in September 1931 (**Figure 1B**). During the 1980s the design was modified further to include a box-shaped double tail-fin that provides better stability when deployed on the faster merchant ships of today (**Figure 1C**). The space within this tail-fin is used in some machines to accommodate physical sensors and flowmeters. The normal maximum tow distance for a CPR is approximately 450 nautical miles (834 km).

By the late 1930s there were seven CPR routes in the North Sea and one in the north-east Atlantic; in 1938 CPRs were towed for over 30 000 miles. After a break for the Second World War, the survey restarted in 1946 and expanded into the eastern North Atlantic. Extension of sampling into the western North Atlantic took place in 1958. The survey reached its greatest extent from 1962 to 1972 when CPRs were towed for at least 120 000 nautical miles annually. Sampling in the western Atlantic, which had been suspended due to funding problems in 1986, recommenced in 1991 and is still ongoing. **Figure 2A** shows the extent of the survey in 1999.

Initially based at the University College of Hull, the survey moved to Leith, Edinburgh in 1950 under the management of the Scottish Marine Biological Association (now the Scottish Association for Marine Science). In 1977 it finally moved to Plymouth as part of the Institute for Marine Environmental Research (now Plymouth Marine Laboratory). After a short period of uncertainty in the late 1980s, when the continuation of the survey was threatened, the Sir Alister Hardy Foundation for Ocean Science (SAHFOS) was formed in November 1990 to operate the survey. Since 1931 more than 200 merchant ships, ocean weather ships, and coastguard cutters – known as 'ships of opportunity' – from many nations have towed CPRs in a voluntary capacity to maintain the survey. The Foundation is greatly indebted to the captains and crews of all these towing ships and their shipping and management companies, without whom the survey could not continue.

During the 1990s CPRs were towed by SAHFOS in several other areas, including the Mediterranean (1998–99), the Gulf of Guinea (1995–99), the Baltic (1998–99), and the Indian Ocean (1999). A separate survey by the National Oceanic and Atmospheric Administration/National Marine Fisheries Service using CPRs along the east coast of the USA off Narragansett has been running since 1974; CPRs are currently towed on two routes in the Middle Atlantic Bight. Following a successful 2000 mile trial tow in the north-east Pacific from Alaska to California in July–August 1997, a 2-year survey by SAHFOS using CPRs in the north-east Pacific started in March 2000. In addition to five tows per year on the Alaska–California route, there is one 3000 mile tow annually east–west from Vancouver to the north-west Pacific (**Figure 2B**). A 'sister' survey, situated in the Southern Ocean south of Australia between 60°E and 160°E, is operated by the Australian Antarctic Division. In this survey CPRs have been deployed since the early 1990s on voyages between Tasmania and stations in the Antarctic.

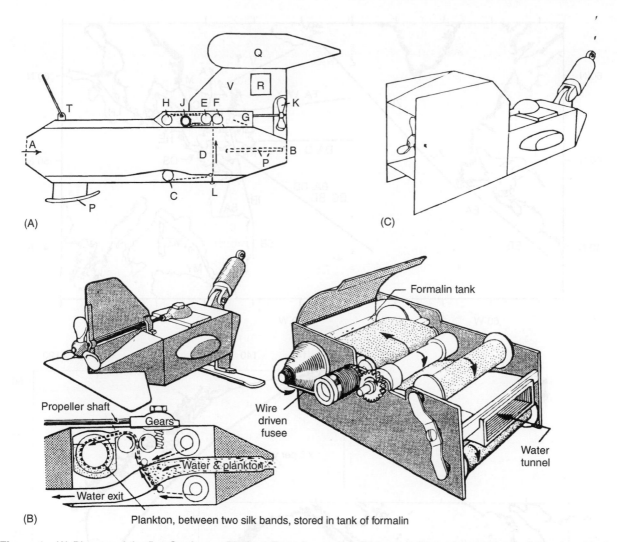

Figure 1 (A) Diagram of the first Continuous Plankton Recorder used on 'Discovery'. (Reproduced with permission from Hardy, 1967). (B) The 'old' CPR, used up to around 1983, showing the internal filtering mechanism. (C) The CPR in current use, with the 'box' tail-fin.

As the operator of a long-term international survey, which has sampled in most of the world's oceans, SAHFOS regularly trains its own staff in plankton identification. In recent years SAHFOS has also trained scientists from the following 10 countries: Benin, Cameroon, Côte d'Ivoire, France, Finland, Ghana, Italy, Nigeria, Thailand, and the USA.

The Database and Open Access Data Policy

The CPR database is housed on an IBM-compatible PC and stored in a relational Microsoft Access DATABASE system. Spatial and temporal data are stored for every sample analysed by the CPR survey since 1948. This amounts to >175 000 samples, with around 400 more samples added per month. There are more than two million plankton data points in the database, which also contains supporting information, including sample locations, dates and times of samples, a taxon catalog, and analyst details. In the near future it will also hold additional conductivity, temperature, and depth (CTD) data. Routine processing procedures ensure that, despite various operational difficulties, the previous year's data are usually available in the database within 9 months.

In 1999 SAHFOS adopted a new open access data policy, i.e. data are freely available to all users worldwide, although a reasonable payment may be incurred for time taken to extract a large amount of data. The only stipulation is that users have to sign a SAHFOS Data Licence Agreement. Details of the database can be found on the web site: http://www.npm.ac.uk/sahfos/. This site advertises the availability of data and allows requests for data to be made easily.

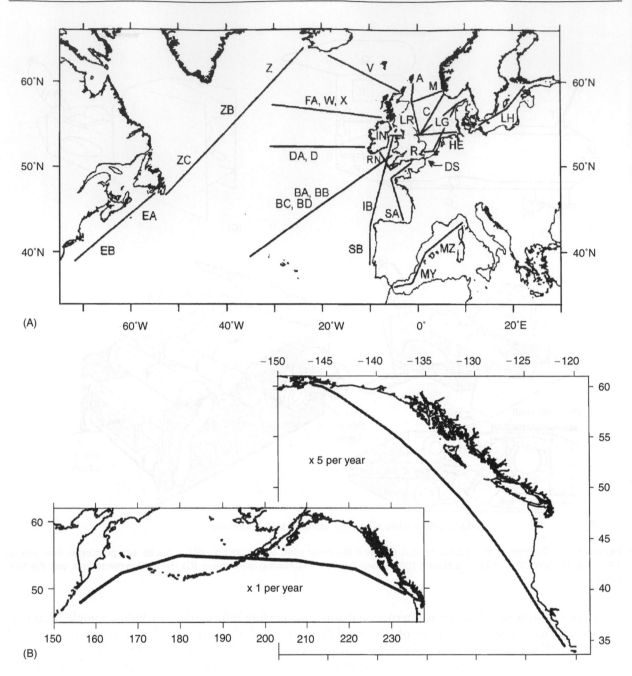

Figure 2 (A) CPR routes in 1999/2000. (B) CPR routes in the Pacific Ocean towed in 2000 and 2001.

The CPR Bibliography, which is available on the SAHFOS web site, lists over 500 references using results from the survey. During the early years many of the papers based on CPR data were published in the 'in-house' journal *Hull Bulletins of Marine Ecology*, which from 1953 onwards became the *Bulletins of Marine Ecology*; this was last published in 1980.

Methods

Merchant ships of many nations tow CPRs each month along 20–25 standard routes (**Figure 2A**) at a depth of 6–10 m. Water enters the CPR through a 12.7 mm square aperture and travels down a tunnel that expands to a cross-section of 50×100 mm, where it passes through a silk filtering mesh with a mesh size of approximately 280 µm. The movement of the CPR through the water turns a propeller that drives a set of rollers and moves the silk across the tunnel. At the top of the tunnel the filtering silk is joined by a covering silk and both are wound onto a spool located in a storage chamber containing formaldehyde solution. The CPRs are then returned to SAHFOS in Plymouth for examination. The green coloration of each silk is visually assessed by reference to a standard color scale; this is known as 'Phytoplankton Color' and gives a crude measure of total phytoplankton biomass. The silks are then cut into sections corresponding to 10 nautical miles (18.5 km) of tow and are distributed randomly to a team of 10–12 analysts. The volume of water filtered per 10-nautical-mile sample is approximately $3\,m^3$. Phytoplankton, small zooplankton (<2 mm in size) and larger zooplankton (>2 mm) are then identified and counted in a three-stage process. Over 400 different taxa are routinely identified during the analysis of samples and the recent expansion of the survey into tropical waters and the Pacific Ocean will certainly increase this figure.

A detailed and thorough quality control examination is carried out by the most experienced analyst on the completed analysis data. Apparently anomalous results are rechecked by the original analyst and the data are altered accordingly where necessary. This system ensures consistency of the data and acts as 'in-service' training for the less experienced analysts.

Instrumentation

On certain routes CPRs carry additional equipment to obtain physical data. In the past temperature has been recorded on certain routes in the North Sea using Braincon™ recording thermographs, prototype electronic packages, and Aquapacks™. Aquapacks record temperature, conductivity, depth, and chlorophyll fluorescence. These are now deployed on CPR routes off the eastern coast of the USA, in the southern Bay of Biscay and, until November 1999, in the Gulf of Guinea. Vemco™ minilogger temperature sensors are used on routes from the UK to Iceland, and from Iceland to Newfoundland. In order to measure flow rate through the CPR, electromagnetic flowmeters are used on some routes. Such recording of key physical and chemical variables simultaneously with abundance of plankton enhances our ability to interpret observed changes in the plankton.

Results and Applications of the Data

The long-term time-series of CPR data acts as a baseline against which to measure natural and anthropogenic changes in biogeography, biodiversity, seasonal variation, inter-annual variation, long-term trends, and exceptional events. The results have applications to studies of eutrophication and are increasingly being applied in statistical analysis of plankton populations and modeling. Some examples are given below.

Another possible application, in the context of the new Pacific CPR programme, is an inter-comparison with data from the CalCOFI Program, the only other existing decadal-scale survey in the world sampling marine plankton. This survey has taken monthly or quarterly net samples from 1949 to the present over an extensive grid of stations off the west coast of California. In the majority of samples the zooplankton has been measured only as displacement volume, rather than being identified to species, but concurrently measured physical and chemical data are more extensive.

Biogeography of Marine Plankton

Much of the early work of the survey focused on biogeography. Using Principal Component Analysis, Colebrook was able to distinguish five main geographical distribution patterns in the plankton – northern oceanic, southern oceanic, northern intermediate, southern intermediate, and neritic. Two closely related species of calanoid copepod – *Calanus finmarchicus* and *C. helgolandicus* – which co-occur in the North Atlantic and are morphologically very similar, show very different distributions (**Figure 3**). *C. finmarchicus* is a cold-water species whose center of distribution lies in the north-west Atlantic gyre and the Norwegian Sea ('northern oceanic'). In contrast, *C. helgolandicus* is a warm–temperate water species occurring in the Gulf Stream, the Bay of Biscay and the North Sea ('southern intermediate'). These different distribution patterns are reflected in their life histories; *C. finmarchicus* overwinters in deep waters off the shelf edge, whereas *C. helgolandicus* overwinters in shelf waters.

A new species of marine diatom, *Navicula planamembranacea* Hendey, was first described from CPR samples taken in 1962. The species was found to have a wide distribution in the western North Atlantic from Newfoundland to Iceland.

An atlas of distribution of 255 species or groups (taxa) of plankton recorded by the CPR survey between 1958 and 1968 was published by the Edinburgh Oceanographic Laboratory in 1973. An updated version of this atlas, covering more than 40 years of CPR data and over 400 taxa, is in preparation.

Phytoplankton, Zooplankton, Herring, Kittiwake Breeding Data, and Weather

A study in the north-eastern North Sea found that patterns of four time-series of marine data and weather showed similar long-term trends. Covering

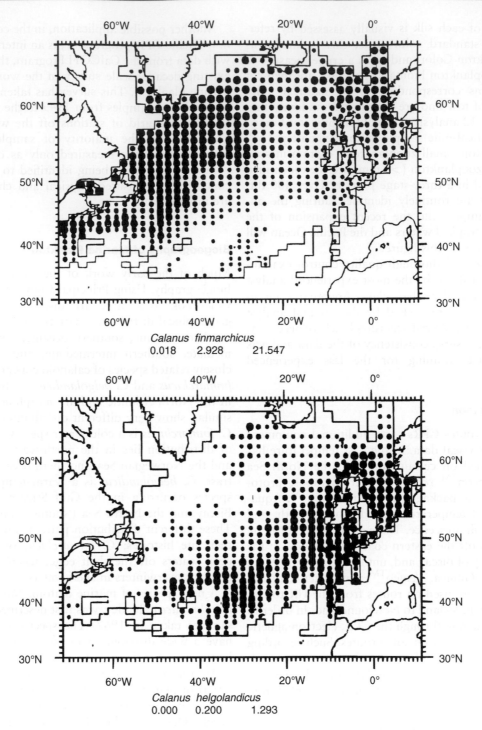

Calanus finmarchicus
0.018   2.928   21.547

Calanus helgolandicus
0.000   0.200   1.293

Figure 3 Distribution of *Calanus finmarchicus* and *C. helgolandicus* recorded in CPR samples from 1958 to 1994.

the period 1955–87, these trends were found in the abundance of phytoplankton and zooplankton (as measured by the CPR), herring in the northern North Sea, kittiwake breeding success (laying date, clutch size, and number of chicks fledged per pair) at a colony on the north-east coast of England, and the frequency of westerly weather (**Figure 4**).

The mechanisms behind the parallelism in these data over the 33-year period are still not fully understood.

Calanus and the North Atlantic Oscillation

The North Atlantic Oscillation (NAO) is a large-scale alternation of atmospheric mass between

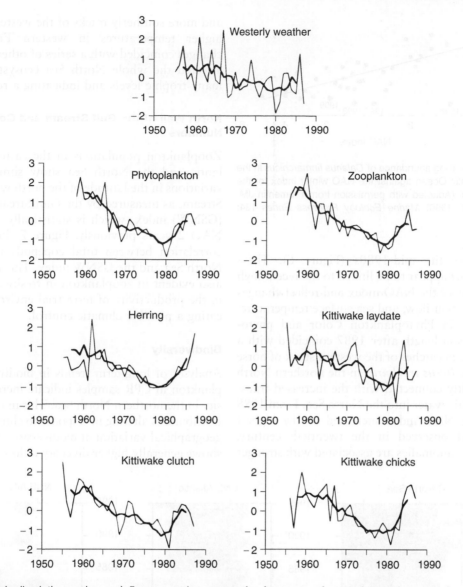

Figure 4 Standardized time-series and 5-year running means for frequency of westerly weather, and for abundances of phytoplankton, zooplankton, herring, and three parameters of kittiwake breeding (laying date, clutch size, and number of chicks fledged per pair), from 1955 to 1987. (Reproduced with permission from Aebischer NJ *et al.* (1990) *Nature* 347: 753–755.)

subtropical high surface pressure, centred on the Azores, and subpolar low surface pressures, centred on Iceland. The NAO determines the speed and direction of the westerly winds across the North Atlantic, as well as winter sea surface temperature. The NAO index is the difference in normalized sea level pressures between Ponta Delgadas (Azores) and Akureyri (Iceland). There is a close association between the abundance of *Calanus finmarchicus* and *C. helgolandicus* in the north-east Atlantic and this index (**Figure 5**). At times of heightened pressure difference between the Azores and Iceland, i.e. a high, positive NAO index, there is low abundance of *C. finmarchicus* and high abundance of *C. helgolandicus*; during a low, negative NAO index the

reverse is true. However, since 1995 this strong *Calanus*/NAO relationship has broken down and the causes of this are presently unknown. It suggests a change in the nature of the link between climate and plankton in the north-east Atlantic.

North Sea Ecosystem Regime Shift

Recent studies have shown changes in CPR Phytoplankton Color, a visual assessment of chlorophyll, for the north-east Atlantic and the North Sea. In the central North Sea and the central north-east Atlantic an increased season length was strikingly evident after the mid-1980s. In contrast, in the north-east Atlantic north of 59°N Phytoplankton Color

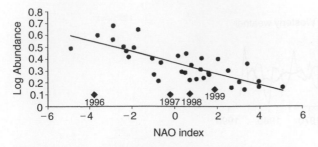

Figure 5 Annual log abundance of *Calanus finmarchicus* in the north-east Atlantic Ocean against the NAO winter index for the period 1962–99. (Adapted with permission from Fromentin JM, and Planque B (1996) *Marine Ecology Progress Series* 134: 111–118.)

declined after the mid-1980s (**Figure 6**). These changes in part appear to be linked to the recent high positive phase of the NAO index and reflect changes in mixing, current flow, and sea surface temperature. The increase in Phytoplankton Color and phytoplankton season length after 1987 coincided with a large increase in catches of the western stock of horse mackerel *Trachurus trachurus* in the northern North Sea, apparently connected with the increased transport of Atlantic water into the North Sea. From 1988 onwards the NAO index increased to the highest positive level observed in the twentieth century. Positive NAO anomalies are associated with stronger

and more southerly tracks of the westerly winds and higher temperatures in western Europe. These changes coincided with a series of other changes that affected the whole North Sea ecosystem, affecting many trophic levels and indicating a regime shift.

North Wall of the Gulf Stream and Copepod Numbers

Zooplankton populations in the eastern North Atlantic and the North Sea show similar trends to variations in the latitude of the north wall of the Gulf Stream, as measured by the Gulf Stream North Wall (GSNW) index, which is statistically related to the NAO 2 years previously. **Figure 7** shows the close correlation between total copepods in the central North Sea and the GSNW index. This relationship is also evident in zooplankton in freshwater lakes and in the productivity of terrestrial environments, indicating a possible climatic control.

Biodiversity

Analyses of long-term trends in biodiversity of zooplankton in CPR samples indicate increases in diversity in the northern North Sea. This may be related to distributions altering in response to climatic change as geographical variation in biodiversity of the plankton shows generally higher diversity at low latitudes than

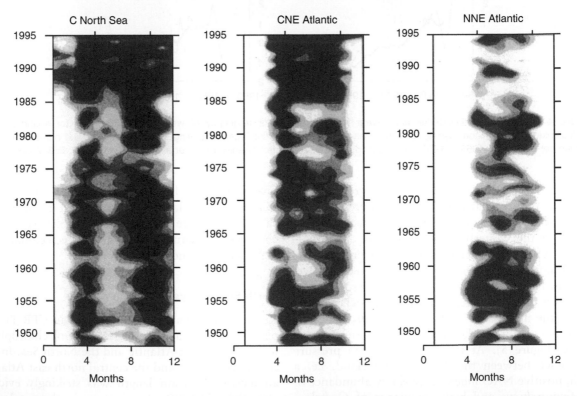

Figure 6 Contour plots of mean monthly Phytoplankton Color during 1948–95 for the central North Sea, and for the central and northern north-east Atlantic. (Reproduced with permission from Reid PC *et al.* (1998) *Nature* 391: 546.)

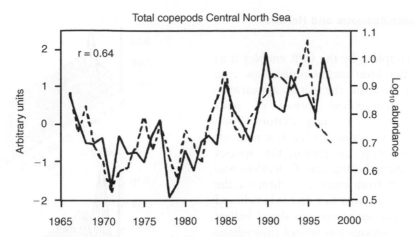

Figure 7 The latitude of the Gulf Stream (the GSNW index 'arbitrary units', broken line) compared with the abundance of total copepods in the central North Sea (solid line). Adapted with permission from Taylor AH *et al.* (1992) *Journal of Mar. Biol. Ass.* UK 72: 919–921.

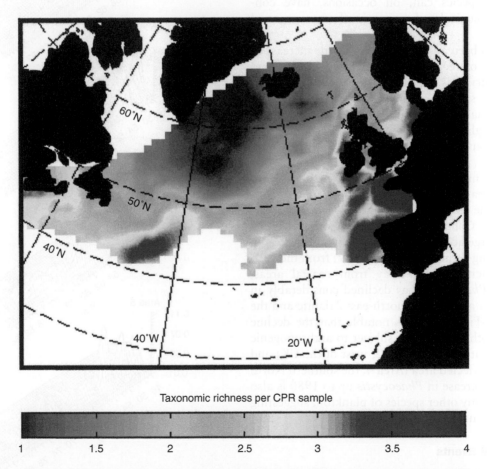

Taxonomic richness per CPR sample

Figure 8 The biodiversity (taxonomic richness) of calanoid copepods in the CPR sampling area. (Adapted with permission from Beaugrand *et al.* (2000) *Marine Ecology Progress Series* 204: 299–303.)

at high latitudes. Calanoid copepods are the dominant zooplankton group in the North Atlantic and the large data set from the CPR survey has been used to map their diversity. This has demonstrated a pronounced local spatial variability in biodiversity. Higher diversity was found in the Gulf Stream extension, the Bay of Biscay, and along the southern part of the European shelf. Cold water south of Greenland, east of Canada, and west of Norway was found to have the lowest diversity (**Figure 8**).

Monitoring for Nonindigenous and Harmful Algal Blooms

The regularity of sampling by the CPR enables it to detect changes in plankton communities. Few case histories exist that describe the initial appearance and subsequent geographical spread of nonindigenous species. In 1977 the large diatom *Coscinodiscus wailesii* was recorded for the first time off Plymouth, when mucilage containing this species was found to be clogging fishing nets. *C. wailesii* was previously known only from southern California, the Red Sea, and the South China Sea and it is believed that it arrived in European waters via ships' ballast water. Since then the species has spread throughout north-west European waters and has become an important contributor to North Sea phytoplankton biomass, particularly in autumn and winter. Such introduced species can, on occasions, have considerable ecological and economic effects on regional ecosystems.

There has been an apparent worldwide increase in the number of recorded harmful algal blooms and the CPR survey is ideally placed to monitor such events. The serious outbreak of paralytic shellfish poisoning that occurred in 1968 on the north-east coast of England was shown by CPR sampling to have been caused by the dinoflagellate *Alexandrium tamarense*.

Increased nutrient inputs into the North Sea since the 1950s have been linked with an apparent increase in the haptophycean alga *Phaeocystis*, particularly in Continental coastal regions where it produces large accumulations of foam on beaches. In contrast, long-term records (1946–87) from the CPR survey, which samples away from coastal areas, show that *Phaeocystis* has declined considerably in the open-sea areas of the north-east Atlantic and the North Sea (**Figure 9**). It is notable that the decline occurred both in areas not subject to anthropogenic nutrient inputs (Areas 1 and 2, west of the UK) and in the most affected area (Area 4, the southern North Sea). This decrease in *Phaeocystis* up to 1980 is also shown by many other species of plankton, suggesting a common causal relationship.

Exceptional Events

Dolioles are indicators of oceanic water and in CPR samples are normally found to the west and southwest of the British Isles; they occur only sporadically in the North Sea and are rarely recorded in the central or southern North Sea. On two occasions in recent years, in October–December 1989 and September–October 1997, the doliolid *Doliolum nationalis* was recorded in CPR samples taken in the

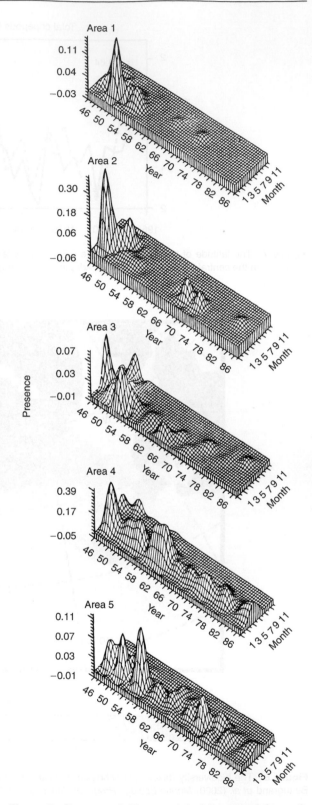

Figure 9 Presence of *Phaeocystis* in five areas of the northeast Atlantic Ocean and the North Sea. Data are plotted for each month for 1946–87 inclusive. (Reproduced with permission from Owens NJP *et al.* (1989) *Journal of Mar. Biol. Ass. UK* 69: 813–821.)

German Bight, accompanied by other oceanic indicator species, suggesting a strong influx of north-east Atlantic water into the North Sea. Both these occasions coincided with higher than average sea surface temperature and salinities.

Summary and the Future

The long-term time-series of CPR data have been used in many different ways:

- mapping the geographical distribution of plankton
- a baseline against which to measure natural and anthropogenically forced change, including eutrophication and climate change
- linking of plankton and environmental forcing
- detecting exceptional events in the sea
- monitoring for newly introduced and potentially harmful species.

In the future new applications of CPR data may include:

- use as 'sea-truthing' for satellites
- regional assessment of plankton biodiversity
- regional studies of responses to climate change
- as input variables to predictive modeling for fish stock and ecosystem management
- for construction and validation of new models comparing ecosystems of different regional seas.

The CPR survey has gathered nearly 70 years of data on marine plankton throughout the North Atlantic Ocean, and has recently extended into the North Pacific Ocean. Alister Hardy's simple concept in the 1920s has succeeded in providing us with a unique and valuable long-term data set. There is increasing worldwide concern about anthropogenic effects on the marine ecosystem, including eutrophication, overfishing, pollution, and global warming. The data in the CPR time-series is being used more and more widely to investigate these problems and now plays a significant role in our understanding of global ocean and climate change.

See also

Plankton. Plankton and Climate. Satellite Remote Sensing: Ocean Color. Satellite Remote Sensing of Sea Surface Temperatures.

Further Reading

Colebrook JM (1960) Continuous Plankton Records: methods of analysis, 1950–59. *Bulletins of Marine Ecology* 5: 51–64.

Gamble JC (1994) Long-term planktonic time series as monitors of marine environmental change. In: Leigh RA and Johnston AE (eds.) *Long-term Experiments in Agricultural and Ecological Sciences*, pp. 365–386. Wallingford: CAB International.

Glover RS (1967) The continuous plankton recorder survey of the North Atlantic. *Symp. Zoological Society of London* 19: 189–210.

Hardy AC (1939) Ecological investigations with the Continuous Plankton Recorder: object, plan and methods. *Hull Bulletins of Marine Ecology* 1: 1–57.

Hardy AC (1956) *The Open Sea: Its Natural History. Part 1: The World of Plankton*. London: Collins.

Hardy AC (1967) *Great Waters*. London: Collins.

IOC and SAHFOS (1991) *Monitoring the Health of the Ocean: Defining the Role of the Continuous Plankton Recorder in Global Ecosystem Studies*. Paris: UNESCO.

Oceanographic Laboratory, Edinburgh (1973) Continuous plankton records: a plankton atlas of the North Atlantic and the North Sea. *Bulletins of Marine Ecology* 7: 1–174.

Reid PC, Planque B, and Edwards M (1998) Is observed variability in the observed long-term results of the Continuous Plankton Recorder survey a response to climate change? *Fisheries Oceanography* 7: 282–288.

Warner AJ and Hays GC (1994) Sampling by the Continuous Plankton Recorder survey. *Progress in Oceanography* 34: 237–256.

MARINE MESOCOSMS

J. H. Steele, Woods Hole Oceanographic Institution, MA, USA

Controlled experiments are the basis of the scientific method. There are obvious difficulties in using this technique when dealing with natural communities or ecosystems, given the great spatial and temporal variability of their environment. On land the standard method is to divide an area of ground, say a field, into a large number of equal plots. Then with a randomized treatment, such as nutrient addition, it is possible to replicate growth of plants and animals over a season.

It is apparent that this approach is not possible in the open sea because of continuous advection and dispersion of water and the organisms in it. Bottom-living organisms are an exception, especially these living near shore, so there have been a wide range of experiments on rocky shores, salt marshes, and sea grasses. But even there, the critical reproductive period for most animals involves dispersion of the larvae in a pelagic phase. Also these experiments require continuous exchange of sea water.

For the completely pelagic plants and animals, short-term experiments – usually a few days – on single species are used to study physiological responses. There can be 24-hour experimental measurements of the rates of grazing of copepods on phytoplankton in liter bottles. But for studies of longer-term interactions, much larger volumes of

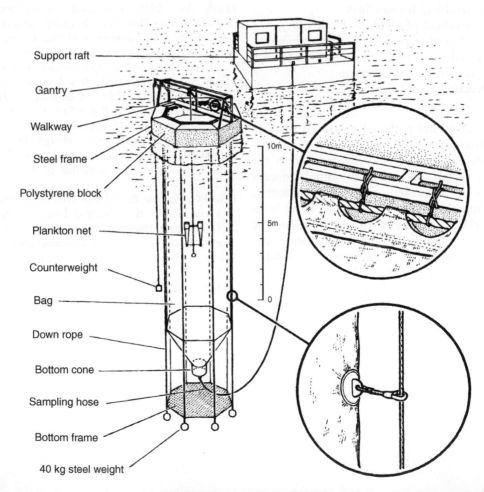

Figure 1 The design of a mesocosm used in Loch Ewe, Scotland for studies of the dynamics of plankton communities and of fish larval growth and mortality (adapted from Davies and Gamble, 1979).

water are necessary, to contain whole communities and to minimize wall effects of the containers.

To this end 'mesocosms' – containers much larger than can fit into the normal laboratory – have been used in a variety of designs and for a diversity of purposes. The first choice is whether to construct these on land, at the sea's edge, or to immerse them in the sea. The former has advantages in durability, ease of access, and re-use. There are constraints on the volumes that can be contained, difficulties in temperature control, and, especially, problems in transferring representative marine communities from the sea to the tanks. This approach was used originally in tall relatively narrow tanks to study populations of copepods and fish larvae; in particular to experiment on factors such as light that control vertical migration. Another use of such large tanks is to study the effect of pollutants on communities of pelagic and benthic organisms.

These shore-based tanks are limited by the weight of water, usually to volumes of 10–30 m^3. Enclosures immersed in the sea do not have this constraint. Instead the problems concern the strength of the flexible materials used for the walls in relation to currents and, especially, wind-induced waves. For this reason, such enclosures are placed in sheltered semienclosed places such as fiords. Nylon-reinforced polythene or vinyl reinforced with fabric have been used for these large 'test-tubes' containing 300–3000 m^3 (Figure 1). A column of water containing the natural plankton is captured by drawing up the bag from the bottom and fastening it in a rigid frame. The water and plankton can then be sampled by normal oceanographic methods.

It is possible to maintain at least three trophic levels – phytoplankton, copepods, and fish larvae – for 100 days or more. The only necessary treatment is addition of nutrients to replace those in the organic matter that sinks out. Such mesocosms can also be used for study of the fates and effects of pollutants.

These mesocosms have the obvious advantages associated with their large volumes – numerous animals for sampling, minimal wall effects. Temperature is regulated by exchange of heat through the walls. But they have various drawbacks. Not only is advection suppressed but vertical mixing decreases so that the outside physical conditions are not reproduced. The greatest disadvantage, however, is lack of adequate replication. There have been only three to six of these mesocosms available for any experiment and pairs did not often agree closely. Thus each tube represents an ecosystem on its own rather than a replicate of a larger community.

The need for experimental results at the community level represents an unresolved problem in biological oceanography. There are smaller-scale experiments continuing. Open mesh containers through which water and plankton pass can be a compromise for the study of small fish and fish larvae. It is now possible to mark a body of water with very sensitive tracers and follow the effects on plankton of the addition of nutrients, specifically iron, for several weeks. The concatenation of these results may have to depend on computer simulations.

See also

Iron Fertilization. Population Dynamics Models.

Further Reading

Cowan JH and Houde ED (1990) Growth and survival of bay anchovy in mesocosm enclosures. *Marine Ecology Progress Series* 68: 47–57.

Davies JM and Gamble JC (1979) *Experiments with large enclosed ecosystems. Philosophical Transcations of the Royal Society, B. Biological Sciences* 286: 523–544.

Gardner RH, Kemp WM, Kennedy VS, and Peterson JS (eds.) (2001) *Scaling Relations in Experimental Ecology.* New York: Columbia University Press.

Grice GD and Reeve MR (eds.) (1982) *Marine Mesocosms.* New York: Springer-Verlag.

Lalli CM (ed.) (1990) *Enclosed Experimental Marine Ecosystems: A Review and Recommendations. Coastal and Estuarine Studies 37.* New York: Springer-Verlag.

Underwood AJ (1997) *Experiments in Ecology.* Cambridge: Cambridge University Press.

MARINE SNOW

R. S. Lampitt, University of Southampton, Southampton, UK

Introduction

Marine snow is loosely defined as inanimate particles with a diameter greater than 0.5 mm. These particles sink at high rates and are thought to be the principal vehicles by which material sinks in the oceans. In addition to this high sinking rate they have characteristic properties in terms of the microenvironments within them, their chemical composition, the rates of bacterial activity and the fauna associated with them. These properties make such particles important elements in influencing the structure of marine food webs and biogeochemical cycles throughout the world's oceans.

Such an apparently simple definition, however, belies the varied and complex processes which exist in order to produce and destroy these particles (**Figure 1**). Similarly it gives no clue as to the wide variety of particulate material which, when aggregated together, falls into this category and to the significance of the material in the biogeochemistry of the oceans.

Marine snow particles are found throughout the world's oceans from the surface to the great depths although with a wide range in concentration reaching the highest levels in the sunlit euphotic zone where production is fastest. The principal features which render this particular class of material so important are:

1. high sinking rates such that they are probably the principal vehicles by which material is transported to depth;
2. a microenvironment which differs markedly from the surrounding water such that they provide a specialized niche for a wide variety of faunal groups and a chemical environment which is different from the surrounding water;
3. Elevated biogeochemical rates within the particles over that in the surrounding water;
4. Provision of a food source for organisms swimming freely outside the snow particles.

Historical Developments

The debate about how material is transported to the deep seafloor has been going on for over a century but at the beginning of the last century some surprisingly modern calculations by Hans Lohmann

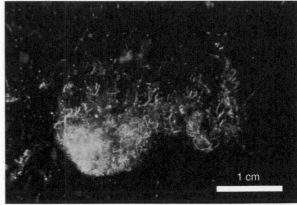

(A)

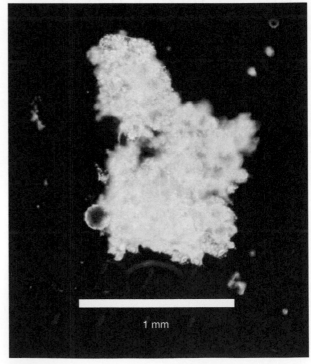

(B)

Figure 1 Examples of marine snow particles. (A) Aggregate comprising living chain-forming diatoms. Scale bar = 1 cm. (B) Aggregate containing a variety of types of material including phytoplankton cells rich in plant pigment (brown colored).

came to the conclusion that large particles must be capable of transporting material from the sunlit surface zone to the abyssal depths. This was based on his observations that near bottom water above the abyssal seabed sometimes contained a surprising range of thin-shelled phytoplankton species, some still in chains and with their fine spines well preserved. He thought that the fecal pellets from some larger members of the plankton (doliolids, salps, and pteropods) were the likely vehicles and in many cases he was entirely correct. The process of aggregation is now thought to involve a variety of different mechanisms of which fecal production is only one.

In 1951 Rachael Carson described the sediments of the oceans as the material from the most stupendous snowfall on earth and this prompted a group of Japanese oceanographers to describe as 'marine snow' the large particles they could see from the submersible observation chamber *Kuroshio* (**Figure 2**). This submersible was a cumbersome device and did not permit anything but the simplest of observations to be made. The scientists did, however, manage to collect some of the material and reported that its main components were the remains of diatoms, although with terrestrial material appearing to provide nuclei for formation.

In spite of these observations and the outstanding questions surrounding material cycles in the oceans, it was, until the late 1970s, a widely held belief that the deep-sea environment received material as a fine 'rain' of small particles. These, it was assumed, would take many months or even years to reach their ultimate destination on the seafloor. The separation of a few kilometers between the top and bottom of the ocean was thought sufficient to decouple the two ecosystems in a substantial way such that any seasonal variation in particle production at the surface would be lost by the time the settling particles reached the seabed. This now seems to have been a fundamental misconception. Part of the reason for this has been lack of understanding of the role of marine snow aggregates.

Methods of Examination

As stated above, the first use of the term was from submersible observations and *in situ* visual observation still serves a valuable role (**Figure 2**). This may be from manned submersibles or the rapidly evolving class of remotely operated vehicles or by subaqua divers. Photographic techniques have developed fast over the past decade and the standard photography systems which provided much of the currently available data on distribution are being replaced by

(A)

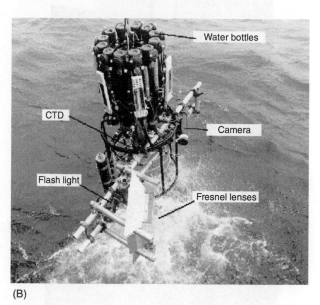

(B)

Figure 2 Devices for observing snow particles. (A) The submersible *Kuroshio* as used in the 1950s. (B) A photographic system incorporating a variety of other sensors and water bottles.

high definition video systems linked to fast computers which can categorize particles in near real time. Holographic techniques are also being developed for *in situ* use and these provide very high resolution images along with the three-dimension coordinates of the particles and organisms surrounding them.

An important goal in any of these studies is to be able to obtain undamaged samples of marine snow particles so that they can be examined under the microscope, chemically analysed, and used for experimentation. Some types of marine snow are, however, very fragile and the devices which are used to collect water often destroy their structure unless special precautions are taken in the design such that the water intake is very large and turbulence around it is reduced (**Figure 3A**). *In situ* pumping systems (**Figure 3B**) have been used for several years to collect material but in this case separation of the large marine snow particles which have distinctive characteristics of sinking rate, chemistry and biology from the far more abundant smaller particles is very difficult and may

lead to erroneous conclusions as a result of this forced aggregation of different types of particle.

Sediment traps are the principal means by which direct measurement can be made of the downward flux of material in the oceans and it may, therefore, be supposed that this provides a means to collect undisturbed marine snow particles (**Figure 3C**). Although the material collected in sediment traps may be considered as that which is sinking fast and is

(A)

(B)

(C)

(D)

Figure 3 Methods of collecting marine snow particles. (A) A 100 l water bottle 'The Snatcher'; (B) a large volume filtration system; (C) a sediment trap; (D) a subaqua diver.

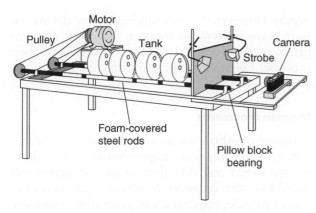

Figure 4 Device for producing marine snow particles in the laboratory.

certainly the material which mediates downward flux, the characteristics of individual marine snow particles or even particles with features in common can not be determined as the boundaries between particles are not retained in the sediment trap sample jars. Furthermore the preservatives and poisons which are usually employed in such devices (formaldehyde, mercuric chloride, etc.) prevent experimentation on the recovered material and some types of chemical analysis.

Laboratory production of marine snow particles was first achieved in the 1970s (**Figure 4**) enabling researchers to produce a plentiful supply of material under controlled and hence repeatable laboratory conditions. This is done by enclosing water samples which have high concentrations of phytoplankton in rotating water bottles for upwards of several hours. After a while the sheer forces encourage aggregation of the solid material. Much progress has been made with these particles and insights have been gained into the ways in which they are formed and the factors which cause their destruction.

We should now ask the most basic of questions about this important class of material: what is marine snow? how is it distributed in time and space? and why is it of such significance?

Characteristics of Marine Snow

Microscopic Composition

The microscopic composition of marine snow particles reflects to a major degree the processes responsible for their creation, and at any one location, the composition varies rather little; in some places it is dominated by diatoms or dinoflagelates and in other places by the mucus webs from larvaceans. Various staining techniques have been used to reveal the presence of transparent exopolymer particles in snow particles dominated by some groups of

phytoplankton and these are thought to be crucial components for the creation of some snow particles. Bacteria are invariably present in large numbers in snow particles.

Sinking Rate, Density, and Porosity

As illustrated in **Figure 5**, sinking rates increase with particle size and this extends throughout the range of particle sizes considered to be marine snow. The rates measured are consistent with other observations on the time delay between phytoplankton blooms at the surface of the ocean and the arrival of fresh material on the abyssal seafloor. Although the dry weight of individual particles increases with increasing size, the density decreases and porosity increases (**Figure 5**). The relationships between size and these physical characteristics is invariably a poor one with much scatter in the data points, a factor which introduces considerable uncertainty in trying to deduce important rates such as downward particulate flux from the size distribution of particles. This variability is due in large part to the variety of sources of material which comprise the final aggregated snow particle, but also to the different processes which contribute to the aggregation mechanism. The processes occurring within the snow particles, such as microbial degradation and photosynthesis, will also have an effect on the physical properties.

Chemical Composition

From the perspective of the biogeochemical processes occurring in the ocean, a knowledge of the chemical composition of snow particles and its variation with region, depth, and time are essential. The basic elemental composition in terms of carbon and nitrogen shows an increase, as expected, with particle size, but the proportion of the dry weight which is carbon or nitrogen tends to show a slight decrease reflecting the incorporation of lithogenic material with time as the aggregate becomes larger (**Figure 5**). In general, the composition of snow particles is not very different from that of the smaller particles found in the same body of water and from which the snow particles have been formed and to which they contribute when they are fragmented. This reflects the frequent and rapid transformations between the different size classes in the sea.

Biological Processes Within the Marine Snow Particles

Biological processes all occur at higher rates within marine snow particles than in comparable volumes of water outwith the snow. Furthermore, in the case of microbial activity, the rates of growth are often

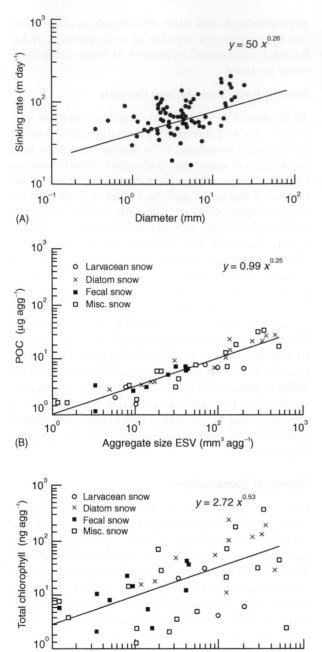

Figure 5 Relationships between marine snow particle size and (A) sinking rate, (B) organic carbon content and (C) chlorophyll content.

depths. However, the high sinking rates of the marine snow particles tend to increase the depth of remineralization and the balance between these two processes can not be described with confidence at the present time.

Microenvironments

Diffusion of solutes within any particulate entity will depend on the physical composition of the particle to a large extent and the effect of such diffusion will depend on the distances involved. With increasing size of particle, the chances of generating distinctive microenvironments within the particle increase. There are several examples in the literature where anoxic conditions have been found within snow particles as a result of enhanced oxygen consumption within the body of the particle and diffusion rates being insufficiently high to restore the concentrations outside the particle. These experiments have been done using laboratory-made snow particles or naturally collected specimens, but in both cases exclusion of metazoans which may fragment snow particles introduces some doubts about the frequency of anoxia in the natural environment. If proved to be common in 'the wild', the effect is to create microenvironments within the particle which will have a profound effect on the chemical and biological processes which take place within it.

Adsorption and Scavenging

During descent through the water column, sinking particles accumulate smaller particles in their path and this scavenging is probably an important means by which small particles with very low sinking rates are transported to depth. Layers of fine particles commonly referred to as nepheloid layers may be reduced in intensity by the descent through them of sticky marine snow particles. Similarly, as with any solid surface, dissolved chemicals may be adsorbed onto marine snow particles and hence be drawn down in the water column at a much faster rate than would be possible by diffusion or downwelling.

Food Source for Grazers, Enrichment Factors

Not only do marine snow particles provide an attractive habitat in which smaller fauna and flora can live, but they are also a food source for a variety of planktonic organisms and fish. The species most commonly found associated with snow particles tend to be copepods, particularly cyclopoids, but there are also examples of heterotrophic dinoflagellates, polychaete larvae, euphausiids, and amphipods. In

substantially higher than for comparable weights of smaller particles. Flagellate and ciliate populations also tend to be enhanced, presumably taking advantage of the elevated levels of bacterial activity. The effect of this is for snow particles to be sites where nutrient regeneration is enhanced. This does not, however, imply that formation of marine snow encourages nutrient regeneration at shallower

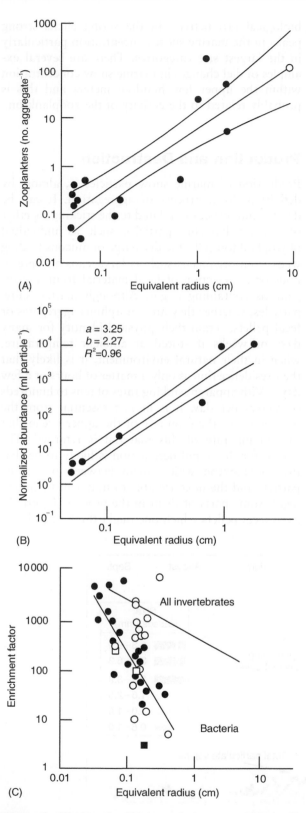

Figure 6 (A) Abundance of zooplankters associated with marine snow particles. (B) Abundance of all invertebrates after normalization to the size of the individual particles. (C) Enrichment factors for bacteria (•), ciliates (○), heterotrophic flagellates (□), and all invertebrates.

the case of the last of these groups one species, *Themisto compressa*, which was previously thought to be an obligate carnivore, was found to feed voraciously on marine snow particles. Recent experimental work has demonstrated the tracking behavior of zooplankton whereby they follow the odor trail left in the water by a sinking snow particle. Such sophisticated behavior suggests that snow particles are important food sources for some species of zooplankton. Enhancement of the concentration of bacteria or zooplankton associated with snow particles can be expressed as an 'enrichment factor' and, as shown in **Figure 6**, this factor appears to decrease with increasing size but at a different rate for the different faunal groups. The effect of this is that with increasing size, the metazoan zooplankton appear to become more important.

Distribution in Time and Space

Marine snow particles are found throughout the world's oceans in all parts of the water column. They are not uniformly distributed either in space or time but are usually found in higher concentrations in the upper water column and in the more productive regions of the oceans. Although it had been suspected since the early observations that their concentration decreases with increasing depth, this has been confirmed only recently. The profiles now becoming available do not however suggest a simple decrease. There is considerable structure, undoubtedly related to the processes of production, destruction, and sinking, which are related to the physics, biology, and chemistry of the water column and of the particles themselves. **Figure 7** shows an example of a profile from the north-east Atlantic. Bearing in mind the strong seasonal variation which can occur even well below the upper mixed layer and the different techniques employed by different researchers to obtain profiles, a common story seems to be emerging. Apart from profiles near to the continental slope where snow concentrations tend to increase near the seabed due to resuspension, there is generally a rapid fall in concentration over the top 100 m. Peak concentrations are not, however, found throughout the upper mixed layer but are located at its base, a feature which is directly related to the rates of production and loss of the marine snow particles in this highly dynamic part of the water column. Sinking rates may well decrease significantly in this part of the water column as the particles sink into water of higher density.

As might be expected for material which is so intimately related to the biological cycles of the oceans,

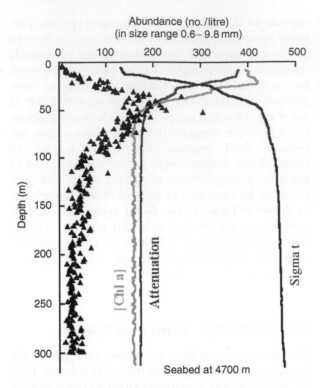

Figure 7 Vertical profile of marine snow concentrations (▲) in the north-east Atlantic (48°N 17°W).

biological productivity in the spring elicits strong peaks in the marine snow concentration particularly in the largest size categories. There are several examples of diel changes in marine snow concentration within the upper few hundred meters and this is probably related to the activity of the zooplankton.

Production and Destruction

Production of marine snow particles is, almost by definition, by a process of aggregation. It can be divided into processes related to the sticking together of smaller biogenic particles such as individual phytoplankton cells, the discharge of mucous feeding webs from organisms such as larvaceans and pteropods or the ejection of fecal material from any organisms containing a gut. Although marine snow particles, whether they are amorphous aggregates or fecal pellets, retain their physical identity for many days or weeks if stored at ambient temperature, when in their natural environment, it is likely that their residence time is only a matter of hours or a few days. With apparent sinking rates of tens to hundreds of meters per day, it is in fact essential from the perspective of the economy of the upper ocean that the sinking rate of this material is retarded. The reason for this rapid destruction is likely to be that the zooplankton which swim between one snow particle and the next are able to fragment them and ingest some parts of them in the process (**Figure 9**). Even fecal pellets which one might think of as being

a strong seasonal cycle in marine snow concentration is usually found even at depths below the euphotic zone. **Figure 8** shows the concentration of marine snow at 270 m depth in the open ocean of the northeast Atlantic. It can be seen that the period of highest

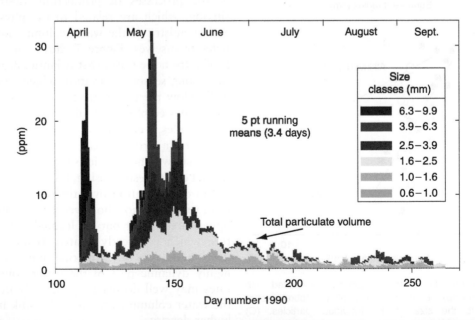

Figure 8 Seasonal change in the concentration of marine snow particles at 270 m depth in the north-east Atlantic (48°N 20°W). This is expressed as a volume concentration to emphasize the importance of the largest size category which are rare but contribute significantly.

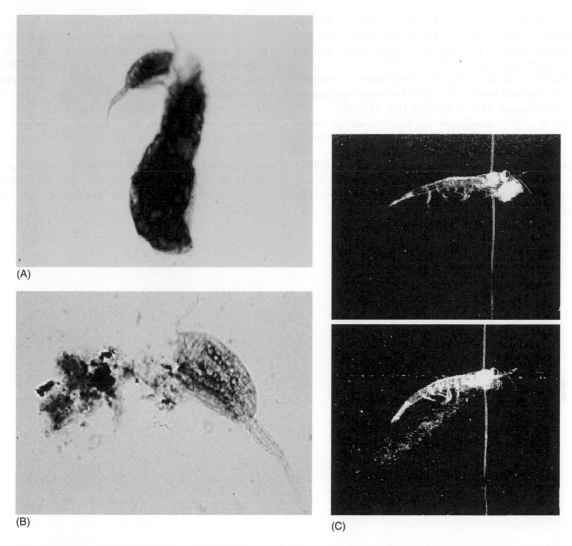

Figure 9 Zooplankton interactions with large particles: (A) feeding on a fecal pellet; (B) feeding on marine snow particle; (C) fragmenting marine snow particle.

an unattractive food resource, are readily broken up (**Figure 9**). Fragmentation by wave action seemed at one stage to be another likely mechanism but even for the more fragile particles, this now seems to be of minor importance as the shear rates are not usually adequate to break a significant number of particles.

Conclusion

Marine snow particles play a crucial role in regulating the supply of material to the deep sea due to their high sinking rates. They provide a microenvironment in which most rates of biogeochemical processes are enhanced and in which many heterotrophic organisms obtain a living. They are a food source for a variety of free-living organisms some of which have already been shown to adopt complex behaviour in order to locate them.

Further Reading

Alldredge A (1998) The carbon, nitrogen and mass content of marine snow as a function of aggregate size. *Deep-Sea Research(Part 1)* 45(4–5): 529–541.

Alldredge AL and Gotschalk C (1988) In situ settling behaviour of marine snow. *Limnology and Oceanography* 33: 339–351.

Alldredge AL, Granata TC, Gotschalk CC, and Dickey TD (1990) The physical strength of marine snow and its implications for particle disaggregation in the ocean. *Limnology and Oceanography* 35: 1415–1428.

Dilling L and Alldredge AL (2000) Fragmentation of marine snow by swimming macrozooplankton: a new process impacting carbon cycling in the sea. *Deep-Sea Research Part 1* 47(7): 1227–1245.

Dilling L, Wilson J, Steinberg D, and Alldredge A (1998) Feeding by the euphausiid *Euphausia pacifica* and the copepod *Calanus pacificus* on marine snow. *Marine Ecology Progress Series* 170: 189–201.

Gotschalk CC and Alldredge AL (1989) Enhanced primary production and nutrient regeneration within aggregated marine diatoms. *Marine Biology* 103: 119–129.

Kiorboe T (1997) Small-scale turbulence, marine snow formation, and planktivorous feeding. *Scientia Marina (Barcelona)* 61(suppl. 1): 141–158.

Lampitt RS, Hillier WR, and Challenor PG (1993) Seasonal and diel variation in the open ocean concentration of marine snow aggregates. *Nature* 362: 737–739.

Lampitt RS, Wishner KF, Turley CM, and Angel MV (1993) Marine snow studies in the northeast Atlantic: distribution, composition and role as a food source for migrating plankton. *Marine Biology* 116: 689–702.

Lohmann H (1908) On the relationship between pelagic deposits and marine plankton. *Int. Rev. Ges. Hydrobiol. Hydrogr* 1(3): 309–323 (in German).

Simon M, Alldredge AL, and Azam F (1990) Bacterial carbon dynamics on marine snow. *Marine Ecology Progress Series* 65(3): 205–211.

Suzuki N and Kato K (1953) Studies on suspended materials. marine snow in the sea. Part 1, sources of marine snow. *Bulletin of the Faculty of Fisheries of Hokkaido University* 4: 132–135.

ICE

SEA ICE: OVERVIEW

W. F. Weeks, Portland, OR, USA

Introduction

Sea ice, any form of ice found at sea that originated from the freezing of sea water, has historically been among the least-studied of all the phenomena that have a significant effect on the surface heat balance of the Earth. Fortunately, this neglect has recently lessened as the result of improvements in observational and operational capabilities in the polar ocean areas. As a result, considerable information is now available on the nature and behavior of sea ice as well as on its role in affecting the weather, the climate, and the oceanography of the polar regions and possibly of the planet as a whole.

Extent

Although the majority of Earth's population has never seen sea ice, in area it is extremely extensive: 7% of the surface of the Earth is covered by this material during some time of the year. In the northern hemisphere the area covered by sea ice varies between 8×10^6 and 15×10^6 km^2, with the smaller number representing the area of multiyear (MY) ice remaining at the end of summer. In summer this corresponds roughly to the contiguous area of the United States and to twice that area in winter, or to between 5% and 10% of the surface of the northern hemisphere ocean. At maximum extent, the ice extends down the western side of the major ocean basins, following the pattern of cold currents and reaching the Gulf of St. Lawrence (Atlantic) and the Okhotsk Sea off the north coast of Japan (Pacific). The most southerly site in the northern hemisphere where an extensive sea ice cover forms is the Gulf of Bo Hai, located off the east coast of China at 40°N. At the end of the summer the perennial MY ice pack of the Arctic is primarily confined to the central Arctic Ocean with minor extensions into the Canadian Arctic Archipelago and along the east coast of Greenland.

In the southern hemisphere the sea ice area varies between 3×10^6 and 20×10^6 km^2, covering between 1.5% and 10% of the ocean surface. The amount of MY ice in the Antarctic is appreciably less than in the Arctic, even though the total area affected by sea ice in the Antarctic is approximately a third larger than in the Arctic. These differences are largely caused by differences in the spatial distributions of land and ocean. The Arctic Ocean is effectively landlocked to the south, with only one major exit located between Greenland and Svalbard. The Southern Ocean, on the other hand, is essentially completely unbounded to the north, allowing unrestricted drift of the ice in that direction, resulting in the melting of nearly all of the previous season's growth.

Geophysical Importance

In addition to its considerable extent, there are good reasons to be concerned with the health and behavior of the world's sea ice covers. Sea ice serves as an insulative lid on the surface of the polar oceans. This suppresses the exchange of heat between the cold polar air above the ice and the relatively warm sea water below the ice. Not only is the ice itself a good insulator, but it provides a surface that supports a snow cover that is also an excellent insulator. In addition, when the sea ice forms with its attendant snow cover, it changes the surface albedo, α (i.e., the reflection coefficient for visible radiation) of the sea from that of open water ($\alpha = 0.15$) to that of newly formed snow ($\alpha = 0.85$), leading to a 70% decrease in the amount of incoming short-wave solar radiation that is absorbed. As a result, there are inherent positive feedbacks associated with the existence of a sea ice cover. For instance, a climatic warming will presumably reduce both the extent and the thickness of the sea ice. Both of these changes will, in turn, result in increases in the temperature of the atmosphere and of the sea, which will further reduce ice thickness and extent. It is this positive feedback that is a major factor in producing the unusually large increases in arctic temperatures that are forecast by numerical models simulating the effect of the accumulation of greenhouse gases.

The presence of an ice cover limits not only the flux of heat into the atmosphere but also the flux of moisture. This effect is revealed by the common presence of linear, local clouds associated with individual leads (cracks in the sea ice that are covered with either open water or thinner ice). In fact, sea ice exerts a significant influence on the radiative energy balance of the complete atmosphere–sea ice–ocean system. For instance, as the ice thickness increases in the range between 0 and 70 cm, there is an increase in the radiation absorption in the ice and a decrease in the ocean. There is also a decrease in the radiation adsorption by the total atmosphere–ice–ocean system. It is also known that the upper 10 cm of the ice can absorb over 50% of the total solar radiation, and that decreases in ice extent produce increases in atmospheric moisture or cloudiness, in turn altering the surface radiation budget and increasing the amount of precipitation. Furthermore, all the ultraviolet and infrared radiation is absorbed in the upper 50 cm of the ice; only visible radiation penetrates into the lower portions of thicker ice and into the upper ocean beneath the ice. Significant changes in the extent and/or thickness of sea ice would result in major changes in the climatology of the polar regions. For instance, recent computer simulations in which the ice extent in the southern hemisphere was held constant and the amount of open water (leads) within the pack was varied showed significant changes in storm frequencies, intensities and tracks, precipitation, cloudiness, and air temperature.

However, there are even less obvious but perhaps equally important air–ice and ice–ocean interactions. Sea ice drastically reduces wave-induced mixing in the upper ocean, thereby favoring the existence of a 25–50 m thick, low-salinity surface layer in the Arctic Ocean that forms as the result of desalination processes associated with ice formation and the influx of fresh water from the great rivers of northern Siberia. This stable, low-density surface layer prevents the heat contained in the comparatively warm (temperatures of up to $+3°C$) but more saline denser water beneath the surface layer from affecting the ice cover. As sea ice rejects roughly two-thirds of the salt initially present in the sea water from which the ice forms, the freezing process is equivalent to distillation, producing both a low-salinity component (the ice layer itself) and a high-salinity component (the rejected brine). Both of these components play important geophysical roles. Over shallow shelf seas, the rejected brine, which is dense, cold, and rich in CO_2, sinks to the bottom, ultimately feeding the deep-water and the bottom-water layers of the world ocean. Such processes are particularly effective in regions where large polynyas exist (semipermanent open water and thin-ice areas at sites where climatically much thicker ice would be anticipated).

The 'fresh' sea ice layer also has an important geophysical role to play in that its exodus from the Arctic Basin via the East Greenland Drift Stream represents a fresh water transport of 2366 km^3 y^{-1} (c. 0.075 Sv). This is a discharge equivalent to roughly twice that of North America's four largest rivers combined (the Mississippi, St. Lawrence, Columbia, and Mackenzie) and in the world is second only to the Amazon. This fresh surface water layer is transported with little dispersion at least as far as the Denmark Strait and in all probability can be followed completely around the subpolar gyre of the North Atlantic. Even more interesting is the speculation that during the last few decades this fresh water flux has been sufficient to alter or even stop the convective regimes of the Greenland, Iceland and Norwegian Seas and perhaps also of the Labrador Sea. This is a sea ice-driven, small-scale analogue of the so-called halocline catastrophe that has been proposed for past deglaciations, when it has been argued that large fresh water runoff from melting glaciers severely limited convective regimes in portions of the world ocean. The difference is that, in the present instance, the increase in the fresh water flux that is required is not dramatic because at near-freezing temperatures the salinity of the sea water is appreciably more important than the water temperature in controlling its density. It has been proposed that this process has contributed to the low near-surface salinities and heavy winter ice conditions observed north of Iceland between 1965 and 1971, to the decrease in convection described for the Labrador Sea during 1968–1971, and perhaps to the so-called 'great salinity anomaly' that freshened much of the upper North Atlantic during the last 25 years of the twentieth century. In the Antarctic, comparable phenomena may be associated with freezing in the southern Weddell Sea and ice transport northward along the Antarctic Peninsula.

Sea ice also has important biological effects at both ends of the marine food chain. It provides a substrate for a special category of marine life, the ice biota, consisting primarily of diatoms. These form a significant portion of the total primary production and, in turn, support specialized grazers and species at higher trophic levels, including amphipods, copepods, worms, fish, and birds. At the upper end of the food chain, seals and walruses use ice extensively as a platform on which to haul out and give birth to young. Polar bears use the ice as a platform while hunting. Also important is the fact that in shelf seas such as the Bering and Chukchi, which are well mixed in the winter, the melting of the ice cover in the spring lowers the surface salinity, increasing the stability of the water column. The reduced mixing concentrates phytoplankton in the near-surface photic zone, thereby enhancing the overall intensity of the spring bloom. Finally, there are the direct effects of sea ice on human activities. The most important of these are its barrier action in limiting the use of otherwise highly advantageous ocean routes between the northern Pacific regions and Europe and its contribution to the numerous operational difficulties that must be overcome to achieve the safe extraction of the presumed oil and gas resources of the polar shelf seas.

Properties

Because ice is a thermal insulator, the thicker the ice, the slower it grows, other conditions being equal. As sea ice either ablates or stops growing during the summer, there is a maximum thickness of first-year (FY) ice that can form during a specific year. The exact value is, of course, dependent upon the local climate and oceanographic conditions, reaching values of slightly over 2 m in the Arctic and as much as almost 3 m at certain Antarctic sites. It is also clear that during the winter the heat flux from areas of open water into the polar atmosphere is significantly greater than the flux through even thin ice and is as much as 200 times greater than the flux through MY ice. This means that, even if open water and thin ice areas comprise less than 1–2% of the winter ice pack, lead areas must still be considered in order to obtain realistic estimates of ocean–atmosphere thermal interactions.

If an ice floe survives a summer, during the second winter the thickness of the additional ice that is added is less than the thickness of nearby FY ice for two reasons: it starts to freeze later and it grows slower. Nevertheless, by the end of the winter, the second-year ice will be thicker than the nearby FY ice. Assuming that the above process is repeated in subsequent years, an amount of ice is ablated away each summer (largely from the upper ice surface) and an amount is added each winter (largely on the lower ice surface). As the year pass, the ice melted on top each summer remains the same (assuming no change in the climate over the ice), while the ice forming on the bottom becomes less and less as a result of the increased insulating effect of the thickening overlying ice. Ultimately, a rough equilibrium is reached, with the thickness of the ice added in the winter becoming equal to the ice ablated in the summer. Such steady-state MY ice floes can be layer cakes of ten or more annual layers with total thicknesses in the range 3.5–4.5 m. Much of the uncertainty in estimating the equilibrium thickness of such floes is the result of uncertainties in the oceanic heat flux. However, in sheltered fiord sites in the Arctic where the oceanic heat flux is presumed to be near zero, MY fast ice with thicknesses up to roughly 15–20 m is known to occur. Another important factor affecting MY ice thickness is the formation of melt ponds on the upper ice surface during the summer in that the thicknesses

and areal extent of these shallow-water bodies is important in controlling the total amount of short-wave radiation that is absorbed. For instance, a melt pond with a depth of only 5 cm can absorb nearly half the total energy absorbed by the whole system. The problem here is that good regional descriptive characterizations of these features are lacking as the result of the characteristic low clouds and fog that occur over the Arctic ice packs in the summer. Particularly lacking are field observations on melt pond depths as a function of environmental variables. Also needed are assessments of how much of the meltwater remains ponded on the surface of the ice as contrasted with draining into the underlying sea water. Thermodynamically these are very different situations.

Conditions in the Antarctic are, surprisingly, rather different. There, surface melt rates within the pack are small compared to the rates at the northern boundary of the pack. The stronger winds and lower humidities encountered over the pack also favor evaporation and minimize surface melting. The limited ablation that occurs appears to be controlled by heat transfer processes at the ice–water interface. As a result, the ice remains relatively cold throughout the summer. In any case, as most of the Antarctic pack is advected rapidly to the north, where it encounters warmer water at the Antarctic convergence and melts rapidly, only small amounts of MY ice remain at the end of summer.

Sea ice properties are very different from those of lake or river ice. The reason for the difference is that when sea water freezes, roughly one-third of the salt in the sea water is initially entrapped within the ice in the form of brine inclusions. As a result, initial ice salinities are typically in the range 10–12‰. At low temperatures (below $-8.7°C$), solid hydrated salts also form within the ice. The composition of the brine in sea ice is a unique function of the temperature, with the brine composition becoming more saline as the temperature decreases. Therefore, the brine volume (the volumetric amount of liquid brine in the ice) is determined by the ice temperature and the bulk ice salinity. Not only is the temperature of the ice different at different levels in the ice sheet but the salinity of the ice decreases further as the ice ages ultimately reaching a value of $\sim 3‰$ in MY ice. Brine volumes are usually lower in the colder upper portions of the ice and higher in the warmer, lower portions. They are particularly low in the above-sea-level part of MY ice as the result of the salt having drained almost completely from this ice. In fact, the upper layers of thick MY ice and of aged pressure ridges produce excellent drinking water when melted. As brine volume is the single most important parameter controlling the thermal, electrical, and mechanical properties of sea ice, these properties show associated large changes both vertically in the same ice sheet and between ice sheets of differing ages and histories. To add complexity to this situation, exactly how the brine is distributed within the sea ice also affects ice properties.

There are several different structural types of sea ice, each with characteristic crystal sizes and preferred crystal orientations and property variations. The two most common structural types are called congelation and frazil. In congelation ice, large elongated crystals extend completely through the ice sheet, producing a structure that is similar to that found in directionally solidified metals. In the Arctic, large areas of congelation ice show crystal orientations that are so similar as to cause the ice to have directionally dependent properties in the horizontal plane as if the ice were a giant single crystal. Frazil, on the other hand, is composed of small, randomly oriented equiaxed crystals that are not vertically elongated. Congelation is more common in the Arctic, while frazil is more common in the Antarctic, reflecting the more turbulent conditions characteristically found in the Southern Ocean.

Two of the more unusual sea ice types are both subsets of so-called 'underwater ice.' The first of these is referred to as platelet ice and is particularly common around margins of the Antarctic continent at locations where ice shelves exist. Such shelves not only comprise 30% of the coastline of Antarctica, they also can be up to 250 m thick. Platelet ice is composed of a loose open mesh of large platelets that are roughly triangular in shape with dimensions of 4–5 cm. In the few locations that have been studied, platelet ice does not start to develop until the fast ice has reached a thickness of several tens of centimeters. Then the platelets develop beneath the fast ice, forming a layer that can be several meters thick. The fast ice appears to serve as a superstrate that facilitates the initial nucleation of the platelets. Ultimately, as the fast ice thickens, it incorporates some of the upper platelets. In the McMurdo Sound region, platelets have been observed forming on fish traps at a depth of 70 m. At locations near the Filchner Ice Shelf, platelets have been found in trawls taken at 250 m. This ice type appears to be the result of crystal growth into water that has been supercooled a fraction of a degree. The mechanism appears to be as follows. There is evidence that melting is occurring on the bottom of some of the deeper portions of the Antarctic ice shelves. This results in a water layer at the ice–water interface that is not only less saline and therefore less dense than the underlying seawater, but also is exactly at its freezing point at that

depth because it is in direct contact with the shelf ice. When this water starts to flow outward and upward along the base of the shelf, supercooling develops as a result of adiabatic decompression. This in turn drives the formation of the platelet ice.

The second unusual ice type is a special type of frazil that results from what has been termed suspension freezing. The conditions necessary for its formation include strong winds, intense turbulence in an open water area of a shallow sea and extreme sub-freezing temperatures. Such conditions are characteristically found either during the initial formation of an ice cover in the fall or in regions where polynya formation is occurring, typically by newly formed ice being blown off of a coast or a fast ice area leaving in its wake an area of open water. When such conditions occur, the water column can become supercooled, allowing large quantities of frazil crystals to form and be swept downward by turbulence throughout the whole water column. Studies of benthic microfossils included in sea ice during such events suggest that supercooling commonly reaches depths of 20–25 m and occasionally to as much as 50 m. The frazil ice crystals that form occur in the form of 1–3 mm diameter discoids that are extremely sticky. As a result, they are not only effective in scavenging particulate matter from the water column but they also adhere to material on the bottom, where they continue to grow fed by the supercooled water. Such so-called anchor ice appears to form selectively on coarser material. The resulting spongy ice masses that develop can be quite large and, when the turbulence subsides, are quite buoyant and capable of floating appreciable quantities of attached sediment to the surface. There it commonly becomes incorporated in the overlying sea ice. In rivers, rocks weighing as much as 30 kg have been observed to be incorporated into an ice cover by this mechanism. Recent interest in this subject has been the result of the possibility that this mechanism has been effective in incorporating hazardous material into sea ice sheets, which can then serve as a long-distance transport mechanism.

Drift and Deformation

If sea ice were motionless, ice thickness would be controlled completely by the thermal characteristics of the lower atmosphere and the upper ocean. Such ice sheets would presumably have thicknesses and physical properties that change slowly and continuously from region to region. However, even a casual examination of an area of pack ice reveals striking local lateral changes in ice thicknesses and characteristics. These changes are invariably caused by ice movements produced by the forces exerted on the ice by winds and currents. Such motions are rarely uniform and lead to the build-up of stresses within ice sheets. If these stresses become large enough, cracks may form and widen, resulting in the formation of leads. Such features can vary in width from a few meters to several kilometers and in length from a few hundred meters to several hundred kilometers. As mentioned earlier, during much of the year in the polar regions, once a lead forms it is immediately covered with a thin skim of ice that thickens with time. This is an ever-changing process associated with the movement of weather systems as one lead system becomes inactive and is replaced by another system oriented in a different direction. As lead formation occurs at varied intervals throughout the ice growth season, the end result is an ice cover composed of a variety of thicknesses of uniform sheet ice.

However, when real pack ice thickness distributions are examined (**Figure 1**), one finds that there is a significant amount of ice thicker than the 4.5–5.0 m maximum that might be expected for steady-state MY ice floes. This thicker ice forms by the closing of leads, a process that commonly results in the piling of broken ice fragments into long, irregular features referred to as pressure ridges. There are many small ridges and large ridges are rare.

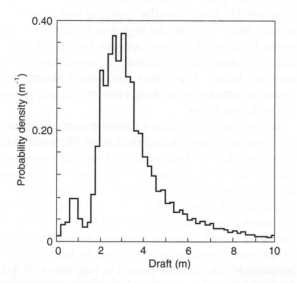

Figure 1 The distribution of sea ice drafts expressed as probability density as determined via the use of upward-looking sonar along a 1400 km track taken in April 1976 in the Beaufort Sea. All ice thicker than ~4 m is believed to be the result of deformation. The peak probabilities that occur in the range between 2.4 and 3.8 m represent the thicknesses of undeformed MY ice, while the values less than 1.2 m come from ice that ice that recently formed in newly formed leads.

Nevertheless, the large ridges are very impressive, the largest free-floating sail height and keel depth reported to date in the Arctic being 13 and 47 m, respectively (values not from the same ridge). Particularly heavily deformed ice commonly occurs in a band of ~150 km running between the north coast of Greenland and the Canadian Arctic Islands and the south coast of the Beaufort Sea. The limited data available on Antarctic ridges suggest that they are generally smaller and less frequent than ridges in the Arctic Ocean. The general pattern of the ridging is also different in that the long sinuous ridges characteristic of the Arctic Ocean are not observed. Instead, the deformation can be better described as irregular hummocking accompanied by the extensive rafting of one floe over another. Floe sizes are also smaller as the result of the passage of large-amplitude swells through the ice. These swells, which are generated by the intense Southern Ocean storms that move to the north of the ice edge, result in the fracturing of the larger floes while the large vertical motions facilitate the rafting process.

Pressure ridges are of considerable importance for a variety of reasons. First, they change the surface roughness at the air–ice and water–ice interfaces, thereby altering the effective surface tractions exerted by winds and currents. Second, they act as plows, forming gouges in the sea floor up to 8 m deep when they ground and are pushed along by the ungrounded pack as it drifts over the shallower (<60 m) regions of the polar continental shelves. Third, as the thickest sea ice masses, they are a major hazard that must be considered in the design of offshore structures. Finally, and most importantly, the ridging process provides a mechanical procedure for transferring the thinner ice in the leads directly and rapidly into the thickest ice categories.

Considerable information on the drift and deformation of sea ice has recently become available through the combined use of data buoy and satellite observations. This information shows that, on the average, there are commonly two primary ice motion features in the Arctic Basin. These are the Beaufort Gyre, a large clockwise circulation located in the Beaufort Sea, and the Trans-Polar Drift Stream, which transports ice formed on the Siberian Shelf over the Pole to Fram Strait between Greenland and Svalbard. The time required for the ice to complete one circuit of the gyre averages 5 years, while the transit time for the Drift Stream is roughly 3 years, with about 9% of the sea ice of the Arctic Basin (919 000 km^2) moving south through Fram Strait and out of the basin each year. There are many interesting features of the ice drift that exist over shorter time intervals. For instance, recent observations show that the Beaufort Gyre may run backward (counter-clockwise) over appreciable periods of time, particularly in the summer and fall. There have even been suggestions that such reversals can occur on decadal timescales. Typical pack ice velocities range from 0 to 20 cm s^{-1}, although extreme velocities of up to 220 cm s^{-1} (4.3 knots) have been recorded during storms. During winter, periods of zero ice motion are not rare. During summers, when considerable open water is present in the pack, the ice appears to be in continuous motion. The highest drift velocities are invariably observed near the edge of the pack. Not only are such locations commonly windy, but the floes are able to move toward the free edge with minimal inter-floe interference. Ice drift near the Antarctic continent is generally westerly, becoming easterly further to the north, but in all cases showing a consistent northerly diverging drift toward the free ice edge.

Trends

Considering the anticipated geophysical consequences of changes in the extent of sea ice, it is not surprising that there is considerable scientific interest in the subject. Is sea ice expanding and thickening, heralding a new glacial age, or retreating and thinning before the onslaught of a greenhouse-gas-induced heatwave? One thing that should be clear from the preceding discussion is that the ice is surprisingly thin and variable. Small changes in meteorological and oceanographic forcing should result in significant changes in the extent and state of the ice cover. They could also produce feedbacks that might have significant and complex climatic consequences.

Before we examine what is known about sea ice variations, let us first examine other related observations that have a direct bearing on the question of sea ice trends. Land station records for 1966–1996 show that the air temperatures have increased, with the largest increases occurring winter and spring over both north-west North America and Eurasia, a conclusion that is supported by increasing permafrost temperatures. In addition, meteorological observations collected on Russian NP drifting stations deployed in the Arctic Basin show significant warming trends for the spring and summer periods. It has also recently been suggested that when proxy temperature sources are considered, they indicate that the late twentieth-century Arctic temperatures were the highest in the past 400 years.

Recent oceanographic observations also relate to the above questions. In the late 1980s the balance between the Atlantic water entering the Arctic Basin and the Pacific water appears to have changed, resulting in an increase in the areal extent of the more saline, warmer Atlantic water. In addition, the Atlantic water is shallower than in the past, resulting in temperature increases of as much as 2°C and salinity increases of up to 2.5‰ at depths of 200 m. The halocline, which isolates the cold near-surface layer and the overlying sea ice cover from the underlying warmer water, also appears to be thinning, a fact that could profoundly affect the state of the sea ice cover and the surface energy budget in the Arctic. Changes revealed by the motions of data buoys placed on the ice show that there has been a weakening of the Beaufort Sea Gyre and an associated increased divergence of the ice peak. There are also indications that the MY ice in the center of the Beaufort Gyre is less prevalent and thinner than in the past and that the amount of surface melt increased from ∼0.8 m in the mid-1970s to ∼2 m in 1997. This conclusion is supported by the operational difficulties encountered by recent field programs such as SHEBA that attempted to maintain on-ice measurements. The increased melt is also in agreement with observed decreases in the salinity of the near-surface water layer.

It is currently believed that these changes appear to be related to atmospheric changes in the Polar Basin where the mean atmospheric surface pressure is decreasing and has been below the 1979–95 mean every year since 1988. Before about 1988–99 the Beaufort High was usually centered over 180° longitude. After this time the high was both weaker and typically confined to more western longitudes, a fact that may account for lighter ice conditions in the western Arctic. There also has been a recent pronounced increase in the frequency of cyclonic storms in the Arctic Basin.

So are there also direct measurements indicating decreases in ice extent and thickness? Historical data based on direct observations of sea ice extent are rare, although significant long-term records do exist for a few regions such as Iceland where sea ice has an important effect on both fishing and transportation. In monitoring the health of the world's sea ice covers the use of satellite remote sensing is essential because of the vast remote areas that must be surveyed. Unfortunately, the satellite record is very short. If data from only microwave remote sensing systems are considered, because of their all-weather capabilities, the record is even shorter, starting in 1973. As there was a 2-year data gap between 1976 and 1978, only 25 years of data are available to date. The imagery shows that there are definitely large seasonal, interannual and regional variations in ice extent. For instance, a decrease in ice extent in the Kara and Barents Seas contrasts with an increase in the Baffin Bay/Davis Strait region and out-of-phase fluctuations occur between the Bering and the Okhotsk Seas. The most recent study, which examined passive microwave data up to December 1996, concludes that the areal extent of Arctic sea ice has decreased by 2.9% ± 0.4% per decade. In addition, record or near-record minimum areas of Arctic sea ice have been observed in 1990, 1991, 1993, 1995, and 1997. A particularly extreme recession of the ice along the Beauford coast was also noted in the fall of 1998. Russian ice reconnaissance maps also show that a significant reduction in ice extent and concentration has occurred over much of the Russian Arctic Shelf since 1987.

Has a systematic variation also been observed in ice thickness? Unfortunately there is, at present, no satellite-borne remote sensing technique that can measure sea ice thicknesses effectively from above. There is also little optimism about the possibilities of developing such techniques because the extremely lossy nature of sea ice limits penetration of electromagnetic signals. Current ice thickness information comes from two very different techniques: *in situ* drilling and upward-looking, submarine-mounted sonar. Although drilling is an impractical technique for regional studies, upward-looking sonar is an extremely effective procedure. The submarine passes under the ice at a known depth and the sonar determines the distance to the underside of the ice by measuring the travel times of the sound waves. The result is an accurate, well-resolved under-ice profile from which ice draft distributions can be determined and ice thickness distributions can be estimated based on the assumption of isostacy. Although there have been a large number of under-ice cruises starting with the USS *Nautilus* in 1958, to date only a few studies have been published that examine temporal variations in ice thickness in the Arctic. The first compared the results of two nearly identical cruises: that of the USS *Nautilus* in 1958 with that of the USS *Queenfish* in 1970. Decreases in mean ice thickness were observed in the Canadian Basin (3.08–2.39 m) and in the Eurasian Basin (4.06–3.57 m). The second study has compared the results of two Royal Navy cruises made in 1976 and 1987, and obtained a 15% decrease in mean ice thickness for a 300 000 km² area north of Greenland. Although these studies showed similar trends, the fact that they each only utilized two years' data caused many scientists to feel that a conclusive trend had not been established. However, a recent study has been able to examine this problem in more detail by comparing data from three submarine cruises made in the 1990s (1993,

1996, 1997) with the results of similar cruises made between 1958 and 1976. The area examined was the deep Arctic Basin and the comparisons used only data from the late summer and fall periods. It was found that the mean ice draft decreased by about 1.3 m from 3.1 m in 1958–76 to 1.8 m in the 1990s, with a larger decrease occurring in the central and eastern Arctic than in the Beaufort and Chukchi Seas. This is a very large difference, indicating that the volume of ice in the region surveyed is down by some 40%. Furthermore, an examination of the data from the 1990s suggests that the decrease in thickness is continuing at a rate of about 0.1 m y^{-1}.

Off the Antarctic the situation is not as clear. One study has suggested a major retreat in maximum sea ice extent over the last century based on comparisons of current satellite data with the earlier positions of whaling ships reportedly operating along the ice edge. As it is very difficult to access exactly where the ice edge is located on the basis of only ship-board observations, this claim has met with some skepticism. An examination of the satellite observations indicates a very slight increase in areal extent since 1973. As there are no upward-looking sonar data for the Antarctic Seas, the thickness database there is far smaller than in the Arctic. However, limited drilling and airborne laser profiles of the upper surface of the ice indicate that in many areas the undeformed ice is very thin (60–80 cm) and that the amount of deformed ice is not only significantly less than in the Arctic but adds roughly only 10 cm to the mean ice thickness (**Figure 2**).

What is one to make of all of this? It is obvious that, at least in the Arctic, a change appears to be under way that extends from the top of the atmosphere to depths below 100 m in the ocean. In the middle of this is the sea ice cover, which, as has been shown, is extremely sensitive to environmental changes. What is not known is whether these changes are part of some cycle or represent a climatic regime change in which the positive feedbacks associated with the presence of a sea ice cover play an important role. Also not understood are the interconnections between what is happening in the Arctic and other changes both inside and outside the Arctic. For instance, could changes in the Arctic system drive significant lower-latitude atmospheric and oceanographic changes or are the Arctic changes driven by more dynamic lower-latitude processes? In the Antarctic the picture is even less clear, although changes are known to be underway, as evidenced by the recent breakup of ice shelves along the eastern coast of the Antarctic Peninsula. Not surprisingly, the scientific community is currently devoting considerable energy to attempting to answer these

Figure 2 (A) Ice gouging along the coast of the Beaufort Sea. (B) Aerial photograph of an area of pack ice in the Arctic Ocean showing a recently refrozen large lead that has developed in the first year. The thinner newly formed ice is probably less than 10 cm thick. (C) A representative pressure ridge in the Arctic Ocean. (D) A rubble field of highly deformed first-year sea ice developed along the Alaskan coast of the Beaufort Sea. The tower in the far distance is located at a small research station on one of the numerous off-shore islands located along this coast. (E) Deformed sea ice along the NW Passage, Canada. (F) Aerial photograph of pack ice in the Arctic Ocean.

questions. One could say that a cold subject is heating up.

See also

Icebergs.

Further Reading

Cavelieri DJ, Gloersen P, Parkinson CL, Comiso JC, and Zwally HJ (1997) Observed hemispheric asymmetry in global sea ice changes. *Science* 278 5340: 1104–1106.

Dyer I and Chryssostomidis C (eds.) (1993) *Arctic Technology and Policy.* New York: Hemisphere.

Leppäranta M (1998) *Physics of Ice-covered Seas, 2 vols.* Helsinki: Helsinki University Printing House.

McLaren AS (1989) The underice thickness distribution of the Arctic basin as recorded in 1958 and 1970. *Journal of Geophysical Research* 94(C4): 4971–4983.

Morison JH, Aagaard K and Steele M (1998) *Study of the Arctic Change Workshop*. (Report on the Study of the Arctic Change Workshop held 10–12 November 1997, University of Washington, Seattle, WA). Arctic System Science Ocean–Atmosphere–Ice Interactions Report No. 8 (August 1998).

Rothrock DA, Yu Y and Maykut G (1999) Thinning of the Arctic sea ice cover. *Geophysical Research Letters* 26.

Untersteiner N (ed.) (1986) *The Geophysics of Sea Ice. NATO Advanced Science Institutes Series B, Physics 146*. New York: Plenum Press.

SEA ICE DYNAMICS

M. Leppäranta, University of Helsinki, Helsinki, Finland

Introduction

Sea ice grows, drifts, and melts under the influence of solar, atmospheric, oceanic, and tidal forcing. Most of it lies in the polar seas but seasonally freezing, smaller basins exist in lower latitudes. In sizeable basins, solid sea ice lids are statically unstable and break into fields of ice floes, undergoing transport as well as opening and ridging which altogether create the exciting sea ice landscape as it appears to the human eye.

The history of sea ice dynamics science initiates from the drift of Nansen's ship *Fram* in 1893–96 in the Arctic Ocean. The average drift speed was 2% of the wind speed and the drift direction deviated 30° to the right from the wind. Much data were collected in Soviet Union North Pole Drifting Stations program commenced in 1937, where science camps drifted from the North Pole to the Greenland Sea. The closure of the sea ice dynamics problem was completed in the 1950s as rheology and conservation laws were established. In the 1970s, two major findings came out in the Arctic Ice Dynamics Joint Experiment (AIDJEX) program: plastic rheology and thickness distribution. Recent progress contains granular flow models, ice kinematics mapping with microwave satellite imagery, anisotropic rheologies, and scaling laws. Methods for remote sensing of sea ice thickness have been slowly progressing, which is the most critical point for the further development of the theory and models of sea ice dynamics.

Drift Ice Medium

Ice State

Drift ice is a peculiar geophysical medium (**Figure 1**). It is granular (ice floes are the basic elements, grains) and the motion takes place on the sea surface plane as a two-dimensional system. The compactness of floe fields may easily change, that is, the medium is compressible, the rheology shows highly nonlinear properties, and by freezing and melting an ice source/sink term exists. The full ice drift problem includes the following unknowns: ice state (a set of relevant material properties), ice velocity, and ice stress. The system is closed by the equations for ice conservation, motion, and rheology.

A sea ice landscape consists of ice floes with ridges and other morphological features, and leads and polynyas. It can be divided into zones of different dynamic character. The central pack is free from immediate influence from the boundaries, and the length scale is the size of the basin. Land fast or fast ice is the immobile coastal sea ice zone extending from the shore to about 10–20-m depths (in Antarctic, grounded icebergs may act as tie points and extend the fast ice zone to deeper waters). Next to fast ice is the shear zone (width 10–200 km), where the mobility of the ice is restricted by the geometry of the boundary and strong deformation takes place. Marginal ice zone (MIZ) lies along the boundary to open sea. It is loosely characterized as the zone, which 'feels the presence of the open ocean' and extends to a distance of 100 km from the ice edge. Well-developed MIZs are found along the oceanic ice edge of the polar oceans. They influence the mesoscale ocean dynamics resulting in ice edge eddies, jets, and upwelling/downwelling.

The horizontal structure of a sea ice cover is well revealed by optical satellite images (**Figure 1(a)**). Ice floes are described by their thickness h and diameter d, and we may examine the drift of an individual floe or a field of floes. For continuum models to be valid for a floe field, the size of continuum material particles D must satisfy $d \ll D \ll \Lambda$, where Λ is the gradient length scale. The ranges are in nature $d \sim 10^1$–10^4 m, $D \sim 10^3$–10^5 m, and $\Lambda \sim 10^4$–10^6 m. As $D \to \Lambda$, discontinuities build up, and as $D \to d$ we have a set of n individual floes.

In the continuum approach, an ice state J is defined for the material description, $\dim(J)$ being the number of levels. The first attempt was $J = \{A, H\}$, where A is ice compactness and H is mean thickness. Three-level ice states $J = \{A, H_\mathrm{u}, H_\mathrm{d}\}$ decomposing the ice into undeformed ice thickness H_u and deformed ice thickness H_d have been used, and the fine-resolution approach is to take the thickness distribution $p(h)$ for the ice state. The thickness classes are fixed, arbitrarily spaced, and their histogram contains the state as the class probabilities π_k:

$$J = \{\pi_0, \pi_1, \pi_2, \ldots\}, \pi_k$$
$$= \mathrm{Prob}\{h = h_k\}, \quad \sum_k \pi_k = 1 \qquad [1]$$

(a)

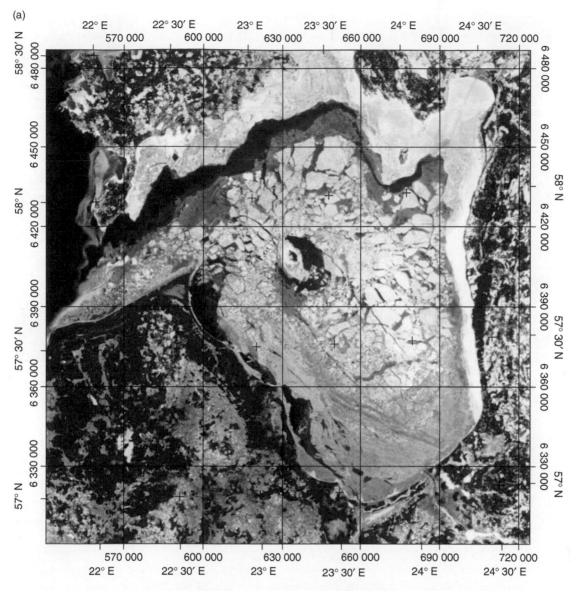

Figure 1 (a) A moderate-resolution imaging spectroradiometer (MODIS) image (NASA's *Terra/Aqua* satellite) of the sea ice cover in the Gulf of Riga, 3 Mar. 2003. The width of the basin is 120 km. (b) Sea ice landscape of the heavy pack ice in the Arctic Ocean, north of Svalbard. (c) Sea ice landscape from the Weddell Sea, Antarctica, showing a first-year ice floe field.

Kinematics

Sea ice kinematics has been mapped using drifters and sequential remote-sensing imagery. In the Arctic Ocean, the long-term drift pattern consists of the Transpolar Drift Stream on the Eurasian side and the Beaufort Sea Gyre on the American side, as illustrated by the historical data in **Figure 2**. In the Antarctica, the governing feature is two annuluses rotating in opposite directions forced by the East Wind and West Wind zones, with meridional movements in places to interchange ice floes between the annuluses, in particular, northward along the Antarctic Peninsula in the Weddell Sea.

Figure 3 shows a typical 1-week time series of sea ice velocity together with wind data. The ice followed the wind with essentially no time lag but sometimes the ice made 'unexpected' steps due to its internal friction. In general, frequency spectra of ice velocity reach highest levels at the synoptic time-scales, and a secondary peak appears at the inertial period. Exceptionally very-high ice velocities of more than $1 \, \mathrm{m \, s^{-1}}$ have been observed in transient currents in straits and along coastlines. An extreme case

(b)

(c)

Figure 1 *Continued.*

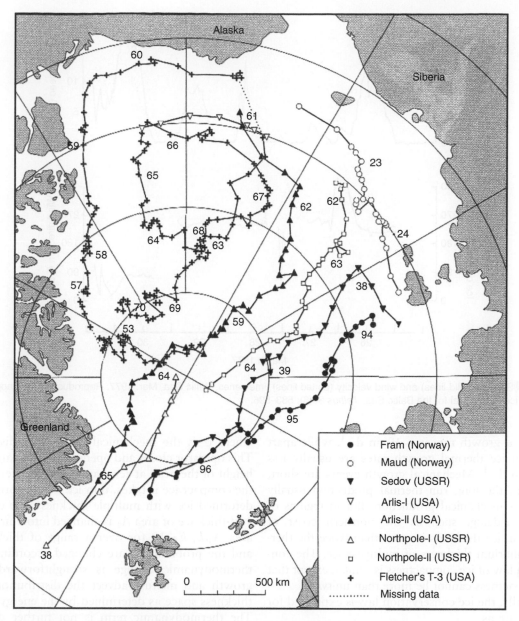

Figure 2 Paths of drifting stations in the Arctic Ocean. The numbers show the year (1894–1970) and marks between are at monthly intervals. Reproduced from Hibler WD, III (1980) Sea ice growth, drift and decay. In: Colbeck S (ed.) *Dynamics of Snow and Ice Masses*, pp. 141–209. New York: Academic Press.

is an 'ice river' phenomenon where a narrow (≈ 0.5 km) band in close ice moves at the speed of up to $3 \, \mathrm{m \, s^{-1}}$. Long-term data of sea ice deformation was obtained in AIDJEX in 1975. The magnitude of strain rate and rotation was $0.01 \, \mathrm{d^{-1}}$ and their levels were higher by 50–100% in summer. In the MIZ, the level is up to $0.05 \, \mathrm{h^{-1}}$ in short, intensive periods. Leads open in divergent directions while ridges form in convergent directions, and both these processes may occur in pure two-dimensional shear deformation. Also, drift ice has a particular asymmetry in the deformation: leads open and close, pressure

(ridged) ice forms, but cannot 'unform' since there is no restoring force.

The ice conservation law tells how ice-state components are changed by advection, mechanical deformation, and thermodynamics. The conservation of ice volume is always required:

$$\frac{\partial H}{\partial t} + \boldsymbol{u} \cdot \nabla H = -H \nabla \cdot \boldsymbol{u} + \phi(H) \qquad [2]$$

where $\boldsymbol{u}$ is ice velocity and $\phi(H)$ is the thermal growth and melt rate. If $H \sim 1$ m and $\nabla \cdot \boldsymbol{u} \sim -0.1 \, \mathrm{d^{-1}}$, the

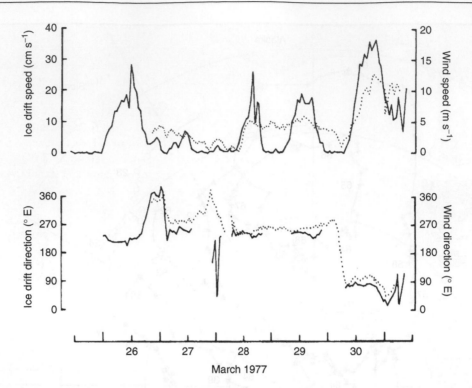

Figure 3 Sea ice (solid lines) and wind velocity (dotted lines) time series, Baltic Sea, Mar. 1977. Reproduced from Leppäranta M (1981) An ice drift model for the Baltic Sea. *Telbus* 33(6): 583–596.

mechanical growth rate is $\sim 10\,\mathrm{cm}\,\mathrm{d}^{-1}$, while apart from thin ice thermal growth rates are usually less than $1\,\mathrm{cm}\,\mathrm{d}^{-1}$. Mechanical growth events are short, so that in the long run thermal production usually overcomes mechanical production. But in regions of intensive ridging, such as the northern coast of Greenland, ice thickness is more than twice the thermal equilibrium thickness of multiyear ice. The conservation law of ice compactness is similar, except that the compactness cannot be more than unity.

Formally, the ice conservation law is expressed for the ice state as

$$\frac{\partial J}{\partial t} + \boldsymbol{u} \cdot \nabla J = \Psi + \Phi \qquad [3]$$

where Ψ and Φ are the change of ice state due to mechanics and thermodynamics, respectively. In multilevel cases, the question is that how deformed ice is produced. In the three-level case $J = \{A,\ H_{\mathrm{u}},\ H_{\mathrm{d}}\}$, all ridging adds on the thickness of deformed ice, but when using the thickness distribution additional assumptions are needed for the redistribution. Formally, one may take the limit $\Delta h_k \to 0$ to obtain the conservation law for the spatial density of thickness:

$$\frac{\partial \pi}{\partial t} + \boldsymbol{u}\cdot\nabla\pi = \psi - \pi\nabla\cdot\boldsymbol{u} + \pi\frac{\partial\phi(h)}{\partial h} \qquad [4]$$

where ψ is the mechanical thickness redistributor. This operator closes and opens leads by changing the height of the peak at zero, and under convergence of the compact ice it takes, thin ice is transformed into deformed ice with multiple thickness; for example, 1-m-thick ice of area A_1 is changed into k m ice with area A_1/k, $k \sim 10$ (or over a range of thicknesses), and the probabilities are changed accordingly. The thermodynamic change is straightforward as ice growth and melting advect the distribution in the thickness space as determined by the energy budget. The thermodynamic term is not further discussed below.

Rheology

The mechanisms behind the rheology are floe collisions, floe breakage, shear friction between floes, and friction between ice blocks and potential energy production in ridging. By observational evidence, it is known that (1) stress level ≈ 0 for $A < 0.8$, (2) yield strength > 0 for $A \approx 1$, and (3) tensile strength $\approx 0 \ll$ shear strength $<$ compressive strength. The rheological law is given in its general form as

$$\boldsymbol{\sigma} = \boldsymbol{\sigma}(J, \boldsymbol{\varepsilon}, \dot{\boldsymbol{\varepsilon}}) \qquad [5]$$

where $\boldsymbol{\sigma}$ is stress, $\boldsymbol{\varepsilon}$ is strain, and $\dot{\boldsymbol{\varepsilon}}$ is strain rate (**Figure 4**). The stress is very low for compactness less

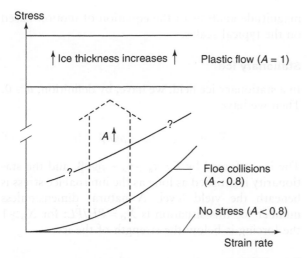

Figure 4 Schematic presentation of the sea ice rheology as a function of ice compactness A and thickness h. The cut in the ordinate axis tells of a jump of several orders of magnitude. Reproduced from Leppäranta M (2005) *The Drift of Sea Ice*, 266pp. Heidelberg: Springer-Praxis.

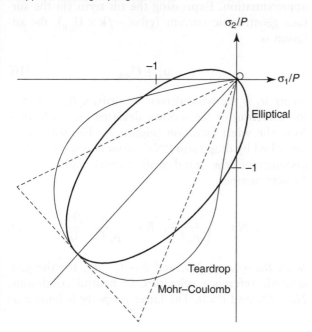

Figure 5 Plastic yield curves for drift ice. Reproduced from Leppäranta M (2005) *The Drift of Sea Ice*, 266pp. Heidelberg: Springer-Praxis.

than 0.8. At higher ice concentrations, the significance of floe collisions and shear friction between ice floes increases, and finally with ridge formation a plastic flow results. How the rheology changes from the superlinear collision rheology to a plastic law is not known. In the plastic regime, the yield strength increases with increasing ice thickness.

Two-dimensional yielding is specified using a yield curve $F(\sigma_1, \sigma_2) = 0$ (**Figure 5**). In numerical modeling,

lower stresses are taken linear elastic (AIDJEX model) or linear viscous (Hibler model). Drucker's postulate for stable materials states that the yield curve serves as the plastic potential, and consequently, the failure strain is directed perpendicular to the yield curve. Drift ice is strain hardening in compression, and therefore, ridging may proceed only to a certain limit. Hibler model is an excellent tool for numerical modeling since it allows an explicit solution for the stress as a function of the strain rate:

$$\boldsymbol{\sigma} = \frac{P}{2}\left(\frac{\dot{\varepsilon}_{\mathrm{I}}}{\max(\Delta, \Delta_{\mathrm{o}})} - 1\right)\mathbf{I} + \frac{P}{2e^2\max(\Delta, \Delta_{\mathrm{o}})}\dot{\boldsymbol{\varepsilon}}' \quad [6]$$

where $P = P^*h \exp[-C(1-A)]$ is the compressive strength, P^* is the strength level constant and C is the strength reduction for lead opening, $\Delta = \sqrt{\dot{\varepsilon}_{\mathrm{I}}^2 + (\dot{\varepsilon}_{\mathrm{II}}/e)^2}$, $\dot{\varepsilon}_{\mathrm{I}}$ and $\dot{\varepsilon}_{\mathrm{II}}$ are strain-rate invariants equal to divergence and twice the maximum rate of shear, Δ_{o} is the maximum viscous creep rate, e is the aspect ratio of the yield ellipse, and $\dot{\boldsymbol{\varepsilon}}'$ is the deviatoric strain rate. The normal parameter values are $P^* = 25\,\mathrm{kPa}$, $C = 20$, $e = 2$, and $\Delta_{\mathrm{o}} = 10^{-9}\,\mathrm{s}^{-1}$.

Equation of Motion

The equation of motion is first integrated through the thickness of the ice for the two-dimensional system (**Figure 6**). This is straightforward. Integration of the divergence of internal ice stress brings the air and ocean surface stresses (τ_{a} and τ_{w}) and internal friction, and the other terms are just multiplied by the mean thickness of ice. The result is

$$\rho H\left[\frac{\partial \boldsymbol{u}}{\partial t} + \boldsymbol{u} \cdot \nabla \boldsymbol{u} + f\mathbf{k} \times \boldsymbol{u}\right]$$
$$= \nabla \cdot \boldsymbol{\sigma} + \boldsymbol{\tau}_{\mathrm{a}} + \boldsymbol{\tau}_{\mathrm{w}} - \rho Hg\boldsymbol{\beta} \quad [7]$$

where ρ is ice density, f is the Coriolis parameter, g is acceleration due to gravity, and $\boldsymbol{\beta}$ is the sea surface slope. The air and water stresses are written as

$$\boldsymbol{\tau}_{\mathrm{a}} = \rho_{\mathrm{a}}C_{\mathrm{a}}U_{\mathrm{ag}}(\cos\theta_{\mathrm{a}} + \sin\theta_{\mathrm{a}}\mathbf{k}\times)U_{\mathrm{ag}} \quad [8a]$$

$$\boldsymbol{\tau}_{\mathrm{w}} = \rho_{\mathrm{w}}C_{\mathrm{w}}|U_{\mathrm{wg}} - \boldsymbol{u}|(\cos\theta_{\mathrm{w}} + \sin\theta_{\mathrm{w}}\mathbf{k}\times)$$
$$(U_{\mathrm{wg}} - \boldsymbol{u}) \quad [8b]$$

where ρ_{a} and ρ_{w} are air and water densities, C_{a} and C_{w} are air and water drag coefficients, θ_{a} and θ_{w} are the air- and water-boundary layer angles, and U_{ag} and U_{wg} are the geostrophic wind and current velocities. Representative Arctic – the best-known region – parameters are $C_{\mathrm{a}} = 1.2 \times 10^{-3}$, $\theta_{\mathrm{a}} = 25°$,

$C_w = 5 \times 10^{-3}$, and $\theta_w = 25°$. In a stratified fluid, the drag parameters also depend on the stability of the stratification; in very stable conditions, $C_a \approx 10^{-4}$ and $\theta_a \approx 35°$ while in unstable conditions $C_a \approx 1.5 \times 10^{-3}$ and $\theta_a \to 0$.

The sea ice dynamics problem can be divided into three categories: (1) stationary ice, (2) free drift, and (3) drift in the presence of internal friction. There are three timescales: local acceleration $T_I = H/(C_w U)$, Coriolis period f^{-1}, and adjustment of the thickness field $T_D = L/U$. These are well separated, $T_I \ll f^{-1} \ll T_D$. **Table 1** shows the result of the magnitude analysis of the equation of motion based on the typical scales.

Stationary Ice

In a stationary ice field, we have, by definition, $u \equiv 0$. Then we have

$$\nabla \cdot \sigma + \tau_a + \tau_w - \rho g h \beta = 0 \qquad [9]$$

The ice is forced by $F = \tau_a + \tau_w - \rho g h \beta$, and the stationarity is satisfied as long as the internal ice stress is beneath the yield level. A natural dimensionless number for this situation is $X_o = PH/FL$: for $X_o > 1$ the forcing is below the strength of the ice.

Free Drift

In free drift, by definition $\nabla \cdot \sigma \equiv 0$. Since the momentum advection is very small and $T_I \sim 1$ h, an algebraic steady-state equation results as a very good approximation. Expressing the tilt term via the surface geostrophic current ($g\beta = -f k \times U_{wg}$), the solution is

$$u = u_a + U_{wg} \qquad [10]$$

where u_a is the wind-driven ice drift, $u_a/U_a = \alpha$, and the direction of ice motion deviates the angle of θ from the wind direction (**Figure 7**). The solution is described by drag ratio ('Na', from Nansen who first documented the wind drift factor) and ice drift Rossby number:

$$Na = \sqrt{\frac{\rho_a C_a}{\rho_w C_w}}, \quad Ro = \frac{\rho}{\rho_w C_w} \frac{fH}{U} \qquad [11a]$$

With $Ro \to 0$, $\alpha \to Na$, and $\theta \to \theta_a - \theta_w$, for the geostrophic reference velocities, in neutral conditions, $Na \approx 2\%$ and $\theta \approx 0$. The latter property is known as

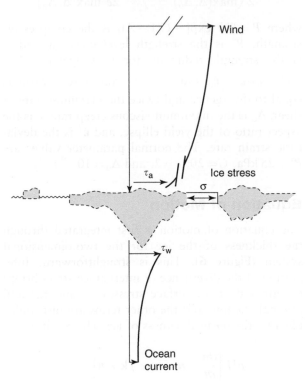

Figure 6 The ice drift problem. Reproduced from Leppäranta M (2005) *The Drift of Sea Ice*, 266pp. Heidelberg: Springer-Praxis.

Table 1 Scaling of the equation of motion of drift ice

Term	Scale	Value	Comments
Local acceleration	$\rho HU/T$	-3	-1 for rapid changes ($T = 10^3$ s)
Advective acceleration	$\rho HU^2/L$	-4	Long-term effects may be significant
Coriolis term	ρHfU	-2	Mostly less than -1
Internal friction	PH/L	$-1 (-\infty)$	Compact ice, $A > 0.9$ ($A < 0.8$)
Air stress	$\rho_a C_a U_{ag}^2$	-1	Mostly significant
Water stress	$\rho_w C_w |U - U_{wg}|^2$	-1	Mostly significant
Sea surface tilt	ρHfU_{wg}	-2	Mostly less than or equal to -2

The representative elementary scales are: $H = 1$ m, $U = 10$ cm s^{-1}, $P = 10$ kPa, $U_a = 10$ m s^{-1}, $U_{wg} = 5$ cm s^{-1}, $T = 1$ day, and $L = 100$ km. The 'Value' column gives the ten-based logarithm of the scale in pascals.
Reproduced from Leppäranta M (2005) *The Drift of Sea Ice*, 266pp. Heidelberg: Springer-Praxis.

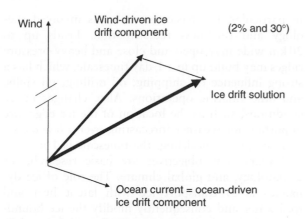

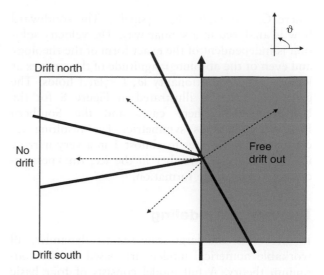

Figure 7 The free drift solution as the vector sum of wind-driven ice drift and geostrophic ocean current.

Figure 8 Steady-state solution of wind-driven zonal flow, Northern Hemisphere. Wind vector is drawn from the crosspoint of the thick lines as indicated by the dashed vectors and the resulting ice velocity is free drift, northward or southward boundary flow, or fast ice, depending on the ice strength and wind direction.

the Zubov's isobaric drift rule. With increasing Rossby number, the wind factor decreases and deviation angle turns more to the right (left) in the Northern (Southern) Hemisphere.

Ice Drift in the Presence of Internal Friction

In compact ice the forcing may become large and break the yield criterion. The ice starts motion, and importance of the internal friction is then described by the friction number:

$$X = \frac{1}{\rho_w C_w U^2} \frac{PH}{L} \qquad [11b]$$

Sea ice rheology has a distinct asymmetry in that during closing the stress may be very high but during opening it is nearly zero. In consequence, the forcing frequencies show up unchanged in ice velocity spectra.

In a channel with a closed end, the full steady-state solution is a stationary ice field. The functional form of the compressive strength, $P = P^* h \exp[-C(1-A)]$, results in a very sharp ice edge, and beyond the edge zone the thickness increases by $dh/dx = \tau_a/P^*$. In the spin-down the ice flows as long as the internal stress overcomes the yield level. Longitudinal boundary zone flow offers a more general frame, representing coastal shear zone or marginal ice zone. The y-axis is aligned along the longitudinal direction, and the system of equations is

$$\rho h\left(\frac{\partial u}{\partial t} + u\frac{\partial u}{\partial x} - fv\right) = \frac{\partial \sigma_{xx}}{\partial x}$$
$$+ \tau_{ax} + \tau_{wx} - \rho h g \beta_x \qquad [12a]$$

$$\rho h\left(\frac{\partial v}{\partial t} + u\frac{\partial v}{\partial x} + fu\right) = \frac{\partial \sigma_{xy}}{\partial x}$$
$$+ \tau_{ay} + \tau_{wy} - \rho h g \beta_y \qquad [12b]$$

$$\sigma = \sigma(h, A, \dot{\varepsilon}), \dot{\varepsilon}_{xx} = \frac{\partial u}{\partial x}, \dot{\varepsilon}_{yx}$$
$$= \dot{\varepsilon}_{yx} = \frac{1}{2}\frac{\partial v}{\partial x}, \dot{\varepsilon}_{yy} = 0 \qquad [12c]$$

$$\frac{\partial\{A, h\}}{\partial t} + \frac{\partial u\{A, h\}}{\partial x} = 0 \quad (0 \le A \le 1) \quad [12d]$$

Consider the full steady state in the Northern Hemisphere (**Figure 8**). The free drift solution results when the direction of the wind stress (ϑ) satisfies $-90° + \theta < \vartheta < 90° + \theta$. Since $\theta \sim 30°$, the angle ϑ must be between $-60°$ and $120°$. Otherwise the ice will stay in contact with the coast influenced by the coastal friction, and the land boundary condition implies $u \equiv 0$. Also, in quite general conditions, $|\sigma_{xy}/\sigma_{xx}| = \gamma = \text{constant} \approx 2$, and an algebraic equation can be obtained for the longitudinal velocity v. The ice may drift north if $\tau_{ay} + \gamma\tau_{ax} > 0$, which means that $90° + \theta < \vartheta < 180° - \arctan(\gamma) \approx 150°$, and then $\sigma_{xy} > 0$. Equations [12] give

$$v = \sqrt{\frac{\tau_{ay} + \gamma\tau_{ax}}{C_N} + \left(\frac{\gamma\rho h f}{2C_N}\right)^2} - \frac{\gamma\rho h f}{2C_N} \qquad [13]$$

where $C_N = \rho_w C_w(\cos\theta_w - \gamma\sin\theta_w)$. The southward flow is analyzed in a similar way. The velocity solution is independent of the exact form of the rheology and even of the absolute magnitude of the stresses as long as the proportionality $|\sigma_{xy}| = \gamma|\sigma_{xx}|$ holds. The general solution is illustrated in **Figure 8** for the Northern Hemisphere case, and the Southern Hemisphere case is symmetric. The resulting ice compactness increases to almost 1 in a very narrow ice edge zone, and further in the ice, thickness increases due to ridge formation.

Numerical Modeling

In mesoscale and large-scale sea ice dynamics, all workable numerical models are based on the continuum theory. A full model consists of four basic elements: (1) ice state J, (2) rheology, (3) equation of motion, and (4) conservation of ice. The elements (1) and (2) constitute the heart of the model and are up to the choice of the modeler: one speaks of a three-level ($\dim(J) = 3$) viscous-plastic sea ice model, etc. The unknowns are ice state, ice velocity, and ice stress, and the number of independent variables is $\dim(J) + 2 + 3$. Any proper ice state has at least two levels.

The model parameters can be grouped into those for (a) atmospheric and oceanic drag, (b) rheology, (c) ice redistribution, and (d) numerical design. The primary geophysical parameters are the drag coefficients and compressive strength of ice. The drag coefficients together with the Ekman angles tune the free drift velocity, while the compressive strength tunes the length scale in the presence of internal friction. The secondary geophysical parameters come from the rheology (other than the compressive strength) and the ice state redistribution scheme. The redistribution parameters would be probably very important but the distribution physics lacks good data. The numerical design parameters include the choice of the grid; also since the system is highly nonlinear, the stability of the solution may require smoothing techniques. Since the continuum particle size D is fairly large, the grid size can be taken as $\Delta x \sim D$.

Because the inertial timescale of sea ice is quite small, the initial ice velocity can be taken as zero. At solid boundary, the no-slip condition is employed, while in open boundary the normal stress is zero (a practical way is to define open water as ice with zero thickness and avoid an explicit open boundary).

In short-term modeling, the timescale is 1 h–10 days. The objectives are basic research of the dynamics of drift ice and coupled ice–ocean system, ice forecasting, and applications for marine technology.

In particular, the basic research has involved rheology and thickness redistribution. Leads up to 20 km wide may open and close and heavy-pressure ridges may build up in a 1-day timescale, which has a strong influence on shipping, oil drilling, oil spills, and other marine operations. Also, changes in ice conditions, such as the location of the ice edge, are important for weather forecasting over a few days.

In long-term modeling, the timescale is 1 month–100 years. The objectives are basic research, ice climatology, and global climate. The role of ice dynamics is to transport ice with latent heat and freshwater and consequently modify the ice boundary and air–sea interaction. Differential ice drift opens and closes leads which means major changes to the air–sea heat fluxes. Mechanical accumulation of ice blocks, like ridging, adds large amount to the total volume of ice.

An example of long-term simulations in the Weddell Sea is shown in **Figure 9**. The drag ratio Na and Ekman angle were 2.4% and 10°, both a bit low, for the Antarctica, but it can be explained that the upper layer water current was the reference in the water stress and not the geostrophic flow. The compressive strength constant was taken as $P^* = 20$ kPa. The grid size was about 165 km, and the model was calibrated with drift buoy data for the ice velocity and upward-looking sonar data for the ice thickness. There is a strong convergence region in the southwest part of the basin, and advection of the ice shows up in larger ice thickness northward along the Antarctic peninsula. The width of the compressive region east of the peninsula is around 500 km.

The key areas of modeling research are now ice thickness distribution and its evolution, and use of satellite synthetic aperture radars (SARs) for ice kinematics. The scaling problem and, in particular, the downscaling of the stress from geophysical to local (engineering) scale is examined for combining scientific and engineering knowledge and developing ice load calculation and forecasting methods. The physics of drift ice is quite well represented in short-term ice forecasting models, in the sense that other questions are more critical for their further development, and the user interface is still not very good. Data assimilation methods are coming into sea ice models which give promises for the improvement of both the theoretical understanding and applications.

Concluding Words

The sea ice dynamics problem contains interesting basic research questions in geophysical fluid dynamics. But perhaps, the principal science motivation is connected to the role of sea ice as a dynamic

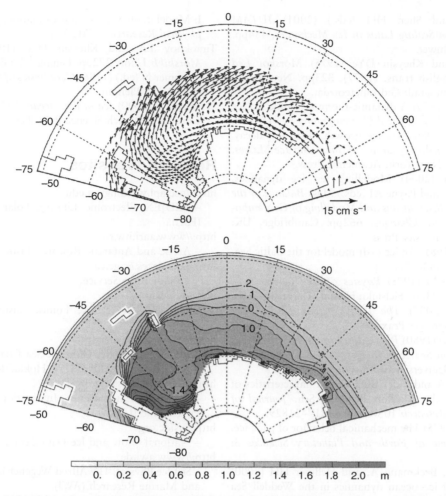

Figure 9 Climatological (a) sea ice velocity and (b) sea ice thickness in the Weddell Sea according to model simulations. Reproduced from Timmermann R, Beckmann A, and Hellmer HH (2000) Simulation of ice–ocean dynamics in the Weddell Sea I: Model configuration and validation. *Journal of Geophysical Research* 107(C3): 10, with permission from the Americal Geophysical Union.

air–ocean interface. The transport of ice takes sea ice (with latent heat and fresh water) to regions, where it would not be formed by thermodynamic processes, and due to differential drift leads open and close and hummocks and ridges form. Sea ice has an important role in environmental research. Impurities are captured into the ice sheet from the seawater, sea bottom, and atmospheric fallout, and they are transported with the ice and later released into the water column. The location of the ice edge is a fundamental boundary condition for the marine biology in polar seas. A recent research line for sea ice dynamics is in paleoclimatology and paleoceanography. Data archive of drift ice and icebergs exists in marine sediments, and via its influence on ocean circulation, the drift ice has been an active agent in the global climate history.

In the practical world, three major questions are connected with sea ice dynamics. Sea ice models have been applied for tactical navigation to provide short-term forecasts of the ice conditions. Ice forcing on ships and fixed structures are affected by the dynamical behavior of the ice. Sea ice information service is an operational routine system to support shipping and other marine operations such as oil drilling in ice-covered seas. In risk assessment for oil spills and oil combating, proper oil transport and dispersion models for ice-covered seas are needed.

See also

Coupled Sea Ice–Ocean Models. Icebergs. Ice-Ocean Interaction. Satellite Passive-Microwave Measurements of Sea Ice. Sea Ice: Overview.

Further Reading

Coon MD, Knoke GS, Echert DC, and Pritchard RS (1998) The architecture of an anisotropic elastic-plastic sea ice mechanics constitutive law. *Journal of Geophysical Research* 103(C10): 21915–21925.

Dempsey JP and Shen HH (eds.) (2001) *IUTAM Symposium on Scaling Laws in Ice Mechanics*, 484pp. Dordrecht: Kluwer.

Doronin YuP and Kheysin DYe (1975) *Morskoi Led (Sea Ice)*, (English trans. (1977), 323pp. New Delhi: Amerind). Leningrad: Gidrometeoizdat.

Hibler WD, III (1979) A dynamic-thermodynamic sea ice model. *Journal of Physical Oceanography* 9: 815–846.

Hibler WD, III (1980) Sea ice growth, drift and decay. In: Colbeck S (ed.) *Dynamics of Snow and Ice Masses*, pp. 141–209. New York: Acamemic Press.

Hibler WD, III (2004) Modelling sea ice dynamics. In: Bamber JL and Payne AJ (eds.) *Mass Balance of the Cryosphere: Observations and Modelling of Contemporary and Future Changes*, 662pp. Cambridge, UK: Cambridge University Press.

Leppäranta M (1981) An ice drift model for the Baltic Sea. *Telbus* 33(6): 583–596.

Leppäranta M (ed.) (1998) *Physics of Ice-Covered Seas*, vols. 1 and 2, 823pp. Helsinki: Helsinki University Press.

Leppäranta M (2005) *The Drift of Sea Ice*, 266pp. Heidelberg: Springer-Praxis.

Pritchard RS (ed.) (1980) *Proceedings of the ICSI/AIDJEX Symposium on Sea Ice Processes and Models*, 474pp. Seattle, WA: University of Washington Press.

Richter-Menge JA and Elder BC (1998) Characteristics of ice stress in the Alaskan Beaufort Sea. *Journal of Geophysical Research* 103(C10): 21817–21829.

Rothrock DA (1975) The mechanical behavior of pack ice. *Annual Review of Earth and Planetary Sciences* 3: 317–342.

Timmermann R, Beckmann A, and Hellmer HH (2000) Simulation of ice–ocean dynamics in the Weddell Sea I: Model configuration and validation. *Journal of Geophysical Research* 107(C3): 10.

Timokhov LA and Kheysin DYe (1987) *Dynamika Morskikh L'dov*, 272pp. Leningrad: Gidrometeoizdat.

Untersteiner N (ed.) (1986) *Geophysics of Sea Ice*, 1196pp. New York: Plenum.

Wadhams P (2000) *Ice in the Ocean*, 351pp. Amsterdam: Gordon & Breach Science Publishers.

Relevant Websites

http://psc.apl.washington.edu
– AIDJEX Electronic Library, Polar Science Center (PSC).

http://www.aari.nw.ru
– Arctic and Antarctic Research Institute (AARI).

http://ice-glaces.ec.gc.ca
– Canadian Ice Service.

http://www.fimr.fi
– Finnish Ice Service, Finnish Institute of Marine Research.

http://www.hokudai.ac.jp
– Ice Chart Off the Okhotsk Sea Coast of Hokkaido, Sea Ice Research Laboratory, Hokkaido University.

http://IABP.apl.washington.edu
– Index of Animations, International Arctic Buoy Programme (IABP).

http://nsidc.org
– National Snow and Ice Data Center (NSIDC).

http://www.awi.de
– Sea Ice Physics, The Alfred Wegener Institute for Polar and Marine Research (AWI).

POLYNYAS

S. Martin, University of Washington, Seattle, WA, USA

Introduction

Polynyas are large, persistent regions of open water and thin ice that occur within much thicker pack ice, at locations where climatologically, thick pack ice would be expected. Polynyas have a rectangular or oval aspect ratio with length scales of order 100 km; they persist with intermittent openings and closings at the same location for up to several months, and recur over many years. In contrast to polynyas, leads – another open water feature – are long, linear transient features associated with the pack ice deformation, are not restricted to a particular location, and generally have a much smaller area than polynyas. Polynyas occur in both winter and summer. Given that their physical behavior in winter is more complicated than in summer, we begin with the winter case, then follow with a shorter description of their transition to summer.

Polynyas can be classified into coastal and open-ocean polynyas. Costal polynas form where the winter winds advect the adjacent pack ice away from the coast, so that sea water at temperatures close to the freezing point is directly exposed to a large negative heat flux, with the resultant rapid formation of new ice. This new ice is advected away from the coast as fast as it forms. For these polynyas, a typical alongshore length is 100–500 km; a typical offshore length 10–100 km. In contrast, the less common open-ocean polynyas have characteristic diameters of 100 km and are driven by the upwelling of warm ocean water, which maintains a large opening in the pack ice. Because the atmospheric heat loss from the open-ocean polynyas goes into cooling of the water column, they are sometimes called 'sensible heat' polynyas; because the heat loss from coastal polynyas goes into ice growth, they are called 'latent heat' polynyas. Finally, some polynyas, notably the North Water polynya in Baffin Bay, are maintained by both upwelling and ice advection.

Figure 1 shows the locations of some of these polynyas for both hemispheres. In the Northern Hemisphere, most of the polynyas are coastal, with the Kashevarov Bank polynya in the Okhotsk Sea being the only purely open-ocean polynya. The largest coastal polynyas occur in the marginal seas, where the adjacent ice edge provides room for ice divergence; this means that large polynyas occur in the Bering, Okhotsk, and Barents Seas. In the Barents Sea, prominent polynyas occur around Novaya Zemlya, Franz Josef Land, and at occasional coastal sites in the Barents and adjacent Kara Seas. In the Laptev Sea, polynyas occur in early winter along Severnaya Zemlya. There is also a long flaw lead, sometimes referred to as a polynya, which occurs between the shorefast and the pack ice north of the Laptev River delta.

In the vicinity of the Bering Strait, polynyas occur in the Chukchi, Beaufort, and Bering Seas. In the Chukchi Sea, an important polynya occurs along the Alaskan coast; in the Beaufort Sea, polynyas form in early winter along the Alaskan coast and in Amundsen Gulf. In the Bering Sea, large polynyas occur along the Alaskan coast and south of the islands, where the most investigated of these occurs south of St. Lawrence Island. Prominent polynyas also occur along the Siberian coast south of the Bering Strait and in Anadyr Gulf. The Canadian islands are also sites of several polynyas, the largest and most studied being the North Water polynya in Baffin Bay. In north-east Greenland, the North-east Water (NEW) polynya is large in summer, and has also been the subject of a recent field study. Finally, the Okhotsk Sea with the adjacent Tatarskiy Strait is the site of several coastal polynyas and the open-ocean Kashevarov polynya. The Kashevarov polynya occurs over the 200 m deep Kashevarov Bank, where the turbulence associated with a strong tidal resonance generates a heat flux to the surface that creates a region of reduced ice cover with a characteristic diameter of about 100 km.

The Southern Hemisphere is characterized by many coastal polynyas and by two or three open-ocean polynyas. The open-ocean polynyas include the Weddell Sea polynya, which only occurred in the 1970s, and the smaller Maud Rise and Cosmonaut Sea polynyas. The coastal polynyas occur at different locations along the Ronne, Amery, and Ross Ice Shelves, where the Ross polynya has recently been studied, and at many locations along the coast. The coastal polynyas form in the lee of headlands, islands, ice shelves, and grounded icebergs and downwind of ice tongues such as the Mertz, Dibble, and Dalton tongues, which extend as much as 100 km off the coast. These polynyas are created by a

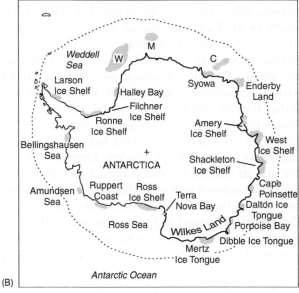

Figure 1 Geographic distribution of polynyas. (A) The Arctic, where SLIP is the St. Lawrence Island Polynya, NEW is the North-east Water, and K is the Kashevarov Bank polynya. (B) The Antarctic, where W is the Weddell polynya, M is the Maud Rise polynya and C is the Cosmonaut Sea polynya. On both figures the dashed line indicates the position of the maximum ice edge. Polar stereographic map projection courtesy of the National Snow and Ice Data Center (NSIDC).

combination of the dominant easterly winds and the very cold powerful katabatic winds that sweep down the glacial drainage basins and across the pack ice for a few kilometers before slowing. The response to

these winds is a series of large polynyas; for example, the overwinter average area of the Mertz Ice Tongue polynya is about 20 000 km². Given the prevalence of coastal polynyas compared with mid-ocean polynyas, most of the water mass modification takes place in the former.

Although these polynyas occupy only a small fraction of the areal winter pack ice extent, because the polar pack ice is a good insulator, very large atmospheric heat losses occur from both polynya types. Given these heat losses, the coastal polynyas are regions where large amounts of ice are generated, the ocean is cooled and salt is added to the underlying waters, while the open-ocean polynyas cool the upwelled water, and in the Southern Ocean are suspected to contribute to modification of the Antarctic Bottom Water. As winter progresses into spring, the polynya regions remain important. Because the predominant winds sweep the polynyas free of ice, their ice cover at the end of winter consists of either open water or thin ice. This means that as spring approaches, when the air temperature rises above freezing and the incident solar radiation increases, the polynya ice is either swept away without replacement, or melts away faster than the surrounding pack ice. In the Okhotsk and Bering Seas, for example, the onset of open water in spring occurs both from melting at the offshore ice edge, and from the disappearance of ice at the coastal polynya sites. Because the other Arctic and Antarctic coastal polynyas behave similarly, the coastal polynyas become seasonally open approximately one month earlier than the interior pack. The open ocean polynyas, such as the Kashevarov, Maud, and Cosmonaut, also serve as regions for initiation of the spring melt. As is shown below, the early spring melt of the coastal polynyas has biological consequences.

Physical Processes within the Two Polynya Types

Coastal Polynyas

Because of their relative accessibility and more frequent occurrence, we know much more about coastal than about open-ocean polynyas. **Figure 2**, a schematic drawing of a coastal polynya, shows that as the winds advect the pack ice away from the coast, open water is exposed to the cold winds. This generates a wind-wave field on the open water surface, where the wave amplitudes and wavelengths increase away from the coast. As the polynya width increases, and if the wind speed is greater than about 5–10 m s⁻¹, the interaction of the waves with the wind stress creates Langmuir circulation within the water column. This

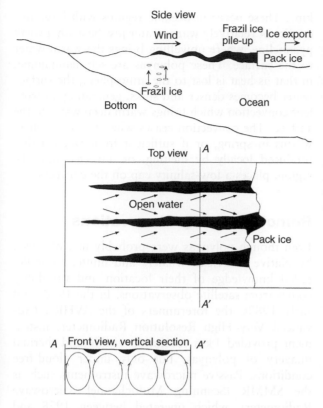

Figure 2 A schematic drawing of a coastal polynya and top and vertical views.

Figure 3 A RADARSAT image of the St. Lawrence Island polynya on 9 January 1999. The length of the island is approximately 150 km. On the image, (a) is the island, (b) is the fast ice south of the island, (c) is open water, (d) is the Langmuir plumes, (e) is the pack ice, (f) is the piled-up grease ice, and (g) is the thick first-year ice north of the island. The wind is blowing from top right to bottom left. Image courtesy of the Alaska SAR Facility, with processing courtesy Harry Stern and copyright the Canadian Space Agency, 1999.

circulation consists of rotating vortices with the rotor axes approximately parallel to the surface winds, where adjacent rotors turn in opposite directions and the rotor diameter is approximately equal to either the bottom depth in well-mixed waters or to the halocline depth. Because of the wave and Langmuir mixing, the initial ice formation occurs as follows. If the sea water temperature is above freezing, the combination of the surface heat loss with the mixing cools the entire column to the freezing point, and sometimes even causes a slight supercooling. This means that once freezing begins, ice formation occurs throughout the water column as small millimeter-scale crystals, called frazil crystals, which float slowly to the surface. As the crystals form, they reject salt to the underlying water column, leading to an oceanic brine flux.

Once these crystals reach the surface, the circulation herds them into slurries taking the form of long bands or plumes of floating ice crystals located at the Langmuir convergence zones. The slurries have thicknesses of order 10 cm, are highly viscous, and damp out the incident short waves. This damping gives the slurries a greasy appearance, so that following old whaling terminology, they are sometimes called grease ice. As the plumes grow downwind, they become wider and increase in thickness. As their thicknesses increase, their surface begins to freeze.

The longer ocean swell propagating through the ice, breaks the surface ice into floes with diameters of 0.3 to 0.5 m, called pancake ice. Because of wave-induced collisions, the swell also causes the growth of raised rims around the pancakes, which increase both the wind-drag on the ice and the radar reflectivity. As these ice growth processes proceed, the ice is advected downwind by the wind stress, where it piles up against the edge of the solid pack ice. As time goes on, the width of this region of piled-up ice grows slowly upwind. As an example, **Figure 3** shows a 100 m resolution RADARSAT image of the St. Lawrence Island polynya. The figure shows the region of open water and Langmuir plumes surrounded by pack ice south of the island, where the plumes are approximately parallel to the wind, and the polynya area is about 5000 km². Within the plumes, the figure also shows the downwind increase in brightness associated with the growth of pancake ice.

If the wind speeds are slow enough that the Langmuir circulation either does not occur or is not strong enough to circulate the ice crystals the polynya behavior is less well understood. Observations suggest that the downwind transport of ice continues to occur, but with either frazil or thin ice forming immediately adjacent to the coast. Given the offshore transport of ice in both the Langmuir and non-Langmuir cases, the question arises of what determines the winter polynya size. The crosswind polynya scale is set by the coastline configuration over which the ice divergence occurs. The downwind scale is set by a balance between the production of new ice within the polynya, its export downwind to the pack ice edge, and the subsequent upwind growth of the piled up new ice. All else remaining constant, a greater polynya area leads to more ice production and a faster upwind growth of the new ice. An equilibrium size occurs when the retreat of the pack ice edge equals the advance of the piled-up new ice. This balance between production and advection sets the polynya offshore length scale, typically 10–100 km. When the winds stop, the export stops, and the frazil ice freezes into a solid ice cover. For the same wind speed, very cold temperatures yield smaller polynyas because the ice production is so much greater; relatively warm temperatures correspond to large polynyas with less ice production.

Open-ocean Polynyas

The open-ocean polynyas generally occur away from the coast, and are driven by the upwelling of warm sea water (**Figure 4**). In the Northern Hemisphere, the Kashevarov polynya is driven by a tidal resonance, and the North Water and NEW are maintained by a combination of wind-induced ice advection and oceanic upwelling. In the Southern Hemisphere, there have been three open-ocean polynyas. The most famous, intriguing, and mysterious of these occurred in the Weddell Sea during 1973–76, and was only observed by remote sensing. The Weddell polynya had an open water and thin ice area of about 2–3×10^5 km^2, which is about the area of Oregon, USA. Because this polynya first occurred over Maud Rise, its subsequent evolution may have been caused by a large eddy that separated from the rise and migrated westward across the Weddell Sea. For the other open-ocean polynyas, both the Cosmonaut and Maud polynyas have maximum areas of about 10^5

km^2. These occur in oceanic regions with large reservoirs of relatively warm water just beneath a weak pycnocline, where upwelling brings the warm water to the surface. These polynyas are self-maintaining, in that as heat is lost to the atmosphere, the surface water becomes denser and sinks, generating a turbulent convection which brings warm deep water to the surface. The convection ceases when the atmosphere warms in spring, or if sufficient fresh water, either produced locally by melting, or advected into the region, places a low-salinity cap on the convection.

Remote Sensing Observations

Even though polynyas were probably first observed by Native American whalers and hunters, our detailed knowledge of their location and variability comes from satellite observations. In the 1970s and early 1980s, the forerunners of the AVHRR (Advanced Very High Resolution Radiometer) instrument provided 1 km resolution visible and thermal imagery of polynyas, but only under cloud-free conditions. Passive microwave instruments such as the SMMR (Scanning Multichannel Microwave Radiometer), which operated between 1978 and 1987, and the SSM/I (Special Sensor Microwave/Imager, operating between 1987 and the present, made it possible to obtain cloud-independent low-resolution imagery of the entire polar pack at intervals of every other day for the SMMR and daily intervals for the SSM/I. These observations led to the discovery of all of the mid-ocean polynyas, and many of the coastal polynyas, and provided time series of their area versus time. The major problem with the SMMR and SMM/I is their low spatial resolution; the 37 GHz SSM/I channel has a 25 km resolution, and the more water vapor-sensitive channel at 85 GHz has a 12.5 km resolution. The planned AMSR (Advanced Microwave Scanning Radiometer), which is scheduled for launch in 2000, has twice the resolution of the SSM/I and should greatly improve our polynya studies. The high-resolution active microwave satellite instruments that are just beginning to be used in polynya research include the 100–500 km swath width synthetic aperture radars (SAR) with their approximately 3-day repeat cycle, and resolutions of 12.5–100 m (**Figure 3**).

Physical Importance

The physical importance of the coastal polynyas is that, because the new ice is being constantly swept away, the polynyas serve as large heat sources to the atmosphere, and as powerful ice and brine factories.

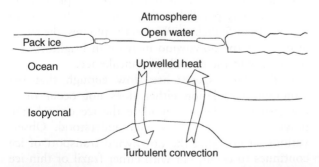

Figure 4 A schematic drawing of an open-ocean polynya, showing mean circulation and mixing.

For example, while a typical Bering Sea ice thickness is about 0.5 m, the seasonal ice growth in the St. Lawrence polynya is about 5 m, or an order of magnitude greater. Combined with the persistent north-easterly winds, this means that the Bering ice cover can be approximated as a conveyor belt, where the ice forms in the coastal polynyas, then is advected to the ice edge where it melts. This simple conveyor-belt model may also apply to the ice covers of the Weddell and Okhotsk Seas. The brine generated by the polynyas also contributes to the oceanic water masses. The large shallow continental shelves where most of the Arctic coastal polynyas occur provide a dynamic constraint on the brine-generated dense water that permits its density and volume to increase. Because of this constraint, when the dense water eventually drains off the shelves into the deeper basins, it is dense enough to contribute to the intermediate and deep water masses. In this way, the polynyas in the Chukchi, Bering, Beaufort, and Barents Sea polynyas contribute to the cold halocline layer of the Arctic Ocean. (The Bering Sea dense water is transported through the Bering Strait into the Arctic by the pressure difference between the Bering Sea and the Arctic Ocean.) In the Okhotsk Sea, the polynya water contributes to the Okhotsk Intermediate Water and to the North Pacific Intermediate Water. The dynamics and fate of this dense water are topics of current scientific investigation.

In the Antarctic, estimates of the polynya ice growth rates are $0.1–0.2\,\mathrm{m\,d^{-1}}$, or about 10 m per season. Although the Antarctic shelves are not as broad as the Arctic shelves, the Antarctic coastal polynyas also generate dense shelf water. The drainage of this water contributes to the Antarctic bottom water in the western Weddell Sea, the Ross Sea, and along the Wilkes Land Coast. In the Weddell Sea, the dense water flows beneath the Filchner-Ronne Ice Shelf, causing melting of the glacial ice, and creating a lower-salinity form of the bottom water. The Antarctic open-ocean polynyas cool the warm upwelled deep ocean water, which leads to modification of the intermediate-depth water into cold bottom water.

The Arctic polynyas also contribute to sediment transport. Because freezing occurs at all depths during the initiation of ice formation in coastal polynyas, nucleating ice crystals adhere to rocks and sediments on the bottom, forming what is called 'anchor ice.' Although this has never been directly observed in polynyas because of the hazardous conditions, observations by Alaskan divers following severe autumnal storms have found sediment-laden frazil ice on the surface, and anchor ice on the bottom. Downwind of these polynya regions, there are also many observations of pack ice containing layers with large sediment concentrations. These observations suggest that as the amount of anchor ice increases, the buoyancy of the sediment/ice mixture lifts the material to the surface, where it accumulates downwind. At river deltas such as the Laptev River, this means that sediments carried by the river into the delta can be incorporated into the frazil ice, for later export across the Arctic. Laboratory studies also suggest that the Langmuir circulation may also directly mix bottom sediments into the water column, where these sediments are then incorporated into the frazil ice and carried to the surface. By a combination of these routes, the coastal polynyas probably serve as a source of the observed sediments in the polar ice. This mechanism also provides a route for sediments laden with contaminants or radionuclides carried into the polynya regions by the rivers, to be incorporated into the pack ice, and then be transported across the Arctic.

Biological Importance

In the winter Canadian Arctic, because the marine mammals living under the ice need breathing holes, these mammals tend to concentrate in the polynya regions. For example, the North Water contains large concentrations of white whales, narwhals, walruses, and seals, with polar bears foraging along the coast. Also, the major winter bird colonies in the Canadian islands are located adjacent to polynyas, especially the North Water, and archeological evidence shows that the coastal region adjacent to the NEW was the site of human settlements. All of this suggests that polynyas are vital for the overwinter survival of arctic species. In spring and summer, the polynya regions continue to have large concentrations of marine mammals. This is because, as the polynya regions become ice-free earlier than the rest of the pack, they annually absorb more solar radiation than an adjacent region covered with thick pack ice. As a result, the polynya regions in summer have a much greater primary productivity than regions with heavy winter pack ice, so that these regions are important feeding areas for whales. Also, in the US and Canadian Arctic, most of the whales migrate into the region through passages generated by the spring melt of polynyas. For example, the opening of the polynya region along the Alaskan Chukchi coast provides a whale migration route to the Beaufort Sea, and in the Canadian Arctic the early opening of the polynya regions provides migration routes for narwhals in Lancaster Strait.

Conclusions

Polynyas are persistent openings in the ice cover that in winter ventilate the warm ocean directly to the cold atmosphere. The major physical importance of the coastal polynyas is due to their large production of ice and brine, where the resultant dense water contributes to various Arctic, Antarctic, and North Pacific water masses. In the Southern Hemisphere, the winter convection within the mid-ocean polynyas also leads to cooling of the upwelled ocean water and its modification into Antarctic bottom water. Also, although this has not been elaborated on because of space considerations, the polynyas may enhance the exchange of gases such as carbon dioxide between the atmosphere and ocean. The biological importance of the polynyas is that at least in the US and Canadian Arctic, they serve as an important winter and summer habitat for marine birds and mammals. The polynyas also play a role in spring and summer as regions of early open water formation, and as sites of significant increases in primary productivity associated with the early absorption of incident solar radiation in the water column.

Acknowledgment

This material is based upon work supported by the National Science Foundation under Grant No. 9811097.

See also

Langmuir Circulation and Instability. Satellite Passive-Microwave Measurements of Sea Ice. Satellite Remote Sensing Sar. Sea Ice: Overview. Sub Ice-Shelf Circulation and Processes. Surface Gravity and Capillary Waves.

Further Reading

Comiso JC and Gordon AL (1998) Interannual variability in summer sea ice minimum, coastal polynyas and bottom water formation in the Weddell Sea. In: Jeffries MO (ed.) *Antarctic Sea Ice, Physical Processes, Interactions and Variability*. Antarctic Research Series 74, pp. 293–315. Washington, DC: American Geophysical Union.

Gordon AL and Comiso JC (1988) Polynyas in the Southern Ocean. *Scientific American* 256(6): 90–97.

Martin S, Steffen K, Comiso JC, *et al.* (1992) Microwave remote sensing of polynyas. In: Carsey F (ed.) *Microwave Remote Sensing of Sea Ice*. Geophysical Monograph 68, pp. 303–312. Washington, DC: American Geophysical Union.

Overland JE, Curtin TB, and Smith WO (eds.) (1995) Special Section: Leads and Polynyas. *Journal of Geophysical Research* 100: 4267–4843.

Smith SD, Muench RD, and Pease CH (1990) Polynyas and leads: an overview of physical processes and environment. *Journal of Geophysical Research* 95(C6): 9461–9479.

Stirling I (1997) The importance of polynyas, ice edges, and leads to marine mammals and birds. *Journal of Marine Systems* 10: 9–21.

ICEBERGS

D. Diemand, Coriolis, Shoreham, VT, USA

Introduction

Icebergs are large blocks of freshwater ice that break away from marine glaciers and floating ice shelves of glacial origin. Although they originate on land, they are often included in discussions of sea ice because they are commonly found surrounded by it. However, unlike sea ice, they are composed of fresh water; therefore their origin, crystal structure, and chemical composition, as well as the hazards they pose, are different.

They are found in both polar regions, their sizes and numbers generally being greater at higher latitudes. They pose a hazard both to shipping and to seabed structures.

Origins and Spatial Distribution

The great ice sheets of Greenland and Antarctica, which produce by far the greatest number of the world's icebergs, flow off the land and into the sea through numerous outlet glaciers. In many cases, especially in Antarctica, the ice spreads out on the sea surface, staying connected to land and forming a floating ice shelf of greater or lesser extent.

There are two major differences between the calving fronts in Greenland and those in Antarctica. First, most of the Greenland icebergs are calved directly from the parent glaciers into the sea, while Antarctic icebergs are mostly calved from the edges of the huge ice shelves that fringe much of the continent. The result is that southern icebergs at the time of calving tend to be very large and tabular, while the northern ones are not so large and have a more compact configuration.

Second, the equilibrium line of the Greenland ice sheet is above 1000 m. Therefore, the entire volume of a Greenland iceberg is composed of ice. In Antarctica, on the other hand, the equilibrium line is at or near the edge of the ice shelves, so that icebergs are commonly calved with an upper layer of permeable firn (see Ice Properties below) of varying thickness that influences later deterioration rates and complicates estimates of draft and mass.

The drift of icebergs is largely governed by ocean currents, although wind may exert some influence. Since ocean currents at depth may differ in speed and direction from surface currents, a large iceberg may move in a direction different from that of the surrounding sea ice, creating a patch of open water behind it. Since its speed and direction are heavily dependent on the depth and shape of the keel, which is usually unknown, trajectory predictions are seldom reliable, even when local current profiles are known.

Because the Antarctic continent is surrounded by oceans while the Arctic is an ocean surrounded by continents, the drift patterns of the icebergs from these areas are very different.

Baffin Bay to North Atlantic Region

About 95% of icebergs in northern latitudes originate on Greenland. Most of these are from western Greenland where they calve directly into Baffin Bay, but a few are produced in eastern Greenland. Many of these remain trapped in the fiords where they originated, deteriorating to a great degree before they reach the sea. Those from eastern Greenland that do reach the sea drift south in the East Greenland Current, a small number continuing south into the North Atlantic where they rapidly dwindle, others being carried around the southern tip of Greenland and then north in the warm West Greenland Current, where the few that survive the long trip join the great numbers of bergs originating from Disko Bay north. **Figure 1** shows some of the most active glaciers on Greenland and the drift paths generally followed by icebergs.

Icebergs may remain in Baffin Bay for several years, circulating north along the Greenland coast and then south along the Canadian arctic islands. Since the water temperature in Baffin Bay remains consistently low throughout the year, little deterioration takes place. However, many do escape southward through the Davis Strait and drift down the Labrador coast in the cold Labrador Current until they break free of the annual pack ice and reach the Grand Banks off Newfoundland.

Icebergs have been sighted as far south as Bermuda and the Azores.

Arctic Ocean

The remaining 5% of northern icebergs are calved from numerous glaciers on Ellesmere Island in the Canadian Arctic, the many islands in or bordering the Barents, Kara, and Laptev Seas, and Alaska (see **Figure 2**). Many of these, especially those calved from ice

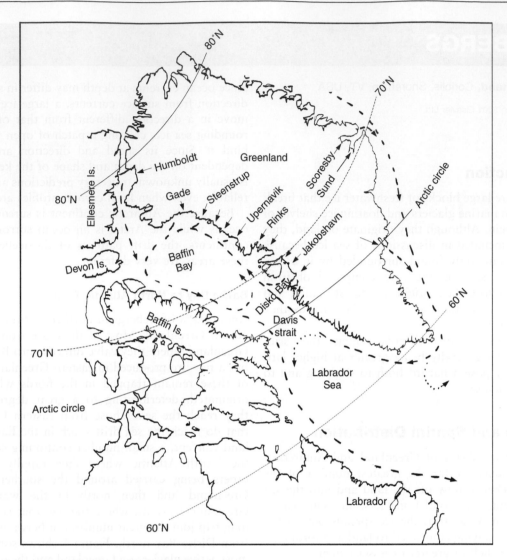

Figure 1 Sources and drift paths of North Atlantic icebergs.

shelves on Ellesmere Island, are tabular in form. When they were first discovered drifting among the Arctic pack, they were referred to as 'ice islands', and the name stays with them. Once they have become incorporated into the pack, they tend to stay there indefinitely, although occasionally one may escape and join the southbound flux through Davis Strait or the east coast of Greenland. The sources and trajectories of these icebergs and ice islands are shown in **Figure 2**.

Southern Regions

Since there is no significant runoff from Antarctica, iceberg production accounts for most mass loss from the continent. Most of these icebergs are calved from the massive ice shelves, such as the Ross, Filchner, Ronne, Larsen, and Amery. About 60–80% by volume are calved from the ice shelves, the remainder from outlet glaciers that empty directly into the sea

or from active ice tongues. Once free of the ice front, they drift with the prevailing current along the coast, in some places westward, in others eastward as shown in **Figure 3**. They may remain close to the coast, where their concentration is the greatest, for periods up to 4 years, protected by the sea ice and the cold water. There are several localized places around the coast where icebergs turn north away from the continent. Once they drift beyond the northern limit of the pack ice, about 60°S, they are carried east and north into ever warmer waters until they deteriorate.

The most northerly reported sighting was at 26°S near the Tropic of Capricorn. Few pass 55°S.

Numbers and Size Distribution

Our knowledge of the numbers and size distribution of icebergs is based on visual observations from

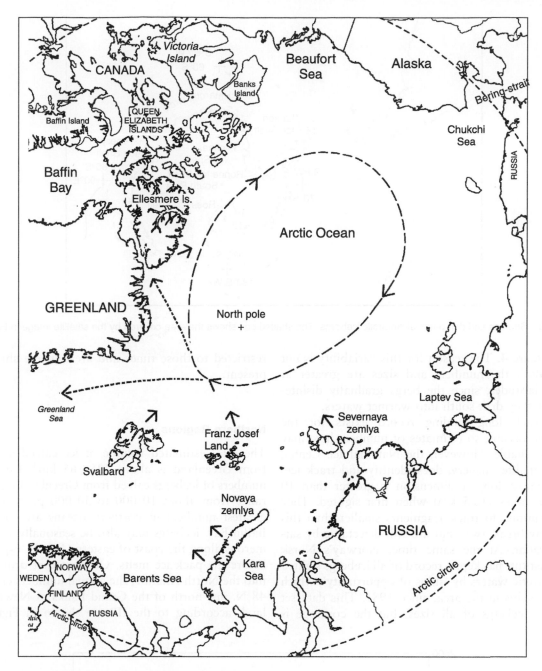

Figure 2 Sources and drift paths of icebergs in the Arctic Ocean.

ships and aircraft; from radar data from ships, shore, and aircraft; and from satellite imagery. Each method has its advantages and shortcomings. For example, satellite imagery covers very large areas and all times of the year, but will not detect small bergs; ship's radar will pick up most icebergs within its range but may miss rounded bergs or small bergs in heavy seas; visual observation will catch all sizes of bergs, but only within a limited area in good weather when someone is looking. Thus, any iceberg census will be slanted toward the size and

shape categories favored by the observation method used.

The most detailed records of iceberg numbers and sizes in a single location have been kept by the US Coast Guard's International Ice Patrol (IIP) which was formed in the aftermath of the sinking of the *Titanic*. The IIP began patrolling the Grand Banks in 1914 and reporting iceberg locations to ships in the area. Since that time the IIP has kept a detailed record of all icebergs crossing 48°N. These numbers are highly variable from year to year as is apparent

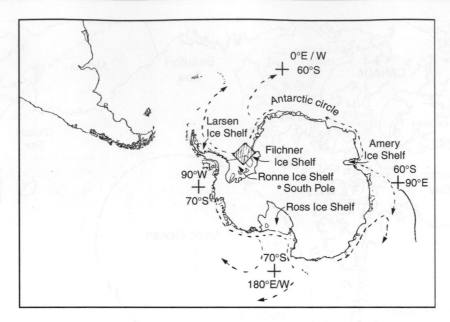

Figure 3 Sources and drift paths of Antarctic icebergs. The shaded box shows the area covered by the satellite image in **Figure 5**

from **Figure 4**. The reason for this variability is not clear. Both the numbers and sizes are greater at higher latitudes, since the bergs gradually disintegrate as they drift south into warmer waters.

No such long-standing record exists for the southern oceans, so estimates of numbers here may be less reliable. However, the National Ice Center, using satellite imagery, does identify and track icebergs whose longest dimension is greater than 10 nautical miles (18.5 km) when first sighted. They also continue to track fragments smaller than this that may break away but are still detectable by satellite radar. At the same time, Norway's Norsk Polarinstitutt has kept a record of all icebergs sighted in Antarctic waters by 'ships of opportunity', which is most ships in the area, since 1981. This data set includes icebergs of all sizes, but the coverage is restricted to those times and areas where ships are present.

Northern Regions

The total estimated volume of ice calved annually from Greenland is about 225 ± 65 km^3. Estimated numbers of icebergs calved from Greenland's glaciers range from about 10 000 to 30 000 per year. The greatest numbers in northern oceans are found in Baffin Bay. Icebergs may also be seasonally very numerous along the coast of eastern Canada, especially before the pack ice melts. Of those that drift south into the North Atlantic, the annual numbers crossing 48°N, just north of the Grand Banks of Newfoundland, according to the IIP are shown in **Figure 4**.

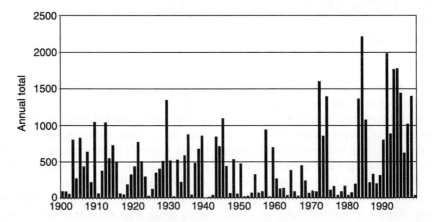

Figure 4 Total numbers of icebergs crossing 48°N each year from 1900 through 1999. Note: The figures for the years of World War I and II are incomplete. (Data courtesy of the US Coast Guard International Ice Patrol.)

Once they encounter the warm Gulf Stream waters, they rapidly deteriorate.

Southern Regions

The total estimated volume of ice calved annually from Antarctica ranges from 750 to 3000 km³ per year. The occasional release of extremely large icebergs has a major impact on annual estimates of Antarctic mass loss. The iceberg shown in **Figure 5.** represents as much ice as the total annual mass loss from a 'normal' year. The shaded box shown in **Figure 3** is the area covered by this image. Estimated numbers of icebergs calved range from 5000 to 10 000 each year.

Shapes and Sizes

The range of sizes of icebergs is enormous, spanning about eight orders of magnitude, from small

fragments with a mass around 1000 tonnes to the immense Antarctic tabular begs with masses in excess of 10^{10} t. **Table 1** shows the normal range of iceberg sizes in the Labrador Sea.

In terms of shape, no two icebergs are the same. However, as bergs deteriorate they do tend to assume characteristic forms (**Figures 6–10**).

The shape classification in common use is given in **Table 2**. Specialized terms used in classification and description are defined in the Glossary. It should be borne in mind that the shape or extent of the 'sail' does not necessarily reflect the shape of the entire iceberg. **Figure 11A** and **B** show a photograph of an iceberg and a computer-generated image of its underwater configuration. The nearly spherical shape of this medium-sized berg suggests that it had rolled, probably recently and probably several times. Any horizontal tongues of ice, or 'rams', would have broken away during this energetic process. **Figure 12A** and **B** show a larger iceberg that had only tilted from its original in the water. Extensive rams are visible extending outward underwater far beyond the extent of the sail, remnants of a far greater mass that has been lost since the berg moved into relatively warm waters. While these two icebergs have roughly similar shapes above water, their underwater configurations are very different.

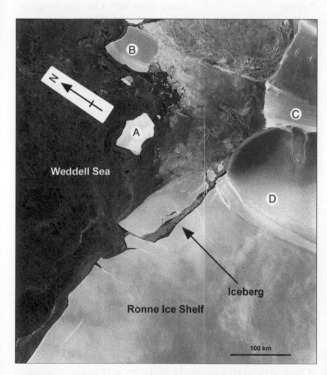

Figure 5 Satellite image showing an extremely large iceberg breaking away from the Ronne Ice Shelf in 1998. The area covered by this image is indicated in **Figure 3**. A and B are the remnants of two other very large icebergs, both of which broke free in 1986 and were grounded at the time of this image. C is a rapidly moving stream of ice moving off the continent through the Filchner Ice Shelf and past Berkner Island (D) into the Weddell Sea. (Radarsat data © 1998 Canadian Space Agency/Agence spatiale canadienne. Received by the Canada Centre for Remote Sensing (CCRS). Processed by Radarsat International (RSI) and the Alaska SAR facility (ASF). Image enhancement and interpretation by CCRS. Provided courtesy of RSI, CCRS, ASF, and the National Ice Center.)

Table 1 Iceberg size categories

Designation	Height (m)	Length (m)	Approximate mass (Mt)
Growler	<1	<5	0.001
Bergy Bit	1–5	5–15	0.01
Small	5–15	15–60	0.1
Medium	16–45	61–120	2
Large	46–75	121–200	10
Very Large	>75	>200	>10

Figure 6 Tabular iceberg. (Photograph by Deborah Diemand.)

Figure 7 Wedge iceberg. (Photograph by Deborah Diemand.)

Figure 10 Domed iceberg. (Photograph by Deborah Diemand.)

Figure 8 Pinnacle iceberg. (Photograph by Deborah Diemand.)

Table 2 Iceberg shape categories

Designation	Description
Tabular types	
Tabular	Steep sides with a flat top; length to height ratio >5:1 (**Figure 6**)
Blocky	Similar to tabular, but length to height ratio <5:1
Nontabular types	
Wedge	One flat side sloping gradually to the water; the opposite side sloping steeply, the two meeting at the peak as a spine (**Figure 7**)
Pinnacle	With one or more sharp peaks (**Figure 8**)
Drydock	With two or more peaks separated by a water-filled channel (**Figure 9**)
Dome	Small with rounded top (**Figure 10**)

Figure 9 Drydock iceberg. (Photograph by Deborah Diemand.)

Because of this uncertainty in the underwater shape, it is impossible to calculate iceberg draft and mass accurately using the sail dimensions. However, certain rules of thumb have emerged from empirical studies. To approximate the draft in meters, the relationship of eqn [1] can be used.

$$\text{Draft} = 49.4 \times \left(\text{height}^{0.2}\right) \qquad [1]$$

To approximate the mass in tonnes, that of eqn [2] can be used.

$$\begin{aligned}\text{Mass} = 3.01 \times [&(\text{longest sail dimension(m)}) \\ &\times(\text{orthogonal width(m)}) \qquad [2] \\ &\times(\text{maximum height(m)})]\end{aligned}$$

These calculations apply only to icebergs composed entirely of ice, with no firn layer.

Northern Regions

In general, the mean size of icebergs in Baffin Bay is about 60 m height, 100 m width and 100 m draft. Mean mass is about 5–10 Mt. The sizes of icebergs in this area are constrained by the water depth near the calving fronts, which is less than 200 m. Icebergs with a mass greater than 20 Mt are extremely rare, and for those found south of 60°N, a mass greater

(A)

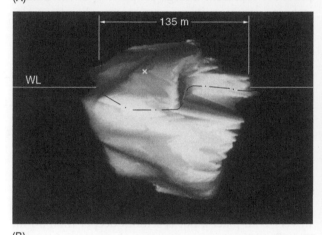

(B)

Figure 11 (A) Medium-sized iceberg grounded in 107 m of water. (B) Computer-generated image showing the underwater profile of this berg. The shape and size of the sail were established through stereophotography; those of the keel were determined using acoustic profiling techniques. These two datasets were joined electronically to create this image as well as **Figure 12B**. WL, water line. ((A) Photograph by Deborah Diemand; (B) courtesy of Dr. James H. Lever.)

than 10 Mt is seldom found. The maximum sail height on record for an iceberg in the North Atlantic is 168 m.

The ice islands in the Arctic Ocean extend about 5 m above sea level. They have a thickness of 30–50 m and an area from a few thousand square metres to 500 km² or more.

Southern Regions

In general, the thickness of the ice shelves at the calving fronts is about 200–250 m, increasing away from the front. Since the ice edge is very long, and often seaward of seabed obstructions, Antarctic icebergs may be extremely large.

The iceberg shown in **Figure 5** is more than 5800 km², larger than the US state of Rhode Island. The

(A)

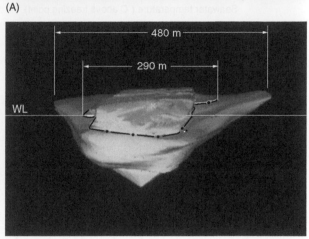

(B)

Figure 12 (A) Large iceberg grounded in 134 m of water. (B) Computer-generated image showing the underwater profile of this berg. WL, water line. ((A) Photograph by Deborah Diemand; (B) courtesy of Dr. James H. Lever.)

largest iceberg ever reported was about 180 km long, with an estimated volume of 1000 km³.

Deterioration

Deterioration begins as soon as an iceberg calves from its parent glacier. Even icebergs locked into sea ice over the winter show signs of mass loss. However, significant deterioration does not usually begin until after the berg breaks free from the pack ice and is exposed to warmer surface water and wave action. The major causes of ice loss are melting, calving, and splitting and ram loss.

Melting

Ice loss due to melting alone is hard to quantify, but is thought to fall within the shaded region shown in **Figure 13**. It is highly dependent on water temperature, but is also influenced by wave action, water currents, and bubble release. Melting rate at the

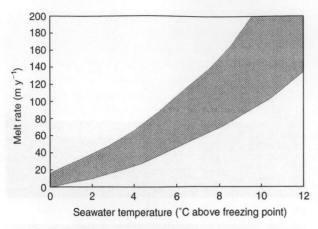

Figure 13 Dependence of the melt rate of icebergs on the temperature above the freezing point of sea water.

water line is far greater than that over the rest of the ice surface. This causes a groove to form, undercutting the ice cliffs and creating sometimes extensive underwater rams.

The importance of melting in the overall mass loss depends on the surface/volume ratio, being more significant for small bergs and growlers than for large ones.

One of the side effects of the rapid side melting of icebergs is the vertical mixing of the surrounding seawater. Driven mostly by the release of air from the bubbles in the ice and partly by the lower density of the fresh water of the melted ice, water flows upward near the berg, drawing deep water to the surface. The combination of nutrients brought up from depth by this process and the decreased salinity of the meltwater surrounding the berg results in a specialized community of plankton and fish in the vicinity of icebergs.

Calving of Cliff Faces

Small pieces of ice are constantly breaking off the sides of icebergs, mostly owing to waterline undercutting. Such calving events may produce only a few small pieces, or a great number, especially in warm water. Usually the individual pieces are quite small and are quickly melted, but the total mass loss can be considerable and the resulting imbalance can cause the berg to roll, causing further ice loss. Once the berg has rolled and stabilized, waterline erosion begins anew. This is probably the major cause of mass loss in medium-sized bergs.

Splitting and Ram Loss

Splitting occurs when a large iceberg breaks into two or more pieces, each of which is an iceberg in its own right. This is a common occurrence for very large

Antarctic tabular bergs, and the resulting fragments can still be extremely large. In this case, probably the main cause of breakup is flexure due to ocean waves, although grounding or collision, or both, may contribute. It is likely that grounding is the main cause of splitting for bergs smaller than 1 km^2.

While undercut cliffs on the sides of icebergs tend to shed many small fragments, the corresponding underwater rams remain intact until they are sufficiently large that buoyancy forces alone cause them to break away, or the berg grounds. These rounded fragments, which may be of considerable size, probably represent a large proportion of the domed icebergs common in warmer waters. For example, the calving of the ram extending to the right in **Figure 12B** would create a new iceberg weighing roughly 100 000 tonnes.

Splitting and ram loss are the major cause of size reduction in extremely large bergs.

Ice Properties

Glacial ice is formed by the gradual accumulation of snow over many centuries. As the snow compacts and recrystallizes, it forms firn, a granular, permeable material. The firn layer may reach as deep as 100 m in very cold places, but is seldom deeper than 50 m. This firn layer is not present on the calving fronts of Greenland, but is present on Antarctica's ice shelves, and in the icebergs calved from them. When the firn reaches a density of 830 kg m^{-3} the pores close off, trapping any air that remains. At this point the ice contains about 10% air by volume. Further densification is a result of compression of the air in the bubbles. The bubbles become smaller and may become incorporated into the crystals through recrystallization. The pressure inside them may be as high as 2 MPa (20 bars).

Acoustics

Melting icebergs in the open ocean make a characteristic sound sometimes referred to as 'bergy seltzer'. This is probably created by the explosion or implosion of bubbles as the ice melts. The frequency range of audible sound produced is quite wide, and is largely masked by ambient ocean noise at frequencies below 6 kHz. The sound seems to vary from berg to berg and is undoubtedly influenced by environmental conditions. Estimated detection distances at frequencies above 6 kHz range from 2 to 150 km.

Ice Temperature

Since ice is a good insulator, the original temperature of a large berg at the time of calving will be retained in its central core, and may be as low as $-22°C$. After a year or more in cold water, where little or no ablation takes place, the surface ice will warm to about 0°C. In relatively warm waters, however, these outer layers of ice are removed more rapidly than the inner cold core warms up, leaving much colder ice near the iceberg's surface. Since the strength of ice is greater at lower temperatures, the result of a collision with such a berg could be more severe than with one that had not undergone significant melting.

Color

In small quantities, ice appears both colorless and transparent. However, because ice selectively transmits light in the blue portion of the visible spectrum while it absorbs light of other frequencies, sufficiently large pieces of clear, bubble-free ice can appear blue. However, this color is frequently masked in glacial ice by the scattering of light of all wavelengths by the bubbles included in the ice, causing the ice to appear white. Blue bands, commonly present within the greater white mass, are caused when cracks form on the parent glacier or later on the iceberg itself and are filled with meltwater that then freezes relatively bubble free. These blue bands range in size from hairline cracks to a meter or more in width.

Green icebergs are fairly common in certain regions. This has variously been attributed to copper or iron compounds, the incorporation of dissolved organic compounds, or to an optical trick caused by red light of the sun near the horizon causing the apparent green color. It is likely that there is no single cause and that all of these factors may make the ice appear green. In some cases the trace substances may originate on land, as in the blue cracks mentioned above. In others they may result from sea water freezing to the underside of an ice shelf. Unlike in ice formed at the water surface, most salts and bubbles are rejected, but certain compounds may be trapped in trace amounts, causing the green appearance of otherwise clear ice.

Icebergs may also have bands of brown or black; these are caused by morainal or volcanic material deposited while the ice was still part of the parent glacier.

Economic Importance

Hazard to Shipping

A large iceberg poses little threat to shipping on the whole. It will not normally exceed a speed of 1 m s^{-1} (2 knots) and it can be detected with normal marine radar at a considerable distance, allowing the ship to alter course. Ironically, a greater danger is posed by much smaller bergs. These 'small' bergs may weigh in excess of 100 000 t, but owing to their small above-water size and frequently rounded shape they may not be detected by radar until they are dangerously close to the ship, especially in storm conditions when the radar return from the rough sea surface (sea clutter) will tend to mask the weak radar return from the iceberg. In such conditions the peril is further increased because these relatively small ice masses, tossed by the heavy seas, may reach maximum instantaneous velocities 4–5 times larger than hourly drift speeds. A 4000 t bergy bit moving at the maximum fluid particle velocity of 4.5 m s^{-1} in typical Grand Banks storm waves could have about a third of the kinetic energy of a 1 Mt iceberg drifting at 0.5 m s^{-1} (~ 1 knot). While the influence of waves on ice movement decreases for larger bergs, icebergs as massive as 1 Mt may still exhibit significantly higher maximum instantaneous velocities than their hourly drift values.

Seabed Damage

While small icebergs pose a serious threat to structures at the sea surface, seabed structures such as well-heads, pipelines, cables, and mooring systems are endangered by large icebergs, which may possess a deep enough draft to collide with the seafloor.

Marine navigators have long known that the keels of icebergs drifting south over the relatively shallow banks of Canada's eastern continental shelf may touch the seabed and become grounded.

Modern iceberg scours appear in the form of linear to curvilinear scour marks and as pits, and occur from the Baffin Bay/Davis Strait region to the Grand Banks of Newfoundland. They are present at water depths up to about 200 m. Seabed scouring has also been documented in Antarctica, but to date has generated little interest because of the absence of seabed structures.

A single scour may be as wide as 30 m, as deep as 10 m, and longer than 100 km. An iceberg may also produce pitting when its draft is suddenly increased through splitting or rolling. It may then remain anchored to the seafloor, rocking and twisting, and may produce a pit deeper than the maximum scour depth.

Usage of Icebergs

In the past there has been considerable interest in the possibility of transporting icebergs, representing as they do an essentially unlimited supply of fresh water, to arid areas such as Saudi Arabia, Western Australia, and South America. The two seemingly insurmountable problems that need to be solved are propulsion and prevention of in-transit breakup in warm seas. Proposed means of moving a sufficiently large ice mass over such long distances have ranged from conventional towing to use of a nuclear submarine to wind power. None has proven feasible.

Destruction of Icebergs

Attempts at destroying icebergs have been numerous and varied. Perhaps the most-studied technique has involved the use of explosives, which have been extensively tested on glacial ice in the form of glaciers and ice islands. Both crater blasting and bench blasting have been attempted. The results of this testing suggest that ice is as difficult to blast as typical hard rock, and that therefore the use of explosives for its destruction is impractical.

Other methods tested include spreading carbon black on the berg's surface to accelerate melting, and introducing various gases into the ice to create holes or cavities that can then be filled with explosives of choice. Attempts have also been made to cut through the ice using various means. There is little evidence that any great success was achieved with any of these methods.

The only report of a successful attempt to break up an iceberg involved the use of thermit, a welding compound that reacts at very high temperatures ($\sim 3000°C$). The explanation was that the very high heat produced by the thermit caused massive thermal shock within the mass of ice that ultimately resulted in its disintegration, much as glass can be fragmented by extreme temperature changes.

Conclusions

There is a great deal of uncertainty surrounding iceberg properties, behavior, drift, and other aspects relating to individual icebergs as opposed to laboratory samples or intact glaciers. This is mostly because of the high cost of expeditions to the remote areas where icebergs are most numerous, and the inherent dangers of hands-on measurement and sampling.

Glossary

Calving The breaking away of an iceberg from its parent glacier or ice shelf. Also the subsequent loss of ice from the iceberg itself.

Equilibrium line On a glacier, the line above which there is a net gain due to snow accumulation and below which there is a net loss due to melt.

Firn Permeable, partially consolidated snow with density between $400 \, kg \, m^{-3}$ and $830 \, kg \, m^{-3}$.

Growler A small fragment of glacial ice extending less than a meter above the sea surface and having a horizontal area of about $20 \, m^2$.

Keel The underwater portion of an iceberg.

Ram Lobe of the underwater portion of an iceberg that extends outward, horizontally, beyond the sail.

Sail The above-water portion of an iceberg.

See also

Sea Ice: Overview. Sub Ice-Shelf Circulation and Processes. Wind Driven Circulation.

Further Reading

Colbeck SC (ed.) (1980) *Dynamics of Snow and Ice Masses.* New York: Academic Press.

Husseiny AA (ed.) (1978) *First International Conference on Iceberg Utilization for Fresh Water Production, Weather Modification, and Other Applications.* Iowa State University, Ames, 1977. New York: Pergamon Press.

Vaughan D (1993) Chasing the rogue icebergs. *New Scientist*, 9 January.

International Ice Patrol (IIP): http://www.uscg.mil/lantarea/iip/home.html

Library of Congress Cold Regions bibliography: http://lcweb.loc.gov/rr/scitech/coldregions/welcome.html

National Ice Center (NIC): http://www.natice.noaa.gov/

ICE–OCEAN INTERACTION

J. H. Morison, University of Washington, Seattle, WA, USA

M. G. McPhee, McPhee Research Company, Naches, WA, USA

Introduction

The character of the sea ice cover greatly affects the upper ocean and vice versa. In many ways ice-covered seas provide ideal examples of the planetary boundary layer. The under-ice surface may be uniform over large areas relative to the vertical scale of the boundary layer. The absence of surface waves simplifies the boundary layer processes. However, thermodynamic and mechanical characteristics of ice–ocean interaction complicate the picture in unique ways. We discuss a few of those unique characteristics.

We deal first with how momentum is transferred to the water and introduce the structure of the boundary layer. This will lead to a discussion of the processes that determine the fluxes of heat and salt. Finally, we discuss some of the unique characteristics imposed on the upper ocean by the larger-scale features of a sea ice cover.

Drag and Characteristic Regions of the Under-ice Boundary Layer

To understand the interaction of the ice and water, it is useful to consider three zones of the boundary layer: the molecular sublayer, surface layer, and outer layer (**Figure 1**). Under a reasonably smooth and uniform ice boundary, these can be described on the basis of the influence of depth on the terms of the equation for a steady, horizontally homogeneous boundary layer (eqn [1]).

$$ifV = \frac{\partial}{\partial z}\left(v\frac{\partial V}{\partial z}\right) + \frac{\partial}{\partial z}\left(K\frac{\partial V}{\partial z}\right) - \rho^{-1}\nabla_b p \quad [1]$$

The coordinate system is right-handed with z positive upward and the origin at the ice under-surface. V is the horizontal velocity vector in complex notation ($V = u + iv$), ρ is water density, and p is pressure. An eddy diffusivity representation is used for turbulent shear stress, $K(\partial V/\partial z) = \overline{V'w'}$, where K is the eddy diffusivity. The term $v(\partial V/\partial z)$ is the viscous shear stress, where v is the kinematic molecular viscosity. The pressure gradient term, $\rho^{-1}\nabla_b p$ is equal to $\rho^{-1}(\partial p/\partial x + i\partial p/\partial y)$.

The stress gradient term due to molecular viscosity is of highest inverse order in z. It varies as z^{-2}, and therefore dominates the stress balance in the molecular sublayer (**Figure 1**) where z is vanishingly small. As a result the viscous stress, $v(\partial V/\partial z)$, is effectively constant in the molecular sublayer, and the velocity profile is linear.

The next layer away from the boundary is the surface layer. Here the relation between stress and velocity depends on the eddy viscosity, which is proportional to the length scale and velocity scale of turbulent eddies. The length scale of the turbulent eddies is proportional to the distance from the boundary, $|z|$. Therefore, the turbulent stress term varies as z^{-1} and becomes larger than the viscous term beyond z greater than $(1/k)(v/u_{*0})$, typically a fraction of a millimeter. The velocity scale in the surface layer is u_{*0}, where ρu_{*0}^2 is equal to τ_0, the average shear stress at the top of the boundary layer. Thus, K is equal to $ku_{*0}|z|$, where Von Kármän's constant, k, is equal to 0.4. Because the turbulent stress term dominates the equations of motion, the stress is roughly constant with depth in the surface layer. This and the linear z dependence of the eddy coefficient result in the log-layer solution or 'law of the wall' (eqn [2]).

$$\frac{u}{u_*} = \frac{1}{k}\ln z + C = \frac{1}{k}\ln\frac{z}{z_0} \quad [2]$$

$C = -(\ln z_0)/k$ is a constant of integration. Under sea ice the surface layer is commonly 1–3 m thick.

The surface layer is where the influence of the boundary roughness is imposed on the planetary boundary layer. In the presence of under-ice roughness, the average stress the ice exerts on the ocean, τ_0, is composed partly of skin friction due to shear and partly of form drag associated with pressure disturbances around pressure ridge keels and other roughness elements. Observations under very rough ice have shown a decrease in turbulent stress toward the surface, presumably because more of the momentum transfer is taken up by pressure forces on the rough surface. The details of this drag partition are not known. Drag partition is complicated further for cases in which stratification exists at depths shallow compared to the depth of roughness elements.

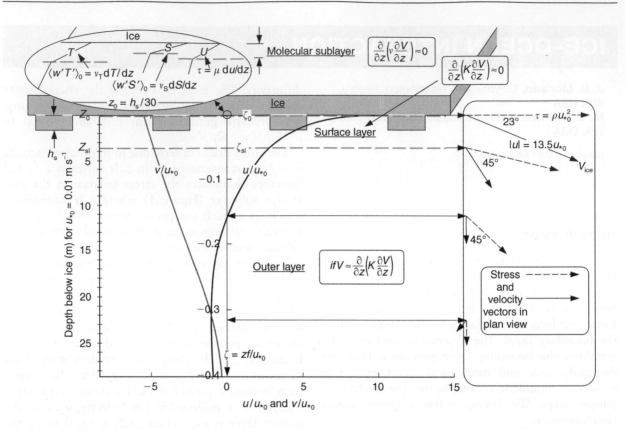

Figure 1 Illustration of three regions of the planetary boundary layer under sea ice: molecular sublayer, surface layer, and outer layer. The velocity profiles are from the Rossby similarity solution (eqns [8], [9], [10] and [11]) for $u_{*0} = 0.01 \text{ m s}^{-1}$, $z_0 = 0.06 \text{ m}$, $\eta_* = 1$. The stress and velocity vector comparisons are from the same solution.

Then it also becomes possible to transfer momentum by internal wave generation. However, for many purposes τ_0 is taken as the turbulent stress evaluated at z_0. Laboratory studies of turbulent flow over rough surfaces suggest that z_0 may be taken equal to $h_s/30$, where h_s is the characteristic height of the roughness elements.

In rare situations the ice surface may be so smooth that bottom roughness and form drag are not factors in the drag partition. In such a hydrodynamically smooth situation, the turbulence is generated by shear induced instability in the flow. The surface length scale, z_0, is determined by the level of turbulent stress and is proportional to the molecular sublayer thickness according to the empirically derived relation $z_0 = 0.13(\nu/u_{*0})$.

In the outer layer farthest from the boundary, the Coriolis and pressure gradient terms in eqn [1], which have no explicit z dependence, are comparable to the turbulent stress terms. The presence of the Coriolis term gives rise to a length scale, h, for the outer boundary layer equal to u_{*0}/f under neutral stratification. This region is far enough from the boundary so that the turbulent length scale becomes

independent of depth and in neutral conditions has been found empirically to be $\lambda = \xi_n u_{*0}/f$, where ξ_n is 0.05. For neutral stratification, u_* and h are the independent parameters that define the velocity profile over most of the boundary layer. The ratio of the outer length scale to the surface region length scale, z_0, is the surface friction Rossby number, $R_0 = u_{*0}/(z_0 f)$.

Solutions for the velocity in the outer layer can be derived for a wide range of conditions if we non-dimensionalize the equations with these Rossby similarity parameters, u_{*0}/f and u_{*0}. However, the growth and melt of the ice produce buoyancy flux that strongly affects mixing. Melting produces a stabilizing buoyancy flux that inhibits turbulence and contracts the boundary layer. Freezing causes a destabilizing buoyancy flux that enhances turbulence and thickens the boundary layer. We can account for the buoyancy flux effect by adjusting the Rossby parameters dealing with length scale. We define the scale of the outer boundary layer as $h_m = u_{*0}\eta_*/f$. If the mixing length of the turbulence in the outer layer is $\lambda_m = \xi_n u_{*0}\eta_*^2/f$, it interpolates in a reasonable way between known values of λ_m for neutral stratification

$(\xi_n u_{*0}/f)$ and stable stratification $(R_c L)$ if η_* is given as eqn [3].

$$\eta_* = \left(1 + \frac{\xi_n u_*}{f}\frac{1}{R_c L}\right)^{1/2} \qquad [3]$$

R_c is the critical Richardson number; the Obukhov length, L, is the ratio of shear and buoyant production of turbulent energy, $\rho u_{*0}^3/kg\langle\rho'w'\rangle$; and $-\langle\rho'w'\rangle g/\rho$ is the turbulent buoyancy flux. With this Rossby similarity normalization of the equations of motion, we can derive analytical expressions for the under-ice boundary layer profile that are applicable to a range of stratification.

For large $|z|$, V will approach the free stream geostrophic velocity, $\bar{V}_g = U_g + iV_g = f^{-1}\rho^{-1}\nabla_h p$. Here we will assume this is zero. However, surface stress-driven absolute velocity solutions can be superimposed on any geostrophic current. We also ignore the time variation and viscous terms and define a normalized stress equal to $\Sigma = (K\partial V/\partial z)/u_{*0}^2$. The velocity is nondimensionalized by the friction velocity and the boundary layer thickness, $U = Vfh_m/u_{*0}^2$, and depth is nondimensionalized by the boundary layer thickness scale, $\zeta = z/h_m$. With these changes eqn [1] becomes eqn [4].

$$iU = \partial\Sigma/\partial\zeta \qquad [4]$$

In terms of nondimensional variables the constitutive law is given by eqn [5].

$$\Sigma = K_*\partial U/\partial\zeta \qquad [5]$$

The nondiemensional eddy coefficient is given by eqn [6].

$$K_* = ku_{*0}\lambda_m/fh_m^2 = k\xi_n \qquad [6]$$

Eqn [6] is the Rossby similarity relation that is the key to providing similarity solutions for stable and neutral conditions. It even provides workable results for slightly unstable conditions.

Eqns [4] and [5] can be combined in an equation for nondimensionalized stress (eqn [7]).

$$(i/K_*)\Sigma = d\Sigma d/\zeta \qquad [7]$$

This has the solution eqns [8].

$$\Sigma = e^{\hat{\delta}\zeta} \qquad [8]$$

$$\hat{\delta} = (i/K_*)^{1/2} \qquad [9]$$

Eqn [8] attenuates and rotates (to the right in the Northern Hemisphere) with depth. It duplicates the salient features found in data and sophisticated numerical models.

In the outer layer, eqns [5] and [8] are satisfied for nondimensional velocity given by eqn [10].

$$U = -i\hat{\delta}e^{\hat{\delta}\zeta} \quad \text{for } \zeta \leq \zeta_0 \qquad [10]$$

Thus the velocity is proportional to stress but rotated $45°$ to the right.

As we see in the derivation of the law of the wall [2], the surface layer variation of the eddy viscosity with depth is critical to the strong shear present there. Thus eqn [10] will not give a realistic profile in the surface layer. We define the nondimensional surface layer thickness, ζ_{sl}, as the depth where the surface layer mixing length, $|z|$, becomes equal to the outer layer mixing length, $\lambda_m = \xi_n u_{*0}\eta_*^2/f$. We find ζ_{sl} is equal to $-\eta_*\xi_n$ and applying the definition [6] gives K_* as $K_{*sl} = -k\zeta/\eta_*$ in the surface layer. If we approximate the stress profile [8] by a Taylor series, we can integrate [5] with K_{*sl} substituted for K_* to obtain the velocity profile in the surface layer.

$$U(\zeta) - U(\zeta_{sl}) = \frac{\eta_*}{k}\left[\ln\left(\frac{\zeta_{sl}}{\zeta_0}\right) + \hat{\delta}(\zeta_{sl} - \zeta)\right] \quad \text{for } \zeta \geq \zeta_0 \qquad [11]$$

Eqn [11] is analogous to [2] except for the introduction of the $\hat{\delta}(\zeta_{sl} - \zeta)$ term. This is the direct result of accounting for the stress gradient in the surface layer. This term is small compared to the logarithmic gradient.

Figure 1 illustrates the stress and velocity vectors at various points in the boundary layer as modeled by eqns [8] through [11]. For neutral conditions the nondimensional boundary layer thickness is typically 0.4 (dimensional thickness is $0.4u_*/f$). Through the outer layer, the velocity vector is $45°$ to the right of the stress vector as a consequence of the $-i\hat{\delta} \propto e^{-i(45°)}$ multiplier in [10]. As the ice surface is approached through the surface layer, the stress vector rotates $10–20°$ to the left to reach the surface direction. However, in the surface layer the velocity shear in the direction of the surface stress is great because of the logarithmic profile. Thus, as the surface is approached, the velocity veers to the left twice as much as stress. At the surface the velocity is about $23°$ to the right of the surface stress.

It is commonly useful to relate the stress on under-ice surface to the relative velocity between ice and water a neutral-stratification drag coefficient,

$\rho u_{*0}^2 = \rho C_z V_{(z)}^2$ where C_z is the drag coefficient for depth z. If z is in the log-layer, eqn [2] can be used to derive the relation between ice roughness and the drag coefficient. We find that $C_z = k^2 [\ln(z/z_0)]^{-2}$. Clearly values of the drag coefficient can vary widely depending on the under-ice roughness. Typical values of z_0 range from 1 to 10 cm under pack ice. A commonly referenced value for the Arctic is 6 cm, which produces a drag coefficient at the outer edge of the log layer of 9.4×10^{-3} (**Figure 1**).

If the reference depth is outside the log layer, the drag coefficient formulation is poorly posed because of the turning in the boundary layer. For neutral conditions, eqns [10] and [11] can be used to obtain a Rossby similarity drag law that yields the non-dimensional surface drift relative to the geostrophic current for unit nondimensional surface stress (eqn [12]).

$$U_0 = \frac{V_0}{u_{*0}} = \frac{1}{k}([\ln(R_0) - A] - iB) \quad [12]$$

Here

$$A = \left(1 - \ln \xi_n - \sqrt{\frac{k}{2\xi_n}} + \sqrt{\frac{\xi_n}{2k}}\right) \cong 2.2$$

$$B = \sqrt{\frac{k}{2\xi_n}} + \sqrt{\frac{\xi_n}{2k}} \cong 2.3 \quad [13]$$

This Rossby similarity drag law for outside the surface layer results in a surface stress that is proportional to $V^{1.8}$ rather than V^2, a result that is supported by observational evidence, and can be significant at high velocities.

Heat and Mass Balance at the Ice–Ocean Interface: Wintertime Convection

The energy balance at the ice–ocean interface not only exerts major influence over the ice mass balance but also dictates the seasonal evolution of upper ocean salinity and temperature structure. At low temperature, water density is controlled mainly by salinity. Salt is rejected during freezing, so that buoyancy flux from basal growth (or ablation), combined with turbulent mixing during storms, determines the depth of the well-mixed layer.

Vertical motion of the ice–ocean interface depends on isostatic adjustment as the ice melts or freezes. The interface velocity is $w_0 + w_i$ where $w_0 = -(\rho_{ice}/\rho)\hbar_b$, $\hbar_b$ is the basal growth rate, and w_i represents isostatic adjustment to runoff of surface melt and percolation of water through the ice cover. In an infinitesimal control volume following the ice–ocean interface, conservation of heat and salt may be expressed in kinematic form as eqns [14] and [15].

$$\dot{q} = \langle w'T' \rangle_0 - w_0 Q_L \quad \text{(with units K m s}^{-1}) \quad [14]$$

$$(w_0 + w_i)(S_0 - S_{ice}) = \langle w'S' \rangle_0 \quad \text{(with units psu m s}^{-1}) \quad [15]$$

where $\dot{q} = H_{ice}/(\rho c_p)$ is flux (H_{ice}) conducted away from the interface in the ice; ρ is water density; c_p is specific heat of seawater; $\langle w'T' \rangle_0$ is the kinematic turbulent heat flux from the ocean; Q_L is the latent heat of fusion (adjusted for brine volume) divided by c_p; S_0 is salinity in the control volume, S_{ice} is ice salinity, and $\langle w'S' \rangle_0$ is turbulent salinity flux. Fluid in the control volume is assumed to be at its freezing temperature, approximated by the freezing line (eqn [16]).

$$T_0 = -mS_0 \quad [16]$$

By standard closure, turbulent fluxes are expressed in terms of mean flow properties (eqns [17] and [18]).

$$\langle w'T' \rangle_0 = c_h u_{*0} \delta T \quad [17]$$

$$\langle w'S' \rangle_0 = c_S u_{*0} \delta S \quad [18]$$

u_{*0} is the square root of kinematic turbulent stress at the interface (friction velocity); $\delta T = T - T_0$ and $\delta S = S - S_0$ are differences between far-field and interface temperature and salinity; and c_h and c_S are turbulent exchange coefficients termed Stanton numbers.

The isostatic basal melt rate, w_0 is the key factor in interface thermodynamics, and in combination with w_i it determines the salinity flux. A first-order approach to calculating w_0 that is often sufficiently accurate (relative to uncertainties in forcing parameters) when melting or freezing is slow, is to assume that $S_0 = S$, the far-field salinity, and that c_h is constant. Combining [14], [16], and [17] gives eqn [19].

$$w_0 = \frac{c_h u_{*0}(T + mS) - \dot{q}}{Q_L} \quad [19]$$

Salinity flux is determined from [15]. Note the c_S is not used, and that this technique fixes (unrealistically) the temperature at the interface to be the mixed layer freezing temperature.

A more sophisticated approach is required when melting or freezing is intense. Manipulation of [14] through [18] produces a quadratic equation for w_0 (eqn [20]).

$$\frac{S_L}{u_{*0}}w_0^2 + (S_T + S_L c_S - S_{ice})w_0 + (u_{*0}c_S + w_i)S_T$$
$$+ u_{*0}c_S S - w_i S_{ice} = 0 \qquad [20]$$

$$S_T = \left(\frac{\dot{q}}{c_h u_{*0}} - T\right)/m \quad \text{and} \quad S_L = Q_L/(mc_h) \quad [21]$$

Here c_h and c_S (turbulent Stanton numbers for heat and salt) are both important and not necessarily the same. Melting or freezing will decrease or increase S_0 relative to far-field salinity, with corresponding changes in T_0.

The Marginal Ice Zone Experiments (MIZEX) in the 1980s showed that existing ice–ocean turbulent transfer models overestimated melt rates by a wide factor. It became clear that the rates of heat and mass transfer were less than momentum transfer (by an order of magnitude or more), and were being controlled by molecular effects in thin sublayers adjacent to the interface. If it is assumed that the extent of the sublayers is proportional to the bottom roughness scale, z_0, then dimensional analysis suggests that the Stanton numbers (nondimensional heat and salinity flux) should depend mainly on two other dimensionless groups, the turbulent Reynolds number, $Re_* = u_{*0}z_0/v$, where v is molecular viscosity, and the Prandtl (Schmidt) numbers, $v/v_{T(S)}$, where v_T and v_S are molecular diffusivities for heat and salt. Laboratory studies of heat and mass transfer over hydraulically rough surfaces suggested approximate expressions for the Stanton numbers of the form shown in eqn [22].

$$c_{h(S)} = \frac{\langle w'T(S)'\rangle_0}{u_{*0}\delta T(S)} \propto (Re_*)^{-1/2}\left(\frac{v}{v_{T(S)}}\right)^{-2/3} \quad [22]$$

The Stanton number, c_h, has been determined in several turbulent heat flux studies since the original MIZEX experiment, under differing ice types with z_0 values ranging from less than a millimeter (eastern Weddell Sea) to several centimeters (Greenland Sea MIZ). According to [22], c_h should vary by almost a factor of 10. Instead, it is surprisingly constant, ranging from about 0.005 to 0.006, implying that the Reynolds number dependence from laboratory results cannot be extrapolated directly to sea ice.

If the Prandtl number dependence of [14] holds, the ratio $c_h/c_S = (v_h/v_S)^{2/3}$ is approximately 30. Under conditions of rapid freezing, the solution of [20] with this ratio leads to significant supercooling of the water column, because heat extraction far outpaces salt injection in what is called double diffusion. This result has caused some concern. Because the amount of heat represented by this supercooling is substantial, it has been hypothesized that ice may spontaneously form in the supercooled layer and drift upward in the form of frazil ice crystals. This explanation has not been supported by ice core sampling, which shows no evidence of widespread frazil ice formation beyond that at the surface of open water.

The physics of the freezing process suggest that the seeming paradox of the supercooled boundary layer may be realistic without spontaneous frazil formation. When a parcel of water starts to solidify into an ice crystal, energy is released in proportion to the volume of the parcel. At large scales this manifests itself as the latent heat of fusion. However, as the parcel solidifies, energy is also required to form the surface of the solid. This surface energy penalty is proportional to the surface area of the parcel and depends on other factors including the physical character of any nucleating material. In any event, if the parcel is very small the ratio of parcel volume to surface area will be so small that the energy released as the volume solidifies is less than the energy needed to create the new solid surface. For this reason, ice crystals cannot form even in supercooled water without a nucleating site of sufficient size and suitable character. In the clean waters of the polar regions, the nearest suitable site may only be at the underside of the ice cover where the new ice can form with no nucleation barrier. Therefore, it is possible to maintain supercooled conditions in the boundary layer without frazil ice formation.

Furthermore, recent results suggest that supercooling in the uppermost part of the boundary layer may be intrinsic to the ice formation process. Sea ice is a porous mixture of pure ice and high-salinity liquid water (i.e., brine). The bottom surface of a growing ice floe consists of vertically oriented pure ice platelets separated by vertical layers of concentrated brine. This platelet–brine sandwich (on edge) structure is on the scale of a fraction of a millimeter, and its formation is controlled by molecular diffusion of heat and salt. The low solid solubility of the salt in the ice lattice results in an increase of the salinity of water in the layer above the advancing freezing interface. Because heat diffuses more rapidly than salt at these scales, the cold brine tends to supercool the water below the ice–water interface. With this local supercooling, any disturbance of the ice bottom will tend to grow spontaneously. The conditions of sea ice growth are such that this instability is always present. Continued growth results in additional rejection of salt, some fraction of which is trapped in the brine layers, and consequently the interfacial region of the ice sheet continues to experience constitutional supercooling. Also, anisotropy in the molecular attachment efficiency intrinsic to the crystal structure of the ice platelets creates an additional supercooling in the interfacial region. The net result is that heat is extracted from the top of the water column at the rate needed to maintain its temperature near but slightly below the equilibrium freezing temperature as salt is added. This and the convective processes in the growing ice may imply that $c_h/c_S = 1$ during freezing. The situation with melting may be quite different, since the physical properties of the interface change dramatically.

Observations to date suggest that c_h remains relatively unchanged with variable ice type and mixed layer temperature elevation above freezing. A value of 5.5×10^{-3} is representative. c_S is not so well known, since direct measurements of $\langle w'S' \rangle$ are relatively rare. The dependence of the exchange coefficients on Prandtl and Schmidt numbers is not clear, and will only be resolved with more research.

Effects of Horizontal Inhomogeneity: Wintertime Buoyancy Flux

Although the under-ice surface may be homogeneous over ice floes hundreds of meters in extent, the key fluxes of heat and salt are characteristically nonuniform. As ice drifts under the action of wind stress, the ice cover is deformed. Some areas are forced together, producing ridging and thick ice, and some areas open in long, thin cracks called leads. In special circumstances the ice may form large, unit-aspect-ratio openings called polynyas. In winter the openings in the ice expose the sea water directly to cold air without an intervening layer of insulating sea ice. This results in rapid freezing. As the ice forms, it rejects salt and results in unstable stratification of the boundary layer beneath open water or thin ice. These effects are so important that, even though such areas may account for less than 10% of the ice cover, they may account for over half the total ice growth and salt flux to the ocean. Thus the dominant buoyancy flux is not homogeneous but is concentrated in narrow bands or patches. Similarly, in the summer solar radiation is reflected from the ice but is nearly completely absorbed by open water. Fresh water from summertime surface melt tends to drain into leads, making them sources of fresh water flux as well.

The effect of wintertime convection in leads is illustrated in **Figure 2**. It shows two extremes in the upper ocean response. **Figure 2A** shows what we might expect in the case of a stationary lead. As the surface freezes, salt is rejected and forms more dense water that sinks under the lead. This sets up a circulation with fresh water flowing in from the sides near the surface and dense water flowing away from the lead at the base of the mixed layer.

Figure 2B illustrates the case in which the lead is embedded in ice moving at a velocity great enough to produce a well-developed turbulent boundary layer (e.g. $0.2\,\mathrm{m\,s^{-1}}$). If the mixed layer is fully turbulent, the cellular convection pattern may not occur; rather, the salt rejected at the surface may simply mix into the surface boundary layer.

The impact of nonhomogeneous surface buoyancy flux on the boundary layer can also be characterized by the equations of motion. The viscous terms in eqn [1] can be neglected at the scales we discuss here, but the possibility of vertical motion associated with large-scale convection requires that we include the vertical component of velocity. For steady state we have eqn [23].

$$\bar{V} \cdot \nabla \bar{V} + \bar{f} \times \bar{V} = \frac{\partial}{\partial z}\left(K \frac{\partial \bar{V}}{\partial}\right) - \rho^{-1}\nabla p \qquad [23]$$

$\bar{V}$ is the velocity vector including the mean vertical velocity w; $\bar{f}$ is the Coriolis parameter times the vertical unit vector. The advective acceleration term, $\bar{V} \cdot \nabla \bar{V}$, and pressure gradient term are necessary to account for the horizontal inhomogeneity that is caused by the salinity flux at the lead surface.

The condition that separates the free convection regime of **Figure 2A** and the forced convection regime of **Figure 2B** is expressed by the relative magnitude of the pressure gradient, $\rho^{-1}\nabla_b p$, and turbulent stress, $\partial/\partial z(K \partial V/\partial z)$, terms in [23]. This ratio can be derived with addition of mass conservation and salt conservation equations, and if we assume the vertical equation is hydrostatic, $\partial p/\partial z = -g\rho = -gMS$, where M is the sensitivity of density to salinity. If we nondimensionalize the equations by the ice velocity U_i, mixed-layer depth, d, average salt flux at the lead surface, F_S, and friction velocity, u_{*0}, the ratio of the pressure gradient term to the turbulent stress term scales as eqn [24].

$$L_0 = \frac{gMF_S d}{\rho_0 U_i u_{*0}^2} \qquad [24]$$

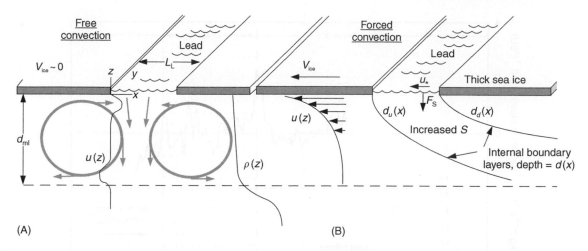

Figure 2 Modes of lead convection. (A) The free convection pattern that results when freezing and salt flux are strong, and the relative velocity of the ice is low. Cellular patterns of convective overturning are driven by pressure gradients that arise from the salinity distrurbance due to ice formation. (B) The forced convection regime that exists when ice motion is strong. The salinity flux and change in surface stress in the lead cause a change in the character of the boundary layer that grows deeper downstream. The balance of forces is primarily Coriolis and turbulent diffusion of momentum. (From Morison JH, MCPhee MG, Curtin T and Paulson CA (1992) The oceanography of winter leads. *Journal of Geophysical Research* 97: 11199–11218.)

If this lead number is small because the ice is moving rapidly or the salt flux is small, the pressure gradient term is not significant in [23]. In this forced convection case, illustrated in **Figure 2B**, the boundary layer behaves as in the horizontally homogeneous case except that salt is advected and diffused away from the lead in the turbulent boundary layer.

If the lead number is large because the ice is moving slowly or the salt flux is large, the pressure gradient term is significant. In this free convention case the salinity disturbance is not advected away, but builds up under the lead. This creates pressure imbalances that can drive the type of cellular motion shown in **Figure 2A**.

Figure 3 shows conditions for which the lead number is unity for a range of ice thickness. Here the salt flux has been parametrized in terms of the air–sea temperature difference, and stress has been parametrized in terms of U_i. The figure shows the locus of points where L_0 is equal to unity. For typical winter and spring conditions, L_0 is close to 1, indicating that a mix of free and forced convection is common. Conditions where lead convection features have been observed are also shown in **Figure 3**. Most of these are in the free convection regime, probably because they are more obvious during quiet conditions.

There have been several dedicated efforts to study the effects of wintertime lead convection. The most recent example was the 1992 Lead Experiment (LeadEx) in the Beaufort Sea. **Figure 4** illustrates the average salinity profile at 9 m under a nearly stationary lead. The data was gathered with an

autonomous underwater vehicle. Using the vehicle vertical motion as a proxy for vertical water velocity, it is also possible to estimate the salt flux $w'S'$. The lead was moving at 0.04 m s^{-1}, and estimates of salt flux put L_0 between 4 and 11 (free convection in **Figure 3**). Salinity increased in the downstream direction across the lead and reached a sharp maximum

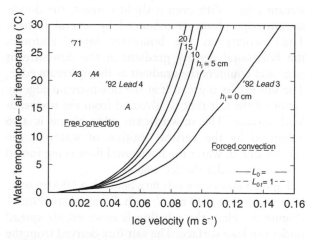

Figure 3 Air–water temperature difference versus U_i for L_0 equal to 1 for various ice thicknesses, h_i. Also shown are the temperature difference and ice velocity values for several observations of lead convection features such as underice plumes. Most of these are in the free convection regime: '71 denotes the AIDJEX pilot study; A3 denotes the 1974 AIDJEX Lead Experiment – lead 3 (ALEX3); A4 denotes ALEX4; A denotes the 1976 Arctic Mixed Layer Experiment; and '92 Lead 4 denotes the 1992 LeadEx lead 4. LeadEx lead 3 ('92 Lead 3) was close to $L_0 = 1$. (From Morison JH, McPhee MG, Curtin T and Paulson CA (1992) The oceanography of winter leads. *Journal of Geophysical Research* 97: 11199–11218.)

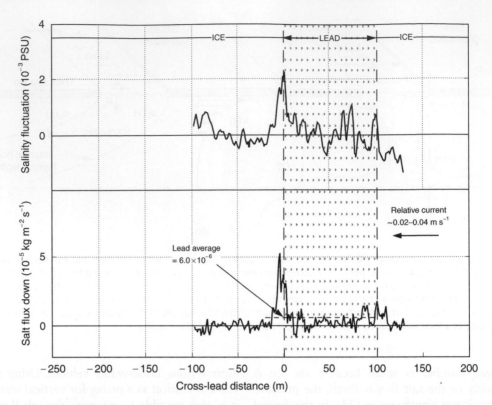

Figure 4 Composites of S' and $w'S'$ at 9 m depth measured with an autonomous underwater vehicle during four runs under lead 4 at the 1992 Lead Experiment. The horizontal profile data have been collected in 1 m bins. (From Morison JH, McPhee MG (1998) Lead convection measured with an autonomous underwater vechicle, *Journal of Geophysical Research* 103: 3257–3281.)

at the downstream edge. The salt flux was highest near the lead edges, but particularly at the downstream edge. With even a slight current, the downstream edge plume is enhanced by several factors. The vorticity in the boundary layer reinforces the horizontal density gradient at the downstream edge and counters the gradient at the upstream edge. The salt excess is greatest at the downstream edge by virtue of the salt that is advected from the upstream lead surface. The downstream edge plume is also enhanced by the vertical motion of water at the surface due to water the horizontal flow being forced downward under the ice edge.

Figure 5 shows the salt flux beneath a 1000 m wide lead moving at 0.14 m s^{-1} with L_0 equal to about 1 (**Figure 3**). Here the salt flux is more evenly spread under the lead surface. The salt flux derived from the direct $w'S'$ correlation method does show some enhancement at the lead edge. This may be partly due to the influence of pressure gradient forces and the reasons cited for the free convection case described above. The other factor that influences the convective pattern is the lead width. In the case of the 100 m lead in even a weak current, the convection may not be fully developed until the downstream edge is reached. For the 1000 m lead of the second case, the convection under the downstream portion of the

lead was a fully developed unstable boundary layer. The energy-containing eddies filled the mixed layer and their dominant horizontal wavelength was equal to about twice the mixed layer depth.

Effects of Horizontal Inhomogeneity: Summertime Buoyancy Flux

The behavior of the boundary layer under summer leads is relatively unknown compared to the winter lead process. Because of the important climate consequences, it is a subject of increasing interest. Summertime leads are thought to exhibit a critical climate-related feature of air–sea–ice interaction, ice-albedo feedback. This is because leads are windows that allow solar radiation to enter the ocean. The proportion of radiation that is reflected (albedo) from sea ice and snow is high (0.6–0.9) while that from open water is low (0.1). The fate of the heat that enters summer leads is important. If it penetrates below the draft of the ice, it warms the boundary layer and is available to melt the bottom of the ice over a large area. If most of the heat is trapped in the lead above the draft of the ice, it will be available to melt small pieces of ice and the ice

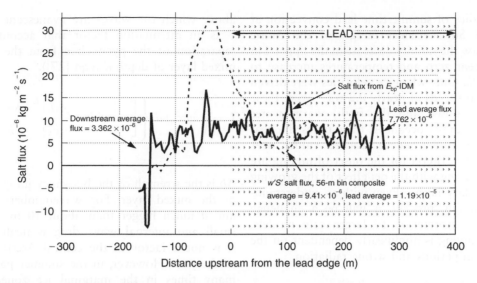

Figure 5 Composite average for autonomous underwater vehicle runs 1 to 5 at lead 3 of the 1992 Lead Experiment. The salinity is band-passed at 1 rad m^{-1} and is indicative of the turbulence level and is used to estimate the salt flux by the E_{bp} – IDM method of Morison and McPee (1998). The salt flux is elevated in the lead and decreases beyond about 72 m downstream of the lead edge. The composite average of $w'S'$ in 56 m bins for the same runs is also shown in the center panel. The average flux and the decrease downstream are about the same as given by the E_{bp} – IDM method, but $w'S'$ suggests elevated fluxes near the lead edge. (From Morison JH, McPhee MG (1998) Lead convection measured with an autonomous underwater vechicle, *Journal of Geophysical Research* 103: 3257–3281.)

floe edges. In the latter case the area of ice will be reduced and the area of open water increased. This allows even more solar radiation to enter the upper ocean, resulting in a positive feedback. This process may greatly affect the energy balance of an ice-covered sea. The critical unknown is the partition of heating between lateral melt of the floe edges and bottom melt.

There are fundamental similarities between the summertime and wintertime lead problems. The equations of motion ([15]–[24]) are virtually identical. Only the sign of the buoyancy flux is opposite. The heat flux is important to summer leads and tends to decrease the density of the surface waters. However, as with winter leads, the buoyancy flux is controlled mainly by salt. As the top surface of the ice melts, much of the water that does not collect in melt ponds on the ice surface instead runs into the leads. If the ambient ice velocity is low, ice melt from the bottom surface will tend to flow upward and collect in the leads as well. Thus leads are the site of a concentrated flux of fresh water accumulated over large areas of ice. If this flux, F_S, into the lead is negative enough relative to the momentum flux represented by u_{*0}, the lead number, L_0, will be a large negative number and shear production of turbulent energy will not be able to overcome the stabilizing buoyant production. This means turbulent mixing will be weak beneath the lead surface and a layer of fresh water will accumulate near the surface of the lead. The stratification

at the bottom of this fresh water layer may be strong enough to prevent mixing until a storm produces a substantial stress. This will be made even more difficult than in the winter situation because of the effect of stabilizing buoyancy flux on the boundary layer generally. The only way the fresh water will be mixed downward is by forced convection; there is no analogue to the wintertime free convection regime.

When there is sufficient stress to mix out a summertime lead, the pattern must resemble that of the forced convection regime in **Figure 2A**. At the upstream edge of the lead, fresh warm water will be mixed downward in an internal boundary layer that increases in thickness downstream until it reaches the steady-state boundary layer thickness appropriate for that buoyancy flux or the ambient mixed layer depth. The rate of growth should scale with the local value of u_{*0} (or perhaps $u_{*0}\eta_*$). At the downstream edge, another boundary layer conforming to the under-ice buoyancy flux and surface stress will begin to grow at a rate roughly scaling with the local u_{*0}. In spite of the generally stabilizing buoyancy flux, this process has the effect of placing colder, more saline water from under the ice on top of fresher and warmer (consequently lighter) water drawn from the lead. Thus, even embedded in the stable summer boundary layer, the horizontal inhomogeneity due to leads may create pockets of instability and more rapid mixing than might be expected on the basis of average conditions.

Recent studies of summertime lead convection at the 1997–98 Surface Heat Budget of the Arctic experiment saw the salinity decrease in the upper 1 m of leads to near zero and temperatures increase to more than 0°C. Only when ice velocities were driven by the wind to speeds of nearly 0.2 m s^{-1} were these layers broken down and the fresh, warm water mixed into the upper ocean. At these times the heat flux measured at 5 m depth reached values over 100 W m^{-2}. The criteria for the onset of mixing are being studied along with the net effect of the growing internal boundary layers. Even with an understanding of the mixing process, it will be a challenge to apply this information to larger-scale models, because the mixing is nonlinearly dependent on the history of calm periods and strong radiation.

Internal Waves and Their Interaction with the Ice Cover

One of the first studies of internal waves originated with observations made by Nansen during his 1883 expedition. It did not actually involve interaction with the ice cover, but with his ship the *Fram*. He found that while cruising areas of the Siberian shelf covered with a thin layer of brackish water, the *Fram* had great difficulty making any headway. It was hypothesized by V. Bjerknes and proved by Ekman that this 'dead water' phenomenon was caused by the drag of the internal wave wake produced by the ship's hull as it passed through the shallow surface layer. This suggests that internal wave generation by deep keels may cause drag on moving ice. Evidence of internal wave generation by keels has been observed by several authors, but estimates of the amount of drag vary widely. This is due mainly to wide differences in the separation of the stratified pycnocline and the keels.

The drag produced by under-ice roughness of amplitude h_0 with horizontal wavenumber β moving at velocity V_i (magnitude v_i) over a pycnocline with stratification given by Brunt–Vaisala frequency, N, a depth d below the ice–ocean interface, can be expressed as an effective internal wave stress (eqn [25]), where C_{wd} (eqn [26]) accounts for the drag that would exist if there were no mixed layer between the ice and the pycnocline.

$$\Sigma_{iw} = -\Gamma C_{wd} V_i \qquad [25]$$

$$C_{wd} = \tfrac{1}{2}\beta_x^2 h_0 [(\beta_c^2/\beta_x^2) - 1]^{1/2} \qquad [26]$$

The wavenumber in the direction of the relative ice velocity, V_i, is β_x, and β_c is the critical wave number above which the waves are evanescent ($\beta_c = N/v_i$). Γ is an attenuation factor that accounts for the separation of the pycnocline from the ice by the mixed layer of depth d (eqn [27]).

$$\Gamma = \left(\sinh^2(\beta d) \left\{ \left[\coth(\beta d) - \frac{\beta \Delta b}{v_i^2 \beta_x^2} \right]^2 + \frac{N^2}{v_i^2 \beta_x^2} - 1 \right\} \right)^{-1}$$

$$[27]$$

Δb is the strength of the buoyancy jump at the base of the mixed layer. For wavenumbers of interest and d much bigger than about 10 m, Γ becomes small and internal wave drag is negligible. Thus it is not a factor in the central Arctic over most of the year. However, in the summer pack ice, and many times in the marginal ice zone, stratification will extend to or close to the surface. Then internal wave drag can be at least as important as form drag.

The ice cover also uniquely affects the ambient internal wave field. In most of the world ocean the internal wave energy level, when normalized for stratification, is remarkably uniform. It has been established by numerous studies that the internal wave energy in the Arctic Ocean is typically several times lower. In part this may be due to the absence of surface gravity waves. The other likely reason is that friction on the underside of the ice damps internal waves. Decomposing the internal wave field into vertical modes, one finds the mode shapes for horizontal velocity are a maximum at the surface. This is perfectly acceptable in the open water situation. However, at the horizontal scales of most internal waves, an ice cover imposes a surface boundary condition of zero horizontal velocity. The effect of this can be estimated by assuming that a time-varying boundary layer is associated with each spectral component of the internal wave field. This is not rigorously correct because all the modes interact in the same nonlinear boundary layer, and are thereby coupled. However, in the presence of a dominant, steady current due to ice motion, the effect on the internal wave modes can be linearized and considered separately. The near-surface internal wave velocity can be approximated as a sum of rotary components (eqn [28]).

$$V(z) = \sum_{n=0}^{M} D_n(z) e^{i\omega_n t} = \sum_{n=0}^{M} [A_n(z) + iB_n(z)] e^{i\omega_n t}$$

$$[28]$$

The internal wave motion away from the boundary $D_{\infty n}$ can be subtracted from the linear time-varying boundary layer equation (eqn [1] with the addition of the time variation acceleration, $\partial V/\partial t$). This yields an equation for each rotary component of velocity in the boundary layer (eqn [29]).

$$i(\omega_n + f)(D_n - D_{\infty n}) = \frac{\partial}{\partial z} K \frac{\partial D_n}{\partial z} \qquad [29]$$

$$D_n = 0 \quad \text{at } z = z_0$$
$$D_n = D_{\infty n} \quad \text{at } z = d$$

This oscillating boundary layer equation can be solved for K of the form $K = ku_{*0}z\exp(-6l + f|z u_{*0})$. When we do this for representative internal wave conditions in the Arctic and compute the energy dissipation, we find the timescale required to dissipate the internal wave energy through under-ice friction is 32 days. This is a factor of 3 smaller than is typical for open ocean conditions. Assuming a steady state with internal wave forcing and other dissipation mechanisms in place, the under-ice boundary layer will result in a 75% reduction in steady-state internal wave energy. This suggests the effect of the under-ice boundary layer is critical to the unique character of internal waves in ice-covered seas.

Outstanding Issues

The outstanding issue of ice–ocean interaction is how the small-scale processes in the ice and at the interface affect the exchange between the ice and water. This is arguably most urgent in the case of heat and salt exchange during ice growth. When we apply laboratory-derived concepts for the diffusion of heat and salt to the ice–ocean interface, we get results that are not supported by observation, such as spontaneous frazil ice formation and large ocean heat flux under thin ice.

These results are causing significant errors in large-scale models. They stem from a molecular sublayer model of the ice–ocean interface (**Figure 1**) and the difference between the molecular diffusivities of heat and salt. What seems to be wrong is the molecular sublayer model. Recent results in the microphysics of ice growth reveal that the structure and thermodynamics of the growing ice produce instabilities and convection within the ice and extending into the water. The ice surface is thus not a passive, smooth surface covered with a thin molecular layer. Rather it is field of jets emitting plumes of supercooled, high-salinity water at a very small scale. This type of

unstable convection likely tends to equalize the diffusion of heat and salt relative to the apparently unrealistic parameterizations we are using now.

Similarly, we do not really understand how the turbulent stress we might measure in the surface layer is converted to drag on the ice. Certainly a portion of this is through viscous friction in the molecular sublayer. However, in most cases the underside of the ice is not hydrodynamically smooth, which suggests that pressure force acting on the bottom roughness elements are ultimately transferring a large share of the momentum. Understanding this will require perceptual breakthroughs in our view of how turbulence and the mean flow interact with a rough surface buried in a boundary layer. Achieving this understanding is complicated greatly by a lack of contemporaneous measurements of turbulence and under-ice topography at the appropriate scales. This drag partition problem is general and not limited to the under-ice boundary layer. However, the marvelous laboratory that the under-ice boundary layer provides may be the place to solve it.

See also

Coupled Sea Ice–Ocean Models. Sea Ice: Overview. Sub Ice-Shelf Circulation and Processes. Under-Ice Boundary Layer.

Further Reading

Johannessen OM, Muench RD, and Overland JE (eds.) (1994) *The Polar Oceans and Their Role in Shaping the Global Environment: The Nansen Centennial Volume.* Washington, DC: American Geophysical Union.

McPhee MG (1994) On the turbulent mixing length in the oceanic boundary layer. *Journal of Physical Oceanography* 24: 2014–2031.

Morison JH, McPhee MG, and Maykutt GA (1987) Boundary layer, upper ocean and ice observations in the Greenland Sea marginal ice zone. *Journal of Geophysical Research* 92(C7): 6987–7011.

Morison JH and McPhee MG (1998) Lead convection measured with an autonomous underwater vehicle. *Journal of Geophysical Research* 103(C2): 3257–3281.

Smith WO (ed.) (1990) *Polar Oceanography.* San Diego, CA: Academic Press.

Wettlaofer JS (1999) Ice surfaces: macroscopic effects of microscopic structure. *Philosphical Transactions of the Royal Society of London A* 357: 3403–3425.

UNDER-ICE BOUNDARY LAYER

M. G. McPhee, McPhee Research Company, Naches,
WA, USA
J. H. Morison, University of Washington, Seattle, WA,
USA

Copyright © 2001 Elsevier Ltd.

Introduction

Sea ice is in almost constant motion in response to
wind, ocean currents, and forces transmitted within
the ice cover itself, thus there is nearly always a
zone of sheared flow between the ice and underlying,
undisturbed ocean where turbulence transports momentum, heat, salt, and other contaminants vertically. The zone in which these turbulent fluxes
occur, which can span from a few to hundreds of
meters, is the under-ice boundary layer (UBL). This
article describes general characteristics of the UBL,
with emphasis on the physics of vertical turbulent
transfer, specifically turbulent mixing length and
eddy diffusivity. Extensive measurements of turbulence in the UBL, not available elsewhere, have not
only made these ideas concrete, but have also provided quantitative guidance on how external forcing
controls the efficiency of vertical exchange. Here we
stress features that the UBL has in common with
ocean boundary layers everywhere. The article on
ice–ocean interaction emphasizes unique aspects of
the interaction between sea ice and the ocean (*see*
Ice–Ocean Interaction).

While largely responsible for the relative paucity
of oceanographic data from polar regions, sea ice
also serves as an exceptionally stable platform, often
moving with the maximum velocity in the water
column. In effect, it provides a rotating geophysical
laboratory with unique opportunities for directly
measuring turbulent fluxes of momentum, heat, and
salt at multiple levels in the oceanic boundary layer –
measurements that are extremely difficult in the
open ocean. Examples of important oceanographic
boundary-layer processes first observed from sea ice
include: (1) the Ekman spiral of velocity with depth;
(2) Reynolds stress through the entire boundary
layer, and its associated spiral with depth; (3) direct
measurements of turbulent heat flux and salinity
flux; (4) direct measurements of eddy viscosity and
diffusivity in the ocean boundary layer; (5) the impact of surface buoyancy, both negative and positive,
on boundary layer turbulence, and (6) internal wave

drag ('dead water') as an important factor in the
surface momentum and energy budgets.

The UBL differs from temperate open ocean
boundary layers by the absence of strong diurnal
forcing and of high frequency, wind-driven surface
waves. It thus lacks the near surface zone of intense
turbulence and dissipation associated with wave
breaking, and organized Langmuir circulation due to
the nonlinear interaction between waves and currents (e.g., the interaction of Stokes drift with near
surface vorticity) (*see* Langmuir Circulation and Instability). On the other hand, quasi-organized roll
structures associated with sheared convective cells
have been observed under freezing ice, and are apparently a ubiquitous feature of freezing leads and
polynyas. Large inertial-period oscillations in UBL
horizontal velocity are observed routinely, especially
in summer when the ice pack is relaxed. The annual
cycle of buoyancy flux from freezing and melting
mimics in some respects the diurnal cycle of heating
and cooling, as well as the annual evolution of
temperate ocean boundary layers. The range of surface forcing, with observations of surface stress
ranging up to 1 Pa, and buoyancy flux magnitudes
as high as $10^{-6} W kg^{-1}$, is comparable to that encountered in open oceans. All these factors suggest
that similarities between the UBL and the open ocean
boundary layer far outweigh the differences.

History and Basic Concepts

Rotational Physics and the Ekman Layer

From 1893 to 1896, the Norwegian research vessel
Fram drifted with the Arctic pack ice north of Eurasia in one of the most productive oceanographic
cruises ever conducted. Among other important
discoveries was the observation by Fridtjof Nansen,
the great scientist–explorer–statesman, that the
drift was consistently to the right of the surface wind.
Nansen surmised that this effect arose from the differential acceleration in a rotating reference frame
(the earth) on the sheared turbulent flow beneath
the ice, and interested the young Swedish scientist,
V.W. Ekman, in the problem. Ekman discovered an
elegantly simple solution to the coupled differential
equations describing the steady-state boundary layer,
which exhibited attenuated circular rotation with
depth (spirals) in both velocity and stress (momentum flux). The solution includes a constant phase
difference between velocity and stress, resulting in a
45° clockwise deflection of surface velocity with

respect to surface stress in the Northern Hemisphere, roughly comparable to the 20–40° deflection Nansen observed. In his classic 1905 paper, Ekman extended his findings with remarkable insight to predict inertial oscillations, large circular currents superimposed on the mean current, and even derived credible estimates of eddy viscosity in the ocean from surface-drift-to-wind-speed ratios. Ekman postulated the eddy viscosity should vary as the square of the surface wind speed, with kinematic values of order $0.04\,\mathrm{m^2\,s^{-1}}$ for typical wind speeds of $10\,\mathrm{m\,s^{-1}}$.

After nearly a century, it is tempting to dismiss Ekman's solution as not adequately accounting for vertical variation of eddy viscosity in the boundary layer. Surface (ice) velocity, for example, is strongly influenced by a zone of intense shear near the ice–ocean interface where eddy viscosity varies linearly with distance from the ice. For typical under-ice conditions, this approximately halves the angle between interfacial stress and velocity and significantly increases the ratio of surface speed to surface stress. The Ekman approach also ignores potentially important effects from density gradients in the water column, or from buoyancy flux at the interface. Nevertheless, measurements from the UBL show that with slight modification, Ekman theory does indeed provide a very useful first-order description of turbulent stress in the UBL. Turbulent stress is not much affected by either variation in eddy viscosity in the near surface layer (across which the stress magnitude varies by only about 10%), or by horizontal gradients in density of the boundary layer ('thermal wind'). Both can have large impact on the mean velocity profile.

The Ekman solution for turbulent stress is derived as follows. Using modern notation, the equations of motion in a noninertial reference frame rotating with the earth include an apparent acceleration resulting in the Coriolis force, with horizontal vector component $\rho f k \times V$, where ρ is density, V is the horizontal velocity vector, k is the vertical unit vector, and f is the Coriolis parameter (positive in the Northern Hemisphere). Ekman postulated that eddy viscosity, K, which behaves similarly to molecular viscosity but is several orders of magnitude larger, relates stress to velocity shear: $\hat{\tau} = K \partial V / \partial z$ where $\hat{\tau}$ is a traction vector combining the horizontal components of stress in the water. Expressing horizontal vectors as complex numbers, e.g., $V = u + iv$, the steady-state, horizontally homogeneous equation for horizontal velocity in an otherwise quiescent ocean forced by stress at the surface is then given by:

$$if V = K \frac{\partial^2 V}{\partial z^2} \qquad [1]$$

Implicit in eqn [1] is that k does not vary with depth, so differentiation of eqn [1] with respect to z and substituting $\hat{\tau}/K$ for $\partial V / \partial z$ yields a second-order differential equation for $\hat{\tau}$ subject to boundary conditions that $\hat{\tau}t$ vanish at depth and that it match the applied interfacial stress, $\hat{\tau}_0$ at $z = 0$.

The solution is simply:

$$\hat{\tau}(z) = \hat{\tau}_0 e^{\hat{\delta} z} \qquad [2]$$

where $\hat{\delta} = (f/|f|)(if/K)^{1/2}$ is a complex extinction coefficient that both attenuates and rotates stress with increasing depth, clockwise in the Northern Hemisphere, counterclockwise in the Southern Hemisphere.

The practical differences between Ekman spirals in velocity and stress are illustrated by measurements of mean velocity and Reynolds stress during a period of rapid ice drift at Ice Station Weddell near 65°S, 50°W (**Figure 1**). The mean current in a reference frame drifting with the ice velocity (i.e., the negative of the dashed vector labeled 'Bot' in **Figure 1A**) shows the characteristic leftward turning with depth, but

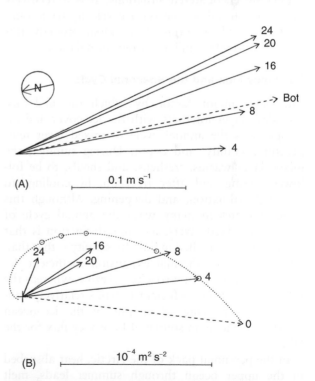

Figure 1 (A) Plan view of mean velocity averaged over a period of steady drift at Ice Station Weddell (1992). Numbers indicate meters from the ice–ocean interface. The vector labeled 'Bot' is the apparent velocity of the seafloor in the drifting reference frame. (B) Horizontal Reynolds stress. The dotted stress hodograph is from a similarity model, with boundary stress (dashed vector) inferred from the model solution that matches observed stress at 4 m. (Reproduced from McPhee MG and Martinson DG (1994) *Science* 263: 218–221.)

also includes a region of strong shear between 4 m and the ice–ocean interface, as well as an apparent eastward geostrophic current of several centimeters per second. The last may include its own vertical shear unrelated to UBL dynamics. None of these complicating factors has much impact on the Reynolds stress (**Figure 1B**), which shows (in a general sense) the depth attenuation and rotation predicted by a simple complex exponential (2) with vertically invariant eddy viscosity. The latter derives from a similarity based value for K, proportional to $|\hat{\tau}_0/f|$, with a magnitude of about $0.02 \, \text{m}^2 \, \text{s}^{-1}$. Since the interfacial stress is approximately proportional to wind speed squared, this is indeed similar to Ekman's development,[1] with the magnitude implied by the observations within a factor of about two of Ekman's prediction. Although the profile of **Figure 1(B)** is especially 'clean,' numerous other examples of spirals in Reynolds stress profiles exist from under-ice measurements, most consistent with the neutral scaling implied by $K \propto |\hat{\tau}_0/f|$. Thus despite its simplicity, the Ekman approach provides a remarkably accurate account of momentum flux in the UBL for many commonly encountered situations. It is a relatively minor step to adjust the surface velocity to account for the variable K surface layer. Done properly, this leads to a Rossby similarity drag formulation.

Buoyancy Flux and the Seasonal Cycle

The other major factor by which the under-ice boundary layer interacts with the ice cover and atmosphere is the annual cycle of mixed layer temperature, salinity, and depth. During summer, the mixed layer warms, freshens, and shoals, to be followed during and after freezeup, by cooling (to freezing), salination, and deepening. Although this cycle emulates in many ways the annual cycle of temperate mixed layers, a major distinction is that buoyancy is controlled mainly by salinity rather than temperature (the thermal expansion coefficient decreases rapidly as T approaches freezing, whereas the saline contraction coefficient remains relatively constant), thus freezing or melting at the ice–ocean interface is the main source of buoyancy flux for the UBL.

In the perennial pack of the Arctic, heat absorbed in the upper ocean through summer leads, melt ponds, and thin ice contributes to bottom melting and is an important part of both the ice mass balance

and the total summer buoyancy increase for the UBL. Away from the continental shelves and ice margins, heat exchange with the deep ocean tends to be small, limited by the cold halocline that separates water of Atlantic origin from the surface. In the eastern Arctic, the marginal ice zone of Fram Strait, and in the vast seasonal sea ice zone surrounding the Antarctic continent, the UBL interacts directly with warmer deep ocean, and oceanic heat mixed into the boundary layer from below often controls the ice mass balance, and exerts major influence on overall ocean stability.

Buoyancy plays a major role in these exchange processes and is not adequately represented by treating eddy viscosity as dependent solely on surface stress. Most of the UBL research in recent years has been devoted to understanding how buoyancy influences turbulent fluxes.

Turbulence in the Under-ice Boundary Layer

Reynolds Flux

When ice is in motion relative to the underlying water, there is a net flux of momentum in the underlying boundary layer, most of which is carried by turbulent fluctuations arising from relatively small, chaotic instabilities in the flow, motions which will also induce fluxes of scalar properties (e.g., T, S) if a mean gradient in the property exists. The turbulent transport process is best demonstrated by considering the advective part of the material derivative. Consider, for example, the simplest form of the heat equation: horizontally homogeneous, with no internal sources or sinks of heat. In a Eulerian reference frame, this reduces to a simple balance between the material derivative of temperature and the vertical gradient of the molecular heat diffusion

$$\frac{dT}{dt} = \frac{\partial T}{\partial t} + \boldsymbol{u} \cdot \nabla T = \frac{\partial}{\partial z} \left(v_T \frac{\partial T}{\partial z} \right) \quad [3]$$

where v_T is the molecular thermal diffusivity. Turbulent flux of temperature variations arises from the advective term, $\boldsymbol{u} \cdot \Delta T$. If velocity and temperature are expressed as the sum of mean and turbulent (fluctuating) parts: $\boldsymbol{u} = \bar{U} + \boldsymbol{u}'$ and $T = \bar{T} + T'$, and the flow is incompressible and horizontally homogeneous with no mean vertical velocity,

$$\boldsymbol{u} \cdot \nabla T = \frac{\partial}{\partial z} \langle w'T' \rangle \quad [4]$$

---

[1] Ekman suggested that the 'depth of frictional influence' $D = \pi \sqrt{(2K/f)}$ varied as wind speed divided by $\sqrt{\sin \Phi}$, where Φ is latitude. This implies no f dependence for k. At high latitudes, this has minor impact.

Normally this term completely dominates the molecular flux and eqn [3] is approximated by

$$\frac{\partial \bar{T}}{\partial t} = -\frac{\partial}{\partial z}\langle w'T' \rangle \qquad [5]$$

In a strict sense, the angle brackets represent an ensemble Reynolds average over many independent realizations of the flow, but for practical applications it is assumed that the large-scale, 'mean' properties of the flow and its turbulent fluctuations respond in different and separable wavenumber bands (so that the local time derivative in eqn [5] has meaning), and that a suitable average in time is representative of the Reynolds flux.

A similar analysis of du/dt leads to the divergence of the Reynolds stress tensor formed from the velocity covariance matrix of the three fluctuating velocity components. Under the same simplifications as above, the advective term in the mean horizontal velocity equation becomes

$$\frac{\partial}{\partial z}(\langle u'w' \rangle + i\langle v'w' \rangle)$$

where the horizontal vector quantity $\tau = \langle u'w' \rangle + i\langle v'w' \rangle$ is traditionally called Reynolds stress. A second important turbulence property associated with the Reynolds stress tensor is its trace

$$q^2 = \langle u'u' \rangle + \langle v'v' \rangle + \langle w'w' \rangle \qquad [6]$$

which is twice the turbulent kinetic energy (TKE) per unit mass.

The connection between turbulence and eddy viscosity becomes apparent when the horizontal velocity equation is written with the simplifying (but often reasonable) assumptions of horizontal homogeneity, no mean vertical velocity, and negligible impact of molecular viscosity:

$$\frac{\partial V}{\partial t} + ifV = -\frac{\partial}{\partial z}(\langle u'w' \rangle + i\langle v'w' \rangle)$$
$$= \frac{\partial}{\partial z}\left(u_\tau \lambda \frac{\partial V}{\partial z}\right) \qquad [7]$$

The last term in eqn [7] represents the mixing-length hypothesis, essentially a scaling argument that Reynolds stress is uniquely related to the mean velocity shear by the product of velocity and length scales characterizing the largest, energy-containing eddies in the flow. Eddy viscosity is $K = u_\tau \lambda$. The steady version of eqn [7] differs from eqn [1] in that K may depend on z and remains within the scope of the outer derivative.

Scales of Turbulence

A reasonable choice for the turbulence velocity scale (u_τ) is the friction speed $u_* = \sqrt{|\hat{\tau}|}$. In exceptional cases where destabilizing buoyancy flux ($\langle w'b' \rangle = (g/\rho)\langle \rho'w' \rangle$) from rapid freezing is the main source of turbulence, a more appropriate choice is the convective scale velocity $w_* = (\lambda|\langle w'b' \rangle_0|)^{1/3}$ where λ is the length scale of the dominant eddies. An alternative scale is q given by eqn [6]; however, observations in the UBL show the ratio q/u_* to be relatively constant (~ 3) in shear-dominated flows; the distinction may therefore be academic until a clear connection between q and w_* is demonstrated.

Mixing length is the distance over which the 'energy-containing' eddies are effective at diffusing momentum. Several observational studies in the UBL have shown a robust relationship between a length scale λ_{peak} inversely proportional to the wavenumber at the maximum of the weighted spectrum of vertical velocity, and λ inferred by other methods. Since the spectrum of vertical velocity is relatively easy to measure, λ_{peak} provides a useful proxy for estimating λ simultaneously at several levels in the UBL.

A diagram of governing turbulence scales in the UBL is presented in **Figure 2**, developed by combining simple boundary-layer similarity theory with numerous observations from drifting sea ice ranging from the marginal ice zone of the Greenland Sea, to the central Arctic ocean under thick ice and at the edges of freezing leads, and in the Weddell Sea. **Figure 2(A)** shows neutral stratification in the bulk of the UBL, when surface buoyancy flux (melt rate) is too small to have appreciable impact on turbulence. This is a common condition for perennial pack ice, which grows or melts slowly most of the year. Working from the interface down, mixing length increases approximately linearly with depth through the surface layer, until it reaches a limiting value proportional to the planetary length scale $\lambda_{max} = \Lambda_* u_{*0}/f$, where $\Lambda_* \sim 0.03$. Usually, the surface layer extends 5 m or less. From there the mixing length holds relatively constant through the extent of the Ekman (or outer) part of the UBL, to the depth of the pycnocline (typically 35–50 m in the western Arctic; 75–150 m in the Weddell Sea). If the neutral layer is very deep, stress decreases more or less exponentially, following approximately the Ekman solution (see the discussion of **Figure 3B** below); however, if the pycnocline is shallow, a finite stress will exist at z_p (indicated in **Figure 2** by u_{*p}) instigating upward mixing of pycnocline water with associated buoyancy flux, $\langle w'b' \rangle_p$. Mixing length in the highly stratified fluid just below the mixed-layer–pycnocline interface is estimated from the turbulent kinetic

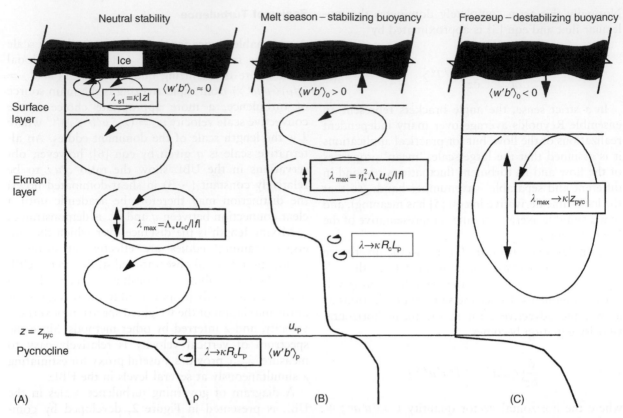

Figure 2 Schematic diagram of mixing length distributions in the UBL under conditions of (A) dynamically negligible surface buoyancy flux (neutral stratification in the well mixed layer), (B) upward buoyancy flux from summer melting, with formation of a seasonal pycnocline and a negative density gradient in the 'well mixed' layer, and (C) downward buoyancy flux from rapid freezing, with positive density gradient to the pycnocline. u_*, Friction velocity; $\langle w'b' \rangle$, buoyancy flux; κ, Kármán's constant, 0.4; Λ_*, similarity constant, 0.028; R_c, critical flux Richardson number, 0.2; f, Coriolis parameter; $L = u_*^3/(\kappa \langle w'b' \rangle)$, Obukhov length; $\eta_* = (1 + \Lambda_* u_*/\kappa R_c |f|L))^{-1/2}$, stability parameter.

energy equation, which is dominated by three terms: production of TKE by shear ($P_S = \hat{\tau} \cdot \partial U/\partial z$), production by buoyancy ($P_b = -\langle w'b' \rangle$), and dissipation by molecular forces (ε). Relating stress and shear by the mixing-length hypothesis, the balance of TKE production with dissipation is

$$u_*^3/\lambda - \langle w'b' \rangle = \varepsilon \qquad [8]$$

The negative ratio of buoyancy production to shear production is the flux Richardson number:

$$-P_b/P_S = \frac{\lambda \langle w'b' \rangle}{u_*^3} = \frac{\lambda}{\kappa L} \qquad [9]$$

where $L = u_*^3/(\kappa \langle w'b' \rangle)$ is known as the Obukhov length. Studies of turbulence in stratified flows have shown that the ratio eqn [9] does not exceed a limiting value (the critical flux Richardson number, R_c) of about 0.2. This establishes a limit for mixing length in stratified flow: $\lambda \le R_c \kappa L$, and it is assumed that in the pycnocline this limit is approached, where

L is based on pycnocline fluxes of momentum and buoyancy.

Estimates of mixing length in a near neutral UBL from the Ice Station Weddell data (**Figure 1**) are illustrated in **Figure 3(A)**. Points marked λ_{peak} were taken from the inverse of the wavenumber at the peak in the vertical velocity spectra (averaged over all 1-h flow realizations), as described above. Values marked λ_ε were obtained using eqn [8] assuming negligible buoyancy flux, with measured values for u_* and ε (obtained from spectral levels in the inertial subrange). They show clearly that the 'wall layer' scaling, $\lambda = \kappa |z|$ does not hold for depths greater than about 4 m.

Rapid melting reduces the extent of the surface layer and the maximum mixing length (**Figure 2B**). The stability factor $\eta_* = (1 + \Lambda_* u_*/(\kappa R_c |f|L))^{-1/2}$ derives from similarity theory and ensures that the mixing length varies smoothly from the neutral limit ($\lambda_{max} \to \Lambda_* u_{*0}/|f|$) to the stable limit ($\lambda_{max} \to \kappa R_c L_0$) for increasing stability. A consequence of reduced scales during melting is formation of a seasonal

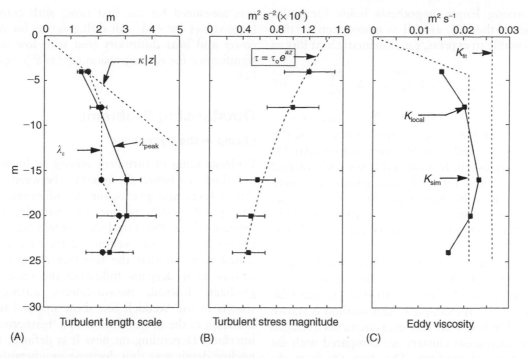

Figure 3 (A) Mixing length determined from the TKE equation (λ_ε) and from the inverse of the wavenumber at the peak in the weighted w spectrum (λ_{peak}). Error bars indicate twice the standard deviation from the spectra calculated from 1-h segments of data. (B) Average Reynolds stress magnitude, with a least-squares fitted exponential decay with depth. Fit coefficients are $\tau_0 = 1.44 \times 10^{-4}\,m^2\,s^{-2}$ and $a = 0.051\,m^{-1}$. (C) Eddy viscosity estimated by three methods as described in the text. (Reproduced from McPhee MG and Martinson DG (1994) *Science* 263, 218–221.)

pycnocline, above a 'trapped' layer with properties indicative of the mixed layer that existed before the freshwater influx.

Rapid ice growth produces negative buoyancy via enhanced salinity at the interface, increasing TKE by the buoyancy production term in eqn [8]. The result is that mixing length and eddy viscosity increase in the UBL, sometimes dramatically. During the 1992 Lead Experiment, turbulent flux and dissipation measured from the edge of a freezing lead in a forced convective regime showed that, compared with the neutral UBL, there was a tenfold increase in mixing length (based on w spectral peaks) and in eddy heat and salt diffusivity (based on measured fluxes and gradients). The Obukhov length was -12 m, about 40% of the mixed layer extent, indicating relatively mild convection, yet the turbulence was greatly altered, apparently by the generation of quasi-organized roll structures in the lead, reminiscent of Langmuir circulations (a thin ice cover precluded any surface waves at the time of the measurements). Mixing length inferred from the lead measurements increased away from the surface following Monin–Obukhov similarity (adapted from atmospheric boundary layer studies), reaching a maximum value roughly comparable to the pycnocline depth scaled by von Kármán's constant.

The density profiles in **Figure 2(B)** and **(C)** are drawn schematically with slight gradients in the so-called mixed layer. This is at odds with conceptual models of the upper ocean which treat the boundary layer as completely mixed, but is consistent with measurements in the UBL. Wherever scalar fluxes of temperature and salinity are measurable, vertical gradients (albeit small) of mean temperature and salinity are found in the fully turbulent UBL, including statically unstable profiles as in **Figure 2(C)**.

Effective Eddy Viscosity and Diffusivity

Figure 3(C) illustrates different methods for estimating bulk eddy viscosity in the UBL. The distribution labeled K_{sim} is from the similarity model used to construct the stress profile of **Figure 1(B)** by matching observed stress at 4 m. The vertical distribution labeled K_{local} is the product $\lambda_{peak}u*$ at each level (**Figure 3A and B**). Its vertical average value is $0.019\,m^2\,s^{-1}$. Finally, the dashed line labeled K_{fit} in **Figure 3(B)** is from the least-squares fitted extinction coefficient $(\mathrm{Re}\{\hat{\delta}\})$ for the Ekman stress solution eqn [2]. The last method is sensitive to small stress values at depth: if the bottommost cluster is ignored, $K_{fit} = 0.020\,m^2\,s^{-1}$.

The mixing length hypothesis holds for scalar properties of the UBL as well as momentum, so that it is reasonable to express, e.g., kinematic heat flux as

$$\langle w'T' \rangle = -u_* \lambda_T \frac{\partial T}{\partial z} = -K_H \frac{\partial T}{\partial z} \qquad [10]$$

In flows where turbulence is fully developed with large eddies and a broad inertial subrange, scalar eddy diffusivity and eddy viscosity are comparable (Reynold's analogy). In stratified flows with internal wave activity and relatively low turbulence levels, momentum may be transferred by pressure forces that have no analog in scalar conservation equations, hence scalar mixing length may be considerably less than λ.

By measuring turbulent heat flux and the mean thermal gradient, it is possible to derive an independent estimate of eddy diffusivity in the UBL from eqn [10]. An example of this method is shown in **Figure 4**, where heat flux measurements averaged over five instrument clusters are compared with the negative thermal gradient. The data are from the same Ice Station Weddell storm as the other turbulence measurements of **Figures 1** and **3**. The mean thermal diffusivity, $K_H = 0.018 \, \text{m}^2 \, \text{s}^{-1}$, is similar to the eddy viscosity (**Figure 3C**). Close correspondence between eddy viscosity and heat diffusivity was also found during the 1989 CEAREX drift north of Fram Strait, and during the 1992 LEADEX project. In the forced convective regime of the latter, salinity flux

was measured for the first time, with comparably large values for eddy salt diffusivity as for eddy viscosity and heat diffusivity (but with low statistical significance for the regression of $\langle w'S' \rangle$ against $\partial S/\partial z$).

Outstanding Problems

Mixing in the Pycnocline

Understanding of turbulent mixing in highly stratified fluid just below the interface between the well-mixed layer and pycnocline is rudimentary. Many conceptual models assume, for example, that fluid 'entrained' at the interface immediately assumes the properties of the well-mixed layer (i.e., is mixed completely), so that the interface sharpens during storms as it deepens following the mean density gradient. Instead, measurements during severe storms in the Weddell Sea show upward turbulent diffusion of the denser fluid with a 'feathering' of the interface. Depending on how it is defined, the pycnocline depth may thus decrease significantly during extreme mixing events. Where the bulk stability of the mixed layer is low and there is large horizontal variability in pycnocline depth (as in the Weddell Sea), advection of horizontal density gradients may have large impact on mixing, both by changing turbulence scales and by conditioning the water column for equation-of-state related effects like cabbeling and thermobaric instability.

Even with the advantage of the stable ice platform, observations in the upper pycnocline are hampered by the small turbulence scales, by the difficulty of separating turbulence from high frequency internal wave velocities, and by rapid migration of the interface in response to internal waves or horizontal advection.

Convection in the Presence of Sea Ice

The cold, saline water that fills most of the abyssal world ocean originates from deep convection at high latitudes. Sea ice formation is a (geophysically) very efficient distillation process and may play a critical role in deep convection in areas like the Greenland, Labrador, and Weddell Seas where the bulk stability of the water column is low. By the same token, melting sea ice is a strong surface stabilizing influence that can rapidly shut down surface driven convection as soon as warm water reaches the well mixed layer from below.

Understanding the physics of turbulent transfer in highly convective regimes is a difficult problem both from theoretical and observational standpoints, complicated not only by uncertainty about how

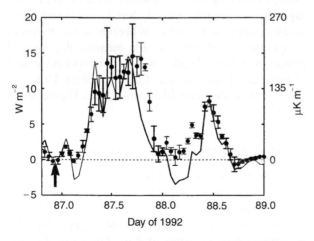

Figure 4 Time series of turbulent heat flux, $\rho c_p \overline{\langle w'T' \rangle}$ (W m^{-2}, circles) and temperature gradient $-\partial T/\partial z (\mu K \, m^{-1}$ curve). The overbar indicates a vertical average over five turbulence clusters from 4 to 24 m. Error bars are twice the sample standard deviation. The temperature gradient was calculated by linear regression, after the calibration of each thermometer was adjusted by a constant amount so that the gradient was zero at time 86.95 when heat flux was zero (heavy arrow). (Reproduced from McPhee MG and Martinson DG (1994) *Science* 263: 218–221.)

large-scale eddies interact with the stably stratified pycnocline fluid, but also by the possibility of frazil ice, small crystals that form within the water column. Depending on where it nucleates, frazil can represent a distributed internal source of buoyancy and heat in the UBL.

Zones of intense freezing tend to be highly heterogeneous, concentrated in lead systems or near the ice margins, and require specialized equipment for studying horizontal structure. Measuring difficulties increase greatly in the presence of frazil ice or supercooled water, because any intrusive instruments present attractive nucleation sites.

In addition to questions of UBL turbulence and surface buoyancy flux, factors related to nonlinearities in the equation of state for sea water may have profound influence on deep convection triggered initially by ice growth and UBL convection. Recent studies have shown, for example, that certain regions of the Weddell Sea are susceptible to thermobaric instability, arising from nonlinearity of the thermal expansion coefficient with increasing pressure. The importance of thermobaric instability for an ice-covered ocean is that once triggered, the potential energy released and converted in to turbulence as the water column overturns thermobarically, may be sufficient to override the surface buoyancy flux that would result from rapid melting as warm water reaches the surface.

Symbols used

f	Coriolis parameter
g	acceleration of gravity
K	eddy viscosity
K_H	scalar eddy diffusivity
i	imaginary number
L	Obukhov length, $u_*^3/(\kappa\langle w'b'\rangle)$
P_b	production rate of turbulent kinetic energy by buoyancy, $-\langle w'b'\rangle$
P^S	production rate of turbulent kinetic energy by shear, u_*^3/λ
q	turbulent kinetic energy scale velocity
R_c	critical flux Richardson number (~ 0.2)
S	salinity
T	temperature
u	three-dimensional velocity vector (u, v, w components)
u_*	friction velocity, square root of kinematic stress
u_τ	turbulence scale velocity
V	horizontal velocity vector
w_*	convective turbulence scale velocity
$\langle w'b'\rangle$	turbulent buoyancy flux, $(g/\rho)\langle w'\rho'\rangle$
$\langle w'T'\rangle$	kinematic turbulent heat flux
$\langle w'S'\rangle$	turbulent salinity flux
$\hat\delta$	complex attenuation coefficient
ε	dissipation rate of turbulent kinetic energy
η_*	stability factor, $(1+\Lambda_* u_*/(\kappa R_c|f|L))^{-1/2}$
κ	von Kármán's constant (0.4)
Λ_*	similarity constant (~ 0.03)
λ	turbulent mixing length scale
λ_T	turbulent scalar mixing length scale
ν	kinematic molecular viscosity, units $m^2\,s^{-1}$
ν_T	molecular scalar (thermal) diffusivity, units $m^2 s^{-1}$
τ	Reynolds stress: $\langle u'w'\rangle+i\langle v'w'\rangle$
Φ	latitude

See also

Ice–Ocean Interaction. Langmuir Circulation and Instability. Wind- and Buoyancy-Forced Upper Ocean.

Further Reading

Ekman VW (1905) On the influence of the earth's rotation on ocean currents. *Ark. Mat. Astr. Fys* 2: 1–52.

Gill AE (1982) *Atmosphere–Ocean Dynamics.* New York: Academic Press.

Johannessen OM, Muench RD, and Overland JE (eds.) (1994) *The Polar Oceans and Their Role in Shaping the Global Environment: The Nansen Centennial Volume.* Washington DC: American Geophysical Society.

McPhee MG (1994) On the turbulent mixing length in the oceanic boundary layer. *Journal of Physical Oceanography* 24: 2014–2031.

Pritchard RS (ed.) (1980) *Sea Ice Processes and Models.* Seattle, WA: University of Washington Press.

Smith WO (ed.) (1990) *Polar Oceanography.* San Diego, CA: Academic Press.

Untersteiner N (ed.) (1986) *The Geophysics of Sea Ice.* New York: Plenum Press.

COUPLED SEA ICE–OCEAN MODELS

A. Beckmann and G. Birnbaum, Alfred-Wegener-Institut für Polar und Meeresforschung, Bremerhaven, Germany

Introduction

Oceans and marginal seas in high latitudes are seasonally or permanently covered by sea ice. The understanding of its growth, movement, and decay is of utmost importance scientifically and logistically, because it affects the physical conditions for air–sea interaction, the large-scale circulation of atmosphere and oceans and ultimately the global climate (e.g., the deep and bottom water formation) as well as human activities in these areas (e.g., ship traffic, offshore technology).

Coupled sea ice–ocean models have become valuable tools in the study of individual processes and the consequences of ice–ocean interaction on regional to global scales. The sea ice component predicts the temporal evolution of the ice cover, thus interactively providing the boundary conditions for the ocean circulation model which computes the resulting water mass distribution and circulation.

A number of important feedback processes between the components of the coupled system can be identified which need to be adequately represented (either resolved or parameterized) in coupled sea ice–ocean models (see **Figure 1**):

- ice growth through freezing of sea water, the related brine release and water mass modification;
- polynya maintenance by continuous oceanic upwelling;
- lead generation by lateral surface current shear and divergence;
- surface buoyancy loss causing oceanic convection; and
- pycnocline stabilization in melting regions.

Not all of these are equally important everywhere, and it is not surprising that numerous variants of coupled sea ice–ocean models exist, which differ in physical detail, parameterizational sophistication, and numerical formulation. Models for studies with higher resolution usually require a higher level of complexity.

The main regions for applying coupled sea ice–ocean models are the Arctic Ocean, the waters surrounding Antarctica, and marginal seas of the Northern Hemisphere (e.g., Baltic Sea, Hudson Bay). A universally applicable model needs to include (either explicitly or by adequate parameterization) the

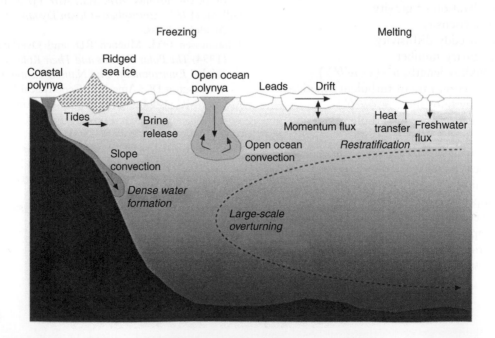

Figure 1 Cartoon of coupled sea ice–ocean processes. Effects in the ocean are in *italics*.

specific mechanisms of each region, e.g., mainly seasonal variations of the ice cover or the presence of thick, ridged multiyear ice.

This article describes the philosophy and design of large- and mesoscale prognostic dynamic–thermo-dynamic sea ice models which are coupled to primitive equation ocean circulation models. The conservation principles, the most widely used para-meterizations, several numerical and coupling as-pects, and model evaluation issues are addressed.

Basics

Sea water and sea ice as geophysical media are quite different; whereas the liquid phase is continuous, three-dimensional, and largely incompressible, the solid phase can be best characterized as granular, two-dimensional and compressible. Both share a high degree of nonlinearity, and many direct feed-backs between oceanic and sea ice processes exist.

Today's coupled sea ice–ocean models are Euler-ian; granular Lagrangian models, which consider the floe–floe interaction explicitly, exist but have so far not been fully coupled to ocean circulation models. Originally designed for use with large-scale coarse resolution ocean models, sea ice in state-of-the-art models is treated as a continuous medium. Following the continuum hypothesis only the average effect of a large number of ice floes is considered, assuming that averaged ice volume and velocities are continuous and differentiable functions of space and time.

Thus, very similar numerical methods are being applied to both sea water and sea ice, which greatly facilitates the coupling of models of these two com-ponents of the climate system. Conceptually, the continuum approach limits the applicability of sea ice models to grid spacings that are much larger than typical floes, i.e., several to several tens of kilometers. Typical values are 100–300 km for global climate studies, 10–100 km for regional climate simulations and about 2 km for process studies and quasi-oper-ational forecasts. The latter cases are stretching the continuum concept for sea ice quite a bit, but still give reasonable results.

State-of-the-art coupled sea ice–ocean models are based on two principles for the description of sea ice, the conservation of mass and momentum, covering its thermodynamics and dynamics. Mass conser-vation for snow is also taken into account. A snow layer modifies the thermal properties of the ice cover through an increased albedo and reduced conductiv-ity. This leads to delayed surface melting and lower basal freezing rates. In the following description sea ice is frozen sea water, and ice is the sum of sea ice and snow.

The temporal change of sea ice and snow volume due to local sources/sinks and drift is described by the mass conservation equation

$$\frac{\partial h_{(i,s)}}{\partial t} + \nabla \cdot \left(\vec{v}_i h_{(i,s)}\right) = S_{h_{(i,s)}}^{thdyn} \qquad [1]$$

where $h_{(i,s)}$ is the sea ice and snow volume per unit area (with $h = h_i + h_s$), respectively, v_i is the two-dimensional ice velocity vector and $S_{h_{(i,s)}}^{thdyn}$ denotes the thermodynamic sources and sinks for sea ice and snow. The corresponding ice velocities are obtained from the momentum equation

$$(\rho_i h_i + \rho_s h_s)\left[\frac{\partial \vec{v}_i}{\partial t} + \vec{v}_i \cdot \nabla \vec{v}_i + f\vec{k} \times \vec{v}_i + g\nabla \mathcal{H}\right]$$
$$= \vec{\tau}_{ai} - \vec{\tau}_{iw} + \vec{\mathcal{F}}_i \qquad [2]$$

where the local time rate of change, advection of momentum and the accelerations due to Coriolis are included. The wind stress $\vec{\tau}_{ai}$ is external to the cou-pled sea ice–ocean model; the ocean surface current stress $\vec{\tau}_{iw}$ and the sea surface height $\mathcal{H}$ is part of the coupling to the ocean. The so-called ice stress term $\vec{\mathcal{F}}_i$ summarizes all internal forces generated by floe–floe interactions.

Subgridscale Parameterizations

The granular nature of the medium, combined with the strong sensitivity of both thermodynamics and dynamics on the number, size, and thickness of in-dividual ice floes requires the inclusion of a sub-gridscale structure of the modeled ice.

Ice Classes

An obvious assumption is that of a subgridscale ice thickness distribution. In this widely used approach the predicted ice volume h is thought to be the average of several compartments, the thermo-dynamic ice classes, which represent both thinner and thicker ice and possibly include open water. The relative contribution of an ice class is fixed (e.g., uniform between 0 and twice the average ice thick-ness). For each ice class, a separate thermodynamic balance is computed; the resulting fluxes are then averaged according to their relative areal coverage.

Ice Categories

As the subgridscale distribution of ice thickness will change with time and location, an even more sophisticated approach considers time-evolution of the volume of each compartment, which is then called an ice category. The form of these prognostic equations follows the conservation eqn [1]. Different subgridscale ice velocities are not taken into account; the advection of all compartments takes place with the resolved velocity field.

Models with several ice categories are in use. A minimum requirement, however, has been identified in the discrimination between ice-covered areas and open water. Then the prognostic equation for ice volume [1] is accompanied by a formally similar equation for ice concentration, i.e., the percentage of ice-covered area per unit area A,

$$\frac{\partial A}{\partial t} + \nabla \cdot (\vec{v}_i A) = S_A^{thdyn} + S_A^{thdyn} \quad [3]$$

Thermodynamic sources and sinks S_A^{thdyn} for ice concentration are chosen empirically and involve parameterizations of subgridscale thermodynamic melting and freezing. The conceptual *ansatz*:

$$S_A^{thdyn} = \frac{1-A}{h_{cls}} \max\left[0, \frac{\partial h}{\partial t}\right] + \frac{A}{h_{opn}} \min\left[\frac{\partial h}{\partial t}, 0\right] \quad [4]$$

describes the formation of new ice between the ice floes with the first term on the right hand side; here h_{cls} is the so-called lead closing parameter. The second term parameterizes basal melting of sea ice with a similar approach involving h_{opn}. The coefficients are often derived from the assumed internal structure of the ice-covered portion of the grid cell, the ice classes. The dynamic source/sink term S_A^{thdyn} follows from the constitutive law (see section on dynamics below). Conceptually, the ice concentration (or compactness) A has to lie between 0 and 1, which has to be enforced separately.

With the introduction of an ice concentration, the ice volume h has to be replaced by the actual ice thickness

$$h^* = \frac{h}{A} \quad [5]$$

i.e., the mean value of individual floe height, such that the total ice volume per grid box is not affected by this approach.

An approach using two categories, open water and ice-covered areas with an internal structure (ice classes), has been proven highly adequate for a large number of situations.

Thermodynamics

Mass conservation for ice is closely tied to the heat balance at its surfaces. Sea ice forms, if the freezing temperature of sea water is reached. The surface freezing point T_f (in K) is a function of salinity, estimated by the polynomial approximation

$$T_f = 273.15 - 0.0575 S_w + 1.710523 \times 10^{-3} S_w^{3/2} \\ - 2.154996 \times 10^{-4} S_w^2 \quad [6]$$

where S_w is the sea surface salinity.

The majority of today's sea ice thermodynamics models is based on a one-dimensional (vertical) heat diffusion equation, which for sea ice (without snow cover) reads

$$\rho_i c_{pi} \frac{\partial T_i}{\partial t} = \frac{\partial}{\partial z}\left(k_i \frac{\partial T_i}{\partial z}\right) + K_i I_{oi} \exp[-K_i z] \quad [7]$$

Here, T_i, ρ_i, c_{pi} and k_i are the sea ice temperature, density, specific heat, and thermal conductivity, respectively. The net short-wave radiation at the sea ice surface is I_{oi} and K_i is the bulk extinction coefficient.

If a snow cover is present, the penetrating short wave radiation is neglected and a second prognostic equation for the snow is solved:

$$\rho_s c_{ps} \frac{\partial T_s}{\partial t} = \frac{\partial}{\partial z}\left(k_s \frac{\partial T_s}{\partial z}\right) \quad [8]$$

At the snow–sea ice interface, the temperatures and fluxes have to match.

Assuming that ice exists, the local time rate of change of ice thickness due to freezing of sea water or melting of ice is the result of the energy fluxes at the surface and the base of the ice. At the surface, the ice temperature and thickness change is determined from the energy balance equation:

$$Q_{a(i,s)} = \left(1 - \alpha_{(i,s)}\right)\left(1 - I_{oi}\right)\mathcal{R}_{SW}^{\downarrow} + \mathcal{R}_{LW}^{\downarrow} \\ - \epsilon_{(i,s)}\sigma_0 T_{0(i,s)}^4 + Q_l + Q_s + Q_c \\ = \begin{cases} 0 & \text{if } T_{o(i,s)} < T_{m(i,s)} \\ -\rho_{(i,s)} L_{(i,s)} \frac{\partial h^*}{\partial t} & \text{if } T_{o(i,s)} = T_{m(i,s)} \end{cases} \quad [9]$$

where $T_{o(i,s)}$ and $T_{m(i,s)}$ are the surface and melting temperatures, respectively, $\rho_{(i,s)}$ and $L_{(i,s)}$ are the density and heat of fusion for sea ice and snow. Besides the conductive heat flux in the ice $Q_c = k_{(i,s)} \partial T_{(i,s)} / \partial t$ the following atmospheric fluxes are considered: downward short-wave radiation $\mathcal{R}_{SW}^{\downarrow}(\phi, \lambda, A_{cl}$, net long-wave radiation $\mathcal{R}_{LW}^{\downarrow}(T_a, A_{cl}) - \epsilon_{(i,s)}\sigma_o T_{o(i,s)}^4$, as well as sensible $Q_s(\vec{v}_a, T_a, T_{o(i,s)})$ and latent $Q_l(\vec{v}_a, q_a, q_{o(i,s)})$ heat fluxes. The albedos $\alpha_{(i,s)}$ and emissivities $\epsilon_{(i,s)}$ are dependent on the surface structure of the medium (sea ice, snow). The atmospheric forcing data are:

- near-surface wind velocity $\vec{v}_a$;
- near-surface atmospheric temperature T_a;
- near-surface atmospheric dew point temperature T_d, or specific humidity q_a;
- cloudiness A_{cl};
- precipitation P and evaporation E (needed for the sea ice and snow mass balance).

Note that the wind velocity is also needed for the atmospheric forcing in the momentum [2].

At the base of the ice (the sea surface), an imbalance of the conductive heat flux in the sea ice (Q_c) and the turbulent heat flux from the ocean

$$Q_{sw} = \rho_w c_{pw} c_h u_\star \left(T_f - T_{ml} \right) \qquad [10]$$

leads to a change in ice thickness:

$$Q_{iw} = -Q_{ow} - Q_c = -\rho_i L_i \frac{\partial h_i^*}{\partial t} \qquad [11]$$

Here, T_{ml} is the ocean surface and mixed layer temperature, c_h is the heat transfer coefficient and $u_\star$ is the friction velocity. In general, the main sink for ice volume is basal melting due to above-freezing temperatures in the oceanic mixed layer. The source for snow is a positive rate of $P–E$, if the air temperature is below the freezing point of fresh water. The main source for sea ice is basal freezing. However, the formation of additional sea ice on the upper ice surface is possible through a process called flooding. This conversion of snow into sea ice takes place when the weight of the snow exceeds the buoyancy of the ice and sea water intrudes laterally.

Often, the vertical structure of temperature is approximated by simple zero-, one- or two-layer formulations, with the resulting internal temperature profile being piecewise linear. The most simple approach, the zero-layer model, eliminates the capacity of the ice to store heat. However, it has been used successfully in areas where sea ice is mostly seasonal and thus relatively thin ($<1\,\mathrm{m}$).

The specifics of the brine-related processes in the sea ice are difficult to implement in models. As a consequence, sea ice models usually assume constant sea ice salinity S_i of about 5 PSU (practical salinity units) to calculate the heat of fusion and the vertical heat transfer coefficient. The errors arising from this assumption are largest during the early freezing processes, when salt concentrations are considerably higher.

The open water part of each grid cell, where the atmosphere is in direct contact with the ocean, is treated like any other air–sea interface. The thermodynamic eqns [9] and [11] are modified to the radiative and heat fluxes between ocean and atmosphere. In the case of heat loss resulting in an ocean

temperature below the freezing point T_f, new ice is formed:

$$
\begin{aligned}
Q_{aw} &= (1-\alpha)(1-I_{ow})\mathscr{R}_{SW}^{\downarrow} + \mathscr{R}_{LW}^{\downarrow} - \varepsilon_w \sigma_0 T_{0w}^4 \\
&\quad + Q_l + Q_s + Q_{sw} \\
&= \begin{cases} 0 & \text{if } T_{ow} > T_f \\ -\rho_i L_i \frac{\partial h^*}{\partial t} & \text{if } T_{ow} = T_f \end{cases}
\end{aligned}
\qquad [12]
$$

In the case of above-freezing ocean surface temperatures, Q_{sw} follows from [10] with T_f replaced by T_{ow}. An illustration summarizing the fluxes is given in **Figure 2**.

The solution of eqns [7], [8], [9] and [11] is conceptually straightforward but algebraically complicated in that it involves iterative solution of the energy balance equation to obtain the surface temperature.

Dynamics

Driven by wind and surface ocean currents, sea ice grown locally is advected horizontally. Free drift (the absence of internal ice stresses) is a good approximation for individual ice floes. In a compact ice cover, however, internal stresses will resist further compression and react to shearing stresses. These internal ice forces are expressed as the divergence of the isotropic two-dimensional internal stress tensor

$$\vec{\mathscr{F}}_i = \nabla \cdot \sigma \qquad [13]$$

which depends on the stress–strain relationship, where the deformation rates are proportional to the spatial derivatives of ice velocities. A general form of the constitutive law is

$$
\sigma = \begin{pmatrix} \eta\left(\frac{\partial u_i}{\partial x} - \frac{\partial v_i}{\partial y}\right) + \zeta\left(\frac{\partial u_i}{\partial x} + \frac{\partial v_i}{\partial y}\right) - \frac{1}{2}P_i & \eta\left(\frac{\partial u_i}{\partial y} + \frac{\partial v_i}{\partial x}\right) \\ \eta\left(\frac{\partial u_i}{\partial y} + \frac{\partial v_i}{\partial x}\right) & \eta\left(\frac{\partial v_i}{\partial y} - \frac{\partial u_i}{\partial x}\right) + \zeta\left(\frac{\partial u_i}{\partial x} + \frac{\partial v_i}{\partial y}\right) - \frac{1}{2}P_i \end{pmatrix} \qquad [14]
$$

here, ζ and η are nonlinear viscosities for compression and shear. P_i is the ice strength.

The most widely used sea ice rheology is based on the viscous–plastic approach. Introduced as the result of the ice dynamics experiment AIDJEX, it has proven to be a universally applicable rheology. It treats the ice as a linear viscous fluid for small deformation rates and as a rigid plastic medium for larger deformation rates. Simpler rheologies have been tested but could not reproduce the observed ice distributions and thicknesses nearly as well as the viscous–plastic approach. The viscosities are then

$$\zeta = e^2 \eta = \frac{P_i}{2\Delta} \qquad [15]$$

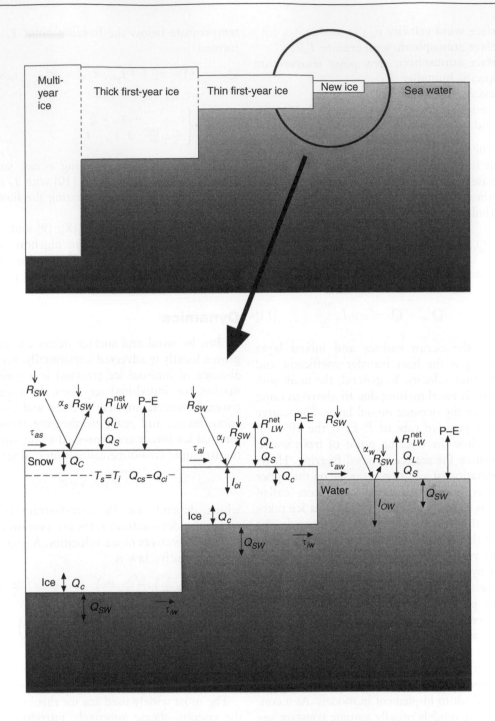

Figure 2 Schematic of the concept of ice classes/categories, surface energy balance and the ice–ocean flux coupling. The upper panel shows a grid cell covered with several classes/categories of ice, including open water. The lower panel is a detailed view at the fluxes between atmosphere, ice, and ocean, in the three cases: snow-covered sea ice, pure sea ice, and open water.

where an elliptic yield curve of ellipticity e is assumed, with the deformation rate given by

$$\Delta = \left[\left\{\left(\frac{\partial u_i}{\partial x}\right)^2 + \left(\frac{\partial v_i}{\partial y}\right)^2\right\}(1+e^{-2}) + \left(\frac{\partial u_i}{\partial y} + \frac{\partial v_i}{\partial x}\right)^2 e^{-2}\right.$$
$$\left. + 2\frac{\partial u_i}{\partial x}\frac{\partial v_i}{\partial y}(1-e^{-2})\right]^{1/2} \qquad [16]$$

In particular, internal forces are only important for densely packed ice floe fields, i.e., for ice concentrations exceeding 0.8. Most sea ice models take this into account by assuming that the ice strength is

$$P_i = P^*h\exp[-C^*(1-A)] \qquad [17]$$

where P^* and C^* are empirical parameters. The same functional dependence is also used successfully to describe the generation of open water areas through shear deformation, which is parameterized by

$$S_A^{dyn} = -0.5\left(\Delta - \left|\nabla \cdot \vec{v}_i\right|\right)\exp[-C^*(1-A)] \quad [18]$$

Subgridscale processes like ridging and rafting can be successfully parameterized this way.

Coupling

Numerical ocean circulation models are described in article . A schematic illustration of the interactions in a coupled sea ice–ocean model is given in **Figure 3**. The coupling between the sea ice and ocean components is done via fluxes of heat, salt, and momentum. They enter the ocean model through the surface boundary conditions to the vertically diffusive/viscous terms. Given the relative (to the depth of the ocean) small draught of the ice, it is assumed not to deform the sea surface, all boundary conditions are applied at the air–sea interface.

Due to the presence of subgridscale ice categories, the fluxes have to be weighed with the areal coverage of open water, and ice of different thickness. The resulting boundary conditions for the simplest two-category (ice and open water) case are:

$$A_v^M \frac{\partial \vec{v}_w}{\partial z}\bigg|_{z=0} = A\vec{\tau}_{iw} + (1-A)\vec{\tau}_{aw} \quad [19]$$

$$A_v^T \frac{\partial T_w}{\partial z}\bigg|_{z=0} = -\frac{1}{\rho_w c_p}(AQ_{iw} + (1-A)Q_{aw}) \quad [20]$$

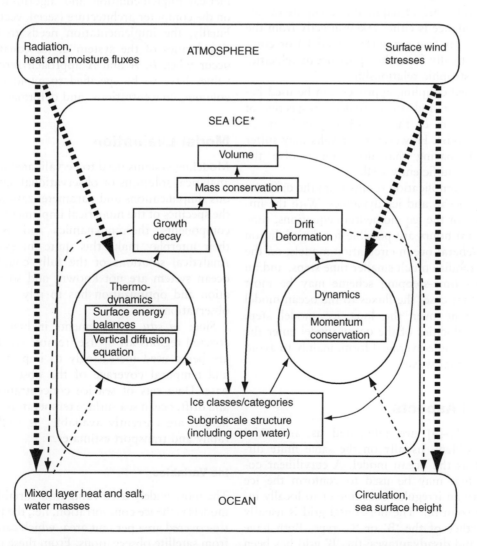

Figure 3 Concept of a coupled sea ice–ocean numerical model with prescribed atmospheric forcing. Thick broken arrows represent the atmospheric forcing, thin dashed arrows represent the coupling pathways between ice and ocean, and thin arrows indicate how the ice volume is computed from thermodynamical and dynamical principles (mass and momentum conservation), with the assumption of a subgridscale ice distribution. * Includes both sea ice and snow.

$$A_v^s \frac{\partial S_w}{\partial z}\bigg|_{z=0} = (S_w - S_i)\frac{\rho_i}{\rho_w}\frac{\partial h_i}{\partial t}$$

$$+ S_w \begin{cases} P - E & \text{if } h = 0 \\ P - E & \text{if } h > 0, T_a > 0 \\ (1 - A)(P - E) & \text{if } h > 0, T_a < 0 \\ \frac{\rho_s}{\rho_w}\frac{\partial h^s}{\partial t} & \text{if } h_s > 0, T_a > 0 \end{cases} \quad [21]$$

where the freshwater flux is converted into a salt flux, and the momentum exchange is parameterized (like the atmospheric wind forcing $\vec{\tau}_{aw}$ and $\vec{\tau}_{ai}$) in the form of the usual quadratic drag law:

$$\vec{\tau}_{iw} = \rho_w C_w |\vec{v}_i - \vec{v}_w|$$
$$\left[(\vec{v}_i - \vec{v}_w)\cos\theta_{iw} + \vec{k} \times (\vec{v}_i - \vec{v}_w)\sin\theta_{iw} \right] \quad [22]$$

Here, C_w is the drag coefficient and θ_{iw} the rotation angle. The vertical viscosities and diffusivities (A_v^M, A_v^T, A_v^S) are ocean model parameters.

The sea surface height required to compute the ice momentum balance is either taken directly from the ocean model (for free sea surface models) or computed diagnostically from the upper ocean velocities using the geostrophic relationship.

The described coupling approach can be used between any sea ice and ocean model, irrespective of vertical resolution, or the use of a special surface mixed layer model. However, the results may suffer from a grid spacing that does not resolve the boundary layer sufficiently well.

A technical complication results from the different timescales in ocean and ice dynamics. With the implicit solution of the ice momentum equations, time steps of several hours are possible for the evolution of the ice. General ocean circulation models, on the other hand, require much smaller time steps, and an asynchronous time-stepping scheme may be most efficient. In that case, the fluxes to the ocean model remain constant over the long ice model step, whereas the velocities of the ocean model enter the fluxes only in a time-averaged form, mainly to avoid aliasing of inertial waves.

Numerical Aspects

Equations [1]–[3] are integrated as an initial boundary problem, usually on the same finite difference grid as the ocean model. A curvilinear coordinate system may be used to conform the ice model grid to an irregular coastline or to locally increase the resolution. The horizontal grid is usually staggered, either of the 'B' or 'C' type. Both have advantages and disadvantages: the 'B' grid has been favored because of the more convenient formulation of the stress terms and the better representation of the Coriolis term; the 'C' grid avoids averaging for the advection and pressure gradient terms. The treatment of coastal boundaries and the representation of flow through passages is also different.

Due to the large nonlinear viscosities in the viscous plastic approach, an explicit integration of the momentum equations would require time steps of the order of seconds, whereas the thickness equations can be integrated with time steps of the order of hours. Therefore, the momentum equations are usually solved implicitly. This leads to a nonlinear elliptic problem, which is solved iteratively. An explicit alternative has been developed for elastic–viscous–plastic rheology.

Other general requirements for numerical fluid dynamics models also apply: a positive definite and monotonic advection scheme is desired to avoid negative ice volume and concentration with the numerical implementation and algorithms depending on the computer architecture (serial, vector, parallel). Finally, the implementation needs to observe the singularities of the system of ice equations, which occur when h, A and Δ approach zero. Minimum values have to be specified to avoid vanishing ice volumes, concentrations, and deformation rates.

Model Evaluation

Modeling systems need to be validated against either analytical solutions or observational data. The various simplifications and parameterizations, as well as the specifics of the numerical implementation of both components' thermodynamics and dynamics and their interplay make this quite an extensive task. Analytical solutions of the fully coupled sea ice–ocean system are not known, and so model validation and optimization has to rely on geophysical observations.

Since *in situ* measurements in high latitude ice-covered regions are sparse, remote sensing products are being used increasingly to improve the spatial and temporal coverage of the observational database. Data sets of sea ice concentration, thickness and drift, ocean sea surface temperature, salinity, and height, are currently available, as well as hydrography and transport estimates.

Ice Variables

The most widely used validation variable for sea ice models is the ice concentration, i.e., the percentage of ice-covered area per unit area, which can be obtained from satellite observations. From these observations, maps of monthly mean sea ice extent are constructed, and compared to model output (see **Figure 4**). It

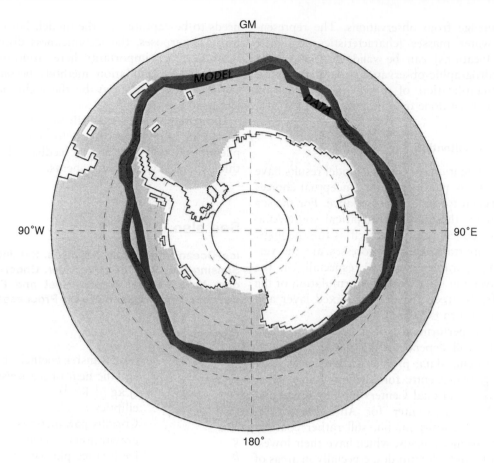

Figure 4 September 1987 simulated (blue) and remotely observed (red) sea ice edge around Antarctica. After Timmermann *et al.*, 2001.

should be noted though that the modeled ice concentrations represent a subgridscale parameterization with an empirically determined source/sink relationship such that an optimization of a model with respect to this quantity may be misleading.

A more rigorous model evaluation focuses on sea ice thickness or drift, which is a conserved quantity and more representative of model performance. Unfortunately, 'ground truth' values of these variables are available in few locations and over relatively short periods only (e.g., upward looking sonar, ice buoys) and comparison of point measurements with coarse resolution model output is always problematic. A successful example is shown in **Figure 5**. The routine derivation of ice thickness estimates from satellite observations will be a major step in the validation attempts. The evaluation of ice motion is done through comparison between satellite-tracked and modeled ice buoys. Thickness and motion of first-year ice in free drift is usually represented well in the models, as long as atmospheric fields resolving synoptic weather systems are used to drive the system.

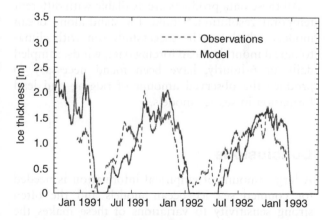

Figure 5 Time series of simulated (blue) and ULS (upward looking sonar) measured (red) sea ice thickness at 15°W, 70°S in the Weddell Sea. After Timmermann *et al.*, 2001.

Ocean Variables

The success of the coupled system also depends on the representation of oceanic quantities. The model's temperature and salinity distribution as well as the corresponding circulation need to be consistent with prior knowledge from observations. The representation of water masses (characteristics, volumes, formation locations) can be validated against the existing hydrographic observational database, but a rigorous quantification of water mass formation products has been done only in few cases.

Parameter Sensitivities

Systematic evaluations of coupled model results have shown that a few parameter and conceptual choices are most crucial for model performance. For the sea ice component, these are the empirical source/sink terms for ice concentration, as well as the rheology. The most important oceanic processes are the vertical mixing (parameterizations), especially in the case of convection in general, the formulation of the heat transfer between the oceanic mixed layer and the ice is central to the coupled system.

Finally, the performance of a coupled sea ice–ocean model will depend on the quality of the atmospheric forcing data; products from the weather centers (European Centre for Medium Range Weather Forecasts, National Centers for Environmental Prediction/National Center for Atmospheric Research) provide consistent, but still rather coarsely resolved atmospheric fields, which have their lowest overall quality in high latitudes, especially in areas of highly irregular terrain and for the P–E and cloudiness fields. Some errors, even systematic ones, are presently unavoidable.

All these data products are available with different temporal resolution. Unlike for stand-alone ocean models, which can be successfully run with climatological monthly mean forcing data, winds, sampled daily or 6-hourly, have been found necessary to produce the observed amount of ridging and lead formation in sea ice models.

Conclusions

A large amount of empirical information is needed for coupled sea ice–ocean models and the often strong sensitivity to variations of these makes the optimization of such modeling systems a difficult task. Yet, several examples of successful simulation of fully coupled ice–ocean interaction exist, which qualitatively and quantitatively compare well with the available observations.

Coupled sea ice–ocean modeling is an evolving field, and much needs to be done to improve parameterizations of vertical (and lateral) fluxes at the ice–ocean interfaces. For climate studies, water mass variability on seasonal and interannual timescales

needs to be captured by the model. For operational forecast purposes, the ice thickness distribution in itself is most important; here atmospheric data quality and assimilation methods become crucial. Obvious next steps may be the inclusion of tides, icebergs, and ice shelves.

Ultimately, however, a fully coupled atmosphere–ice–ocean model is required for the simulation of phenomena that depend on feedback between the three climate system components.

See also

Ice–Ocean Interaction. Polynyas. Satellite Remote Sensing Sar. Sea Ice: Overview. Under-Ice Boundary Layer. Upper Ocean Heat and Freshwater Budgets. Upper Ocean Mixing Processes.

Glossary

c_h	heat transfer coefficient
$c_{p(i,s,w)}$	specific heat of sea ice/snow/water ($\text{J kg}^{-1}\,\text{K}^{-1}$)
e	ellipticity
f	Coriolis parameter (s^{-1})
g	gravitational acceleration (m s^{-2})
h	ice (sea ice plus snow) volume per unit area (m)
$h_{(i,s)}$	sea ice/snow volume per unit area (m)
h^*	actual ice thickness (m)
$h^*_{(i,s)}$	actual sea ice/snow thickness (m)
h_{cls}, h_{opn}	lead closing/opening parameter (m)
$k_{(i,s)}$	thermal conductivity of sea ice/snow ($\text{W m}^{-1}\,\text{K}^{-1}$)
$\vec{k}$	vertical unit vector
q_a	atmospheric specific humidity
$q_{o(i,s)}$	surface specific humidity of sea ice/snow
t	time (s)
$\vec{v_i} = (u_i, v_i)$	horizontal ice velocity (m s^{-1})
$u_\star$	friction velocity (m s^{-1})
$\vec{v_a} = (u_a, v_a)$	wind velocity (m s^{-1})
$\vec{v}_w = (u_{w_1}, v_w)$	ocean surface velocity (m s^{-1})
x, y, z	spatial directions (m)
A	ice concentration
A_{cl}	cloudiness
A_v^M, A_v^T, A_v^S	oceanic vertical mixing coefficients ($\text{m}^2\,\text{s}^{-1}$)
C^*	empirical parameter
C_w	oceanic drag coefficient
$I_{o(i,w)}$	short wave radiation penetrating sea ice/water (W m^{-2})

E	evaporation (m s^{-1})
K_i	bulk extinction coefficient (m^{-1})
$L_{(i,s)}$	heat of fusion (J kg^{-1})
P	precipitation (m s^{-1})
P_i	ice strength (N m^{-1})
$P*$	ice strength parameter (N m^{-2})
$Q_{a(i,s,w)}$	net energy flux between atmosphere and sea ice/snow/water (W m^{-2})
Q_c	conductive heat flux in the ice (W m^{-2})
Q_l, Q_s	atmospheric latent/sensible heat flux (W m^{-2})
Q_{iw}	turbulent heat flux at the ocean surface (W m^{-2})
Q_{sw}	oceanic sensible heat flux (W m^{-2})
$\mathscr{R}^{\downarrow}_{SW}, \mathscr{R}^{\downarrow}_{LW}$	downward short/long wave radiation (W m^{-2})
$S_{(i,w)}$	sea ice/sea water salinity (PSU)
T_a, T_i, T_w	air/ice/water temperature (K)
T_d	dew point temperature (K)
T_f	freezing temperature of sea water (K)
T_{ml}	oceanic mixed layer temperature (K)
$T_{m(i,s)}$	melting temperature of sea ice/snow at the surface (K)
$T_{o(i,s,w)}$	sea ice/snow/water surface temperature (K)
$\mathscr{F} \rightarrow$	internal ice forces (N m^{-2})
$\mathscr{H}$	sea surface elevation (m)
$S_A^{(thdyn,dyn)}$	source/sink terms for ice concentration (s^{-1})
$S_{h(i,s)}^{thdyn}$	source/sink terms for sea ice/snow volume per unit area (m s^{-1})
$\alpha_{(i,sw)}$	sea ice/snow/sea water albedo
Δ	ice deformation rate (s^{-1})
$\varepsilon_{(i,sw)}$	sea ice/snow/sea water emissivity
η, ζ	nonlinear viscosities (kg s^{-1})
$\nabla = (\frac{\partial}{\partial x}, \frac{\partial}{\partial y})$	horizontal gradient operator
λ, ϕ	geographical longitude/latitude (deg)
$\rho_{(a,i,s,w)}$	densities of air/sea ice/snow/water (kg m^{-3})
σ	two-dimensional stress tensor (N m^{-1})
σ_o	Stefan-Boltzmann constant (W m^{-2} K^{-4})
θ_{iw}	turning angle (deg)
$\overrightarrow{\tau_{ai}}, \overrightarrow{\tau_{iw}}, \overrightarrow{\tau_{aw}}$	air-ice/ice-water/air-water stress (N m^{-2})

Appendix

A typical parameter set for simulations with coupled dynamic–thermodynamic ice--ocean models (e.g.,

Timmermann et al., 2001), as shown in **Figures 4 and 5**, are

$\rho_a = 1.3$ kg m-3
$\rho_i = 910$ kg m-3
$\rho_s = 290$ kg m-3
$\rho_w = 1027$ kg m-3
e$=2$
C$^*=20$
P$^*=2000$ N m^{-2}
$h_{cls}=1$ m
$h_{opn}=2$ h*
$C_w = 3 \times 10^{-3}$
$c_h = 1.2 \times 10^{-3}$
$\alpha_w = 0.1$
$\alpha_i = 0.75$
$\alpha_i = 0.65$ (melting)
$\alpha_s = 0.8$
$\alpha_w = 0.7$ (melting)
$K_i = 0.04$ m^{-1}
$S_i = 5$ PSU
$\theta = 10$ degrees
$c_{pi} = 2000$ J K^{-1} kg^{-1}
$c_{pw} = 4000$ J K^{-1} kg^{-1}
$c_{pa} = 1004$ J K^{-1} kg^{-1}
$L_i = 3.34 \times 10^5$ J kg^{-1}
$L_s = 1.06 \times 10^5$ J kg^{-1}
$k_i = 2.1656$ W m^{-1} K^{-1}
$k_s = 0.31$ W m^{-1} K^{-1}

Further Reading

Curry JA and Webster PJ (1999) *Thermodynamics of Atmospheres and Oceans*. London: Academic Press. International Geophysics Series..

Fichefet T, Goosse H, and Morales Maqueda M (1998) On the large-scale modeling of sea ice and sea ice–ocean interaction. In: Chassignet EP and Verron J (eds.) *Ocean Modeling and Parameterization*, pp. 399–422. Dordrecht: Kluwer Academic.

Haidvogel DB and Beckmann A (1999) *Numerical Ocean Circulation Modeling*. London: Imperial College Press.

Hibler WD III (1979) A dynamic-thermodynamic sea ice model. *Journal of Physical Oceanography* 9: 815–846.

Kantha LH and Clayson CA (2000) *Numerical Models of Oceans and Oceanic Processes*. San Diego: Academic Press.

Leppäranta M (1998) The dynamics of sea ice. In: Leppäranta M (ed.) *Physics of Ice-Covered Seas*, vol. 1, 305–342.

Maykut GA and Untersteiner N (1971) Some results from a time-dependent thermodynamic model of sea ice. *Journal of Geophysical Research* 76: 1550–1575.

Mellor GL and Häkkinen S (1994) A review of coupled ice–ocean models. In: Johannessen OM, Muench RD and Overland JE (eds) *The Polar Oceans and Their*

Role in Shaping the Global Environment. AGU Geophysical Monograph, 85, 21–31.

Parkinson CL and Washington WM (1979) A large-scale numerical model of sea ice. *Journal of Geophysical Research* 84: 311–337.

Timmermann R, Beckmann A and Hellmer HH (2001) Simulation of ice–ocean dynamics in the Weddell Sea. Part I: Model description and validation. *Journal of Geophysical Research* (in press).

SUB ICE-SHELF CIRCULATION AND PROCESSES

K. W. Nicholls, British Antarctic Survey,
Cambridge, UK

Introduction

Ice shelves are the floating extension of ice sheets. They extend from the grounding line, where the ice sheet first goes afloat, to the ice front, which usually takes the form of an ice cliff dropping down to the sea. Although there are several examples on the north coast of Greenland, the largest ice shelves are found in the Antarctic where they cover 40% of the continental shelf. Ice shelves can be up to 2 km thick and have horizontal extents of several hundreds of kilometers. The base of an ice shelf provides an intimate link between ocean and cryosphere. Three factors control the oceanographic regime beneath ice shelves: the geometry of the sub-ice shelf cavity, the oceanographic conditions beyond the ice front, and tidal activity. These factors combine with the thermodynamics of the interaction between sea water and the ice shelf base to yield various glaciological and oceanographic phenomena: intense basal melting near deep grounding lines and near ice fronts; deposition of ice crystals at the base of some ice shelves, resulting in the accretion of hundreds of meters of marine ice; production of sea water at temperatures below the surface freezing point, which may then contribute to the formation of Antarctic Bottom Water; and the upwelling of relatively warm Circumpolar Deep Water.

Although the presence of the ice shelf itself makes measurement of the sub-ice shelf environment difficult, various field techniques have been used to study the processes and circulation within sub-ice shelf cavities. Rates of basal melting and freezing affect the flow of the ice and the nature of the ice–ocean interface, and so glaciological measurements can be used to infer the ice shelf's basal mass balance. Another indirect approach is to make ship-based oceanographic measurements along ice fronts. The properties of in-flowing and out-flowing water masses give clues as to the processes needed to transform the water masses. Direct measurements of oceanographic conditions beneath ice shelves have been made through natural access holes such as rifts, and via access holes created using thermal (mainly hot-water) drills. Numerical models of the sub-ice shelf regime have been developed to complement the field measurements. These range from simple one-dimensional models following a plume of water from the grounding line along the ice shelf base, to full three-dimensional models coupled with sea ice models, extending out to the continental shelf-break and beyond.

The close relationship between the geometry of the sub-ice shelf cavity and the interaction between the ice shelf and the ocean implies a strong dependence of the ice shelf/ocean system on the state of the ice sheet. During glacial climatic periods the geometry of ice shelves would have been radically different to their geometry today, and ice shelves probably played a different role in the climate system.

Geographical Setting

By far the majority of the world's ice shelves are found fringing the Antarctic coastline (**Figure 1**). Horizontal extents vary from a few tens to several hundreds of kilometers, and maximum thickness at the grounding line varies from a few tens of meters to 2 km. By area, the Ross Ice Shelf is the largest at around 500 000 km². The most massive, however, is the very much thicker Filchner-Ronne Ice Shelf in the southern Weddell Sea. Ice from the Antarctic Ice Sheet flows into ice shelves via fast-moving ice streams (**Figure 2**). As the ice moves seaward, further nourishment comes from snowfall, and, in some cases, from accretion of ice crystals at the ice shelf base. Ice is lost by melting at the ice shelf base and by calving of icebergs at the ice front. Current estimates suggest that basal melting is responsible for around 25% of the ice loss from Antarctic ice shelves; most of the remainder calves from the ice fronts as icebergs.

Over central Antarctica the weight of the ice sheet depresses the lithosphere such that the seafloor beneath many ice shelves deepens towards the grounding line. The effect of the lithospheric depression has probably been augmented during glacial periods by the scouring action of ice on the seafloor: at the glacial maxima the grounding line would have been much closer to the continental shelf-break. Since ice shelves become thinner towards the ice front and float freely in the ocean, a typical sub-ice shelf cavity has the shape of a cavern that dips downwards towards the grounding line (**Figure 2**).

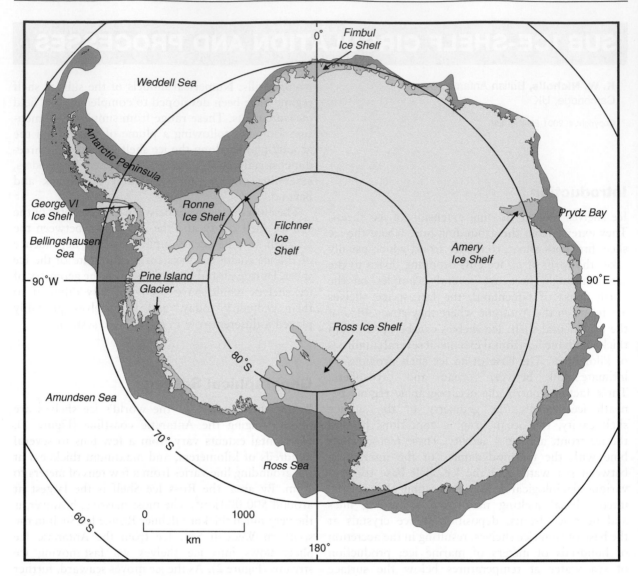

Figure 1 Map showing ice shelves (blue) covering about 40% of the continental shelf (dark gray) of Antarctica.

This geometry has important consequences for the ocean circulation within the cavity.

Oceanographic Setting

The oceanographic conditions over the Antarctic continental shelf depend on whether relatively warm, off-shelf water masses are able to cross the continental shelf-break.

For much of Antarctica a dynamic barrier at the shelf-break prevents advection of circumpolar deep water (CDW) onto the continental shelf itself. In these regions the principal process determining the oceanographic conditions is production of sea ice in coastal polynyas and leads, and the water column is largely dominated by high salinity shelf water (HSSW). Long residence times over some of the broader continental shelves, for example in the Ross and southern Weddell seas, enable HSSW to attain salinities of over 34.8 PSU. HSSW has a temperature at or near the surface freezing point (about − 1.9°C), and is the densest water mass in Antarctic waters. Conditions over the continental shelves of the Bellingshausen and Amundsen seas (**Figure 1**) represent the other extreme. There, the barrier at the shelf-break appears to be either weak or absent. At a temperature of about 1°C, CDW floods the continental shelf.

Between these two extremes there are regions of continental shelf where tongues of modified warm deep water (MWDW) are able to penetrate the shelf-break barrier (**Figure 3**), in some cases reaching as far as ice fronts. MWDW comes from above the warm core of CDW: the continental shelf effectively skims

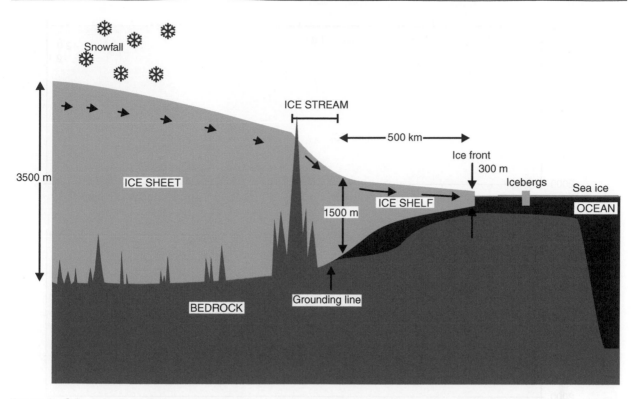

Figure 2 Schematic cross-section of the Antarctic ice sheet showing the transition from ice sheet to ice stream to ice shelf. Also shown is the depression of the lithosphere that results in the deepening of the seabed towards the continental interior.

off the shallower and cooler part of the water column.

What Happens When Ice Shelves Melt into Sea Water?

The freezing point of fresh water is 0°C at atmospheric pressure. When the water contains dissolved salts, the freezing point is depressed: at a salinity of around 34.7 PSU the freezing point is −1.9°C. Sea water at a temperature above −1.9°C is therefore capable of melting ice. The freezing point of water is also pressure dependent. Unlike most materials, the pressure dependence for water is negative: increasing the pressure decreases the freezing point. The freezing point T_f of sea water is approximated by:

$$T_f = aS + bS^{3/2} - cS^2 - dp$$

where $a = -5.75 \times 10^{-2}$°C PSU^{-1}, $b = 1.710523 \times 10^{-3}$°C PSU$^{-3/2}$, $c = -2.154996 \times 10^{-4}$°C PSU^{-2} and $d = -7.53 \times 10^{-4}$°C dbar^{-1}. S is the salinity in PSU, and p is the pressure in dbar. Every decibar increase in pressure therefore depresses the freezing point by 0.75 m°C. The depression of the freezing point with pressure has important consequences for the interaction between ice shelves and the ocean. Even though HSSW is already at the surface freezing point, if it can

be brought into contact with an ice shelf base, melting will take place. As the freezing point at the base of deep ice shelves can be as much as 1.5°C lower than the surface freezing point, the melt rates can be high.

When ice melts into sea water the effect is to cool and freshen. Consider unit mass of water at temperature T_0, and salinity S_0 coming into contact with the base of an ice shelf where the *in situ* freezing point is T_f. The water first warms m kg of ice to the freezing point, and then supplies the latent heat necessary for melting. The resulting mixture of melt and sea water has temperature T and salinity S. If the initial temperature of the ice is T_i, the latent heat of melting is L, the specific heat capacity of sea water and ice, c_w and c_i, then heat and salt conservation requires that:

$$(T - T_f)(1 + m)c_w + m(c_i(T_f - T_i) + L)$$
$$= (T_o - T_f)c_w$$
$$S(1 + m) = S_o$$

Eliminating m, and then expressing T as a function of S reveals the trajectory of the mixture in T–S space as a straight line passing through (S_0, T_0), with a gradient given by:

$$\frac{dT}{dS} = \frac{L}{S_o c_w} + \frac{(T_f - T_i)c_i}{S_o c_w} + \frac{(T_o - T_f)}{S_o}$$

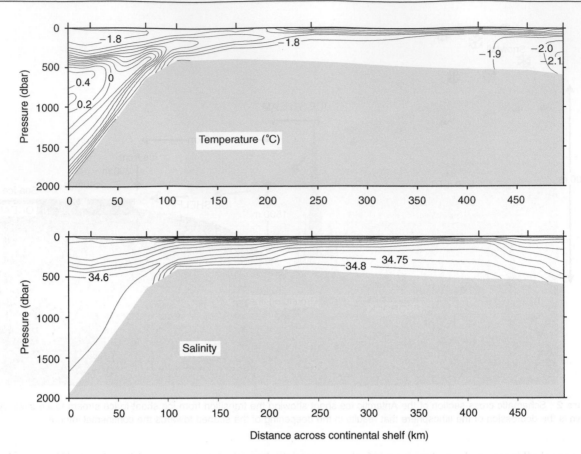

Figure 3 Hydrographic section over the continental slope and across the open continental shelf in the southern Weddell Sea, as far as the Ronne Ice Front. Water below the surface freezing point (−1.9°C) can be seen emerging from beneath the ice shelf. The majority of the continental shelf is dominated by HSSW, although in this location a tongue of warmer MWDW penetrates across the shelf-break. The station locations are shown by the heavy tick marks along the upper axes.

The gradient is dominated by the first term, which evaluates to about 2.4°C PSU^{-1}. In polar waters the third term is two orders of magnitude lower than the first; the second term results from the heat needed to warm the ice, and, at about a tenth the size of the first term, makes a measurable contribution to the gradient. This relationship allows the source water for sub-ice shelf processes to be found by inspection of the T–S properties of the resultant water masses. Examples of T–S plots from beneath ice shelves in warm and cold regimes are shown in **Figure 4**.

Two important passive tracers are introduced into sea water when glacial ice melts. When water evaporates from the ocean, molecules containing the lighter isotope of oxygen, ^{16}O, evaporate preferentially. Compared with sea water the snow that makes up the ice shelves is therefore low in ^{18}O. By comparing the ^{18}O/^{16}O ratios of the outflowing and inflowing water it is possible to calculate the concentration of melt water, provided the ratio is known for the glacial ice. Helium contained in the air bubbles in the ice is also introduced into the sea water when the ice melts. As helium's solubility in

water increases with increasing water pressure, the concentration of dissolved helium in the melt water can be an order of magnitude greater than in ambient sea water, which has equilibrated with the atmosphere at surface pressure.

Modes of Sub-ice Shelf Circulation

Various distinguishable modes of circulation appear to be possible within a sub-ice shelf cavity. Which mode is active depends primarily on the oceanographic forcing from seaward of the ice front, but also on the geometry of the sub-ice shelf cavity. Thermohaline forcing drives three modes of circulation, although the tidal activity is thought to play an important role by supplying energy for vertical mixing. Another mode results from tidal residual currents.

Thermohaline Modes

Cold regime external ventilation Over the parts of the Antarctic continental shelf dominated by the

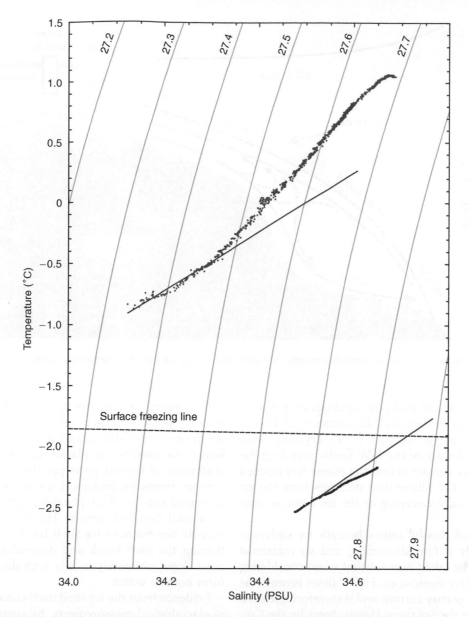

Figure 4 Temperature and salinity trajectories from CTD stations through the George VI Ice Shelf (red) and Ronne Ice Shelf (blue). The cold end of each trajectory corresponds to the base of the ice shelf. The straight lines are at the characteristic gradient for ice melting into sea water. For the Ronne data, as the source water will be HSSW at the surface freezing point, the intersection of the characteristic with the broken line gives the temperature and salinity of the source water. The isopycnals (gray lines) are referenced to sea level.

production of HSSW, such as in the southern Weddell Sea, the Ross Sea, and Prydz Bay, the circulation beneath large ice shelves is driven by the drainage of HSSW into the sub-ice shelf cavities. The schematic in **Figure 5** illustrates the circulation mode. HSSW drains down to the grounding line where tidal mixing brings it into contact with ice at depths of up to 2000 m. At such depths HSSW is up to 1.5°C warmer than the freezing point, and relatively rapid melting ensues (up to several meters of ice per year). The HSSW is cooled and diluted,

converting it into ice shelf water (ISW), which is defined as water with a temperature below the surface freezing point.

ISW is relatively buoyant and rises up the inclined base of the ice shelf. As it loses depth the *in situ* freezing point rises also. If the ISW is not entraining sufficient HSSW, which is comparatively warm, the reduction in pressure will result in the water becoming *in situ* supercooled. Ice crystals are then able to form in the water column and possibly rise up and accrete at the base of the ice shelf. This 'snowfall' at

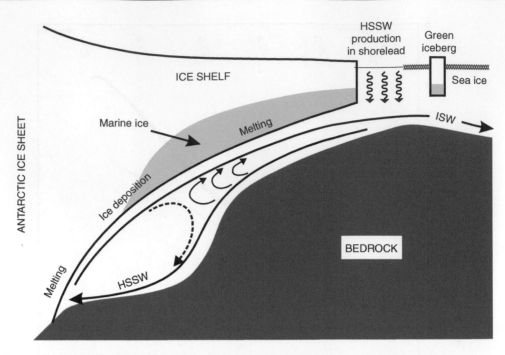

Figure 5   Schematic of the two thermohaline modes of sub-ice shelf circulation for a cold regime ice shelf.

the ice shelf base can build up hundreds of meters of what is termed 'marine ice'. Entrainment of HSSW, and the possible production of ice crystals, often result in the density of the ISW finally matching the ambient water density before the plume has reached the ice front. The plume then detaches from the ice shelf base, finally emerging at the ice front at mid-water depths.

The internal Rossby radius beneath ice shelves is typically only a few kilometers, and so rotational effects must be taken into account when considering the flow in three dimensions. HSSW flows beneath the ice shelf as a gravity current and is therefore gathered to the left (in the Southern Hemisphere) by the Coriolis force. As an organized flow, it then follows bathymetric contours. Once converted into ISW, the flow is again gathered to the left, following either the coast, or topography in the ice base. If the ISW plume fills the cavity, conservation of potential vorticity would demand that it follow contours of constant water column thickness. The step in water column thickness caused by the ice front then presents a topographic obstacle for the outflow of the ISW. However, the discontinuity can be reduced by the presence of trenches in the seafloor running across the ice front. This has been proposed as the mechanism that allows ISW to flow out from beneath the Filchner Ice Shelf, in the southern Weddell Sea (**Figure 1**).

Initial evidence for this mode of circulation came from ship-based oceanographic observations along the ice front of several of the larger ice shelves. Water

with temperatures up to 0.3°C below the surface freezing point indicated interaction with ice at a depth of at least 400 m, and the $^{18}O/^{16}O$ ratio confirmed the presence of glacial melt water at a concentration of several parts per thousand. Nets cast near ice fronts for biological specimens occasionally recovered masses of ice platelets, again from depths of several hundred meters. The ISW flowing from beneath the Filchner Ice Shelf has been traced overflowing the shelf-break and descending the continental slope, ultimately to mix with deep waters and form bottom water.

Evidence from the ice shelf itself comes in the form of glaciological measurements. By assuming a steady state (the ice shelf neither thickening nor thinning with time at any given point) conservation arguments can be used to derive the basal mass balance at individual locations. The calculation needs measurements of the local ice thickness variation, the horizontal spreading rate of the ice as it flows under its own weight, the horizontal speed of the ice, and the surface accumulation rate. This technique has been applied to several ice shelves, but is time-consuming, and has rarely been used to provide a good areal coverage of basal mass balance. However, it has demonstrated that high basal melt rates do indeed exist near deep grounding lines; that the melt rates reduce away from the grounding line; that further still from the grounding line, melting frequently switches to freezing; and that the balance usually returns to melting as the ice front is approached.

One-dimensional models have been to study the development of ISW plumes from the grounding line to where they detach from the ice shelf base. The most sophisticated includes frazil ice dynamics, and suggests that the deposition of ice at the base depends not only on its formation in the water column, but also on the flow regime being quiet enough to allow the ice to settle at the ice base. As the flow regime usually depends on the basal topography, the deposition is often highly localized. For example, a reduction in basal slope reduces the forcing on the buoyant plume, thereby slowing it down and possibly allowing any ice platelets to be deposited.

Deposits of marine ice become part of the ice shelf itself, flowing with the overlying meteoric ice. This means that, although the marine ice is deposited in well-defined locations, it moves towards the ice front with the flow of the ice and may or may not all be melted off by the time it reaches the ice front. Icebergs that have calved from Amery Ice Front frequently roll over and reveal a thick layer of marine ice. Impurities in marine ice result in different optical properties, and these bergs are often termed 'green icebergs'.

Ice cores obtained from the central parts of the Amery and Ronne ice shelves have provided other direct evidence of the production of marine ice. The interface between the meteoric and marine ice is clearly visible – the ice changes from being white and bubbly, to clear and bubble-free. Unlike normal sea ice, which typically has a salinity of a few PSU, the salinity of marine ice was found to be below 0.1 PSU. The salinity in the cores is highest at the interface itself, decreasing with increasing depth. A different type of marine ice was found at the base of the Ross Ice Shelf. There, a core from near the base showed 6 m of congelation ice with a salinity of between 2 and 4 PSU. Congelation ice differs from marine ice in its formation mechanism, growing at the interface directly rather than being created as an accumulation of ice crystals that were originally formed in the water column.

Airborne downward-looking radar campaigns have mapped regions of ice shelf that are underlain by marine ice. The meteoric (freshwater) ice/marine ice interface returns a characteristically weak echo, but the return from marine ice/ocean boundary is generally not visible. By comparing the thickness of meteoric ice found using the radar with the surface elevation of the freely floating ice shelf, it is possible to calculate the thickness of marine ice accreted at the base. In some parts of the Ronne Ice Shelf basal accumulation rates of around $1 \, m \, a^{-1}$ result in a marine ice layer over 300 m thick, out of a total ice column depth of 500 m. Accumulation rates of that magnitude would be expected to be associated with high ISW fluxes. However, cruises along the Ronne Ice Front have been unsuccessful in finding commensurate ISW outflows.

Internal recirculation   Three-dimensional models of the circulation beneath the Ronne Ice Shelf have revealed the possibility of an internal recirculation of ISW. This mode of circulation is driven by the difference in melting point between the deep ice at the grounding line, and the shallower ice in the central region of the ice shelf. The possibility of such a recirculation is indicated in **Figure 5** by the broken line. Intense deposition of ice in the freezing region salinifies the water column sufficiently to allow it to drain back towards the grounding line. In three dimensions, the recirculation consists of a gyre occupying a basin in the topography of water column thickness. The model predicts a gyre strength of around one Sverdrup ($10^6 \, m^3 \, s^{-1}$).

This mode of circulation is effectively an 'ice pump' transporting ice from the deep grounding line regions to the central Ronne Ice Shelf. The mechanism does not result in a loss or gain of ice overall. The heat used to melt the ice at the grounding line is later recovered in the freezing region. The external heat needed to maintain the recirculation is therefore only the heat to warm the ice to the freezing point before it is melted. Ice leaves the continent at a temperature of around $-30 \, °C$, and has a specific heat capacity of around $2010 \, J \, kg^{-1} °C^{-1}$. As the latent heat of ice is $335 \, kJ \, kg^{-1}$, the heat required for warming is less than 20% of that required for melting. To support an internal redistribution of ice therefore requires a small fraction of the external heat that would be needed to melt and remove the ice from the system entirely. A corollary is that a recirculation of ISW effectively decouples much of the ice shelf base from external forcings that might be imposed, for example, by climate change.

Apart from the lack of a sizable ISW outflow from beneath the Ronne Ice Front, evidence in support of an ISW recirculation deep beneath the ice shelf is scarce, as it would require observations beneath the ice. Direct measurements of conditions beneath ice shelves are limited to a small number of sites. Fissures through George VI and Fimbul ice shelves (**Figure 1**) have allowed instruments to be deployed with varying degrees of success. The more important ice shelves, such as the Ross, Amery and Filchner-Ronne system have no naturally occurring access points. Instead, access holes have to be created using hot water, or other thermal-type drills. In the late 1970s researchers used various drilling techniques to gain access to the cavity at one location beneath the

Ross Ice Shelf before deploying various instruments. During the 1990s several access holes were made through the Ronne Ice Shelf, and data from these have lent support both to the external mode of circulation, and most recently, to the internal recirculation mode first predicted by numerical models.

Warm regime external ventilation The flooding of the Bellingshausen and Amundsen seas' continental shelf by barely modified CDW results in very high basal melt rates for the ice shelves in that sector. The floating portion of Pine Island Glacier (**Figure 1**) has a mean basal melt rate estimated to be around $12 \, \mathrm{m \, a^{-1}}$, compared with estimates of a few tens of centimeters per year for the Ross and Filchner-Ronne ice shelves. Basal melt rates for Pine Island Glacier are high even compared with other ice shelves in the region. George VI Ice Shelf on the west coast of the Antarctic Peninsula, for example, has an estimated mean basal melt rate of $2 \, \mathrm{m \, a^{-1}}$. The explanation for the intense melting beneath Pine Island Glacier can be found in the great depth at the grounding line. At over $1100 \, \mathrm{m}$, the ice shelf is $700 \, \mathrm{m}$ thicker than George VI Ice Shelf, and this results in not only a lower freezing point, but also steeper basal slopes. The steep slope provides a stronger buoyancy forcing, and therefore greater turbulent heat transfer between the water and the ice.

The pattern of circulation in the cavities beneath warm regime ice shelves is significantly different to its cold regime counterpart. Measurements from ice front cruises show an inflow of warm CDW ($+1.0°C$), and an outflow of CDW mixed with glacial melt water. **Figure 6** shows a two-dimensional schematic of this mode of circulation. Over the open continental shelf the ambient water column consists of CDW overlain by colder, fresher water left over from sea ice production during the previous winter. Although the melt water-laden outflow is colder, fresher, and of lower density than the inflow, it is typically warmer and saltier than the overlying water, but of similar density. Somewhat counter-intuitively, therefore, the products of sub-glacial melt are often detected over the open continental shelf as relatively warm and salty intrusions in the upper layers. Again, measurements of oxygen isotope ratio, and also helium, provide the necessary confirmation that the upwelled CDW contains melt water from the base of ice shelves. In the case of warm regime ice shelves, melt water concentrations can be as high as a few percent.

Tidal Forcing

Except for within a few ice thicknesses of grounding lines, ice shelves float freely in the ocean, rising and falling with the tides. Tidal waves therefore propagate through the ice shelf-covered region, but are modified by three effects of the ice cover: the ice shelf base provides a second frictional surface, the draft of the ice shelf effectively reduces the water column thickness, and the step change in water column thickness at the ice front presents a topographic feature that has significant consequences for the generation of residual tidal currents and the propagation of topographic waves along the ice front.

Conversely, tides modify the oceanographic regime of sub-ice shelf cavities. Tidal motion helps transfer heat and salt beneath the ice front. This is a result

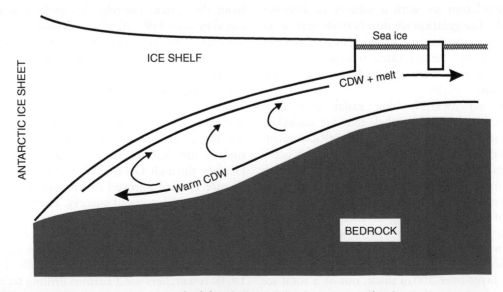

Figure 6 Schematic of the thermohaline mode of sub-ice shelf circulation for a warm regime ice shelf.

both of the regular tidal excursions, which take water a few kilometers into the cavity, and of residual tidal currents which, in the case of the Filchner-Ronne Ice Shelf, help ventilate the cavity far from the ice front. The effect of the regular advection of potentially seasonally warmed water from seaward of the ice shelf is to cause a dramatic increase in basal melt rates in the vicinity of the ice front. Deep beneath the ice shelf, tides and buoyancy provide the only forcing on the regime. Tidal activity contributes energy for vertical mixing, which brings the warmer, deeper waters into contact with the base of the ice shelf. **Figure 7A** shows modeled tidal ellipses for the M_2 semidiurnal tidal constituent for the southern Weddell Sea, including the sub-ice shelf domain. A map of the modeled residual currents for the area of the ice shelf is shown in **Figure 7B**. Apart from the activity near the ice front itself, a residual flow runs along the west coast of Berkner Island, deep under the ice shelf. However, this flow probably makes only a minor contribution to the ventilation of the cavity.

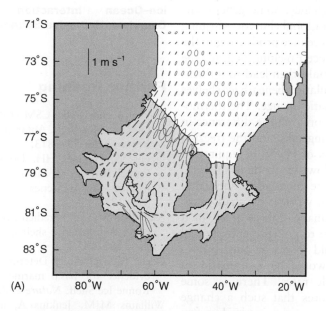

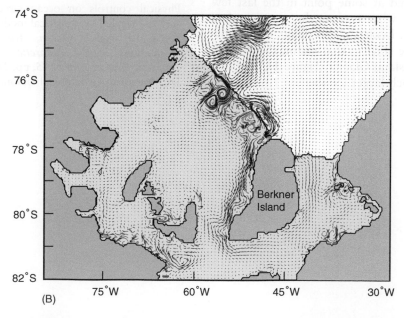

Figure 7 Results from a tidal model of the southern Weddell Sea, in the vicinity of the Ronne Ice Shelf. (A) The tidal ellipses for the dominant M_2 species. (B) Tidally induced residual currents.

How Does the Interaction between Ice Shelves and the Ocean Depend on Climate?

The response to climatic changes of sub-ice shelf circulation depends on the response of the oceanographic conditions over the open continental shelf. In the case of cold regime continental shelves, a reduction in sea ice would lead to a reduction in HSSW production. Model results, together with the implications of seasonality observed in the circulation beneath the Ronne Ice Shelf, suggest that drainage of HSSW beneath local ice shelves would then reduce, and that the net melting beneath those ice shelves would decrease as a consequence. Some general circulation models predict that global climatic warming would lead to a reduction in sea ice production in the southern Weddell Sea. Reduced melting beneath the Filchner-Ronne Ice Shelf would then lead to a thickening of the ice shelf. Recirculation beneath ice shelves is highly insensitive to climatic change. The thermohaline driving is dependent only on the difference in depths between the grounding lines and the freezing areas. A relatively small flux of HSSW is required to warm the ice in order to allow this mode to operate.

The largest ice shelves are in a cold continental shelf regime. If intrusions of warmer off-shelf water were to become more dominant in these areas, or if the shelf-break barrier were to collapse entirely and the regime switch from cold to warm, then the response of the ice shelves would be a dramatic increase in their basal melt rates. There is some evidence from sediment cores that such a change might have occurred at some point in the last few thousand years in what is now the warm regime Bellingshausen Sea. Evidence also points to the possibility that one ice shelf in that sector, the floating extension of Pine Island Glacier (**Figure 1**), might be a remnant of a much larger ice shelf.

During glacial maxima the Antarctic ice sheet thickens and the ice shelves become grounded. In many cases they ground as far as the shelf-break. There are two effects. The continental shelf becomes very limited in extent, and so there is little possibility for the production of HSSW; and where the ice shelves overhang the continental shelf-break, the only possible mode of circulation will be the warm regime mode. Substantial production of ISW during glacial conditions is therefore unlikely.

See also

Ice–Ocean   Interaction.   Polynyas.   SeaIce: Overview. Under-Ice Boundary Layer.

Further Reading

Jenkins A and Doake CSM (1991) Ice–ocean interactions on Ronne Ice Shelf, Antarctica. *Journal of Geophysical Research* 96: 791–813.

Jacobs SS, Hellmer HH, Doake CSM, Jenkins A, and Frolich RM (1992) Melting of ice shelves and the mass balance of Antarctica. *Journal of Glaciology* 38: 375–387.

Nicholls KW (1997) Predicted reduction in basal melt rates of an Antarctic ice shelf in a warmer climate. *Nature* 388: 460–462.

Oerter H, Kipfstuhl J, Determann J, *et al.* (1992) Ice-core evidence for basal marine shelf ice in the Filchner-Ronne Ice Shelf. *Nature* 358: 399–401.

Williams MJM, Jenkins A, and Determann J (1998) Physical controls on ocean circulation beneath ice shelves revealed by numerical models. In: Jacobs SS and Weiss RF (eds.) *Ocean, Ice, and Atmosphere: Interactions at the Antarctic Continental Margin, Antarctic Research Series* 75, pp. 285–299. Washington DC: American Geophysical Union.

NOBLE GASES AND THE CRYOSPHERE

M. Hood, Intergovernmental Oceanographic Commission, Paris, France

Introduction

Ice formation and melting strongly influence a wide range of water properties and processes, such as dissolved gas concentrations, exchange of gases between the atmosphere and the ocean, and dense water formation. As water freezes, salt and gases dissolved in the water are expelled from the growing ice lattice and become concentrated in the residual water. As a result of the increased salt content, this residual water becomes more dense than underlying waters and sinks to a level of neutral buoyancy, carrying with it the dissolved gas load. Dense water formation is one of the primary mechanisms by which atmospheric and surface water properties are transported into the interior and deep ocean, and observation of the effects of this process can answer fundamental questions about ocean circulation and the ocean–atmosphere cycling of biogeochemically important gases such as oxygen and carbon dioxide. Because it is not possible to determine exactly when and where dense water formation will occur, it is not an easy process to observe directly, and thus information about the rates of dense water formation and circulation is obtained largely through the observation of tracers. However, when dense water formation is triggered by ice formation, interaction of surface water properties with the ice and the lack of full equilibration between the atmosphere and the water beneath the growing ice can significantly modify the concentrations of the tracers in ways that are not yet fully understood. Consequently, the information provided by tracers in these ice formation areas is often ambiguous.

A suite of three noble gases, helium, neon, and argon, have the potential to be excellent tracers in the marine cryosphere, providing new information about the interactions of dissolved gases and ice, the cycling of gases between the atmosphere and ocean, and mixing and circulation pathways in high latitude regions of the world's oceans and marginal seas. The physical chemistry properties of these three gases span a wide range of values, and these differences cause them to respond to varying degrees to physical processes such as ice formation and melting or the transfer of gas between the water and air. By observing the changes of the three tracers as they respond to these processes, it is possible to quantify the effect the process has on the gases as a function of the physical chemistry of the gases. Subsequently, this 'template' of behavior can be used to determine the physical response of any gas to the process, using known information about the physical chemistry of the gas. Although this tracer technique is still being developed, results from laboratory experiments and field programs have demonstrated the exciting potential of the nobel gases to provide unique, quantitative information on a range of processes that it is not possible to obtain using conventional tracers.

Noble Gases in the Marine Environment

The noble gases are naturally occurring gases found in the atmosphere. **Table 1** shows the abundance of the noble gases in the atmosphere as a percentage of the total air composition, and the concentrations of the gases in surface sea water when in equilibrium with the atmosphere.

Other sources of these gases in sea water include the radioactive decay of uranium and thorium to helium-4 (^{4}He), and the radioactive decay of potassium (^{40}K) to argon (^{40}Ar). For most areas of the surface ocean, these radiogenic sources of the noble gases are negligible, and thus the only significant source for these gases is the atmosphere.

The noble gases are biogeochemically inert and are not altered through chemical or biological reactions, making them considerably easier to trace and quantify as they move through a system than other gases whose concentrations are modified through reactions. The behavior of the noble gases is largely determined by the size of the molecule of each gas and the natural affinity of each gas to reside in a gaseous or liquid state. The main physical chemistry

Table 1 Noble gases in the atmosphere and sea water

Gas	Abundance in the atmosphere (%)	Concentration in seawater ($cm^3 g^{-1}$)
Helium	0.0005	3.75×10^{-8}
Neon	0.002	1.53×10^{-7}
Argon	0.9	2.49×10^{-4}

parameters of interest are the solubility of the gas in liquid, the temperature dependence of this solubility, and the molecular diffusivity of the gas. The suite of noble gases have a broad range of these properties, and the behavior of the noble gases determined by these properties, can serve as a model for the behavior of most other gases.

One unique characteristic of the noble gases that makes them ideally suited as tracers of the interactions between gases and ice is that helium and neon are soluble in ice as well as in liquids. It has been recognized since the mid-1960s that helium and neon, and possibly hydrogen, should be soluble in ice because of the small size of the molecules, whereas gases having larger atomic radii are unable to reside in the ice lattice. These findings, however, were based on theoretical treatises and carefully controlled laboratory studies in idealized conditions. It was not until the mid-1980s that this process was shown to occur on observable scales in nature, when anomalies in the concentrations of helium and neon were observed in the Arctic.

The solubility of gases in ice can be described by the same principles governing solubility of gases in liquids. Solubility of gases in liquids or ice occurs to establish equilibrium, where the affinities of the gas to reside in the gaseous, liquid, and solid state are balanced. The solubility process can be described by two principle mechanisms:

1. creation of a cavity in the solvent large enough to accommodate a solute molecule;
2. introduction of the solute molecule into the liquid or solid surface through the cavity.

In applying this approach to the solubility of gases in ice, it follows that if the atomic radius of the solute gas molecule is smaller than the cavities naturally present in the lattice structure of ice, then the energy required to make a cavity in the solvent is zero, and the energy required for the solubility process is then only a function of the energy required to introduce the solute molecule into the cavity. For this reason, the solubility of a gas molecule capable of fitting in the ice lattice is greater than its solubility in a liquid. The solubilities of helium and neon in ice have been determined in two separate laboratory studies, and although the values agree for the solubility of helium in ice, the values for neon disagree. The size of neon is very similar to the size of a cavity in the ice lattice, and the discrepancies between the two reported values for the solubility of neon in ice may result from small differences in the experimental procedure.

During ice formation, most gases partition between the water and air phases to try to establish equilibrium under the changing conditions, whereas

Table 2 Noble gas partitioning in three phases

Partition phases	Helium	Neon	Argon
Bubble to water	106.8	81.0	18.7
Bubble to ice	56.9	90.0, 56.3	∞
Ice to water	1.9	0.9, 1.4	0

helium and neon additionally partition into the ice phase. As water freezes, salt and gases are rejected from the growing ice lattice, increasing the concentrations of salt and gas in the residual water. Helium and neon partition between the water and ice reservoirs according to their solubility in water and ice. The concentrations of the gases in the residual water that have been expelled from the ice lattice, predominantly oxygen and nitrogen, can become so elevated through this process that the pressure of the dissolved gases in the water exceeds the *in situ* hydrostatic pressure and gas bubbles form. The gases then partition between the water, the gas bubble, and the ice according to the solubilities of the gases in each phase. This three-phase partitioning process can occur either at the edge of the growing ice sheet at the ice–water interface, or in small liquid water pockets, called 'brine pockets' in salt water systems, entrained in the ice during rapid ice formation.

Table 2 quantitatively describes how the noble gases partition between the three phases when a system containing these three phases is in equilibrium in fresh water at 0°C. The numbers represent the amount of the gas found in one phase relative to the other. For example, the first row describes the amount of each gas that would reside in the gaseous bubble phase relative to the liquid phase; thus for helium, there would be 106.8 times more helium present in the bubble than in the water. This illustrates the small solubility of helium in water and its strong affinity for the gas phase. Because helium is 1.9 times more soluble in ice than in water, helium partitions less strongly between the bubble and ice phases compared to the partition between the bubble and water phases. The two numbers shown for neon represent the two different estimates for the solubility of neon in the ice phase. One estimate suggests that neon is less soluble in ice than in water, whereas the other suggests that it is more soluble in ice.

Application of the Noble Gases as Tracers

The noble gases have been used as tracers of air–sea gas exchange processes for more than 20 years. Typically, the noble gases are observed over time at a single location in the ocean along with other meteorological and hydrodynamic parameters to characterize and quantify the behavior of each of the gases in response to the driving forces of gas exchange such as water temperature, wind speed, wave characteristics, and bubbles injected from breaking waves. Because both the amount and rate of a gas transferred between the atmosphere and ocean depend on the solubility and diffusivities of the gas, the noble gases have long been recognized as ideal tracers for these processes. In addition, argon and oxygen have very similar molecular diffusivities and solubilities, making argon an excellent tracer of the physical behavior of oxygen. By comparing the relative concentration changes of argon and oxygen over time, it is possible to account for the relative contributions of physical and biological processes (such as photosynthesis by phytoplankton in the surface ocean) to the overall concentrations, thus constraining the biological signal and allowing for estimates of the biological productivity of the surface ocean.

The observations of anomalous helium and neon concentrations in ice formation areas and the suggestion that these anomalies could be the result of solubility of these gases in the ice were made in 1983, and since that time, a number of laboratory and field studies have been conducted to characterize and quantify these interactions. The partitioning of the noble gases among the three phases of gas, water, and ice creates a very distinctive 'signature' of the noble gas concentrations left behind in the residual water. Noble gas concentrations are typically expressed in terms of 'saturation', which is the concentration of a gas dissolved in the water relative to its equilibrium with the atmosphere at a given temperature. For example, a parcel of water at standard temperature and pressure containing the concentrations of noble gases shown in column 2 of **Table 1** would be said to have a saturation of 100%. Saturations that deviate from this 100% can arise when equilibration with the atmosphere is incomplete, either because the equilibration process is slow relative to some other dynamic process acting on the system (for example, rapid heating or cooling, or injection of bubbles from breaking waves), or because full equilibration between the water and atmosphere is prevented, as in the case of ice formation.

Typical saturations for the noble gases in the surface ocean range from 100 to 110% of atmospheric equilibrium, due mostly to the influx of gas from bubbles. Ice formation, however, can lead to quite striking saturations of -70 to -60% for helium and neon and $+230$% for argon in the relatively undiluted residual water. Ice melting can also lead to

large anomalous saturations of the noble gases, showing the reverse of the freezing pattern for the gas saturations, where helium and neon are supersaturated while argon is undersaturated with respect to the atmosphere.

The interactions of noble gases and ice have been well-documented and quantified in relatively simple freshwater systems. Observations of large noble gas anomalies in a permanently ice-covered antarctic lake were quantitatively explained using the current understanding of the solubility of helium and neon in ice and the partitioning of the gases in a three-phase system. Characteristics of ice formed from salt water are more complex than ice formed from fresh water, and the modeling of the system more complex. Using a set of equations developed in 1983 and measurements of the ice temperature, salinity, and density, it is possible to calculate the volume of the brine pockets in the ice and the volume of bubbles in the ice. With this type of information, a model of the ice and the dissolved gas balance in the various phases in the ice and residual water can be constructed. Such an ice model was developed during a field study of gas–ice interactions in a seasonally ice-covered lagoon, and the model predicted the amount of argon, nitrogen, and oxygen measured in bubbles in similar types of sea ice. No measurements are available for the amount of helium and neon in the bubbles of sea ice to verify the results for these gases. It is also possible to predict the relative saturations of the noble gases in the undiluted residual water at the ice–water interface, and this unique fingerprint of the noble gases can then serve as a tracer of the mixing and circulation of this water parcel as it leaves the surface and enters the interior and deep ocean. In this manner, the supersaturations of helium from meltwater have been successfully used as a tracer of water mass mixing and circulation in the Antarctic, and the estimated sensitivity of helium as a tracer for these processes is similar to the use of the conventional tracer, salinity, for these processes.

As an illustration of the ways in which the noble gases can be used to distinguish between the effects of ice formation, melting, injection of air bubbles from breaking waves, or temperature changes on dissolved gases, **Figure 1** shows a vector diagram of the characteristic changes of helium compared to argon resulting from each of these processes.

From a starting point of equilibrium with the atmosphere (100% saturation), both helium and argon saturations increase as a result of bubbles injected from breaking waves. Ice formation increases the saturation of argon and decreases the saturation of helium, whereas ice melting has the opposite effect. Changes in temperature with no gas exchange with

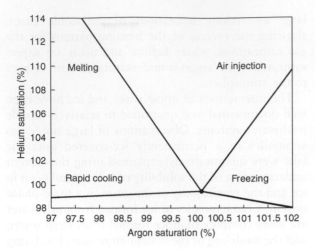

Figure 1 Vector diagram of helium and argon saturation changes in response to upper ocean processes.

the atmosphere to balance this change can lead to modest changes in the saturations of the gases, where the gas saturations decrease with decreasing temperature and increase with increasing temperature. The trends presented here are largely qualitative indicators, since quantitative assessment of the changes depend on the exact nature of the system being studied. However, this diagram does illustrate the general magnitude of the changes that these processes have on the noble gases and conversely, the ability of the noble gases to differentiate between these effects.

Conclusions

The use of the noble gases as tracers in the marine cryosphere is in its infancy. Our understanding of the interactions of the noble gases and ice have progressed from controlled, idealized laboratory conditions to natural freshwater systems and simple salt water systems, and the initial results from these studies are extremely encouraging. This technique is currently being developed more fully to provide quantitative information about the interactions of dissolved gases and ice, and to utilize the resulting effects of these interactions to trace water mass mixing and circulation in the range of dynamic ice formation environments. Water masses in the interior and deep ocean originating in ice formation and melting areas have been shown to have distinct noble gas ratios, which are largely imparted to the water mass at the time of its formation in the surface ocean. By understanding and quantifying the processes responsible for these distinct ratios, we will be

able to learn much about where and how the water mass was formed and the transformations it has experience since leaving the surface ocean. These issues are important for our understanding of the global cycling of gases between the atmosphere and the ocean and for revealing the circulation pathways of water in the Arctic, Antarctic, and high latitude marginal seas. The noble gases could represent a significant addition to the set of tracers typically used to study these processes.

See also

Air–Sea Gas Exchange. Bubbles. CFCs in the Ocean. Ice–Ocean Interaction. Polynyas. Sea Ice: Overview. Sub Ice-Shelf Circulation and Processes.

Further Reading

Bieri RH (1971) Dissolved noble gases in marine waters. *Earth and Planetary Science Letters* 10: 329–333.

Cox GFN and Weeks WF (1982) Equations for determining the gas and brine volumes in sea ice samples, *USA Cold Regions Research and Engineering Laboratory Report* 82-30, Hanover, New Hampshire.

Craig H and Hayward T (1987) Oxygen supersaturations in the ocean: biological vs. physical contributions. *Science* 235: 199–202.

Hood EM, Howes BL, and Jenkins WJ (1998) Dissolved gas dynamics in perennially ice-covered Lake Fryxell, Antarctica. *Limnology and Oceanography* 43(2): 265–272.

Hood EM (1998) *Characterization of Air–sea Gas Exchange Processes and Dissolved Gas/ice Interactions Using Noble Gases*. PhD thesis, MIT/WHOI, 98–101.

Kahane A, Klinger J, and Philippe M (1969) Dopage selectif de la glace monocristalline avec de l'helium et du neon. *Solid State Communications* 7: 1055–1056.

Namoit A and Bukhgalter EB (1965) Clathrates formed by gases in ice. *Journal of Structural Chemistry* 6: 911–912.

Schlosser P (1986) Helium: a new tracer in Antarctic oceanography. *Nature* 321: 233–235.

Schlosser P, Bayer R, Flodvik A, *et al.* (1990) Oxygen-18 and helium as tracers of ice shelf water and water/ice interaction in the Weddell Sea. *Journal of Geophysical Research* 95: 3253–3263.

Top Z, Martin S, and Becker P (1988) A laboratory study of dissolved noble gas anomaly due to ice formation. *Geophysical Research Letters* 15: 796–799.

Top Z, Clarke WB, and Moore RM (1983) Anomalous neon–helium ratios in the Arctic Ocean. *Geophysical Research Letters* 10: 1168–1171.

MEASUREMENT TECHNIQUES INCLUDING REMOTE SENSING

SENSORS FOR MEAN METEOROLOGY

K. B. Katsaros, Atlantic Oceanographic and Meteorological Laboratory, NOAA, Miami, FL, USA

Introduction

Basic mean meteorological variables include the following: pressure, wind speed and direction, temperature, and humidity. These are measured at all surface stations over land and from ships and buoys at sea. Radiation (broadband solar and infrared) is also often measured, and sea state, swell, wind sea, cloud cover and type, and precipitation and its intensity and type are evaluated by an observer over the ocean. Sea surface temperature and wave height (possibly also frequency and direction of wave trains) may be measured from a buoy at sea; they are part of the set of parameters required for evaluating net surface energy flux and momentum transfer. Instruments for measuring the quantities described here have been limited to the most common and basic. Precipitation is an important meteorological variable that is measured routinely over land with rain gauges, but its direct measurement at sea is difficult because of ship motion and wind deflection by ships' superstructure and consequently it has been measured routinely over the ocean only from ferry boats. However, it can be estimated at sea by satellite techniques, as can surface wind and sea surface temperature. Satellite methods are included in this article, since they are increasing in importance and provide the only means for obtaining complete global coverage.

Pressure

Several types of aneroid barometers are in use. They depend on the compression or expansion of an evacuated metal chamber for the relative change in atmospheric pressure. Such devices must be compensated for the change in expansion coefficient of the metal material of the chamber with temperature, and the device has to be calibrated for absolute values against a classical mercury in glass barometer, whose vertical mercury column balances the weight of the atmospheric column acting on a reservoir of mercury. The principle of the mercury barometer was developed by Evangilista Toricelli in the 17th century, and numerous sophisticated details were worked out over a period of two centuries. With modern manufacturing techniques, the aneroid has become standardized and is the commonly used device, calibrated with transfer standards back to the classical method. The fact that it takes a column of about 760 mm of the heavy liquid metal mercury (13.6 times as dense as water) illustrates the substantial weight of the atmosphere. Corrections for the thermal expansion or contraction of the mercury column must be made, so a thermometer is always attached to the device. Note that the word 'weight' is used, which implies that the value of the earth's gravitational force enters the formula for converting the mercury column's height to a pressure (force/unit area). Since gravity varies with latitude and altitude, mercury barometers must be corrected for the local value of the acceleration due to gravity.

Atmospheric pressure decreases with altitude. The balancing column of mercury decreases or the expansion of the aneroid chamber increases as the column of air above the barometer has less weight at higher elevations. Conversely, pressure sensors can therefore be used to measure or infer altitude, but must be corrected for the variation in the atmospheric surface pressure, which varies by as much as 10% of the mean (even more in case of the central pressure in a hurricane). An aneroid barometer is the transducer in aircraft altimeters.

Wind Speed and Direction

Wind speed is obtained by two basic means, both depending on the force of the wind to make an object rotate. This object comprises either a three- or four-cup anemometer, half-spheres mounted to horizontal axes attached to a vertical shaft (**Figure 1A**). The cups catch the wind and make the shaft rotate. In today's instruments rotations are counted by the frequency of the interception of a light source to produce a digital signal.

Propeller anemometers have three or four blades that are turned by horizontal wind (**Figure 1B**). The propeller anemometer must be mounted on a wind vane that keeps the propeller facing into the wind. For propeller anemometers, the rotating horizontal shaft is inserted into a coil. The motion of the shaft generates an electrical current or a voltage difference that can be measured directly. The signal is large enough that no amplifiers are needed.

Both cup and propeller anemometers, as well as vanes, have a threshold velocity below which they do not turn and measure the wind. For the propeller

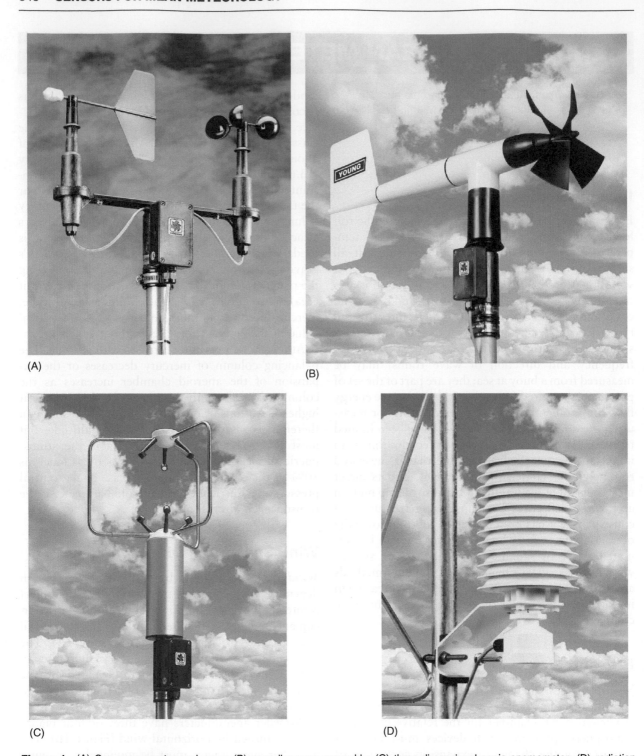

Figure 1 (A) Cup anemometer and vane; (B) propeller vane assembly; (C) three-dimensional sonic anemometer; (D) radiation shield for temperature and humidity sensors. (Photographs of these examples of common instruments were provided courtesy of R.M. Young Company.)

anemometer, the response of the vane is also crucial, for the propeller does not measure wind speed off-axis very well. These devices are calibrated in wind tunnels, where a standard sensor evaluates the speed in the tunnel. Calibration sensors can be fine cup anemometers or pitot tubes.

The wind direction is obtained from the position of a wind vane (a vertical square, triangle, or otherwise shaped wind-catcher attached to a horizontal shaft, **Figure 1A** and **B**). The position of a sliding contact along an electrical resistance coil moved by the motion of the shaft gives the wind direction relative to the zero position of the coil. The position is typically a fraction of the full circle (minus a small gap) and must be calibrated with a compass for absolute direction with respect to the Earth's north.

Other devices such as sonic anemometers can determine both speed and direction by measuring the modification of the travel time of short sound pulses between an emitter and a receiver caused by the three-dimensional wind. They often have three sound paths to allow evaluation of the three components of the wind (**Figure 1C**). These devices have recently become rugged enough to be used to measure mean winds routinely, and have a high enough frequency response to also determine the turbulent fluctuations. The obvious advantage is that the instrument has no moving parts. Water on the sound transmitter or receiver causes temporary difficulties, so a sonic anemometer is not an all-weather instrument. The sound paths can be at arbitrary angles to each other and to the natural vertical. Processing of the data transforms the measurements into an Earth-based coordinate system. The assumptions of zero mean vertical velocity and zero mean cross-wind velocity allow the relative orientation between the instrument axes and the Earth-based coordinate system to be found. Difficulties arise if the instrument is experiencing a steady vertical velocity at its location due to flow distortion around the measuring platform, for instance.

Cup and propeller anemometers are relatively insensitive to rain. However, snow and frost are problematic to all wind sensors, particularly the ones described above with moving parts. Salt contamination over the ocean also causes deterioration of the bearings in cup and propeller anemometers. Proper exposure of wind sensors on ships is problematic because of severe flow distortion by increasingly large ships. One solution has been to have duplicate sensors on port and starboard sides of the ship and selecting the valid one on the basis of the recording of the ship's heading and the relative wind direction.

Temperature

The measurements of both air and water temperature will be considered here, since both are important in air–sea interaction. Two important considerations for measuring temperature are the exposure of the sensor and shielding from solar radiation. The axiom that a 'thermometer measures its own temperature' is a good reminder. For the thermometer to represent the temperature of the air, it must be well ventilated, which is sometimes assured by a protective housing and a fan pulling air past the sensor. Shielding from direct sunlight has been done traditionally over land

and island stations by the use of a 'Stephenson screen,' a wooden-roofed box with slats used for the sides, providing ample room for air to enter. Modern devices have individual housings based on the same principles (**Figure 1D**).

The classic measurements of temperature were done with mercury in glass or alcohol in glass thermometers. For sea temperature, such a thermometer was placed in a canvas bucket of water hauled up on deck. Today, electronic systems have replaced most of the glass thermometers. **Table 1** lists some of these sensors (for details see the Further Reading section).

The sea surface temperature (SST) is an important aspect of air–sea interaction. It enters into bulk formulas for estimating sensible heat flux and evaporation. The temperature differences between the air at one height and the SST is also important for determining the atmospheric stratification, which can modify the turbulent fluxes substantially compared with neutral stratification.

The common measure of SST is the temperature within the top 1 or 2 m of the interface, obtained with any of the contact temperature sensors described in **Table 1**. On ships, the sensor is typically placed in the ship's water intake, and on buoys it may even be placed just inside the hull on the bottom, shaded side of the buoy. Because the heat losses to the air occur at the air–sea interface, while solar heating penetrates of the order of tens of meters (depth depending on sun angle), a cool skin, 1–2 mm in depth and 0.1–0.5°C cooler than the lower layers, is often present just below the interface. Radiation thermometers are sometimes used from ships or piers to measure the skin temperature directly (*see* Radiative Transfer in the Ocean).

Humidity

The Classical Sling Psychrometer

An ingenious method for evaluating the air's ability to take up water (its deficit in humidity with respect to the saturation value, see 68 on Evaporation and Humidity) is the psychrometric method. Two thermometers (of any kind) are mounted side by side, and one is provided with a cotton covering (a wick) that is wetted with distilled water. The sling psychrometer (**Figure 2**) is vigorously ventilated by swinging it in the air. The air passing over the sensors changes their temperatures to be in equilibrium with the air; the dry bulb measures the actual air temperature, the wet bulb adjusts to a temperature that is intermediate between the dew point and air temperature. As water from the wick is evaporated, it takes heat out of the air passing over the wick until

Table 1 Electronic devices for measuring temperature in air or water

Name	Principle	Typical use
Thermocouple	Thermoelectric junctions between two wires (e.g. Copper-Constantan) set up a voltage in the circuit, if the junctions are at different temperatures. The reference junction temperature must be measured as well	Good for measuring differences of temperature
Resistance thermometer	$R = R_{Ref}(1 + \alpha^T)$ Where R is the electrical resistance, R_{Ref} is resistance at a reference temperature, and α is the temperature coefficient of resistance	Platinum resistance thermometers are used for calibration and as reference thermometers
Thermistor	$R = a \exp (b/T)$ Where R is resistance, T is absolute temperature, and a and b are constants	Commonly used in routine sensor systems
Radiation thermometer	Infrared radiance in the atmospheric window, 8–12 μm, is a measure of the equivalent black body temperature	Usually used for measuring water's skin temperature

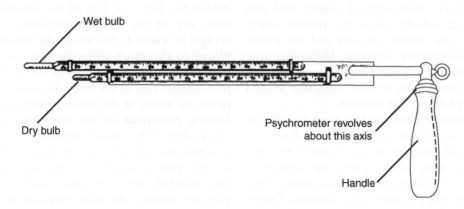

Figure 2 Sling pyschrometer. (Reproduced with permission from Parker, 1977.)

an equilibrium is reached between the heat supplied to the wet bulb by the air and the heat lost due to evaporation of water from the wick. This is the wet bulb temperature. The *Smithsonian Tables* provide the dew point temperature (and equivalent saturation humidity) corresponding to the measured 'wet bulb temperature depression,' i.e. the temperature difference between the dry bulb and the wet bulb thermometers at the existing air temperature.

Resistance Thermometer Psychrometer

?>A resistance thermometer psychrometer consists of stainless steel-encased platinum resistance thermometers housed in ventilated cylindrical shields. Ventilation can be simply due to the natural wind (in which case errors at low wind speeds may develop), or be provided by a motor and a fan (typically an air

speed of $3 \, \mathrm{m \, s^{-1}}$ is required). A water reservoir must be provided to ensure continuous wetting of the wet bulb. The reservoir should be mounted below the psychrometer so that water is drawn onto the wet bulb with a long wick. (This arrangement assures that the water has had time to equilibrate to the wet bulb temperature of the air.)

If these large wet bulbs collect salt on them over time, the relative humidity may be in error. This is not a concern for short-term measurements. A salt solution of 3.6% on the wet bulb would result in an overestimate of the relative humidity of approximately 2%.

Capacitance Sensors of Humidity

The synoptic weather stations often use hygrometers based on the principle of capacitance change as the small transducer absorbs and desorbs water vapor. To avoid contamination of the detector, special filters cover the sensor. Dirty filters (salt or other contaminants) may completely mask the atmospheric effects. Even the oil from the touch of a human hand is detrimental. Two well-known sensors go under the names of Rotronic and Humicap. Calibration with mercury in glass psychrometers is useful.

Exposure to Salt

As for wind and temperature devices, the humidity sensors are sensitive to flow distortion around ships and buoys. Humidity sensors have an additional problem in that salt crystals left behind by evaporating spray droplets, being hygroscopic, can modify the measurements by increasing the local humidity around them. One sophisticated, elegant, and expensive device that has been used at sea without success is the dew point hygrometer. It depends on the cyclical cooling and heating of a mirror. The cooling continues until dew forms, which is detected by changes in reflection of a light source off the mirror, and the temperature at that point is by definition the dew point temperature. The problem with this device is that during the heating cycle sea salt is baked onto the mirror and cannot be removed by cleaning.

Several attempts to build devices that remove the spray have been tried. Regular Stephenson screen-type shields provide protection for some time, the length of which depends both on the generation of spray in the area, the height of the measurement, and the size of the transducer (i.e. the fraction of the surface area that may be contaminated). One of the protective devices that was successfully used in the Humidity Exchange over the Sea (HEXOS) experiment is the so-called 'spray flinger.'

Spray-removal Device

The University of Washington 'spray flinger' (**Figure 3**) was designed to minimize flow modification on scales important to the eddy correlation calculations of evaporation and sensible heat flux employing data from temperature and humidity sensors inside the housing. The design aims to ensure that the droplets removed from the airstream do not remain on the walls of the housing or filter where they could evaporate and affect the measurements. The device has been tested to ensure that there are no thermal effects due to heating of the enclosure, but this would be dependent on the meteorological conditions encountered, principally insolation. The housing should be directed upwind.

Although there is a slow draw of air through the unit by the upwind and exit fans ($1-2 \ m \ s^{-1}$), it is mainly a passive device with respect to the airflow. Inside the tube, wet and dry thermocouples or other temperature and humidity sensors sample the air for mean and fluctuating temperature and humidity. Wind tunnel and field tests showed the airflow inside the unit to be steady and about one-half the ambient wind speed for wind directions $<40°$ off the axis. Even in low wind speeds there is adequate ventilation for the wet bulb sensor. Comparison between data from shielded and unshielded thermocouples (respectively, inside and outside the spray flinger) show

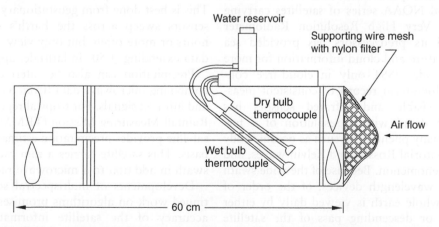

Figure 3 Sketch of aspirated protective housing, the 'spray flinger', used for the protection of a thermocouple psychrometer by the University of Washington group. The system is manually directed upwind. The spray flinger is a 60 cm long tube, 10 cm in diameter, with a rotating filter screen and fan on the upwind end, and an exit fan and the motor at the downwind end. The filter is a single layer of nylon stocking, which is highly nonabsorbent, supported by a wire mesh. Particles and droplets are intercepted by the rotating filter and flung aside, out of the airstream entering the tube. The rotation rate of the filter is about 625 rpm. Inspection of the filter revealed that this rate of rotation prevented build-up of water or salt. The nylon filter needs to be replaced at least at weekly intervals. (Reproduced with permission from Katsaros et al., 1994.)

that the measurements inside are not noticeably affected by the housing.

A quantitative test of the effectiveness of the spray flinger in removing aerosols from the sample airstream was performed during HEXMAX (the HEXOS Main Experiment) using an optical particle counter to measure the aerosol content with diameters between 0.5 and 32 μm in the environmental air and at the rear of the spray flinger. Other devices have been constructed, but have had various difficulties, and the 'spray flinger' is not the final answer. Intake tubes that protect sensors have also been designed for use on aircraft.

Satellite Measurements

With the global ocean covering 70% of the earth's surface, large oceanic areas cannot be sampled by *in situ* sensors. Most of the meteorological measurements are taken by Voluntary Observing Ships (VOS) of the merchant marine and are, therefore, confined to shipping lanes. Research vessels and military ships may be found in other areas and have contributed substantially to our knowledge of conditions in areas not visited by VOS. The VOS report their observations on a 3 hour or 6 hour schedule. Some mean meteorological quantities such as SST and wind speed and direction are observable by satellites directly, while others can be inferred from less directly related measurements. Surface insolation and precipitation depend on more complex algorithms for evaluation. Satellite-derived surface meteorological information over the ocean are mostly derived from polar-orbiting, sun-synchronous satellites. The famous TIROS and NOAA series of satellites carrying the Advanced Very High Resolution Radiometer (AVHRR) and its predecessors has provided sea surface temperature and cloud information for more than three decades (SST only in cloud-free conditions). This long-term record of consistent measurements by visible and infrared sensors has provided great detail with a resolution of a few kilometers of many phenomena such as oceanic eddy formation, equatorial Rossby and Kelvin waves, and the El Niño phenomenon. Because of the wide swath of these short wavelength devices, of the order of 2000 km, the whole earth is viewed daily by either the ascending or descending pass of the satellite overhead, once in daytime and once at night.

Another mean meteorological variable observable from space is surface wind speed, with microwave radiometers and the wind vector from scatterometers, active microwave instruments. Both passive and active sensors depend on the changing roughness of the sea as a function of wind speed for their ability to 'sense' the wind. The first scatterometer was launched on the Seasat satellite in 1978, operating for 3 months only. The longest record is from the European Remote Sensing (ERS) satellites 1 and 2 beginning in 1991 and continuing to function well in 2000. Development of interpretation of the radar returns in terms of both speed and direction depends on the antennae viewing the same ocean area several times at different incidence angles relative to the wind direction. A recently launched satellite (QuikSCAT in 1999) carries a new design with a wider swath, the SeaWinds instrument. Scatterometers are providing surface wind measurements with accuracy of $\pm 1.6 \, \mathrm{m \, s^{-1}}$ approximately in speed and $\pm 20°$ in direction at 50 km resolution for ERS and 25 km for SeaWinds. They view all of the global ocean once in 3 days for ERS and in approximately 2 days for QuikSCAT. Microwave radiometers such as the Special Sensor Microwave/Imager (SSM/I), operational since 1987 on satellites in the US Defense Meteorological Satellite Program, have wider swaths covering the globe daily, but they are not able to sense the ocean surface in heavy cloud or rainfall areas and do not give direction. They can be assimilated into numerical models where the models provide an initial guess of the wind fields, which are modified to be consistent with the details of the radiometer-derived wind speeds.

Surface pressure and atmospheric surface air temperature are not yet amenable to satellite observations, but surface humidity can be inferred from total column water content. From the satellite-observed cloudiness, solar radiation at the surface can be inferred by use of radiative transfer models. This is best done from geostationary satellites whose sensors sweep across the Earth's surface every 3 hours or more often, but only view a circle of useful data extending $\pm 50°$ in latitude, approximately.

Precipitation can also be inferred from satellites combining microwave data (from SSM/I) with visible and infrared signals. For tropical regions, the Tropical Rainfall Measuring Mission (TRMM) on a low-orbit satellite provides precipitation estimates on a monthly basis. This satellite carries a rain radar with 500 km swath in addition to a microwave radiometer.

Developments of multispectral sensors and continued work on algorithms promises to improve the accuracy of the satellite information on air–sea interaction variables. Most satellite programs depend on the simple *in situ* mean meteorological measurements described above for calibration and validation. A good example is the important SST record provided by the US National Weather Service and used by all weather services. The analysis procedure employs surface data on SST from buoys, particularly small, inexpensive, free-drifting buoys that are spread over the global oceans to 'tie-down' the correction for atmospheric interference for the satellite estimates of SST. The satellite-observed infrared radiances are modified by the transmission path from the sea to the satellite, where the unknown is the aerosol that can severely affect the interpretation. The aerosol signal is not directly observable yet by satellite, so the surface-measured SST data serve an important calibration function.

Future Developments

New measurement programs are being developed by international groups to support synoptic definition of the ocean's state similarly to meteorological measurements and to provide forecasts. The program goes under the name of the Global Ocean Observing System (GOOS). It includes new autonomous buoys cycling in the vertical to provide details below the interface, a large surface drifter component, and the VOS program, as well as certain satellite sensors. The GOOS is being developed to support a modeling effort, the Global Ocean Data Assimilation Experiment (GODAE), which is an experiment in forecasting the oceanic circulation using numerical models with assimilation of the GOOS data.

See also

Air–Sea Gas Exchange. Breaking Waves and Near-Surface Turbulence. Evaporation and Humidity. Heat and Momentum Fluxes at the Sea Surface. Sensors for Micrometeorological and Flux Measurements. Wind Driven Circulation.

Further Reading

Atlas RS, Hoffman RN, Bloom SC, Jusem JC, and Ardizzone J (1996) A multiyear global surface wind velocity data set using SSM/I wind observations. *Bulletin of the American Meteorological Society* 77: 869–882.

Bentamy A, Queffeulou P, Quilfen Y, and Katsaros KB (1999) Ocean surface wind fields estimated from satellite active and passive microwave instruments. *IEEE Transactions Geosci Remote Sens* 37: 2469–2486.

de Leeuw G (1990) Profiling of aerosol concentrations, particle size distributions, and relative humidity in the atmospheric surface layer over the North Sea. *Tellus* 42B: 342–354.

Dobson F, Hasse L, and Davies R (eds.) (1980) *Instruments and Methods in Air–Sea Interaction*, pp. 293–317. New York: Plenum.

Geernaert GL and Plant WJ (eds.) (1990) *Surface Waves and Fluxes*, 2, pp. 339–368. Dordrecht: Kluwer Academic Publishers.

Graf J, Sasaki C, *et al.* (1998) NASA Scatterometer Experiment. *Asta Astronautica* 43: 397–407.

Gruber A, Su X, Kanamitsu M, and Schemm J (2000) The comparison of two merged rain gauge-satellite precipitation datasets. *Bulletin of the American Meteorological Society* 81: 2631–2644.

Katsaros KB (1980) Radiative sensing of sea surface temperatures. In: Dobson F, Hasse L, and Davies R (eds.) *Instruments and methods in Air–Sea Interaction*, pp. 293–317. New York: Plenum Publishing Corp.

Katsaros KB, DeCosmo J, Lind RJ, *et al.* (1994) Measurements of humidity and temperature in the marine environment. *Journal of Atmospheric and Oceanic Technology* 11: 964–981.

Kummerow C, Barnes W, Kozu T, Shiue J, and Simpson J (1998) The tropical rainfall measuring mission (TRMM) sensor package. *Journal of Atmospheric and Oceanic Technology* 15: 809–817.

Liu WT (1990) Remote sensing of surface turbulence flux. In: Geenaert GL and Plant WJ (eds.) *Surface Waves and Fluxesyyy*, 2, pp. 293–309. Dordrecht: Kluwer Academic Publishers.

Parker SP (ed.) (1977) *Encyclopedia of Ocean and Atmospheric Science*. New York: McGraw-Hill.

Pinker RT and Laszlo I (1992) Modeling surface solar irradiance for satellite applications on a global scale. *Journal of Applied Meteorology* 31: 194–211.

Reynolds RR and Smith TM (1994) Improved global sea surface temperature analyses using optimum interpolation. *Journal of Climate* 7: 929–948.

List RJ (1958) *Smithsonian Meteorological Tables* 6th edn. City of Washington: Smithsonian Institution Press.

van der Meulen JP (1988) On the need of appropriate filter techniques to be considered using electrical humidity sensors. In: *Proceedings of the WMO Technical Conference on Instruments and Methods of Observation (TECO-1988)*, pp. 55–60. Leipzig, Germany: WMO.

Wentz FJ and Smith DK (1999) A model function for the ocean-normalized radar cross-section at 14 GHz derived from NSCAT observations. *Journal of Geophysical Research* 104: 11 499–11 514.

SENSORS FOR MICROMETEOROLOGICAL AND FLUX MEASUREMENTS

J. B. Edson, Woods Hole Oceanographic Institution, Woods Hole, MA, USA

Introduction

The exchange of momentum, heat, and mass between the atmosphere and ocean is the fundamental physical process that defines air–sea interactions. This exchange drives ocean and atmospheric circulations, and generates surface waves and currents. Marine micrometeorologists are primarily concerned with the vertical exchange of these quantities, particularly the vertical transfer of momentum, heat, moisture, and trace gases associated with the momentum, sensible heat, latent heat, and gas fluxes, respectively. The term flux is defined as the amount of heat (i.e., thermal energy) or momentum transferred per unit area per unit time.

Air–sea interaction studies often investigate the dependence of the interfacial fluxes on the mean meteorological (e.g., wind speed, degree of stratification or convection) and surface conditions (e.g., surface currents, wave roughness, wave breaking, and sea surface temperature). Therefore, one of the goals of these investigations is to parametrize the fluxes in terms of these variables so that they can be incorporated in numerical models. Additionally, these parametrizations allow the fluxes to be indirectly estimated from observations that are easier to collect and/or offer wider spatial coverage. Examples include the use of mean meteorological measurements from buoys or surface roughness measurements from satellite-based scatterometers to estimate the fluxes.

Direct measurements of the momentum, heat, and moisture fluxes across the air–sea interface are crucial to improving our understanding of the coupled atmosphere–ocean system. However, the operating requirements of the sensors, combined with the often harsh conditions experienced over the ocean, make this a challenging task. This article begins with a description of desired measurements and the operating requirements of the sensors. These requirements involve adequate response time, reliability, and survivability. This is followed by a description of the sensors used to meet these requirements, which includes examples of some of the obstacles that marine researchers have had to overcome. These obstacles include impediments caused by environmental conditions and engineering challenges that are unique to the marine environment. The discussion is limited to the measurement of velocity, temperature, and humidity. The article concludes with a description of the state-of-the-art sensors currently used to measure the desired fluxes.

Flux Measurements

The exchange of momentum and energy a few meters above the ocean surface is dominated by turbulent processes. The turbulence is caused by the drag (i.e., friction) of the ocean on the overlying air, which slows down the wind as it nears the surface and generates wind shear. Over time, this causes faster-moving air aloft to be mixed down and slower-moving air to be mixed up; the net result is a downward flux of momentum. This type of turbulence is felt as intermittent gusts of wind that buffet an observer looking out over the ocean surface on a windy day.

Micrometeorologists typically think of these gusts as turbulent eddies in the airstream that are being advected past the observer by the mean wind. Using this concept, the turbulent fluctuations associated with these eddies can be defined as any departure from the mean wind speed over some averaging period (eqn [1]).

$$u(t) = U(t) - \bar{U} \qquad [1]$$

In eqn [1], $u(t)$ is the fluctuating (turbulent) component, $U(t)$ is the observed wind, and the overbar denotes the mean value over some averaging period. The fact that an observer can be buffeted by the wind indicates that these eddies have some momentum. Since the eddies can be thought to have a finite size, it is convenient to consider their momentum per unit volume, given by $\rho_a U(t)$, where ρ_a is the density of air. In order for there to be an exchange of momentum between the atmosphere and ocean, this horizontal momentum must be transferred downward by some vertical velocity. The mean vertical velocity associated with the turbulent flux is normally assumed to be zero. Therefore, the turbulent transfer of this momentum is almost exclusively via the turbulent vertical velocity, $w(t)$, which we associate with overturning air.

The correlation or covariance between the fluctuating vertical and horizontal wind components is the most direct estimate of the momentum flux. This approach is known as the eddy correlation or direct covariance method. Computation of the covariance involves multiplying the instantaneous vertical velocity fluctuations with one of the horizontal components. The average of this product is then computed over the averaging period.

Because of its dependence on the wind shear, the flux of momentum at the surface is also known as the shear stress defined by eqn [2], where $\hat{i}$ and $\hat{j}$ are unit vectors, and v is the fluctuating horizontal component that is orthogonal to u.

$$\tau_0 = \hat{i}\overline{uw} - \hat{j}\overline{vw} \qquad [2]$$

Typically, the coordinate system is rotated into the mean wind such that u, v, and w denote the longitudinal, lateral, and vertical velocity fluctuations, respectively. Representative time series of longitudinal and vertical velocity measurements taken in the marine boundary layer are shown in **Figure 1**. The velocities in this figure exhibit the general trend that downward-moving air (i.e., $w < 0$) is transporting eddies with higher momentum per unit mass (i.e., $\rho_a u > 0$) and vice versa. The overall correlation is therefore negative, which is indicative of a downward flux of momentum.

Close to the surface, the wave-induced momentum flux also becomes important. At the interface, the turbulent flux actually becomes negligible and the

momentum is transferred via wave drag and viscous shear stress caused by molecular viscosity. The ocean is surprisingly smooth compared to most land surfaces. This is because the dominant roughness elements that cause the drag on the atmosphere are the wind waves shorter than 1 m in length. Although the longer waves and swell give the appearance of a very rough surface, the airflow tends to follow these waves and principally act to modulate the momentum flux supported by the small-scale roughness. Therefore, the wave drag is mainly a result of these small-scale roughness elements.

Turbulence can also be generated by heating and moistening the air in contact with the surface. This increases the buoyancy of the near-surface air, and causes it to rise, mix upward, and be replaced by less-buoyant air from above. The motion generated by this convective process is driven by the surface buoyancy flux (eqn [3]).

$$B_0 = \rho_a c_p \overline{w\theta_v} \qquad [3]$$

In eqn [3] c_p is the specific heat of air at constant pressure and θ_v is the fluctuating component of the virtual potential temperature defined in eqn [4], where Θ and θ are the mean and fluctuating components of the potential temperature, respectively, and q is the specific humidity (i.e., the mass of water vapor per unit mass of moist air).

$$\theta_v = \theta + 0.61\bar{\Theta}q \qquad [4]$$

These quantities also define the sensible heat (eqn [5]) and latent heat (eqn [6]) fluxes.

$$H_0 = \rho_a c_p \overline{w\theta} \qquad [5]$$

$$E_0 = L_e \overline{w\rho_a q} \qquad [6]$$

where L_e is the latent heat of evaporation. The parcels of air that are heated and moistened via the buoyancy flux can grow into eddies that span the entire atmospheric boundary layer. Therefore, even in light wind conditions with little mean wind shear, these turbulent eddies can effectively mix the marine boundary layer.

Conversely, when the air is warmer than the ocean, the flow of heat from the air to water (i.e., a downward buoyancy flux) results in a stably stratified boundary layer. The downward buoyancy flux is normally driven by a negative sensible heat flux. However, there have been observations of a downward latent heat (i.e., moisture) flux associated with the formation of fog and possibly condensation at the ocean surface. Vertical velocity fluctuations have

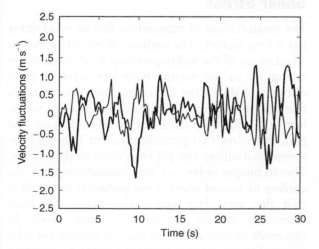

Figure 1 Time-series of the longitudinal (thick line) and vertical velocity (thin line) fluctuations measured from a stable platform. The mean wind speed during the sampling period was $10.8\,\mathrm{m\,s^{-1}}$.

to work to overcome the stratification since upward-moving eddies are trying to bring up denser air and vice versa. Therefore, stratified boundary layers tend to dampen the turbulent fluctuations and reduce the flux compared to their unstable counterpart under similar mean wind conditions. Over the ocean, the most highly stratified stable boundary layers are usually a result of warm air advection over cooler water. Slightly stable boundary conditions can also be driven by the diurnal cycle if there is sufficient radiative cooling of the sea surface at night.

Sensors

Measurement of the momentum, sensible heat, and latent heat fluxes requires a suite of sensors capable of measuring the velocity, temperature, and moisture fluctuations. Successful measurement of these fluxes requires instrumentation that is rugged enough to withstand the harsh marine environment and fast enough to measure the entire range of eddies that transport these quantities. Near the ocean surface, the size of the smallest eddies that can transport these quantities is roughly half the distance to the surface; i.e., the closer the sensors are deployed to the surface, the faster the required response. In addition, micrometeorologists generally rely on the wind to advect the eddies past their sensors. Therefore, the velocity of the wind relative to a fixed or moving sensor also determines the required response; i.e., the faster the relative wind, the faster the required response. For example, planes require faster response sensors than ships but require less averaging time to compute the fluxes because they sample the eddies more quickly.

The combination of these two requirements results in an upper bound for the required frequency response (eqn [7]).

$$\frac{fz}{U_r} \approx 2 \qquad [7]$$

Here f is the required frequency response, z is the height above the surface, and U_r is the relative velocity. As a result, sensors used on ships, buoys, and fixed platforms require a frequency response of approximately 10–20 Hz, otherwise some empirical correction must be applied. Sensors mounted on aircraft require roughly an order of magnitude faster response depending on the sampling speed of the aircraft.

The factors that degrade sensor performance in the marine atmosphere include contamination, corrosion, and destruction of sensors due to sea spray and salt water; and fatigue and failure caused by long-term operation that is accelerated on moving

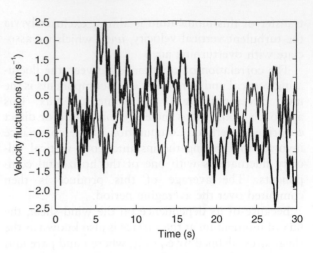

Figure 2 Time-series of the longitudinal (thick line) and vertical velocity (thin line) fluctuations measured from a 3 m discus buoy. The mean wind speed during the sampling period was 10.9 m s^{-1}. The measured fluctuations are a combination of turbulence and wave-induced motion of the buoy.

platforms. Additionally, if the platform is moving, the motion of the platform will be sensed by the instrument as an additional velocity and will contaminate the desired signal (**Figure 2**). Therefore, the platform motion must be removed to accurately measure the flux. This requires measurements of the linear and angular velocity of the platform. The alternative is to deploy the sensors on fixed platforms or to reduce the required motion correction by mounting the sensors on spar buoys, SWATH vessels, or other platforms that are engineered to reduced the wave-induced motion.

Shear Stress

The measurement of momentum flux or shear stress has a long history. The earliest efforts attempted to adapt many of the techniques commonly used in the laboratory to the marine boundary layer. A good example of this is the use of hot-wire anemometers that are well-suited to wind tunnel studies of turbulent flow. Hot-wire anemometry relies on very fine platinum wires that provide excellent frequency response and satisfy eqn [7] even close to the surface. The technique relies on the assumption that the cooling of heated wires is proportional to the flow past the wire. Hot-wire anemometers are most commonly used in constant-temperature mode. In this mode of operation, the current heating the wire is varied to maintain a constant temperature using a servo loop. The amount of current or power required to maintain the temperature is a measure of the cooling of the wires by the wind.

Unfortunately, there are a number of problems associated with the use of these sensors in the marine environment. The delicate nature of the wires (they are typically 10 μm in diameter) makes them very susceptible to breakage. Hot-film anemometers provide a more rugged instrument with somewhat slower, but still excellent, frequency response. Rather than strands of wires, a hot-film anemometer uses a thin film of nickel or platinum spread over a small cylindrical quartz or glass core. Even when these sensors are closely monitored for breakage, aging and corrosion of the wires and films due to sea spray and other contaminants cause the calibration to change over time. Dynamic calibration in the field has been used but this requires additional sensors. Therefore, substantially more rugged anemometers with absolute or more stable calibrations have generally replaced these sensors in field studies.

Another laboratory instrument that meets these requirements is the pitot tube; this uses two concentric tubes to measure the difference between the static pressure of the inner tube, which acts as a stagnation point, and the static pressure of the air flowing past the sensor. The free stream air also has a dynamic pressure component. Therefore, the difference between the two pressure measurements can be used to compute the dynamic pressure of the air flow moving past the sensor using Bernoulli's equation (eqn [8]).

$$\Delta p = \frac{1}{2}\rho_a \alpha U^2 \qquad [8]$$

Here α is a calibration coefficient that corrects for departures from Bernoulli's equations due to sensor geometry. A calibrated pitot tube can then be used to measure the velocity.

The traditional design is most commonly used to measure the streamwise velocity. However, three-axis pressure sphere (or cone) anemometers have been used to measure fluxes in the field. These devices use a number of pressure ports that are referenced against the stagnation pressure to measure all three components of the velocity. This type of anemometer has to be roughly aligned with the relative wind and its ports must remain clear of debris (e.g., sea spray and other particulates) to operate properly. Consequently, it has been most commonly used on research aircraft where the relative wind is large and particulate concentrations are generally lower outside of clouds and fog.

The thrust anemometer has also been used to directly measure the momentum flux in the marine atmosphere. This device measures the frictional drag of the air on a sphere or other objects. The most successful design uses springs to attach a sphere and its supporting structure to a rigid mount. The springs allow the sphere to be deflected in both the horizontal and vertical directions. The deflection due to wind drag on the sphere is sensed by proximity sensors that measure the displacement relative to the rigid mount. Carefully calibrated thrust anemometers have been used to measure turbulence from fixed platforms for extended periods. They are fairly rugged and low-power, and have adequate response for estimation of the flux. The main disadvantages of these devices are the need to accurately calibrate the direction response of each sensor and sensor drift due to aging of the springs.

A very robust sensor for flux measurements, particularly for use on fixed platforms, relies on a modification of the standard propellor vane anemometer used to measure the mean wind. The modification involves the use of two propellers on supporting arms set at 90° to each other. The entire assembly is attached to a vane that keeps the propeller pointed into the wind. The device is known as a K-Gill anemometer from the appearance of the twin propeller-vane configuration (**Figure 3**). The twin propellers are capable of measuring the instantaneous vertical and streamwise velocity, and the vane reading allows the streamwise velocity to be broken down into its u and v components. This device is also very robust and low power. However, it has a complicated inertial response on moving

Figure 3 The instrument at the far right is K-Gill anemometer shown during a deployment on a research vessel. The instrument on the left is a sonic anemometer that is shown in more detail in **Figure 4**. Photograph provided by Olc Persson (CIRES/NOAA/ETL).

platforms and is therefore most appropriate for use on fixed platforms. Additionally, the separation between the propellers (typically 0.6) acts as a spatial filter (i.e., it cannot detect eddies smaller than the separation), so it cannot be used too close to the surface and still satisfy eqn [7]. This is generally not a problem at the measurement heights used in most field deployments.

Over the past decade, sonic anemometers have become the instrument of choice for most investigations of air–sea interaction. These anemometers use acoustic signals that are emitted in either a continuous or pulsed mode. At present, the pulse type sonic anemometers are most commonly used in marine research. Most commercially available devices use paired transducers that emit and detect acoustic pulses (**Figure 4**). One transducer emits the pulse and the other detects it to measure the time of flight between them. The functions are then reversed to measure the time of flight in the other direction. The basic concept is that in the absence of any wind the time of flight in either direction is the same.

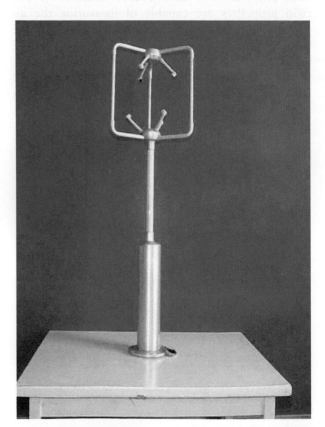

Figure 4 A commercially available pulse-type sonic anemometer. The three sets of paired transducers are cabable of measuring the three components of the velocity vector. This type of device produced the time-series in **Figure 1** and **Figure 2**. The sonic anemometer measures 0.75 m for top to bottom.

However, the times of flight differ if there is a component of the wind velocity along the path between the transducers. The velocity is directly computed from the two time of flight measurements, t_1 and t_2, using eqn [9], where L is the distance between the transducers.

$$U = \frac{L}{2}\left(\frac{1}{t_1} - \frac{1}{t_2}\right) \qquad [9]$$

Three pairs of transducers are typically used to measure all three components of the velocity vector. These devices have no moving parts and are therefore far less susceptible to mechanical failure. They can experience difficulties when rain or ice covers the transducer faces or when there is a sufficient volume of precipitation in the sampling volume. However, the current generation of sonic anemometers have proven themselves to be remarkably reliable in long-term deployments over the ocean; so much so that two-axis versions of sonic anemometers are also beginning to replace cup and propellor/vane anemometers for mean wind measurements over the ocean.

Motion Correction

The measurement of the fluctuating velocity components necessary to compute the fluxes is complicated by the platform motion on any aircraft, sea-going research vessel, or surface mooring. This motion contamination must be removed before the fluxes can be estimated. The contamination of the signal arises from three sources: instantaneous tilt of the anemometer due to the pitch, roll, and yaw (i.e., heading) variations; angular velocities at the anemometer due to rotation of the platform about its local coordinate system axes; and translational velocities of the platform with respect to a fixed frame of reference. Therefore, motion sensors capable of measuring these quantities are required to correct the measured velocities. Once measured, these variables are used to compute the true wind vector from eqn [10].

$$\mathbf{U} = T(\mathbf{U}_m + \mathbf{\Omega}_m \times \mathbf{R}) + \mathbf{U}_p \qquad [10]$$

Here $\mathbf{U}$ is the desired wind velocity vector in the desired reference coordinate system (e.g., relative to water or relative to earth); $\mathbf{U}_m$ and $\mathbf{\Omega}_m$ are the measured wind and platform angular velocity vectors respectively, in the platform frame of reference; T is the coordinate transformation matrix from the platform coordinate system to the reference coordinates; $\mathbf{R}$ is the position vector of the wind sensor with respect to the motion sensors; and U_p is the translational velocity vector of the platform measured at the location of the motion sensors.

A variety of approaches have been used to correct wind sensors for platform motion. True inertial navigation systems are standard for research aircraft. These systems are expensive, so simpler techniques have been sought for ships and buoys, where the mean vertical velocity of the platform is unambiguously zero. These techniques generally use the motion measurements from either strapped-down or gyro-stabilized systems.

The strapped-down systems typically rely on a system of three orthogonal angular rate sensors and accelerometers, which are combined with a compass to get absolute direction. The high-frequency component of the pitch, roll, and yaw angles required for the transformation matrix are computed by integrating and highpass filtering the angular rates. The low-frequency component is obtained from the lowpass accelerometer signals or, more recently, the angles computed from differential GPS. The transformed accelerometers are integrated and highpass filtered before they are added to lowpass filtered GPS or current meter velocities for computation of U_p relative to earth or the sea surface, respectively. The gyro-stabilized system directly computes the orientation angles of the platform. The angular rates are then computed from the time-derivative of the orientation angles.

Heat Fluxes

The measurement of temperature fluctuations over the ocean surface has a similar history to that of the velocity measurements. Laboratory sensors such as thermocouples, thermistors, and resistance wires are used to measure temperature fluctuations in the marine environment.

Thermocouples rely on the Seebeck effect that arises when two dissimilar materials are joined to form two junctions: a measuring junction and a reference junction. If the temperature of the two junctions is different, then a voltage potential difference exists that is proportional to the temperature difference. Therefore, if the temperature of the reference junction is known, then the absolute temperature at the junction can be determined. Certain combinations of materials exhibit a larger effect (e.g., copper and constantan) and are thus commonly used in thermocouple design. However, in all cases the voltage generated by the thermoelectric effect is small and amplifiers are often used along with the probes.

Thermistors and resistance wires are devices whose resistance changes with temperature. Thermistors are semiconductors that generally exhibit a large negative change of resistance with temperature (i.e., they have a large negative temperature coefficient of resistivity). They come in a variety of different forms including beads, rods, or disks. Microbead thermistors are most commonly used in turbulence studies; in these the semiconductor is situated in a very fine bead of glass. Resistance wires are typically made of platinum, which has a very stable and well-known temperature–resistance relationship. The trade-off is that they are less sensitive to temperature change than thermistors. The probe supports for these wires are often similar in design to hot-wire anemometers, and they are often referred to as cold-wires.

All of these sensors can be deployed on very fine mounts (**Figure 5**), which greatly reduces the adverse effects of solar heating but also exposes them to harsh environments and frequent breaking. Additionally, the exposure invariably causes them to become covered with salt from sea spray. The coating of salt causes spurious temperature fluctuations due to condensation and evaporation of water vapor on these hygroscopic particles. These considerations generally require more substantial mounts and some

Figure 5 A thermocouple showing the very fine mounts used for turbulence applications. The actual thermocouple is situated on fine wires between the probe supports and is too small to be seen in this photograph.

sort of shielding from the radiation and spray. While this is acceptable for mean temperature measurements, the reduction in frequency response caused by the shields often precludes their use for turbulence measurements.

To combat these problems, marine micrometeorologists have increasingly turned to sonic thermometry. The time of flight measurements from sonic anemometers can be used to measure the speed of sound c along the acoustic path (eqn [11]).

$$c = \frac{L}{2}\left(\frac{1}{t_1} + \frac{1}{t_2}\right) \qquad [11]$$

The speed of sound is a function of temperature and humidity and can be used to compute the sonic temperature T_s defined by eqn [12], where U_N is velocity component normal to the transducer path.

$$T_s = T(1 + 0.51q) = \frac{c^2 + U_N^2}{403} \qquad [12]$$

The normal wind term corrects for lengthening of the acoustic path by this component of the wind. This form of velocity crosstalk has a negligible effect on the actual velocity measurements, but has a measurable effect on the sonic temperature.

Sonic thermometers share many of the positive attributes of sonic anemometers. Additionally, they suffer the least from sea salt contamination compared to other fast-response temperature sensors. The disadvantage of these devices is that velocity crosstalk must be corrected for and they do not provide the true temperature signal as shown by eqn [12]. Fortunately, this can be advantageous in many investigations because the sonic temperature closely approximates the virtual temperature in moist air $T_v = T(1 + 0.61q)$. For example, in many investigations over the ocean an estimate of the buoyancy flux is sufficient to account for stability effects. In these investigations the difference between the sonic and virtual temperature is often neglected (or a small correction is applied), and the sonic anemometer/thermometer is all that is required. However, due to the importance of the latent heat flux in the total heat budget over the ocean, accurate measurement of the moisture flux is often a crucial component of air–sea interaction investigations.

The accurate measurement of moisture fluctuations required to compute the latent heat flux is arguably the main instrumental challenge facing marine micrometeorologists. Sensors with adequate frequency response generally rely on the ability of water vapor in air to strongly absorb certain wavelengths of radiation. Therefore, these devices require a narrowband source for the radiation and a detector to measure the reduced transmission of that radiation over a known distance.

Early hygrometers of this type generated and detected ultraviolet radiation. The Lyman α hygrometer uses a source tube that generates radiation at the Lyman α line of atomic hydrogen which is strongly absorbed by water vapor. This device has excellent response characteristics when operating properly. Unfortunately, it has proven to be difficult to operate in the field due to sensor drift and contamination of the special optical windows used with the source and detector tubes. A similar hygrometer that uses krypton as its source has also been used in the field. Although the light emitted by the krypton source is not as sensitive to water vapor, the device still has more than adequate response characteristics and generally requires less maintenance than the Lyman α. However, it still requires frequent calibration and cleaning of the optics. Therefore, neither device is particularly well suited for long-term operation without frequent attention.

Commercially available infrared hygrometers are being used more and more in marine micrometeorological investigations (**Figure 6**). Beer's law

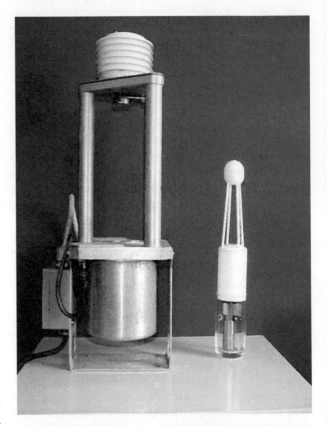

Figure 6 Two examples of commercially available infrared hygrometers. The larger hygrometer is roughly the height of the sonic anemometer shown in **Figure 4**.

(eqn [13]) provides the theoretical basis for the transmission of radiation over a known distance.

$$T = e^{-(\gamma+\delta)D} \qquad [13]$$

Here T is the transmittance of the medium, D is the fixed distance, and γ and δ are the extinction coefficients for scattering and absorption, respectively. This law applies to all of the radiation source described above; however, the use of filters with infrared devices allows eqn [13] to be used more directly. For example, the scattering coefficient has a weak wavelength dependence in the spectral region where infrared absorption is strongly wavelength dependent. Filters can be designed to separate the infrared radiation into wavelengths that exhibit strong and weak absorption. The ratio of transmittance of these two wavelengths is therefore a function of the absorption (eqn [14]), where the subscripts s and w identify the variables associated with the strongly and weakly absorbed wavelengths.

$$\frac{T_s}{T_w} \approx e^{-\delta_s D} \qquad [14]$$

Calibration of this signal then provides a reliable measure of water vapor due to the stability of current generation of infrared sources.

Infrared hygrometers are still optical devices and can become contaminated by sea spray and other airborne contaminants. To some extent the use of the transmission ratio negates this problem if the contamination affects the two wavelengths equally. Obviously, this is not the case when the optics become wet from rain, fog, or spray. Fortunately, the devices recover well once they have dried off and are easily cleaned by the rain itself or by manual flushing with water. Condensation on the optics can also be reduced by heating their surfaces. These devices require longer path lengths (0.2–0.6 m) than Lyman α or krypton hygrometers to obtain measurable absorption (**Figure 6**). This is not a problem as long as they are deployed at heights $\gg D$.

Conclusions

The state of the art in sensor technology for use in the marine surface layer includes the sonic anemometer/thermometer and the latest generation of infrared hygrometers (**Figure 7**). However, the frequency response of these devices, mainly due to spatial averaging, precludes their use from aircraft. Instead, aircraft typically rely on gust probes for measurement of the required velocity fluctuations, thermistors for temperature fluctuations, and Lyman α hygrometers

Figure 7 A sensor package used to measure the momentum, sensible heat, and latent heat fluxes from a moving platform. The cylinder beneath the sonic anemometer/thermometer holds 3-axis angular rate sensors and linear accelerometers, as well as a magnetic compass. Two infrared hygrometers are deployed beneath the sonic anemometer. The radiation shield protects sensors that measure the mean temperature and humidity. Photograph provided by Wade McGillis (WHOI).

for moisture fluctuations. Hot-wire and hot-film anemometers along with the finer temperature and humidity devices are also required to measure directly the viscous dissipation of the turbulent eddies that occurs at very small spatial scales.

Instruments for measuring turbulence are generally not considered low-power when compared to the mean sensors normally deployed on surface moorings, so past deployments of these sensors were mainly limited to fixed platforms or research vessels with ample power. Recently, however, sensor packages mounted on spar and discus buoys have successfully measured motion-corrected momentum and buoyancy fluxes on month- to year-long deployments with careful power management. The use of these sensor packages is expected to continue owing to the desirability of these measurements and technological advances leading to improved power sources and reduced power consumption by the sensors.

See also

Air–Sea Gas Exchange. Breaking Waves and Near-Surface Turbulence. Heat and Momentum Fluxes at the Sea Surface. Sensors for Mean Meteorology. Surface Gravity and Capillary Waves. Upper Ocean Heat and Freshwater Budgets. Upper Ocean Mixing Processes. Wave Generation by Wind. Wind- and Buoyancy-Forced Upper Ocean. Wind Driven Circulation.

Further Reading

Ataktürk SS and Katsaros KB (1989) The K-Gill, a twin propeller-vane anemometer for measurements of atmospheric turbulence. *Journal of Atmospheric and Oceanic Technology* 6: 509–515.

Buck AL (1976) The variable path Lyman-alpha hygrometer and its operating characteristics. *Bulletin of the American Meteorological Society* 57: 1113–1118.

Crawford TL and Dobosy RJ (1992) A sensitive fast-response probe to measure turbulence and heat flux from any airplane. *Boundary-Layer Meteorology* 59: 257–278.

Dobson FW, Hasse L, and Davis RE (1980) *Air–Sea Interaction; Instruments and Methods*. New York: Plenum Press.

Edson JB, Hinton AA, Prada KE, Hare JE, and Fairall CW (1998) Direct covariance flux estimates from mobile platforms at sea. *Journal of Atmospheric and Oceanic Technology* 15: 547–562.

Fritschen LJ and Gay LW (1979) *Environmental Instrumentation*. New York: Springer-Verlag.

Kaimal JC and Gaynor JE (1991) Another look at sonic thermometry. *Boundary-Layer Meteorology* 56: 401–410.

Larsen SE, Højstrup J, and Fairall CW (1986) Mixed and dynamic response of hot wires and measurements of turbulence statistics. *Journal of Atmospheric and Oceanic Technology* 3: 236–247.

Schmitt KF, Friehe CA, and Gibson CH (1978) Humidity sensitivity of atmospheric temperature sensors by salt contamination. *Journal of Physical Oceanography* 8: 141–161.

Schotanus P, Nieuwstadt FTM, and de Bruin HAR (1983) Temperature measurement with a sonic anemometer and its application to heat and moisture fluxes. *Boundary-Layer Meteorology* 26: 81–93.

IR RADIOMETERS

C. J. Donlon, Space Applications Institute,
Ispra, Italy

Introduction

Measurements of sea surface temperature (SST) are most important for the investigation of the processes underlying heat and gas exchange across the air–sea interface, the surface energy balance, and the general circulation of both the atmosphere and the oceans. Complementing traditional subsurface contact temperature measurements, there is a wide variety of infrared radiometers, spectroradiometers, and thermal imaging systems that can be used to determine the SST by measuring thermal emissions from the sea surface. However, the SST determined from thermal emission can be significantly different from the subsurface temperature ($> \pm 1\,K$) because the heat flux passing through the air–sea interface typically results in a strong temperature gradient. Radiometer systems deployed on satellite platforms provide daily global maps of SSST (sea surface temperature) at high spatial resolution ($\sim 1\,km$) whereas those deployed from ships and aircraft provide data at small spatial scales of centimeters to meters. In particular, the development of satellite radiometer systems providing a truly synoptic view of surface ocean thermal features has been pivotal in the description and understanding of the global oceans.

This article reviews the infrared properties of water and some of the instruments developed to measure thermal emission from the sea surface. It focuses on *in situ* radiometers although the general principles described are applicable to satellite sensors treated elsewhere in this volume.

Infrared Measurement Theory

Infrared (IR) radiation is heat energy that is emitted from all objects that have a temperature above 0 K ($-273.16°C$). It includes all wavelengths of the electromagnetic spectrum between $0.75\,\mu m$ and $\sim 100\,\mu m$ (**Figure 1**) and has the same optical properties as visible light, being capable of reflection, refraction, and forming interference patterns.

The following total quantities, conventional symbols and units provide the theoretical foundation for the measurement of IR radiation and are schematically shown in **Figure 2**. Spectral quantities can be represented by restricting each to a specific waveband.

- Radiant energy Q, is the total energy radiated from a point source in all directions in units of joules (J).
- Radiant flux $\phi = dQ/dt$ is the flux of all energy radiated in all directions from a point source in units of watts (W).
- Emittance $M = d\phi/dA$ is the radiant flux density from a surface area A in units of $W\,m^{-2}$. This is an integrated flux (i.e., independent of direction) and will therefore vary with orientation relative to a nonuniform source.
- Radiant intensity $I = d\phi/d\omega$ is the radiant flux of a point source per solid angle ω (steradian, sr) and is a directional flux in units of $W\,sr^{-1}$.
- Radiance $L = dI/d(A\cos\theta)$ is the radiant intensity of an extended source per unit solid angle in a given direction θ, per unit area of the source projected in the same θ. It has units of $W\,sr^{-1}\,m^{-2}$.

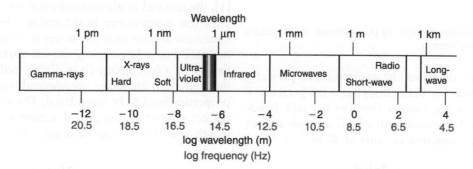

Figure 1 Schematic diagram of the electromagnetic spectrum showing the location and interval of the infrared waveband.

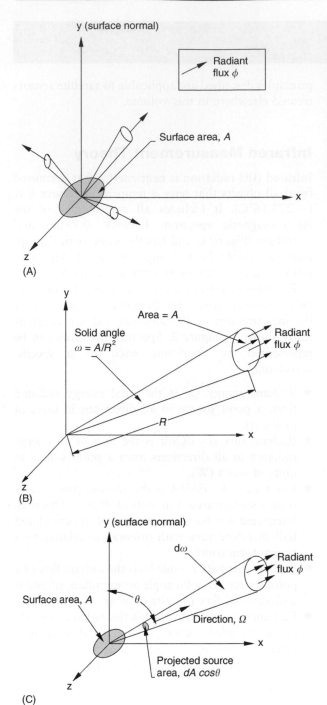

(A)

(B)

(C)

Figure 2 Schematic definition of (A) Emittance, E; (B) radiant intensity, I; (C) radiance, L.

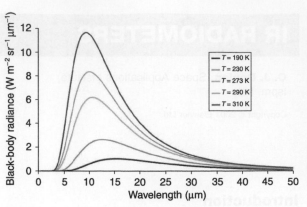

Figure 3 Black-body radiance as a function of wavelength computed using eqns [1] and [2] for different target temperatures.

where $h = 6.626 \times 10^{-34}$ J s is Planck's constant, $c = 2.998 \times 10^{8}$ m s^{-1} is the speed of light and $k = 1.381 \times 10^{-23}$ J K^{-1} is Boltzmann's constant. The sea surface is considered a Lambertian source (i.e., uniform radiance in all directions) so that the spectral radiance L_λ is related to M_λ by

$$L_\lambda = \frac{M_\lambda}{\pi} \qquad [2]$$

Figure 3 shows L_λ computed for several temperatures as a function of wavelength. Considering temperatures of 273–310 K as representative of the global ocean, maximum emission occurs at a wavelength of 9.3–10.7 µm. Atmospheric attenuation is minimal at ~3.5 µm, 9.0 µm and 11.0 µm that are the spectral intervals often termed atmospheric 'windows'. Instruments operating within these intervals are optimal for sea surface measurements – especially in the case of satellite deployment where atmospheric attenuation can be significant. In the 3–5 µm spectral region. L_λ is a strong function of temperature (**Figure 3**) highlighting the possibility to increase in radiometer sensitivity by utilizing this spectral interval.

By measuring L_λ using eqn [2] and inverting eqn [1], the spectral brightness temperature, $B_{(T,\lambda)}$, rather than the temperature is determined because these equations assume that sea water is a perfect emitter or black body. In practice, the sea surface does not behave as a black body (it is slightly reflective in the infrared) and therefore its spectral and geometric properties need to be considered. The emittance of a perfect emitter at the actual temperature T, wavelength λ, and view angle θ is given by

Planck's law describes the emittance of a perfectly emitting surface (or black body) at a temperature T in Kelvin. It is the radiant flux (ϕ) per unit bandwidth centered at wavelength λ leaving a unit area of surface in any direction in units of W m^{-2} m^{-1}.

$$M_{\lambda,T} = \frac{2\pi h c^{2}}{\lambda^{5}(e^{hc/\lambda kT} - 1)} \qquad [1]$$

$$M_{(T,\lambda,\theta)} = \frac{\pi L_{(T,\lambda,\theta)}}{\varepsilon_{(\lambda,\theta)}} \qquad [3]$$

where the emissivity, $\varepsilon_{(\lambda,\theta)}$, can be calculated using

$$\varepsilon_{(\lambda,\theta)} = \frac{M_{(T,\lambda,\theta)} measured}{M_{(T,\lambda,\theta)} blackbody} \qquad [4]$$

which has a strong dependence on wavelength and viewing geometry. The effective emissivity, ε, integrates $\varepsilon_{(\lambda,\theta)}$ over all wavelengths of interest for radiometer view angle θ. **Figure 4A** shows the calculated normal reflectivity, ρ, of pure water as a function of wavelength for the spectral region 1–100 µm. Pure water differs only slightly from sea water in this context. Note that ρ is minimal at a wavelength of

~11 µm ($\rho \approx 0.0015$) and following Kirchoff's law

$$\rho_\lambda + \tau_\lambda + \varepsilon_\lambda = 1 \qquad [5]$$

where ρ_λ is the spectral reflectivity and τ_λ is the spectral transmissivity. The e-folding penetration depth (i.e., the depth of 63% emission) or optical depth at a typical wavelength of 11 µm is ≈ 10 µm, τ_λ can be neglected and ε_λ can be calculated using

$$\varepsilon_\lambda = 1 - \rho_\lambda \qquad [6]$$

Although the actual optical depth is wavelength dependent, it is clear that IR radiometers determine

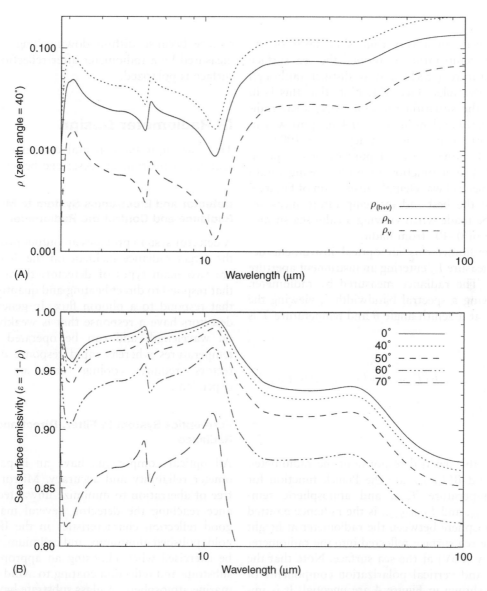

Figure 4 (A) The normal reflectivity, ρ, of the pure water as a function of wavelength (full line). Also shown are the horizontal polarized (ρ_h) and the vertically polarized (ρ_v) components of ρ. (B) The spectral emissivity of pure water as a function of viewing zenith angle. h, height above sea surface.

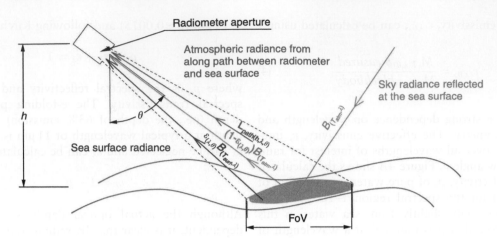

Figure 5   Schematic diagram showing the radiance components measured by an IR radiometer viewing the sea surface. Fov, Field of view.

the temperature of a very thin 'skin' layer of the ocean. This temperature is termed the sea surface skin temperature (SSST) and is distinct (although related) to the subsurface SST. Note that this is in contrast to the situation for short-wave solar radiation (having wavelengths of ~0.4–0.7 μm) which, for clear water, penetrates to a depth of ~100 m.

Figure 4B shows the ε_λ of pure water computed from eqn [6] as a function of both viewing zenith angle and spectral wavelength. Inspection of **Figure 4** reveals that the best radiant temperature measurement will be made when viewing a calm sea surface at an angle of 0–40° from nadir.

An IR radiometer is an optical instrument designed to measure L_λ entering an instrument aperture (**Figure 5**). The radiance measured by radiometer, $L_{(T,\lambda,\theta)}$, having a spectral bandwidth λ, viewing the sea surface at a zenith angle θ and temperature T is given by:

$$L_{(T,\lambda,\theta)} = \int_0^\alpha \xi_\lambda [\varepsilon_{(\lambda,\theta)} B(T_{surf},\lambda) + (1 - \varepsilon_{(\lambda,\theta)}) B(T_{atm},\lambda) + L_{path(h,\lambda,\theta)}] d\lambda$$

[7]

where ξ_λ is the spectral response of the radiometer, $B(T_{surf},\lambda$ and $B(T_{atm},\lambda$ are the Planck function for surface temperature T_{surf}, and atmospheric temperature T_{atm}, and $L_{path(h,\lambda,\theta)}$ is the radiance emitted by the atmosphere between the radiometer at height h above the sea surface reflected into the radiometer field-of-view (FoV) at the sea surface. Note that the horizontal and vertical polarization components of reflectivity shown in **Figure 4** are unequal. It is important to consider the polarization of surface reflectance when making measurements of the sea

surface because diffuse downwelling sky radiance measured by a radiometer after reflection at the sea surface is polarized.

IR Radiometer Design

There are four fundamental components to all IR radiometer instruments described below.

Detector and Electronics System to Measure Radiance and Control the Radiometer

A detector system provides an output proportional to the target radiance incident on the detector. There are two main types of detector: thermal detectors that respond to direct heating and quantum detectors that respond to a photon flux. In general, thermal detectors have a response that is weakly dependent on wavelength and can be operated at ambient temperatures whereas rapid response quantum detectors require cooling and are wavelength dependent.

Fore-optics System to Filter, Direct and Focus Radiance

All optical components have an impact on radiometer reliability and accuracy. Mirrors should be free of aberration to minimize unwanted stray radiance reaching the detector. Several materials have good reflection characteristics in the IR including, gold, polished aluminum, and cadmium. Care should be exercised when choosing an appropriate mirror substrate and reflection coating to avoid decay in the marine atmosphere. A glass substrate having a 'hard' scratch-resistant polished gold surface provides >98% reflectance and good environmental wear.

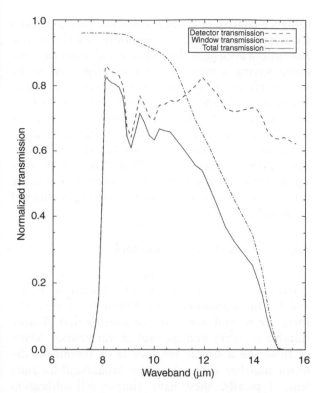

Figure 6 Normalized spectral transmission of an IR window and detector shown together with the combined total response.

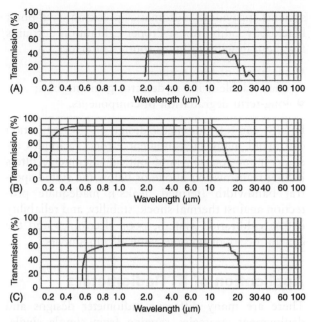

Figure 7 Spectral transmission for common IR window materials. (A) Germanium, (B) sodium chloride, (C) zinc selenide.

Spectral filter windows and lenses require spectral properties that, together with the detector characteristics, define the overall spectral characteristics of a radiometer. **Figure 6** shows the combined spectral response for a broadband radiometer together with the component spectral response of the window and detector.

Environmental System to Protect and Thermally Stabilize the Radiometer

For any optical instrument intended for use in the harsh marine environment, adequate environmental protection is critical. Rain, sea-water spray, and high humidity can destroy a poorly protected instrument rapidly and components such as electrical connections and fore-optics should be resistant to these effects. *In situ* radiometer windows are particularly important in this context. They should not significantly reduce the incoming signal or render it noisy, and be strong enough to resist mechanical, thermal and chemical degradation. **Figure 7** describes several common materials. Germanium (Ge) windows have good transmission characteristics but are very brittle. Sodium chloride (NaCl) is a low-cost, low-absorption material but is of little use in the marine environment because it is water-soluble. Zinc selenide (ZnSe) has high transmission, is nonhygroscopic

and resistant to thermal shock but is soft and requires a protective 'hard' surface finish coating.

Certain window materials achieve better performance when antireflection (AR) coatings are used to minimize reflection from the window. For example, when an AR coating is used on a ZnSe window the transmission increases from ∼70% to ∼90%. Other coatings provide windows that polarize the incident radiance signal such as optically thin interference coatings and wire grid diffraction polarizers.

Deposition of marine NaCl on all optical components (especially calibration targets) presents an unavoidable problem. Although NaCl itself has good infrared transmission properties (**Figure 7**), contaminated surfaces may become decoupled from temperature sensors and the noise introduced by the optical system will increase. Finally, adequate thermal control using reflective paint together with substantial instrument mass is required so that instruments are not sensitive to thermal shock. In higher latitudes, it may be necessary to provide an extensive antifreeze capability.

Calibration System to Quantify the Radiometer Output

The role of a calibration system is to quantify the instrument output in terms of the measured radiance incident on the detector. Calibration techniques are specific to the particular design of radiometer and vary considerably from simple bias corrections to systems providing automatic precision two-point blackbody calibrations. Proper calibration accounts for the following primary sources of error:

- the effect of fore-optics;
- unavoidable drifts in detector gain and bias;
- long-term degradation of components.

Finally, careful radiometer design and configuration can avoid many measurement errors. Examples include: poor focusing and optical alignment; filters having transmission above or below the stated spectral bandwidth (termed 'leaks'); inadequate protection against thermal shock, stability, and reliability of the calibration system; poor electronics and the general decay of optomechanical components.

Application of IR Radiometers

There are many different radiometer designs and deployment strategies ranging from simple single-channel hand-held devices to complex spectro-radiometers. Instrument accuracy, sensitivity and stability depends on both the deployment scenario and radiometer design. Accordingly, the following sections describe several different examples.

Broadband Pyrgeometers

A pyrgeometer is an integrating hemispheric radiometer which, by definition, measures the band-limited spectral emittance, E, so that the angle dependency in eqn [7] is redundant. They are used to determine the long-wave heat flux at the air–sea interface by measuring the difference between atmospheric and sea surface radiance either using two individual sensors (**Figure 8A**) or as a single combined sensor (**Figure 8B**). A thermopile detector is often used which is a collection of thermocouple detectors composed of two dissimilar metallic conductors connected together at two 'junctions.' The measurement junction is warmed by incident

radiance relative to a stable reference junction and a μV signal is produced. The response of a thermopile has little wavelength dependence and a hemispheric dome having a filter ($\sim 3–50\,\mu m$) deposited on its inner surface typically defines the spectral response. Direct compensation for thermal drift using a temperature sensor located at the thermopile reference junction is sometimes used but regular calibration using an independent laboratory blackbody is mandatory. An accuracy of $<10\ \mathrm{W\ m^{-2}}$ is possible after significant correction for instrument temperature drift and stray radiance contribution using additional on-board temperature sensors.

Narrow Beam Filter Radiometers

Narrow beam filter radiometers are often used to determine the SSST for air–sea interaction studies and for the validation of satellite derived SSST and there are several low-cost instruments that provide suitable accuracy and spectral characteristics. Many of these use a simple thermopile or thermistor detector together with a low-cost broadband-focusing lens. Typically, they have simple self-calibration techniques based on the temperature of the instrument and/or detector. Consequently they have poor resistance to thermal shock and have fore-optics that readily degrade in the marine atmosphere. However, handled with care, these devices are accurate to ± 0.1 K, albeit with limited sensitivity.

Precision narrow beam filter radiometers often use pyroelectric detectors that produce a small electrical current in response to changes in detector temperature forced by incident radiation. They have a fast response at ambient temperatures but require a modulated signal to operate. Modulation is accomplished by using an optical chopper having high reflectivity 'vanes' to alternately view a reference radiance source by reflection and a free path to the target radiance. The most common chopper systems are rotary systems driven by a small electric motor phase locked to the

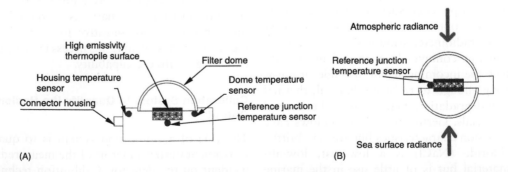

Figure 8 (A) A typical design of a long-wave pyrgeometer. (B) A net-radiation pyrgeometer for determination of the net long-wave flux at the sea surface.

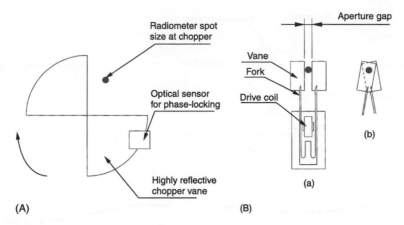

Figure 9 (A) Schematic layout of a rotary chopper; (B) schematic layout of a tuning fork chopper; (a) open; (b) closed.

detector output by an optoelectronic sensor **Figure 9A**. An alternative design driven by a small oscillating electromagnetic coil called a tuning fork chopper is shown in **Figure 9B**. As the coil resonates, the reflective vanes of the chopper oscillate alternately opening and closing an aperture 'gap.'

Dynamic detector bias compensation is inherent when using an optical chopper. The detector alternately measures radiance from the sea surface L_{src} and a reference blackbody, L_{bb} (sometimes this is the detector itself) reflected by the chopper vanes resulting in two signals

$$S_1 = L_{bb} + \delta \qquad [8]$$

$$S_2 = L_{src} + \delta \qquad [9]$$

Assuming L_{bb} remains constant during a short chopping cycle, the bias term δ in eqns [8] and [9] is eliminated

$$\Delta S = S_1 - S_2 = L_{bb} - L_{src} \qquad [10]$$

It is important to recognize the advantages to this technique, which is widely used:

- There is minimal thermal drift of the detector;
- The detector is dynamically compensated for thermal shock;
- A precise modulated signal is generated well suited to selective filtering providing excellent noise suppression and signal stability.

However, in order to compensate for instrument gain changes, an additional blackbody sources(s) is required. These are periodically viewed by the detector to provide a mechanism for absolute calibration. Either the black body is moved into the detector FoV or an adjustable mirror reflects radiance from the black body on to the detector.

Calibration cycles should be made at regular intervals so that gain changes can be accurately monitored and calibration sources need to be viewed using the same optical path as that used to view the sea surface.

A basic 'black-body' calibration strategy uses an external bath of sea water as a high ε (>0.95) reference as shown in **Figure 10**. In this scheme, the radiometer periodically views the water bath that is stirred vigorously to prevent the development of a thermal skin temperature deviation. The view geometry for the water bath and the sea surface are assumed to be identical and, by measuring the temperature of the water bath the radiometer can be absolutely calibrated. An advantage of this technique is that $\varepsilon_{(\lambda,\theta)}$ is not required to determine the SSST. However, in practice, it is difficult to continuously operate a water bath at sea and surface roughness differences between the bath and sea surface are ignored.

On reflection at the sea surface, diffuse sky radiance is polarized and, at Brewster's angle ($\sim 50°$ from nadir at a wavelength of 11 μm), the vertical v-polarization is negligible for a given wavelength (**Figure 11**). Only the horizontal h-polarization component remains so that if the radiometer filter response is v-polarized (i.e., only passes v-polarized radiance), negligible reflected sky radiance is measured by the radiometer. In practice, because Brewster's angle is very sensitive to the geometry of a particular deployment (approximately $\pm 2°$) this technique is only applicable to deployments from fixed platforms and when the sea surface is relatively calm. Further, the use of a polarizing filter will significantly reduce the signal falling on the detector increasing the signal-to-noise ratio.

The use of fabricated black-body cavities (**Figure 12A**) provides an accurate, versatile and, compact calibration system. Normally, two

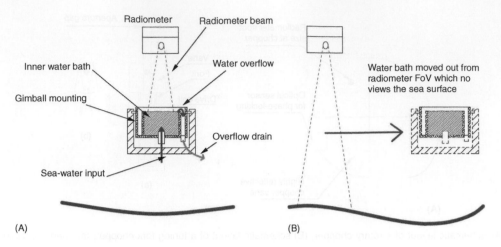

Figure 10 Schematic diagram showing the stirred water bath calibration scheme. (A) The radiometer in calibration mode, (B) the radiometer viewing the sea surface after the water bath has been moved out of the field of view (FoV).

black-body cavities are used, one of which follows the ambient temperature of the instrument and a second is heated to a nominal temperature above this. High ε (>0.99) is attained by a combination of specialized surface finish and black-body geometry. The cavity radiance is determined as a function of the black body temperature that is easily measured. **Figure 12B** shows a schematic outline of a typical black-body radiometer design using a rotary chopper

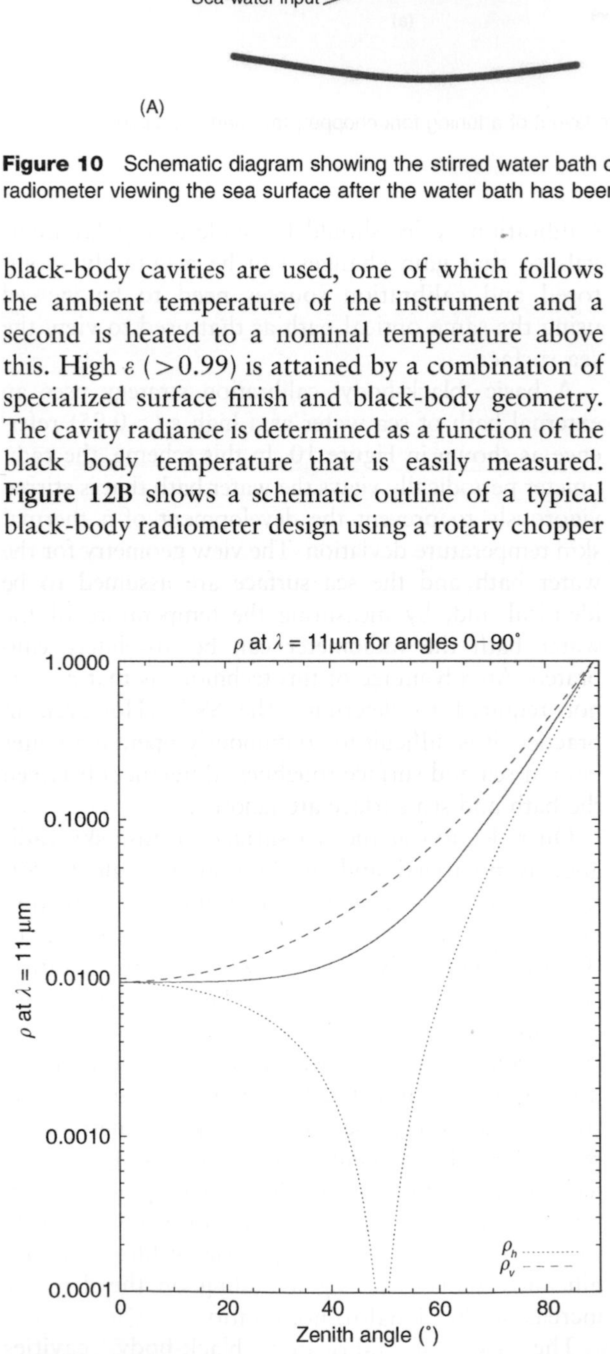

and **Figure 12C** provides a schematic diagram of a typical output signal.

Note that for all calibration schemes, larger errors are expected beyond the calibrated temperature range which can be a problem for sky radiance measurements where clear sky temperatures of <200 K are common.

Multichannel Radiometers

The terms in eqn [7] are directly influenced by the height, h, of the radiometer above the sea surface and are different in magnitude for *in situ* and spacecraft deployments. In the case of a sea surface *in situ* radiometer deployment, $L_{path(h,\lambda,\theta)}$ is typically neglected because h is normally <10 m unless the atmosphere has a heavy water vapor loading (e.g., $>90\%$) or an aircraft deployment is considered. However, for a spacecraft deployment, this is a significant term requiring explicit correction. Conversely, the $B(T_{atm},\lambda)$ term is critical to the accuracy of an *in situ* radiometer deployment but of little impact (except perhaps at the edge of clouds) for a satellite instrument deployment because $L_{path(h,\lambda,\theta)}$ dominates the signal. A multispectral capability can be used to explicitly account for $L_{path(h,\lambda,\theta)}$ in eqn [7] because of unequal atmospheric attenuation for different spectral wavebands. Multichannel radiometers are exclusively used on satellite platforms for this reason. Many *in situ* multichannel radiometers are designed primarily for the radiant calibration or validation of specific satellite radiometers and the development of satellite radiometer atmospheric correction algorithms. They often have several selectable filters matched to those of the satellite sensor.

ρ at $\lambda = 11\,\mu$m for angles 0–90°

ρ at $\lambda = 11\,\mu$m

1.0000

0.1000

0.0100

0.0010

0.0001

0   20   40   60   80

Zenith angle (°)

ρ_h ....
ρ_v - - - -

Figure 11 Polarization of sea surface reflection at 11 μm as a function of view angle. Total polarization is shown as a solid line.

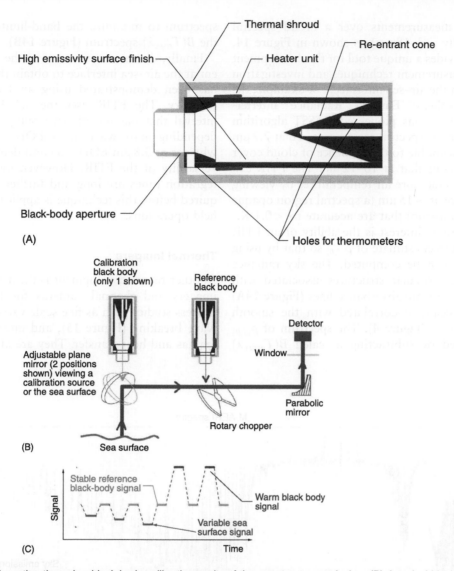

Figure 12 (A) A section through a black-body calibration cavity of the re-entrant cone design. (B) A typical black-body calibration radiometer using a rotary optical chopper. (C) A schematic diagram of a typical detector output signal, showing sea, reference, and calibration signals.

It is worth noting that multiangle view radiometers are also capable of providing an explicit correction for atmospheric attenuation. Often operated from satellite and aircraft, these instruments provide a direct measure of atmospheric attenuation by making two views of the same sea surface area at different angles using a geometry that doubles the atmospheric pathlength (**Figure 13**). The assumption is made that atmospheric and oceanic conditions are stationary in the time between each measurement.

Spectroradiometers

A recent development is the use of Fourier transform infrared spectrometers (FTIR) that are capable of accurate ($\sim 0.05\,K$). High spectral resolution

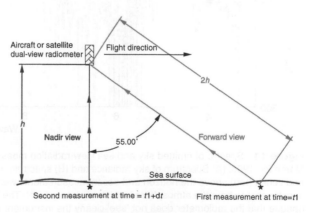

Figure 13 Schematic diagram of dual view, double atmospheric path radiometer deployment geometry.

(~ 0.5 cm^{-1}) measurements over a broad spectral range (typically ~ 3–18 μm) as shown in **Figure 14**. The FTIR provides a unique tool for the development of new IR measurement techniques and investigation of processes at the air–sea interface. For example, the Marine-Atmospheric Emitted Radiance Interferometer (M-AERI) has pioneered a SSST algorithm that uses a narrow spectral region centered at 7.7 μm that is less susceptible to the influence of cloud cover and sky emissions that at 10–12 μm. The FTIR can also be used to measure air temperatures by viewing the atmosphere at ~ 15 μm (a spectral region opaque due to CO$_2$ emission) that are accurate to <0.1 K.

Of considerable interest is the ability of an FTIR provide an indirect estimate of $\rho_{(\lambda,\theta)}$ so that by using eqn [5], $\varepsilon_{(\lambda,\theta)}$ can be computed. The sky radiance spectrum has particular structures associated with atmospheric emission–absorbance lines (**Figure 14A**) that are physically uncorrelated with the smooth spectrum of $\rho_{(\lambda,\theta)}$ (**Figure 4**). The spectrum of $\rho_{(\lambda,\theta)}$ can be derived by subtracting a scaled $B(T_{arm}, \lambda)$

spectrum to minimize the band-limited variance of the $B(T_{sea}, \lambda)$ spectrum (**Figure 14B**).

Finally, direct measurement of the thermal gradient at the air–sea interface to obtain the net heat flux has been demonstrated using an FTIR in the laboratory. The FTIR uses the 3.3–4.1 μm spectral interval that has an effective optical depth (EOD) depending on the wavelength (EOD $= 0$ μm at 3.3 μm whereas at 3.8 μm EOD $= 65$ μm) demonstrating the versatility of the FTIR. However, measurement integration times are long and further progress is required before this technique is applicable for normal field operations.

Thermal Imagers

Another recent development is the application of IR imagers and thermal cameras for high-resolution process studies such as fine-scale variability of SSST, wave breaking (**Figure 15**), and understanding air–sea gas and heat transfer. They are also used during

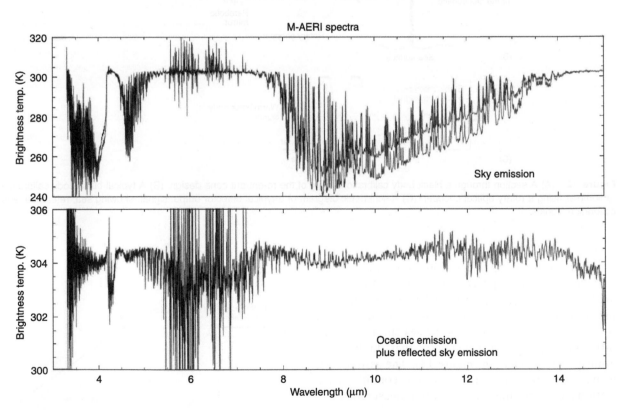

Figure 14 Spectra of emitted sky and sea view radiation measured by the M-AERI FTIR in the tropical Western Pacific Ocean on March 24, 1996. (A) Spectrum of sky radiance and (B) spectrum of corresponding sea radiance Sky measurements were made at 45° and zenith (red) above the horizon and ocean measurements were made at 45° below the horizon. The cold temperatures in the sky spectra show where the atmosphere is relatively transparent. The 'noise' in the 5.5–7 μm range is caused by the atmosphere being so opaque that the radiometer does not 'see' clearly the instrument internal black-body targets and calibration is void. The spectrum of upwelling radiation (B) consists of emission from the sea surface, reflected sky emission and emission from the atmospheric pathlength between the sea surface and the radiometer.

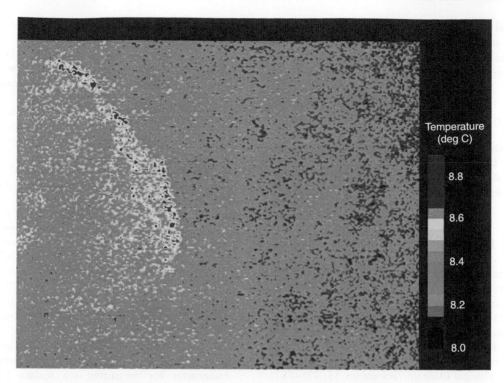

Figure 15   A thermal image of a breaking wave. Each pixel is ∼25 × 25 cm and the wavelength of the camera is 8–12 μm. (Courtesy of D. Woolf.)

air–sea rescue operations providing a nighttime capability for detecting warm objects such as a life raft or survivors.

These instruments use a focal plane array (FPA) detector (a matrix of individual detectors, e.g., 256 × 256) located at the focal point of incoming radiance together with a charge couple device that is used to 'read' the FPA. This type of 'staring array' system generates a two-dimensional image either by using a mechanical scanning system (in the case of a small FPA array) or as an instantaneous image. Larger FPA arrays are much more power efficient, lighter and smaller than more elaborate mechanical scanning systems. Rapid image acquisition (>15 frames s⁻¹) is typical of these instruments that are available in a wide variety of spectral configurations and a typical accuracy of ∼ ±0.1 K. However, considerable problems are encountered when obtaining sky radiance data due to the difficulty of geometrically matching sea and sky radiance data.

The major problem with FPA detector technology is nonuniformity between FPA detector elements and drifts in detector gain and bias. Many innovative self-calibration methods which range in quality are used to correct for these problems. For example, a small heated calibration plate assumed to be at an isothermal temperature is periodically viewed by the detector to provide an absolute calibration. However, further development of this technology will

eventually provide extremely versatile instrumentation for the investigation of fine-scale sea surface emission.

Future Direction and Conclusions

In the last 10 years, considerable progress has been made in the development and application of IR sensors to study the air–sea interface. The continued development and use of FTIR sensors will provide the capability to accurately investigate the spectral characteristics of the sea surface in order to optimize the spectral intervals used by space sensors to determine SSST. It can be expected that in the near future, new algorithms will emerge for the direct measurement of the air–sea heat flux using multispectral sounding techniques and the accurate *in situ* determination of sea surface emissivity. Although still in their infancy, the development and use of thermal cameras will provide valuable insight into the fine-resolution two-dimensional spatial and temporal variability of the ocean surface. These data will be useful in developing and understanding the sampling limitations of large footprint satellite sensors and in the refinement of validation protocols. Finally, as satellite radiometers are now providing consistent and accurate observations of the SSST (e.g., ATSR), there is a need for autonomous operational *in situ* radiometer systems for ongoing validation of their data. Such intelligent systems that are extremely robust against the harsh realities of the marine environment are currently being developed.

See also

Air–Sea Gas Exchange. Heat and Momentum Fluxes at the Sea Surface. Radiative Transfer in the Ocean. Satellite Remote Sensing of Sea Surface Temperatures.

Further Reading

Bertie JE and Lan ZD (1996) Infrared intensities of liquids: the intensity of the OH stretching band revisited, and the best current values of the optical constants H_2O (1) at 25°C between 15,000 and 1 cm^{-1}. *Applied Spectroscopy* 50: 1047–1057.

Donlon CJ, Keogh SJ, Baldwin DJ, *et al.* (1998) Solid state measurements of sea surface skin temperature. *Journal of Atmospherical Oceanic Technology* 15: 775–787.

Donlon CJ and Nightingale TJ (2000) The effect of atmospheric radiance errors in radiometric sea surface skin temperature measurements. *Applied Optics* 39: 2392–2397.

Jessup AT, Zappa CJ, and Yeh H (1997) Defining and quantifying micro-scale wave breaking with infrared imagery. *Journal of Geophysical Research* 102: 23145–23153.

McKeown W and Asher W (1997) A radiometric method to measure the concentration boundary layer thickness at an air–water interface. *Journal of Atmospheric and Oceanic Technology* 14: 1494–1501.

Shaw JA (1999) Degree of polarisation in spectral radiances from water viewing infrared radiometers. *Applied Optics* 15: 3157–3165.

Smith WL, Knuteson RO, Rivercombe HH, *et al.* (1996) Observations of the infrared radiative properties of the ocean – implications for the measurement of sea surface temperature via satellite remote sensing. *Bulletin of the American Meteorological Society* 77: 41–51.

Suarez MJ, Emery WJ, and Wick GA (1997) The multi-channel infrared sea truth radiometric calibrator (MISTRC). *Journal of Atmospheric and Oceanic Technology* 14: 243–253.

Thomas JP, Knight RJ, Roscoe HK, Turner J, and Symon C (1995) An evaluation of a self-calibrating infrared radiometer for measuring sea surface temperature. *Journal of Atmospheric and Oceanic Technology* 12: 301–316.

MANNED SUBMERSIBLES, SHALLOW WATER

T. Askew, Harbor Branch Oceanographic
Institute, Ft Pierce, FL, USA

Introduction

Early man's insatiable curiosity to look beneath the surface of the sea in search of natural treasures that were useful in a primitive, comfortless mode of living were a true test of his limits of endurance. The fragile vehicle of the human body quickly discovered that most of the sea's depths were unapproachable without some form of protection against the destructive hostilities of the ocean.

Modern technology has paved the way for man to conquer the hostile marine environment by creating a host of manned undersea vehicles. Called submersibles, these small engineering marvels carry out missions of science, exploration, and engineering. The ability to conduct science and other operations under the sea rather than from the surface has stimulated the submersible builder/operator to further develop the specialized tools and instruments which provide humans with the opportunity to be present and perform tasks in relative comfort in ocean bottom locations that would otherwise be destructive to human life.

Over the past fifty years a depth of 1000 m has surfaced as the transition point for shallow vs deep water manned submersibles. During the prior 100 years any device that enabled man to explore the ocean depths beyond breath-holding capabilities would have been considered deep.

History

While it is difficult to pinpoint the advent of the first submersible it is thought that in 1620 Cornelius van Drebel constructed a vehicle under contract to King James I of England. It was operated by 12 rowers with leather sleeves, waterproofing the oar ports. It is said that the craft navigated the Thames River for several hours at a depth of 4 m and carried a secret substance that purified the air, perhaps soda lime?

In 1707, Dr Edmund Halley built a diving bell with a 'lock-out' capability. It had glass ports above to provide light, provisions for replenishing its air, and crude umbilical-supplied diving helmets which permitted divers to walk around outside. In 1776, Dr David Bushnell built and navigated the first submarine employed in war-like operations. Bushnell's Turtle was built of wood, egg-shaped with a conning tower on top and propelled horizontally and vertically by a primitive form of screw propeller after flooding a tank which allowed it to submerge.

In the early 1800s, Robert Fulton, inventor of the steamship, built two iron-formed copper-clad submarines, *Nautilus* and *Mute*. Both vehicles carried out successful tests, but were never used operationally.

The first 'modern submersible' was Simon Lake's *Argonaut I*, a small vehicle with wheels and a bottom hatch that could be opened after the interior was pressurized to ambient. While there are numerous other early submarines, the manned submersible did not emerge as a useful and functional means of accomplishing underwater work until the early 1960s.

It was during this same period of time that the French-built *Soucoupe*, sometimes referred to as the 'diving saucer', came into being. Made famous by Jacques Cousteau on his weekly television series, the *Soucoupe* is credited with introducing the general population to underwater science. Launched in 1959, the diving saucer was able to dive to 350 m.

The *USS Thresher* tragedy in 1963 appears to have spurred a movement among several large corporations such as General Motors (Deep Ocean Work Boat, DOWB), General Dynamics (*Star I, II, III, Sea Cliff*, and *Turtle*), Westinghouse (*Deepstar 2000, 4000*), and General Mills (*Alvin*), along with numerous other start-up companies formed solely to manufacture submersibles. Perry Submarine Builders, a Florida-based company, started manufacturing small shallow-water, three-person submersibles in 1962, and continued until 1980 (**Figures 1** and **2**). International Hydrodynamics Ltd (based in Vancouver, BC, Canada) commenced building the Pisces series of submersibles in 1962. The pressure hull material in the 1960s was for the most part steel with one or more view ports. The operating depth ranged from 30 m to 600 m, which was considered very deep for a free-swimming, untethered vehicle.

The US Navy began design work on *Deep Jeep* in 1960 and after 4 years of trials and tribulations it was commissioned with a design depth of 609 m. A two-person vehicle, *Deep Jeep* included many features incorporated in today's submersibles, such as a dropable battery pod, electric propulsion motors that operate in silicone oil-filled housings, and shaped resin blocks filled with glass micro-balloons

Figure 1 Perry-Link *Deep Diver*, 1967. Owned by Harbor Branch Oceanographic Institution. Length 6.7 m, beam 1.5 m, height 2.6 m, weight 7485 kg, crew 1, observers 2, duration 3–5 hours.

used to create buoyancy. *Deep Jeep* was eventually transferred to the Scripps Institution of Oceanography in 1966 after a stint searching for a lost 'H' bomb off Palomares, Spain. Unfortunately, *Deep Jeep* was never placed into service as a scientific submersible due to a lack of funding. The missing bomb was actually found by another vehicle, *Alvin*. *Alvin* did get funding and proved to be useful as a scientific tool.

The Nekton series of small two-person submersibles appeared in 1968, 1970, and 1971. The *Alpha*, *Beta*, and *Gamma* were the brainchildren of Doug Privitt, who started building small submersibles for recreation in the 1950s. The Nektons had a depth capability of 304 m. The tiny submersibles conducted hundreds of dives for scientific purposes as well as for military and oilfield customers. In 1982, the *Nekton Delta*, a slightly larger submersible with a depth rating of 365 m was

unveiled and is still operating today with well over 3000 dives logged.

A few of the submersibles were designed with a diver 'lock-out' capability. The first modern vehicle was the Perry-Link 'Deep Diver' built in 1967 and able to dive to 366 m. This feature enabled a separate compartment carrying divers to be pressurized internally to the same depth as outside, thus allowing the occupants to open a hatch and exit where they could perform various tasks while under the supervision of the pilots. Once the work was completed, the divers would re-enter the diving compartment, closing the outer and inner hatches; thereby maintaining the bottom depth until reaching the surface, where they could decompress either by remaining in the compartment or transferring into a larger, more comfortable decompression chamber via a transfer trunk.

Acrylic plastic was tested for the first time as a new material for pressure hulls in 1966 by the US Navy.

Figure 2   Perry PC 1204 *Clelia*, 1976. Owned and operated by Harbor Branch Oceanographic Institution. Length 6.7 m, beam 2.4 m, height 2.4 m, weight 8160 kg, crew 1, observers 2, duration 3–5 hours.

The *Hikino*, a unique submersible that incorporated a 142 cm diameter and a 0.635 cm thick hull, was only able to dive to 6 m. This experimental vehicle was used to gain experience with plastic hulls, which eventually led to the development of *Kumukahi*, *Nemo* (Naval Experimental Manned Observatory), *Makakai*, and *Johnson-Sea-Link*.

The *Kumukahi*, launched in 1969, incorporated a unique 135 cm acrylic plastic sphere formed in four sections. It was 3.175 cm thick and could dive to 92 m.

The *Nemo*, launched in 1970, and *Makakai*, launched in 1971, both utilized spheres made of 12 curved pentagons formed from a 6.35 cm flat sheet of Plexiglas™. The pentagons were bonded together to make one large sphere capable of diving to 183 m.

The *Johnson-Sea-Link*, designed by Edwin Link and built by Aluminum Company of America (ALCOA), utilized a *Nemo*-style Plexiglas™ sphere, 167.64 cm in diameter, 10.16 cm thick, and made of 12 curved pentagons formed from flat sheet and bonded together. This new thicker hull had an operational depth of 304 m.

Present Day Submersibles

The submersibles currently in use today are for the most part classified as either shallow-water or deep-water vehicles, the discriminating depth being approximately 1000 m. This is where the practicality of using compressed gases for ballasting becomes

impractical. The deeper diving vehicles utilize various drop weight methods; most use two sets of weights (usually scrap steel cut into uniform blocks). One set of weights is released upon reaching the bottom, allowing the vehicle to maneuver, travel, and perform tasks in a neutral condition. The other set of weights is dropped to make the vehicle buoyant, which carries it back to the surface once the dive is complete.

The shallow vehicles use their thrusters and/or water ballast to descend to the bottom and some of the more sophisticated submersibles have variable ballast systems which allow the pilot to achieve a neutral condition by varying the water level in a pressure tank.

One advantage of the shallow vehicles is that the view ports (commonly called windows) can be much larger both in size and numbers, and where an acrylic plastic sphere is used the entire hull becomes a window.

Since the late 1960s and early 1970s acrylic plastic pressure hulls have emerged as an ideal engineering solution to create a strong, transparent, corrosion-resistant, nonmagnetic pressure hull. The limiting factor of the acrylic sphere is its ability to resist implosion from external pressure at great depths. Its strength comes from the shape and wall thickness. Therefore, the greater the depth the operator aims to reach, the thicker the sphere must be, which results in a hull that is much too heavy to be practical for use on a small submersible designed to go deeper than 1000 m.

These shallow-water submersibles, once quite numerous because of their usefulness in the offshore oilfield industry, are now limited to a few operators and mostly used for scientific investigations.

The *Johnson-Sea-Links* (J-S-Ls) stand out as two of the most advanced manned submersibles (**Figure 3**). *J-S-L I*, commissioned by the Smithsonian Institution in January 1971, was named for designer and donors Edwin A. Link and J. Seward Johnson.

Edwin Link, responsible for the submersible's unique design and noted for his inventions in the aviation field, turned his energies to solving the problems of undersea diving, a technology then still in its infancy. One of his objectives was to carry out scientific work under water for lengthy periods.

The *Johnson-Sea-Link* was the most sophisticated diving craft he had created for this purpose, and it promised to be one of the most effective of the new generation of small submersible vehicles that were being built to penetrate the shallow depths of the continental shelf (183 m or 100 fathoms).

Originally designed for a depth of 304 m, the vessel's unique features include a two-person transparent acrylic sphere, 1.82 m in diameter and 10.16 cm thick, that provides panoramic underwater visibility to a pilot and a scientist/observer. Behind the sphere, there is a separate 2.4 m long cylindrical, welded, aluminum alloy lock-out/lock-in compartment that will enable scientists to exit from its bottom and collect specimens of undersea flora and fauna. The acrylic sphere and the aluminum cylinder are enclosed within a simple jointed aluminum tubular frame, a configuration that makes the vessel resemble a helicopter rather than a conventionally shaped submarine. Attached to the frame are the vessel's ballast tanks, thrusters, compressed air, mixed gas flasks, and battery pod.

The aluminum alloy parts of the submersible, lightweight and strong, along with the acrylic capsule which was patterned after the prototype used by the US Navy on the *Nemo*, had extraordinary advantages over traditional materials like steel. They were most of all immune to the corrosive effects of sea water.

The emphasis in engineering of the submersible was on safety. Switches, connectors, and all operating gear were especially designed to avoid possible safety hazards. The rear diving compartment allows one diver to exit for scientific collections while tethered for communications and breathing air supply, while the other diver/tender remains inside as a safety backup. Once the dive is completed and the submersible is recovered by a special deck-mounted crane on the support ship, the divers can transfer into a larger decompression chamber via a transfer trunk which is bolted to the lock-out/lock-in compartment.

Now, 30 years later, the *Johnson-Sea-Links* with a 904 m depth rating, remain state of the art underwater vehicles. Sophisticated hydraulic manipulators work in conjunction with a rotating bin collection platform which allow 12 separate locations to be sampled and simultaneously documented by digital color video cameras mounted on electric pan and tilt mechanisms and aimed with lasers. Illumination is provided by a variety of underwater lighting systems utilizing zenon arc lamps, metal halide, and halogen bulbs. Acoustic beacons provide real time position

Figure 3 *Johnson-Sea-Link I* and *II*, 1971 and 1976. Owned and operated by Harbor Branch Oceanographic Institution. Length 7.2 m, beam 2.5 m, height 3.1 m, weight 10400 kg, crew 2, observers 2, duration 3–5 hours.

and depth information to shipboard computer tracking systems that not only show the submersible's position on the bottom, but also its relationship to the ship in latitude and longitude via the satellite-based global positioning system (GPS). The lock-out/lock-in compartment is now utilized as an observation and instrumentation compartment, which remains at one atmosphere.

Today's shallow-water submersibles (average dive 3–5 h) require a support vessel to provide the necessities that are not available due to their relatively small size. The batteries must be charged, compressed air and oxygen flasks must be replenished. Carbon dioxide removal material, usually soda lime or lithium hydroxide, is also replenished so as to provide maximum life support in case of trouble. Most submersibles today carry 5 days of life support, which allows time to effect a rescue should it become necessary.

The support vessel also must have a launch/recovery system capable of safely handling the submersible in all sea conditions. Over the last 30 years, the highly trained crews that operate the ship's handling systems and the submersibles, have virtually made the shallow-water submersibles an everyday scientific tool where the laboratory becomes the ocean bottom.

Operations

The two *Johnson-Sea-Links* have accumulated over 8000 dives for science, engineering, archaeology, and training purposes since 1971. They have developed into highly sophisticated science tools. Literally thousands of new species of marine life have been photographed, documented by video camera and collected without disturbing the surrounding habitat. Behavioral studies of fish, marine mammals and invertebrates as well as sampling of the water column and bottom areas for chemical analysis and geological studies are everyday tasks for the submersibles. In addition, numerous historical shipwrecks from galleons to warships like the *USS Monitor* have been explored and documented, preserving their legacy for future generations.

Johnson-Sea-Links I and *II* (J-S-Ls) were pressed into service to assist in locating, identifying, and ultimately recovering many key pieces of the ill-fated Space Shuttle Challenger. This disaster, viewed by the world via television, added a new dimension to the J-S-Ls' capabilities. Previously only known for their pioneering efforts in marine science, they proved to be valuable assets in the search and recovery operation. The J-S-Ls completed a total of 109 dives, including mapping a large area of the right solid rocket booster debris at a depth of 365 m. The vehicles proved their worth throughout the operation by consistently performing beyond expectations. They were launched and recovered easily and quickly. They could work on several contacts per day, taking NASA engineers to the wreckage for first-hand detailed examination of debris while video cameras recorded what was being seen and said. Significant pieces were rigged with lifting bridles for recovery. The autonomous operation of the J-S-Ls, a dedicated support vessel, and highly trained operations personnel made for a successful conclusion to an operation that had a significant impact on the future of the US Space Program.

Summary

There is no question that the manned submersible has earned its place in history. Much of what are now cataloged as new species were discovered in the last 30 years with the aid of submersibles. The ability to conduct marine science experiments *in situ* led to the development of intricate precision instruments, sampling devices for delicate invertebrates and gelatinous organisms that previously were only seen in blobs or pieces due to the primitive methods used to collect them.

While some suggest that remotely operated vehicles (ROVs) could, and have replaced the manned submersible, in reality they are complementary. There is no substitute for the autonomous, highly maneuverable submersible that can approach and collect without contact delicate zooplankton, while observing behavior and measuring the levels of bioluminescence, or probing brine pools and cold seep regions in the Gulf of Mexico for specialized collections of biological, geological, and geochemical samples. Tubeworms are routinely marked for growth rate studies and collected individually, along with other biological species that thrive in these chemosynthetic communities. Sediments and methane ice (gas hydrates) are also selectively retrieved for later analysis.

Some new vehicles are still being produced, but have limited payloads, which restricts them to specific tasks such as underwater camera platforms or observation. Some are easily transportable but are small and restricted to one occupant; they can be carried by smaller support vessels and are more economical to operate. Man's desire to explore the lakes, oceans, and seas has not diminished. New technology will only enable, not reduce, the need for man's presence in these hostile environments.

Further Reading

Askew TM (1980) *JOHNSON-SEA-LINK Operations Manual*. Fort Pierce: Harbor Branch Foundation.

Busby F (1976, 1981) *Undersea Vechicles Directory*. Arlington: Busby & Associates.

Forman WR (1968) *KUMUKAHI Design and Operations Manual*. Makapuu, HI: Oceanic Institute.

Forman W (1999) *The History of American Deep Submersible Operations*. Flagstaff, AZ: Best Publishing Co.

Link MC (1973) *Windows in the Sea*. Washington, DC: Smithsonian Institute Press.

Stachiw JD (1986) *The Origins of Acrylic Plastic Submersibles*. American Society of Mechanical Engineers, Asme Paper 86-WA/HH-5.

Van Hoek S and Link MC (1993) *From Sky to Sea; A Story of Ed Link*. Flagstaff, AZ: Best Publishing Co.

AIRCRAFT REMOTE SENSING

L. W. Harding, Jr and W. D. Miller, University of
Maryland, College Park, MD, USA
R. N. Swift, and C. W. Wright, NASA Goddard
Space Flight Center, Wallops Island, VA, USA

Introduction

The use of aircraft for remote sensing has steadily
grown since the beginnings of aviation in the early
twentieth century and today there are many applications in the Earth sciences. A diverse set of remote
sensing uses in oceanography developed in parallel
with advances in aviation, following increased aircraft capabilities and the development of instrumentation for studying ocean properties. Aircraft
improvements include a greatly expanded range of
operational altitudes, development of the Global
Positioning System (GPS) enabling precision navigation, increased availability of power for instruments, and longer range and duration of missions.
Instrumentation developments include new sensor
technologies made possible by microelectronics,
small, high-speed computers, improved optics, and
increased accuracy of digital conversion of electronic
signals. Advances in these areas have contributed
significantly to the maturation of aircraft remote
sensing as an oceanographic tool.

Many different types of aircraft are currently used
for remote sensing of the oceans, ranging from balloons to helicopters, and from light, single engine
piston-powered airplanes to jets. The data and information collected on these platforms are commonly used to enhance sampling by traditional
oceanographic methods, giving increased spatial and
temporal resolution for a number of important
properties. Contemporary applications of aircraft
remote sensing to oceanography can be grouped into
several areas, among them ocean color, sea surface
temperature (SST), sea surface salinity (SSS), wave
properties, near-shore topography, and bathymetry.
Prominent examples include thermal mapping using
infrared (IR) sensors in both coastal and open ocean
studies, lidar and visible radiometers for ocean color
measurements of phytoplankton distributions and
'algal blooms', and passive microwave sensors to
make observations of surface salinity structure of
estuarine plumes in the coastal ocean. These topics
will be discussed in more detail in subsequent

sections. Other important uses of aircraft remote
sensing are to test instruments slated for deployment
on satellites, to calibrate and validate space-based
sensors using aircraft-borne counterparts, and to
make 'under-flights' of satellite instruments and assess the efficacy of atmospheric corrections applied
to data from space-based observations.

Aircraft have some advantages over satellites for
oceanography, including the ability to gather data
under cloud cover, high spatial resolution, flexibility
of operations that enables rapid responses to 'events',
and less influence of atmospheric effects that complicate the processing of satellite data. Aircraft remote sensing provides nearly synoptic data and
information on important oceanographic properties
at higher spatial resolution than can be achieved by
most satellite-borne instruments. Perhaps the greatest advantage of aircraft remote sensing is the ability
to provide consistent, high-resolution coverage at
larger spatial scales and more frequent intervals than
are practical with ships, making it feasible to use
aircraft for monitoring change.

Disadvantages of aircraft remote sensing include
the relatively limited spatial coverage that can be
obtained compared with the global coverage available from satellite instruments, the repeated expense
of deploying multiple flights, weather restrictions on
operations, and lack of synopticity over large scales.
Combination of the large-scale, synoptic data that are
accessible from space with higher resolution aircraft
surveys of specific locations is increasingly recognized
as an important and useful marriage that takes advantages of the strengths of both approaches.

This article begins with a discussion of sensors that
use lasers (also called active sensors), including airborne laser fluorosensors that have been used to
measure chlorophyll (chl-a) and other properties;
continues with discussions of lidar sensors used for
topographic and bathymetric mapping; describes
passive (sensors that do not transmit or illuminate,
but view naturally occurring reflections and emissions) ocean color remote sensing directed at quantifying phytoplankton biomass and productivity;
moves to available or planned hyper-spectral aircraft
instruments; briefly describes synthetic aperture
radar applications for waves and wind, and closes
with a discussion of passive microwave measurements of salinity. Readers are directed to the Further
Reading section if they desire additional information
on individual topics.

Active Systems

Airborne Laser Fluorosensing

The concentrations of certain waterborne constituents, such as chl-a, can be measured from their fluorescence, a relationship that is exploited in shipboard sensors such as standard fluorometers and flow cytometers that are discussed elsewhere in this encyclopedia. NASA first demonstrated the measurement of laser-induced chl-a fluorescence from a low-flying aircraft in the mid-1970s. Airborne laser fluorosensors were developed shortly thereafter in the USA, Canada, Germany, Italy, and Russia, and used for measuring laser-induced fluorescence of a number of marine constituents in addition to chl-a. Oceanic constituents amenable to laser fluorosensing include phycoerythrin (photosynthetic pigment in some phytoplankton taxa), chromophoric dissolved organic matter (CDOM), and oil films. Airborne laser fluorosensors have also been used to follow dyes such as fluorescein and rhodamine that are introduced into water masses to trace their movement.

The NASA Airborne Oceanographic Lidar (AOL) is the most advanced airborne laser fluorosensor. The transmitter portion features a dichroic optical device to spatially separate the temporally concurrent 355 and 532 nm pulsed-laser radiation, followed by individual steering mirrors to direct the separated beams to respective oceanic targets separated by ~1 m when flown at the AOL's nominal 150 m operational altitude. The receiver focal plane containing both laser-illuminated targets is focused onto the input slits of the monochromator. The monochromator output focal planes are viewed by custom-made optical fibers that transport signal photons from the focal planes to the photo-cathode of each photo-multiplier module (PMM) where the conversion from photons to electrons takes place with a substantial gain. Time-resolved waveforms are collected in channels centered at 404 nm (water Raman) and 450 nm (CDOM) from 355 nm laser excitation, and at 560 nm and 590 nm (phycoerythrin), 650 (water Raman), and 685 nm (chl-a) from the 532 nm laser excitation. The water Raman from the respective lasers is the red-shifted emission from the OH^- bonds of water molecules resulting from radiation with the laser pulse. The strength of the water Raman signal is directly proportional to the number of OH^- molecules accessed by the laser pulse. Thus, the water Raman signal is used to normalize the fluorescence signals to correct for variations in water attenuation properties in the surface layer of the ocean. The AOL is described in more detail on http://lidar.wff.nasa.gov.

The AOL has supported major oceanographic studies throughout the 1980s and 1990s, extending the usefulness of shipboard measurements over wide areas to permit improved interpretation of the ship-derived results. Examples of data from the AOL show horizontal structure of laser-induced fluorescence converted to chl-a concentration (**Figure 1**). Prominent oceanographic expeditions that have benefited from aircraft coverage with the AOL include the North Atlantic Bloom Experiment (NABE) of the Joint Global Ocean Flux Study (JGOFS), and the Iron Enrichment Experiment (IRONEX) of the Equatorial Pacific near the Galapagos Islands. This system has been flown on NASA P-3B aircraft in open-ocean missions that often exceeded 6 h in duration and collected hundreds of thousands of spectra. Each 'experiment' is able to generate both active and passive data in 'pairs' that are used for determining ocean color and recovering chl-a and other constituents, and that are also useful in the development of algorithms for measuring these constituents from oceanic radiance spectra.

Airborne Lidar Coastal Mapping

The use of airborne lidar (light detection and ranging) sensors for meeting coastal mapping requirements is a relatively new and promising application of laser-ranging technology. These applications include high-density surveying of coastline and beach morphology and shallow water bathymetry. The capability to measure distance accurately with lidar sensors has been available since the early 1970s, but their application to airborne surveying of terrestrial features was seriously hampered by the lack of knowledge of the position of the aircraft from which the measurement was made. The implementation of the Department of Defense GPS constellation of satellites in the late 1980s, coupled with the development of GPS receiver technology, has resulted in the capability to provide the position of a GPS antenna located on an aircraft fuselage in flight to an accuracy approaching 5 cm using kinematic differential methodology. These methods involve the use of a fixed receiver (generally located at the staging airport) and a mobile receiver that is fixed to the aircraft fuselage. The distance between mobile and fixed receivers, referred to as the baseline, is typically on the order of tens of kilometers, and can be extended to hundreds of kilometers by using dual frequency survey grade GPS receivers aided by tracking the phase code of the carrier from each frequency.

Modern airborne lidars are capable of acquiring 5000 or more discrete range measurements per second. Aircraft attitude and heading information are

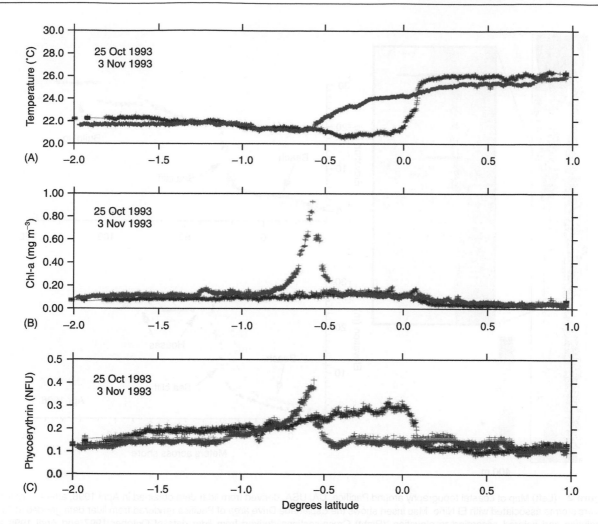

Figure 1 Cross-section profiles flown across a large oceanic front west of the Galapagos Islands on 25 October and 3 November 1993. (A) A dramatic horizontal displacement of the front as measured with an infrared radiometer; (B) and (C) show corresponding changes in laser-induced fluorescence of chl-a and phycoerythrin. This flight was made as part of the original IRONEX investigation in late 1993. NFU, normalized fluorescence units.

used along with the GPS-determined platform position to locate the position of the laser pulse on the Earth's surface to a vertical accuracy approaching 10 cm with some highly accurate systems and <30 cm for most of these sensors. The horizontal accuracy is generally 50–100 cm. Depending on the pulse repetition rate of the laser transmitter, the off-nadir pointing angle, and the speed of the aircraft platform, the density of survey points can exceed one sample per square meter.

NASA's Airborne Topographic Mapper (ATM) is an example of a topographic mapping lidar used for coastal surveying applications. An example of data from ATM shows shoreline features off the east coast of the USA (**Figure 2**). ATM was originally developed to measure changes in the elevation of Arctic ice sheets in response to global warming. The sensor was applied to measurements of changes in coastal morphology beginning in 1995. Presently, baseline

topographic surveys exist for most of the Atlantic and Gulf coasts between central Maine and Texas and for large sections of the Pacific coast. Affected sections of coastline are re-occupied following major coastal storms, such as hurricanes and 'Nor'easters', to determine the extent of erosion and depositional patterns resulting from the storms. Additional details on the ATM and some results of investigations on coastal morphology can be found on websites (http://lidar.wff.nasa.gov and http://aol.wff.nasa.gov/aoltm/projects/beachmap/98results/).

Other airborne lidar systems have been used to survey coastal morphology, including Optec lidar systems by Florida State University and the University of Texas at Austin. Beyond these airborne lidar sensors, there are considerably more instruments with this capability that are currently in use in the commercial sector for surveying metropolitan areas, flood plains, and for other terrestrial

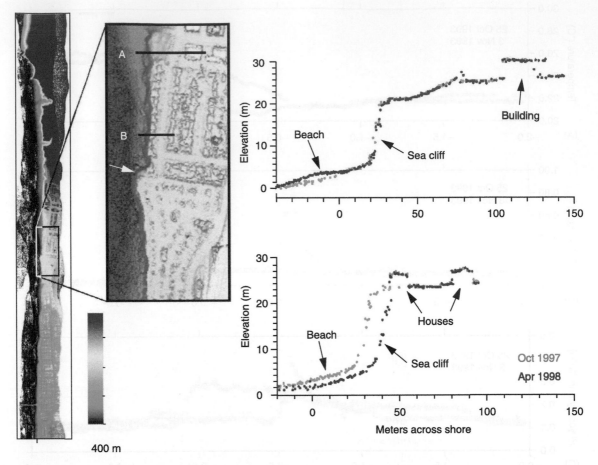

Figure 2   (Left) Map of coastal topography around Pacifica, CA, USA, derived from lidar data obtained in April 1998, after a winter of severe storms associated with El Niño. Map insert shows the Esplanade Drive area of Pacifica rendered from lidar data gridded at 2 m resolution and colored according to elevation. (Right) Cross-sections derived from lidar data of October 1997 and April 1998 at locations marked in the map inset. The profiles in (A) show a stable cliff and accreting beach, whereas about 200 m to the south the profiles in (B) show erosion of the sea cliff and adjacent beach resulting in undermining of houses. Each profile shows individual laser spot elevations that fall within a 2 m wide strip oriented approximately normal to shoreline. (Reproduced with permission from Sallenger *et al.*, 1999.)

applications. At the last count (early 2000) there were approximately 60 airborne lidars in operation worldwide, with most engaged in a variety of survey applications generally outside the field of coastal mapping.

Pump and Probe Fluorometry

Several sections in this chapter describe recoveries of phytoplankton biomass as chl-a by active and passive measurements. Another recent accomplishment is an active, airborne laser measurement intended to aid in remote detection of photosynthetic performance, an important ingredient of primary productivity computations. Fluorometric techniques, such as fast repetition rate (FRR) fluorometry (explained in another article of this encyclopedia), have provided an alternative approach to ^{14}C assimilation and O_2 evolution in the measurement of primary productivity. This

technology has matured with the commercial availability of FRR instruments that can give vertical profiles or operate in a continuous mode while underway. There have been several attempts to develop airborne lidar instruments to determine phytoplankton photosynthetic characteristics from aircraft in the past decade. NASA scientists have deployed a pump and probe fluorometer from aircraft, wherein the AOL laser (described above) acts as the pump and a second laser with variable power options and rapid pulsing capabilities (10 ns) functions as the probe.

Passive Systems

Multichannel Ocean Color Sensor (MOCS)

Passive ocean color measurements using visible radiometers to measure reflected natural sunlight from the ocean have been made with a number of instruments in the past two decades. These instruments include the Multichannel Ocean Color Sensor (MOCS) that was flown in studies of Nantucket Shoals in the early 1980s, the passive sensors of the AOL suite that have been used in many locations around the world, and more recently, simple radiometers that have been deployed on light aircraft in regional studies of Chesapeake Bay (see below). MOCS was one of the earliest ocean color sensors used on aircraft. It provided mesoscale data on shelf and slope chl-a in conjunction with shipboard studies of physical structure, nutrient inputs, and phytoplankton primary productivity.

Ocean Data Acquisition System (ODAS) and SeaWiFS Aircraft Simulator (SAS)

Few aircraft studies have obtained long time-series sufficient to quantify variability and detect secular trends. An example is ocean color measurements made from light aircraft in the Chesapeake Bay region for over a decade, providing data on chl-a and SST from >250 flights. Aircraft over-flights of the Bay using the Ocean Data Acquisition System (ODAS) developed at NASA's Goddard Space Flight Center commenced in 1989. ODAS was a nadir-viewing, line-of-flight, three-band radiometer with spectral coverage in the blue-green region of the visible spectrum (460–520 nm), a narrow $1.5°$ field-of-view, and a 10 Hz sampling rate. The ODAS instrument package included an IR temperature sensor (PRT-5, Pyrometrics, Inc.) for measuring SST. The system was flown for ∼7 years over Chesapeake Bay on a regular set of tracks to determine chl-a and SST. Over 150 flights were made with ODAS between 1989 and 1996, coordinated with *in situ* observations from a multi-jurisdictional monitoring program and other cruises of opportunity.

ODAS was flown together with the SeaWiFS Aircraft Simulator (SAS II, III, Satlantic, Inc., Halifax, Canada) beginning in 1995 and was retired soon thereafter and replaced with the SAS units. SAS III is a multi-spectral (13-band, 380–865 nm), line-of-flight, nadir viewing, 10 Hz, passive radiometer with a $3.5°$ field-of-view that has the same wavebands as the SeaWiFS satellite instrument, and several additional bands in the visible, near IR, and UV. The SAS systems include an IR temperature sensor (Heimann Instruments, Inc.). Chl-a estimates are obtained using a curvature algorithm applied to water-leaving radiances at wavebands in the blue-green portion of the visible spectrum with validation from concurrent shipboard measurements. Flights are conducted at ∼50–60 m s^{-1} (100–120 knots), giving an along-track profile with a resolution of 5–6 m averaged to

50 m in processing, and interpolated to 1 km^2 for visualization. Imagery derived from ODAS and SAS flights is available on a web site of the NOAA Chesapeake Bay Office for the main stem of the Bay (http://noaa.chesapeakebay.net), and for two contrasting tributaries, the Choptank and Patuxent Rivers on a web site of the Coastal Intensive Sites Network (CISNet) (http://www.cisnet-choptank.org).

Data from ODAS and SAS have provided detailed information on the timing, position, and magnitude of blooms in Chesapeake Bay, particularly the spring diatom bloom that dominates the annual phytoplankton cycle. This April–May peak of chl-a represents the largest accumulation of phytoplankton biomass in the Bay and is a proximal indicator of over-enrichment by nutrients. Data from SeaWiFS for spring 2000 show the coast-wide chl-a distribution for context, while SAS III data illustrate the high-resolution chl-a maps that are obtained regionally (**Figure 3**). A well-developed spring bloom corresponding to a year of relatively high freshwater flow from the Susquehanna River, the main tributary feeding the estuary, is apparent in the main stem Bay chl-a distribution. Estimates of primary productivity are now being derived from shipboard observations of key variables combined with high-resolution aircraft measurements of chl-a and SST for the Bay.

Hyper-spectral Systems

Airborne Visible/Infrared Imaging Spectrometer (AVIRIS)

The Airborne Visible/Infrared Imaging Spectrometer (AVIRIS) was originally designed in the late 1980s by NASA at the Jet Propulsion Laboratory (JPL) to collect data of high spectral and spatial resolution, anticipating a space-based high-resolution imaging spectrometer (HIRIS) that was planned for launch in the mid-1990s.

Because the sensor was designed to provide data similar to satellite data, flight specifications called for both high altitude and high speed. AVIRIS flies almost exclusively on a NASA ER-2 research aircraft at an altitude of 20 km and an airspeed of 732 km h^{-1}. At this altitude and a $30°$ field of view, the swath width is almost 11 km. The instantaneous field of view is 1 mrad, which creates individual pixels at a resolution of 20 m^2. The sensor samples in a whisk-broom fashion, so a mirror scans back and forth, perpendicular to the line-of-flight, at a rate of 12 times per second to provide continuous spatial coverage. Each pixel is then sent to four separate spectrometers by a fiber optic cable. The spectrometers are arranged so that they each cover a part of the

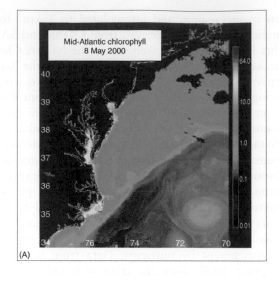

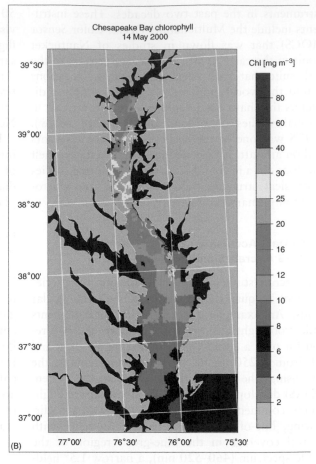

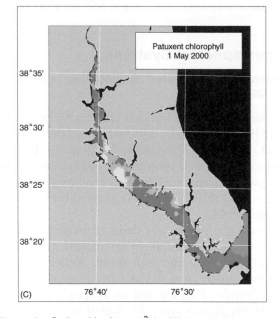

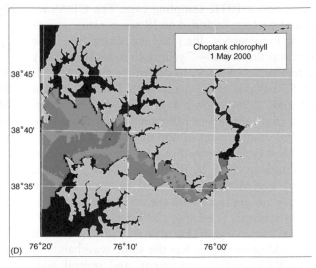

Figure 3 Spring chl-a (mg m⁻³) in: (A) the mid-Atlantic region from SeaWiFS; (B) Chesapeake Bay; (C) Patuxent R; (D) Choptank R from SAS III.

spectrum from 0.40 to 2.4 μm, providing continuous spectral coverage at 10 nm intervals over the entire spectrum from visible to near IR. Data are recorded to tape cassettes for storage until rectification, atmospheric correction, and processing at JPL. A typical AVIRIS 'scene' is a 40 min flight line. At ER-2 flight parameters, this creates an image roughly 500 km long and 11 km wide. Data are encoded at 12-bits for a high degree of discrimination. The physical dimensions of AVIRIS are quite large, 84 cm wide × 160 cm long × 117 cm tall at a weight of 720 pounds.

Data collected with AVIRIS have been used for terrestrial, marine, and atmospheric applications. Accomplishments of AVIRIS include separation of the chl-a signature from bottom reflectance for clear lake waters of Lake Tahoe and turbid waters near Tampa Bay, interpretation of spectral signals from resuspended sediment and dissolved organic materials in W. Florida, and of suspended sediment and kelp beds in S. California. Recent efforts have focused on improving atmospheric correction procedures for both AVIRIS and satellite data, providing inputs for bio-optical models which determine inherent optical properties (IOPs) from reflectance, algorithm development, and sporadic attempts at water quality monitoring (e.g., chl-a, suspended sediment, diffuse attenuation coefficient, k_d). AVIRIS data have recently been used as an input variable to a neural network model developed to estimate water depth. The model was able to separate the contributions of different components to the total water-leaving radiance and to provide relatively accurate estimates of depth (rms error $= 0.48$ m).

Compact Airborne Spectrographic Imager (CASI)

The Compact Airborne Spectrographic Imager (CASI) is a relatively small, lightweight hyper-spectral sensor that has been used on a variety of light aircraft. CASI was developed by Itres Research Ltd (Alberta, Canada) in 1988 and was designed for a variety of remote sensing applications in forestry, agriculture, land-use planning, and aquatic monitoring. By allowing user-defined configurations, the 12-bit, push-broom-type sensor (333 scan lines per second) using a charge-coupled detector (CCD) can be adapted to maximize either spatial (37.8° across track field of view, 0.077° along-track, 512 pixels – pixel size varies with altitude) or spectral resolution (288 bands at 1.9 nm intervals between 400 and 1000 nm). Experiments have been conducted using CASI to determine bottom type, benthic cover, submerged aquatic vegetation, marsh type, and in-water constituents such as suspended sediments, chl-a, and other algal pigments.

Portable Hyper-spectral Imager for Low-Light Spectroscopy (PHILLS)

The Portable Hyper-spectral Imager for Low-Light Spectroscopy (PHILLS) has been constructed by the US Navy (Naval Research Laboratory) for imaging the coastal ocean. PHILLS uses a backside-illuminated CCD for high sensitivity, and an all-reflective spectrograph with a convex grating in an Offner configuration to produce a distortion-free image. The instrument benefits from improvements in large-format detector arrays that have enabled increased spectral resolution and higher signal-to-noise ratios for imaging spectrographs, extending the use of this technology in low-albedo coastal waters. The ocean PHILLS operates in a push-broom scanned mode whereby cross-track ground pixels are imaged with a camera lens onto the entrance slit of the spectrometer, and new lines of the along-track ground pixels are attained by aircraft motion. The Navy's interest in hyper-spectral imagers for coastal applications centers on the development of methods for determining shallow water bathymetry, topography, bottom type composition, underwater hazards, and visibility. PHILLS precedes a planned hyper-spectral satellite instrument, the Coastal Ocean Imaging Spectrometer (COIS) that is planned to launch on the Naval Earth Map Observer (NEMO) spacecraft.

Radar Altimetry

Ocean applications of airborne radar altimetry systems include several sensors that retrieve information on wave properties. Two examples are the Radar Ocean Wave Spectrometer (ROWS), and the Scanning Radar Altimeter (SRA), systems designed to measure long-wave directional spectra and near-surface wind speed. ROWS is a K_u-band system developed at NASA's Goddard Space Flight Center in support of present and future satellite radar missions. Data obtained from ROWS in a spectrometer mode are used to derive two-dimensional ocean spectral wave estimates and directional radar backscatter. Data from the pulse-limited altimeter mode radar yield estimates of significant wave height and surface wind speed.

Synthetic Aperture Radar (SAR)

Synthetic Aperture Radar (SAR) systems emit microwave radiation in several bands and collect the reflected radiation to gain information about sea surface conditions. Synthetic aperture is a technique that is used to synthesize a long antenna by combining signals, or echoes, received by the radar as it moves along a flight track. Aperture refers to the opening that is used to collect reflected energy and form an image. The analogous feature of a camera to the aperture would be the shutter opening. A synthetic aperture is constructed by moving a real aperture or antenna through a series of positions along a flight track.

NASA's Jet Propulsion Laboratory and Ames Research Center have operated the Airborne SAR (AIRSAR) on a DC-8 since the late 1980s. The radar of AIRSAR illuminates the ocean at three microwave wavelengths: C-band (6 cm), L-band (24 cm), and P-band (68 cm). Brightness of the ocean (the amount of energy reflected back to the antenna) depends on the roughness of the surface at the length scale of the microwave (Bragg scattering). The primary source of roughness, and hence brightness, at the wavelengths used is capillary waves associated with wind. Oceanographic applications derive from the responsiveness of capillary wave amplitude to factors that affect surface tension, such as swell, atmospheric stability, and the presence of biological films. For example, the backscatter characteristics of the ocean are affected by surface oil and slicks can be observed in SAR imagery as a decrease of radar backscatter; SAR imagery appears dark in an area affected by an oil spill, surface slick, or biofilm, as compared with areas without these constituents.

Microwave Salinometers

Passive microwave radiometry (L-band) has been tested for the recovery of SSS from aircraft, and it may be possible to make these measurements from space. Salinity affects the natural emission of EM radiation from the ocean, and the microwave signature can be used to quantify SSS. Two examples of aircraft instruments that have been used to measure SSS in the coastal ocean are the Scanning Low-Frequency

Microwave Radiometer (SLFMR), and the Electronically Thinned Array Radiometer (ESTAR).

SLFMR was used recently in the estuarine plume of Chesapeake Bay on the east coast of the USA to follow the buoyant outflow that dominates the near-shore density structure and constitutes an important tracer of water mass movement. SLFMR is able to recover SSS at an accuracy of about 1 PSU (**Figure 4**). This resolution is too coarse for the open ocean, but is quite suitable for coastal applications where significant gradients occur in regions influenced by freshwater inputs. SLFMR has a bandwidth of 25 MHz, a frequency of 1.413 GHz, and a single antenna with a beam width of approximately 16° and six across-track positions at ±6°, ±22° and ±39°. Tests of SLFMR off the Chesapeake Bay demonstrated its effectiveness as a 'salinity mapper' by characterizing the trajectory of the Bay plume from surveys using light aircraft in joint operations with ships. Flights were conducted at an altitude of 2.6 km, giving a resolution of about 1 km. The accuracy of SSS in this example is ~0.5 PSU.

ESTAR is an aircraft instrument that is the prototype of a proposed space instrument for measuring SSS. This instrument relies on an interferometric technique termed 'aperture synthesis' in the across-track dimension that can reduce the size of the antenna aperture needed to monitor SSS from space. It has been described as a 'hybrid of a real and a synthetic aperture radiometer.' Aircraft surveys of

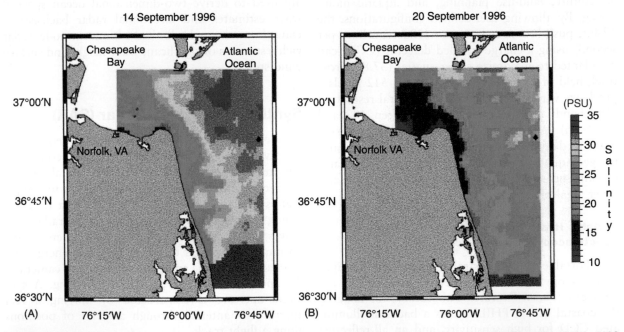

Figure 4 Sea surface salinity from an airborne microwave salinity instrument for (A) 14 September 1996; (B) 20 September 1996. Images reveal strong onshore-offshore gradients in salinity from the mouth of Chesapeake Bay to the plume and shelf, and the effect of high rainfall and freshwater input on the salinity distribution over a 1-week interval. (Adapted with permission from Miller *et al.*, 1998.)

SSS using ESTAR in the coastal current off Maryland and Delaware showed good agreement with thermosalinograph measurements from ships in the range of 29–31 PSU.

See also

IR Radiometers. Iron Fertilization. Satellite Oceanography, History, and Introductory Concepts. Satellite Remote Sensing: Ocean Color. Satellite Remote Sensing of Sea Surface Temperatures. Satellite Remote Sensing: Salinity Measurements. Upper Ocean Time and Space Variability.

Further Reading

Blume H-JC, Kendall BM, and Fedors JC (1978) Measurement of ocean temperature and salinity via microwave radiometry. *Boundary Layer Meteorology* 13: 295–308.

Campbell JW and Esaias WE (1985) Spatial patterns in temperature and chlorophyll on Nantucket Shoals from airborne remote sensing data, May 7–9, 1981. *Journal of Marine Research* 43: 139–161.

Harding LW Jr, Itsweire EC, and Esaias WE (1994) Estimates of phytoplankton biomass in the Chesapeake Bay from aircraft remote sensing of chlorophyll concentrations, 1989–92. *Remote Sensing Environment* 49: 41–56.

Le Vine DM, Zaitzeff JB, D'Sa EJ, *et al.* (2000) Sea surface salinity: toward an operational remote-sensing system. In: Halpern D (ed.) *Satellites, Oceanography and Society*, pp. 321–335. Elsevier Science.

Miller JL, Goodberlet MA, and Zaitzeff JB (1998) Airborne salinity mapper makes debut in coastal zone. *EOS Transactions of the American Geophysical Union* 79: 173–177.

Sallenger AH Jr, Krabill W, Brock J, *et al.* (1999) *EOS Transactions of the American Geophysical Union* 80: 89–93.

Sandidge JC and Holyer RJ (1998) Coastal bathymetry from hyper-spectral observations of water radiance. *Remote Sensing Environment* 65: 341–352.

SATELLITE OCEANOGRAPHY, HISTORY, AND INTRODUCTORY CONCEPTS

W. S. Wilson, NOAA/NESDIS, Silver Spring, MD, USA
E. J. Lindstrom, NASA Science Mission Directorate, Washington, DC, USA
J. R. Apel[†], Global Ocean Associates, Silver Spring, MD, USA

Published by Elsevier Ltd.

Oceanography from a satellite – the words themselves sound incongruous and, to a generation of scientists accustomed to Nansen bottles and reversing thermometers, the idea may seem absurd.

Gifford C. Ewing (1965)

Introduction: A Story of Two Communities

The history of oceanography from space is a story of the coming together of two communities – satellite remote sensing and traditional oceanography.

For over a century oceanographers have gone to sea in ships, learning how to sample beneath the surface, making detailed observations of the vertical distribution of properties. Gifford Ewing noted that oceanographers had been forced to consider "the class of problems that derive from the vertical distribution of properties at stations widely separated in space and time."

With the introduction of satellite remote sensing in the 1970s, traditional oceanographers were provided with a new tool to collect synoptic observations of conditions at or near the surface of the global ocean. Since that time, there has been dramatic progress; satellites are revolutionizing oceanography. (Appendix 1 provides a brief overview of the principles of satellite remote sensing.)

Yet much remains to be done. Traditional subsurface observations and satellite-derived observations of the sea surface – collected as an integrated set of observations and combined with state-of-the-art models – have the potential to yield estimates of the three-dimensional, time-varying distribution of properties for the global ocean. Neither a satellite nor an *in situ* observing system can do this on its own. Furthermore, if such observations can be collected over

the long term, they can provide oceanographers with an observational capability conceptually similar to that which meteorologists use on a daily basis to forecast atmospheric weather.

Our ability to understand and forecast oceanic variability, how the oceans and atmosphere interact, critically depends on an ability to observe the three-dimensional global oceans on a long-term basis. Indeed, the increasing recognition of the role of the ocean in weather and climate variability compels us to implement an integrated, operational satellite and *in situ* observing system for the ocean now – so that it may complement the system which already exists for the atmosphere.

The Early Era

The origins of satellite oceanography can be traced back to World War II – radar, photogrammetry, and the V-2 rocket. By the early 1960s, a few scientists had recognized the possibility of deriving useful

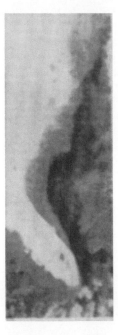

Figure 1 Thermal infrared image of the US southeast coast showing warmer waters of the Gulf Stream and cooler slope waters closer to shore taken in the early 1960s. While the resolution and accuracy of the TV on *Tiros* were not ideal, they were sufficient to convince oceanographers of the potential usefulness of infrared imagery. The advanced very high resolution radiometer (AVHRR) scanner (see text) has improved images considerably. Courtesy of NASA.

[†] Deceased

oceanic information from the existing aerial sensors. These included (1) the polar-orbiting meteorological satellites, especially in the 10–12-μm thermal infrared band; and (2) color photography taken by astronauts in the Mercury, Gemini, and Apollo manned spaceflight programs. Examples of the kinds of data obtained from the National Aeronautics and Space Administration (NASA) flights collected in the 1960s are shown in **Figures 1** and **2**.

Such early imagery held the promise of deriving interesting and useful oceanic information from space, and led to three important conferences on space oceanography during the same time period.

In 1964, NASA sponsored a conference at the Woods Hole Oceanographic Institution (WHOI) to examine the possibilities of conducting scientific research from space. The report from the conference, entitled *Oceanography from Space*, summarized findings to that time; it clearly helped to stimulate a number of NASA projects in ocean observations and sensor development. Moreover, with the exception of the synthetic aperture radar (SAR), all instruments flown through the 1980s used techniques described in this report. Dr. Ewing has since come to be justifiably regarded as the father of oceanography from space.

A second important step occurred in 1969 when the Williamstown Conference was held at Williams College in Massachusetts. The ensuing Kaula report set forth the possibilities for a space-based geodesy mission to determine the equipotential figure of the Earth using a combination of (1) accurate tracking of satellites and (2) the precision measurement of satellite elevation above the sea surface using radar altimeters. Dr. William Von Arx of WHOI realized the possibilities for determining large-scale oceanic currents with precision altimeters in space. The requirements for measurement precision of 10-cm height error in the elevation of the sea surface with respect to the geoid were articulated. NASA scientists and engineers felt that such accuracy could be achieved in the long run, and the agency initiated the Earth and Ocean Physics Applications Program, the first formal oceans-oriented program to be established within the organization. The required accuracy was not to be realized until 1992 with *TOPEX/Poseidon*, which was reached only over a 25-year

Figure 2 Color photograph of the North Carolina barrier islands taken during the Apollo-Soyuz Mission (AS9-20-3128). Capes Hatteras and Lookout, shoals, sediment- and chlorophyll-bearing flows emanating from the coastal inlets are visible, and to the right, the blue waters of the Gulf Stream. Cloud streets developing offshore the warm current suggest that a recent passage of a cold polar front has occurred, with elevated air–sea evaporative fluxes. Later instruments, such as the coastal zone color scanner (CZCS) on *Nimbus-7* and the SeaWiFS imager have advanced the state of the art considerably. Courtesy of NASA.

period of incremental progress that saw the flights of five US altimetric satellites of steadily increasing capabilities: *Skylab*, *Geos-3*, *Seasat*, *Geosat*, and *TOPEX/Poseidon* (see **Figure 3** for representative satellites).

A third conference, focused on sea surface topography from space, was convened by the National Oceanic and Atmospheric Administration (NOAA), NASA, and the US Navy in Miami in 1972, with 'sea surface topography' being defined as undulations of the ocean surface with scales ranging from

approximately 5000 km down to 1 cm. The conference identified several data requirements in oceanography that could be addressed with space-based radar and radiometers. These included determination of surface currents, Earth and ocean tides, the shape of the marine geoid, wind velocity, wave refraction patterns and spectra, and wave height. The conference established a broad scientific justification for space-based radar and microwave radiometers, and it helped to shape subsequent national programs in space oceanography.

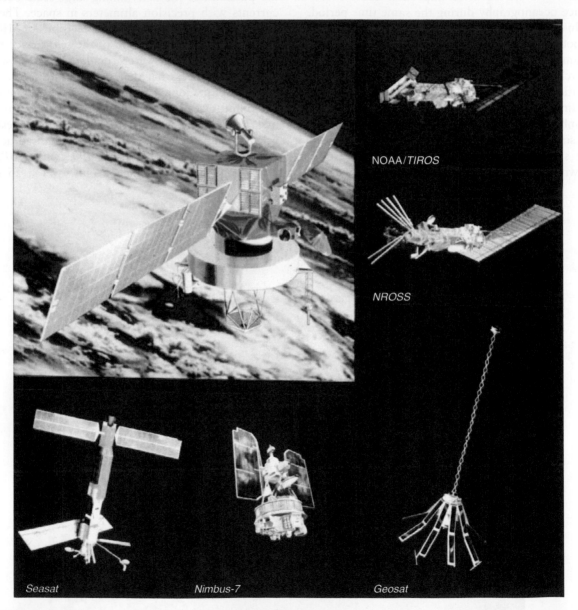

Figure 3 Some representative satellites: (1) *Seasat*, the first dedicated oceanographic satellite, was the first of three major launches in 1978; (2) the *Tiros* series of operational meteorological satellites carried the advanced very high resolution radiometer (AVHRR) surface temperature sensor; *Tiros-N*, the first of this series, was the second major launch in 1978; (3) *Nimbus-7*, carrying the CZCS color scanner, was the third major launch in 1978; (4) *NROSS*, an oceanographic satellite approved as an operational demonstration in 1985, was later cancelled; (5) *Geosat*, an operational altimetric satellite, was launched in 1985; and (6) this early version of *TOPEX* was reconfigured to include the French *Poseidon*; the joint mission *TOPEX/Poseidon* was launched in 1992. Courtesy of NASA.

The First Generation

Two first-generation ocean-viewing satellites, *Skylab* in 1973 and *Geos-3* in 1975, had partially responded to concepts resulting from the first two of these conferences. *Skylab* carried not only several astronauts, but a series of sensors that included the S-193, a radar-altimeter/wind-scatterometer, a long-wavelength microwave radiometer, a visible/infrared scanner, and cameras. S-193, the so-called Rad/Scatt, was advanced by Drs. Richard Moore and Willard Pierson. These scientists held that the scatterometer could return wind velocity measurements whose accuracy, density, and frequency would revolutionize marine meteorology. Later aircraft data gathered by NASA showed that there was merit to their assertions. *Skylab*'s scatterometer was damaged during the opening of the solar cell panels, and as a consequence returned indeterminate results (except for passage over a hurricane), but the altimeter made observations of the geoid anomaly due to the Puerto Rico Trench.

Geos-3 was a small satellite carrying a dual-pulse radar altimeter whose mission was to improve the knowledge of the Earth's marine geoid, and coincidentally to determine the height of ocean waves via the broadening of the short transmitted radar pulse upon reflection from the rough sea surface. Before the end of its 4-year lifetime, *Geos-3* was returning routine wave height measurements to the National Weather Service for inclusion in its Marine Waves Forecast. Altimetry from space had become a clear possibility, with practical uses of the sensor immediately forthcoming. The successes of *Skylab* and *Geos-3* reinforced the case for a second generation of radar-bearing satellites to follow.

The meteorological satellite program also provided measurements of sea surface temperature using far-infrared sensors, such as the visible and infrared scanning radiometer (VISR), which operated at wavelengths near 10 μm, the portion of the terrestrial spectrum wherein thermal radiation at terrestrial temperatures is at its peak, and where coincidentally the atmosphere has a broad passband. The coarse, 5-km resolution of the VISR gave blurred temperature images of the sea, but the promise was clearly there. **Figure 1** is an early 1960s TV image of the southeastern USA taken by the NASA *TIROS* program, showing the Gulf Stream as a dark signal. While doubts were initially held by some oceanographers as to whether such data actually represented the Gulf Stream, nevertheless the repeatability of the phenomenon, the verisimilitude of the positions and temperatures with respect to conventional wisdom, and their own objective judgment finally convinced most workers of the validity of the data. Today, higher-resolution, temperature-calibrated infrared imagery constitutes a valuable data source used frequently by ocean scientists around the world.

During the same period, spacecraft and aircraft programs taking ocean color imagery were delineating the possibilities and difficulties of determining sediment and chlorophyll concentrations remotely. **Figure 2** is a color photograph of the North Carolina barrier islands taken with a hand-held camera, with Cape Hatteras in the center. Shoals and sediment- and chlorophyll-bearing flows emanating from the coastal inlets are visible, and to the right, the blue waters of the Gulf Stream. Cloud streets developing offshore the warm stream suggest a recent passage of a cold polar front and attendant increases in air–sea evaporative fluxes.

The Second Generation

The combination of the early data and advances in scientific understanding that permitted the exploitation of those data resulted in spacecraft sensors explicitly designed to look at the sea surface. Information returned from altimeters and microwave radiometers gave credence and impetus to dedicated microwave spacecraft. Color measurements of the sea made from aircraft had indicated the efficacy of optical sensors for measurement of near-surface chlorophyll concentrations. Infrared radiometers returned useful sea surface temperature measurements. These diverse capabilities came together when, during a 4-month interval in 1978, the USA launched a triad of spacecraft that would profoundly change the way ocean scientists would observe the sea in the future. On 26 June, the first dedicated oceanographic satellite, *Seasat*, was launched; on 13 October, *TIROS-N* was launched immediately after the catastrophic failure of *Seasat* on 10 October; and on 24 October, *Nimbus-7* was lofted. Collectively they carried sensor suites whose capabilities covered virtually all known ways of observing the oceans remotely from space. This second generation of satellites would prove to be extraordinarily successful. They returned data that vindicated their proponents' positions on the measurement capabilities and utility, and they set the direction for almost all subsequent efforts in satellite oceanography.

In spite of its very short life of 99 days, *Seasat* demonstrated the great utility of altimetry by measuring the marine geoid to within a very few meters, by inferring the variability of large-scale ocean surface currents, and by determining wave heights. The wind scatterometer could yield oceanic surface wind velocities equivalent to 20 000 ship observations per day. The scanning multifrequency radiometer also provided wind speed and atmospheric water content data; and the SAR penetrated clouds to show features on the surface of the sea, including surface and internal waves, current boundaries, upwellings, and rainfall patterns. All of these measurements could be extended to basin-wide scales, allowing oceanographers a view of the sea never dreamed of before. *Seasat* stimulated several subsequent generations of ocean-viewing satellites, which outline the chronologies and heritage for the world's ocean-viewing spacecraft. Similarly, the early temperature and color observations have led to successor programs that provide large quantities of quantitative data to oceanographers around the world.

The Third Generation

The second generation of spacecraft would demonstrate that variables of importance to oceanography could be observed from space with scientifically useful accuracy. As such, they would be characterized as successful concept demonstrations. And while both first- and second-generation spacecraft had been exclusively US, international participation in demonstrating the utility of their data would lead to the entry of Canada, the European Space Agency (ESA), France, and Japan into the satellite program during this period. This article focuses on the US effort. Additional background on US third-generation missions covering the period 1980–87 can be found in the series of *Annual Reports for the Oceans Program* (NASA Technical Memoranda 80233, 84467, 85632, 86248, 87565, 88987, and 4025).

Partnership with Oceanography

Up to 1978, the remote sensing community had been the prime driver of oceanography from space and there were overly optimistic expectations. Indeed, the case had not yet been made that these observational techniques were ready to be exploited for ocean science. Consequently, in early 1979, the central task was establishing a partnership with the traditional oceanographic community. This meant involving them in the process of evaluating the performance of *Seasat* and *Nimbus-7*, as well as building an ocean science program at NASA headquarters to complement the ongoing remote sensing effort.

National Oceanographic Satellite System

This partnership with the oceanographic community was lacking in a notable and early false start on the part of NASA, the US Navy, and NOAA – the National Oceanographic Satellite System (NOSS). This was to be an operational system, with a primary and a backup satellite, along with a fully redundant ground data system. NOSS was proposed shortly after the failure of *Seasat*, with a first launch expected in 1986. NASA formed a 'science working group' (SWG) in 1980 under Francis Bretherton to define the potential that NOSS offered the oceanographic community, as well as to recommend sensors to constitute the 25% of its payload allocated for research. However, with oceanographers essentially brought in as junior partners, the job of securing a new start for NOSS fell to the operational community – which it proved unable to do. NOSS was canceled in early 1981. The prevailing and realistic view was that the greater community was not ready to implement such an operational system.

Science Working Groups

During this period, SWGs were formed to look at each promising satellite sensing technique, assess its potential contribution to oceanographic research, and define the requirements for its future flight. The notable early groups were the TOPEX SWG formed in 1980 under Carl Wunsch for altimetry, Satellite Surface Stress SWG in 1981 under James O'Brien for scatterometry, and Satellite Ocean Color SWG in 1981 under John Walsh for color scanners. These SWGs were true partnerships between the remote sensing and oceanographic communities, developing consensus for what would become the third generation of satellites.

Partnership with Field Centers

Up to this time, NASA's Oceans Program had been a collection of relatively autonomous, in-house activities run by NASA field centers. In 1981, an overrun in the space shuttle program forced a significant budget cut at NASA headquarters, including the Oceans Program. This in turn forced a reprioritization and refocusing of NASA programs. This was a blessing in disguise, as it provided an opportunity to initiate a comprehensive, centrally led program – which would ultimately result in significant funding for the oceanographic as well as remote-sensing communities. Outstanding relationships with individuals like Mous Chahine in senior management at the Jet Propulsion Laboratory (JPL) enabled the partnership between NASA headquarters and the two prime ocean-related field centers (JPL and the Goddard Space Flight Center) to flourish.

Partnerships in Implementation

A milestone policy-level meeting occurred on 13 July 1982 when James Beggs, then Administrator of NASA, hosted a meeting of the Ocean Principals Group – an informal group of leaders of the ocean-related agencies. A NASA presentation on opportunities and prospects for oceanography from space was received with much enthusiasm. However, when asked how NASA intended to proceed, Beggs told the group that – while NASA was the sole funding agency for space science and its missions – numerous agencies were involved in and support oceanography. Beggs said that NASA was willing to work with other agencies to implement an ocean satellite program, but that it would not do so on its own. Beggs' statement defined the approach to be pursued in implementing oceanography from space, namely, a joint approach based on partnerships.

Research Strategy for the Decade

As a further step in strengthening its partnership with the oceanographic community, NASA collaborated with the Joint Oceanographic Institutions Incorporated (JOI), a consortium of the oceanographic institutions with a deep-sea-going capability. At the time, JOI was the only organization in a position to represent and speak for the major academic oceanographic institutions. A JOI satellite planning committee (1984) under Jim Baker examined SWG reports, as well as the potential synergy between the variety of oceanic variables which could be measured from space; this led to the idea of understanding the ocean as a system. (From this, it was a small leap to understanding the Earth as a system, the goal of NASA's Earth Observing System (EOS).)

The report of this Committee, *Oceanography from Space: A Research Strategy for the Decade, 1985–1995*, linked altimetry, scatterometry, and ocean color with the major global ocean research programs being planned at that time – the World Ocean Circulation Experiment (WOCE), Tropical Ocean Global Atmosphere (TOGA) program, and Joint Global Ocean Flux Study (JGOFS). This strategy, still being followed today, served as a catalyst to engage the greater community, to identify the most important missions, and to develop an approach for their prioritization. Altimetry, scatterometry, and ocean color emerged from this process as national priorities.

Promotion and Advocacy

The *Research Strategy* also provided a basis for promoting and building an advocacy for the NASA program. If requisite funding was to be secured to pay for proposed missions, it was critical that government policymakers, the Congress, the greater oceanographic community, and the public had a good understanding of oceanography from space and its potential benefits. In response to this need, a set of posters, brochures, folders, and slide sets was designed by Payson Stevens of Internetwork Incorporated and distributed to a mailing list which grew to exceed 3000. These award-winning materials – sharing a common recognizable identity – were both scientifically accurate and esthetically pleasing.

At the same time, dedicated issues of magazines and journals were prepared by the community of involved researchers. The first example was the issue of *Oceanus* (1981), which presented results from the second-generation missions and represented a first step toward educating the greater oceanographic community in a scientifically useful and balanced way about realistic prospects for satellite oceanography.

Implementation Studies

Given the SWG reports taken in the context of the *Research Strategy*, the NASA effort focused on the following sensor systems (listed with each are the various flight opportunities which were studied):

- altimetry – the flight of a dedicated altimeter mission, first *TOPEX* as a NASA mission, and then *TOPEX/Poseidon* jointly with the French Centre Nationale d'Etudes Spatiales (CNES);
- scatterometry – the flight of a NASA scatterometer (NSCAT), first on NOSS, then on the *Navy Remote Ocean Observing Satellite* (*NROSS*), and finally on the *Advanced Earth Observing Satellite* (*ADEOS*) of the Japanese National Space Development Agency (NASDA);
- visible radiometry – the flight of a NASA color scanner on a succession of missions (NOSS, *NOAA-H/-I*, *SPOT-3* (*Système Pour l'Observation de la Terre*), and *Landsat-6*) and finally the purchase of ocean color data from the Sea-viewing Wide Field-of-view Sensor (SeaWiFS) to be flown by the Orbital Sciences Corporation (OSC);
- microwave radiometry – a system to utilize data from the series of special sensor microwave imager (SSMI) radiometers to fly on the Defense Meteorological Satellite Program satellites;
- SAR – a NASA ground station, the Alaska SAR Facility, to enable direct reception of SAR data from the *ERS-1/-2*, *JERS-1*, and *Radarsat* satellites of the ESA, NASDA, and the Canadian Space Agency, respectively.

New Starts

Using the results of the studies listed above, the Oceans Program entered the new start process at NASA headquarters numerous times attempting to secure funds for implementation of elements of the third generation. *TOPEX* was first proposed as a NASA mission in 1980. However, considering limited prospects for success, partnerships were sought and the most promising one was with the French. CNES initially proposed a mission using a *SPOT* bus with a US launch. However, NASA rejected this because *SPOT*, constrained to be Sun-synchronous, would alias solar tidal components. NASA proposed instead a mission using a US bus capable of flying in a non-Sun-synchronous orbit with CNES providing an *Ariane* launch. The NASA proposal was accepted for study in fiscal year (FY) 1983, and a new start was finally secured for the combined *TOPEX/Poseidon* in FY 1987.

In 1982 when the US Navy first proposed *NROSS*, NASA offered to be a partner and provide a scatterometer. The US Navy and NASA obtained new starts for both *NROSS* and NSCAT in FY 1985. However, *NROSS* suffered from a lack of strong support within the navy, experienced a number of delays, and was finally terminated in 1987. Even with this termination, NASA was able to keep NSCAT alive until establishing the partnership with NASDA for its flight on their *ADEOS* mission.

Securing a means to obtain ocean color observations as a follow-on to the coastal zone color scanner (CZCS) was a long and arduous process, finally coming to fruition in 1991 when a contract was signed with the OSC to purchase data from the flight of their SeaWiFS sensor. By that time, a new start had already been secured for NASA's EOS, and ample funds were available in that program for the Sea-WiFS data purchase.

Finally, securing support for the Alaska SAR Facility (now the Alaska Satellite Facility to reflect its broader mission) was straightforward; being small in comparison with the cost of flying space hardware, its funding had simply been included in the new start that NSCAT obtained in FY 1985. Also, funding for utilization of SSMI data was small enough to be covered by the Oceans Program itself.

Implementing the Third Generation

With the exception of the US Navy's *Geosat*, these third-generation missions would take a very long time to come into being. As seen in **Table 1**, *TOPEX/Poseidon* was launched in 1992 – 14 years after *Seasat*; NSCAT was launched on *ADEOS* in 1996 – 18 years after *Seasat*; and SeaWiFS was launched in 1997 – 19 years after *Nimbus-7*. (In addition to the missions mentioned in **Table 1**, the Japanese *ADEOS-1* included the US NSCAT in its sensor complement, and the US *Aqua* included the Japanese advanced microwave scanning radiometer (AMSR); the United States provided a launch for the Canadian *RADARSAT-1*.) In fact, these missions came so late that they had limited overlap with the field phases of the major ocean research programs (WOCE, TOGA, and JGOFS) they were to complement. Why did it take so long?

Understanding and Consensus

First, it took time to develop a physically unambiguous understanding of how well the satellite sensors actually performed, and this involved learning to cope with the data – satellite data rates being orders of magnitude larger than those encountered in traditional oceanography. For example, it was not until 3 years after the launch of *Nimbus-7* that CZCS data could be processed as fast as collected by the satellite. And even with only a 3-month data set from *Seasat*, it took 4 years to produce the first global maps of variables such as those shown in **Figure 4**.

In evaluating the performance of both *Seasat* and *Nimbus-7*, it was necessary to have access to the data. *Seasat* had a free and open data policy; and after a very slow start, the experiment team concept (where team members had a lengthy period of exclusive access to the data) for the *Nimbus-7* CZCS was replaced with that same policy. Given access to the data, delays were due to a combination of sorting out the algorithms for converting the satellite observations into variables of interest, as well as being constrained by having limited access to raw computing power.

In addition, the rationale for the third-generation missions represented a major paradigm shift. While earlier missions had been justified largely as demonstrations of remote sensing concepts, the third-generation missions would be justified on the basis of their potential contribution to oceanography. Hence, the long time it took to understand sensor performance translated into a delay in being able to convince traditional oceanographers that satellites were an important observational tool ready to be exploited for ocean science. As this case was made, it was possible to build consensus across the remote sensing and oceanographic communities.

Space Policy

Having such consensus reflected at the highest levels of government was another matter. The *White House*

Table 1 Some major ocean-related missions

Year	USA	Russia	Japan	Europe	Canada	Other
1968		Kosmos 243				
	Nimbus-3					
1970	Nimbus-4	Kosmos 384				
1972	Nimbus-5					
1974	Skylab					
	Nimbus-6, Geos-3					
1976						
1978	Nimbus-7, Seasat					
		Kosmos 1076				
1980		Kosmos 1151				
1982						
		Kosmos 1500				
1984		Kosmos 1602				
	Geosat					
1986		Kosmos 1776				
		Kosmos 1870	MOS 1A			
1988		OKEAN 1				
1990		OKEAN 2	MOS 1B			
		Almaz-1,		ERS-1		
		OKEAN 3				
1992	Topex/Poseidon[a]		JERS-1			
1994		OKEAN 7				
		OKEAN 8		ERS-2	RADARSAT-1	
1996			ADEOS-1			
	SeaWiFS		TRMM[b]			
1998	GFO					
	Terra, QuikSCAT	OKEAN-O #1				OCEANSAT-1[i]
2000				CHAMP[c]		
		Meteor-3 #1		Jason-1[e]		
2002	Aqua; GRACE[d]			ENVISAT		HY-1A[j]
	WINDSAT		ADEOS-2			
2004	ICESat	SICH-1M[f]				
				CRYOSat		
2006			ALOS	GOCE, MetOp-1		HY-1B[j]
		Meteor-3M #2		SMOS	RADARSAT-2	OCEANSAT-2[i]
2008				OSTM/Jason-2[g]		
	NPP; Aquarius[h]			CryoSat-2		HY-1C, HY-2A[j]
2010		GCOM-W		Sentinel-3, MetOp-2		OCEANSAT-3[i]

[a]US/France TOPEX/Poseidon.
[b]Japan/US TRMM.
[c]German CHAMP.
[d]US/German GRACE.
[e]France/US Jason-1.
[f]Russia/Ukraine Sich-1M.
[g]France/US Jason-2/OSTM.
[h]US/Argentina Aquarius.
[i]India OCEANSAT series.
[j]China HY series.
Updated version of similar data in Wilson WS, Fellous JF, Kawamura H, and Mitnik L (2006) A history of oceanography from space. In: Gower JFR (ed.) Manual of Remote Sensing, Vol. 6: Remote Sensing of the Environment, pp. 1–31. Bethesda, MD: American Society for Photogrammetry and Remote Sensing.

Fact Sheet on US Civilian Space Policy of 11 October 1978 states, "... emphasizing space applications ... will bring important benefits to our understanding of earth resources, climate, weather, pollution ... and provide for the private sector to take an increasing responsibility in remote sensing and other applications." Landsat was commercialized in 1979 as part of this space policy. As Robert Stewart explains, "Clearly the mood at the presidential level was that earth remote sensing, including the oceans, was a practical space application more at home outside the scientific community. It took almost a decade to get an

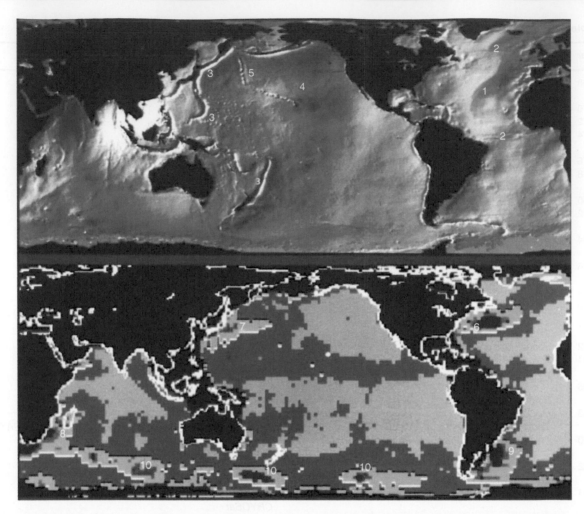

Figure 4 Global sea surface topography *c.* 1983. This figure shows results computed from the 70 days of *Seasat* altimeter data in 1978. Clearly visible in the mean sea surface topography, the marine geoid (upper panel), are the Mid-Atlantic Ridge (1) and associated fracture zones (2), trenches in the western Pacific (3), the Hawaiian Island chain (4), and the Emperor seamount chain (5). Superimposed on the mean surface is the time-varying sea surface topography, the mesoscale variability (lower panel), associated with the variability of the ocean currents. The largest deviations (10–25 cm), yellow and orange, are associated with the western boundary currents: Gulf Stream (6), Kuroshio (7), Agulhas (8), and Brazil/Falkland Confluence (9); large variations also occur in the West Wind Drift (10). Courtesy of NASA.

understanding at the policy level that scientific needs were also important, and that we did not have the scientific understanding necessary to launch an operational system for climate." The failures of NOSS, and later *NROSS*, were examples of an effort to link remote sensing directly with operational applications without the scientific underpinning.

The view in Europe was not dissimilar; governments felt that cost recovery was a viable financial scheme for ocean satellite missions, that is, the data have commercial value and the user would be willing to pay to help defray the cost of the missions.

Joint Satellite Missions

It is relatively straightforward to plan and implement missions within a single agency, as with NASA's

space science program. However, implementing a satellite mission across different organizations, countries, and cultures is both challenging and time-consuming. An enormous amount of time and energy was invested in studies of various flight options, many of which fizzled out, but some were implemented. With the exception of the former Soviet Union, NASA's third-generation missions would be joint with each nation having a space program at that time, as well as with a private company.

The *Geosat* Exception

Geosat was the notable US exception, having been implemented so quickly after the second generation. It was approved in 1981 and launched in 1985 in order to address priority operational needs on the part of the US Navy. During the second half of its mission, data would become available within 1–2 days. As will be discussed below, *Geosat* shared a number of attributes with the meteorological satellites: it had a specific focus; it met priority operational needs for its user; experience was available for understanding and using the observations; and its implementation was done in the context of a single organization.

Challenges Ahead

Scientific Justification

As noted earlier, during the decade of the 1980s, there was a dearth of ocean-related missions in the United States, it being difficult to justify a mission based on its contribution to ocean science. Then later in that decade, NASA conceived of the EOS and was able to make the case that Earth science was sufficient justification for a mission. Also noted earlier, ESA initially had no appropriate framework for Earth science missions, and a project like *ERS-1* was pursued under the assumption that it would help develop commercial and/or operational applications of remote sensing of direct societal benefit. Its successor, *ERS-2*, was justified on the basis of needing continuity of SAR coverage for the land surface, rather than the need to monitor ocean currents. And the *ENVISAT* was initially decided by ESA member states as part of the Columbus program of the International Space Station initiative. The advent of an Earth Explorer program in 1999 represented a change in this situation. As a consequence, new Earth science missions – *GOCE*, *CryoSat*, and *SMOS* – all represent significant steps forward.

This ESA program, together with similar efforts at NASA, are leading to three sets of ground-breaking scientific missions which have the potential to significantly impact oceanography. *GOCE* and *GRACE* will contribute to an improved knowledge of the Earth's gravity field, as well as the mass of water on the surface of the Earth. *CryoSat-2* and *ICESat* will contribute to knowledge of the volume of water locked up in polar and terrestrial ice sheets. Finally, *SMOS* and *Aquarius* will contribute to knowledge of the surface salinity field of the global oceans. Together, these will be key ingredients in addressing the global water cycle.

Data Policy

The variety of missions described above show a mix of data policies, from full and open access without any period of exclusive use (e.g., *TOPEX/Poseidon*, *Jason*, and *QuikSCAT*) to commercial distribution (e.g., real-time SeaWiFS for nonresearch purposes, *RADARSAT*), along with a variety of intermediate cases (e.g., *ERS*, *ALOS*, and *ENVISAT*). From a scientific perspective, full and open access is the preferred route, in order to obtain the best understanding of how systems perform, to achieve the full potential of the missions for research, and to lay the most solid foundation for an operational system. Full and open access is also a means to facilitate the development of a healthy and competitive private sector to provide value-added services. Further, if the international community is to have an effective observing system for climate, a full and open data policy will be needed, at least for that purpose.

In Situ Observations

Satellites have made an enormous contribution enabling the collection of *in situ* observations from *in situ* platforms distributed over global oceans. The Argos (plural spelling; not to be confused with *Argo* profiling floats) data collection and positioning system has flown on the NOAA series of polar-orbiting operational environmental satellites continuously since 1978. It provides one-way communication from data collection platforms, as well as positioning of those platforms. While an improved Argos capability (including two-way communications) is coming with the launch of *MetOp-1* in 2006, oceanographers are looking at alternatives – Iridium being one example – which offer significant higher data rates, as well as two-way communications.

In addition, it is important to note that the Intergovernmental Oceanographic Commission and the World Meteorological Organization have established the Joint Technical Commission for Oceanography and Marine Meteorology (JCOMM) to bring a focus to the collection, formatting, exchange, and archival of data collected at sea, whether they be oceanic or atmospheric. JCOMM has established a center, JCOMMOPS, to serve as the specific institutional focus to harmonize the national contributions of *Argo* floats, surface-drifting buoys, coastal tide gauges, and fixed and moored buoys. JCOMMOPS will play an important role helping contribute to 'integrated observations' described below.

Integrated Observations

To meet the demands of both the research and the broader user community, it will be necessary to focus, not just on satellites, but 'integrated' observing systems. Such systems involve combinations of satellite and *in situ* systems feeding observations into data-processing systems capable of delivering a comprehensive view of one or more geophysical variables (sea level, surface temperature, winds, etc.).

Three examples help illustrate the nature of integrated observing systems. First, consider global sea level rise. The combination of the *Jason-1* altimeter, its precision orbit determination system, and the suite of precision tide gauges around the globe allow scientists to monitor changes in volume of the oceans. The growing global array of *Argo* profiling floats allows scientists to assess the extent to which those changes in sea level are caused by changes in the temperature and salinity structure of the upper ocean. *ICESat* and *CryoSat* will provide estimates of changes in the volume of ice sheets, helping assess the extent to which their melting contributes to global sea level rise. And *GRACE* and *GOCE* will provide estimates of the changes in the mass of water on the Earth's surface. Together, systems such as these will enable an improved understanding of global sea level rise and, ultimately, a reduction in the wide range of uncertainty in future projections.

Second, global estimates of vector winds at the sea surface are produced from the scatterometer on *QuikSCAT*, a global array of *in situ* surface buoys, and the Seawinds data processing system. Delivery of this product in real time has significant potential to improve marine weather prediction. The third example concerns the *Jason-1* altimeter together with *Argo*. When combined in a sophisticated data assimilation system – using a state-of-the-art ocean model – these data enable the estimation of the physical state of the ocean as it changes through time. This information – the rudimentary 'weather map' depicting the circulation of the oceans – is a critical component of climate models and provides the fundamental context for addressing a broad range of issues in chemical and biological oceanography.

Transition from Research to Operations

The maturing of the discipline of oceanography includes the development of a suite of global oceanographic services being conducted in a manner similar to what exists for weather services. The delivery of these services and their associated informational products will emerge as the result of the successes in ocean science ('research push'), as well as an increasing demand for ocean analyses and forecasts from a variety of sectors ('user pull').

From the research perspective, it is necessary to 'transition' successfully demonstrated ('experimental') observing techniques into regular, long-term, systematic ('operational') observing systems to meet a broad range of user requirements, while maintaining the capability to collect long-term, 'research-quality'

observations. From the operational perspective, it is necessary to implement proven, scientifically sound, cost-effective observing systems – where the uninterrupted supply of real-time data is critical. This is a big challenge to be met by the space systems because of the demand for higher reliability and redundancy, at the same time calling for stringent calibration and accuracy requirements. Meeting these sometimes competing, but quite complementary demands will be the challenge and legacy of the next generation of ocean remote-sensing satellites.

Meteorological Institutional Experience

With the launch in 1960 of the world's first meteorological satellite, the polar-orbiting *Tiros-1* carrying two TV cameras, the value of the resulting imagery to the operational weather services was recognized immediately. The very next year a National Operational Meteorological System was implemented, with NASA to build and launch the satellites and the Weather Bureau to be the operator. The feasibility of using satellite imagery to locate and track tropical storms was soon demonstrated, and by 1969 this capability had become a regular part of operational weather forecasting. In 1985, Richard Hallgren, former Director of the National Weather Service, stated, "the use of satellite information simply permeates every aspect of the [forecast and warning] process and all this in a mere 25 years." In response to these operational needs, there has been a continuing series of more than 50 operational, polar-orbiting satellites in the United States alone!

The first meteorological satellites had a specific focus on synoptic meteorology and weather forecasting. Initial image interpretation was straightforward (i.e., physically unambiguous), and there was a demonstrated value of resulting observations in meeting societal needs. Indeed, since 1960 satellites have ensured that no hurricane has gone undetected. In addition, the coupling between meteorology and remote sensing started very near the beginning. An 'institutional mechanism' for transition from research to operations was established almost immediately. Finally, recognition of this endeavor extended to the highest levels of government, resulting in the financial commitment needed to ensure success.

Oceanographic Institutional Issue

Unlike meteorology where there is a National Weather Service in each country to provide an institutional focus, ocean-observing systems have multiple performer and user institutions whose interests must be reconciled. For oceanography, this is a significant challenge working across 'institutions', where the *in situ* research is in one or more agencies, the space research and development is in another, and operational activities in yet another – with possibly separate civil and military systems. In the United States, the dozen agencies with ocean-related responsibilities are using the National Oceanographic Partnership Program and its Ocean.US Office to provide a focus for reconciling such interests. In the United Kingdom, there is the Interagency Committee on Marine Science and Technology.

In France, there is the Comité des Directeurs d'Organismes sur l'Océanographie, which gathers the heads of seven institutions interested in the development of operational oceanography, including CNES, meteorological service, ocean research institution, French Research Institute for Exploitation of the Sea (IFREMER), and the navy. This group of agencies has worked effectively over the past 20 years to establish a satellite data processing and distribution system (AVISO), the institutional support for a continuing altimetric satellite series (*TOPEX/Poseidon, Jason*), the framework for the French contribution to the *Argo* profiling float program (CORIOLIS), and to create a public corporation devoted to ocean modeling and forecasting, using satellite and *in situ* data assimilation in an operational basis since 2001 (Mercator). This partnership could serve as a model in the effort to develop operational oceanography in other countries. Drawing from this experience working together within France, IFREMER is leading the European integrated project, MERSEA, aimed at establishing a basis for a European center for ocean monitoring and forecasting.

Ocean Climate

If we are to adequately address the issue of global climate change, it is essential that we are able to justify the satellite systems required to collect the global observations 'over the long term'. Whether it be global sea level rise or changes in Arctic sea ice cover or sea surface temperature, we must be able to sustain support for the systems needed to produce climate-quality data records, as well as ensure the continuing involvement of the scientific community.

Koblinsky and Smith have outlined the international consensus for ocean climate research needs and identified the associated observational requirements. In addition to their value for research, we are compelled by competing interests to demonstrate the value of such observations in meeting a broad range of societal needs. Climate observations pose

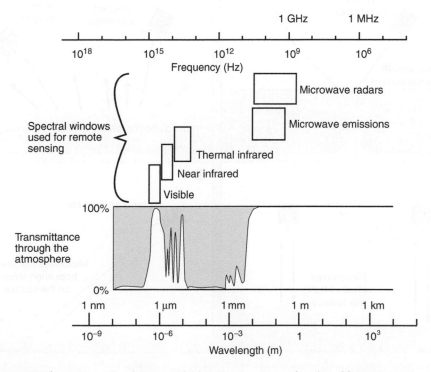

Figure 5 The electromagnetic spectrum showing atmospheric transmitance as a function of frequency and wavelength, along with the spectral windows used for remote sensing. Microwave bands are typically defined by frequency and the visible/infrared by wavelength. Adapted from Robinson IS and Guymer T (1996) Observing oceans from space. In: Summerhayes CP and Thorpe SA (eds.) *Oceanography: An Illustrated Guide*, pp. 69–87. Chichester, UK: Wiley.

challenges, since they require operational discipline to be collected in a long-term systematic manner, yet also require the continuing involvement of the research community to ensure their scientific integrity, and have impacts that may not be known for decades (unlike observations that support weather forecasting whose impact can be assessed within a matter of hours or days). Together, the institutional and observational challenges for ocean climate have been difficult to surmount.

International Integration

The paper by the Ocean Theme Team prepared under the auspices of the Integrated Global Observing Strategy (IGOS) Partnership represents how the space-faring nations are planning for the collection of global ocean observations. IGOS partners include the major global research program sponsors, global observing systems, space agencies, and international organizations.

On 31 July 2003, the First Earth Observations Summit – a high-level meeting involving ministers from over 20 countries – took place in Washington, DC, following a recommendation adopted at the G-8 meeting held in Evian the previous month; this summit proposed to 'plan and implement' a Global Earth Observation System of Systems (GEOSS). Four additional summits have been held, with participation having grown to include 60 nations and 40 international organizations; the GEOSS process provides the political visibility – not only to implement the plans developed within the IGOS Partnership – but to do so in the context of an overall Earth observation framework. This represents a remarkable opportunity to develop an improved understanding of the oceans and their influence on the Earth system, and to contribute to the delivery of improved oceanographic products and services to serve society.

Appendix 1: A Brief Overview of Satellite Remote Sensing

Unlike the severe attenuation in the sea, the atmosphere has 'windows' in which certain electromagnetic (EM) signals are able to propagate. These windows, depicted in **Figure 5**, are defined in terms

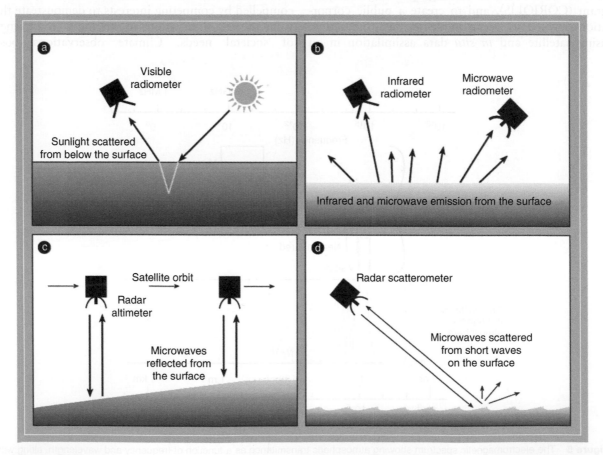

Figure 6 Four techniques for making oceanic observations from satellites: (a) visible radiometry, (b) infrared and microwave radiometry, (c) altimetry, and (d) scatterometry. Adapted from Robinson IS and Guymer T (1996) Observing oceans from space. In: Summerhayes CP and Thorpe SA (eds.) *Oceanography: An Illustrated Guide*, pp. 69–87. Chichester, UK: Wiley.

	Passive sensors (radiometers)			Active sensors (microwave radars)		
Sensor type	Visible	Infrared	Microwave	Altimetry	Scatterometry	SAR
Measured physical variable	Solar radiation backscattered from beneath the sea surface	Infrared emission from the sea surface	Microwave emission from the sea surface	Travel time, shape, and strength of reflected pulse	Strength of return pulse when illuminated from different directions	Strength and phase of return pulse
Applications	Ocean color; chlorophyll; primary production; water clarity; shallow-water bathymetry	Surface temperature; ice cover	Ice cover, age and motion; sea surface temperature; wind speed	Surface topography for geostrophic currents and tides; bathymetry; oceanic geoid; wind and wave conditions	Surface vector winds; ice cover	Surface roughness at fine spatial scales; surface and internal wave patterns; bathymetric patterns; ice cover and motion

Figure 7 Measured physical variables and applications for both passive and active sensors, expressed as a function of sensor type.

of atmospheric transmittance – the percentage of an EM signal which is able to propagate through the atmosphere – expressed as a function of wavelength or frequency.

Given a sensor on board a satellite observing the ocean, it is necessary to understand and remove the effects of the atmosphere (such as scattering and attenuation) as the EM signal propagates through it. For passive sensors (**Figures 6(a)** and **6(b)**), it is then possible to relate the EM signals collected by the sensor to the associated signals at the bottom of the atmosphere, that is, the natural radiation emitted or reflected from the sea surface. Note that passive sensors in the visible band are dependent on the Sun for natural illumination.

Active sensors, microwave radar (**Figures 6(c)** and **6(d)**), provide their own source of illumination and have the capability to penetrate clouds and, to a certain extent, rain. Atmospheric correction must be done to remove effects for a round trip from the satellite to the sea surface.

With atmospheric corrections made, measurements of physical variables are available: emitted radiation for passive sensors, and the strength, phase, and/or travel time for active sensors. **Figure 7** shows typical measured physical variables for both types of sensors in their respective spectral bands, as well as applications or derived variables of interest – ocean

color, surface temperature, ice cover, sea level, and surface winds. The companion articles on this topic address various aspects of **Figure 7** in more detail, so only this general overview is given here.

Acknowledgments

The authors would like to acknowledge contributions to this article from Mary Cleave, Murel Cole, William Emery, Michael Freilich, Lee Fu, Rich Gasparovic, Trevor Guymer, Tony Hollingsworth[†], Hiroshi Kawamura, Michele Lefebvre, Leonid Mitnik, Jean-François Minster, Richard Moore, William Patzert, Willard Pierson[†], Jim Purdom, Keith Raney, Payson Stevens, Robert Stewart, Ted Strub, Tasuku Tanaka, William Townsend, Mike Van Woert, and Frank Wentz.

See also

Aircraft Remote Sensing. IR Radiometers. Satellite Passive-Microwave Measurements of Sea Ice. Satellite Remote Sensing: Ocean Color. Satellite Remote Sensing of Sea Surface Temperatures. Satellite Remote Sensing: Salinity Measurements. Satellite Remote Sensing Sar. Upper Ocean Time and Space Variability.

Further Reading

Apel JR (ed.) (1972) Sea surface topography from space. *NOAA Technical Reports: ERL No. 228, AOML No. 7.* Boulder, CO: NOAA.

Cherny IV and Raizer VY (1998) *Passive Microwave Remote Sensing of Oceans.* Chichester, UK: Praxis.

Committee on Earth Sciences (1995) *Earth Observations from Space: History, Promise, and Reality.* Washington, DC: Space Studies Board, National Research Council.

Ewing GC (1965) Oceanography from space. *Proceedings of a Conference held at Woods Hole, 24–28 August 1964. Woods Hole Oceanographic Institution Ref. No. 65-10.* Woods Hole, MA: WHOI.

Fu L-L, Liu WT, and Abbott MR (1990) Satellite remote sensing of the ocean. In: Le Méhauté B (ed.) *The Sea, Vol. 9: Ocean Engineering Science*, pp. 1193–1236. Cambridge, MA: Harvard University Press.

Guymer TH, Challenor PG, and Srokosz MA (2001) Oceanography from space: Past success, future challenge. In: Deacon M (ed.) *Understanding the Oceans: A Century of Ocean Exploration*, pp. 193–211. London: UCL Press.

JOI Satellite Planning Committee (1984) *Oceanography from Space: A Research Strategy for the Decade, 1985–1995*, parts 1 and 2. Washington, DC: Joint Oceanographic Institutions.

Kaula WM (ed.) (1969) The terrestrial environment: solid-earth and ocean physics. *Proceedings of a Conference held at William College, 11–21 August 1969, NASA CR-1579.* Washington, DC: NASA.

Kawamura H (2000) Era of ocean observations using satellites. *Sokko-Jiho* 67: S1–S9 (in Japanese).

Koblinsky CJ and Smith NR (eds.) (2001) *Observing the Oceans in the 21st Century.* Melbourne: Global Ocean Data Assimilation Experiment and the Bureau of Meteorology.

Masson RA (1991) *Satellite Remote Sensing of Polar Regions.* London: Belhaven Press.

Minster JF and Lefebvre M (1997) *TOPEX/Poseidon* satellite altimetry and the circulation of the oceans. In: Minster JF (ed.) *La Machine Océan*, pp. 111–135 (in French). Paris: Flammarion.

Ocean Theme Team (2001) *An Ocean Theme for the IGOS Partnership.* Washington, DC: NASA. http://www.igospartners.org/docs/theme_reports/IGOS-Oceans-Final-0101.pdf (accessed Mar. 2008).

Purdom JF and Menzel WP (1996) Evolution of satellite observations in the United States and their use in meteorology. In: Fleming JR (ed.) *Historical Essays on Meteorology: 1919–1995*, pp. 99–156. Boston, MA: American Meteorological Society.

Robinson IS and Guymer T (1996) Observing oceans from space. In: Summerhayes CP and Thorpe SA (eds.) *Oceanography: An Illustrated Guide*, pp. 69–87. Chichester, UK: Wiley.

Victorov SV (1996) *Regional Satellite Oceanography.* London: Taylor and Francis.

Wilson WS (ed.) (1981) *Special Issue: Oceanography from Space. Oceanus* 24: 1–76.

Wilson WS, Fellous JF, Kawamura H, and Mitnik L (2006) A history of oceanography from space. In: Gower JFR (ed.) *Manual of Remote Sensing, Vol. 6: Remote Sensing of the Environment*, pp. 1–31. Bethesda, MD: American Society for Photogrammetry and Remote Sensing.

Wilson WS and Withee GW (2003) A question-based approach to the implementation of sustained, systematic observations for the global ocean and climate, using sea level as an example. *MTS Journal* 37: 124–133.

Relevant Websites

http://www.aviso.oceanobs.com
– AVISO.

http://www.coriolis.eu.org
– CORIOLIS.

http://www.eohandbook.com
– Earth Observation Handbook, CEOS.

http://www.igospartners.org
– IGOS.

http://wo.jcommops.org
– JCOMMOPS.

http://www.mercator-ocean.fr
– Mercator Ocean.

SATELLITE REMOTE SENSING OF SEA SURFACE TEMPERATURES

P. J. Minnett, University of Miami, Miami, FL, USA

Introduction

The ocean surface is the interface between the two dominant, fluid components of the Earth's climate system: the oceans and atmosphere. The heat moved around the planet by the oceans and atmosphere helps make much of the Earth's surface habitable, and the interactions between the two, that take place through the interface, are important in shaping the climate system. The exchange between the ocean and atmosphere of heat, moisture, and gases (such as CO_2) are determined, at least in part, by the sea surface temperature (SST). Unlike many other critical variables of the climate system, such as cloud cover, temperature is a well-defined physical variable that can be measured with relative ease. It can also be measured to useful accuracy by instruments on observation satellites.

The major advantage of satellite remote sensing of SST is the high-resolution global coverage provided by a single sensor, or suite of sensors on similar satellites, that produces a consistent data set. By the use of onboard calibration, the accuracy of the time-series of measurements can be maintained over years, even decades, to provide data sets of relevance to research into the global climate system. The rapid processing of satellite data permits the use of the global-scale SST fields in applications where the immediacy of the data is of prime importance, such as weather forecasting – particularly the prediction of the intensification of tropical storms and hurricanes.

Measurement Principle

The determination of the SST from space is based on measuring the thermal emission of electromagnetic radiation from the sea surface. The instruments, called radiometers, determine the radiant energy flux, B_λ, within distinct intervals of the electromagnetic spectrum. From these the brightness temperature (the temperature of a perfectly emitting 'black-body' source that would emit the same radiant flux) can be calculated by the Planck equation:

$$B_\lambda(T) = 2hc^2\lambda^{-5}\left(e^{hc/(\lambda kT)} - 1\right)^{-1} \qquad [1]$$

where h is Planck's constant, c is the speed of light in a vacuum, k is Boltzmann's constant, λ is the wavelength and T is the temperature. The spectral intervals (wavelengths) are chosen where three conditions are met: (1) the sea emits a measurable amount of radiant energy, (2) the atmosphere is sufficiently transparent to allow the energy to propagate to the spacecraft, and (3) current technology exists to build radiometers that can measure the energy to the required level of accuracy within the bounds of size, weight, and power consumption imposed by the spacecraft. In reality these constrain the instruments to two relatively narrow regions of the infrared part of the spectrum and to low-frequency microwaves. The infrared regions, the so-called atmospheric windows, are situated between wavelengths of $3.5-4.1\mu m$ and $8-12\mu m$ (**Figure 1**); the microwave measurements are made at frequencies of 6–12 GHz.

As the electromagnetic radiation propagates through the atmosphere, some of it is absorbed and scattered out of the field of view of the radiometer, thereby attenuating the original signal. If the attenuation is sufficiently strong none of the radiation from the sea reaches the height of the satellite, and such is the case when clouds are present in the field of view of infrared radiometers. Even in clear-sky conditions a significant fraction of the sea surface emission is absorbed in the infrared windows. This energy is re-emitted, but at a temperature characteristic of that height in the atmosphere. Consequently the brightness temperatures measured through the clear atmosphere by a spacecraft radiometer are cooler than would be measured by a similar device just above the surface. This atmospheric effect, frequently referred to as the temperature deficit, must be corrected accurately if the derived sea surface temperatures are to be used quantitatively.

Infrared Atmospheric Correction Algorithms

The peak of the Planck function for temperatures typical of the sea surface is close to the longer wavelength infrared window, which is therefore well suited to SST measurement (**Figure 1**). However, the

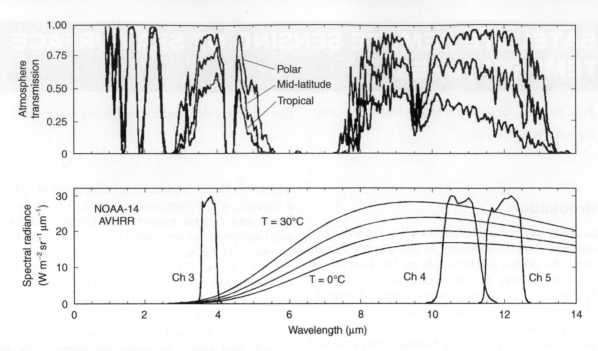

Figure 1 Spectra of atmospheric transmission in the infrared (wavelengths 1–14 μm) calculated for three typical atmospheres from diverse parts of the ocean; polar, mid-latitude and tropical with integrated water vapor content of 7 kg m^{-2} (polar), 29 kg m^{-2} (mid-latitude) and 54 kg m^{-2} (tropical). Regions where the transmission is high are well suited to satellite remote sensing of SST. The lower panel shows the electromagnetic radiative flux for four sea surface temperatures (0, 10, 20, and 30°C) with the relative spectral response functions for channels 3, 4, and 5 of the AVHRR on the NOAA-14 satellite. The so-called 'split-window' channels, 4 and 5, are situated where the sea surface emission is high, and where the atmosphere is comparatively clear but exhibits a strong dependence on atmospheric water vapor content.

main atmospheric constituent in this spectral interval that contributes to the temperature deficit is water vapor, which is very variable both in space and time. Other molecular species that contribute to the temperature deficit are quite well mixed throughout the atmosphere, and therefore inflict a relatively constant temperature deficit that is simple to correct.

The variability of water vapor requires an atmospheric correction algorithm based on the information contained in the measurements themselves. This is achieved by making measurements at distinct spectral intervals in the windows when the water vapor attenuation is different. These spectral intervals are defined by the characteristics of the radiometer and are usually referred to as bands or channels (**Figure 1**). By invoking the hypothesis that the difference in the brightness temperatures measured in two channels, i and j, is related to the temperature deficit in one of them, the atmospheric correction algorithm can be formulated thus:

$$SST_{ij} - T_i = f(T_i - T_j) \qquad [2]$$

where SST_{ij} is the derived SST and T_i, T_j are the brightness temperatures in channels i, j.

Further, by assuming that the atmospheric attenuation is small in these channels, so that the radiative transfer can be linearized, and that the channels are spectrally close so that Planck's function can be linearized, the algorithm can be expressed in the very simple form:

$$SST_{ij} = a_o + a_i T_i + a_j T_j \qquad [3]$$

where are a_o, a_i, and a_j are coefficients. These are determined by regression analysis of either coincident satellite and *in situ* measurements, mainly from buoys, or of simulated satellite measurements derived by radiative transfer modeling of the propagation of the infrared radiation from the sea surface through a representative set of atmospheric profiles.

The simple algorithm has been applied for many years in the operational derivation of the sea surface from measurements of the Advanced Very High Resolution Radiometer (AVHRR, see below), the product of which is called the multi-channel SST (MCSST), where i refers to channel 4 and j to channel 5.

More complex forms of the algorithms have been developed to compensate for some of the shortcomings of the linearization. One such widely applied algorithm takes the form:

$$SST_{ij} = b_o + b_1 T_i + b_2(T_i - T_j)SST_r$$
$$+ b_3(T_i - T_j)(\sec\theta - 1) \qquad [4]$$

where SST_r is a reference SST (or first-guess temperature), and θ is the zenith angle to the satellite radiometer measured at the sea surface. When applied to AVHRR data, with i and j referring to channels 4 and 5 derived SST is called the nonlinear SST (NLSST). A refinement is called the Pathfinder SST (PFSST) in the research program designed to post-process AVHRR data over a decade or so to provide a consistent data set for climate research. In the PFSST, the coefficients are derived on a monthly basis for two different atmospheric regimes, distinguished by the value of the T_4–T_5 differences being above or below 0.7 K, by comparison with measurements from buoys.

The atmospheric correction algorithms work effectively only in the clear atmosphere. The presence of clouds in the field of view of the infrared radiometer contaminates the measurement so that such pixels must be identified and removed from the SST retrieval process. It is not necessary for the entire pixel to be obscured, even a portion as small as 3–5%, dependent on cloud type and height, can produce unacceptable errors in the SST measurement. Thin, semi-transparent cloud, such as cirrus, can have similar effects to subpixel geometric obscuration by optically thick clouds. Consequently, great attention must be paid in the SST derivation to the identification of measurements contaminated by even a small amount of clouds. This is the principle disadvantage to SST measurement by spaceborne infrared radiometry. Since there are large areas of cloud cover over the open ocean, it may be necessary to composite the cloud-free parts of many images to obtain a complete picture of the SST over an ocean basin.

Similarly, aerosols in the atmosphere can introduce significant errors in SST measurement. Volcanic aerosols injected into the cold stratosphere by violent eruptions produce unmistakable signals that can bias the SST too cold by several degrees. A more insidious problem is caused by less readily identified aerosols at lower, warmer levels of the atmosphere that can introduce systematic errors of a much smaller amplitude.

Microwave Measurements

Microwave radiometers use a similar measurement principle to infrared radiometers, having several spectral channels to provide the information to correct for extraneous effects, and black-body calibration targets to ensure the accuracy of the measurements. The suite of channels is selected to include sensitivity to the parameters interfering with the SST measurements, such as cloud droplets and surface wind speed, which occurs with microwaves at higher frequencies. A simple combination of the brightness temperature, such as eqn [2], can retrieve the SST.

The relative merits of infrared and microwave radiometers for measuring SST are summarized in Table 1.

Characteristics of Satellite-derived SST

Because of the very limited penetration depth of infrared and microwave electromagnetic radiation in sea water the temperature measurements are limited to the sea surface. Indeed, the penetration depth is typically less than 1 mm in the infrared, so that temperature derived from infrared measurements is characteristic of the so-called skin of the ocean. The skin temperature is generally several tenths of a degree cooler than the temperature measured just below, as a result of heat loss from the ocean to atmosphere. On days of high insolation and low wind speed, the absorption of sunlight causes an increase in near surface temperature so that the water just below the skin layer is up to a few degrees warmer than that measured a few meters deeper, beyond the influence of the diurnal heating. For those people interested in a temperature characteristic of a depth of a few meters or more, the decoupling of the skin and deeper, bulk temperatures is perceived as a disadvantage of using satellite SST. However, algorithms generated by comparisons between satellite

Table 1 Relative merits of infrared and microwave radiometers for sea surface temperature measurement

Infrared	Microwave
Good spatial resolution (~1 km)	Poor spatial resolution (~50 km)
Surface obscured by clouds	Clouds largely transparent, but measurement perturbed by heavy rain
No side-lobe contamination	Side-lobe contamination prevents measurements close to coasts or ice
Aperture is reasonably small; instrument can be compact for spacecraft use	Antenna is large to achieve spatial resolution from polar orbit heights (~800 km above the sea surface)
4 km resolution possible from geosynchronous orbit; can provide rapid sampling data	Distance to geosynchronous orbit too large to permit useful spatial resolution with current antenna sizes

Table 2 Spectral characteristics of current and planned satellite-borne infrared radiometers

AVHRR		ATSR		MODIS		OCTS		GLI	
λ (μm)	NEΔT (K)	λ (μm)	NEΔT (K)	λ (μm)	NEΔT (K)	λ (μm)	NEΔT (K)	λ (μm)	NEΔT (K)
3.75	0.12	3.7	0.019	3.75	0.05	3.7	0.15	3.715	<0.15
				3.96	0.05				
				4.05	0.05				
				8.55	0.05	8.52	0.15	8.3	<0.1
10.5	0.12	10.8	0.028	11.03	0.04	10.8	0.15	10.8	<0.1
11.5	0.12	12.0	0.025	12.02	0.04	11.9	0.15	12	<0.1

and *in situ* measurements from buoys include a mean skin effect masquerading as part of the atmospheric effect, and so the application of these results in an estimate of bulk temperatures.

The greatest advantage offered by satellite remote sensing is, of course, coverage. A single, broad-swath, imaging radiometer on a polar-orbiting satellite can provide global coverage twice per day. An imaging radiometer on a geosynchronous satellite can sample much more frequently, once per half-hour for the Earth's disk, or smaller segments every few minutes, but the spatial extent of the data is limited to that part of the globe visible from the satellite.

The satellite measurements of SST are also reasonably accurate. Current estimates for routine measurements show absolute accuracies of ± 0.3 to ± 0.5 K when compared to similar measurements from ships, aircraft, and buoys.

Spacecraft Instruments

All successful instruments have several attributes in common: a mechanism for scanning the Earth's surface to generate imagery, good detectors, and a mechanism for real-time, in-flight calibration. Calibration involves the use of one or more black-body calibration targets, the temperatures of which are accurately measured and telemetered along with the imagery. If only one black-body is available a measurement of cold dark space provides the equivalent of a very cold calibration target. Two calibration points are needed to provide in-flight calibration; nonlinear behavior of the detectors is accounted for by means of pre-launch instrument characterization measurements.

The detectors themselves inject noise into the data stream, at a level that is strongly dependent on their temperature. Therefore, infrared radiometers require cooled detectors, typically operating from 78 K ($-195°$C) to 105 K ($-168°$C) to reduce the noise

equivalent temperature difference (NEΔT) to the levels shown in **Table 2**.

The Advanced Very High Resolution Radiometer (AVHRR)

The satellite instrument that has contributed the most to the study of the temperature of the ocean surface is the AVHRR that first flew on TIROS-N launched in late 1978. AVHRRs have flown on successive satellites of the NOAA series from NOAA-6 to NOAA-14, with generally two operational at any given time. The NOAA satellites are in a near-polar, sun-synchronous orbit at a height of about 780 km above the Earth's surface and with an orbital period of about 100 min. The overpass times of the two NOAA satellites are about 2.30 a.m. and p.m. and about 7.30 a.m. and p.m. local time. The AVHRR has five channels: 1 and 2 at ~ 0.65 and ~ 0.85 μm are responsive to reflected sunlight and are used to detect clouds and identify coastlines in the images from the daytime part of each orbit. Channels 4 and 5 (**Table 2** and **Figure 1**) are in the atmospheric window close to the peak of the thermal emission from the sea surface and are used primarily for the measurement of sea surface temperature. Channel 3, positioned at the shorter wavelength atmospheric window, is responsive to both surface emission and reflected sunlight. During the nighttime part of each orbit, measurements of channel 3 brightness temperatures can be used with those from channels 4 and 5 in variants of the atmospheric correction algorithm to determine SST. The presence of reflected sunlight during the daytime part of the orbit prevents much of these data from being used for SST measurement. Because of the tilting of the sea surface by waves, the area contaminated by reflected sunlight (sun glitter) can be quite extensive, and is dependent on the local surface wind speed. It is limited to the point of specular reflection only in very calm seas.

The images in each channel are constructed by scanning the field of view of the AVHRR across the

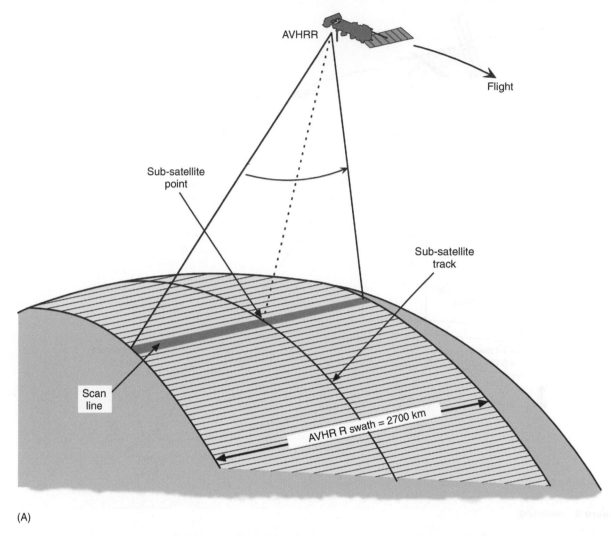

Figure 2 Scan geometries of AVHRR (A) and ATSR (B). The continuous wide swath of the AVHRR is constructed by linear scan lines aligned across the direction of motion of the subsatellite point. The swaths of the ATSR are generated by an inclined conical scan, which covers the same swath through two different atmospheric path lengths. The swath is limited to 512 km by geometrical constraints. Both radiometers are on sun-synchronous, polar-orbiting satellites.

Earth's surface by a mirror inclined at 45° to the direction of flight (**Figure 2A**). The rate of rotation, 6.67 Hz, is such that successive scan lines are contiguous at the surface directly below the satellite. The width of the swath (~2700 km) means that the swaths from successive orbits overlap so that the whole Earth's surface is covered without gaps each day.

The Along-Track Scanning Radiometer (ATSR)

An alternative approach to correcting the effects of the intervening atmosphere is to make a brightness temperature measurement of the same area of sea surface through two different atmospheric path lengths. The pairs of such measurements must be made in quick succession, so that the SST and atmospheric conditions do not change in the time interval. This approach is that used by the ATSR, two of which have flown on the European satellites ERS-1 and ERS-2.

The ATSR has infrared channels in the atmospheric windows comparable to those of AVHRR, but the rotating scan mirror sweeps out a cone inclined from the vertical by its half-angle (**Figure 2B**). The field of view of the ATSR sweeps out a curved path on the sea surface, beginning at the point directly below the satellite, moving out sideways and forwards. Half a mirror revolution later, the field of view is about 900 km ahead of the sub-satellite track in the center of the 'forward view'. The path of the field of view returns to the sub-satellite point, which,

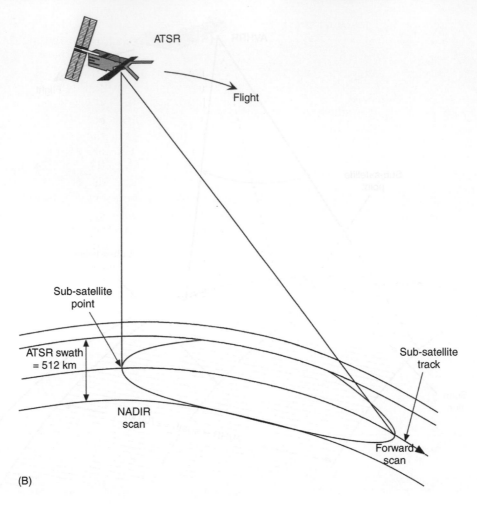

(B)

Figure 2 continued

during the period of the mirror rotation, has moved 1 km ahead of the starting point. Thus the pixels forming the successive swaths through the nadir point are contiguous. The orbital motion of the satellite means that the nadir point overlays the center of the forward view after about 2 min. The atmospheric path length of the measurement at nadir is simply the thickness of the atmosphere, whereas the slant path to the center of the forward view is almost double that, resulting in colder brightness temperatures. The differences in the brightness temperatures between the forward and nadir swaths are a direct measurement of the effect of the atmosphere and permit a more accurate determination of the sea surface temperature. The atmospheric correction algorithm takes the form:

$$SST = c_o + \sum_i c_{n,i} T_{n,i} + \sum_i c_{f,i} T_{f,i} \qquad [5]$$

where the subscripts n and f refer to measurements from the nadir and forward views, i indicates two or three atmospheric window channels and the set of c are coefficients. The coefficients, derived by radiative transfer simulations, have an explicit latitudinal dependence.

Accurate calibration of the brightness temperatures is achieved by using two onboard black-body cavities, situated between the apertures for the nadir and forward views such that they are scanned each rotation of the mirror. One calibration target is at the spacecraft ambient temperature while the other is heated, so that the measured brightness temperatures of the sea surface are straddled by the calibration temperatures.

The limitation of the simple scanning geometry of the ATSR is a relatively narrow swath width of 512 km. The ERS satellites have at various times in their missions been placed in orbits with repeat patterns of 3, 35, and 168 days, and given the narrow ATSR swath, complete coverage of the globe has been possible only for the 35 and 186 day cycles. This disadvantage is offset by the intended improvement in absolute accuracy of the atmospheric correction, and of its better insensitivity to aerosol effects.

The Moderate Resolution Imaging Spectroradiometer (MODIS)

The MODIS is a 36-band imaging radiometer on the NASA Earth Observing System (EOS) satellites *Terra*, launched in December 1999, and *Aqua*, planned for launch by late 2001. MODIS is much more complex than other radiometers used for SST measurement, but uses the same atmospheric windows. In addition to the usual two bands in the 10–12 μm interval, MODIS has three narrow bands in the 3.7–4.1 μm windows, which, although limited by sun-glitter effects during the day, hold the potential for much more accurate measurement of SST during the night. Several of the other 31 bands of MODIS contribute to the SST measurement by better identification of residual cloud and aerosol contamination.

The swath width of MODIS, at 2330 km, is somewhat narrower than that of AVHRR, with the result that a single day's coverage is not entire, but the gaps from one day are filled in on the next. The spatial resolution of the infrared window bands is 1 km at nadir.

The GOES Imager

SST measurements from geosynchronous orbit are made using the infrared window channels of the GOES Imager. This is a five-channel instrument that remains above a given point on the Equator. The image of the Earth's disk is constructed by scanning the field of view along horizontal lines by an oscillating mirror. The latitudinal increments of the scan line are done by tilting the axis of the scan mirror. The spatial resolution of the infrared channels is 2.3 km (east–west) by 4 km (north–south) at the subsatellite point. There are two imagers in orbit at the same time on the two GOES satellites, covering the western Atlantic Ocean (GOES-East) and the eastern Pacific Ocean (GOES-West). The other parts of the global oceans visible from geosynchronous orbit are covered by three other satellites operated by Japan, India, and the European Meteorological Satellite organization (Eumetsat). Each carries an infrared imager, but with lesser capabilities than the GOES Imager.

TRMM Microwave Imager (TMI)

The TMI is a nine-channel microwave radiometer on the Tropical Rainfall Measuring Mission satellite, launched in 1997. The nine channels are centered at five frequencies: 10.65, 19.35, 21.3, 37.0, and 85.5 GHz, with four of them being measured at two polarizations. The 10.65 GHz channels confer a sensitivity to SST, at least in the higher SST range found in the tropics, that has been absent in microwave radiometers since the SMMR (Scanning Multifrequency Microwave Radiometer) that flew on the short-lived Seasat in 1978 and on Nimbus-7 from 1978 to 1987. Although SSTs were derived from SMMR measurements, these lacked the spatial resolution and absolute accuracy to compete with those of the AVHHR. The TMI complements AVHRR data by providing SSTs in the tropics where persistent clouds can be a problem for infrared retrievals. Instead of a rotating mirror, TMI, like other microwave imagers, uses an oscillating parabolic antenna to direct the radiation through a feed-horn into the radiometer.

The swath width of TMI is 759 km and the orbit of TRMM restricts SST measurements to within 38.5° of the equator. The beam width of the 10.65 GHz channels produces a footprint of 37×63 km, but the data are over-sampled to produce 104 pixels across the swath.

Applications

With absolute accuracies of satellite-derived SST fields of ∼0.5 K or better, and even smaller relative uncertainties, many oceanographic features are resolved. These can be studied in a way that was hitherto impossible. They range from basin-scale perturbations to frontal instabilities on the scales of tens of kilometers. SST images have revealed the great complexity of ocean surface currents; this complexity was suspected from shipboard and aircraft measurements, and by acoustically tacking neutrally buoyant floats. However, before the advent of infrared imagery the synoptic view of oceanic variability was elusive, if not impossible.

El Niño

The El Niño Southern Oscillation (ENSO) phenomenon has become a well-known feature of the coupled ocean–atmosphere system in terms of perturbations that have a direct influence on people's lives, mainly by altering the normal rainfall patterns causing draughts or deluges – both of which imperil lives, livestock, and property.

The normal SST distribution in the topical Pacific Ocean is a region of very warm surface waters in the west, with a zonal gradient to cooler water in the east; superimposed on this is a tongue of cool surface water extending westward along the Equator. This situation is associated with heavy rainfall over the western tropical Pacific, which is in turn associated with lower level atmospheric convergence and deep atmospheric convection. The atmospheric convergence and convection are part of the large-scale global circulation. The warm area of surface water, enclosed by the 28°C isotherm, is commonly referred to as the 'Warm Pool' and in the normal situation is confined to the western part of the tropical Pacific. During an El Niño event the warm surface water, and associated convection and rainfall, migrate eastward perturbing the global atmospheric circulation. El Niño events occur up to a few times per decade and are of very variable intensity. Detailed knowledge of

the shape, area, position, and movement of the Warm Pool can be provided from satellite-derived SST to help study the phenomenon and forecast its consequences.

Figure 3 shows part of the global SST fields derived from the Pathfinder SST algorithm applied to AVHRR measurements. The tropical Pacific SST field in the normal situation (December 1993) is shown in the upper panel, while the lower panel shows the anomalous field during the El Niño event of 1997–98. This was one of the strongest El Niños on record, but also the best documented and forecast. Seasonal

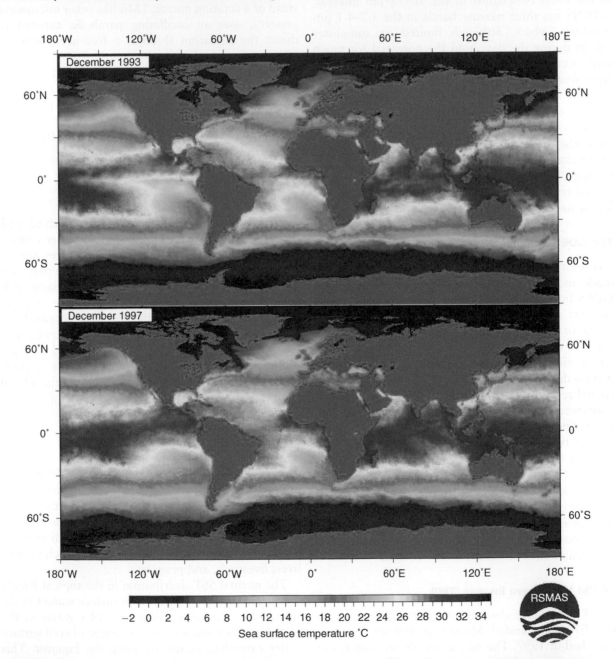

Figure 3 Global maps of SST derived from the AVHRR Pathfinder data sets. These are monthly composites of cloud-free pixels and show the normal situation in the tropical Pacific Ocean (above) and the perturbed state during an El Niño event (below).

predictions of disturbed patterns of winds and rain-fall had an unprecedented level of accuracy and provided improved useful forecasts for agriculture in many affected areas. Milder than usual hurricane and tropical cyclone seasons were successfully fore-cast, as were much wetter winters and severe coastal erosion on the Pacific coasts of the Americas.

Hurricane Intensification

The Atlantic hurricane season in 1999 was one of the most damaging on record in terms of land-falling storms in the eastern USA, Caribbean, and Central America. Much of the damage was not a result of high winds, but of torrential rainfall. Accurate fore-casting of the path and intensity of these land-falling storms is very important, and a vital component of this forecasting is detailed knowledge of SST patterns in the path of the hurricanes. The SST is indicative of the heat stored in the upper ocean that is available to intensify the storms, and SSTs of >26°C appear to be necessary to trigger the intensification of the hurricanes. Satellite-derived SST maps are used in the prediction of the development of storm propagation across the Atlantic Ocean from the area off Cape Verde where atmospheric disturbances spawn the nascent storms. Closer to the USA and Caribbean, the SST field is important in determining the sudden intensification that can occur just before landfall. After the hurricane has passed, they sometimes leave a wake of cooler water in the surface that is readily identifiable in the satellite-derived SST fields.

Frontal Positions

One of the earliest features identified in infrared images of SST were the positions of ocean fronts, which delineate the boundaries between dissimilar surface water masses. Obvious examples are western boundary currents, such as the Gulf Stream in the Atlantic Ocean (**Figure 4**) and the Kuroshio in the Pacific Ocean, both of which transport warm surface water poleward and away from the western coast-lines. In the Atlantic, the path of the warm surface water of the Gulf Stream can be followed in SST images across the ocean, into the Norwegian Sea, and into the Arctic Ocean. The surface water loses heat to the atmosphere, and to adjacent cooler waters on this path from the Gulf of Mexico to the Arctic, producing a marked zonal difference in the climates of the opposite coasts of the Atlantic and Greenland-Norwegian Seas. Instabilities in the fronts at the sides of the currents have been revealed in great detail in the SST images. Some of the large-scale instabilities can lead to loops on scales of a few tens to hundreds of kilometers that can become

'pinched off' from the flow and evolve as independ-ent features, migrating away from the currents. When these occur on the equator side of the current these are called 'Warm Core Rings' and can exist for many months; in the case of the Gulf Stream these can propagate into the Sargasso Sea.

Figure 5 shows a series of instabilities along the boundaries of the Equatorial current system in the Pacific Ocean. The extent and structure of these features were first described by analysis of satellite SST images.

Coral Bleaching

Elevated SSTs in the tropics have adverse influences on living coral reefs. When the temperatures exceed the local average summertime maximum for several days, the symbiotic relationship between the coral polyps and their algae breaks down and the reef-building animals die. The result is extensive areas where the coral reef is reduced to the skeletal struc-ture without the living and growing tissue, giving the reef a white appearance. Time-series of AVHRR-de-rived SST have been shown to be valuable predictors of reef areas around the globe that are threatened by warmer than usual water temperatures. Although it is not possible to alter the outcome, SST maps have been useful in determining the scale of the problem and identifying threatened, or vulnerable reefs.

The 'Global Thermometer'

Some of the most pressing problems facing environ-mental scientists are associated with the issue of global climate change: whether such changes are natural or anthropogenic, whether they can be forecast accurately on regional scales over decades, and whether undesirable consequences can be avoi-ded. The past decade has seen many air temperature records being surpassed and indeed the planet ap-pears to be warming on a global scale. However, the air temperature record is rather patchy in its distri-bution, with most weather stations clustered on Northern Hemisphere continents.

Global SST maps derived from satellites provide an alternative approach to measuring the Earth's temperature in a more consistent fashion. However, because of the very large thermal inertia of the ocean (it takes as much heat to raise the temperature of only the top meter of the ocean through one degree as it does for the whole atmosphere), the SST changes indicative of global warming are small. Climate change forecast models indicate a rate of temperature increase of only a few tenths of a degree per decade, and this is far from certain because of our incomplete understanding of how the climate

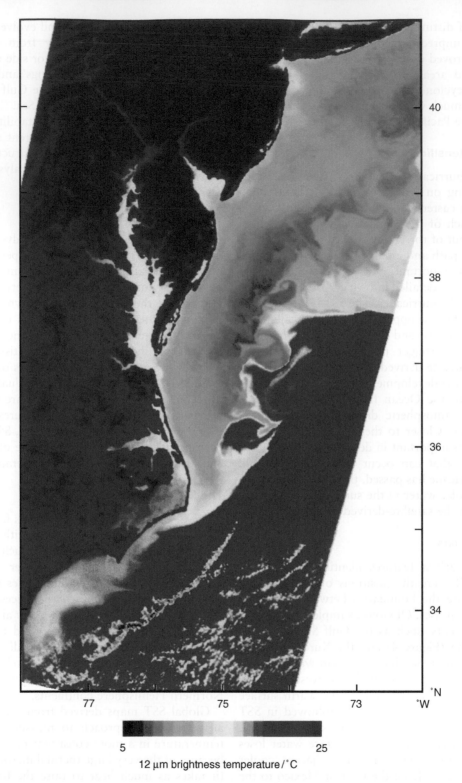

Figure 4 Brightness temperature image derived from the measurements of the ATSR on a nearly cloud-free day over the eastern coast of the USA. The warm core of the Gulf Stream is very apparent; it departs from the coast at Cape Hatteras. The cool, shelf water from the north entrains the warmer outflows from the Chesapeake and Delaware Bays. The north wall of the Gulf Stream reveals very complex structure associated with frontal instabilities that lead to exchanges between the Gulf Stream and inshore waters. The small-scale multicolored patterns over the warm Gulf Stream waters to the south indicate the presence of cloud. This image was taken at 15.18 UTC on 21 May 1992, and is derived from nadir view data from the 12 μm channel. (Generated from data © NERC/ESA/RAL/BNSC, 1992)

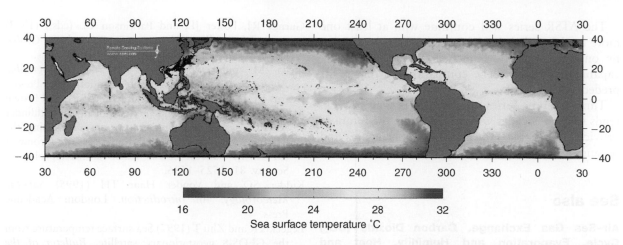

Figure 5 Tropical SSTs produced by microwave radiometer measurements from the TRMM (Tropical Rainfall Measuring Mission) Microwave Imager (TMI). This is a composite generated from data taken during the week ending December 22, 1999. The latitudinal extent of the data is limited by the orbital geometry of the TRMM satellite. The measurement is much less influenced by clouds than those in the infrared, but the black pixels in parts of the oceans where there are no islands indicate areas of heavy rainfall. The image reveals the cold tongue of surface water along the Equator in the Pacific Ocean and cold water off the Pacific coast of South America, indicating a non-El Niño situation. Note that the color scale is different from that used in **Figure 3**. The image was produced by Remote Sensing Systems, sponsored in part by NASA's Earth Science Information Partnerships (ESIP) (a federation of information sites for Earth science); and by the NOAA/NASA Pathfinder Program for early EOS products; principal investigator: Frank Wentz.

system functions, especially in terms of various feedback factors such as those involving changes in cloud and aerosol properties. Such a rate of temperature increase will require SST records of several decades length before the signal, if present, can be unequivocally identified above the uncertainties in the accuracy of the satellite-derived SSTs. Furthermore, the inherent natural variability of the global SST fields tends to mask any small, slow changes. Difficult though this task may be, global satellite-derived SSTs are an important component in climate change research.

Air-sea Exchanges

The SST fields play further indirect roles in the climate system in terms of modulating the exchanges of heat and greenhouse gases between the ocean and atmosphere. Although SST is only one of several variables that control these exchanges, the SST distributions, and their evolution on seasonal timescales can help provide insight into the global patterns of the air–sea exchanges. An example of this is the study of tropical cloud formation over the ocean, a consequence of air–sea heat and moisture exchange, in terms of SST distributions.

Future Developments

Over the next several years continuing improvement of the atmospheric correction algorithms can be anticipated to achieve better accuracies in the derived

SST fields, particularly in the presence of atmospheric aerosols. This will involve the incorporation of information from additional spectral channels, such as those on MODIS or other EOS era satellite instruments. Improvements in SST coverage, at least in the tropics, can be expected in areas of heavy, persistent cloud cover by melding SST retrievals from high-resolution infrared sensors with those from microwave radiometers, such as the TMI.

Continuing improvements in methods of validating the SST retrieval algorithms will improve our understanding of the error characteristics of the SST fields, guiding both the appropriate applications of the data and also improvements to the algorithms.

On the hardware front, a new generation of infrared radiometers designed for SST measurements will be introduced on the new operational satellite series, the National Polar-Orbiting Environmental Satellite System (NPOESS) that will replace both the civilian (NOAA-n) and military (DMSP, Defense Meteorological Satellite Program) meteorological satellites. The new radiometer, called VIIRS (the Visible and Infrared Imaging Radiometer Suite), will replace the AVHRR and MODIS. The prototype VIIRS will fly on the NPP (NPOESS Preparatory Program) satellite scheduled for launch in late 2005. At present, the design details of the VIIRS are not finalized, but the physics of the measurement constrains the instrument to use the same atmospheric window channels as previous and current instruments, and have comparable, or better, measurement accuracies.

The ATSR series will continue with at least one more model, called the Advanced ATSR (AATSR) to fly on Envisat to be launched in 2001. The SST capability of this will be comparable to that of its predecessors.

Thus, the time-series of global SSTs that now extends for two decades will continue into the future to provide invaluable information for climate and oceanographic research.

See also

Air–Sea Gas Exchange. Carbon Dioxide (CO₂) Cycle. Evaporation and Humidity. Heat and Momentum Fluxes at the Sea Surface. IR Radiometers. Penetrating Shortwave Radiation. Radiative Transfer in the Ocean. Satellite Oceanography, History, and Introductory Concepts. Satellite Remote Sensing Sar. Upper Ocean Time and Space Variability.

Further Reading

Barton IJ (1995) Satellite-derived sea surface temperatures: Current status. *Journal of Geophysical Research* 100: 8777–8790.

Gurney RJ, Foster JL, and Parkinson CL (eds.) (1993) *Atlas of Satellite Observations Related to Global Change.* Cambridge: Cambridge University Press.

Ikeda M and Dobson FW (1995) *Oceanographic Applications of Remote Sensing.* London: CRC Press.

Kearns EJ, Hanafin JA, Evans RH, Minnett PJ, and Brown OB (2000) An independent assessment of Pathfinder AVHRR sea surface temperature accuracy using the Marine-Atmosphere Emitted Radiance Interferometer (M-AERI). Bulletin of the American Meteorological Society. 81: 1525–1536.

Kidder SQ and Vonder Haar TH (1995) *Satellite Meteorology: An Introduction.* London: Academic Press.

Legeckis R and Zhu T (1997) Sea surface temperature from the GEOS-8 geostationary satellite. *Bulletin of the American Meteorological Society* 78: 1971–1983.

May DA, Parmeter MM, Olszewski DS, and Mckenzie BD (1998) Operational processing of satellite sea surface temperature retrievals at the Naval Oceanographic Office. *Bulletin of the American Meteorological Society,* 79: 397–407.

Robinson IS (1985) *Satellite Oceanography: An Introduction for Oceanographers and Remote-sensing Scientists.* Chichester: Ellis Horwood.

Stewart RH (1985) *Methods of Satellite Oceanography.* Berkeley, CA: University of California Press.

Victorov S (1996) *Regional Satellite Oceanography.* London: Taylor and Francis.

SATELLITE REMOTE SENSING SAR

A. K. Liu and S. Y. Wu, NASA Goddard
Space Flight Center, Greenbelt, MD, USA

Introduction

Synthetic aperture radar (SAR) is a side-looking imaging radar usually operating on either an aircraft or a spacecraft. The radar transmits a series of short, coherent pulses to the ground producing a footprint whose size is inversely proportional to the antenna size, its aperture. Because the antenna size is generally small, the footprint is large and any particular target is illuminated by several hundred radar pulses. Intensive signal processing involving the detection of small Doppler shifts in the reflected signals from targets to the moving radar produces a high resolution image that is equivalent to one that would have been collected by a radar with a much larger aperture. The resulting larger aperture is the 'synthetic aperture' and is equal to the distance traveled by the spacecraft while the radar antenna is collecting information about the target. SAR techniques depend on precise determination of the relative position and velocity of the radar with respect to the target, and on how well the return signal is processed.

SAR instruments transmit radar signals, thus providing their own illumination, and then measure the strength and phase of the signals scattered back to the instrument. Radar waves have much longer wavelengths compared with light, allowing them to penetrate clouds with little distortion. In effect, radar's longer wavelengths average the properties of air with the properties and shapes of many individual water droplets, and are only affected while entering and exiting the cloud. Therefore, microwave radar can 'see' through clouds.

SAR images of the ocean surface are used to detect a variety of ocean features, such as refracting surface gravity waves, oceanic internal waves, wind fields, oceanic fronts, coastal eddies, and intense low pressure systems (i.e. hurricanes and polar lows), since they all influence the short wind waves responsible for radar backscatter. In addition, SAR is the only sensor that provides measurements of the directional wave spectrum from space. Reliable coastal wind vectors may be estimated from calibrated SAR images using the radar cross-section. The ability of a SAR to provide valuable information on the type, condition, and motion of the sea ice and surface signatures of swells, wind fronts, and eddies near the ice edge has also been amply demonstrated.

With all-weather, day/night imaging capability, SAR penetrates clouds, smoke, haze, and darkness to acquire high quality images of the Earth's surface. This makes SAR the frequent sensor of choice for cloudy coastal regions. Space agencies from the USA, Canada, and Europe use SAR imagery on an operational basis for sea ice monitoring, and for the detection of icebergs, ships, and oil spills. However, there can be considerable ambiguity in the interpretation of physical processes responsible for the observed ocean features. Therefore, the SAR imaging mechanisms of ocean features are briefly described here to illustrate how SAR imaging is used operationally in applications such as environmental monitoring, fishery support, and marine surveillance.

History

The first spaceborne SAR was flown on the US satellite Seasat in 1978. Although Seasat only lasted 3 months, analysis of its data confirmed the sensitivity of SAR to the geometry of surface features. On March 31, 1991 the Soviet Union became the next country to operate an earth-orbiting SAR with the launch of Almaz-1. Almaz-1 returned to earth in 1992 after operating for about 18 months. The European Space Agency (ESA) launched its first remote sensing satellite, ERS-1, with a C-band SAR on July 17, 1991. Shortly thereafter, the JERS-1 satellite, developed by the National Space Development Agency of Japan (NASDA), was launched on February 11, 1992 with an L-band SAR. This was followed a few years later by Radarsat-1, the first Canadian remote sensing satellite, launched in November 1995. Radarsat-1 has a ScanSAR mode with a 500 km swath and a 100 m resolution, an innovative variation of the conventional SAR (with a swath of 100 km and a resolution of 25 m). ERS-2 was launched in April 1995 by ESA, and Envisat-1 with an Advanced SAR is underway with a scheduled launch date in July 2001. The Canadian Space Agency (CSA) has Radarsat-2 planned for 2002, and NASDA has Advanced Land Observing Satellite (ALOS) approved for 2003. **Table 1** shows all major ocean-oriented spaceborne SAR missions worldwide from 1978 to 2003.

Table 1 Major ocean-oriented spaceborne SAR missions

Platform	Nation	Launch	Band[a]	Status
Seasat	USA	1978	L	Ended
Almaz-1	USSR	1991	S	Ended
ERS-1	Europe	1991	C	Standby
JERS-1	Japan	1992	L	Ended
ERS-2	Europe	1995	C	Operational
Radarsat-1	Canada	1995	C	Operational
Envisat-1	Europe	2001	C	Launch scheduled
Radarsat-2	Canada	2002	C	Approved
ALOS	Japan	2003	L	Approved

[a]Some frequently used radar wavelengths are: 3.1 cm for X-band, 5.66 cm for C-band, 10.0 cm for S-band, and 23.5 cm for L-band.

Aside from these free-flying missions, a number of early spaceborne SAR experiments in the USA were conducted using shuttle imaging radar (SIR) systems flown on NASA's Space Shuttle. The SIR-A and SIR-B experiments, in November 1981 and October 1984, respectively, were designed to study radar system performance and obtain sample data of the land using various incidence angles. The SIR-B experiment provided a unique opportunity for studying ocean wave spectra due to the relatively lower orbit of the Shuttle as compared with satellites. The low orbital altitude increases the frequency range of ocean waves that could be reliably imaged, because blurring of the detected waves caused by the motion of ocean surface during the imaging process is reduced. The final SIR mission, SIR-C in April and October 1994, simultaneously recorded SAR data at three wavelengths (L-, C-, and X-bands). These multiple-frequency data from SIR-C improved our understanding of the radar scattering properties of the ocean surface.

Imaging Mechanism of Ocean Features

For a radar with an incidence angle of 20°–50°, such as all spaceborne SARs, backscatter from the ocean

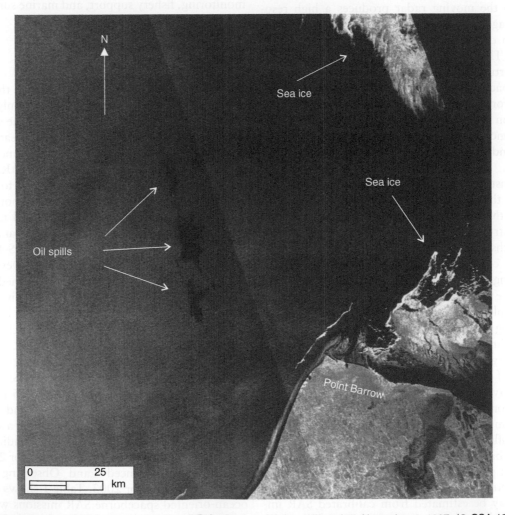

Figure 1 Radarsat ScanSAR image of oil spills off Point Barrow, Alaska, collected on November 2, 1997. (© CSA 1997.)

surface is produced primarily by the Bragg resonant scattering mechanism. That is, surface waves traveling in the radar range (across-track) direction with a wavelength of $\lambda/(2\sin\theta)$, called the Bragg resonant waves, account for most of the backscattering. In this formula, λ is the radar wavelength, and θ is the incidence angle. In general, the Bragg resonant waves are short gravity waves with wavelengths in the range of 3–30 cm, depending on the radar wavelength or band, as shown in **Table 1**. Because SAR is most sensitive to waves of this wavelength, or roughness of this scale, any ocean phenomenon or process that produces modulation in these particular wavelengths is theoretically detectable by SAR. The radar cross-section of the ocean surface is affected by any geophysical variable, such as wind stress, current shear, or surface slicks, that can modulate the ocean surface roughness at

Bragg-scattering scales. Thus, SARs have proven to be an excellent means of mapping ocean features.

For ocean current features, the essential element of the surface manifestation is the interaction between the current field and the wind-driven ocean surface waves. The effect of the surface current is to alter the short-wave spectrum from its equilibrium value, while the natural processes of wave energy input from the wind restores the ambient equilibrium spectrum. A linear SAR system is one for which the variation of the SAR image intensity is proportional to the gradient of the surface velocity. The proportionality depends on radar wavelength, radar incidence angle, angle between the radar look direction and the current direction, azimuth angle, and the wind velocity. Under high wind condition, large wind waves may overwhelm the weaker current feature. When current flows in the cross-wind direction, the

Figure 2 ERS-1 SAR image of shallow water bathymetry at Taiwan Tan acquired on July 27, 1994. (© ESA 1994.)

wave–current interaction is relatively weak, causing a weak radar backscattering signal.

For ocean frontal features, the change in surface brightness across a front in a SAR image is caused by the change in wind stress exerted onto the ocean surface. The wind stress in turn depends on wind speed and direction, air–sea temperature difference, and surface contamination. The effects of wind stress upon surface ripples and therefore upon radar cross-section, have been modeled and demonstrated as shown in the example below. In the high wind stress area, the ocean surface is rougher and appears as a brighter area in a SAR image. On the other side of the front where the wind is lower, the surface is smoother and appears as a darker area in a SAR image.

The reason why surface films are detectable on radar images is that oil films have a dampening effect on short surface waves. Radar is remarkably sensitive to small changes in the roughness of sea surface. Oil slicks also have a dark appearance in radar images and are thus similar to the appearance of areas of low winds. The distinctive shape and sharp boundary of localized surface films allows them to be distinguished from the relatively large regions of low wind.

Examples of Ocean Features from SAR Applications

A number of important SAR applications have emerged recently, particularly since ERS-1/2, and Radarsat-1 data became available and the ability to process SAR data has improved. In the USA, the National Oceanic and Atmospheric Administration (NOAA) and the National Ice Center use SAR imagery on an operational basis for sea ice monitoring, iceberg detection, fishing enforcement, oil spill detection, wind and storm information. In Canada, sea ice surveillance is now a proven near-real-time operation, and new marine and coastal applications for

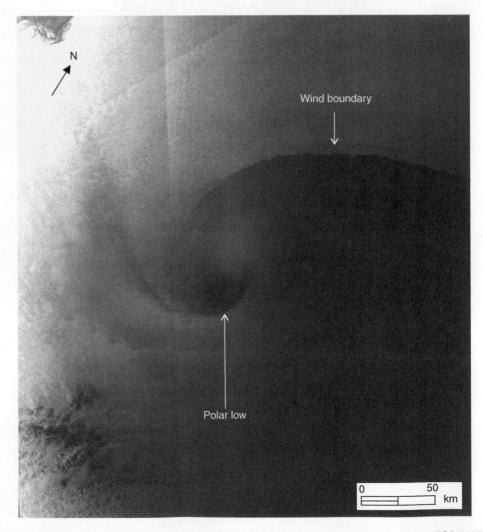

Figure 3 Radarsat ScanSAR image of a polar low in the Bering Sea collected on February 5, 1998. (© CSA 1998.)

SAR imagery are still emerging. In Europe, research on SAR imaging of ocean waves has received great attention in the past 10 years, and has contributed to better global ocean wave forecasts. However, the role of SAR in the coastal observing system still remains at the research and development stage. For reference, examples of some typical ocean features from SAR applications are provided below.

For marine environmental monitoring, features such as oil spills, bathymetry, and polar lows are important for tracking and can often be identified easily with SAR. In early November 1997, Radarsat's SAR sensor captured an oil spill off Point Barrow, Alaska. The oil slicks showed up clearly on the ScanSAR imagery on November 2, 3 and 9. The oil spill is suspected to be associated with the Alaskan Oil Pipeline. **Figure 1** shows a scene containing the oil slicks cropped out from the original ScanSAR image for a closer look. Tracking oil spills using SAR is useful for planning clean-up activities. Early

detection, monitoring, containment, and clean up of oil spills are crucial to the protection of the environment.

Under favorable wind conditions with strong tidal current, the surface signature of bottom topography in shallow water has often been observed in SAR images. **Figure 2**, showing an ERS-1 SAR image of the shallow water bathymetry of Taiwan Tan collected on July 27, 1994, is such an example. Taiwan Tan is located south west of Taiwan in the Taiwan Strait. Typical water depth there is around 30 m with a valley in the middle and the continental shelf break to the south. An extensive sand wave field (**Figure 2**) is developed at Taiwan Tan regularly by wind, tidal current, and surface waves. Monitoring the changes of bathymetry is critical to ship navigation, especially in the areas where water is shallow and ship traffic is heavy.

Intense low pressure systems in polar regions, often referred to as polar lows, may develop above

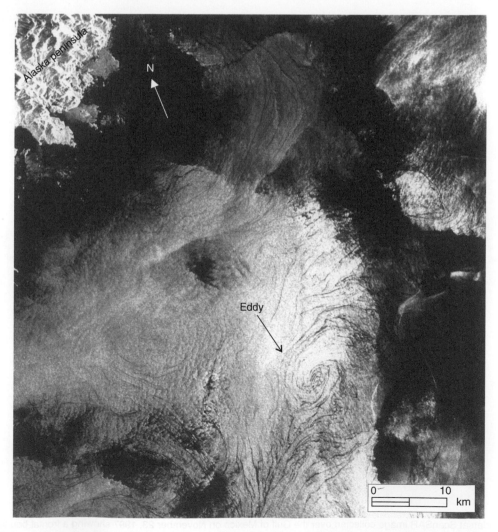

Figure 4   ERS-1 SAR image of lower Shelikof Strait, acquired on October 23, 1991 showing a spiral eddy. (© ESA 1991.)

regions between colder ice/land and warmer ocean during cold air outbreaks. These intense polar lows are formed off major jet streams in cold air masses. Since they usually occur near polar regions where data are sparse, SAR images have been a useful tool for studying these phenomena. **Figure** 3 shows a Radarsat ScanSAR image of a polar low in the Bering Sea (centered at 58.0°N, 174.9°E) collected on February 5, 1998. It has a wind boundary to the north spiraling all the way to the center of the storm that separates the high wind (bright) area from the low wind (dark) area. The rippled character along the wind boundary indicates the presence of an instability disturbance induced by the shear flow, which in turn is caused by the substantial difference in wind speed across the boundary.

Ocean features such as eddies, fronts, and ice edges can result in changes in water temperature, turbulence, or transport and may be the primary determinant of recruitment to fisheries. The survival of larvae is enhanced if they remain on the continental shelf and ultimately recruit to nearshore nursery areas. Features such as fronts and eddies can retain larval patches within the shelf zone. **Figure** 4 shows an ERS-1 SAR image acquired on October 23, 1991 (centered at 56.69°N, 156.07°W) in lower Shelikof Strait, the Gulf of Alaska. In this image, an eddy with a diameter of approximately 20 km is visible due to low wind conditions. The eddy is characterized by spiraling curvilinear lines which are most likely associated with current shears, surface films, and to a lesser extent temperature contrasts. SAR has the potential to locate these eddies over extensive areas in coastal oceans.

A Radarsat ScanSAR image over the Gulf of Mexico taken on November 23, 1997 (**Figure** 5) shows a distinct, nearly straight front stretching at least 300 km in length. The center of the front in this scene is in the Gulf of Mexico some 400 km south west of New Orleans. The frontal orientation is about 76° east of north. Closer inspection reveals that there are many surface film-like filaments on the

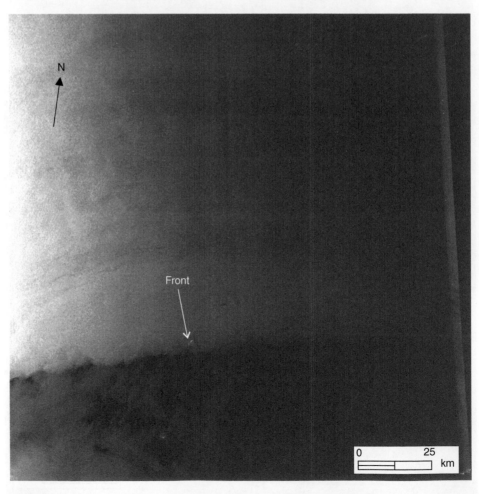

Figure 5   Radarsat ScanSAR image collected over the Gulf of Mexico on November 23, 1997 showing a frontal boundary. (© CSA 1997.)

south side of the front. Concurrent wind data suggest that surface currents converge along the front. Therefore, the formation of this front is probably caused by the accumulation of natural surface films brought about by the convergence around the front. This example highlights SAR's sensitivity to the changes of wind speed and the presence of surface films.

The edge of the sea ice has been found to be highly productive for plankton spring bloom and fishery feeding. In the Bering Sea, fish abundance is highly correlated with yearly ice extent because for their survival many species of fish prefer the cold pools left behind after ice retreat. SAR images are very useful for tracking the movement of the ice edge. **Figure 6** shows a Radarsat ScanSAR image near the ice edge in the Bering Sea collected on February 29, 2000 (centered at 59.6°N and 177.3°W). The sea ice pack with ice bands extending from the ice edge can be clearly seen as the bright area because sea ice surface is rougher than ocean surface (dark area). In the same image, a front is also visible and may or may not be associated with the ice edge to the north. The cold water near the ice edge dampens wave action and appears as a darker area compared with the other side of the front, where it shows up as brighter area due to higher wind and higher sea states.

Information on surface and internal waves, as well as ship wakes, are very important and valuable for marine surveillance and ship navigation. The principal use of SAR for oceanographic studies has been for the detection of ocean waves. The wave direction and height derived from SAR data can be incorporated into models of wind–wave forecast and other applications such as wave–current interaction. **Figure 7** shows an ERS-1 SAR image of long surface waves (or swells) in the lower Shelikof Strait collected on October 17, 1991 (centered at 56.11°N,

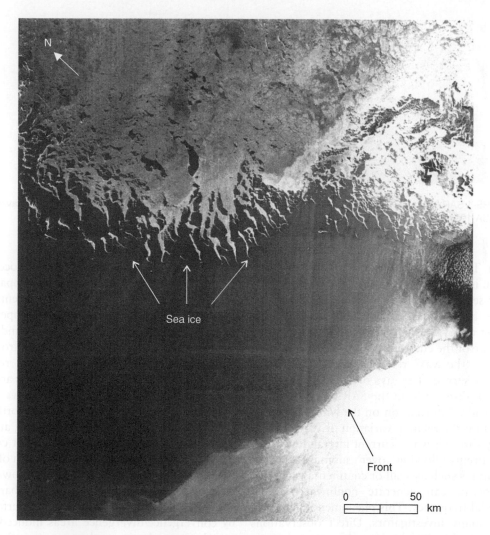

Figure 6   Radarsat ScanSAR image collected over the Bering Sea on February 29, 2000 showing a front near the ice edge. (© CSA 2000.)

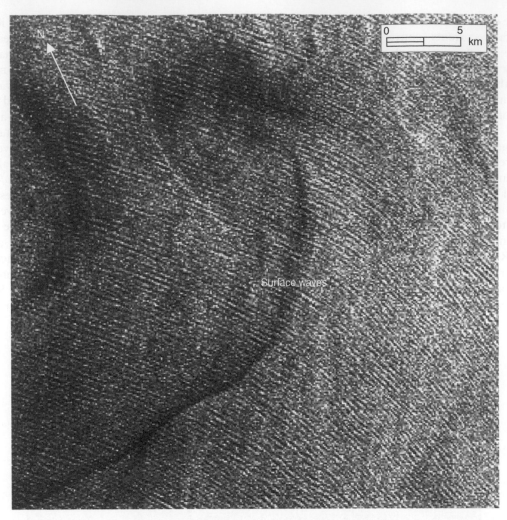

Figure 7   ERS-1 SAR image of lower Shelikof Strait, obtained on October 17, 1991 showing long surface gravity waves refracted by current. (© ESA 1991.)

156.36°W). The location is close to that of the spiral eddy shown in **Figure 4**. Because of the higher winds and higher sea states at the time the image was acquired, the eddy is less conspicuous in this SAR image taken 6 days earlier. Although the direct surface signature of the eddy cannot be discerned clearly in this image, the wave refraction in the eddy area can still be observed. The rays of the wave field can be traced out directly from the SAR image. The ray pattern provides information on the wave refraction pattern and on the relative variation of wave energy along a ray through wave–current interaction.

Tidal currents flowing over submarine topographic features such as a sill or continental shelf in a stratified ocean can generate nonlinear internal waves of tidal frequency. This phenomenon has been studied by many investigators. Direct observations have lent valuable insight into the internal wave generation process and explained the role they play

in the transfer of energy from tides to ocean mixing. These nonlinear internal waves are apparently generated by internal mixing as tidal currents flow over bottom features and propagate in the open ocean. In the South China Sea near DongSha Island, enormous westward propagating internal waves from the open ocean are often confronted by coral reefs on the continental shelf. As a result, the waves are diffracted upon passing the reefs. **Figure 8** shows a Radarsat ScanSAR image collected over the northern South China Sea on April 26, 1998, showing at least three packets of internal waves. Each packet consists of a series of internal waves, and the pattern of each wave is characterized by a bright band followed immediately by a dark band. The bright/dark bands indicate the contrast in ocean rough/smooth surfaces caused by convergence/divergence areas induced by the internal waves. At times, the wave 'crest' as observed by SAR from the length of bands can be over 200 km

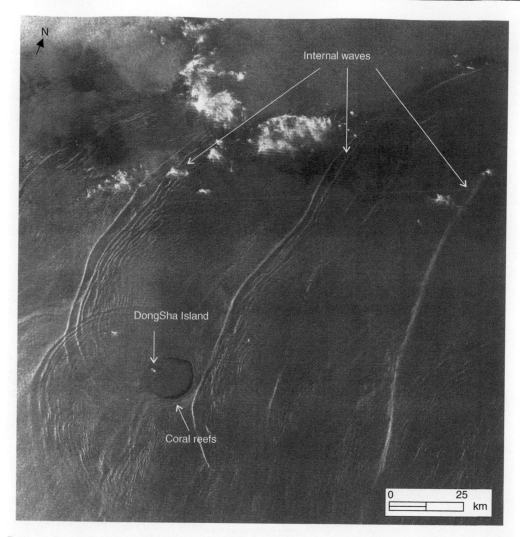

Figure 8 Radarsat ScanSAR image collected over the South China Sea on April 26, 1998 showing three internal wave packets. (© CSA 1998.)

long. After passing the DongSha coral reefs, the waves regroup themselves into two separate packets of internal waves. Later, they interact with each other and emerge as a single wave packet again. SAR can be a very useful tool for studying these shelf processes and the effect of the internal waves on oil drilling platforms, nutrient mixing, and sediment transport.

Ships and their wakes are commonly observable in high-resolution satellite SAR imagery. Detection of ships and ship wakes by means of remote sensing can be useful in the areas of national defense intelligence, shipping traffic, and fishing enforcement. **Figure 9** is an ERS-1 SAR image collected on May 31, 1995 near the northern coast of Taiwan. The image is centered at 25.62°N and 121.15°E, approximately 30 km offshore in the East China Sea. A surface ship heading north east, represented by a bright spot, can be easily identified. Behind this ship, a long dark turbulence

wake is clearly visible. The turbulent wake dampens any short waves, resulting in an area with low backscattering as indicated by the arrow A in **Figure 9**. Near the ship, the dark wake is accompanied by a bright line which may be caused by the vortex shed by the ship into its wake. The ship track follows the busy shipping lane between Hong Kong, Taiwan, and Japan. The ambient dark slicks are natural surface films induced by upwelling on the continental shelf. In the lower part of the image, another ship turbulent wake (long and dark linear feature oriented east–west) can be identified near the location B in **Figure 9**. A faint bright line connects to the end of this turbulent wake, forming a V-shaped wake in the box B. The faintness of this second ship may be caused by very low backscattering of the ship configuration or the wake could have been formed by a submarine. In the latter case, it must have been operating very close to the ocean surface, since the surface wake is

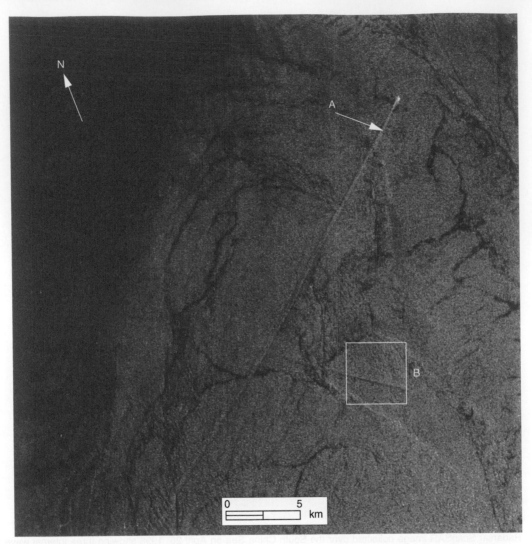

Figure 9 ERS-1 SAR image of East China Sea, obtained on May 31, 1995 showing a surface ship and its wake (arrow A) and a V-shaped wake in box B. (© ESA 1995.)

observable. The ship wake is pointing to the east, indicating that the faint ship was moving from mainland China toward the open ocean.

Discussion

As mentioned earlier, SAR has the unique capability of operating during the day or night and under all weather conditions. With repeated coverage, spaceborne SAR instruments provide the most efficient means to monitor and study the changes in important elements of the marine environment. As demonstrated by the above examples, the use of SAR-derived observations to track eddies, fronts, ice edges, and oil slicks can supply valuable information and can aid in the management of the fishing industry and the protection of the environment. In overcast coastal

areas at high latitudes, the uniformly cold sea surface temperature and persistent cloud cover preclude optical and infrared measurement of surface temperature features, and obscure ocean color observations. The mapping of ocean features by SAR in these challenging coastal regions is, therefore, a potentially major application for satellite-based SAR, particularly for the wider swath ScanSAR mode. Furthermore, SAR data provide unique information for studying the health of the Earth system, as well as critical data for natural hazards and resource assessments.

The prospect of SAR data collection extending well into the twenty-first century gives impetus to current research in SAR applications in ocean science and opens the doors to change detection studies on decadal timescales. The next step is to move into the operational use of SAR data to complement ground measurements. The challenge is to increase cooperation in the scheduling, processing, dissemination, and pricing of SAR data from all SAR satellites between international space agencies. Such cooperation might permit near-real-time high-resolution coastal SAR measurements of sufficient temporal and spatial coverage to impact weather forecasting for selected heavily populated coastal regions. It is necessary to bear in mind that each satellite image is a snapshot and can be complemented with buoy and ship measurements. Ultimately, these data sets should be integrated by numerical models. Such validated and calibrated models will prove extremely useful in understanding a wide variety of oceanic processes.

See also

Aircraft Remote Sensing. Ice–Ocean Interaction. Satellite Oceanography, History, and Introductory Concepts. Satellite Passive-Microwave Measurements of Sea Ice. Satellite Remote Sensing of Sea Surface Temperatures. Surface Films. Wave Generation by Wind.

Further Reading

Alaska SAR Facility User Working Group (1999) *The Critical Role of SAR in Earth System Science.* (http://www.asf.alaska.edu/)

Beal RC and Pichel WG (eds) (2000) *Coastal and Marine Applications of Wide Swath SAR.* Johns Hopkins APL Technical Digest, 21.

European Space Agency (1995) *Scientific Achievements of ERS-1.* ESA SP-1176/I.

Fu L and Holt B (1982) *Seasat Views Oceans and Sea Ice with Synthetic Aperture Radar,* JPL Publication, pp. 81–120. Pasadena, CA: NASA, JPL/CIT.

Hsu MK, Liu AK, and Liu C (2000) An internal wave study in the China Seas and Yellow Sea by SAR. *Continental Shelf Research* 20: 389–410.

Liu AK, Peng CY, and Schumacher JD (1994) Wave–current interaction study in the Gulf of Alaska for detection of eddies by SAR. *Journal of Geophysical Research* 99: 10075–10085.

Liu AK, Peng CY, and Weingartner TJ (1994) Ocean–ice interaction in the marginal ice zone using SAR. *Journal of Geophysical Research* 99: 22391–22400.

Liu AK, Peng CY, and Chang YS (1996) Mystery ship detected in SAR image. *EOS, Transactions, American Geophysical Union* 77: 17–18.

Liu AK, Peng CY, and Chang YS (1997) Wavelet analysis of satellite images for coastal watch. *IEEE Journal of Oceanic Engineering* 22: 9–17.

Tsatsoulis C and Kwok R (1998) *Analysis of SAR Data of the Polar Oceans.* Berlin: Springer-Verlag.

SATELLITE REMOTE SENSING: OCEAN COLOR

C. R. McClain, NASA Goddard Space Flight Center, Greenbelt, MD, USA

Introduction

The term 'ocean color' refers to the spectral composition of the visible light field that emanates from the ocean. The color of the ocean depends on the solar irradiance spectra, atmospheric conditions, solar and viewing geometries, and the absorption and scattering properties of water and the substances that are dissolved and suspended in the water column, for example, phytoplankton and suspended sediments. Water masses whose reflectance is determined primarily by absorption due to water and phytoplankton are generally referred to as 'case 1' waters. In other situations, where scattering is the dominant process or where absorption is dominated by substances other than phytoplankton or their derivatives, the term 'case 2' is applied.

The primary optical variable of interest for remote sensing purposes is the water-leaving radiance, L_w, that is, the subsurface upwelled radiance (light moving upward in the water column) propagating through the air–sea interface, but not including the downwelling irradiance (light moving downward through the atmosphere) reflected at the interface. To simplify the interpretation of ocean color, measurements of the water-leaving radiances are normalized by the surface downwelling irradiances to produce 'remote sensing' reflectances, which provide an unambiguous measure of the ocean's subsurface optical signature. Clear open-ocean reflectances have a spectral peak at blue wavelengths because water absorbs strongly in the near-infrared (NIR) and scatters blue light more effectively than at longer wavelengths. As the concentrations of microscopic green plants (phytoplankton) and suspended materials increase, absorption and scattering reduce the reflectance at blue wavelengths and increase the reflectance at green wavelengths, that is, the color shifts from blue to green and brown. This spectral shift in reflectance can be quantified and used to estimate concentrations of optically active components such as chlorophyll a.

The goal of satellite ocean color analysis is to accurately estimate the water-leaving radiance spectra in order to derive other geophysical quantities, for example, chlorophyll a concentration and the diffuse attenuation coefficient. The motivation for spaceborne observations of this kind lies in the need for frequent high-resolution spatial measurements of these geophysical parameters on regional and global scales for addressing both research and operational requirements associated with marine primary production, ecosystem dynamics, fisheries management, ocean dynamics, and coastal sedimentation and pollution, to name a few. The first proof-of-concept satellite ocean color mission was the Coastal Zone Color Scanner (CZCS) on the *Nimbus*-7 spacecraft which was launched in the summer of 1978. The CZCS was intended to be a 1-year demonstration with very limited data collection, ground processing, and data validation requirements. However, because of the extraordinary quality and unexpected utility of the data for both coastal and open-ocean research, data collection continued until June 1986 when the sensor ceased operating. The entire CZCS data set was processed, archived, and released to the research community by 1990. As a result of the CZCS experience, a number of other ocean color missions have been launched, for example, the Ocean Color and Temperature Sensor (OCTS; Japan; 1996–97), the Sea-viewing Wide Field-of-view Sensor (SeaWiFS, 1997 and continuing; US), and two Moderate Resolution Imaging Spectroradiometer (MODIS on the *Terra* spacecraft, 2000 and continuing; MODIS on the *Aqua* spacecraft, 2002 and continuing; US), with the expectation that continuous global observations will be maintained in the future.

Ocean Color Theoretical and Observational Basis

Reflectance can be defined in a number of ways and the relationship between the various quantities can be confusing. The most common definition is irradiance reflectance, R, just below the surface, as given in eqn [1], where E_u and E_d are the upwelling and downwelling irradiances, respectively, and the superscript minus sign implies the value just beneath the surface:

$$R(\lambda) = \frac{E_u(\lambda, 0^-)}{E_d(\lambda, 0^-)} \qquad [1]$$

In general, irradiance and radiance are functions of depth (or altitude in the atmosphere) and viewing geometry with respect to the sun. R has been theoretically related to the absorption and scattering properties of the ocean as in eqn [2], where Q is the ratio of $E_u(\lambda, 0^-)$ divided by the upwelling radiance, $L_u(\lambda, 0^-)$, $\ell_1 = 0.0949$, $\ell_2 = 0.0794$, $b_b(\lambda)$ is the backscattering coefficient, and $a(\lambda)$ is the absorption coefficient.

$$R(\lambda) = Q(\lambda) \sum_{i=1}^{2} \ell_i \left[\frac{b_b(\lambda)}{a(\lambda) + b_b(\lambda)} \right]^i \qquad [2]$$

Both $b_b(\lambda)$ and $a(\lambda)$ represent the sum of the contributions of the various optical components (water, inorganic particulates, dissolved substances, phytoplankton, etc.) which are often specified explicitly. If the angular distribution of E_u were directionally uniform, that is, Lambertian, Q would equal π. However, the irradiance distribution is not uniform and is dependent on a number of variables. Some experimental results indicate that Q is roughly 4.5 and, to a first approximation, independent of wavelength. In case 1 water, the approximation, $R(\lambda) \sim f[b_b(\lambda)/a(\lambda)]$ is often used and f is assigned a constant value of 0.33. In reality, f is wavelength-dependent. f and Q are also functions of the solar zenith and viewing angles.

When concentrations of substances, for example, chlorophyll a, are measured, the coefficients can be specified in terms of the concentrations, for example, $a_\varphi(\lambda) = a_\varphi^*(\lambda)[\text{chl } a]$ where $a_\varphi(\lambda)$ and $a_\varphi^*(\lambda)$ are the phytoplankton absorption coefficient and specific absorption coefficient, respectively, and $[\text{chl } a]$ is the chlorophyll a concentration. The absorption coefficients are designated for phytoplankton rather than chlorophyll a because the actual absorption by living cells can vary substantially for a fixed amount of chlorophyll a. Thus, $R(\lambda)$ can be expressed in terms of the specific absorption and scattering coefficients and pigment and particle concentrations. When $R(\lambda)$ is observed at a sufficient number of wavelengths across the spectrum, the $R(\lambda)$ values can be inverted to provide estimates of the scattering and absorption coefficients and pigment concentrations.

For satellite applications, the reflectances and radiances just above the surface are more appropriate to use than $R(\lambda)$. Therefore, water-leaving radiance, remote sensing reflectance, $R_{rs}(\lambda)$, and normalized water-leaving radiance, $L_{wn}(\lambda)$, are commonly used. These are defined in eqns [3]–[5], respectively, where ρ is the Fresnel reflectance of the air–sea interface, n is the index of refraction of seawater, $F_o(\lambda)$ is the extraterrestrial solar irradiance, and the plus sign denotes the value just above the surface:

$$L_w(\lambda) = \left[\frac{1 - \rho}{n^2} \right] L_u(\lambda, 0^-) \qquad [3]$$

$$R_{rs}(\lambda) = \frac{L_w(\lambda)}{E_d(\lambda, 0^+)} \qquad [4]$$

and

$$L_{wn}(\lambda) = F_o(\lambda) \frac{L_w(\lambda)}{E_d(\lambda, 0^+)} \qquad [5]$$

The term $(1 - \rho)/n^2$ is approximately equal to 0.54. Furthermore, it can be shown that $L_{wn}(\lambda)$ is related to $R(\lambda)$ by eqn [6] where the $\langle \rho \rangle$ and $\langle r \rangle$ denote the angular mean Fresnel reflectances for downwelling irradiance above the surface and irradiance below the surface, respectively, and depend on the angular distributions of those irradiances:

$$L_{wn}(\lambda) = F_o(\lambda) \left[\frac{1 - \rho}{n^2} \right] \left[\frac{1 - \langle \rho \rangle}{1 - R \langle r \rangle} \right] \frac{R(\lambda)}{Q(\lambda)} \qquad [6]$$

Combining eqns [2] and [6] for case 1 waters yields

$$L_{wn}(\lambda) \approx F_o(\lambda) \mathscr{R}(\lambda) \left[\frac{f(\lambda)}{Q(\lambda)} \right] \left[\frac{b_b(\lambda)}{a(\lambda)} \right] \qquad [7]$$

$\mathscr{R}$ is the product of the two bracketed terms in eqn [6] and is largely dependent on surface roughness (wind speed). f/Q is usually given in tables derived from Monte Carlo simulations.

The actual surface reflectance includes not only L_w, but also the Fresnel reflection of photons not scattered by the atmosphere (the direct component), photons that are scattered by the atmosphere (skylight or the indirect component), and light reflected off whitecaps. The angular distribution of total reflection broadens as wind speed increases and the sea surface roughens, and the total surface whitecap coverage also increases with wind speed. For clear case 1 waters, that is, areas where the chlorophyll a is less than about 0.25 mg chl a/m^3, the L_{wn} spectrum is fairly constant, for example, $L_{wn}(550) \sim 0.28$ mW/cm^2 · μm · sr, and the values are referred to as 'clear water normalized water-leaving radiances'.

Another optical parameter of interest is the diffuse attenuation coefficient for upwelled radiance, K_{L_u}, (eqn [8]):

$$K_{L_u}(\lambda, z) = -\left(\frac{1}{L_u(\lambda, z)} \right) \left(\frac{dL_u(\lambda, z)}{dz} \right) \qquad [8]$$

Near the surface, where optical constituents are relatively uniform and K_{L_u} is constant,

$$L_u(\lambda, z) = L_u(\lambda, 0^-) e^{-K_{L_u}(\lambda) z} \qquad [9]$$

Similar K relationships hold for irradiance. For convenience, $K(\lambda)$ will be used to denote $K_{L_u}(\lambda)$. The various upwelling and downwelling diffuse

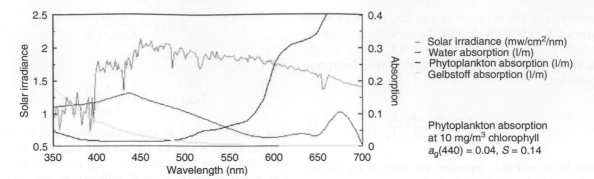

Figure 1 Solar irradiance, chlorophyll specific absorption, yellow substance, and water absorption spectra.

attenuation coefficients for radiance and irradiance are commonly interchanged, but, strictly speaking, they are different quantities and are not equal.

Key to the ocean color measurement technique are the relationships between the solar irradiance, water absorption, and chlorophyll a absorption spectra, chlorophyll a being the primary chemical associated with photosynthesis. The solar spectrum peaks at blue wavelengths which correspond to the maximum transparency of water and the peak in chlorophyll a absorption. Thus, phytoplankton photosynthesis is tuned to the spectral range of maximum light. **Figure 1** provides spectra for $F_o(\lambda)$, $a_\varphi^*(\lambda)$, ocean water absorption ($a_w(\lambda)$), and colored dissolved organic matter ($a_g(\lambda)$; CDOM, also known as yellow substance, gelbstoff, and gilvin). The spectrum for CDOM has the form $a_g(\lambda) = a_g(440) \exp[S(\lambda-440)]$, where S ranges from 0.01 to 0.02 nm^{-1}. There are other optical constituents besides phytoplankton and CDOM, such as sediments, which are particularly important in coastal regions.

Satellite Ocean Color Methodology

In order to obtain accurate estimates of geophysical quantities such as chlorophyll a and $K(490)$ from satellite measurements, a number of radiometric issues must be addressed including (1) sensor design and performance; (2) postlaunch sensor calibration stability; (3) atmospheric correction, that is, the removal of light due to atmospheric scattering, atmospheric absorption, and surface reflection; and (4) bio-optical algorithms, that is, the transformation of $R_{rs}(\lambda)$ or $L_{wn}(\lambda)$ values into geophysical parameter values. Items 2–4 represent developments that progress over time during a mission. For example, over the first 9 years of operation, the radiometric sensitivity of *SeaWiFS* degraded by as much as 18% in one of the NIR bands. Also, as radiative transfer theory develops and additional optical data are

obtained, atmospheric correction and bio-optical algorithms will improve and replace previous versions, and new data products will be defined by the research community. Therefore, flight projects, such as *SeaWiFS*, are prepared to periodically reprocess their entire data set. In fact, the *SeaWiFS* data was reprocessed 5 times in the first 8 years. **Figure 2** provides a graphical depiction of all components of the satellite measurement scenario, each aspect of which is discussed below.

Sensor Design and Performance

Sensor design and performance characteristics encompass many considerations which cannot be elaborated on here, but are essential to meeting the overall measurement accuracy requirements. Radiometric factors include wavelength selection, bandwidth, saturation radiances, signal-to-noise ratios (SNRs), polarization sensitivity, temperature sensitivity, scan angle dependencies (response vs. scan, RVS), straylight rejection, out-of-band contamination, field of view (spatial resolution), band co-registration, and a number of others, all of which must be accurately quantified (characterized) prior to launch and incorporated into the data-processing algorithms. Other design features may include sensor tilting for sunglint avoidance and capacities for tracking the sensor stability on-orbit (e.g., internal lamps, solar diffusers, and lunar views), as instruments generally lose sensitivity over time due to contamination of optical components, filter and detector degradation, etc. Also, there are a variety of spacecraft design criteria including attitude control for accurate navigation, power (solar panel and battery capacities), onboard data storage capacity, telemetry bandwidth (command uplink and downlink data volumes, transmission frequencies, and ground station compatibility and contact constraints), and real-time data broadcast (high-resolution picture transmission (HRPT) station compatibility). Depending on the specifications,

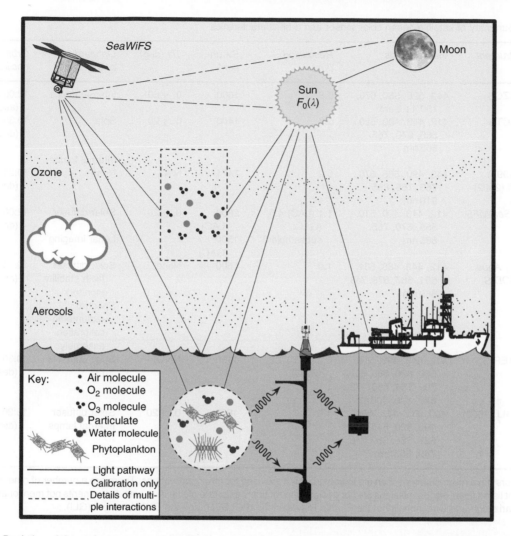

Figure 2 Depiction of the various sensor, atmospheric, and oceanic optical processes which must be accurately accounted for in-satellite ocean color data processing.

which must be accurately described and linked to the final data product accuracy requirements, instruments can be built in a variety of ways to optimize performance. For example, the CZCS used a grating for spectral separation, while *SeaWiFS* and *MODIS* use filters. Along with the specifications, the testing procedures and test equipment must be carefully designed to ensure that the characterization data adequately reflect the true sensor performance. Sensor characterization can be a technically difficult, time-consuming, and expensive process. **Table 1** provides a summary of past, present, and planned ocean color sensors and certain observational attributes. It is important to note that only the two *MODIS* instruments have a common design, although their behaviors on orbit were significantly different.

To date, all ocean color missions, both previous and currently approved, have been designed for low-altitude Sun-synchronous orbits, although sensors on high-altitude geostationary platforms are being considered. Sun-synchronous orbits provide global coverage as the Earth rotates so the data is collected at about the same local time, for example, local noon. Typically, the satellite circles the Earth about every 90–100 min (14–15 orbits/day). Multiple views for a given day are possible only at high latitudes where the orbit tracks converge. **Figure 3** shows the daily global area coverage (GAC) of *SeaWiFS*. The gaps between the swaths are filled on the following day as the ground track pattern progressively shifts. The *SeaWiFS* scan is $\pm 57°$, but the GAC is truncated to $\pm 45°$ and subsampled (every fourth pixel and line) on the spacecraft. The full-resolution local area coverage (LAC, 1.1 km at nadir) is broadcast as HRPT data and includes the entire swath at full resolution. Note that as the scan angle increases, pixels (the area on the surface being viewed) become much larger and the atmospheric corrections are less

Table 1 Summary of satellite ocean color sensor and orbit characteristics

Mission instrument (launch yr)	Visible bands	Resolution (km)	Swath (km)	Tilt (deg)	Onboard calibration	Orbit node (inclination)
Nimbus-7 CZCS (1978)	443, 520, 550, 670, 750 11.5 µm	0.8	1600	0, ±20	Internal lamps	12:00 noon (ascending)
ADEOS-I OCTS (1996)	412, 443, 490, 520, 565, 670, 765, 865 nm	0.7	1400	0, ±20	Solar Internal lamps	10:30 am (descending)
ADEOS-I (1996) ADEOS-II (2002) POLDER	443, 490, 565, 670, 763, 765, 865, 910 nm	6.2	2471	N/A	None	10:30 am (descending)
OrbView-2 SeaWiFS (1997)	412, 443, 490, 510, 555, 670, 765, 865 nm	1.1 (LAC) 4.5 (GAC: subsampled)	2800 (LAC) 1500 (GAC)	0, ±20	Solar diffuser Lunar imaging	12:00 noon (descending)
Terra (2000) Aqua (2002) MODIS	412, 443, 488, 531, 551, 667, 678, 748, 869 nm	1.0	1500	None	Solar diffuser (with stability monitor) Spectral radiometric calibration assembly	10:30 (ascending)
Envisat-1 MERIS (2002)	412, 443, 490, 510, 560, 620, 665, 681, 709, 754, 760, 779, 870, 890, 900 nm	0.3 (LAC) 1.2 (GAC)	1150	None	Solar diffusers (3)	10:30 (descending)
ADEOS-II GLI (2002)	400, 412, 443, 460, 490, 520, 545, 565, 625, 666, 680, 710, 749, 865 nm	1.0	1600	0, ±20	Solar diffuser Internal lamps	10:30 (descending)

Some sensors have more channels than are indicated which are used for other applications. There are a number of other ocean color missions not listed because the missions are not designed to routinely generate global data sets. Instruments not mentioned in the text are the Polarization and Directionality of the Earth's Reflectances (POLDER) and the Global Imager (GLI).

Figure 3 An example of the daily global area coverage (GAC) from SeaWiFS. The SeaWiFS scan extends to ±57° and samples at about a 1-km resolution. This data is the LAC and is continuously broadcast for HRPT station reception. The GAC data are subsampled at every fourth pixel and line over only the center ±45° of the scan. GAC data are stored on the spacecraft and downlinked to the NASA and Orbimage Corporation (the company that owns SeaWiFS) ground stations twice per day.

reliable. MODIS has a swath similar to SeaWiFS LAC, all of which is recorded and broadcast real-time at full resolution. The data gaps in each swath about the subsolar point are where SeaWiFS is tilted from −20° to +20° to avoid viewing into the sunglint. The tilt operation is staggered on successive days in order to ensure every-other-day coverage of the gap. The MODIS and Medium Resolution Imaging Spectrometer (MERIS) do not tilt.

Geostationary orbits, that is, orbits having a fixed subsatellite (nadir) point on the equator, only allow hemispheric coverage with decreased spatial resolution away from nadir, but can provide multiple views each day. Multiple views per day allow for the evaluation of tidal and other diurnal time-dependent biases in sampling to be evaluated, and also provide more complete sampling of a given location as cloud patterns change.

Postlaunch Sensor Calibration Stability

The CZCS sensitivity at 443 nm changed by about 40% during its 7.7 years of operation. Even the relatively small changes noted above for SeaWiFS undermine the mission objectives for global change research because they introduce spurious trends in the derived products. Quantifying changes in the sensor can be very difficult, especially if the changes are gradual. In the case of the CZCS, there was no ongoing comprehensive validation program after its first year of operation, because the mission was a proof-of-concept. Subsequent missions have some level of continuous validation. In the case of *SeaWiFS*, a combination of solar, lunar, and field observations (oceanic and atmospheric) are used. The solar measurements are made daily using a solar diffuser to detect sudden changes in the sensor. The solar measurements cannot be used as an absolute calibration because the diffuser reflectance gradually changes over time. *MODIS* has a solar diffuser stability monitor which accounts for diffuser degradation. *SeaWiFS* lunar measurements are made once a month at a fixed lunar phase angle (7°) using a spacecraft pitch maneuver that allows the sensor to image the Moon through the Earth-viewing optics. This process provides an accurate estimate of the sensor stability relative to the first lunar measurement (**Figure 4**). The data do require a number of corrections, for example, Sun–Moon distance, satellite–Moon distance, and lunar libration variations. The lunar measurements are not used for an absolute calibration because the moon's surface reflectance is not known to a sufficient accuracy.

Once sensor degradation is removed, field measurements can be used to adjust the calibration gain factors so the satellite retrievals of L_{wn} match independent radiometric field observations, the so-called 'vicarious' calibration (meaning that the vicarious calibration replaces the original laboratory-based prelaunch calibration), and involves the atmospheric correction (discussed below). This adjustment is necessary because the prelaunch calibration is only accurate to within about 3%, the sensor can change during launch and orbit raising, and any biases in the atmospheric correction can be removed in this way. The Marine Optical Buoy (MOBY) was developed and located off Lanai, Hawaii, to provide the *SeaWiFS* and *MODIS* vicarious calibration data. Given cloud cover, sun glint contamination, and satellite sampling frequency at Hawaii, it can take up to 3 years to collect enough high-quality comparisons (25–40 depending on the sensor and the wavelength), to derive statistically stable gain correction factors which should be independent of the sensor-viewing geometry and the solar zenith angle. It is important to note that in-water measurements cannot help in accessing biases in the NIR band calibrations. Atmospheric measurements of optical depth and other parameters are needed to evaluate the calibration of these wavelengths.

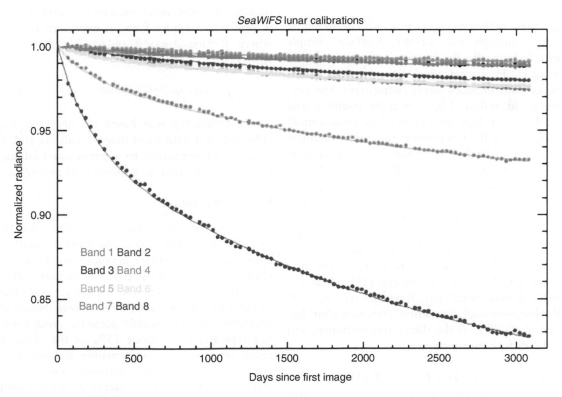

Figure 4 *SeaWiFS* degradation as observed in the monthly lunar calibration data.

Atmospheric Correction

Solar irradiance propagates through the atmosphere where it is attenuated by molecular (Rayleigh) and aerosol scattering and absorption. Rayleigh scattering can be calculated theoretically with a high degree of accuracy. Aerosol scattering and absorption are much more difficult to estimate because their horizontal and vertical distributions are highly variable, as are their absorption and scattering properties. The estimation of the aerosol effects on the upwelling radiance at the top of the atmosphere is one of the most difficult aspects of satellite remote sensing. Ozone (O_3) is the primary absorbing gas that must be considered. Fortunately, ozone is concentrated in a thin band near the top of the atmosphere and its global distribution is mapped daily by sensors such as the Total Ozone Mapping Spectrometer (TOMS), Total Ozone Vertical Sounder (TOVS), and the Ozone Monitoring Instrument (OMI) which have been deployed separately on various spacecraft. Continuous global satellite ozone measurements have been made since 1978 when the first TOMS was flown with the CZCS on *Nimbus*-7. The other absorbing gases of interest are O_2 which has a strong absorption band (the A-band) between 750 and 770 nm and NO_2 which absorbs ultraviolet (UV) and blue light. O_2 is well mixed in the atmosphere and only requires atmospheric pressure data for correction. NO_2 is highly variable and requires global concentration data from satellite sensors such as OMI.

Light that reaches the surface is either reflected or penetrates through the air–sea interface. Simple Fresnel reflection off a flat interface is easily computed theoretically. However, once the surface is wind-roughened and includes foam (whitecaps), the estimation of the reflected light is more complex and empirical relationships must be invoked. Only a small percentage of the light that enters the water column is reflected upward back through the air–sea interface in the general direction of the satellite sensor. Of that light, only a fraction makes its way back through the atmosphere into the sensor. L_w accounts for no more than about 15% of the top-of-the-atmosphere radiance, L_t. Therefore, each process must be accurately accounted for in estimating L_w. The radiances associated with each process are additive, to the first order, and can be expressed as eqn [10] where the subscripts denote 'total' (t), 'Rayleigh' (r), 'aerosol' (a), 'Rayleigh–aerosol interaction' (ra), 'sun glint' (g), and 'foam' (f), $T(\lambda)$ is the direct transmittance, and $t(\lambda)$ is the diffuse transmittance:

$$L_t(\lambda) = L_r(\lambda) + L_a(\lambda) + L_{ra}(\lambda) + T(\lambda)L_g(\lambda) \\ + t(\lambda)[L_f(\lambda) + L_w(\lambda)] \qquad [10]$$

Note that radiance lost to each absorbing gas is estimated separately and added to the satellite radiance to yield L_t. The Rayleigh radiances can be estimated theoretically and the aerosol radiances are usually inferred from NIR wavelengths using aerosol models and the assumption that L_w is negligible. Determining values for all terms on the right side of eqn [10], with the exception of L_w, constitutes the 'atmospheric correction' which allows eqn [10] to be solved for L_w. As mentioned earlier, if L_w is known, then L_t can be adjusted to balance eqn [10] to derive a 'vicarious' calibration.

Bio-optical Algorithms

Bio-optical algorithms are used to define relationships between the water-leaving radiances or reflectances and constituents in the water. The basis for the chlorophyll a algorithm is the change in reflectance spectral slope with increasing concentration, that is, as chlorophyll a increases, the blue end of the spectrum is depressed by pigment absorption and the red end is elevated by increased particle scattering. In case 1 waters, $R(510)$ shows little variation with chlorophyll a and is alluded to as the 'hinge point'. Algorithms can be strictly empirical (statistical regressions) or semi-empirical relationships, that is, based on theoretical expressions using measured values of certain optical variables. For example, the empirical chlorophyll a (chl a) relationship, OC4v4, being used by the *SeaWiFS* mission is depicted in **Figure 5** and is expressed in eqn [11], where $R = \log_{10}[\max(R_{rs}(443), R_{rs}(490), R_{rs}(510)/R_{rs}(555))]$:

$$\text{chl } a = 10^{0.366 - 3.067R + 1.930R^2 + 0.649R^3 - 1.532R^4} \qquad [11]$$

This relationship was based on observed R_{rs} and chlorophyll a data from many locations and, therefore, is not optimized to a particular biological or optical regime, that is, a bio-optical province.

Product Validation

Validation of the derived products can be approached in several ways. The most straightforward approach is to compare simultaneous field and satellite data. Another approach is to make statistical comparisons, for example, frequency distributions, of large *in situ* and satellite data sets. Simultaneous comparisons can provide accurate error estimates, but typically only about 15% of the observations result in valid matchups mainly because of cloud cover, sun glint, spatial inhomogeneities, and time differences. **Figure 6** represents all comparisons for the first 9 years of *SeaWiFS*. Statistical comparisons

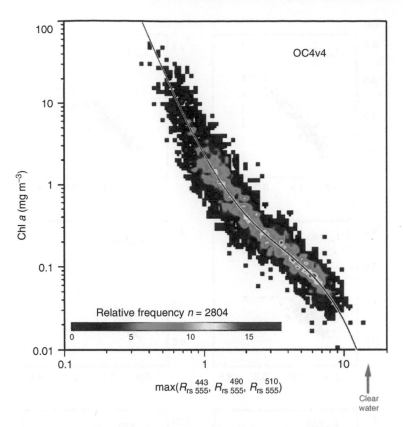

Figure 5 An empirical chlorophyll *a* algorithm, OC4v4, based on a regression of *in situ* chlorophyll *a* vs. R_{rs} ratio observations. In this algorithm, the exponent numerator is the R_{rs} that has the greatest value of the three shown.

of cumulative data sets allow utilization of much more data, but can be subject to sampling biases. Differences in field measurements and satellite estimates can be due to a number of sources including erroneous satellite estimations of L_w, and inaccurate *in situ* values. In the early 1990s, to minimize *in situ* measurement errors, the *SeaWiFS* project initiated a calibration comparison program, the development and documentation of *in situ* measurement protocols, and a number of technology development activities. These were continued under the Sensor Intercomparison and Merger for Biological and Interdisciplinary Oceanic Studies (SIMBIOS) program until 2003. Much of the documentation from *SeaWiFS* and SIMBIOS is available online at http://oceancolor.gsfc.nasa.gov/DOCS/.

An important aspect of satellite data validation is the comparison of data from different missions to assess the level of agreement in the derived products, especially the L_{wn} and chlorophyll *a* values. Because every sensor has a different design and, therefore, different sensitivities to the host of attributes listed earlier, identifying the causes for any differences in the L_{wn} values can be difficult. Good examples are *SeaWiFS* and *MODIS*, which have very different designs. The only design feature these sensors have in common is the use of filters, rather than prisms or gratings, to define the spectral bands. In some situations, the design differences can be used to identify problems in the data products. This point is illustrated by two examples, polarization sensitivity and noise characteristics.

SeaWiFS has a polarization scrambler, which reduces the sensor polarization sensitivity to $\sim 0.25\%$. *SeaWiFS* also incorporates a rotating telescope rather than a mirror which reduces the amount of polarization introduced by the optical surfaces. *MODIS* has a rotating mirror and a polarization sensitivity of several percent. Because the Rayleigh radiance is highly polarized, errors in the polarization characterization can introduce substantial errors in the L_{wn} values. In fact, an error in the *MODIS* polarization tables did initially cause large regional differences ($>50\%$) between the *SeaWiFS* and *MODIS* L_{wn} values. This difference prompted a review of the *MODIS* calibration inputs resulting in the error being corrected, greatly reducing the regional and seasonal differences in the *SeaWiFS* and *MODIS* products.

While the *MODIS* and *SeaWiFS* mean L_{wn} and chlorophyll values are quite comparable, the *MODIS* products have much lower variances (less noise) providing a more reasonable estimate of

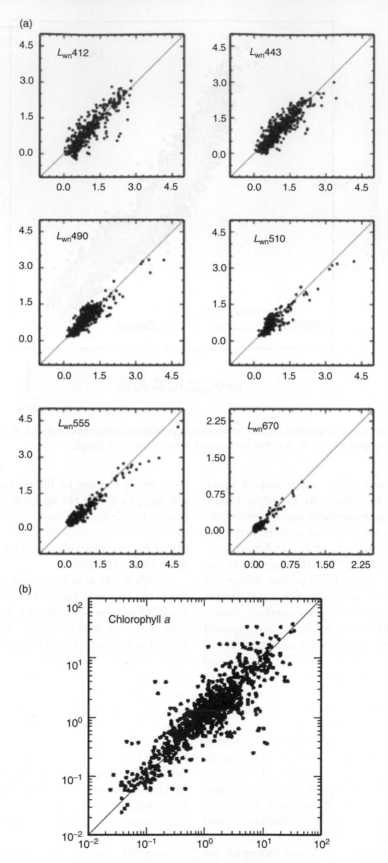

Figure 6 A comparison of simultaneous *SeaWiFS*-derived and *in situ* (a) L_{wn} and (b) chlorophyll *a* data. Only about 15% of the field data collected are used in the comparisons, because many of the data are excluded due to cloud cover, sun glint, time difference, and other rejection criteria. The dashed line is the least squares fit to the data.

sampling variability in the global binned fields, that is, data merged over on a fixed grid for various lengths of time (daily, weekly, monthly, etc.). There are several reasons for *SeaWiFS* having higher variability. First, *MODIS* has higher SNRs. Also, the *MODIS* global data set has about 20 times

more pixels (samples) due to the *SeaWiFS* GAC sub-sampling. The *SeaWiFS* subsampling allows small clouds to escape detection in the GAC processing in which case stray light is uncorrected (stray light is scattered light within the instrument that contaminates measurements in adjacent pixels), thereby elevating

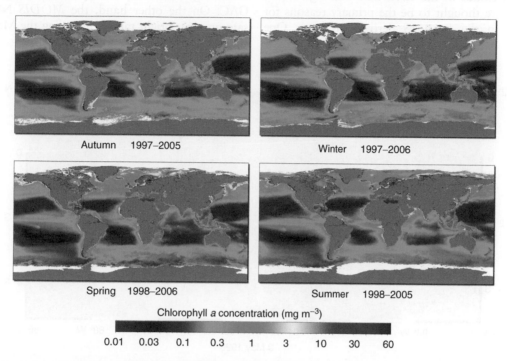

Chlorophyll *a* concentration (mg m^{-3})

0.01   0.03   0.1   0.3   1   3   10   30   60

Figure 7 Seasonal average chlorophyll *a* concentrations from the 9 years of *SeaWiFS* operation. The composites combine all chlorophyll *a* estimates within 9 km square 'bins' obtained during each 3-month period. A variety of quality control exclusion criteria are applied before a sample (pixel) value is included in the average.

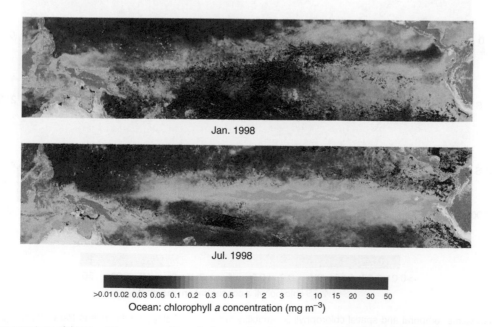

>0.01 0.02 0.03 0.05 0.1   0.2   0.3   0.5   1   2   3   5   10   15   20   30   50

Ocean: chlorophyll *a* concentration (mg m^{-3})

Figure 8 A comparison of the monthly average chlorophyll *a* concentrations at the peaks of the 1997 El Niño and the 1998 La Niña in the equatorial Pacific Ocean.

the L_t values. Finally, the *SeaWiFS* data is truncated from 12 to 10 bits on the data recorder resulting in coarser digitization, especially in the NIR bands where the SNRs are relatively low. Noise can cause 'jitter' in the aerosol model selection which amplifies the variability in visible L_{wn} values via the aerosol correction. Undetected clouds in the GAC data and digitization truncation are thought to be the primary reasons for 'speckling' in the *SeaWiFS* derived products. One

strength of the *SeaWiFS* detector array or focal plane design is the bilinear gain which prevents bright pixels from saturating any band. The prelaunch characterization data provided enough information for a stray light correction algorithm to be derived. This correction works well in the LAC data processing and for correcting the effects of large bright targets in the GAC. On the other hand, the *MODIS* NIR bands saturate over bright targets including the Moon.

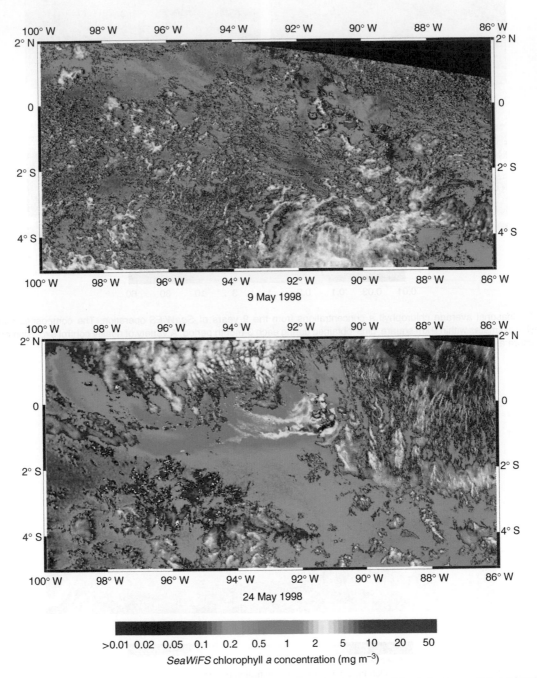

9 May 1998

24 May 1998

>0.01 0.02 0.05 0.1 0.2 0.5 1 2 5 10 20 50

SeaWiFS chlorophyll *a* concentration (mg m^{-3})

Figure 9 Mesoscale temporal and spatial chlorophyll *a* variability around the Galapagos Islands in the eastern equatorial Pacific Ocean before and after the sudden onset of the 1998 La Niña. The sea surface temperature near the islands dropped nearly 10 °C between 9 and 24 May.

Satellite Ocean Color Data Sets and Applications

In this section, examples of satellite ocean color data products are presented to illustrate some research applications. These include (1) an annual seasonal cycle of global chlorophyll a, (2) interannual variability due to the El Niño–Southern Oscillation (ENSO) cycle in the equatorial Pacific, (3) mesoscale variability (scenes from the Galápagos Islands), and (4) blooms of special phytoplankton groups (coccolithophores).

Plant growth in the ocean is regulated by the supply of macronutrients (e.g., nitrate, phosphate, and silicate), micronutrients (iron, in particular), light, and temperature. Light is modulated by cloud cover and time of year (solar zenith angle). Nutrient supply and temperature are determined by ocean circulation and mixing, especially the vertical fluxes, and heat exchange with the atmosphere. **Figure 7** provides seasonal average chlorophyll a concentrations derived from SeaWiFS. Areas such as the North Atlantic show a clear seasonal cycle. The seasonality in the North Atlantic is the result of deep mixing in the winter, which renews the surface nutrient supply because the deeper waters are a reservoir for nitrate and other macronutrients. Once illumination begins to increase in the spring, the mixed layer shallows providing a well-lit, nutrient-rich surface layer ideal for phytoplankton growth. A bloom results and persists into the summer until zooplankton grazing and nutrient depletion curtail the bloom.

Figure 8 depicts the effects of El Niño and La Niña on the ecosystems of the equatorial Pacific during 1997–98. Under normal conditions, the eastern equatorial Pacific is one of the most biologically productive regions in the world ocean as westward winds force a divergent surface flow resulting in upwelling of nutrient-rich subsurface water into the euphotic zone, that is, the shallow illuminated layer where plant photosynthesis occurs. During El Niño, warm, nutrient-poor water migrates eastward from the western Pacific and replaces the nutrient-rich water, resulting in a collapse of the ecosystem. Eventually, the ocean–atmosphere system swings back to cooler conditions, usually to colder than normal ocean temperatures, causing La Niña. The result is an extensive bloom which eventually declines to more typical concentrations as the atmosphere–ocean system returns to a more normal state.

The 1998 transition from El Niño to La Niña occurred very rapidly. **Figure 9** compares the chlorophyll a concentrations around the Galapagos Islands in early and late May. During the time between these two high-resolution scenes, ocean temperatures around the islands dropped by about 10 °C. The chlorophyll a concentrations jumped dramatically as nutrient-rich waters returned and phytoplankton populations could grow unabated in the absence of zooplankton grazers.

The final example illustrates that some phytoplankton have special optical properties which allows them to be uniquely identified. **Figure 10**, a composite of three Rayleigh-corrected SeaWiFS bands, shows an extensive bloom of coccolithophores in the Bering Sea. In their mature stage of development, coccolithophores shed calcite platelets, which turn the water a milky white. Under these conditions, algorithms for reflectance (eqn [2]) and chlorophyll a concentration (eqn [10]) are not valid. However, the anomalously high reflectance allows for their detection and removal from spatial and temporal averages of the satellite-derived products. Since coccolithophores are of interest for a number of ecological and biogeochemistry pursuits, satellite ocean color data can be used to map the temporal and spatial distribution of these blooms. In the case of the Bering Sea, the occurrence of coccolithophores had been rare prior to 1997 when the bloom persisted for c. 6 months. The ecological impact of the blooms in 1997 and 1998, which encompassed the entire western Alaska continental shelf, was dramatic and caused fish to avoid the bloom region resulting in

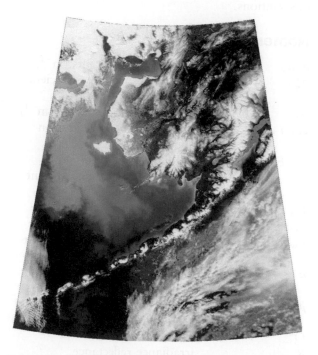

Figure 10 A true color depiction of a coccolithophore, *Emiliania huxleyi*, bloom in the Bering Sea. The true color composite is formed by summing the Rayleigh radiance-corrected 412-, 555-, and 670-nm SeaWiFS images.

extensive starvation of certain marine mammals and seabirds (those with limited foraging range), and formed a barrier for salmon preventing them from spawning in the rivers along that coast. Research is presently being conducted on methods of identifying other types of blooms such as trichodesmium, important for understanding nitrification in the ocean, and red tides (toxic algal blooms).

Conclusions

Satellite ocean color remote sensing combines a broad spectrum of science and technology. In many ways, the CZCS demonstrated that the technique could work. However, to advance the ocean biology and biogeochemistry on a global scale and conduct research on the effects of global warming, many improvements in satellite sensor technology, atmospheric and oceanic radiative transfer modeling, field observation methodologies, calibration metrology, and other areas have been necessary and are continuing to evolve. As these develop, new products and applications will become feasible and will require periodic reprocessing of the satellite data to incorporate these advances. Ultimately, the goal of the international ocean science community, working with the various space agencies, is to develop a continuous long-term global time series of highly accurate and well-documented satellite ocean color observations.

Nomenclature

a, a_{φ}	absorption coefficents (m^{-1})
a^*_{φ}	specific absorption coefficient $(m^2/mg\ Chl\ a)$
b_b	backscattering coefficent (m^{-1})
chl a	chlorophyll a concentration $(mg\ (Chl\ a)/m^3)$
E_d, E_u, F_o	irradiances $(mW/cm^2\ \mu m)$
K	diffuse attenuation coefficient (m^{-1})
$L_a, L_f, L_g, L_D,$	radiances $(mW/cm^2\ \mu m\ sr)$
L_{ra}, L_D, L_u, L_w	
L_{wn}	normalized water-leaving radiance $(mW/cm^2\ \mu m\ sr)$
n	index of refraction (dimensionless)
ρ, r	Fresnel reflectances (dimensionless)
Q	E_u/L_u ratio (sr)
R	irradiance reflectance (dimensionless)
R_{rs}	remote sensing reflectance (sr^{-1})
S	gelvin absorption spectra parameter (nm^{-1})
t	diffuse transmittance (dimensionless)
T	direct transmittance (dimensionless)

See also

Carbon Cycle. Marine Silica Cycle. Nitrogen Cycle. Penetrating Shortwave Radiation. Penetrating Shortwave Radiation. Phosphorus Cycle. Primary Production Processes. Satellite Oceanography, History, and Introductory Concepts. Upper Ocean Mixing Processes. Whitecaps and Foam.

Further Reading

Hooker SB and Firestone ER (eds.) (1996) *SeaWiFS Technical Report Series, NASA Technical Memorandum 104566*, vols. 1–43. Greenbelt, MD: NASA Goddard Space Flight Center.

Hooker SB and Firestone ER (eds.) (2000) *SeaWiFS Postlaunch Technical Report Series, NASA Technical Memorandum Year-206892*, vols. 1–29. Greenbelt, MD: NASA Goddard Space Flight Center.

Jerlov NG (1976) *Marine Optics*, 231pp. New York: Elsevier.

Kirk JTO (1994) *Light and Photosynthesis in Aquatic Ecosystems*, 509pp. Cambridge, UK: Cambridge University Press.

Martin S (2004) *An Introduction to Ocean Remote Sensing*, 426pp. New York: Cambridge University Press.

McClain CR, Feldman GC, and Hooker SB (2004) An overview of the SeaWiFS project and strategies for producing a climate research quality global ocean bio-optical time series. *Deep-Sea Research II* 51(1–3): 5–42.

Mobley CD (1994) *Light and Water, Radiative Transfer in Natural Waters*, 592 pp. New York: Academic Press.

Robinson IS (1985) *Satellite Oceanography*, 455pp. Chichester, UK: Wiley.

Shifrin KS (1988) *Physical Optics of Ocean Water*, 285pp. New York: American Institute of Physics.

Stewart RH (1985) *Methods of Satellite Oceanography*, 360pp. Los Angeles: University of California Press.

Relevant Website

http://oceancolor.gsfc.nasa.gov
 – OceanColor Documentation, OceanColor Home Page.

SATELLITE REMOTE SENSING: SALINITY MEASUREMENTS

G. S. E. Lagerloef, Earth and Space Research, Seattle, WA, USA

Introduction

Surface salinity is an ocean state variable which controls, along with temperature, the density of seawater and influences surface circulation and formation of dense surface waters in the higher latitudes which sink into the deep ocean and drive the thermohaline convection. Although no satellite measurements are made at present, emerging new technology and a growing scientific need for global measurements have stimulated programs now underway to launch salinity-observing satellite sensors within the present decade. Salinity remote sensing is possible because the dielectric properties of seawater which depend on salinity also affect the surface emission at certain microwave frequencies. Experimental heritage extends more than 35 years in the past, including laboratory studies, airborne sensors, and one instrument flown briefly in space on Skylab. Requirements for very low noise microwave radiometers and large antenna structures have limited the advance of satellite systems, and are now being addressed. Science needs, primarily for climate studies, dictate a resolution requirement of approximately 100 km spatial grid, observed monthly, with approximately 0.1 error on the Practical Salinity Scale (or 1 part in 10 000), which demand very precise radiometers and that several ancillary errors be accurately corrected. Measurements will be made in the 1.413 GHz astronomical hydrogen absorption band to avoid radio interference.

Definition and Theory

How Salinity Is Defined and Measured

Salinity represents the concentration of dissolved inorganic salts in seawater (grams salt per kilogram seawater, or parts per thousand, and historically given by the symbol ‰). Oceanographers have developed modern methods based on the electrical conductivity of seawater which permit accurate measurement by use of automated electronic *in situ* sensors. Salinity is derived from conductivity, temperature, and pressure with an international standard set of empirical equations known as the Practical Salinity Scale, established in 1978 (PSS-78), which is much easier to standardize and more precise than previous chemical methods and which numerically represents grams per kilogram. Accordingly, the modern literature often quotes salinity measurements in practical salinity units (psu), or refers to PSS-78. Salinity ranges from near-zero adjacent to the mouths of major rivers to more than 40 in the Red Sea. Aside from such extremes, open ocean surface values away from coastlines generally fall between 32 and 37 (**Figure 1**).

This global mean surface salinity field has been compiled from all available oceanographic observations. A significant fraction of the 1° latitude and longitude cells have no observations, requiring such maps to be interpolated and smoothed over scales of several hundred kilometers. Seasonal to interannual salinity variations can only be resolved in very limited geographical regions where the sampling density is suitable. Data are most sparse over large regions of the Southern Hemisphere. Remote sensing from satellite will be able to fill this void and monitor multiyear variations globally.

Remote sensing theory Salinity remote sensing with microwave radiometry is likewise possible through the electrically conductive properties of seawater. A radiometric measurement of an emitting surface is given in terms of a 'brightness temperature' (T_B), measured in kelvin (K). T_B is related to the true absolute surface temperature (T) through the emissivity coefficient (e):

$$T_B = eT$$

For seawater, e depends on the complex dielectric constant (ε), the viewing angle (Fresnel laws), and surface roughness (due to wind waves). The complex dielectric constant is governed by the Debye equation

$$\varepsilon = \varepsilon_\infty + \frac{\varepsilon_s(S, T) - \varepsilon_\infty}{1 + i2\pi f \tau(S, T)} - \frac{iC(S, T)}{2\pi f \varepsilon_0}$$

and includes electrical conductivity (C), the static dielectric constant ε_s, and the relaxation time τ, which are all sensitive to salinity and temperature (S, T). The equation also includes radio frequency (f)

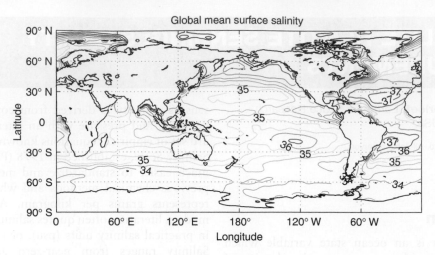

Figure 1 Contour map of the mean global surface salinity field (contour interval 0.5) based on the World Ocean Atlas, 1998 (WOA98). Arctic Ocean salinities <32 are not contoured. Elevated mid-ocean salinities, especially the subtropical Atlantic ones, are caused by excess evaporation. Data were obtained from the US Department of Commerce, NOAA, National Oceanographic Data Center.

and terms for permittivity at infinite frequency (ε_∞) which may vary weakly with T, and permittivity of free space (ε_0, a constant). The relation of electrical conductivity to salinity and temperature is derived from the Practical Salinity Scale. The static dielectric and time constants have been modeled by making laboratory measurements of ε at various frequencies, temperatures, and salinities, and fitting ε_s and τ to polynomial expressions of (S, T) to match the ε data. Different models in the literature show similar variations with respect to (f, S, T).

Emissivity for the horizontal (H) and vertical (V) polarization state is related to ε by Fresnel reflection:

$$e_H = 1 - \left[\frac{\cos \theta - (\varepsilon - \sin^2 \theta)^{1/2}}{\cos \theta + (\varepsilon - \sin^2 \theta)^{1/2}} \right]^2$$

$$e_V = 1 - \left[\frac{\varepsilon \cos \theta - (\varepsilon - \sin^2 \theta)^{1/2}}{\varepsilon \cos \theta + (\varepsilon - \sin^2 \theta)^{1/2}} \right]^2$$

where θ is the vertical incidence angle from which the radiometer views the surface, and $e_H = e_V$ when $\theta = 0$. The above set of equations provides a physically based model function relating T_B to surface S, T, θ, and H or V polarization state for smooth water (no wind roughness). This can be inverted to retrieve salinity from radiometric T_B measurements provided the remaining parameters are known. The microwave optical depth is such that the measured emission originates in the top 1 cm of the ocean, approximately.

The rate at which T_B varies with salinity is sensitive to microwave frequency, achieving levels practical for salinity remote sensing at frequencies below

c. 3 GHz. Considerations for selecting a measurement radio frequency include salinity sensitivity, requisite antenna size (see below), and radio interference from other (mostly man-made) sources. A compromise of these factors, dominated by the interference issue, dictates a choice of about a 27-MHz-wide frequency band centered at 1.413 GHz, which is the hydrogen absorption band protected by international treaty for radio astronomy research. This falls within a frequency range known as L-band. Atmospheric clouds have a negligible effect, allowing observations in all weather except possibly heavy rain. Accompanying illustrations are based on applying $f = 1.413$ GHz in the Debye equation and using a model that included laboratory dielectric constant measurements at the nearby frequency 1.43 GHz. Features of this model function and their influence on measurement accuracy are discussed in the section on resolution and error sources.

Antennas Unusually large radiometer antennas will be required to be deployed on satellites for measuring salinity. Radiometer antenna beam width varies inversely with both antenna aperture and radio frequency. 1.413 GHz is a significantly lower frequency than found on conventional satellite microwave radiometers, and large antenna structures are necessary to avoid excessive beam width and accordingly large footprint size. For example, a 50 km footprint requires about a 6-m-aperture antenna whereas conventional radiometer antennas are around 1–2 m. To decrease the footprint by a factor of 2 requires doubling the antenna size. Various filled and thinned array technologies for large antennas have now reached a development stage where application to salinity remote sensing is feasible.

History of Salinity Remote Sensing

The only experiment to date to measure surface salinity from space took place on the NASA Skylab mission during the fall and winter 1973–74, when a 1.413 GHz microwave radiometer with a 1 m antenna collected intermittent data. A weak correlation was found between the sensor data and surface salinity, after correcting for other influences. There was no 'ground truth' other than standard surface charts, and many of the ambient corrections were not as well modeled then as they could be today. Research leading up to the Skylab experiment began with several efforts during the late 1940s and early 1950s to measure the complex dielectric constant of saline solutions for various salinities, temperatures, and microwave frequencies. These relationships provide the physical basis for microwave remote sensing of the ocean as described above.

The first airborne salinity measurements were demonstrated in the Mississippi River outflow and published in 1970. This led to renewed efforts during the 1970s to refine the dielectric constants and governing equations. Meanwhile, a series of airborne experiments in the 1970s mapped coastal salinity patterns in the Chesapeake and Savannah river plumes and freshwater sources along the Puerto Rico shoreline. In the early 1980s, a satellite concept was suggested that might achieve an ideal precision of about 0.25 and spatial resolution of about 100 km. At that time, space agencies were establishing the oceanic processes remote sensing program around missions and sensors for measuring surface dynamic topography, wind stress, ocean color, surface temperature, and sea ice. For various reasons, salinity remote sensing was then considered only marginally feasible from satellite and lacked a strongly defined scientific need.

Interest in salinity remote sensing revived in the late 1980s with the development of a 1.4 GHz airborne electrically scanning thinned array radiometer (ESTAR) designed primarily for soil moisture measurements. ESTAR imaging is done electronically with no moving antenna parts, thus making large antenna structures more feasible. The airborne version was developed as an engineering prototype and to provide the proof of concept that aperture synthesis can be extrapolated to a satellite design. The initial experiment to collect ocean data with this sensor consisted of a flight across the Gulf Stream in 1991 near Cape Hatteras. The change from 36 in the offshore waters to <32 near shore was measured, along with several frontal features visible in the satellite surface temperature image from the same day. This Gulf Stream transect demonstrated that small salinity variations typical of the open ocean can be detected as well as the strong salinity gradients in the coastal and estuary settings demonstrated previously.

By the mid-1990s, a new airborne salinity mapper scanning low frequency microwave radiometer (SLFMR) was developed for light aircraft and has been extensively used by NOAA and the US Navy to survey coastal and estuary waters on the US East Coast and Florida. A version of this sensor is now being used in Australia, and a second-generation model is presently used for research by the US Navy.

In 1999 a satellite project was approved by the European Space Agency for the measurement of Soil Moisture and Ocean Salinity (SMOS), now with projected launch in 2009. The SMOS mission design emphasizes soil moisture measurement requirements, which is done at the same microwave frequency for many of the same reasons as salinity. The T_B dynamic range is about 70–80 K for varying soil moisture conditions and the precision requirement is therefore much less rigid than for salinity. SMOS will employ a large two-dimensional phased array antenna system that will yield 40–90 km resolution across the measurement swath. The Aquarius/SAC-D mission is being jointly developed by the US (NASA) and Argentina (CONAE) and is due to be launched in 2010. The Aquarius/SAC-D mission design puts primary emphasis on ocean salinity rather than soil moisture, with the focus on optimizing salinity accuracy with a very precisely calibrated microwave radiometer and key ancillary measurements for addressing the most significant error sources.

Requirements for Observing Salinity from Satellite

Scientific Issues

Three broad scientific themes have been identified for a satellite salinity remote sensing program. These themes relate directly to the international climate research and global environmental observing program goals.

Improving seasonal to interannual climate predictions This focuses primarily on El Niño forecasting and involves the effective use of surface salinity data (1) to initialize and improve the coupled climate forecast models, and (2) to study and model the role of freshwater flux in the formation and maintenance of barrier layers and mixed layer heat budgets in the Tropics. Climate prediction models in which satellite altimeter sea level data are assimilated must be adjusted for steric height (sea level change due to ocean density) caused by the variations in upper layer salinity. If not, the adjustment for model heat content is incorrect and the prediction skill is degraded. Barrier layer formation occurs when excessive rainfall creates a shallow, freshwater-stratified, surface layer which effectively isolates the deeper thermocline from exchanging heat with the atmosphere with consequences on the air–sea coupling processes that govern El Niño dynamics.

Improving ocean rainfall estimates and global hydrologic budgets   Precipitation over the ocean is still poorly known and relates to both the hydrologic budget and to latent heating of the overlying atmosphere. Using the ocean as a rain gauge is feasible with precise surface salinity observations coupled with ocean surface current velocity data and mixed layer modeling. Such calculations will reduce uncertainties in the surface freshwater flux on climate timescales and will complement satellite precipitation and evaporation observations to improve estimates of the global water and energy cycles.

Monitoring large-scale salinity events and thermohaline convection   Studying interannual surface salinity variations in the subpolar regions, particularly the North Atlantic and Southern Oceans, is essential to long-time-scale climate prediction and modeling. These variations influence the rate of oceanic convection and poleward heat transport (thermohaline circulation) which are known to have been coupled to extreme global climate changes in the geologic record. Outside of the polar regions, salinity signals are stronger in the coastal ocean and marginal seas than in the open ocean in general, but large footprint size will limit near-shore applications of the data. Many of the larger marginal seas which have strong salinity signals might be adequately resolved nonetheless, such as the East China Sea, Bay of Bengal, Gulf of Mexico, Coral Sea/Gulf of Papua, and the Mediterranean.

Science requirements   From the above science themes, preliminary accuracy and spatial and temporal resolution requirements from satellite observations have been suggested as the minimum to study the following ocean processes: (1) barrier layer effects on tropical Pacific heat flux: 0.2 (PSS-78), 100 km, 30 days; (2) steric adjustment of heat storage from sea level: 0.2, 200 km, 7 days; (3) North Atlantic

thermohaline circulation: 0.1, 100 km, 30 days; and (4) surface freshwater flux balance: 0.1, 300 km, 30 days. Thermohaline circulation and convection in the subpolar seas has the most demanding requirement, and is the most technically challenging because of the reduced T_B/salinity ratio at low seawater temperatures (see below). This can serve as a prime satellite mission requirement, allowing for the others to be met by reduced mission requirements as appropriate. Aquarius/SAC-D is a pathfinder mission capable of meeting the majority of these requirements with a grid scale of 150 km and salinity error less than 0.2 observed monthly.

Resolution and Error Sources

Model function   **Figure 2** shows that the dynamic range of T_B is about 4 K over the range of typical open ocean surface salinity and temperature conditions. T_B gradients are greater with respect to salinity than to temperature. At a given temperature, T_B decreases as salinity increases, whereas the tendency with respect to temperature changes sign. The differential of T_B with respect to salinity ranges from -0.2 to -0.7 K per salinity unit. Corrected T_B will need to be measured to 0.02–0.07 K precision to achieve 0.1 salinity resolution. The sensitivity is strongly affected by temperature, being largest at the highest temperatures and yielding better measurement precision in warm versus cold ocean conditions. Random error can be reduced by temporal and spatial averaging. The degraded measurement precision in higher latitudes will be somewhat compensated by the greater sampling frequency from a polar orbiting satellite.

The T_B variation with respect to temperature falls generally between ± 0.15 K $°C^{-1}$ and near zero over a broad S and T range. Knowledge of the surface temperature to within a few tenths of degrees Celsius will be adequate to correct T_B for temperature effects and can be obtained using data from other satellite systems. The optical depth for this microwave frequency in seawater is about 1–2 cm, and the remotely sensed measurement depends on the T and S in that surface layer thickness. T_B for the H and V polarizations have large variations with incidence angle and spacecraft attitude will need to be monitored very precisely.

Other errors   Several other error sources will bias T_B measurements and must be either corrected or avoided. These include ionosphere and atmosphere effects, cosmic and galactic background radiations, surface roughness from winds, sun glint, solar flux, and rain effects. Cosmic background and lower atmospheric adsorption are nearly constant biases

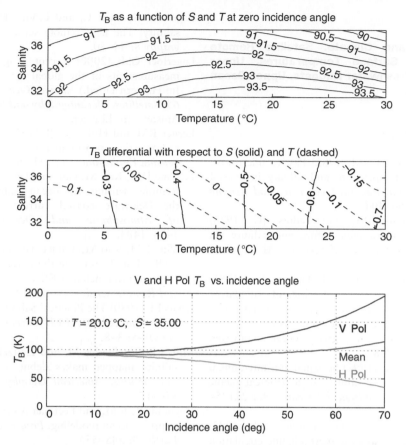

Figure 2 Brightness temperature (T_B) properties as a function of S, T, and incidence angle for typical ocean surface conditions. Top: T_B contours. Middle: T_B derivatives relative to S (solid curves) and T (dashed curves). Bottom: T_B variation vs. incidence angle for H and V polarization. Calculations based on formulas in Klein LA and Swift CT (1977). An improved model for the dielectric constant of sea water at microwave frequencies. *IEEE Transactions on Antennas and Propagation* AP-25(1): 104–111.

easily corrected. An additional correction will be needed when the reflected radiation from the galactic core is in the field of view. The ionosphere and surface winds (roughness) have wide spatial and temporal variations and require ancillary data and careful treatment to avoid T_B errors of several kelvins.

The ionosphere affects the measurement though attenuation and through Faraday rotation of the H and V polarized signal. There is no Faraday effect when viewing at nadir ($\theta = 0$) because H and V emissivities are identical, whereas off-nadir corrections will be needed to preserve the polarization signal. Correction data can be obtained from ionosphere models and analyses but may be limited by unpredictable short-term ionosphere variations. Onboard correction techniques have been developed that require fully polarimetric radiometer measurements from which the Faraday rotation may be derived. Sun-synchronous orbits can be selected that minimize the daytime peak in ionosphere activity as well as solar effects.

The magnitude of the wind roughness correction varies with incidence angle and polarization, and ranges between 0.1 and $0.4\,K/(ms^{-1})$. Sea state conditions can change significantly within the few hours that may elapse until ancillary measurements are obtained from another satellite. Simultaneous wind roughness measurement can be made with an onboard radar backscatter sensor. A more accurate correction can be applied using a direct relationship between the radar backscatter and the T_B response rather than rely on wind or sea state information from other sensors.

Microwave attenuation by rain depends on rain rate and the thickness of the rain layer in the atmosphere. The effect is small at the intended microwave frequency, but for the required accuracy the effect must be either modeled and corrected with ancillary data, or the contaminated data discarded. For the accumulation of all the errors described here, it is anticipated that the root sum square salinity error will be reduced to less than 0.2 with adequate radiometer engineering, correction models, onboard measurements, ancillary data, and spatiotemporal filtering with methods now in development.

See also

Heat Transport and Climate. Satellite Remote Sensing of Sea Surface Temperatures. Upper Ocean Heat and Freshwater Budgets. Upper Ocean Mixing Processes.

Further Reading

Blume HC and Kendall BNM (1982) Passive microwave measurements of temperature and salinity in coastal zones. *IEEE Transactions on Geoscience and Remote Sensing* GE 20: 394–404.

Blume H-JC, Kendall BM, and Fedors JC (1978) Measurement of ocean temperature and salinity via microwave radiometry. *Boundary-Layer Meteorology* 13: 295–380.

Blume H-JC, Kendall BM, and Fedors JC (1981) Multifrequency radiometer detection of submarine freshwater sources along the Puerto Rican coastline. *Journal of Geophysical Research* 86: 5283–5291.

Broecker WS (1991) The great ocean conveyer. *Oceanography* 4: 79–89.

Delcroix T and Henin C (1991) Seasonal and interannual variations of sea surface salinity in the tropical Pacific Ocean. *Journal of Geophysical Research* 96: 22135–22150.

Delworth T, Manabe S, and Stouffer RJ (1993) Interdecadal variations of the thermohaline circulation in a coupled ocean–atmosphere model. *Journal of Climate* 6: 1993–2011.

Dickson RR, Meincke R, Malmberg S-A, and Lee JJ (1988) The 'great salinity anomaly' in the northern North Atlantic, 1968–1982. *Progress in Oceanography* 20: 103–151.

Droppelman JD, Mennella RA, and Evans DE (1970) An airborne measurement of the salinity variations of the Mississippi River outflow. *Journal of Geophysical Research* 75: 5909–5913.

Kendall BM and Blanton JO (1981) Microwave radiometer measurement of tidally induced salinity changes off the Georgia coast. *Journal of Geophysical Research* 86: 6435–6441.

Klein LA and Swift CT (1977) An improved model for the dielectric constant of sea water at microwave frequencies. *IEEE Transactions on Antennas and Propagation* AP-25(1): 104–111.

Lagerloef G, Swift C, and LeVine D (1995) Sea surface salinity: The next remote sensing challenge. *Oceanography* 8: 44–50.

Lagerloef GSE (2000) Recent progress toward satellite measurements of the global sea surface salinity field. In: Halpern D (ed.) *Elsevier Oceaongraphy Series, No. 63: Satellites, Oceanography and Society*, pp. 309–319. Amsterdam: Elsevier.

Lerner RM and Hollinger JP (1977) Analysis of 1.4 GHz radiometric measurements from Skylab. *Remote Sensing Environment* 6: 251–269.

Le Vine DM, Kao M, Garvine RW, and Sanders T (1998) Remote sensing of ocean salinity: Results from the Delaware coastal current experiment. *Journal of Atmospheric and Ocean Technology* 15: 1478–1484.

Le Vine DM, Kao M, Tanner AB, Swift CT, and Griffis A (1990) Initial results in the development of a synthetic aperture radiometer. *IEEE Transactions on Geoscience and Remote Sensing* 28(4): 614–619.

Lewis EL (1980) The Practical Salinity Scale 1978 (PSS-78) and its antecedents. *IEEE Journal of Oceanic Engineering* OE-5: 3–8.

Miller J, Goodberlet M, and Zaitzeff J (1998) Airborne salinity mapper makes debut in coastal zone. *EOS Transactions, American Geophysical Union* 79(173): 176–177.

Reynolds R, Ji M, and Leetmaa A (1998) Use of salinity to improve ocean modeling. *Physics and Chemistry of the Earth* 23: 545–555.

Swift CT and McIntosh RE (1983) Considerations for microwave remote sensing of ocean-surface salinity. *IEEE Transactions on Geoscience and Remote Sensing* GE-21: 480–491.

UNESCO (1981) The Practical Salinity Scale 1978 and the International Equation of State of Seawater 1980. *Technical Papers in Marine Science* 36, 25pp.

Webster P (1994) The role of hydrological processes in ocean–atmosphere interactions. *Reviews of Geophysics* 32(4): 427–476.

Yueh SH, West R, Wilson WJ, Li FK, Njoku EG, and Rahmat-Samii Y (2001) Error sources and feasibility for microwave remote sensing of ocean surface salinity. *IEEE Transactions on Geoscience and Remote Sensing* 39: 1049–1060.

SATELLITE PASSIVE-MICROWAVE MEASUREMENTS OF SEA ICE

C. L. Parkinson, NASA Goddard Space Flight Center, Greenbelt, MD, USA

Published by Elsevier Ltd.

Introduction

Satellite passive-microwave measurements of sea ice have provided global or near-global sea ice data for most of the period since the launch of the *Nimbus 5* satellite in December 1972, and have done so with horizontal resolutions on the order of 25–50 km and a frequency of every few days. These data have been used to calculate sea ice concentrations (percent areal coverages), sea ice extents, the length of the sea ice season, sea ice temperatures, and sea ice velocities, and to determine the timing of the seasonal onset of melt as well as aspects of the ice-type composition of the sea ice cover. In each case, the calculations are based on the microwave emission characteristics of sea ice and the important contrasts between the microwave emissions of sea ice and those of the surrounding liquid-water medium.

The passive-microwave record is most complete since the launch of the scanning multichannel microwave radiometer (SMMR) on the *Nimbus 7* satellite in October 1978; and the SMMR data and follow-on data from the special sensor microwave imagers (SSMIs) on satellites of the United States Defense Meteorological Satellite Program (DMSP) have been used to determine trends in the ice covers of both polar regions since the late 1970s. The data have revealed statistically significant decreases in Arctic sea ice coverage and much smaller magnitude increases in Antarctic sea ice coverage.

Background on Satellite Passive-Microwave Sensing of Sea Ice

Rationale

Sea ice is a vital component of the climate of the polar regions, insulating the oceans from the atmosphere, reflecting most of the solar radiation incident on it, transporting cold, relatively fresh water toward equator, and at times assisting overturning in the ocean and even bottom water formation through its rejection of salt to the underlying water. Furthermore, sea ice spreads over vast distances, globally covering an area approximately the size of North America, and it is highly dynamic, experiencing a prominent annual cycle in both polar regions and many short-term fluctuations as it is moved by winds and waves, melted by solar radiation, and augmented by additional freezing. It is a major player in and indicator of the polar climate state and has multiple impacts on all levels of the polar marine ecosystems. Consequently it is highly desirable to monitor the sea ice cover on a routine basis. In view of the vast areal coverage of the ice and the harsh polar conditions, the only feasible means of obtaining routine monitoring is through satellite observations. Visible, infrared, active-microwave, and passive-microwave satellite instruments are all proving useful for examining the sea ice cover, with the passive-microwave instruments providing the longest record of near-complete sea ice monitoring on a daily or near-daily basis.

Theory

The tremendous value of satellite passive-microwave measurements for sea ice studies results from the combination of the following four factors:

1. Microwave emissions of sea ice differ noticeably from those of seawater, making sea ice generally readily distinguishable from liquid water on the basis of the amount of microwave radiation received by the satellite instrument. For example, **Figure 1** presents color-coded images of the data from one channel on a satellite passive-microwave instrument, presented in units (termed 'brightness temperatures') indicative of the intensity of emitted microwave radiation at that channel's frequency, 19.4 GHz. The ice edge, highlighted by the broken white curve, is readily identifiable from the brightness temperatures, with open-ocean values of 172–198 K outside the ice edge and sea ice values considerably higher, predominantly greater than 230 K, within the ice edge.

2. The microwave radiation received by Earth-orbiting satellites derives almost exclusively from the Earth system. Hence, microwave sensing does not require sunlight, and the data can be collected irrespective of the level of daylight or darkness. This is a major advantage in the polar latitudes, where darkness lasts for months at a time, centered on the winter solstice.

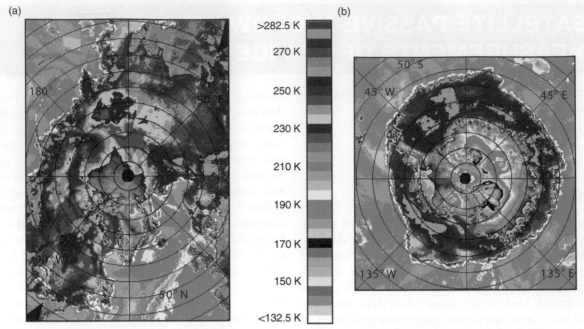

Figure 1 Late-winter brightness temperature images of 19.4-GHz vertically polarized (19.4 V) data from the DMSP SSMI for (a) the north polar region on 15 Mar. 1998 and (b) the south polar region on 15 Sep. 1998, showing near-maximum sea ice coverage in each hemisphere. The broken white curve has been added to indicate the location of the sea ice edge. Black signifies areas of no data; the absence of data poleward of 87.6° latitude results from the satellite's near-polar orbit and is consistent throughout the SSMI data set.

3. Many of the microwave data are largely unaffected by atmospheric conditions, including the presence or absence of clouds. Storm systems can produce atmospheric interference, but, at selected wavelengths, the microwave signal from the ice–ocean surface can pass through most nonprecipitating clouds essentially unhindered. Hence, microwave sensing of the surface does not require cloud-free conditions.

4. Satellite passive-microwave instruments can obtain a global picture of the sea ice cover at least every few days with a resolution of 50 km or better, providing reasonable spatial resolution and extremely good temporal resolution for most large-scale or climate-related studies.

Satellite Passive-Microwave Instruments

The first major satellite passive-microwave imager was the electrically scanning microwave radiometer (ESMR) launched on the *Nimbus 5* satellite of the United States National Aeronautics and Space Administration (NASA) in December 1972, preceded by a nonscanning passive-microwave radiometer launched on the Russian Cosmos satellite in September 1968. The ESMR was a single-channel instrument recording radiation at a wavelength of 1.55 cm and

corresponding frequency of 19.35 GHz. It collected good-quality data for much of the 4-year period from January 1973 through December 1976, although with some major data gaps, including one that lasted for 3 months, from June through August 1975. Being a single-channel instrument, it did not allow some of the more advanced studies that have been done with subsequent instruments, but its flight was a highly successful proof-of-concept mission, establishing the value of satellite passive-microwave technology for observing the global sea ice cover and other variables. The ESMR data were used extensively in the determination and analysis of sea ice conditions in both the Arctic and the Antarctic over the 4 years 1973–76. Emphasis centered on the determination of ice concentrations (percent areal coverages of ice) and, based on the ice concentration results, the calculation of ice extents (integrated areas of all grid elements with ice concentration $\geq 15\%$). This 4-year data set established key aspects of the annual cycles of the polar sea ice covers, including the nonuniformity of the growth and decay seasons and the marked interannual differences even within a 4-year data set.

The *Nimbus 5* ESMR was followed by a less successful ESMR on the *Nimbus 6* satellite and then by the more advanced 10-channel SMMR on board NASA's *Nimbus 7* satellite and a sequence of seven-channel SSMIs on board satellites of the DMSP.

Nimbus 7 was launched in late October 1978, and the SMMR on board was operational through mid-August 1987. The first of the DMSP SSMIs was operational as of early July 1987, providing a welcome data overlap with the *Nimbus 7* SMMR and thereby allowing intercalibration of the SMMR and SSMI data sets. SSMIs continue to operate into the twenty-first century. There was also an SMMR on board the short-lived *Seasat* satellite in 1978; and there was a two-channel microwave scanning radiometer (MSR) on board the Japanese Marine Observation Satellites starting in February 1987. Each of these successor satellite passive-microwave instruments, after the ESMR, has been multichannel, allowing both an improved accuracy in the ice concentration derivations and the calculation of additional sea ice variables, including ice temperature and the concentrations of separate ice types.

The Japanese developed a 12-channel advanced microwave scanning radiometer (AMSR) for the Earth Observing System's (EOS) *Aqua* satellite (formerly named the *PM-1* satellite), launched by NASA in May 2002, and for the Advanced Earth Observing Satellite II (*ADEOS-II*), launched by the Japanese National Space Development Agency in December 2002. *ADEOS-II* prematurely ceased operations in October 2003, but the *Aqua* AMSR, labeled AMSR-E in recognition of its place in the EOS and to distinguish it from the *ADEOS-II* AMSR, has collected a multiyear data record. The AMSR-E has a major advantage over the SMMR and SSMI instruments in allowing sea ice measurements at a higher spatial resolution (12–25 km vs. 25–50 km for the major derived sea ice products). It furthermore has an additional advantage over the SMMR in having channels at 89 GHz in addition to its lower-frequency channels, at 6.9, 10.7, 18.7, 23.8, and 36.5 GHz.

Sea Ice Determinations from Satellite Passive-Microwave Data

Sea Ice Concentrations

Ice concentration is among the most fundamental and important parameters for describing the sea ice cover. Defined as the percent areal coverage of ice, it is directly critical to how effectively the ice cover restricts exchanges between the ocean and the atmosphere and to how much incoming solar radiation the ice cover reflects. Ice concentration is calculated at each ocean grid element, for whichever grid is being used to map or otherwise display the derived satellite products. A map of ice concentrations presents the areal distribution of the ice cover, to the resolution of the grid.

With a single channel of microwave data, taken at a radiative frequency and polarization combination that provides a clear distinction between ice and water, approximate sea ice concentrations can be calculated by assuming a uniform radiative brightness temperature TB_w for water and a uniform radiative brightness temperature TB_I for ice, with both brightness temperatures being appropriate for the values received at the altitude of the satellite, that is, incorporating an atmospheric contribution. Assuming no other surface types within the field of view, the observed brightness temperature TB is given by

$$TB = C_w TB_w + C_I TB_I \qquad [1]$$

C_w is the percent areal coverage of water and C_I is the ice concentration. With only the two surface types, $C_w + C_I = 1$, and eqn [1] can be expressed as

$$TB = (1 - C_I)TB_w + C_I TB_I \qquad [2]$$

This is readily solved for the ice concentration:

$$C_I = \frac{TB - TB_w}{TB_I - TB_w} \qquad [3]$$

Equation [3] is the standard equation used for the calculation of ice concentrations from a single channel of microwave data, such as the data from the ESMR instrument. A major limitation of the formulation is that the polar ice cover is not uniform in its microwave emission, so that the assumption of a uniform TB_I for all sea ice is only a rough approximation, far less justified than the assumption of a uniform TB_w for seawater, although that also is an approximation.

Multichannel instruments allow more sophisticated, and generally more accurate, calculation of the ice concentrations. They additionally allow many options as to how these calculations can be done. To illustrate the options, two algorithms will be described, both of which assume two distinct ice types, thereby advancing over the assumption of a single ice type made in eqns [1]–[3] and being particularly appropriate for the many instances in which two ice types dominate the sea ice cover. For this approximation, the assumption is that the field of view contains only water and two ice types, type 1 ice and type 2 ice (see the section 'Sea ice types' for more information on ice types), and that the three surface types have identifiable brightness temperatures, TB_w, TB_{I1}, and TB_{I2}, respectively. Labeling the concentrations of the two ice types as C_{I1} and C_{I2}, respectively, the percent coverage of water is $1 - C_{I1} - C_{I2}$, and the integrated observed brightness temperature is

$$TB = (1 - C_{I1} - C_{I2})TB_w + C_{I1}TB_{I1} + C_{I2}TB_{I2} \quad [4]$$

With two channels of information, as long as appropriate values for TB_w, TB_{I1}, and TB_{I2} are known for each of the two channels, eqn [4] can be used individually for each channel, yielding two linear equations in the two unknowns C_{I1} and C_{I2}. These equations are immediately solvable for C_{I1} and C_{I2}, and the total ice concentration C_I is then given by

$$C_I = C_{I1} + C_{I2} \quad [5]$$

Although the scheme described in the preceding paragraph is a marked advance over the use of a single-channel calculation (eqn [3]), most algorithms for sea ice concentrations from multichannel data make use of additional channels and concepts to further improve the ice concentration accuracies. A frequently used algorithm (termed the NASA Team algorithm) for the SMMR data employs three of the 10 SMMR channels, those recording horizontally polarized radiation at a frequency of 18 GHz and vertically polarized radiation at frequencies of 18 and 37 GHz. The algorithm is based on both the polarization ratio (PR) between the 18-GHz vertically polarized data (abbreviated 18 V) and the 18-GHz horizontally polarized data (18 H) and the spectral gradient ratio (GR) between the 37-GHz vertically polarized data (37 V) and the 18-V data. PR and GR are defined as:

$$PR = \frac{TB(18\,V) - TB(18\,H)}{TB(18\,V) + TB(18\,H)} \quad [6]$$

$$GR = \frac{TB(37\,V) - TB(18\,V)}{TB(37\,V) + TB(18\,V)} \quad [7]$$

Substituting into eqns [6] and [7] expanded forms of TB(18 V), TB(18 H), and TB(37 V) obtained from eqn [4], the result yields equations for PR and GR in the two unknowns C_{I1} and C_{I2}. Solving for C_{I1} and C_{I2} yields two algebraically messy but computationally straightforward equations for C_{I1} and C_{I2} based on PR, GR, and numerical coefficients determined exclusively from the brightness temperature values assigned to water, type 1 ice, and type 2 ice for each of the three channels (these assigned values are termed 'tie points' and are determined empirically). These are the equations that are then used for the calculation of the concentrations of type 1 and type 2 ice once the observations are made and are used to calculate PR and GR from eqns [6] and [7]. The total ice concentration C_I is then obtained from eqn [5]. The use of PR and GR in this formulation reduces the impact of ice temperature variations on the ice concentration calculations. This algorithm is complemented by a weather filter that sets to 0 all ice concentrations at locations and times with a GR value exceeding 0.07. The weather filter eliminates many of the erroneous calculations of sea ice presence arising from the influence of storm systems on the microwave data.

For the SSMI data, the same basic NASA Team algorithm is used, although 18 V and 18 H in eqns [6] and [7] are replaced by 19.4 V and 19.4 H, reflecting the placement on the SSMI of channels at a frequency of 19.4 GHz rather than the 18-GHz channels on the SMMR. Also, because the data from the 19.4-GHz channels tend to be more contaminated by water vapor absorption/emission and other weather effects than the 18-GHz data, the weather filter for the SSMI calculations incorporates a threshold level for the GR calculated from the 22.2-GHz vertically polarized data and 19.4-V data as well as a threshold for the GR calculated from the 37 V and 19.4-V data. To illustrate the results of this ice concentration algorithm, **Figure 2** presents the derived sea ice concentrations for 15 March 1998 in the Northern Hemisphere and for 15 September 1998 in the Southern Hemisphere, the same dates as used in **Figure 1**.

As mentioned, there are several alternative ice concentration algorithms in use. Contrasts from the NASA Team algorithm just described include: use of different microwave channels; use of regional tie points rather than hemispherically applicable tie points; use of cluster analysis on brightness temperature data, without PR and GR formulations; use of iterative techniques whereby an initial ice concentration calculation leads to refined atmospheric temperatures and opacities, which in turn lead to refined ice concentrations; use of iterative techniques involving surface temperature, atmospheric water vapor, cloud liquid water content, and wind speed; incorporation of higher-frequency data to reduce the effects of snow cover on the computations; and use of a Kalman filtering technique in conjunction with an ice model. The various techniques tend to obtain very close overall distributions of where the sea ice is located, although sometimes with noticeable differences (up to 20%, and on occasion even higher) in the individual, local ice concentrations. The differences can often be markedly reduced by adjustment of tunable parameters, such as the algorithm tie points, in one or both of the algorithms being compared. However, the lack of adequate ground data often makes it difficult to know which tuning is most appropriate or which algorithm is yielding the better results. To help resolve the uncertainties, *in situ* and aircraft measurements are being made in both the Arctic and the

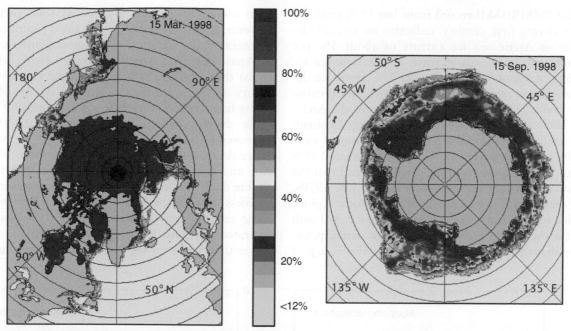

Figure 2 North and south polar sea ice concentration images for 15 Mar. 1998 and 15 Sep. 1998, respectively. The ice concentrations are derived from the data of the DMSP SSMI, including the 19.4-V data depicted in **Figure 1**.

Antarctic to help validate and improve the AMSR-E sea ice products.

Whichever ice concentration algorithm is used, the result provides estimates of ice coverage, not ice thickness. In cases where ice thickness data are also available, the combination of ice concentration and ice thickness allows the calculation of ice volume. Ice thickness data, however, are quite limited both spatially and over time. Furthermore, they are generally not derived from the passive-microwave observations and so are not highlighted in this article. The limited ice thickness data traditionally have come from *in situ* and submarine measurements, although some recent data are also available from satellite radar altimetry and, since the January 2003 launch of the Ice, Cloud and land Elevation Satellite (*ICE-Sat*), from satellite laser altimetry. The laser technique appears promising, although the laser on board ICESat operates only for short periods, and hence the full value of the technique will not be realized until a new laser is launched with a longer operational lifetime. Average sea ice thickness in the Antarctic is estimated to be in the range 0.4–1.5 m, and average sea ice thickness in the Arctic is estimated to be in the range 1.5–3.5 m.

Sea Ice Extents and Trends

Sea ice extent is defined as the total ocean area of all grid cells with sea ice concentration of at least 15% (or, occasionally, an alternative prescribed minimum percentage). Sea ice extents are now routinely calculated for the north polar region as a whole, for the south polar region as a whole, and for each of several subregions within the two polar domains, using ice concentration maps determined from satellite passive-microwave data.

A major early result from the use of satellite passive-microwave data was the detailed determination of the seasonal cycle of ice extents in each hemisphere. Incorporating the interannual variability observed from the 1970s through the early twenty-first century, the Southern Ocean ice extents vary from about 2–4×10^6 km^2 in February to about 17–20×10^6 km^2 in September, and the north polar ice extents vary from about 5–8×10^6 km^2 in September to about 14–16×10^6 km^2 in March. The exact timing of minimum and maximum ice coverage and the smoothness of the growth from minimum to maximum and the decay from maximum to minimum vary noticeably among the different years.

As the data sets lengthened, a major goal became the determination of trends in the ice extents and the placement of these trends in the context of other climate variables and climate change. Because of the lack of a period of data overlap between the ESMR and the SMMR, matching of the ice extents derived from the ESMR data to those derived from the SMMR and SSMI data has been difficult and uncertain. Consequently, most published results regarding trends found from the SMMR and SSMI data sets do not include the ESMR data.

The SMMR/SSMI record from late 1978 until the early twenty-first century indicates an overall decrease in Arctic sea ice extents of about 3% per decade and an overall increase in Antarctic sea ice extents of about 1% per decade. The Arctic ice decreases have received particular attention because they coincide with a marked warming in the Arctic and likely are tied closely to that warming. Although the satellite data reveal significant interannual variability in the ice cover, even as early as the late 1980s it had become clear from the satellite record that the Arctic as a whole had lost ice since the late 1970s. The picture was mixed regionally, with some regions having lost ice and others having gained ice; and with such a short data record, there was a strong possibility that the decreases through the 1980s were part of an oscillatory pattern and would soon reverse. However, although the picture remained complicated by interannual variability and regional differences, the Arctic decreases overall continued (with fluctuations) through the 1990s and into the twenty-first century, and by the middle of the first decade of the twenty-first century no large-scale regions of the Arctic showed overall increases since late 1978. Moreover, by the early twenty-first century the Arctic sea ice decreases were apparent in all seasons of the year, and the decreases in ice extent found from satellite data were complemented by decreases in ice thickness found from submarine and *in situ* data.

The satellite-derived Arctic sea ice decreases are illustrated in **Figure 3** with 26-year March and September time series for the Northern Hemisphere sea

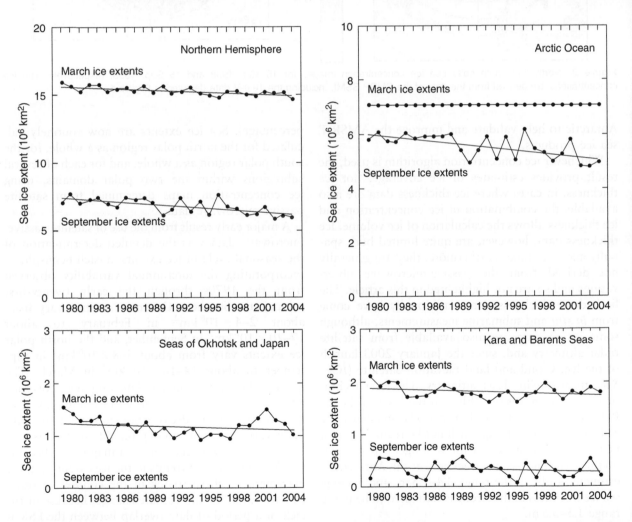

Figure 3   Time series of monthly average 1979–2004 March and September sea ice extents for the Northern Hemisphere and the following three regions within the Northern Hemisphere: the Arctic Ocean, the Seas of Okhotsk and Japan, and the Kara and Barents Seas. All ice extents are derived from the *Nimbus 7* SMMR and DMSP SSMI satellite passive-microwave data. The trend lines are linear least squares fits through the data points, and the slopes of the trend lines for the Northern Hemisphere total are $-29\,500 \pm 5400\,\text{km}^2\,\text{yr}^{-1}$ ($-1.9 \pm 0.35\%$ per decade) for the March values and $-51\,300 \pm 10\,800\,\text{km}^2\,\text{yr}^{-1}$ ($6.9 \pm 1.5\%$ per decade) for the September values. For the Seas of Okhotsk and Japan, all the September ice extents are $0\,\text{km}^2$, as the ice cover fully disappeared from these seas by the end of summer in each of the 26 years.

ice cover as a whole and for three regions within the Northern Hemisphere ice cover, those being the Arctic Ocean, the Seas of Okhotsk and Japan, and the Kara and Barents Seas. March and September are typically the months of maximum and minimum Northern Hemisphere sea ice coverage. Among the regional and seasonal differences visible in the plots, the Arctic Ocean shows little or no variation in March ice extents because of consistently being fully covered with ice in March, to at least 15% ice coverage in each grid square, but shows noticeable fluctuations and overall decreases in September ice coverage. In contrast, the region of the Seas of Okhotsk and Japan has no variability in September, because of having no ice in any of the Septembers, but for March shows marked fluctuations and slight (not statistically significant) overall decreases, interrupted by prominent increases from 1994 to 2001. The Kara and Barents Seas exhibit significant variability in both the March and September values, although with slight (not statistically significant) overall decreases for both months (**Figure 3**). The March and September ice extent decreases for the Northern Hemisphere as a whole are both statistically significant at the 99% level, as are the September decreases for the Arctic Ocean region.

The lack of uniformity in the Arctic sea ice losses (e.g., **Figure 3**) complicates making projections into the future. Nonetheless, in the early twenty-first century, several scientists have offered projections in light of the expectation of continued warming of the climate system. These projections – some based on extrapolation from the data, others on computer modeling – suggest continued Arctic sea ice decreases, with the effects of warming dominating over oscillatory behavior and other fluctuations. Some studies project a totally ice-free late summer Arctic Ocean by the end of the century, the middle of the century, or even, in one projection, by as early as the year 2013. An ice-free Arctic would greatly ease shipping but would also have multiple effects on the Arctic climate (by seasonally eliminating the highly reflective and insulating ice cover) and on Arctic ecosystems (by seasonally removing both the habitat for the organisms that live within the ice and the platform on which a variety of polar animals depend).

In contrast to the Arctic sea ice, the Antarctic sea ice cover as a whole does not show a warming signal over the period from the start of the multichannel satellite passive-microwave record, in late 1978, through the end of 2004. Over this period, some regions in the Antarctic experienced overall ice cover increases and other regions experienced overall ice cover decreases, with the hemispheric 1% per decade ice extent increases incorporating contrasting regional conditions. In **Figure 4**, the 26-year February (generally the month of Antarctic sea ice minimum) and September (generally the month of Antarctic sea ice maximum) ice extent time series are plotted for the Southern Hemisphere total, the Weddell Sea, the Bellingshausen and Amundsen Seas, and the Ross Sea. The region of the Bellingshausen and Amundsen Seas shows statistically significant (99% confidence level) February sea ice decreases but shows very slight (not statistically significant) September sea ice increases. The Weddell Sea instead shows statistically significant (95% confidence level) increases in February ice coverage and slight (not statistically significant) decreases in September ice coverage, while the Ross Sea shows ice increases in both months, with the February increases being statistically significant at the 99% confidence level (**Figure 4**). In line with the mixed pattern of ice extent increases and decreases, the Antarctic has also experienced a mixed pattern of temperature increases and decreases, with the Antarctic Peninsula (separating the Bellingshausen and Weddell seas) being the one region of the Antarctic with a prominent warming signal.

Sea Ice Types

The sea ice covers in both polar regions are mixtures of various types of ice, ranging from individual ice crystals to coherent ice floes several kilometers across. Common ice types include frazil ice (fine spicules of ice suspended in water); grease ice (a soupy layer of ice that looks like grease from a distance); slush ice (a viscous mass formed from a mixture of water and snow); nilas (a thin elastic sheet of ice 0.01–0.1-m thick); pancake ice (small, roughly circular pieces of ice, 0.3–3 m across and up to 0.1-m thick); first-year ice (ice at least 0.3-m thick that has not yet undergone a summer melt period); and multiyear ice (ice that has survived a summer melt).

Because different ice types have different microwave emission characteristics, once these differences are understood, appropriate satellite passive-microwave data can be used to distinguish ice types. The ice types most frequently distinguished with such data are first-year ice and multiyear ice in the Arctic Ocean. In fact, the NASA Team algorithm described earlier was initially developed specifically for first-year and multiyear ice, with the resulting calculations yielding the concentrations, C_{I1} and C_{I2}, of those two ice types. First-year and multiyear ice are distinguishable in their microwave signals because the summer melt process drains some of the salt content downward through the ice, reducing the salinity of the upper layers of the ice and thereby changing the microwave emissions; these

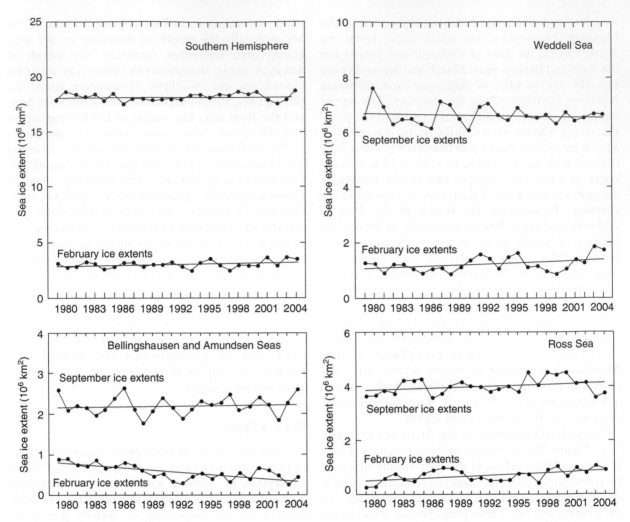

Figure 4 Time series of monthly average 1979–2004 February and September sea ice extents for the Southern Hemisphere and the following three regions within the Southern Hemisphere: the Weddell Sea, the Bellingshausen and Amundsen Seas, and the Ross Sea. All ice extents are derived from the *Nimbus 7* SMMR and DMSP SSMI satellite passive-microwave data. The trend lines are linear least squares fits through the data points, and the slopes of the trend lines for the Southern Hemisphere total are $14\,100\pm8000\,\text{km}^2\,\text{yr}^{-1}$ ($5.0\pm2.8\%$ per decade) for the February values and $7000\pm8200\,\text{km}^2\,\text{yr}^{-1}$ ($0.4\pm0.5\%$ per decade) for the September values.

changes are dependent on the frequency and polarization of the radiation. To illustrate the differences, **Figure 5** presents a plot of the tie points employed in the NASA Team algorithm for the Arctic ice. Tie points were determined empirically and are included for each of the three SSMI channels used in the calculation of ice concentrations prior to the application of the weather filter. The plot shows that while the transition from first-year to multiyear ice lowers the brightness temperatures for each of the three channels, the reduction is greatest for the 37-V data and least for the 19.4-V data. The plot further reveals that the polarization PR (eqn [6], revised for 19.4 GHz rather than 18-GHz data) is larger for multiyear ice than for first-year ice, and larger for

water than for either ice type. Furthermore, the GR (eqn [7], revised for 19.4-GHz data) is positive for water, slightly negative for first-year ice, and considerably more negative for multiyear ice (**Figure 5**). The differences allow the sorting out, either through the calculation of C_{I1} and C_{I2} as described earlier or through alternative algorithms, of the first-year ice and multiyear ice percentages in the satellite field of view.

Other Sea Ice Variables: Season Length, Temperature, Melt, Velocity

Although ice concentrations, ice extents, and, to a lesser degree, ice types have been the sea ice variables

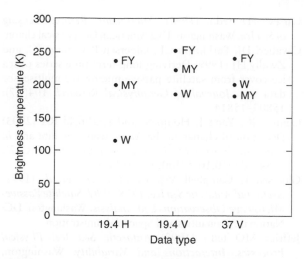

Figure 5 Typical brightness temperatures of first-year ice (FY), multiyear ice (MY), and liquid water (W) at three channels of SSMI data from the DMSP F13 satellite, specifically those for 19.4-GHz horizontally polarized data (19.4 H), 19.4-GHz vertically polarized data (19.4 V), and 37-GHz vertically polarized data (37 V). These are the values used as tie points for the Arctic calculations in the NASA Team algorithm described in the text. Data from Cavalieri DJ, Parkinson CL, Gloersen P, Comiso JC, and Zwally HJ (1999) Deriving long-term time series of sea ice cover from satellite passive-microwave multisensor data sets. *Journal of Geophysical Research* 104(C7): 15803–15814.

most widely calculated and used from satellite passive-microwave data, several additional variables have also been obtained from these data, including the length of the sea ice season, sea ice temperature, sea ice melt, and sea ice velocity. The length of the sea ice season for any particular year is calculated directly from that year's daily maps of sea ice concentrations, by counting, at each grid element, the number of days with ice coverage of at least some predetermined (generally 15% or 30%) ice concentration. Trends in the length of the sea ice season from the late 1970s to the early twenty-first century show coherent spatial patterns in both hemispheres, with a predominance of negative values (shortening of the sea ice season) in the Northern Hemisphere and in the vicinity of the Antarctic Peninsula and a much lesser predominance of positive values in the rest of the Southern Hemisphere's sea ice region, consistent with the respective hemispheric trends in sea ice extents.

The passive-microwave-based ice temperature calculations generally depend on the calculated sea ice concentrations, empirically determined ice emissivities, a weighting of the water and ice temperatures within the field of view, and varying levels of sophistication in incorporating effects of the polar atmosphere and the presence of multiple ice types. The derived temperature is not the surface temperature but the temperature of the radiating portion of the ice, for whichever radiative frequency being used. The ice temperature fields derived from passive-microwave data complement those derived from satellite infrared data, which have the advantages of generally having finer spatial resolution and of more nearly approaching surface temperatures but have the disadvantage of greater contamination by clouds. The passive-microwave and infrared data are occasionally used together, iteratively, for an enhanced ice temperature product.

The seasonal melting of portions of the sea ice and its overlying snow cover generally produces marked changes in microwave emissions, first increasing the emissions as liquid water emerges in the snow, then decreasing the emissions once the snow has melted and meltwater ponds cover the ice. Because of the emission changes, these events on the ice surface frequently become detectable through time series of the satellite passive-microwave data. The onset of melt in particular can often be identified, and hence yearly maps can be created of the dates of melt onset. Melt ponds, however, present greater difficulties, as they can have similar microwave emissions to those of the water in open spaces between ice floes, so that a field of heavily melt-ponded ice can easily be confused in the microwave data with a field of low-concentration ice. The ambiguities can be reduced through analysis of the passive-microwave time series and comparisons with active-microwave, visible, and infrared data. Still, because of these complications under melt conditions, the passive-microwave-derived ice concentrations tend to have larger uncertainties for summertime ice than for wintertime ice.

The calculation of sea ice velocities from satellite data has in general relied upon data with fine enough resolution to distinguish individual medium-sized ice floes, such as visible data and active-microwave data rather than the much coarser resolution passive-microwave data. However, in the 1990s, several groups devised methods of determining ice velocity fields from passive-microwave data, some using techniques based on cross-correlation of brightness temperature fields and others using wavelet analysis. These techniques have yielded ice velocity maps on individual dates for the entire north and south polar sea ice covers. Comparisons with buoy and other data have been quite encouraging regarding the potential of using the passive-microwave satellite data for long-term records and monitoring of ice motions.

Looking toward the Future

Monitoring of the polar sea ice covers through satellite passive-microwave technology is ongoing with the operational SSMI instruments on the DMSP satellites and the Japanese AMSR-E instrument on NASA's Aqua satellite. Both Japan and the United States anticipate launching additional passive-microwave instruments to maintain an uninterrupted satellite passive-microwave data record. The resulting lengthening sea ice records should continue to provide an improved basis with which scientists can examine trends in the sea ice cover and interactions between the sea ice and other elements of the climate system. For instance, the lengthened records will be essential to answering many of the questions raised concerning whether the negative overall trends found in Arctic sea ice extents for the first quarter century of the SMMR/SSMI record will continue and how these trends relate to temperature trends, in particular to climate warming, and to oscillations within the climate system, in particular the North Atlantic Oscillation, the Arctic Oscillation, and the Southern Oscillation. In addition to covering a longer period, other expected improvements in the satellite passive-microwave record of sea ice include further algorithm developments, following additional analyses of the microwave properties of sea ice and liquid water. Such analyses are likely to lead both to improved algorithms for the variables already examined and to the development of techniques for calculating additional sea ice variables from the satellite data.

Glossary

Brightness temperature Unit used to express the intensity of emitted microwave radiation received by the satellite, presented in temperature units (K) following the Rayleigh–Jeans approximation to Planck's law, whereby the radiation emitted from a perfect emitter at microwave wavelengths is proportional to the emitter's physical temperature.

Sea ice concentration Percent areal coverage of sea ice.

Sea ice extent Integrated area of all grid elements with sea ice concentration of at least 15%.

See also

Ice–Ocean Interaction. Polynyas. Satellite Oceanography, History, and Introductory Concepts. Satellite Remote Sensing Sar. Sea Ice: Overview.

Further Reading

Barry RG, Maslanik J, Steffen K, et al. (1993) Advances in sea-ice research based on remotely sensed passive microwave data. *Oceanography* 6(1): 4–12.

Carsey FD (ed.) (1992) *Microwave Remote Sensing of Sea Ice*. Washington, DC: American Geophysical Union.

Cavalieri DJ, Parkinson CL, Gloersen P, Comiso JC, and Zwally HJ (1999) Deriving long-term time series of sea ice cover from satellite passive-microwave multisensor data sets. *Journal of Geophysical Research* 104(C7): 15803–15814.

Comiso JC, Yang J, Honjo S, and Krishfield RA (2003) Detection of change in the Arctic using satellite and *in situ* data. *Journal of Geophysical Research* 108(C12): 3384 (doi:10.1029/2002JC001347).

Gloersen P, Campbell WJ, Cavalieri DJ, et al. (1992) *Arctic and Antarctic Sea Ice, 1978–1987: Satellite Passive-Microwave Observations and Analysis*. Washington, DC: National Aeronautics and Space Administration.

Jeffries MO (ed.) (1998) *Antarctic Sea Ice: Physical Processes, Interactions and Variability*. Washington, DC: American Geophysical Union.

Johannessen OM, Bengtsson L, Miles MW, et al. (2004) Arctic climate change: Observed and modelled temperature and sea-ice variability. *Tellus* 56A(4): 328–341.

Kramer HJ (2002) *Observation of the Earth and Its Environment*, 4th edn. Berlin: Springer.

Lubin D and Massom R (2006) *Polar Remote Sensing, Vol. 1: Atmosphere and Oceans*. Berlin: Springer-Praxis.

Parkinson CL (1997) *Earth from Above: Using Color-Coded Satellite Images to Examine the Global Environment*. Sausalito, CA: University Science Books.

Parkinson CL (2004) Southern Ocean sea ice and its wider linkages: Insights revealed from models and observations. *Antarctic Science* 16(4): 387–400.

Smith WO, Jr. and Grebmeier JM (eds.) (1995) *Arctic Oceanography: Marginal Ice Zones and Continental Shelves*. Washington, DC: American Geophysical Union.

Thomas DN and Dieckmann GS (eds.) (2003) *Sea Ice: An Introduction to Its Physics, Chemistry, Biology and Geology*. Oxford, UK: Blackwell Science.

Ulaby FT, Moore RK, and Fung AK (eds.) (1986) Monitoring sea ice. In: *Microwave Remote Sensing: Active and Passive, Vol. III: From Theory to Applications*, pp. 1478–1521. Dedham, MA: Artech House.

Walsh JE, Anisimov O, Hagen JOM, et al. (2005) Cryosphere and hydrology. In: Symon C, Arris L, and Heal B (eds.) *Arctic Climate Impact Assessment*, pp. 183–242. Cambridge, UK: Cambridge University Press.

Relevant Websites

http://www.awi-bremerhaven.de
– Alfred Wegener Institute for Polar and Marine Research.

http://www.antarctica.ac.uk
– Antarctica, British Antarctic Survey.

http://www.arctic.noaa.gov
– Arctic Change, NOAA Arctic Theme Page.

http://www.aad.gov.au
– Australian Antarctic Division.

http://www.dcrs.dtu.dk
  – Danish Center for Remote Sensing.
http://www.jaxa.jp
  – Japan Aerospace Exploration Agency.
http://www.spri.cam.ac.uk
  – Scott Polar Research Institute.

http://www.nasa.gov
  – US National Aeronautics and Space Administration.
http://www.natice.noaa.gov
  – US National Ice Center.
http://nsidc.org
  – US National Snow and Ice Data Center.

INDEX

Notes

Cross-reference terms in italics are general cross-references, or refer to subentry terms within the main entry (the main entry is not repeated to save space). Readers are also advised to refer to the end of each article for additional cross-references - not all of these cross-references have been included in the index cross-references.

The index is arranged in set-out style with a maximum of three levels of heading. Major discussion of a subject is indicated by bold page numbers. Page numbers suffixed by T and F refer to Tables and Figures respectively. vs. indicates a comparison.

This index is in letter-by-letter order, whereby hyphens and spaces within index headings are ignored in the alphabetization. For example, 'oceanography' is alphabetized before 'ocean optics.' Prefixes and terms in parentheses are excluded from the initial alphabetization.

Where index subentries and sub-subentries pertaining to a subject have the same page number, they have been listed to indicate the comprehensiveness of the text.

Abbreviations used in subentries

AUV - autonomous underwater vehicle
CPR - continuous plankton recorder
DIC - dissolved inorganic carbon
DOC - dissolved organic carbon
ENSO - El Niño Southern Oscillation
MOC - meridional overturning circulation
POC - particulate organic carbon
ROV - remotely operated vehicle
SAR - synthetic aperture radar
SST - sea surface temperature

Additional abbreviations are to be found within the index.

Printed and bound by CPI Group (UK) Ltd, Croydon, CR0 4YY

03/10/2024

01040318-0011